Lecture Notes in Computer Science 2518

Edited by G. Goos, J. Hartmanis, and J. van Leeuwen

Springer
Berlin
Heidelberg
New York
Barcelona
Hong Kong
London
Milan
Paris
Tokyo

Prosenjit Bose   Pat Morin (Eds.)

Algorithms and Computation

13th International Symposium, ISAAC 2002
Vancouver, BC, Canada, November 21-23, 2002
Proceedings

Springer

Series Editors

Gerhard Goos, Karlsruhe University, Germany
Juris Hartmanis, Cornell University, NY, USA
Jan van Leeuwen, Utrecht University, The Netherlands

Volume Editors

Prosenjit Bose
Pat Morin
Carleton University, School of Computer Science
1125 Colonel By Drive, Ottawa, Ontario, Canada, K1S 5B6
E-mail: {jit,morin}@scs.carleton.ca

Cataloging-in-Publication Data applied for

A catalog record for this book is available from the Library of Congress

Bibliographic information published by Die Deutsche Bibliothek
Die Deutsche Bibliothek lists this publication in the Deutsche Nationalbibliographie;
detailed bibliographic data is available in the Internet at <http://dnb.ddb.de>.

CR Subject Classification (1998): F.2, F.1, C.2, G.2-3, I.3.5, F.1

ISSN 0302-9743
ISBN 3-540-00142-5 Springer-Verlag Berlin Heidelberg New York

Springer-Verlag Berlin Heidelberg New York
a member of BertelsmannSpringer Science+Business Media GmbH

http://www.springer.de

© Springer-Verlag Berlin Heidelberg 2002
Printed in Germany

Typesetting: Camera-ready by author, data conversion by PTP-Berlin, Stefan Sossna e.K.
Printed on acid-free paper   SPIN: 10871021   06/3142   5 4 3 2 1 0

Preface

This volume contains the papers selected for presentation at the Thirteenth Annual International Symposium on Algorithms and Computation (ISAAC 2002) to be held in Vancouver, Canada on November 21-23, 2002. This is the first time that ISAAC has ventured into North America, as it is traditionally held in the Asia-Pacific region.

ISAAC is an annual international symposium that covers the very broad areas of algorithms and computing in general. The scope of this symposium includes both theoretical and applied research from these fields. The main aim of the symposium is to promote the exchange of ideas in this active research community. The specific themes targeted for ISAAC 2002 were Computational Geometry, Algorithms and Data Structures, Approximation Algorithms, Randomized Algorithms, Graph Drawing and Graph Algorithms, Combinatorial Optimization, Computational Biology, Computational Finance, Cryptography, and Parallel and Distributed Computing.

In response to our call for papers, we received close to 160 submissions. The challenging task of selecting the papers for presentation was performed by the members of our program committee, their support staff, and referees. All of them deserve special thanks for their hard work in the difficult task of selecting a mere 54 papers from the high-quality submissions. Unfortunately, due to the time constraints imposed by a three-day conference, we were forced to turn away many excellent papers.

John Iacono will receive the traditional *best paper award* for his paper entitled *Key Independent Optimality*. Our three invited speakers are Luc Devroye from McGill University who will speak on *Random Tries*, János Pach from New York University who will speak on *Monotone Drawings of Planar Graphs*, and Nicholas Pippenger from the University of British Columbia who will speak on the *Expected Acceptance Counts for Finite Automata with Almost Uniform Input*.

No conference can succeed without exceptional local organization. For the often overlooked and thankless task of finding a suitable venue and banquet location, dealing with the logistics of registration, organizing lunches and coffee, etc., we would like to thank the the organizing committee, Nadja Rence, and Olga Stachova for all their efforts above and beyond the call of duty. We thank our sponsors Simon Fraser University, the Pacific Institute of Mathematical Sciences, the institute for Mathematics of Information Technology and Complex Systems, and Bajai Inc. for their generous support. Finally, we thank all the authors and participants. We hope this edition of ISAAC will be a success.

August 2002

Prosenjit Bose
Pat Morin

Organization

Organizing Committee

Binay Bhattacharya (Simon Fraser University, Canada)
Prosenjit Bose (Carleton University, Canada)
Arvind Gupta (Simon Fraser University, Canada)
Tiko Kameda (Simon Fraser University, Canada)

Program Committee

Prosenjit Bose, Program Chair (Carleton University, Canada)
Tetsuo Asano (JAIST, Japan)
Binay Bhattacharya (Simon Fraser University, Canada)
Hans Bodlaender (Utrecht University, The Netherlands)
Rudolf Fleischer (HKUST, Hong Kong)
Naveen Garg (IIT Delhi, India)
Arvind Gupta (Simon Fraser University, Canada)
Wen-Lian Hsu (Academia Sinica, Taiwan)
David Kirkpatrick (University of British Columbia, Canada)
Danny Krizanc (Wesleyan University, USA)
Ming Li (University of Waterloo, Canada)
Michael Molloy (University of Toronto, Canada)
Pat Morin (Carleton University, Canada)
Ian Munro (University of Waterloo, Canada)
Lata Narayanan (Concordia University, Canada)
Rajeev Raman (University of Leicester, UK)
Sunil Shende (Rutgers University, USA)
Hisao Tamaki (Meiji University, Japan)
Roberto Tamassia (Brown Univeristy, USA)
Denis Thérien (McGill University, Canada)
Takeshi Tokuyama (Tohoku University, Japan)

Referees

Jochen Alber
Richard Anstee
François Anton
Petra Berenbrink
Anne Berry

Therese Biedl
Patricia Bouyer
Alex Brodsky
David Bryant
Claude Crepeau

Henri Darmon
Will Evans
Paolo Ferragina
Tom Friedetzky
Leszek Gasieniec
Daya Ram Gaur
Dhrubajyoti Goswami
Joachim Gudmundsson
Hovhannes Harutyunyan
Laurie Hendren
Lisa Higham
Dawei Hong
Tsan-sheng Hsu
John Iacono
Klaus Jansen
Juha Kaerkkainen
Michael Kaufmann
Rohit Khandekar
Ralf Klasing
Hirotada Kobayashi
Ekkehard Koehler
Jan Kratochvil
Matthias Krause
Ramesh Krishnamurti
Dietrich Kuske
Stefan Langerman
PeiZong Lee
Hon Fung Li
Chi-Jen Lu
Hsueh-I Lu
Wei-Fu Lu
Cris Moore

Jaroslav Opatrny
Andrzej Proskuroski
Pavel Pudlak
Jaikumar Radhakrishnan
Md. Saidur Rahman
Rajeev Raman
Suneeta Ramaswami
S. S. Ravi
Ran Raz
Udi Rotics
Pranab Sen
Jiri Sgall
Akiyoshi Shioura
Hiroki Shizuya
Michiel Smid
Jack Snoeyink
Ting-Yi Sung
Chaitanya Swamy
Richard B. Tan
Jan Arne Telle
Pascal Tesson
Mikkel Thorup
Yosihito Toyama
Jacques Vaisey
Heribert Vollmer
Alan Wagner
René Weiskircher
Anthony Whitehead
Gerhard Woeginger
David Wood
17 additional anonymous referees

Sponsors

Simon Fraser University
Pacific Institute of Mathematical Sciences (PIMS)
Mathematics of Information Technology and Complex Systems (MITACS)
Bajai Inc.

Table of Contents

Session 5A

Session 5B

Session 6A

Session 6B

Session 7A

Session 7B

Session 8A

Session 8B

Session 9A

Session 9B

Invited Talks

Author Index

Biased Skip Lists

Amitabha Bagchi[1*], Adam L. Buchsbaum[2], and Michael T. Goodrich[3*]

[1] Dept. of Computer Science, Johns Hopkins Univ., Baltimore, MD 21218,
bagchi@cs.jhu.edu
[2] AT&T Labs, Shannon Laboratory, 180 Park Ave., Florham Park, NJ 07932,
alb@research.att.com
[3] Dept. of Info. & Computer Science, Univ. of California, Irvine, CA 92697-3425,
goodrich@ics.uci.edu

Abstract. We design a variation of skip lists that performs well for generally biased access sequences. Given n items, each with a positive weight w_i, $1 \le i \le n$, the time to access item i is $O\left(1 + \log \frac{W}{w_i}\right)$, where $W = \sum_{i=1}^{n} w_i$; the data structure is dynamic. We present deterministic and randomized variations, which are nearly identical; the deterministic one simply ensures the balance condition that the randomized one achieves probabilistically. We use the same method to analyze both.

1 Introduction

The primary goal of data structures research is to design data organization mechanisms that admit fast access and update operations. For a generic n-element ordered data set that is accessed and updated uniformly, this goal is typically satisfied by dictionaries that achieve $O(\log n)$-time search and update performance; e.g., AVL-trees [2], red-black trees [12], and (a, b)-trees [13].

Nevertheless, many dictionary applications involve sets of weighted data items subject to non-uniform access patterns that are *biased* according to the weights. For example, operating systems (e.g., see Stallings [22]) deal with biasing in memory requests. Other recent examples of biased sets include client web server requests [11] and DNS lookups [6]. For such applications, a *biased search structure* is more appropriate—that is, a structure that achieves search times faster than $\log n$ for highly weighted items. Biased searching is also useful in auxiliary structures deployed inside other data structures [5,10,20].

Formally, a *biased dictionary* is a data structure that maintains an ordered set X, each element i of which has a *weight*, w_i; without loss of generality, we assume $w_i \ge 1$. The operations are as follows.

Search(X, i). Determine if i is in X.
Insert(X, i). Add i to X.
Delete(X, i). Delete i from X.
Join(X_L, X_R). Assuming that $i < j$ for each $i \in X_L$ and $j \in X_R$, create a new set $X = X_L \cup X_R$. The operation destroys X_L and X_R.

* Supported by DARPA Grant F30602-00-2-0509 and NSF Grant CCR-0098068.

P. Bose and P. Morin (Eds.): ISAAC 2002, LNCS 2518, pp. 1–13, 2002.
© Springer-Verlag Berlin Heidelberg 2002

Split(X, i). Assuming without loss of generality that $i \notin X$, create $X_L = \{j \in X : j < i\}$ and $X_R = \{j \in X : j > i\}$. The operation destroys X.

FingerSearch(X, i, j). Determine if j is in X, exploiting a handle in the data structure to element $i \in X$.

Reweight(X, i, w_i'). Change the weight of i to w_i'.

In this paper, we study efficient data structures for biased data sets subject to these operations. We desire search times that are asymptotically optimal and update times that are also efficient. For example, consider the case when w_i is the number of times item i is accessed. Define $W = \sum_{i=1}^{n} w_i$. A biased dictionary with $O\left(\log \frac{W}{w_i}\right)$ search time for the i'th item can perform m searches on n items in $O(m(1 - \sum_{i=1}^{n} p_i \log p_i))$ time, where $p_i = \frac{w_i}{m}$, which is asymptotically optimal [1]. We therefore desire $O\left(\log \frac{W}{w_i}\right)$ search times and similar update times for general biased data (with arbitrary weights). We also seek biased structures that would be simple to implement and that do not require major restructuring operations, such as tree rotations, to achieve biasing. Tree rotations, in particular, make structures less amenable to augmentation, for such rotations often require the complete rebuilding of auxiliary structures stored at the affected nodes.

1.1   Related Prior Work

The study of biased data structures is a classic topic in algorithmics. Early work includes a dynamic programming method by Knuth [14,15] for constructing a static biased binary search tree for items weighted by their search frequencies. Subsequent work focuses primarily on achieving asymptotically optimal search times while also admitting efficient updates. Most of the known methods for constructing dynamic biased data structures use search trees, and they differ from one another primarily in their degree of complication and whether or not their resulting time bounds are amortized, randomized, or worst case.

Sleator and Tarjan [21] introduce the theoretically elegant *splay trees*, which automatically adjust themselves to achieve optimal amortized biased access times for access-frequency weights. Splay trees store no balance or weight information, but they perform many tree rotations after every access, which makes them less practically efficient than even AVL-trees in many applications [3]. These rotations can be particularly deleterious when nodes are augmented with auxiliary structures.

Bent, Sleator, and Tarjan [4] and Feigenbaum and Tarjan [9] design biased search trees for arbitrary weights that significantly reduce, but do not eliminate, the number of tree rotations needed. They offer efficient worst-case and amortized performance of biased dictionary operations but do so with complicated implementations.

Seidel and Aragon [19] demonstrate randomized bounds with *treaps*. Like splay trees, treaps perform a large number of rotations after every access. Their data structure is elegant and efficient in practice, but its performance does not achieve bounds that are efficient in a worst-case or amortized sense.

Pugh [18] introduces an alternative *skip list* structure, which efficiently implements an unbiased dictionary without using rotations. Skip lists store the items in series of a linked lists, which are themselves linked together in a leveled fashion. Pugh presents skip lists as a randomized structure that is easily implemented and shows that they are empirically faster than fast balanced search trees, such as AVL-trees. Search and updates take $O(\log n)$ expected time in skip lists, with no rotations or other rebalancing needed for updates. Exploiting the relationship between skip lists and (a, b)-trees, Munro, Papadakis, and Sedgewick [17] show how to implement a deterministic version of skip lists that achieves similar bounds in the worst case using simple promote and demote operations.

For biased skip lists, much less prior work exists. Mehlhorn and Näher [16] anticipated biased skip lists but claimed only a partial result and omitted details and analysis. Recently, Ergun *et al.* [7,8] presented a biased skip list structure that is designed for a specialized notion of biasing, in which access to an item i takes $O(\log r(i))$ expected time, where $r(i)$ is the number of items accessed since the last time i was accessed. Their data structure is incomparable to a general biased dictionary, as each provides properties not present in the other.

1.2   Our Results

We present a comprehensive design of a biased version of skip lists. It combines techniques underlying deterministic skip lists [17] with Mehlhorn and Näher's suggestion [16]. Our methods work for arbitrarily defined item weights and provide asymptotically optimal search times based on these weights. Using skip list technology eliminates tree rotations. We present complete descriptions of all the biased dictionary operations, with time performances that compare favorably with those of the various versions of biased search trees. We give both deterministic and randomized implementations. Our deterministic structure achieves worst-case running times similar to those of biased search trees [4,9] but uses techniques that are arguably simpler. A node in a deterministic biased skip list is assigned an initial level based on its weight, and simple invariants govern promotion and demotion of node levels to ensure desired access times. Our randomized structure achieves expected bounds similar to the respective amortized and randomized bounds of splay trees [21] and treaps [19]. Our randomized structure does not use partial rebuilding and hence does not need any amortization of its own. Table 1 (at the end) juxtaposes our results against biased search trees, splay trees, and treaps.

In Section 2, we define our deterministic biased skip list structure, and in Section 3 we describe how to perform updates efficiently in this structure. In Section 4 we describe a simple, randomized variation of biased skip lists and analyze its performance. We conclude in Section 5.

2   Biased Skip Lists

A *skip list* [18] S is a dictionary data structure, storing an ordered set X, the items of which we number 1 through $|X|$. Each item $i \in X$ has a key, x_i, and a corresponding *node* in the skip list of some integral *height*, $h_i \geq 0$. The *height* of S is $H(S) = \max_{i \in X} h_i$. The *depth*, d_i, of i is $H(S) - h_i$. We use the terms item, node, and key interchangeably; the context clarifies any ambiguity. We assume without loss of generality that the keys in X are unique: $x_i < x_{i+1}, 1 \leq i < |X|$.

Each node i is implemented by a linked list or array of length $h_i + 1$, which we call the *tower* for that node. The *level-j successor* of a node i is the least node $\ell > i$ of height $h_\ell \geq j$. Symmetrically define *level-j predecessor*. For node i and each $0 \leq j \leq h_i$, the j'th element of the node contains pointers to the j'th elements of the level-j successor and predecessor of i. Two distinct nodes $x < y$ are called *consecutive* if and only if $h_z < \min(h_x, h_y)$ for all $x < z < y$. A *plateau* is a maximal sequence of consecutive nodes of equal height.

For convenience we assume sentinel nodes of height $H(S)$ at the beginning (with key $-\infty$) and end (with key ∞) of S; in practice, this assumption is not necessary. We orient the pointers so that the skip list stores items in left-to-right order, and the node levels progress bottom to top. See Figure 1(a).

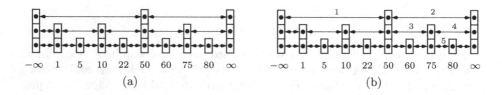

$$\begin{array}{ccccccccc} -\infty & 1 & 5 & 10 & 22 & 50 & 60 & 75 & 80 & \infty \\ & & & & (a) & & & & & \end{array}$$

$$\begin{array}{ccccccccc} -\infty & 1 & 5 & 10 & 22 & 50 & 60 & 75 & 80 & \infty \\ & & & & (b) & & & & & \end{array}$$

Fig. 1. (a) A skip list for the set $X = \{1, 5, 10, 22, 50, 60, 75, 80\}$. (b) Searching for key 80; numbers over the pointers indicate the order in which they are traversed.

To search for an item with key K we start at level $H(S)$ of the left sentinel. When searching at level i from some node we follow the level-i links to the right until we find a key matching K or a pair of nodes j, k such that k is the level-i successor of j and $x_j < K < x_k$. We then continue the search at level $i - 1$ from node j. The search ends with success if we find a node with key K, or failure if we find nodes j and k as above on level 0. See Figure 1(b).

We describe a deterministic, biased version of skip lists. In addition to key x_i each item $i \in X$ has a *weight*, w_i; without loss of generality, assume $w_i \geq 1$. Define the *rank* of item i as $r_i = \lfloor \log_a w_i \rfloor$, where a is a constant parameter.

Definition 1. *For a and b such that $1 < a \leq \lfloor \frac{b}{2} \rfloor$, an (a,b)-biased skip list is one in which each item has height $h_i \geq r_i$ and the following invariants hold.*

(I1) *There are never more than b consecutive items of any height in $[0, H(S)]$.*
(I2) *For each node x and all $r_x < i \leq h_x$, there are at least a nodes of height $i - 1$ between x and any consecutive node of height at least i.*

To derive exact bounds for the case when an item does not exist in the skip list we eliminate redundant pointers. For every pair of adjacent items $i, i + 1$, we ensure that the pointers between them on level $\min(h_i, h_{i+1}) - 1$ are nil; the pointers below this level are undefined. (In Figure 1, for example, the level-0 pointers between $-\infty$ and 1 become nil.) When searching for an item $i \notin X$, we assert failure immediately upon reaching a nil pointer.

Throughout the remainder of the paper, we define $W = \sum_{i \in X} w_i$ to be the weight of S before any operation. For any key i, we denote by i^- the item in X with largest key less than i, and by i^+ the item in X with smallest key greater than i. The main result of our definition of biased skip lists is summarized by the following lemma, which bounds the depth of any node.

Lemma 1 (Depth Lemma). *The depth of any node i in an (a, b)-biased skip list is $O\left(\log_a \frac{W}{w_i}\right)$.*

Before we prove the depth lemma, consider its implication on *access time* for key i: the time it takes to find i in S if $i \in X$ or pair i^-, i^+ in S if $i \notin X$.

Corollary 1 (Access Lemma). *The access time for key i in an (a, b)-biased skip list is $O\left(1 + b \log_a \frac{W}{w_i}\right)$ if $i \in X$ and $O\left(1 + b \log_a \frac{W}{\min(w_{i^-}, w_{i^+})}\right)$ if $i \notin X$.*

Proof. By **(I1)**, at most $b + 1$ pointers are traversed at any level. A search stops upon reaching the first nil pointer, so the Depth Lemma implies the result.

It is important to note that while all the bounds we prove rely on W, the data structure itself need not maintain this value.

To prove the depth lemma, observe that the number of items of any given rank that can appear at higher levels decreases geometrically by level. Define $N_i = |\{x : r_x = i\}|$ and $N_i' = |\{x : r_x \leq i \wedge h_x \geq i\}|$.

Lemma 2. $N_i' \leq \sum_{j=0}^{i} \frac{1}{a^{i-j}} N_j$.

Proof. By induction. The base case, $N_0' = N_0$, is true by definition. For $i > 0$, **(I2)** implies that $N_{i+1}' \leq N_{i+1} + \lfloor \frac{1}{a} N_i' \rfloor \leq N_{i+1} + \frac{1}{a} N_i'$, which, together with the induction hypothesis, proves the lemma.

Intuitively, a node promoted to a higher level is supported by enough weight associated with items at lower levels. Define $W_i = \sum_{r_x \leq i} w_x$.

Corollary 2. $W_i \geq a^i N_i'$.

Proof. By definition, $W_i \geq \sum_{j=0}^{i} a^j N_j = a^i \sum_{j=0}^{i} \frac{1}{a^{i-j}} N_j$. Apply Lemma 2.

Define $R = \max_{x \in X} r_x$. Any nodes with height exceeding R must have been promoted from lower levels to maintain the invariants. **(I2)** thus implies that $H(S) \leq R + \log_a N_R'$, and therefore the maximum possible depth of an item i is $d_i \leq H(S) - r_i \leq R + \log_a N_R' - r_i$.

By Corollary 2, $W = W_R \geq a^R N_R'$. Therefore $\log_a N_R' \leq \log_a W - R$. Hence, $d_i \leq \log_a W - r_i$. The Depth Lemma follows, because $\log_a w_i - 1 < r_i \leq \log_a w_i$.

(I1) and (I2) resemble the invariants defining (a, b)-skip lists [17], but (I2) is stronger than their analogue. Just to prove the Depth Lemma, it would suffice for a node of height h exceeding its rank, r, to be supported by at least a items to each side only at level $h - 1$, not at every level between r and $h - 1$. The update procedures in the next section, however, require support at every level.

3   Updating Deterministic Biased Skiplists

We describe insertion in detail and then sketch implementations for the other operations. All details will be available in the full paper.

The *profile* of an item i captures its predecessors and successors of increasingly greater level. For $h_{i-} \leq j \leq H(S)$, let L_j^i be the level-j predecessor of i; for $h_{i+} \leq j \leq H(S)$, let R_j^i be the level-j successor of i. Define the ordered set

$$PL(i) = \left(j : h_{L_j^i} = j,\ h_{i-} \leq j \leq H(S)\right):$$ the set of distinct heights of the nodes

to the left of i. Symmetrically define $PR(i) = \left(j : h_{R_j^i} = j,\ h_{i+} \leq j \leq H(S)\right)$.

We call the ordered set $\left(L_j^i : j \in PL(i)\right) \cup \left(R_j^i : j \in PR(i)\right)$ the *profile* of i. We call the subset of predecessors the *left profile* and the subset of successors the *right profile* of i. For example, in Figure 1, $PL(60) = (3)$; $PR(60) = (2, 3)$; the left profile of 60 is (50); and the right profile of 60 is $(75, \infty)$.

These definitions assume $i \in S$ but are also precise when $i \notin S$, in which case they apply to the (nonexistent) node that would contain key i. Given node i or, if $i \notin S$, nodes i^- and i^+, we can trace i's profile from lowest-to-highest nodes by starting at i^- (rsp., i^+) and, at any node x, iteratively finding its level-$(h_x + 1)$ predecessor (rsp., successor), until we reach the left (rsp., right) sentinel.

3.1   Inserting an Item

The following procedure inserts a new item with key i into an (a, b)-biased skip list S. If i already exists in the skip list, we discover it in Step 1.

1. Search S for i to discover the pair i^-, i^+.
2. Create a new node of height r_i to store i, and insert it between i^- and i^+ in S, splicing predecessors and successors as in a standard skip list [18].
3. Restore (I2), if necessary. Any node x in the left (sym., right) profile of i might need to have its height demoted, because i might interrupt a plateau of height less than h_x, leaving fewer than a nodes to x's left (sym., right). In this case, x is demoted to the next lower height in the profile (or r_x, whichever is higher). More precisely, for j in turn from h_{i-} up through r_i, if $j \in PL(i)$, consider node $u = L_j^i$. If (I2) is violated at node u, then demote u to height r_u if $u = i^-$ and otherwise to height $\max(j', r_u)$, where j' is the predecessor of j in $PL(i)$; let h_u' be the new height of u. If the demotion violates (I1) at level h_u', then, among the $k \in (b, 2b]$ consecutive items of height h_u', promote the $\lfloor \frac{k}{2} \rfloor$'th node (in order) to height $h_u' + 1$. (See Figure 2.) Iterate at the next j. Symmetrically process right profile of i.

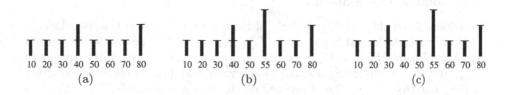

10 20 30 40 50 60 70 80   10 20 30 40 50 55 60 70 80   10 20 30 40 50 55 60 70 80
     (a)   (b)   (c)

Fig. 2. (a) A (2,4)-biased skip list. Nodes are drawn to reflect their heights; hatch marks indicate ranks. Pointers are omitted. (b) After inserting 55 with rank 3, node 40 violates **(I2)**. (c) After demotion of 40 and compensating promotion of 30.

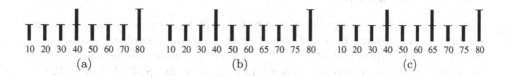

10 20 30 40 50 60 70 80   10 20 30 40 50 60 65 70 75 80   10 20 30 40 50 60 65 70 75 80
     (a)   (b)   (c)

Fig. 3. (a) The (2,4)-biased skip list of Figure 2(a). (b) **(I1)** is violated by the insertion of 65 and 75 with rank 1 each. (c) After promoting node 65.

4. Restore **(I1)**, if necessary. Starting at node i and level $j = r_i$, if node i violates **(I1)** at level j, then, among the $b+1$ consecutive items of height j, promote the $\lfloor \frac{b+1}{2} \rfloor$'th node (in order), u, to height $j+1$, and iterate at node u and level $j+1$. Continue until the violations stop. (See Figure 3.)

Theorem 1. *Inserting an item i in an (a,b)-biased skip list can be done in* $O\left(1 + b\log_a \frac{W+w_i}{\min(w_{i-},w_i,w_{i+})}\right)$ *time.*

Proof. We omit the proof of correctness. By the Depth and Access Lemmas, Steps 1 and 2 take $O\left(1 + b\log_a \frac{W+w_i}{\min(w_{i-},w_i,w_{i+})}\right)$ time. If $\min(h_{i-},h_{i+}) \leq r_i$, Step 3 performs $O(b)$ work at each level between $min(h_{i-},h_{i+})$ and r_i; Step 4 performs $O(b)$ work at each level from r_i through $H(S)$. Again apply the Depth Lemma.

3.2 Deleting an Item

Deletion is the inverse of insertion. After finding i, i^-, and i^+, remove i and splice predecessors and successors as required. Then restore **(I1)**, if necessary, as removing i might unite plateaus into sequences of length exceeding b. This is done analogously to Step 4 of insertion, starting at level $\min(h_{i-},h_{i+})$ and proceeding up through level $h_i - 1$. Finally, restore **(I2)**, if necessary, as removing i might decrease the length of a plateau of height h_i to $a - 1$. This is done analogously to Step 3 of insertion, starting at level h_i. The proof of correctness is analogous to that for insertion, and the time is $O\left(1 + b\log_a \frac{W}{\min(w_{i-},w_i,w_{i+})}\right)$.

3.3   Joining Two Skiplists

Consider biased skip lists S_L and S_R of total weights W_L and W_R, rsp. Denote the item with the largest key in S_L by $L_{\max}$ and that with the smallest key in S_R by $R_{\min}$. Assume $L_{\max} < R_{\min}$. To join S_L and S_R, trace through the profiles of $L_{\max}$ and $R_{\min}$ to splice S_L and S_R together. Restore (I1), if necessary, starting at level $\max(h_{L_{\max}}, h_{R_{\min}})$ and proceeding through level $\max(H(S_L), H(S_R))$, as in Step 4 of insertion. (I2) cannot be violated by the initial splicing, as plateaus never shrink, nor by the promotion of the node in the middle in the restoration of (I1). The time is $O\left(1 + b\log_a \frac{W_L}{w_{L_{\max}}} + b\log_a \frac{W_R}{w_{R_{\min}}}\right)$.

3.4   Splitting a Skiplist

We can split a biased skip list S of total weight W into two biased skip lists, S_L and S_R, containing keys in S less than (rsp., greater than) some $i \notin S$. First insert i into S with weight $w_i = a^{H(S)+1}$. Then disconnect the pointers between i and its predecessors (rsp., successors) to form S_L (rsp., S_R). (I1) and (I2) are true after inserting i by the correctness of insertion. Because i is taller than all of its predecessors and successors, disconnecting the pointers between them and i does not violate either invariant. The time is $O\left(1 + b\log_a \frac{W}{\min(w_{i^-}, w_{i^+})}\right)$.

3.5   Finger Searching

We can search for a key j in a biased skip list S starting at any node i to which we are given an initial pointer (or *finger*). Assume without loss of generality that $j > i$. The case $j < i$ is symmetric.

At any point in the search, we are at some height h of some node u. Initially, $u = i$ and $h = r_i$. In the *up phase*, while $R_h^u < j$, we continually set $h \leftarrow h + 1$ when $h < h_u$ and $u \leftarrow R_h^u$ when $h = h_u$. Once $R_h^u \geq j$, we enter the *down phase*, in which we search from u at height h using the normal search procedure.

The up phase moves up and to the right until we detect a node $u < j$ with some level-h successor $R_h^u > j$. That the procedure finds j (or j^-, j^+ if $j \notin S$) follows from the correctness of the vanilla search procedure and that we enter the down phase at the specified node u and height h.

Define $V(i,j) = \sum_{i \leq u \leq j} w_u$. For any node u and $h \in [r_u, h_u]$, it follows by induction that $V(L_h^u, u) \geq a^h$ and $V(u, R_h^u) \geq a^h$. Using this fact we can show that sufficient weight supports either the link into which u is originally entered during the up phase or the link out of which u is exited during the down phase. It follows that the time is $O\left(1 + b\log_a \frac{V(i,j)}{\min(w_i, w_j)}\right)$ if $j \in X$ and $O\left(1 + b\log_a \frac{V(i,j^+)}{\min(w_i, w_{j^-}, w_{j^+})}\right)$ if $j \notin X$.

3.6   Changing the Weight of an Item

We can change the weight of an item i to w_i' without deleting and reinserting i. Let $r_i' = \lfloor \log_a w_i' \rfloor$. If $r_i' = r_i$, then stop. If $r_i' > r_i$, then stop if $h_i \geq r_i'$.

Otherwise, promote i to height r'_i; restore (I2) as in insertion, starting at height $h_i + 1$; and restore (I1) as in insertion, starting at height r'_i. Finally, if $r'_i < r_i$, then demote i to height r'_i; restore (I1) as in deletion, starting at height r'_i; and restore (I2) as in deletion, starting at the least $j \in PL(i)$ greater than r'_i.

Correctness follows analogously as for insertion (in case $r'_i > r_i$) or deletion (in case $r'_i < r_i$). The time is $O\left(1 + b\log_a \frac{W + w'_i}{\min(w_i, w'_i)}\right)$.

4   Randomized Updates

We can randomize the structure to yield expected optimal access times without any promotions or demotions. Mehlhorn and Näher [16] suggested the following approach but claimed only that the expected maximal height of a node is $\log W + O(1)$. We will show that the expected depth of a node i is $E[d_i] = O\left(\log \frac{W}{w_i}\right)$.

A *randomized biased skip list* S is parameterized by a positive constant $0 < p < 1$. Here we define the *rank* of an item i as $r_i = \lfloor \log_{\frac{1}{p}} w_i \rfloor$. When inserting i into S, we assign its height to be $h_i = r_i + e_i$ with probability $p^{e_i}(1 - p)$ for $e_i \in \mathbb{Z}$, which we call the *excess height* of i. Algorithmically, we start node i at height r_i and continually increment the height by one as long as a biased coin flip returns heads (with probability p).

Reweight is the only operation that changes the height of a node. The new height is chosen as for insertion but based on the new weight, and the tower is adjusted appropriately. The remaining operations (insert, delete, join, split, and (finger) search) perform no rebalancing.

Lemma 3 (Randomized Height Lemma). *The expected height of any item i in a randomized, biased skip list is $\log_{\frac{1}{p}} w_i + O(1)$.*

Proof. $E[h_i] = r_i + E[e_i] = r_i + \sum_{j=0}^{\infty} jp^j(1 - p) = r_i + \frac{p}{1-p} = \lfloor \log_{\frac{1}{p}} w_i \rfloor + O(1)$.

The proof of the Depth Lemma for the randomized structure follows that for the deterministic structure. Recall the definitions $N_i = |\{x : r_x = i\}|$; $N'_i = |\{x : r_x \le i \wedge h_x \ge i\}|$; and $W_i = \sum_{r_x \le i} w_x$.

Lemma 4. $E[N'_i] = \sum_{j=0}^{i} p^{i-j} N_j$.

Proof. By induction. By definition, $N'_0 = N_0$. Since the excess heights are i.i.d. random variables, for $i > 0$, $E[N'_{i+1}] = N_{i+1} + pE[N'_i]$, which, together with the induction hypothesis, proves the lemma.

Corollary 3. $E[N'_i] \le p^i W_i$.

Lemma 5 (Randomized Depth Lemma). *The expected depth of any node i in a randomized, biased skip list S is $O\left(\log_{\frac{1}{p}} \frac{W}{w_i}\right)$.*

Proof. The depth of i is $d_i = H(S) - h_i$. Again define $R = \max_{x \in X} r_x$. By standard skip list analysis [18],

$$E[H(S)] = R + O(E[\log_{\frac{1}{p}} N_R']) \leq R + cE[\log_{\frac{1}{p}} N_R'] \text{ for some constant } c$$
$$\leq R + c \log_{\frac{1}{p}} E[N_R'] \text{ by Jensen's inequality}$$
$$\leq R + c \left(\log_{\frac{1}{p}} W_R - R \right) \text{ by Corollary 3}$$
$$= c \log_{\frac{1}{p}} W - (c-1)R.$$

By the Randomized Height Lemma, therefore, $E[d_i] \leq c \log_{\frac{1}{p}} W - (c-1)R - \log_{\frac{1}{p}} w_i$. The lemma follows by observing that $R \geq \lfloor \log_{\frac{1}{p}} w_i \rfloor$.

Corollary 4 (Randomized Access Lemma). *The expected access time for any key i in a randomized, biased skip list is* $O \left(1 + \frac{1}{p} \log_{\frac{1}{p}} \frac{W}{w_i} \right)$ *if $i \in X$ and* $O \left(1 + \frac{1}{p} \log_{\frac{1}{p}} \frac{W}{\min(w_{i-}, w_{i+})} \right)$ *if $i \notin X$.*

Proof. As $n \to \infty$, the probability that a plateau starting at any given node is of size k is $p(1-p)^{k-1}$. The expected size of any plateau is thus $1/p$.

The operations discussed in Section 3 become simple to implement.

Insert(S, i). Locate i^- and i^+ and create a new node between them to hold i. The expected time is $O \left(1 + \frac{1}{p} \log_{\frac{1}{p}} \frac{W + w_i}{\min(w_{i-}, w_i, w_{i+})} \right)$.

Delete(S, i). Locate and remove node i. The Randomized Depth and Access Lemmas continue to hold, because S is as if i had never been inserted. The expected time is $O \left(1 + \frac{1}{p} \log_{\frac{1}{p}} \frac{W}{\min(w_{i-}, w_i, w_{i+})} \right)$.

Join(S_L, S_R). Trace through the profiles of $L_{\max}$ and $R_{\min}$ to splice the pointers leaving S_L together with the pointers going into S_R. The expected time is $O \left(1 + \frac{1}{p} \log_{\frac{1}{p}} \frac{W_L}{w_{L_{\max}}} + \frac{1}{p} \log_{\frac{1}{p}} \frac{W_R}{w_{R_{\min}}} \right)$.

Split(S, i). Disconnect the pointers that join the left profile of i^- to the right profile of i^+. The expected time is $O \left(1 + \frac{1}{p} \log_{\frac{1}{p}} \frac{W}{\min(w_{i-}, w_{i+})} \right)$.

FingerSearch(S, i, j). Perform **FingerSearch**(S, i, j) as described in Section 3.5. The expected time if $j \in X$ is $O \left(1 + \frac{1}{p} \log_{\frac{1}{p}} \frac{V(i,j)}{\min(w_i, w_j)} \right)$ and if $j \notin X$ is $O \left(1 + \frac{1}{p} \log_{\frac{1}{p}} \frac{V(i,j^+)}{\min(w_i, w_{j-}, w_{j+})} \right)$.

Reweight(S, i, w_i'). Reconstruct the tower for node i. The expected time is $O \left(1 + \frac{1}{p} \log_{\frac{1}{p}} \frac{W + w_i'}{\min(w_i, w_i')} \right)$.

Table 1. Time bounds for biased data structures. In all bounds, W is the total weight of all items before the operation; $V(i,j) = \sum_{k=i}^{j} w_k$. For each table entry, E, the associated time bound is $O(1+E)$.

	Biased Search Trees [4]	Splay Trees [21] amort.	Treaps [19] rand.	Biased Skip Lists w.c. & rand.
Search(X,i)	$\log \dfrac{W}{w_i}$ amort./w.c.	$\log \dfrac{W}{w_i}$	$\log \dfrac{W}{w_i}$	$\log \dfrac{W}{w_i}$
Insert(X,i)	$\log \dfrac{W+w_i}{\min(w_{i^-}+w_{i^+},\, w_i)}$ amort. $\log \dfrac{W}{w_{i^-}+w_{i^+}} + \log \dfrac{W+w_i}{w_i}$ w.c.	$\log \dfrac{W}{\min(w_{i^-},\, w_{i^+})} + \log \dfrac{W+w_i}{w_i}$	$\log \dfrac{W+w_i}{\min(w_{i^-},\, w_i,\, w_{i^+})}$	$\log \dfrac{W+w_i}{\min(w_{i^-},\, w_i,\, w_{i^+})}$
Delete(X,i)	$\log \dfrac{W}{w_i}$ amort. $\log \dfrac{W}{w_i} + \log \dfrac{W-w_i}{w_{i^-}+w_{i^+}}$ w.c.	$\log \dfrac{W}{w_i} + \log \dfrac{W-w_i}{w_{i^-}}$	$\log \dfrac{W+w_i}{\min(w_{i^-},\, w_i,\, w_{i^+})}$	$\log \dfrac{W+w_i}{\min(w_{i^-},\, w_i,\, w_{i^+})}$
Join(X_L, X_R)	$\log \dfrac{W_L+W_R}{w_{L\max}+w_{R\min}}$ w.c.	$\log \dfrac{W_L+W_R}{w_{L\max}}$	$\log \dfrac{W_L}{w_{L\max}} + \log \dfrac{W_R}{w_{R\min}}$	$\log \dfrac{W_L}{w_{L\max}} + \log \dfrac{W_R}{w_{R\min}}$
Split(X,i)	$\log \dfrac{W}{w_{i^-}+w_{i^+}}$ amort./w.c.	$\log \dfrac{W}{\min(w_{i^-},\, w_{i^+})}$	$\log \dfrac{W}{\min(w_{i^-},\, w_{i^+})}$	$\log \dfrac{W}{\min(w_{i^-},\, w_{i^+})}$
Reweight(X,i,w_i')	$\log \dfrac{\max(W,W')}{\min(w_i,\, w_i')}$ amort. $\log \dfrac{W}{w_i} + \log \dfrac{W'}{w_i'}$ w.c.		$\log \dfrac{\max(w_i, w_i')}{\min(w_i, w_i')}$	$\log \dfrac{W'}{\min(w_i, w_i')}$
FingerSearch(X,i,j)			$\log \dfrac{V(i,j)}{\min(w_i, w_j)}$	$\log \dfrac{V(i,j)}{\min(w_i, w_j)}$

5 Conclusion

Open is whether a deterministic biased skip list can be devised that has not only the worst-case times that we provide but also an amortized bound of $O(\log w_i)$ for *updating* node i; i.e., once the location of the update is discovered, inserting or deleting should take $O(\log w_i)$ amortized time.

The following counterexample demonstrates that our initial method of promotion and demotion does not yield this bound. Consider a node i such that $h_i - r_i$ is large and, moreover, that separates two plateaus of size $b/2$ at each level j between $r_i + 1$ and h_i and two plateaus of size $b/2$ and $b/2 + 1$, rsp., at level r_i. Deleting i will cause a promotion starting at level r_i that will percolate to level h_i. Reinserting i with weight a^{r_i} will restore the structural condition before the deletion of i. This pair of operations can be repeated infinitely often; since $h_i - r_i$ is arbitrary, the cost of restoring the invariants cannot be amortized.

We might generalize the promotion operation to split a plateau of size exceeding b into several plateaus of size about b/η each, for some suitable constant η. Above, $\eta = 2$. The counterexample generalizes, however.

References

1. N. Abramson. *Information Theory and Coding*. McGraw-Hill, New York, 1963.
2. G. M. Adel'son-Vel'skii and Y. M. Landis. An algorithm for the organisation of information. *Dokl. Akad. Nauk SSSR*, 146:263–6, 1962. English translation in *Soviet Math. Dokl.* **3**:1259–62, 1962.
3. J. Bell and G. Gupta. Evaluation of self-adjusting binary search tree techniques. *Soft. Prac. Exp.*, 23(4):369–382, 1993.
4. S. W. Bent, D. D. Sleator, and R. E. Tarjan. Biased search trees. *SIAM J. Comp.*, 14(3):545–68, 1985.
5. Y.-J. Chiang and R. Tamassia. Dynamization of the trapezoid method for planar point location in monotone subdivisions. *Int'l. J. Comp. Geom. Appl.*, 2(3):311–333, 1992.
6. E. Cohen and H. Kaplan. Proactive caching of DNS records: Addressing a performance bottleneck. In *Proc. SAINT '01*, pages 85–92. IEEE, 2001.
7. F. Ergun, S. C. Sahinalp, J. Sharp, and R. K. Sinha. Biased dictionaries with fast inserts/deletes. In *Proc. 33rd ACM STOC*, pages 483–91, 2001.
8. F. Ergun, S. C. Sahinalp, J. Sharp, and R. K. Sinha. Biased skip lists for highly skewed access patterns. In *Proc. 3rd ALENEX*, volume 2153 of *LNCS*, pages 216–29. Springer, 2001.
9. J. Feigenbaum and R. E. Tarjan. Two new kinds of biased search trees. *BSTJ*, 62(10):3139–58, 1983.
10. M. T. Goodrich and R. Tamassia. Dynamic ray shooting and shortest paths in planar subdivisions via balanced geodesic triangulations. *J. Alg.*, 23:51–73, 1997.
11. S. D. Gribble and E. A. Brewer. System design issues for internet middleware services: Deductions from a large client trace. In *Proc. 1st USENIX Symp. on Internet Tech. and Syst.*, 1997.
12. L. J. Guibas and R. Sedgewick. A dichromatic framework for balanced trees. In *Proc. 19th IEEE FOCS*, pages 8–21, 1978.

13. S. Huddleston and K. Mehlhorn. A new data structure for representing sorted lists. *Acta Inf.*, 17:157–84, 1982.
14. D. E. Knuth. Optimum binary search trees. *Acta Inf.*, 1:14–25, 1971.
15. D. E. Knuth. *The Art of Computer Programming*, volume 3: *Sorting and Searching*. Addison-Wesley, 1973.
16. K. Mehlhorn and S. Näher. Algorithm design and software libraries: Recent developments in the LEDA project. In *Proc. IFIP '92*, volume 1, pages 493–505. Elsevier, 1992.
17. J. I. Munro, T. Papadakis, and R. Sedgewick. Deterministic skip lists. In *Proc. 3rd ACM-SIAM SODA*, pages 367–75, 1992.
18. W. Pugh. Skip lists: A probabilistic alternative to balanced trees. *C. ACM*, 33(6):668–76, June 1990.
19. R. Seidel and C. R. Aragon. Randomized search trees. *Algorithmica*, 16(4/5):464–97, 1996.
20. D. D. Sleator and R. E. Tarjan. A data structure for dynamic trees. *JCSS*, 26(3):362–91, 1983.
21. D. D. Sleator and R. E. Tarjan. Self-adjusting binary search trees. *J. ACM*, 32(3):652–86, 1985.
22. W. Stallings. *Operating Systems: Internals and Design Principles*. Prentice-Hall, 4th edition, 2001.

Space-Efficient Data Structures for Flexible Text Retrieval Systems

Kunihiko Sadakane

Graduate School of Information Sciences, Tohoku University
Aramaki Aza Aoba09, Aoba-ku, Sendai 980-8579, Japan
sada@dais.is.tohoku.ac.jp

Abstract. We propose space-efficient data structures for text retrieval systems that have merits of both theoretical data structures like suffix trees and practical ones like inverted files. Traditional text retrieval systems use the inverted files and support ranking queries based on the *tf*idf* (term frequency times inverse document frequency) scores of documents that contain given keywords, which cannot be solved by using only the suffix trees. A drawback of the systems is that the scores can be computed for only predetermined keywords. We extend the data structure so that the scores can be computed for any pattern efficiently while keeping the size of the data structures moderate. The size is comparable with the text size, which is an improvement from existing methods using $O(n \log n)$ bit space for a text collection of length n.

1 Introduction

Text retrieval systems are now indispensable to search for important documents from a large collection of text documents such as Web, genome sequence, etc. There are two typical data structures for text retrieval: the suffix tree [15] and the inverted file [3]. The former is mainly used for finding arbitrary patterns, while the latter is for finding text documents containing predetermined patterns. The latter is widely used in Web search systems because of its space efficiency, search speed, easiness of implementation, etc.

A typical text retrieval system stores a collection of text documents, and supports the following queries for a given keyword p:

1. to count the number $tf(p, d)$ of occurrences of p in a document d
2. to count the number $df(p)$ of documents containing p
3. to enumerate all documents containing p in the order of scores of documents

The first and the second ones are necessary to compute the well-known *tf*idf* scores [14] of documents that is used in the third query. These queries can be easily supported by using the inverted file. However, the inverted file does not support the queries for arbitrary pattern. Therefore it is not suitable for genome sequence databases, Japanese or Chinese text databases, etc. On the other hand, the suffix tree supports the first query for arbitrary pattern although it does not support the others.

P. Bose and P. Morin (Eds.): ISAAC 2002, LNCS 2518, pp. 14–24, 2002.

Muthukrishnan [10] proposed an optimal-time algorithm for the following problem:

Problem 1 (Document Listing Problem [10]). We are given a set of text documents $d_1, d_2, \ldots, d_k$, which may be preprocessed. The document listing query list(p) for a pattern p is to return the set of all documents in which p is present. That is, the output is $\{j | d_j[i..i + m - 1] = p \text{ for some } i\}$ where m is the length of p.

This is a basic version of the third query and can be solved in $O(m + q)$ time after $O(n)$ time preprocessing where q is the number of documents containing p. In contrast to the method based on the inverted files, it is impossible to perform the query for arbitrary words efficiently. Therefore Muthukrishnan's algorithm has the merits of both the suffix tree and the inverted file. Unfortunately the data structure is about twice as large as the suffix tree.

In this paper we consider problems related to the document listing problem. First we reduce the size of the data structure for the problem. Though Muthukrishnan's algorithm solved the document listing problem optimally, the size of its data structure is $O(n \log n)$ bits, which is much larger than that of the inverted file. Based on the ideas of Sadakane [12,13], we propose space-efficient data structures for the document listing problem whose size is proportional to the text size, that is, the size is asymptotically optimal. We also propose succinct data structures for the above three queries. We show that the *tf*idf* score can be efficiently computed for arbitrary pattern and documents.

The rest of the paper is organized as follows. In Section 2 we describe some basic data structures used in our method. In Section 3 we show a space-efficient data structure for range minimum queries which is used to solve the above problems. In Section 4 we propose a space-efficient data structure for the document listing problem. In Section 5 we propose succinct data structures to compute *tf*idf* scores for arbitrary patterns. Section 6 summarizes the results.

2   Preliminary Data Structures

2.1   Suffix Trees and Suffix Arrays

Let $T[1..n] = T[1]T[2] \cdots T[n]$ be a text of length n on an alphabet $\mathcal{A}$. The j-th suffix of T is defined as $T[j..n] = T[j]T[j + 1] \ldots T[n]$ and expressed by T_j. A substring $T[j..l]$ is called a prefix of T_j. The suffix array $SA[1..n]$ of T is an array of integers j that represent suffixes T_j. The integers are sorted in lexicographic order of the corresponding suffixes. The suffix tree of a string $T[1..n]$ is a compressed trie built on all suffixes of T. It has n leaves, each of which corresponds to a suffix $T_{SA[i]}$. Any pattern p in T is represented uniquely by a prefix of a path from the root node to a node v of T. Let $leaf(i)$ denote the leaf of the suffix tree that corresponds the suffix $T_{SA[i]}$. Let $lca(v, w)$ be the lowest common ancestor of nodes v and w. Both the suffix tree and the suffix array occupy $O(n \log n)$ bits, which are not linear to the text size $n \log |\mathcal{A}|$.

2.2   Generalized Suffix Trees

If we are given a set of k text documents $d_1, d_2, \ldots, d_k$, we concatenate them into a text T and construct the suffix tree for T. The tree is called *generalized suffix tree* and denoted by GST. To make any leaf have a unique label, we append a unique terminator for each documents, that is, we let $T = d_1\$_1 d_2\$_2 \cdots d_k\$_k$. Figure 1 shows an example of the generalized suffix tree.

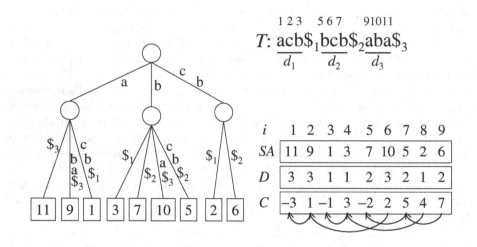

Fig. 1. The generalized suffix tree for "acb$\$_1bcb\$_2aba\$_3$" and data structures for document retrieval problems.

2.3   Succinct Data Structures

We use several basic data structures to reduce the size of the data structure for the document retrieval problems. A basic one is the succinct representation of a tree [8]. An n-node tree is represented by a nested parentheses sequence P of length $2n$. The sequence is defined from the tree as follows. We traverse the tree from the root in a depth-first manner. When we go down an edge we put an open parenthesis '(,' and when we go up an edge we put a close parenthesis ')' to P. Then a traversal on the tree can be simulated by a traversal on the sequence. An example of the sequence is depicted in Fig. 2.

To make the traversal quick, we use auxiliary data structures that support the following functions. The function $rank_p(P, i)$ returns the number of occurrences of pattern p up to the position i in a string P, where p is for example '().' The function $select_p(P, i)$ returns the position of i-th occurrence of pattern p. Both functions take constant time using auxiliary data structures of size $o(n)$ bits [9].

The (generalized) suffix tree and the suffix array can be compressed. We use the compressed suffix array [6]. The suffix array is compressed from $n \log_2 n$ bits to $O(\epsilon^{-1} n |\mathcal{A}|)$ bits where ϵ is a constant between 0 and 1. Therefore the size of the compressed suffix array is proportional to the text size. Each element $SA[i]$ is extracted in $O(\log^\epsilon n)$ time. The size is further reduced to $O(nH_k)$ bits where H_k is the order-k entropy of the text [4,5]. We denote the size of the compressed suffix array by $|CSA|$. By using the compressed suffix array we can compute not only $SA[i]$ but also $SA^{-1}[i]$ in $O(\log^\epsilon n)$ time [12]. The interval $[l, r]$ of the suffix array that corresponds to a pattern p, that is, $T[SA[i]..SA[i] + |p| - 1] = p$ for any $i \in [l, r]$, can be computed in $O(|p|)$ time [4,13].

We also use a succinct data structure for computing *lowest common ancestor* (*lca*) between two leaves of a tree [13], which is based on the algorithm of Bender and Farach-Colton [2]. In the paper an *lca* query is reduced to a range minimum query on an integer array where the difference between two adjacent elements is always 1 or -1. The size of the data structure is $2n + o(n)$ bits and each *lca* query is done in constant time. In this paper we propose a data structure for range minumum queries on arbitrary arrays.

3   Succinct Data Structures for Range Minimum Query

In this section we propose succinct data structures for range minimum queries on arbitrary arrays, which will be used in the proposed algorithms.

Problem 2 (Range Minimum Query). For indices l and r between 1 and n of an array C, the *range minimum query* $\mathrm{RMQ}_C(l, r)$ returns the index of the smallest element in the subarray $C[l..r]$. If there is a tie-breaking we choose the leftmost one.

It is known that a query can be done in constant time using $O(n \log n)$ bits space after $O(n)$ time preprocessing [2]. We reduce the size into $4n + o(n)$ bits. Note that we do not store the array C itself. Therefore we can find only the index i of the minimum element $C[i]$.

The range minimum query is reduced to finding the *lca* between two nodes in a tree [2]. We construct a binary tree and store pairs $(i, C[i])$ in it. The root node of the tree stores the minimum value $C[x]$ in $C[1..n]$ and its index x. The left subtree stores $(i, C[i])$ for $1 \le i \le x - 1$ and the right subtree stores those for $x + 1 \le i \le n$ recursively. The tree is constructed in $O(n)$ time. Then $\mathrm{RMQ}_C(l, r)$ is equal to the index stored in the *lca* between two nodes storing $C[l]$ and $C[r]$.

We use a similar tree to store $C[i]$. Before compressing the data structure we temporally construct a tree M with n internal nodes and n leaves, and store pairs $(i, C[i])$ in it. Each internal node labeled i has exactly one leaf that stores $C[i]$ as a middle child, and has at most two child nodes corresponding to left and right children. In the tree of Bender and Farach-Colton $C[i]$ are stored in internal nodes of the tree, while in our tree they are stored in leaves. The reason is that we do not want to store neither the value $C[i]$ nor the index i explicitly. A node of the tree may not have the left or the right child. Then if we store $C[i]$ in internal nodes and encode the tree to the parentheses sequence, we cannot distinguish between a node having only left child and that having only right child. To solve the problem Munro and Raman [8] use an isomorphism between a binary tree and an ordered tree. In a binary tree we distinguish the left child and the right child of a node, whereas in an ordered tree we do not. Because we cannot distinguish them in the parentheses sequence, the binary tree is converted into an ordered tree. The left child v of a node u in a binary tree M corresponds to the leftmost child of u in an ordered tree M', and the right child w of u in M corresponds to a sibling node of u in M'. Although the tree M' is encoded as a parentheses sequence in $2n$ bits, which is smaller than our method using $4n$ bits, we do not use their original method as it is because we cannot compute the lca of nodes in the sequence.

In this paper we use another representation of a binary tree. We add a leaf to each node that stores $C[i]$ between the left and the right children. Then we can distinguish them and compute lca between nodes. Furthermore, we can compute the inorder of each leaf from the sequence because for each leaf its preorder and inorder coincide.

Sadakane [13] showed that the data structure for lca queries is represented in $2n + o(n)$ bits. In our case the tree M is represented in $4n + o(n)$ bits because it has n internal nodes and n leaves. A node is represented by the position of an open parenthesis in P.

To solve a range minimum query $\mathrm{RMQ}_C(l, r)$ it is necessary to convert an index i to the element $C[i]$ into the position e of the open parenthesis in P that corresponds to a leaf whose parent is labeled i. Because any leaf has no child, it is represented by () in P. Moreover, because leaves appear in P in the order of depth-first search, the leaves are sorted in the order of i. Therefore e and i are converted to each other as follows:

$$e = select_{()}(P, i)$$
$$i = rank_{()}(P, e).$$

To find the lca between two nodes in a parentheses sequence, we consider an imaginary integer array P'. We define $P'[i] = rank_((P, i) - rank_)(P, i) - 1$. In other words, $P'[i]$ is the depth of a node labeled i. Because the array P' is represented by P and an auxiliary data structure of size $o(n)$ bits so that each element can be extracted in constant time, we do not store P' explicitly. Then an lca query is reduced to range minimum query on P'.

The index i of the minimum element $C[i]$ in $C[l..r]$ can be found in constant time as follows:

1. $x = select_{()}(P, l)$, $y = select_{()}(P, r)$
2. $z = \mathrm{RMQ}_{P'}(x, y)$
3. if $P[z+2] = {')'}$ then $f = z + 1$ else $f = z - 1$
4. $i = rank_{()}(P, f)$.

Step 3 finds the position f of the open parenthesis that represents the middle leaf. $P[z+1]$ is the open parenthesis of a child. If it is a leaf $P[z+2]$ is the close

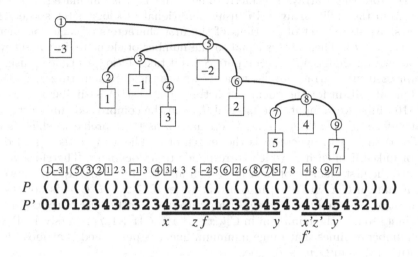

Fig. 2. A tree representation of the array C and its parentheses encoding.

parenthesis, otherwise $P[z+2]$ is the open parenthesis of its child. In the latter case $P[z]$ is the close parenthesis and $P[z-1]$ is the open parenthesis of the leaf.

For example, to find the minimum in $C[4..7]$ we first compute positions x and y of '()' in P corresponding to $C[4]$ and $C[7]$ (see Fig. 2). Then we compute the position z of the minimum element in $P'[x..y]$. In this case $P[z]$ represents the close parenthesis of the left child of $C[5]$ and $P[z+1]$ represents the open parenthesis of the middle child. Therefore the index i of the minimum element $C[i]$ is equal to the number of '()' in $P[1..z-1]$. In another example, to find the minimum in $C[8..9]$, we find the position z' of the minimum element in $P'[x'..y']$. Then $P[z']$ is the close parenthesis of the middle child of $C[8]$.

We have the following:

Theorem 1. *For an array C of n elements, a range minimum query is done in constant time using a data structure of size $4n + o(n)$ bits after $O(n)$ time preprocessing.*

4 Succinct Data Structures for the Document Listing Problem

As in the original paper on the document listing problem [10] we define two integer arrays C and D. We define $D[i] = c$ if the suffix $T_{SA[i]}$ is contained in document d_c, and define $C[i] = j$ where j is the largest index such that $j < i$ and $D[j] = D[i]$. We define $C[i] = -D[i]$ if such j does not exist[1].

---

[1] In the original paper $C[i] = -1$ in this case.

We store these arrays succinctly. The value $D[i]$ is calculated in constant time from the suffix array $SA[i]$ using additional $O(k \log \frac{n}{k})$ bit space [11] as follows. We store a set of positions of the first characters of each document d_i ($i = 1, 2, \ldots, k$). Then $D[i]$ is equal to the number of elements in the set which are no greater than $SA[i]$. To compute $SA[i]$ it takes $O(\log^\epsilon n)$ time by using the compressed suffix array. For the array C we use the data structure in Section 3.

Our algorithm for the document listing query list(p) is similar to the original [10]. First we compute the interval $[l, r]$ on the compressed suffix array such that suffixes $T_{SA[l]}, T_{SA[l+1]}, \ldots, T_{SA[r]}$ have p as their prefixes, which is done in $O(m)$ time [4,13] where m is the length of p. The array C is regarded as a set of linked lists each of which corresponds to a document. Therefore we enumerate the first element of each list which appears in $C[l..r]$. To do so, we find the minimum $C[x]$ in $C[l..r]$ in constant time. If $C[x] \geq l$, $C[x]$ is not the first element of a list and thus the algorithm terminates, otherwise outputs $D[x]$ and continues to find the minimum in $C[l..x-1]$ and $C[x+1..r]$ recursively. Because the number of times that range minimum query is performed is at most $2q$, the algorithm runs in $O(m + q)$ time.

In the original algorithm the array C is used to avoid outputting a duplicate document id. However we cannot use the same algorithm because we do not store the exact values of C. Instead we use a simple marking algorithm. To check the duplication of the output, we use a bit-vector $V[1..k]$. In the preprocess we set $V[i] = 0$ for $i = 1, 2, \ldots, k$, which takes $O(k) = O(n)$ time. In a query list(p) we check whether $V[x] = 1$ or 0 before outputting $D[x]$. If $V[x] = 0$, we output $D[x]$ and set $V[x] = 1$. $D[x]$ is computed in constant time from $SA[x]$ using $O(k \log \frac{n}{k})$ bits space [11]. After outputting all document id's, we set $V[x] = 0$ for each x which was output. Therefore the query is done in $O(m + q \log^\epsilon n)$ time where q is the output size. Note that this algorithm requires temporary $q \log k$ bits space to store the set of document id's which are output.

Let us compute the size of the data structure. The array D is represented by the compressed suffix array and an auxiliary data structure of size $O(k \log \frac{n}{k})$ bits. The array C is represented in $4n + o(n)$ bits. The vector V has size k bits. Therefore the total is $|CSA| + 4n + o(n) + O(k \log \frac{n}{k})$ bits.

Theorem 2. *The* list(p) *query is done in* $O(m + q \log^\epsilon n)$ *time using a data structure of size* $|CSA| + 4n + o(n) + O(k \log \frac{n}{k})$ *bits created by* $O(n)$ *time preprocessing where* q *is the size of output.*

5 How to Compute *tf**idf Scores for Arbitrary Patterns

Many text information retrieval systems use the *tf**idf ranking method, where *tf* means *term frequency* and *idf* means *inverse document frequency* [14]. A score of a document d according to a set of keywords is calculated from *tf**idf values. The term frequency $tf(p, d)$ is the number of occurrences of a keyword p in the document d, the document frequency $df(p)$ is defined by the number of documents which contain p, and the inverse document frequency is defined as $idf(p) = \log \frac{k}{df(p)}$ where k is the number of documents in a database. The score of a document for a set of given keywords p is defined as $\sum_p tf(p,d) \cdot idf(p)$, and ranks of documents are defined in decreasing order of the score. The inverse document frequency is used to decrease the weight of keywords which appear in many documents because such keywords are not important.

The tf and idf scores are stored in the inverted file in addition to document id's. For each keyword p appearing in a set of documents, the $idf(p)$ score and pairs $(d, tf(p,d))$ for all documents d that contain p are stored. In this paper we consider the problem of computing those scores for arbitrary patterns in nearly optimal time and space.

5.1   Data Structures for Computing $tf(p,d)$

We construct the compressed suffix array CSA for all documents and the compressed suffix array CSA_d for each document d. Then the term frequency $tf(p,d)$ is obviously computed in $O(|p|)$ time by using CSA_d as follows. Let $[l,r]$ be the interval that corresponds to p in SA_d. Then $tf(p,d) = r - l + 1$.

In an information retrieval system it is often necessary to compute all $tf(p,d)$ scores for all q documents that contain the pattern p. It takes $O(|p|q)$ time to compute them by using the above naive algorithm for each document. We show an algorithm to compute them in $O(|p| + q\log^\epsilon n)$ time.

We first find the interval $[l,r]$ in CSA in $O(|p|)$ time. Then we find the index i corresponding to the first occurrence of $D[i]$ in $[l,r]$ for each document $d = D[i]$. We also find the index j corresponding to the last occurrence of $D[j]$. To find j we use another data structure which is similar to C. The array C is regarded as a set of linked lists for each document. We define another array C' that represents linked lists in the opposite direction. We define $C'[i] = j$ where j is the smallest index such that $i < j$ and $D[j] = D[i]$. We define $C'[i] = n + D[i]$ if such j does not exist. Then we can enumerate indices j of the last occurrences of $D[j]$ in $[l,r]$ by using range maximum query to C'. The size of C' is at most $4(n+k) + o(n)$ bits.

To compute the number of occurrences of d in $D[l..r]$ we use the compressed suffix array CSA_d of the document d. Let i and j be the positions of the first and the last occurrences of d in $D[l..r]$ computed by using C and C'. We compute $x = SA[i]$ and $y = SA[j]$ using CSA. Because x and y are indices in T we convert them into those in document d, say x' and y'. Then we compute $i' = SA_d^{-1}[x']$ and $j' = SA_d^{-1}[y']$ in $O(\log^\epsilon n)$ time by using CSA_d. Because the lexicographic order of suffixes in document d does not change in either CSA_d or CSA, we have $tf(p,d) = j' - i' + 1$.

The size of the data structure becomes as follows. The compressed suffix array for T is denoted by $|CSA|$. The total of the size of the compressed suffix array for each document is roughly equal to $|CSA|$. The array C and C' has size $4n + o(n)$ bits and $4(n+k) + o(n)$ bits respectively. Therefore the total is $2|CSA| + 8n + 4k + o(n)$ bits.

Note that the order of the output of the range minimum query and that of the range *maximum* query will be different. Therefore we need to sort the document id's. We use an $O(q\log\log q)$ time sorting algorithm of Andersson et al. [1]. Therefore we have the following:

Theorem 3. *The term frequency $tf(p, d)$ is computed in $O(m)$ time, and term frequencies for all q documents d containing a pattern p is computed in $O(m + q \log^\epsilon n)$ time using a data structure of size $2|CSA| + 8n + 4k + o(n)$ bits.*

5.2   Data Structures for Computing $idf(p)$

The inverse document frequency $idf(p)$ is also computed in $O(|p|)$ time using CSA and a data structure of size at most $2n + o(n)$ bits. We use Hui's algorithm [7] and modify its data structure to reduce the size.

Previous Method: In each internal node v of GST the original algorithm stores a number $u(v)$ which represents the number of times that two leaves under the node v come from the same document. More precisely, let $n_d(v)$ be the number of leaves from document d in the subtree rooted at v. Then $u(v) = \sum_{d:n_d(v)>0}(n_d(v)-1)$. Let l and r be the indices of the leftmost and the rightmost leaves in the subtree rooted at v. To compute $u(v)$ we first compute $h(w)$ for all nodes w in the subtree where $h(w)$ is the number of times that $lca(leaf(i), leaf(i+1)) = w$ for some $i \in [l, r]$. Then $u(v)$ is equal to the summation of all $h(w)$ in the subtree. Because each lca query takes constant time after $O(n)$ time processing, $h(w)$'s for all nodes can be computed in $O(n)$ time. Then the values of $u(v)$ for all nodes are computed in linear time by a bottom-up traversal of GST. Finally we have $df(p) = (r - l + 1) - u(v)$ and $idf(p) = \log \frac{k}{df(p)}$.

New Data Structure: The above data structure has size $O(n \log n)$ bits. We reduce the size to $2n + o(n)$ bits. We temporary construct a GST' of T in which all internal nodes have two children, that is, any internal node of GST which have $c > 2$ children is divided into $c - 1$ nodes each of which has two children (see Fig. 3). Then we compute $h(w)$ for each node in GST' in linear time.

Instead of storing $u(v)$ we store $h(v)$ in an array $H[1..n - 1]$ in which the values are arranged in *inorder* of nodes. Let i be the *inorder* of a node v. Then $v = lca(leaf(i), leaf(i + 1))$ holds and we store $h(v)$ in $H[i]$.

The value $u(v)$ is equal to the summation of $h(w)$ for all descendants of v. Because $h(w)$'s are stored in *inorder*, $u(v)$ is equal to the summation of $H[l..r-1]$ where l and r are the lexicographic order of the leftmost and the rightmost leaves of v. If v is the node corresponding to a pattern p, the indices l and r are calculated in $O(m)$ time by using GST or the compressed suffix array of T.

The value $h(v)$ is encoded as unary code in a sequence H', that is, encoded as $h(v)$ zeroes followed by a one, for example 0 is encoded as 1 and 2 as 001. Then $u(v)$ is computed as follows:

$$x = select_1(H', l - 1) + 1$$
$$y = select_1(H', r)$$
$$u(v) = rank_0(H', y) - rank_0(H', x)$$

where l and r are indices defined above. Therefore $u(v)$ is computed in constant time. The size of the bit-vector is at most $2n - d$ bits because there are n ones and at most $n - d$ zeroes.

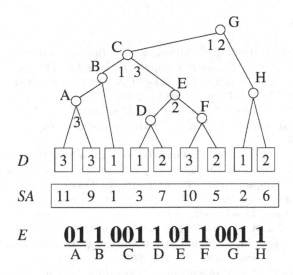

Fig. 3. Data structure for computing $idf(p)$.

Theorem 4. *Given the interval of the suffix array of T that corresponds to a pattern p, the inverse document frequency $idf(p)$ can be computed in constant time using a data structure of size $2n + o(n)$ bits.*

6   Concluding Remarks

We have extended the data structure for the document listing problem so that it can be used to compute *tf*idf* scores. The size of the data structure is proportional to the text size, which is an improvement from the previous algorithm using $O(n \log n)$ bit space. The size of the data structure is $2|CSA| + 10n + O(k \log \frac{n}{k})$ bits where $|CSA|$ is the size of the compressed suffix array for a collection of documents, and it can be smaller than the size of the documents. Therefore the size of the whole data structure is about three times as large as previous methods. The time complexities are at most $O(\log^\epsilon n)$ times of that by using inverted files, while our data structures support queries for any pattern.

Acknowledgment. The author would like to thank Prof. Takeshi Tokuyama of Tohoku University for his valuable comments. The work of the author was supported in part by the Grant-in-Aid of the Ministry of Education, Science, Sports and Culture of Japan.

References

1. A. Andersson, T. Hagerup, S. Nilsson, and R. Raman. Sorting in Linear Time? In *ACM Symposium on Theory of Computing*, pages 427–436, 1995.
2. M. Bender and M. Farach-Colton. The LCA Problem Revisited. In *Proceedings of LATIN2000*, LNCS 1776, pages 88–94, 2000.
3. A. Blumer, J. Blumer, D. Haussler, R. McConnell, and A. Ehrenfeucht. Complete inverted files for efficient text retrieval and analysis. *Journal of the ACM*, 34(3):578–595, 1987.
4. P. Ferragina and G. Manzini. Opportunistic Data Structures with Applications. In *41st IEEE Symp. on Foundations of Computer Science*, pages 390–398, 2000.
5. R. Grossi, A. Gupta, and J. S. Vitter. Higher Order Entropy Analysis of Compressed Suffix Arrays. In *DIMACS Workshop on Data Compression in Networks and Applications*, March 2002.
6. R. Grossi and J. S. Vitter. Compressed Suffix Arrays and Suffix Trees with Applications to Text Indexing and String Matching. In *32nd ACM Symposium on Theory of Computing*, pages 397–406, 2000.
7. L. Hui. Color Set Size Problem with Applications to String Matching. In *Proc. of the 3rd Annual Symposium on Combinatorial Pattern Matching (CPM'92)*, LNCS 644, pages 227–240, 1992.
8. J. I. Munro and V. Raman. Succinct Representation of Balanced Parentheses and Static Trees. *SIAM Journal on Computing*, 31(3):762–776, 2001.
9. J. I. Munro, V. Raman, and S. Srinivasa Rao. Space Efficient Suffix Trees. *Journal of Algorithms*, 39(2):205–222, May 2001.
10. S. Muthukrishnan. Efficient Algorithms for Document Retrieval Problems. In *Proc. ACM-SIAM SODA*, pages 657–666, 2002.
11. R. Raman, V. Raman, and S. Srinivasa Rao. Succinct Indexable Dictionaries with Applications to Encoding k-aray Trees and Multisets. In *Proc. ACM-SIAM SODA*, pages 233–242, 2002.
12. K. Sadakane. Compressed Text Databases with Efficient Query Algorithms based on the Compressed Suffix Array. In *Proceedings of ISAAC'00*, number 1969 in LNCS, pages 410–421, 2000.
13. K. Sadakane. Succinct Representations of *lcp* Information and Improvements in the Compressed Suffix Arrays. In *Proc. ACM-SIAM SODA 2002*, pages 225–232, 2002.
14. G. Salton, A. Wong, and C. S. Yang. A Vector Space Model for Automatic Indexing. *Communications of the ACM*, 18(11):613–620, 1975.
15. P. Weiner. Linear Pattern Matching Algorihms. In *Proceedings of the 14th IEEE Symposium on Switching and Automata Theory*, pages 1–11, 1973.

Key Independent Optimality

John Iacono

Department of Computer and Information Science
Polytechnic University
5 MetroTech Center, Brooklyn NY 11201, USA
jiacono@poly.edu

Abstract. A new form of optimality for comparison based static dictionaries is introduced. This type of optimality, key-independent optimality, is motivated by applications that assign key values randomly. It is shown that any data structure that is key-independently optimal is expected to execute any access sequence where the key values are assigned arbitrarily to unordered data as fast as any offline binary search tree algorithm, within a multiplicative constant. Asymptotically tight upper and lower bounds are presented for key-independent optimality. Splay trees are shown to be key-independently optimal.

1 Introduction

The problem of executing a sequence of operations on a binary search tree is one of the oldest and most fundamental problems in computer science. In this paper examines the simplest form of this problem, where there is a sequence A of m accesses (successful searches) to a tree containing a static set of n items. For simplicity, neither insertions nor deletions are considered, and it is assumed the access sequence is not short (at least $n \log n$). The binary search trees are of the standard type, and allow unit cost edge traversals and rotations.

There has been a long history both of structures for this problem, and of ways of bounding the runtime of a structure as a function of the access sequence. A brief history of runtime characterizations will be presented, along with a description of associated structures. This history will motivate the main result.

First, a few notes on notation and terminology. As previously stated, we are only interested in standard binary search trees that store n items and execute an access sequence A containing m items $a_1 \ldots a_m$. The access sequence A will often be an implicit parameter of the functions that we will define. When we need to make this explicit, a subscript is used. Since insertions and deletions are not under consideration in this work, we assume without loss of generality, that the data stored is simply the integers from 1 to n. By $\log x$ we mean $\max(1, \log_2 x)$. A standard binary search tree structure is a binary search tree storing n items subject to the following conditions. A pointer to one node in the structure is maintained. At unit cost the pointer may be moved to the parent, left child, or right child of the current node. Also at unit cost, a single left or right rotation may be performed. We say that a binary search structure executes an access sequence

P. Bose and P. Morin (Eds.): ISAAC 2002, LNCS 2518, pp. 25–31, 2002.

A if after executing all of the pointer movements and rotations required by the structure, A is a subsequence of the sequence formed by data in all of the nodes touched.

Property 1 *A binary search tree structure has the $O(\log n)$ runtime property if it executes A in time $O(m \log n)$.*

The $O(\log n)$ runtime property is expected to be optimal if the accesses in A are independently drawn at random from a uniform distribution. It is also known from [11] that there exist some access sequences which no binary search structure can execute in time $o(m \log n)$. All data structures discussed in this paper have the $O(\log n)$ runtime property. The simplest such structure would be a static balanced binary tree.

Property 2 *A data structure has the static finger property at finger f if the time to execute A is $O(\sum_{i=1}^{m} \log |f - a_i|)$.*

There exist specialized structures [5] that given f, have the static finger property at f. Splay trees [10] have the static finger property for any given f, without being explicitly initialized with f.

Property 3 *Let $f(x)$ be the number of accesses to x in A. A data structure has the static optimality property if it executes A in time $O(\sum_{i=1}^{n} f(i) \log \frac{m}{f(i)})$.*

Optimal search trees [9] have the static optimality property, however, their construction requires that the access frequencies $\frac{m}{f(i)}$ be known. Splay trees have the static optimality property [10] and do not require any distribution-specific parametrization. Search trees with the static optimality property are expected to be optimal if A consists of a sequence of independent accesses drawn from a fixed (static) distribution. "Optimal" search trees and other structures with the static optimality property are not necessarily optimal for deterministic access sequences, or randomly generated access sequences where one access is dependent on the next. It has been shown that any data structure with the static optimality property also has the static finger property, for any choice of f [8].

Property 4 *Let $l(i, x)$ be j if a_j is the index of the last access to x in the subsequence $a_1, a_2, \ldots a_{i-1}$. If x is not accessed in $a_1, a_2, \ldots a_{i-1}$, we define $l(i, x)$ to be 1. Let $w(i, x)$ be the number of distinct items accessed in the subsequence $a_{l_i(x)+1}, a_{l_i(x)+2} \ldots a_i$. We use $w(a_i)$ to denote $w(i, a_i)$. We say a data structure has the working set property if it can execute A in time $O(\sum_{i=1}^{m} \log w(a_i))$*

Splay trees have the working set property [10]. It has been shown that the working set property implies the static optimality property [6], and thus also the static finger property. A data structure that is not a tree was introduced in [7] that has worst-case $O(\log w(a_i))$ runtime per access. As it was also shown that similar worst-case results are impossible to obtain for the static finger property and the static optimality property [8], the working-set property can be viewed as the most natural of the three previous properties.

Property 5 *A data structure with the dynamic finger property executes A in time* $O(\sum_{i=2}^{m} \log |a_{i-1} - a_i|)$

Level-linked trees of Brown and Tarjan [1] were shown to have the dynamic finger property. However, they do not meet our definition of a standard binary search tree structure, because there are additional pointers that the trees must be augmented with. However, it was shown that splay trees have the dynamic finger property [3,2]. The dynamic finger property implies the static finger property.

Neither the dynamic finger property nor the working set property imply each other, and these are known to be the two best analyses of splay trees, and search trees in general. We now leave the proven properties of binary search trees and continue on to conjectured properties.

Property 6 *A data structure with the unified property executes an access sequence A in time* $O(\sum_{i=1}^{m} \min_j \log(w(i,j) + |a_i - j|))$.

The unified property [7] is an attempt to unify the dynamic finger property and the working set property. It implies the dynamic finger property when $j = a_{i-1}$ and the working set property when $j = a_i$. Splay trees are conjectured to have the unified property, but so far the only structure with the unified property is not a binary search tree. It has been known that the unified property is not a tight bound on the runtime of splay trees [4], as access sequences exist that splay trees execute faster than the unified property would indicate.

Property 7 *We define OPT(A) to be the fastest any binary search tree structure can execute access sequence A. We allow the binary search structure to choose the initial tree. A data structure has the dynamic optimality property if it executes A in time* $O(OPT(A))$

The famous dynamic optimality conjecture [10] simply states that splay trees have the dynamic optimality property. That is, it is conjectured that no binary search tree, even offline and with infinite preprocessing, can execute *any* A by more than a multiplicative constant faster than splay trees. Note that the choice of an initial tree can not asymptotically change $O(OPT(A))$ for $m = \Omega(n)$.

So, in summary, the best binary search tree structure appears to be the splay tree. The splay tree's runtime has not yet been tightly analyzed, and there are currently two "best" bounds on how fast a splay tree executes A, neither of which implies the other. Even the conjectured explicit bound of the unified property is not equal to the dynamic optimality conjecture. So far the only proven forms of optimality for search trees are quite old, and require independent accesses from static distributions to generate entropy-based lower bounds. These lower bounds apply not specifically to trees but to any comparison based structure. On the other hand, dynamic optimality, while powerful remains but a conjecture. The purpose of this paper is to propose and *prove* a new form of optimality for binary search trees.

Property 8 *We define* $KIOPT(A) = E(OPT(b(A)))$, *where b is a random bijection from n to n and* $b(A) = b(a_1), b(a_2) \ldots b(a_m)$. *A data structure is key independently optimal if it executes A in time* $O(KIOPT(A))$.

In this definition we introduce a new form of optimality for binary search trees, *key-independent optimality*. Our definition is motivated by a particular type of application, where there is no correlation between the rank distance between two key values and their relative likelihood of a proximate access. This could occur for several reasons.

One reason could be because the key values were assigned randomly to otherwise unordered data. We will prove (starting in the next paragraph) that if this is the case, no data structure is expected to execute any *fixed* access sequence on unordered data faster than $KIOPT(A)$ after the key values have been assigned. It is important to note that other than picking the total ordering of the key values, the access sequence is fixed. Unlike previous optimality results, there is no requirement that the access sequence be created by independent drawings from a distribution. One can think of many such applications where key values are generated arbitrarily.

We now formally prove the claim of the previous paragraph. Let U be a set of n unordered elements. Let $A' = a'_1, a'_2 \ldots a'_m$ be a fixed sequence of m elements of U, representing an access sequence. Let b' be a random bijection that maps the elements U to the totally ordered set $X = 1, 2 \ldots n$. Let $b'(A') = b'(a'_1), b'(a'_2), \ldots b'(a'_m)$. Thus $b'(A')$ represents the accesses to U after a random ordering has been imposed on the elements, and $OPT(b'(A'))$ is the fastest that such a sequence can be executed on a binary search tree.

Lemma 1. $E(OPT(b'(A'))) = KIOPT(b'(A'))$

By the definition of $KIOPT$, $KIOPT(b'(A')) = E(OPT(b(b'(A))))$. Observing that both b and b' were defined to be random bijections, and noting that composing two random bijections yields a random bijection completes the proof of the lemma.

Now having shown that key-independent optimality is a useful notion, we now present the positive result that we are able to tightly bound $KIOPT(A)$.

Theorem 1 *The working set property and key-independent optimality are asymptotically the same.* $KIOPT(A) = \Theta(\sum_{i=1}^{m} \log w(a_i))$

The proof of this theorem is the subject of Section 2.

Corollary 1 *Splay trees are key-independently optimal*

This is immediate, since splay trees have the working set property. If the key values in a given application are assigned randomly, then no binary search tree can be expected to execute A more than a constant multiplicative factor faster than splay trees, even with complete foreknowledge of the access sequence, and infinite preprocessing time.

2   Proof of Main Result

Theorem 1 *The working set property and key-independent optimality are asymptotically the same.* $KIOPT(A) = \Theta(\sum_{i=1}^{m} \log w(a_i))$.

Proof: We proceed with the upper and lower bounds separately, and state each as a lemma.

Lemma 2. $KIOPT(A) = O(\sum_{i=1}^{m} \log w(a_i))$

As previously stated, splay trees were shown in Sleator and Tarjan [10] to have the working set property. Thus splay trees execute A in time $O(\sum_{i=1}^{m} \log w(a_i))$. Since the definition of $w(x)$ does not take into account the key values at all, $\sum_{i=1}^{m} \log w_i(a_i) = \sum_{i=1}^{m} \log w(b(a_i))$ for every bijection b, including, of course, a randomly chosen one. Thus since splay trees are a binary search tree structure that will execute $b(A)$ in time $O(\sum_{i=1}^{m} \log w(a_i))$, this is an upper bound on $KIOPT(A)$.

Lemma 3. $KIOPT(A) = \Omega(\sum_{i=1}^{m} \log w(a_i))$

Our lower bound is based upon the second lower bound of Wilbur found in [11]. Given a sequence A and $j < i$ we define the untouched region at j of i in A, $u_A(i, j)$ to be the largest interval (x, y) that contains a_j and that no element of (x, y) is in $a_j \ldots a_{i-1}$. We say two regions (x, y) and (x', y') are completely different if $x \neq x'$ and $y \neq y'$. Let $v_A(i)$ be the number of completely different regions $u_A(i, j)$ for all $j < i$. Wilbur's lower bound can be stated as follows:

Theorem 2 *The time to execute A on a binary search tree is $\Omega(\sum_{i=1}^{m} v_A(i))$.*

We make the following two observations: If $j < k$ and $a_j = a_k$ then $u_A(i, j) = u_A(i, j+1)$. If $a_j = a_i$ then $u_A(i, k) = \emptyset$ for all $k \leq j$. We do not count $\emptyset$ in the formulation of $v_A(i)$.

Look at the access sequence $A = a_1, a_2, \ldots a_m$. We now remove some items according two the following two rules: First, if for some $j < i$, $a_j = a_i$, remove $a_1 \ldots a_j$. Secondly, if for some $j < k < i$, if $a_j = a_k$, remove a_j. We refer to this new sequence as the compressed sequence at i, $C_A(i) = c_1, c_2, \ldots c_{|C_A(i)|}$. According to the two observations of the previous paragraph we see that

Fact 1 $v_A(i) = v_{C_A(i)}(|C_A(i)|)$

Next we observe that $C_A(i)$ contains exactly the distinct items accessed since the last access to a_i plus a_i itself. Thus from the definition of w we observe that

Fact 2 $|C_A(i)| = w_A(c_i)$

We now use these facts to make progress in the proof of Theorem 1, starting from Wilbur's bound. We define $R(i)$ to be a random sequence of i distinct values.

$OPT(A) = \Omega(\sum_{i=1}^{m} v_A(i))$ This is Wilbur's bound

$KIOPT(A) = \Omega(E(\sum_{i=1}^{m} v_{b(A)}(b(i))))$ By the definition of key independent optimality

$KIOPT(A) = \Omega(E(\sum_{i=1}^{m} v_{C_{b(A)}(b(i))}(|C_{b(A)}(b(i))|))$ This follows from fact 1.

$KIOPT(A) = \Omega(E(\sum_{i=1}^{m} v_{C_{b(A)}(b(i))}(w_{b(A)}(b(i)))))$ This follows from fact 2.

$KIOPT(A) = \Omega(E(\sum_{i=1}^{m} v_{C_{b(A)}(b(i))}(w(i)))$ Since w is unaffected by a bijection.

$KIOPT(A) = \Omega(E(\sum_{i=1}^{m} v_{R(w(i))}(w(i)))$ Since a sequence of distinct numbers passed through a random bijection is just a random sequence of distinct numbers.

Now assume $E(v_{R(i)}(i)) = \Omega(\log i)$ (this will be proven as a lemma below). We can then conclude

$KIOPT(A) = \Omega(E(\sum_{i=1}^{m} \log w(i)))$

This is Theorem 1. Thus to prove Theorem 1, all that remains is the following Lemma:

Lemma 4. *Given a sequence of randomly distributed distinct numbers $r, r_2, ... r_i$,*
$E(v_{R(i)}(i)) = \Omega(\log i)$

We will refer to r_i as $p(R)$, the pivot. Let $S(R)$ be the set of indicies of items smaller than the pivot, $\{i | r_i < p(R)\}$. Let $L(R)$ be the set of indicies greater than the pivot, $\{i | r_i > p(R)\}$. Note that $P(R), L(R) and S(R)$ partition R.

Let the small decreasing sequence of R, $SD(R)$ be $\{i | \forall_{j>i} r_j > r_i\}$. Let the large increasing sequence of R, $LI(R)$ be $\{i | \forall_{j>i}, r_j < r_i\}$. Observe that $SD(R) \subseteq S(R)$ and $LI(R) \subseteq L(R)$.

There is a correlation between $SD(R)$, $LI(R)$ and the notion of an untouched region. given a set S, define $N(S, i)$ to be the smallest $j \in S$ such that $j \geq i$. the untouched region $u_R(|R|, i)$ is $(r_{N(SD(R),i)}, r_{N(LI(R),i)})$. Thus the number of different untouched regions is at least $|SD(R)| + |LI(R)|$. (The "at least" is because there may be one or two different untouched regions at values of i where $N(SD(R), i)$ or $N(LI(R), i)$ is undefined.) Thus we have established that the untouched regions change for values of i in $SD(R)$ and $LI(R)$, each time yielding a different untouched region. However, what we are interested in is not the number of different untouched regions, but the number of totally different untouched regions. This corresponds to how the sets $SD(R)$ and $LI(R)$ are interleaved.

We define the interleave of two sets containing distinct totally ordered items, $I(A, B)$ to be the number of items in $A \cup B$ where the item and its successor come from different sets. We observe that $I(SD(R), LI(R)) = v_R(|R|)$. Thus what we wish to show is that $E(I(SD(R), LI(R))) = \Omega(\log |R|)$.

With probability at least $1/2$, $|S(R)|$ and $|L(R)|$ are at least $|R|/4$. Given $|S(R)| > |R|/4$, we know $E(SD(R)) \geq \log \frac{|R|}{4}$. Given two random sequences A and B of length n, $I(A, B)$ is a binomial distribution with mean n. We then thus conclude that $E(I(SD(R), LI(R))) = \Omega(\log |R|)$.

3   Further Work

As Wilbur's bound seems quite powerful, it would be of interest to see whether or not it is equivalent to the dynamic optimality property. If one were to try to prove non-equivalence, the simplest way would be to come up with a sequence A that splay trees execute faster than Wilbur's lower bound on this sequence.

References

1. M.R. Brown and R.E. Tarjan. Design and analysis of a data structure for representing sorted lists. *SIAM J. Comput.*,9:594–614, 1980.
2. R.Cole. On the dynamic finger conjecture for splay trees. part II: The proof. *SIAM J. Comp.*, 30(1):44–85,2000.
3. R. Cole, B. Mishra, J. Schmidt, and A. Siegel.On the dynamic finger conjecture for splay trees. part I: Splay sorting log n-block sequences.*SIAM J. Comp.*, 30(1):1–43, 2000.
4. M.L. Fredman, 2001. Private Communication.
5. L.J. Guibas, E.M. McCreight, M.F. Plass, and J.R. Roberts. A new representation for linear lists. In *Proc. 9th Ann. ACM Symp. on Theory of Computing*, pages 49–60, 1977.
6. J. Iacono. New upper bounds for pairing heaps. In *Scandinavian Workshop on Algorithm Theory (LNCS 1851)*, pages 32–45, 2000.
7. J. Iacono.Alternatives to splay trees with $o(\log n)$worst-case access times. In *Symposium on Discrete Algorithms*, pages 516–522, 2001.
8. J. Iacono. *Distribution Sensitive Data Structures*. PhD thesis, Rutgers, The State University of New Jersey, Graduate School, New Brunswick, 2001
9. D.E. Knuth Optimum binary search trees. *Acta inf.*, 1:14–25, 1971
10. D.D. Sleator and R.E. Tarjan . Self-adjusting binary trees. *JACM*, 32:652–686, 1985
11. R. Wilbur. Lower bounds for accessing binary search trees with rotation. In *Proc. 27th Symp. on foundation of Computer Science*, pages 61–69, 1986

On the Comparison-Addition Complexity of All-Pairs Shortest Paths*

Seth Pettie

Seth Pettie

Department of Computer Sciences
The University of Texas at Austin
Austin, TX 78712

Abstract. We present an all-pairs shortest path algorithm for arbitrary graphs that performs $O(mn \log \alpha)$ comparison and addition operations, where m and n are the number of edges and vertices, resp., and $\alpha = \alpha(m, n)$ is Tarjan's inverse-Ackermann function. Our algorithm eliminates the sorting bottleneck inherent in approaches based on Dijkstra's algorithm, and for graphs with $O(n)$ edges our algorithm is within a tiny $O(\log \alpha)$ factor of optimal. The algorithm can be implemented to run in polynomial time (though it is not a pleasing polynomial). We leave open the problem of providing an efficient implementation.

1 Introduction

In 1975 Fredman [F76] presented a simple and elegant algorithm for the all-pairs shortest paths problem that performs only $O(n^{2.5})$ comparison and addition operations, rather than the $O(n^3)$ bound of Floyd's algorithm (see [CLRS01]). However, Fredman gave no polynomial-time implementation of this algorithm, illustrating that the notion of *comparison-addition complexity* in shortest paths problems can be studied apart from the usual notion of *algorithmic complexity*, that is, the actual running times of shortest path programs. We present, in the same vein, an APSP algorithm that makes $O(mn \log \alpha(m, n))$ comparisons and additions, where m and n are the number of edges and vertices, resp., and α is the mind-bogglingly slow growing inverse-Ackermann function. For sparse graphs, the best comparison-addition-based algorithm to date was established very recently [Pet02]; it runs in $O(mn + n^2 \log \log n)$ time, improving on the long-standing bound of $O(mn + n^2 \log n)$ [Dij59,FT87,J77]. A trivial lower bound on the APSP problem is $\Omega(n^2)$, implying that our algorithm is tantalizingly close to optimal for edge-density $m/n = O(1)$. For dense graphs, the best implementable algorithm is due to Takaoka [Tak92], running in time $O(n^3 \sqrt{\log \log n / \log n})$. We refer the reader to Zwick's survey [Z01] for a summary of other shortest path algorithms.

It is still an open question whether there are $O(n^2) + o(mn)$ algorithms for APSP when $m = O(n^{1.5})$. Karger et al. [KKP93] have shown that $\Omega(mn)$ is

---

* This work was supported by Texas Advanced Research Program Grant 003658-0029-1999, NSF Grant CCR-9988160, and an MCD Graduate Fellowship.

P. Bose and P. Morin (Eds.): ISAAC 2002, LNCS 2518, pp. 32–43, 2002.
© Springer-Verlag Berlin Heidelberg 2002

a lower bound among algorithms that only compare path-lengths. Fredman's algorithm obviously does not fit into this class, and neither does our algorithm. This raises the interesting possibility that our techniques could be used to obtain $O(n^2) + o(mn)$ APSP algorithms for sparse graphs.

Our APSP algorithm is based on the *component hierarchy* (CH) approach to single source shortest paths invented by Thorup [Tho99] for the special case of undirected graphs, and generalized by Hagerup [Hag00] to directed graphs. The [Tho99,Hag00] algorithms were designed for *integer*-weighted graphs in the RAM model of computation. Their improved running times depended crucially on the ability of RAMs to sort n integers in $o(n \log n)$ time. It was, therefore, not obvious whether these algorithms could be translated into good algorithms for real-weighted graphs in the comparison-addition model. Pettie & Ramachandran [PR02] gave an adaptation of Thorup's algorithm to real-weighted undirected graphs; it solves APSP in $O(mn\alpha)$ time.[1] Pettie et al. implemented a simplified version of [PR02]; in their experiments with real-weighted graphs it consistently outperformed Dijkstra's algorithm. The techniques used in [PR02] are specific to undirected graphs and simply have no analogues in directed graphs. Pettie [Pet02], using a different set of techniques, gave a version of Hagerup's algorithm for real-weighted directed graphs. It solves APSP in $O(mn + n^2 \log \log n)$ time. Table 1 summarizes the state of the art in APSP for real-weighted graphs.

Table 1. The best comparison-addition APSP algorithms to date, both uniform and non-uniform. Excluded from this table are algorithms for integer-weighted graphs and average-case algorithms. See [Z01] for a good survey on shortest path algorithms.

APSP Algorithms for Real-Weighted Graphs

Citation	Complexity	Uniform?
Fredman [F76]	$O(n^{2.5})$	NO
Takaoka [Tak92]	$O(n^3 \sqrt{\frac{\log \log n}{\log n}})$	yes
Pettie-Ramachandran [PR02] (undirected graphs only)	$O(mn\alpha(m,n))$	yes
Pettie [Pet02]	$O(mn + n^2 \log \log n)$	yes
this paper	$O(mn \log \alpha(m,n))$	NO

In this paper we build on the techniques introduced in [Pet02]. Specifically, our algorithm leverages *approximate* shortest path distances in the computation of exact distances, and it uses a novel mechanism for different SSSP computations to share information. In the next Section we give a technical introduction to the component hierarchy approach which focusses on a high-level feature of all CH-type algorithms [Tho99,Hag00,PR02,Pet02] and not on the algorithmic particulars. This high-level characterization also turns out to be useful in *lower bounding* the complexity of the CH approach in a comparison-based model.

---

[1] Actually, it solves the s-sources shortest path problem, $s > \log n$, in $O(sm\alpha)$ time.

1.1 Technical Introduction

One way to characterize Dijkstra's SSSP algorithm [Dij59] is to say that it finds a permutation π_s of the vertices such that

$$\pi_s(u) < \pi_s(v) \quad \Rightarrow \quad d(s, u) \leq d(s, v)$$

where $d(\cdot, \cdot)$ is the distance function and s is the source. We give a similar characterization of the shortest path algorithms based on component hierarchies [Tho99,Hag00,PR02,Pet02].

Suppose for this discussion that the graph is strongly connected. Let $circ(u, v)$ be the set of all cycles containing vertices u and v and let $sep(u, v)$ be defined as

$$sep(u, v) \;=\; \min_{C \in circ(u,v)} \; \max_{e \in C} \; length(e)$$

All component hierarchy-based algorithms [Tho99,Hag00,PR02,Pet02] generate a permutation π_s satisfying Property 1.

Property 1. $\forall u, v : \; d(s, v) \geq d(s, u) + sep(u, v) \quad \Rightarrow \quad \pi_s(u) < \pi_s(v)$

It is not obvious whether there is a sorting bottleneck inherent in Property 1. In [Pet02] it is proved that any directed SSSP algorithm obeying Property 1 must make $\Omega(m + \min\{n \log n, n \log r\})$ operations, where r is the ratio of the maximum to minimum edge length, even if the sep function is already known. Interestingly, these bounds become significantly weaker for undirected graphs. In an upcoming full version of [PR02] it is proved that any Property 1 undirected SSSP algorithm must perform $\Omega(m + \min\{n \log n, n \log \log r\})$ operations (notice the weaker dependence on r); however, if the sep function is already known there is only a trivial $\Omega(m)$ lower bound for undirected graphs.

What conclusions should be made from these lower bounds? First, one should not waste time trying to develop substantially faster SSSP algorithms obeying Property 1: the directed & undirected SSSP algorithms in [PR02] are tight to within α factors. Second, any directed APSP algorithm that first computes a component hierarchy (read: computes the sep function) then performs n *independent* SSSP computations obeying Property 1 must make $\Omega(mn + n^2 \log n)$ operations since each SSSP computation is subject to the lower bound of [Pet02]. The key technique to improving this bound, which was used to a lesser extent in [Pet02], is to make the SSSP computations *dependent*. In the algorithm presented here, we perform a sequence of n SSSP computations in such a way that later SSSP computations learn from the time-consuming mistakes of earlier ones.

2 Preliminaries

The input is a weighted, directed graph $G = (V, E, \ell)$ where $|V| = n, |E| = m$, and $\ell : E \to \mathbb{R}$ assigns a real *length* to every edge. It is well-known [J77] that the shortest path problem is reducible in $O(mn)$ time to one of the same size but having only non-negative edge lengths. We therefore assume that $\ell : E \to \mathbb{R}^+$

assigns only non-negative lengths. We let $d(u,v)$ denote the length of the shortest path from u to v, or ∞ if none exists. The all-pairs shortest path problem is to compute $d(\cdot,\cdot)$ and the single-source shortest paths problem is to compute $d(s,\cdot)$ where the first argument, the *source*, is fixed. Generalizing the d notation, let $d(u,H)$ be the shortest distance from u to H, where H is a subgraph or an object associated with a subgraph.

2.1   The Comparison-Addition Model

In the comparison-addition model real numbers are only subject to comparisons and additions and *comparison-addition complexity* refers to the number of such operations. In order to specify an *implementation* of a comparison-addition-based algorithm one would also need to fix some kind of underlying model governing non-real number computation such as a pointer machine or RAM; implementations, however, are not the focus of this paper. We frequently use subtraction in our algorithms; refer to [PR02] for a simulation of subtraction.

There are several lower bounds for shortest paths in the comparison-addition model though they are all for restricted classes of algorithms. See [PR02] for a summary.

3   The Component Hierarchy Approach

Dijkstra's classic algorithm [Dij59] computes SSSP by visiting vertices by increasing distance from the source s. It can be thought of as simulating a physical process. Suppose the graph-edges represent water pipes and at time zero we begin releasing water from vertex s. Dijkstra's algorithm simulates the flow of water at unit-speed through the graph. Component hierarchy-based algorithms can also be thought of as simulating this process, though in a much coarser way. Instead of maintaining the same simulated time throughout the whole graph, as Dijkstra's algorithm does, CH-based algorithms decompose the graph into a hierarchy of subgraphs (the component hierarchy), where each subgraph maintains its *own* local simulated time. Progress is made by giving a well-selected subgraph, say at simulated time a, permission to advance its clock to simulated time $b > a$. The correctness of this scheme is not obvious, and depends upon the subgraphs and intervals $[a,b)$ being chosen carefully. Due to space constraints we can only sketch the basic component hierarchy algorithm; refer to [Pet02, Pet02b] for a complete description.

The component hierarchy we use is the same CH given in [Pet02,Pet02b]. Below we describe a generalized CH meant solely for understanding our APSP algorithm; it leaves out many important details from [Pet02,Pet02b]. Assume w.l.o.g. that the graph is strongly connected. The CH is defined w.r.t. an increasing sequence of real lengths $(\ell_1,\ldots,\ell_k)$ where ℓ_1 is the minimum edge length in the graph. Let G_{i-1} denote the graph G restricted to edges with length less than ℓ_i, so for instance, G_0 contains no edges. A level i component hierarchy node x corresponds to a strongly connected component C_x of G_i. The notation $diam(C_x)$ refers to the diameter of C_x (the longest shortest path length) and

$norm(x) = \ell_i$ by definition. A node x is an ancestor of y if C_y is a subgraph of C_x. Since we would like to ignore CH nodes with only one child, define the 'parent' of a CH node to be its nearest ancestor with a strictly larger strongly connected component. If $\{x_j\}_j$ is the set of children of x then C_x^c denotes the subgraph derived from C_x by contracting the subgraphs $\{C_{x_j}\}_j$. Because of the nice correspondence between component hierarchy nodes and subgraphs, we frequently treat them as equivalent in our notation, so $d(s,x)$ refers to the distance from s to C_x, and $y \in V(C_x^c)$ is understood to mean y is a child of x.

The basic idea of the component hierarchy approach [Tho99,Hag00,PR02, Pet02] is to compute $d(s,x)$ for all $x \in CH$. Since the leaves of the component hierarchy represent graph vertices, this solves SSSP from source s as well. One can imagine that there is a separate process identified with each CH node where the job of y is to compute $d(s,y)$. If x is the parent of y, y simply waits for x to compute $d(s,x)$, then y computes $d(s,y) - d(s,x)$. The key observation from [Pet02] is that this scheme can be made very efficient in the comparison-addition model if y is supplied with a key piece of information: an integer approximation to $(d(s,y) - d(s,x))/norm(x)$ that is accurate to within some absolute constant. We summarize the important aspects of the high-level CH algorithm [Pet02, Pet02b], from the point of view of the process of a CH node y, child of x.

First, all CH algorithms, like Dijkstra's, maintain a set S of visited vertices whose distance from the source has been fixed. Let $d_S(s,u)$ be the distance from s to u using only intermediate vertices from S. Define $D(y)$ as $\min\{d_S(s,u) : u \in V(C_y)\}$.[2] Note that d_S is simply Dijkstra's tentative distance function; D represents the tentative distance to whole subgraphs represented by CH nodes.

As soon as x (parent of y) discovers $d(s,x)$ it creates a bucket array of at least $diam(C_x)/norm(x) + 1$ buckets, each representing a real interval of width $norm(x)$. The first bucket begins at t_0, a real such that $t_0 \leq d(s,x) < t_0 + norm(x)$. (We will refer to buckets by their place in the array or by their associated interval, whichever is more convenient.) It would be nice to guarantee that y always appears in the correct bucket, namely bucket number $\lfloor (D(y) - t_0)/norm(x) \rfloor$. This "ideal" invariant is maintained in the CH-based algorithms of [Tho99,Hag00]; however, we do not know how to maintain it efficiently in the comparison-addition model. Our solution is to simulate the ideal bucket array with an *actual* bucket array and a heap, denoted H_x.

Invariant 1 *If $d(s,y)$ has not yet been fixed, then either y appears in an actual bucket between $\lfloor \frac{d(s,y)-t_0}{norm(x)} \rfloor - 2$ and $\lfloor \frac{D(y)-t_0}{norm(x)} \rfloor$ inclusive, or in the heap H_x.*

The purpose of the heap H_x is to hold nodes until enough information is available to bucket them in accordance with Invariant 1. The efficiency of this scheme depends on there being relatively few heap insertions, since heap deletion is a non-constant time operation. We will not go into why the "-2" is a tolerable error (see [Pet02b]).

---

[2] Updating and querying D-values is a non-trivial task, requiring $O(m \log \alpha(m,n))$ comparisons per SSSP computation using Gabow's split-findmin structure [G85], as modified in [PR02].

Our algorithm should be thought of as consisting of two levels, the "high-level" component hierarchy algorithm (see [Pet02b]) and a "low-level" algorithm that maintains Invariant 1 behind the scenes, which we give in Section 4. The low-level algorithm uses some simple, though non-obvious properties of shortest paths.

4   Our Algorithm

In Section 4.1 we define a set of new length functions $\{\gamma_x\}_{x \in CH}$ and a set of *relative distance* functions $\{\Gamma_x\}_{x \in CH}$. A relative distance is just the difference between two distances. In Section 4.2 we show that, using *discrete approximations* of the length and relative distance functions, it can be possible to stitch together new shortest paths from previously computed ones, in a manner that is cheaper than computing them from scratch. There is a tradeoff between the accuracy of the discrete approximations and their usefulness; our amortized analysis depends on the degree of accuracy being chosen carefully.

Naming conventions. The letters x, y, z will refer to CH nodes and u, v, w, s to graph vertices. A hat ($\hat{\ }$) or tilde ($\tilde{\ }$) indicates a discrete approximation to a real quantity.

4.1   Approximating Relative Distances

Let $anchor_x(u)$ be some vertex in $V(C_x^c)$ (recall, $V(C_x^c)$ corresponds to the children of x) and $a_x(u) = d(u, anchor_x(u))$. The anchor of u is specified as soon as possible. That is, as soon as $d(u, y)$ is known for *some* $y \in V(C_x^c)$, $anchor_x(u)$ is set to y. The edge-labeling functions γ_x and $\hat{\gamma}_x$ are also calculated as soon as possible. As soon as $a_x(v)$ and $a_x(u)$ are known, $\gamma_x(u, v)$ is set to:

$$\gamma_x(u, v) \stackrel{\text{def}}{=} \ell(u, v) + a_x(v) - a_x(u)$$

It follows that for edges $(u, v) \in E(C_x)$, $\gamma_x(u, v) = \ell(u, v)$ is fixed immediately, since $a_x(u) = a_x(v) = 0$ is known a priori. As far as conserving comparisons & additions, it turns out that γ_x is not as useful as a discrete approximation to γ_x. If $\gamma_x(u, v) > 2 \cdot diam(C_x)$ then $\hat{\gamma}_x(u, v) \stackrel{\text{def}}{=} \infty$. Otherwise, define $\hat{\gamma}_x(u, v)$ as

$$\hat{\gamma}_x(u, v) \stackrel{\text{def}}{=} \epsilon_x \cdot \left\lfloor \frac{\gamma_x(u, v)}{\epsilon_x} \right\rfloor \qquad \text{where} \qquad \epsilon_x \stackrel{\text{def}}{=} \frac{norm(x)}{4 \cdot |V(C_x^c)|}$$

Why is $\hat{\gamma}_x$ better than γ_x? The difference is in how they are represented. We represent γ_x in the natural way, as a real number kept in a real variable. On the other hand $\hat{\gamma}_x$ is represented *implicitly*. That is, the statement "$\hat{\gamma}_x(u, v)$ is known" means the *integer* $\hat{\gamma}_x(u, v)/\epsilon_x$ can be derived from previous comparisons and additions. Clearly two γ_x-values require one operation to be compared or added. Manipulating $\hat{\gamma}_x$-values is really just a mental exercise; there are no comparisons or additions involved in computing functions of the $\hat{\gamma}_x$ values.

Define $\Gamma_x, \hat{\Gamma}_x : V(G) \times V(C_x^c) \to \mathbb{R}$. As above, the integer $\hat{\Gamma}_x/\epsilon_x$ will be represented implicitly.

$$\Gamma_x(u,y) \stackrel{\text{def}}{=} d(u,y) - a_x(u) \qquad \text{and} \qquad \hat{\Gamma}_x(u,y) \stackrel{\text{def}}{=} \epsilon_x \cdot \left\lfloor \frac{\Gamma_x(u,y)}{\epsilon_x} \right\rfloor$$

Lemma 1. *Let* $x, y \in CH$, $y \in V(C_x^c)$ *and* u, v *be vertices.*

(i) *Given* $\gamma_x(u,v)$ *(resp.,* $\Gamma_x(u,y)$), $\hat{\gamma}_x(u,v)$ *(resp.,* $\hat{\Gamma}_x(u,y)$) *can be computed with* $O(\log \frac{|V(C_x^c)| \cdot diam(C_x)}{norm(x)})$ *comparisons and additions.*

(ii) *Computing* $\hat{\gamma}_x(e)$, *over all* x *and edges* e, *takes* $O(mn)$ *comparisons and additions.*

$\hat{\Gamma}_x$-values are very useful for conserving on comparison-addition operations in the component hierarchy algorithm; however, even given Γ_x, $\hat{\Gamma}_x$ is fairly expensive to compute. Lemma 1(ii) illustrates that we can afford to compute all $\hat{\gamma}$-values; however, it will become clear in the analysis that we can only afford to compute a small fraction of the $\hat{\Gamma}$-values Our solution is to introduce another approximation of Γ_x which is significantly less accurate than $\hat{\Gamma}_x$. Define $\tilde{\Gamma}_x : V(G) \times V(C_x^c) \to \mathbb{R}$ to be any function that satisfies:

$$\Gamma_x(u,y) - \tilde{\Gamma}_x(u,y) \in [0, |V(C_x^c)| \epsilon_x) \qquad \text{and} \qquad \frac{\tilde{\Gamma}_x}{\epsilon_x} \text{ is represented as an integer}$$

Notice that $\epsilon_x |V(C_x^c)| = norm(x)/4$ which is just about as accurate as we will ever need. Lemma 2 illustrates the relationship between $\hat{\gamma}_x, \hat{\Gamma}_x, \tilde{\Gamma}_x$, and ϵ_x.

Lemma 2. *Suppose* $\mathcal{P} = \langle u_0, u_1, \ldots, u_k \rangle$, $u_k \in V(C_y)$, $y \in V(C_x^c)$, *is known to be the shortest path from* u_0 *to* C_y, *and* u_h *is the first vertex in* $\mathcal{P}$ *for which* $\hat{\Gamma}_x(u_h, y)$ *is known. Then if* $h < |V(C_x^c)|$, $\tilde{\Gamma}_x(u_i, y)$ *is known as well, for* $i \le h$.

Proof. Because $\mathcal{P}$ is known to be the shortest path to some vertex in $V(C_x^c)$, it follows that $anchor_x(w)$ has been chosen for all vertices $w \in \mathcal{P}$ and that $\gamma_x(e), \hat{\gamma}_x(e)$ are computed for all edges $e \in \mathcal{P}$. We now prove that for $i \le h$

$$\Gamma_x(u_i, y) - \left(\hat{\Gamma}_x(u_h, y) + \sum_{j=i}^{h-1} \hat{\gamma}(u_j, u_{j+1}) \right) \in [0, (h-i+1)\epsilon_x)$$

Hence $\hat{\Gamma}_x(u_h, y) + \sum_{j=i}^{h-1} \hat{\gamma}(u_j, u_{j+1})$ is a good enough approximation to $\Gamma_x(u_i, y)$ to satisfy the constraints put on $\tilde{\Gamma}_x(u_i, y)$, so long as $h - i < |V(C_x^c)|$. Let $[a, b)$ denote some number in that interval. Then, in general, $\hat{\gamma}_x = \gamma_x - [0, \epsilon_x)$ and $\hat{\Gamma}_x = \Gamma_x - [0, \epsilon_x)$. Let $\xi = [0, \epsilon_x(h-i+1))$. Then

$$\Gamma_x(u_i, y) - \left(\hat{\Gamma}_x(u_h, y) + \sum_{j=i}^{h-1} \hat{\gamma}(u_j, u_{j+1}) \right)$$

$$= \Gamma_x(u_i, y) - \left(d(u_h, y) - a_x(u_h) + \sum_{j=i}^{h-1} (\ell(u_j, u_{j+1}) + a_x(u_{j+1}) - a_x(u_j)) \right) + \xi$$

$$= \Gamma_x(u_i, y) - d(u_i, y) + a_x(u_i) + \xi \quad = \quad \xi$$

The above equalities follow directly from the definitions of $\gamma, \hat{\gamma}, \Gamma$, and $\hat{\Gamma}$. $\square$

4.2   The Algorithm

The algorithm is best described as a list of triggers of the form $\mathcal{P} \longrightarrow \mathcal{A}$, where $\mathcal{P}$ is a precondition and $\mathcal{A}$ an action to be performed. The high-level component hierarchy algorithm can only proceed if none of the triggers are applicable. We have already informally defined a few triggers. Let us state them formally.

Trigger 1 $anchor_x(u)$ *is unspecified but* $d(u,y)$ *is known,* $y \in V(C_x^c)$ $\longrightarrow$ *Set* $anchor_x(u) := y,\ a_x(u) := d(u,y)$

Trigger 2 (u,v) *is an edge and both* $a_x(u)$ *and* $a_x(v)$ *are known* $\longrightarrow$ *Compute* $\gamma_x(u,v)$ *and* $\hat{\gamma}_x(u,v)$

Trigger 3 *Some edge is relaxed, decreasing* $D(y)$, *where* $y \in V(C_x^c)$, $d(s,x)$ *is known.* $\longrightarrow$ *If possible, bucket* y *according to Invariant 1*

Lemma 2 suggests another trigger. Let $OUT(u)$ and $IN(u)$ denote the *known* trees of shortest paths out of, and into u, respectively.[3] So, for instance, $IN(y)$, $y \in CH$, initially has no edges because we do not know any non-trivial shortest paths to C_y. After each SSSP computation, from say, source s, one can see that $IN(y)$ grows the minimal amount to incorporate s. It will be important to know the $\tilde{\Gamma}_x(u,y)$ function for vertices $u \in IN(y)$; by Lemma 2 it is enough to compute $\hat{\Gamma}_x(\cdot,y)$-values for a sufficiently large and well-chosen subset of the vertices in $IN(y)$. It can be proved that Trigger 4 fits the bill.

Trigger 4 *The closest ancestor of* u *in* $IN(y)$ *for which* $\hat{\Gamma}_x(\cdot,y)$ *is known is at distance exactly* $|V(C_x^c)|$ $\longrightarrow$ *Compute* $\hat{\Gamma}_x(w,y)$, *where* w *is the ancestor of* u *in* $IN(y)$ *at distance exactly* $\lfloor |V(C_x^c)|/2 \rfloor$

Lemma 3. *For CH nodes* x *and* $y \in V(C_x^c)$
(i)   The $\tilde{\Gamma}_x(\cdot,y)$ *function is known for vertices in* $IN(y)$.
(ii)   At most $2n/|V(C_x^c)|$ *different* $\hat{\Gamma}_x(\cdot,y)$-values *are computed.*
(iii) The total comparison-addition cost of (ii), over all x,y, *is* $O(n^2)$.

Proof. Part (i): Trigger 4 ensures that every vertex in $IN(y)$ has some ancestor at distance at most $|V(C_x^c)| - 1$ whose $\hat{\Gamma}_x(\cdot,y)$-value is known. By Lemma 2 the $\tilde{\Gamma}_x(\cdot,y)$-value of every vertex in $IN(y)$ is also known. Part (ii) is straightforward [Pet02b]. Part (iii) can be shown to follow from Part (ii) and Lemma 1(i) [Pet02b]. □

Let $G^{\hat{\gamma}_x}$ be the subgraph of G consisting of edges whose $\hat{\gamma}_x$-values are known. Every time $G^{\hat{\gamma}_x}$ grows we can better estimate shortest distances. Trigger 5, given below, attempts to bucket nodes residing in the heap as soon as possible.

---
[3] For $y \in CH$, $IN(y)$ is really an in-forest.

Trigger 5 *Edge(s) are added to $G^{\hat{\gamma}_x}$ $\longrightarrow$ If possible, migrate nodes from H_x to the bucket array, consistent with Invariant 1.*

We now clarify exactly what is meant by "if possible" in Triggers 3 and 5, that is, how we decide if it is possible to bucket a node $y \in V(C_x^c)$. Suppose s is the source in the current SSSP computation. The first moment we are concerned about the distance from s to C_y, $y \in V(C_x^c)$, is when $d(s,x)$ becomes known. At this moment the shortest path from s to C_y consists of a *head* in $OUT(s)$, a *bridge*, and a *tail* in $IN(y)$. We show that if the head, bridge, and tail satisfy certain conditions, then a good, discrete approximation of $d(s,x) - d(s,y)$ is known implicitly, allowing us to bucket y in constant time. Every time new edges are added to $G^{\hat{\delta}_x}$ or when $D(y)$ decreases (Triggers 5 and 3, resp.) we have a new opportunity to bucket y.

We now describe the bucketing procedure for $y \in V(C_x^c)$ more carefully. Let $f \in C_z$, $z \in V(C_x^c)$ be such that $d(s,f) = d(s,x)$, that is, f is the closest vertex to s in C_x — Figure 1 diagrams our situation. Because we are attempting to bucket y, $d(s,x)$ must already be known. Let P_{sf} denote the shortest s-to-f path (which is also the shortest s-to-C_x path).

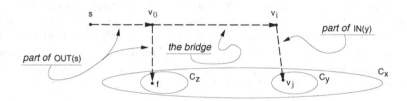

Fig. 1. The path $\langle s, \ldots, v_j \rangle$, divided into a head $\langle s, \ldots, v_0 \rangle$, a bridge $\langle v_0, \ldots, v_i \rangle$, and a tail $\langle v_i, \ldots, v_j \rangle$.

Definition 1. *Let $\mathcal{Q}_y$ be the set of paths $\{\langle v_0, \ldots, v_i, \ldots, v_j \rangle\}$ satisfying*

(i) $v_0 \in P_{sf} \subseteq OUT(s)$
(ii) $i \le |V(C_x^c)|$ and $\langle v_0, \ldots, v_i \rangle \in G^{\hat{\gamma}_x}$
(iii) $v_j \in C_y$ and $\langle v_i, \ldots, v_j \rangle \in IN(y)$

Bucketing y by its D-value is equivalent to estimating $D(y) - d(s,f)$, however we generally will not have enough information to do this. Our solution is not to focus solely on the current path with length $D(y)$, but to estimate the distance of *many* hypothetically shortest paths from s to C_y.

For $Q \in \mathcal{Q}_y$, $Q = \langle v_0, \ldots, v_i, \ldots, v_j \rangle$, define $diff(Q)$ and $diff(\mathcal{Q}_y)$ as

$$diff(Q) \overset{\text{def}}{=} \tilde{\Gamma}_x(v_i, y) + \sum_{k=0}^{i-1} \hat{\gamma}_x(v_k, v_{k+1}) - \tilde{\Gamma}_x(v_0, z)$$

$$diff(\mathcal{Q}_y) \overset{\text{def}}{=} \min_{Q \in \mathcal{Q}_y} diff(Q)$$

Lemma 4. *diff(Q) requires no comparisons or additions to compute, and*
$$diff(Q) = \ell(Q) - d(v_0, x) + (-\tfrac{norm(x)}{2}, \tfrac{norm(x)}{4})$$

Proof. Let $\xi = (-2\epsilon_x |V(C_x^c)|, \epsilon_x |V(C_x^c)|) = \left(-\tfrac{norm(x)}{2}, \tfrac{norm(x)}{4}\right)$

Recall that $f \in C_z$, $z \in V(C_x^c)$ were such that $d(s, f) = d(s, z) = d(s, x)$.

$$diff(Q) = \sum_{k=0}^{i-1} \gamma_x(v_k, v_{k+1}) + \Gamma_x(v_i, y) - \Gamma_x(v_0, z) + \xi \tag{1}$$

$$= \ell(\langle v_0, \dots, v_i \rangle) - a_x(v_0) + a_x(v_i) + \Gamma_x(v_i, y) - \Gamma_x(v_0, z) + \xi \tag{2}$$

$$= \ell(\langle v_0, \dots, v_j \rangle) - a_x(v_0) - \Gamma_x(v_0, z) + \xi \tag{3}$$

$$= \ell(Q) - d(v_0, x) + \xi \tag{4}$$

Line 1 follows from the equalities $\tilde{\Gamma}_x = \Gamma_x - [0, \epsilon_x |V(C_x^c)|)$ and $\hat{\gamma}_x = \gamma_x - [0, \epsilon_x)$, and the bound $i \leq |V(C_x^c)|$. Line 2 is derived by cancelling the terms in the telescoping sum. Line 3 follows from the equality $\Gamma_x(v_i, y) = d(v_i, y) - a_x(v_i) = \ell(\langle v_i, \dots, v_j \rangle) - a_x(v_i)$, and Line 4 from the equality $\Gamma_x(v_0, z) = d(v_0, z) - a_x(v_0) = d(v_0, x) - a_x(v_0)$.

The $\hat{\gamma}_x$ terms in $diff(Q)$ are known from the fact that $Q \in \mathcal{Q}_y$. By Lemma 3 the $\tilde{\Gamma}_x(v_i, y)$ and $\tilde{\Gamma}_x(v_0, z)$ terms are also implicitly known. Therefore, $diff(Q)$ can be computed with no real number operations. $\square$

Our procedure for bucketing $y \in V(C_x^c)$ is as follows. Recall that we denote the beginning of x's bucket array with t_0. Let $[\beta, \beta + norm(x))$ be the bucket s.t. $\beta \leq t_0 + diff(\mathcal{Q}_y) < \beta + norm(x)$. Since $\beta - t_0$, $norm(x)$, and $diff(\mathcal{Q}_y)$ are all known multiples of ϵ_x, this bucket can be identified without real number operations.

---

1. If $D(y) \geq \beta$, put y in bucket $[\beta, \beta + norm(x))$ and stop.
2. If $D(y) \geq \beta - norm(x)$, put y in bucket $[\beta - norm(x), \beta)$ and stop.
3. Otherwise, put y in H_x (or keep y in H_x if it is already there).

Lemma 5. *The bucketing procedure does not violate Invariant 1, and if $\mathcal{Q}_y$ contains a suffix of a shortest s-to-C_y path, then y is bucketed in Line 1 or 2.*

Proof. Recall from Section 3 that t_0 was chosen so that $d(s, x) \in [t_0, t_0 + norm(x))$. Lines 1 and 2 guarantee that y is never bucketed in a higher bucket than $\lfloor \tfrac{D(y) - t_0}{norm(x)} \rfloor$. We only need to show that in Line 1, y is not bucketed before bucket $\lfloor \tfrac{d(s,y) - t_0}{norm(x)} \rfloor - 2$. Because $diff(\mathcal{Q}_y)$ cannot correspond to a path shorter than $d(s, y)$, we have, from Lemma 4, $diff(\mathcal{Q}_y) > d(s, y) - d(s, x) - \tfrac{1}{2} norm(x)$. Using the inequality $d(s, x) < t_0 + norm(x)$, we also have $diff(\mathcal{Q}_y) > d(s, y) - t_0 - \tfrac{3}{2} norm(x)$. So bucketing y according to $diff(\mathcal{Q}_y)$ can put it at most $\lceil \tfrac{3}{2} \rceil = 2$ buckets before bucket $\lfloor \tfrac{d(s,y) - t_0}{norm(x)} \rfloor$. For the last part of the Lemma, assume that some $Q \in \mathcal{Q}_y$ is a suffix of the shortest s-to-C_y path. It follows from Lemma 4 that $diff(\mathcal{Q}_y) < d(s, y) - d(s, x) + \tfrac{1}{4} norm(x)$. This, together with the inequalities $d(s, x) \geq t_0$ and $\beta \leq diff(\mathcal{Q}_y) + t_0$, implies $D(y) > \beta - \tfrac{1}{4} norm(x)$, meaning y must be bucketed in Line 1 or 2. $\square$

We address the efficiency of our bucketing procedure in Lemma 6. Since deleting items from a heap is a non-constant operation, we must show that the percentage of times Step 3 is reached in the bucketing procedure is sufficiently low to counterbalance the cost of the heap operations. Lemma 6 does not depend on any fancy heap implementation. It holds if H_x supports constant time insert and decrease-key operations, and deletion of any subset of H_x in time linear in $|H_x| \le |V(C_x^c)|$. These are very weak assumptions.

Lemma 6. *The bucketing and heap costs over n SSSP computations are $O(mn)$.*

Proof. (sketch) We prove in [Pet02b] that the bucketing procedure above is called only $O(mn)$ times, requiring a constant number of comparisons per invocation, and that the total cost of heap operations is $O(n^2)$. We briefly outline how one would bound the number of heap operations.

Suppose in the SSSP computation from source s, $y \in V(C_x^c)$ is inserted into H_x. Let $P_{sy} = \langle P_1, P_2, P_3 \rangle$ be the shortest s-to-C_y path, where P_1 and P_3 are maximal such that $P_1 \subseteq P_{sx} \subseteq OUT(s)$ and $P_3 \subseteq IN(y)$. From Lemma 5 we know that $\langle P_2, P_3 \rangle \notin Q_y$, otherwise y would have been bucketed properly. Why wasn't $\langle P_2, P_3 \rangle \in Q_y$? From Definition 1 there can be only two reasons: either (a) $|P_2| > |V(C_x^c)|$ or (b) some edge in P_2 is not in $G^{\hat{\gamma}_x}$, which by Trigger 1 means some vertex in P_2 is unanchored. One can easily bound the number of times (a) occurs, since after the current SSSP computation with source s, at least $|V(C_x^c)|$ edges are added to $IN(y)$. A sufficient bound on (b) is n times for each CH node x, since any unanchored vertex in P_2 will, by Trigger 1, be anchored by the end of the current SSSP computation. One can, using other properties of the component hierarchy, prove that the heap costs due to (a) and (b) are $O(n^2)$ [Pet02b]. □

The only costs not covered by Lemma 6 are constructing the component hierarchy, which is $O(m \log n)$ [Pet02b], computing the $\hat{\Gamma}$ and $\hat{\gamma}$ functions, which is $O(mn)$ by Lemma 1 and 3, and maintaining the D-values of CH nodes, which is $O(m \log \alpha(m, n))$ for each SSSP computation [G85,PR02]. Regarding an actual implementation of this algorithm, the tricky part is maintaining shortest distances in the graphs $\{G^{\hat{\gamma}_x}\}_{x \in CH}$ under insertion of new edges. Simply running Bellman-Ford every time a batch of new edges are inserted gives a bound of $O(mn^3)$. However, if we use the dynamic shortest path algorithm from [RR96] the upper bound can be reduced to $\tilde{O}(mn^2)$. Theorem 1 follows.

Theorem 1. *The all-pairs shortest path problem on arbitrarily weighted, directed graphs can be solved with $O(mn \log \alpha(m, n))$ comparisons & additions in $\tilde{O}(mn^2)$ time, where m and n are the number of edges & vertices, resp., and α is the inverse-Ackermann function.*

Acknowledgment. We thank Vijaya Ramachandran for her comments and Camil Demetrescu for references on dynamic shortest paths.

References

[CLRS01]   T. Cormen, C. Leiserson, R. Rivest, C. Stein. *Introduction to Algorithms.* MIT Press, 2001.

[Dij59]   E. W. Dijkstra. A note on two problems in connexion with graphs. In *Numer. Math.*, 1 (1959), 269-271.

[F76]   M. Fredman. New bounds on the complexity of the shortest path problem. *SIAM J. Comput.* 5 (1976), no. 1, 83–89.

[FT87]   M. L. Fredman, R. E. Tarjan. Fibonacci heaps and their uses in improved network optimization algorithms. In *JACM* 34 (1987), 596–615.

[G85]   H. N. Gabow. A scaling algorithm for weighted matching on general graphs. In *26th Ann. Symp. on Foundations of Computer Science (FOCS 1985)*, 90–99.

[Hag00]   T. Hagerup. Improved shortest paths on the word RAM. In *Proceedings 27th Int'l Colloq. on Automata, Languages and Programming (ICALP 2000)*, LNCS volume 1853, 61–72.

[J77]   D. B. Johnson. Efficient algorithms for shortest paths in sparse networks. *JACM* 24 (1977), 1–13.

[KKP93]   D. R. Karger, D. Koller, S. J. Phillips. Finding the hidden path: time bounds for all-pairs shortest paths. *SIAM J. on Comput.* 22 (1993), no. 6, 1199–1217.

[Pet02]   S. Pettie. A faster all-pairs shortest path algorithm for real-weighted sparse graphs. *Proceedings 29th Int'l Colloq. on Automata, Languages and Programming (ICALP 2002)*, LNCS 2380, 85–97.

[Pet02b]   S. Pettie. On the comparison-addition complexity of all-pairs shortest paths. UTCS Technical Report TR-02-21, May 2002.

[PR02]   S. Pettie, V. Ramachandran. Computing shortest paths with comparisons and additions (extended abstract). *Proceedings of the 13th Annual ACM-SIAM Symposium on Discrete Algorithms (SODA)*, 2002, 267–276.

[PRS02]   S. Pettie, V. Ramachandran, S. Sridhar. Experimental evaluation of a new shortest path algorithm. *4th Workshop on Algorithm Engineering and Experiments (ALENEX)*, 2002.

[RR96]   G. Ramalingam, T. Reps. An incremental algorithm for a generalization of the shortest path problem. *J. Algorithms* 21 (1996), 267–305.

[Tak92]   T. Takaoka. A new upper bound on the complexity of the all pairs shortest path problem. *Inform. Process. Lett.* 43 (1992), no. 4, 195–199.

[Tho99]   M. Thorup. Undirected single source shortest paths with positive integer weights in linear time. *JACM* 46 (1999), no. 3, 362–394.

[Z01]   U. Zwick. Exact and approximate distances in graphs – A survey. Updated version at http://www.cs.tau.ac.il/ zwick/ *Proc. of 9th ESA* (2001), 33–48.

On the Clique-Width of Graphs in Hereditary Classes

Rodica Boliac and Vadim Lozin

RUTCOR, Rutgers University, 640 Bartholomew Rd, Piscataway NJ 08854-8003,
USA
{boliac, lozin}@rutcor.rutgers.edu

Abstract. The paper presents several results that can be helpful for deciding whether the clique-width of graphs in a certain class is bounded or not, and applies these results to a number of particular graph classes.

1 Introduction

The notion of clique-width of a graph was introduced in [7] and is defined as the minimum number of labels needed to construct the graph by means of the four graph operations: creation of a new vertex v with label i (denoted $i(v)$), disjoint union of two labeled graphs G and H (denoted $G \oplus H$), connecting vertices with specified labels i and j (denoted $\eta_{i,j}$) and renaming label i to label j (denoted $\rho_{i \to j}$). Every graph can be defined by an algebraic expression using the four operations above. For instance, the graph consisting of two adjacent vertices x and y can be defined by the expression $\eta_{1,2}(1(x) \oplus 2(y))$, and the cycle C_5 on vertices a, b, c, d, e (listed along the cycle) can be defined by the following expression:

$$\eta_{4,1}(\eta_{4,3}(4(e) \oplus \rho_{4 \to 3}(\rho_{3 \to 2}(\eta_{4,3}(4(d) \oplus \eta_{3,2}(3(c) \oplus \eta_{2,1}(2(b) \oplus 1(a))))))))) . \quad (1)$$

Alternatively, any algebraic expression defining G can be represented as a rooted tree, whose leaves correspond to the operations of vertex creation, the internal nodes correspond to the $\oplus$-operations, and the root is associated with G. The operations η and ρ are assigned to the respective edges of the tree. An example of the tree representing the expression (1) is depicted in Figure 1. Notice that any expression defining an n-vertex graph contains exactly $n-1$ $\oplus$-operations, and hence the corresponding tree has exactly $2n$ nodes.

An expression built from the above four operations is called a k-expression if it uses k different labels. The clique-width of a graph G, denoted $cwd(G)$, is the minimum k such that there exists a k-expression defining G. Thus, from the above example we conclude that $cwd(C_5) \le 4$. Moreover, it is not hard to see that the clique-width of any cycle is at most 4. Graphs with bounded clique-width are of special interest, because many problems that are NP-hard in general graphs admit polynomial time solutions when restricted to graphs with bounded clique-width [8].

P. Bose and P. Morin (Eds.): ISAAC 2002, LNCS 2518, pp. 44–54, 2002.

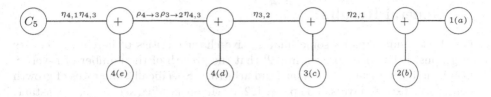

Fig. 1. The tree representing the expression (1) defining a C_5

The objective of this paper is the problem of deciding whether the clique-width of graphs in a certain class is bounded or not. We study *hereditary* classes, i.e. those closed under vertex deletion. It is well known that a class of graphs X is hereditary if and only if it can be characterized by a set Y of forbidden induced subgraphs, in which case we say that graphs in X are Y-free.

Simple arguments show that the clique-width of P_4-free graphs (cographs) is at most 2 and that of trees (forests) is at most 3. A more complicated analysis leads to the conclusion that the clique-width is bounded for some extensions of P_4-free graphs [19] and for distance-hereditary graphs [14] that generalize trees. Also, the clique-width is bounded for the bipartite analog of cographs [13] and some of their extensions [12,17,18]. On the other hand, the clique-width is unbounded in the class of split graphs [19], unit interval, permutation [14] and even bipartite permutation graphs [4].

In the present paper we establish several general results that can be useful to solve the problem in question, and apply them to particular subclasses of bipartite graphs, claw-free graphs and monotone classes, i.e. hereditary classes of graphs closed under deletion of edges.

Most notations we use are customary: $V(G)$ and $E(G)$ denote the vertex set and the edge set of a graph G, respectively. The complement to G is called *co-G* and is denoted $\overline{G}$. Given a subset of vertices $U \subseteq V(G)$, we denote by $G - U$ the subgraph of G induced by $V(G) - U$. If $U = \{v\}$, we write $G - v$ instead of $G - \{v\}$. A cut-point in a graph G is a vertex v such that $G - v$ has strictly more connected components than G. A block in a graph is a connected induced subgraph without cut-points.

As usual, C_n is the chordless cycle, P_n is the chordless path, and K_n is the complete graph on n vertices. Also, $K_{n,m}$ is the complete bipartite graph with parts of size n and m, and $2K_2$ is the disjoint union of two copies of K_2. We denote by $S_{i,j,k}$ a tree with exactly three leaves being at distance i,j,k from the only vertex of degree three. The class of all graphs whose every connected component is either a P_n or a $S_{i,j,k}$ will be denoted by S. We use special names for several particular graphs: a *claw* is $S_{1,1,1} = K_{1,3}$; a *paw* is the graph obtained from a claw by adding an arbitrary edge; a *diamond*, denoted $K_4 - e$, is obtained form a K_4 by deleting an edge.

2   General Results

Our first result exploits some quantitative characteristics of hereditary classes of graphs. It has been proven in [22] that the growth of the number of n-vertex graphs in a hereditary class is far from arbitrary. Specifically, the rates of growth constitute discrete layers. The paper [22] distinguishes five such layers: constant, polynomial, exponential, factorial and superfactorial. Independently, a similar result has been obtained in [1]. Moreover, the latter paper provides the first three layers with complete structural characterizations and describes all minimal factorial classes of graphs. In terms of the clique-width, the graph classes in the first three layers are of no interest, since their structure is rather simple and leads to the immediate conclusion that all of them are of bounded clique-width. In the present paper, our special concern is the classes in the factorial layer.

A class of graphs X is said to be *factorial* if the number X_n of n-vertex graphs in X satisfies the inequalities $n^{c_1 n} \leq X_n \leq n^{c_2 n}$ for some constants c_1 and c_2.

Theorem 1 *For any integer $k > 1$, the class of graphs with clique-width at most k, denoted by $\mathcal{C}(k)$, is factorial.*

Proof. $\mathcal{C}(2)$ contains the class of cographs, which is known to be factorial (see, e.g., [1]). This observation gives the lower bound. To obtain an upper bound on the number of graphs in $\mathcal{C}(k)$, we associate with each n-vertex graph $G \in \mathcal{C}(k)$ a k-expression $F(G)$ of minimum length and the tree $T(F(G))$ representing $F(G)$. The minimality of $F(G)$ ensures that the length of the label of any edge of the tree $T(F(G))$ is bounded by a constant depending on k but not n. Consequently, $T(F(G))$ can be encoded with a binary word of length $O(n \log_2 n)$, which immediately implies the required upper bound n^{cn} on the number of graphs in $\mathcal{C}(k)$.

An encoding of $T(F(G))$ of the desired length can be obtained in different ways. The reader can apply, for instance, the Prüfer code [21,5] for a labeled tree, i.e. a tree T with a given injection $V(T) \to A$, where A is a set of vertex labels. The Prüfer code of an n-vertex labeled tree consists of $n - 2$ vertex labels, and therefore has the length $(n - 2) \log_2 |A|$ in binary representation. The tree $T(F(G))$ can be easily transformed into a labeled tree by assigning the labels of the edges to the respective parent nodes. Since $T(F(G))$ is a binary tree with $2n$ nodes and the length of any edge label is bounded by a constant, we obtain a labeled tree with node labels of length $O(\log_2 n)$ (in binary representation), as required. ∎

As an immediate consequence from Theorem 1 it follows that the clique-width of bipartite, co-bipartite and split graphs is unbounded, since the number of n-vertex graphs in those classes is at least $2^{n^2/4}$ (notice that a direct proof of unboundedness of the clique-width of split graphs has been obtained in [19]).

Our next result deals with graph operations that can be helpful to prove boundedness of the clique-width in certain classes of graphs.

For a class of graphs Y, we denote by $[Y]_k$ the class of graphs G such that $G - U$ belongs to Y for a subset $U \subseteq V(G)$ of cardinality at most k. Also, given a class of graphs Y, $[Y]_B$ denotes the class of graphs whose every block belongs to Y.

Theorem 2 *If Y is a class of graphs of bounded-clique width, then so are $[Y]_k$ and $[Y]_B$.*

Proof. To prove the theorem for $[Y]_k$, it suffices to consider the case $k = 1$. Let $G \in [Y]_1$ and $G - v \in Y$. Given a p-expression F for $G - v$, we first construct a $2p$-expression F' for $G - v$ in such a way that every labeled vertex $l(u)$ in F changes either to $l'(u)$ or to $l''(u)$ in F' depending on adjacency of u to v. Now we need only one additional label to obtain a $(2p+1)$-expression for G from F'. Thus, $cwd(G) \leq 2cwd(G - v) + 1$ and the proposition follows for $[Y]_k$.

To prove the theorem for $[Y]_B$, consider a graph G in $[Y]_B$. By assumption, the clique-width of every block of G is bounded by p. We show by induction on the number of blocks that $cwd(G) \leq p + 2$. Two additional labels needed to construct a $(p + 2)$-expression for G will be denoted α and β. Let first G be a block and v an arbitrary vertex in G. Any p-expression $F(G)$ defining G can trivially be modified into a $(p + 1)$-expression $F'_v(G)$ in which v is the only vertex labeled with α. We then transform $F'_v(G)$ into a $(p+2)$-expression $F_v(G)$ by re-labeling every vertex different from v with β. Now let H be a graph with $b > 1$ blocks and G be a block in H with a single cut-point v. Deleting from H all the vertices of G except v, we obtain a graph with $b - 1$ blocks. Such a graph can be defined by a $(p + 2)$-expression T due to the inductive hypothesis. Assume the vertex v is created in T with label j. Then substituting $j(v)$ with $\rho_{\alpha \to j}(F_v(G))$ in T we obtain a $(p + 2)$-expression defining H. To see this, it is enough to notice that the label β is never renamed or used in any η-operation in T by the inductive hypothesis. ∎

3   Monotone Graph Classes

A hereditary graph class is called *monotone* if it is closed under edge deletion. The classes of bipartite graphs and forests constitute two well-known examples of monotone classes. For a set of graphs Y, let us denote by $Free_m(Y)$ the class of graphs containing no graph in Y as a subgraph (not necessarily induced). It is well known that a class of graphs X is monotone if and only if $X = Free_m(Y)$ for some set Y. A monotone graph class of particular interest in this section is $L_k := Free_m(\{C_j \mid j \geq k\})$.

Lemma 1 *For each $k \geq 3$, the clique width of graphs in L_k is bounded by a constant.*

Proof. For $k = 3$, the proposition follows from the fact that L_3 is the class of forests. For $k > 3$, we use the induction on k and Theorem 2.

Let G be a graph in L_{k+1} and H a block in G. We will show that $H \in [L_k]_k$, which is a trivial observation in the case that H contains at most one cycle of length k. Now let C^1 and C^2 be two cycles of length k in H. Assume they are vertex disjoint. Consider two edges $e_1 \in C^1$ and $e_2 \in C^2$. Since H is connected and has no cut-points, there is a cycle in H containing both e_1 and e_2. In this cycle, one can distinguish two disjoint paths P' and P'', each of which contains the endpoints in C^1 and C^2, and the remaining vertices outside the cycles. The endpoints of the paths P' and P'' partition each of the cycles C^1 and C^2 into two parts. The larger parts in both cycles together with paths P' and P'' form a cycle of length at least $k + 2$, contradicting the assumption that $G \in L_{k+1}$. This contradiction shows that any two cycles of length k in H have a vertex in common. Therefore, removing the vertices of any cycle of length k from H results in a graph in L_k, as required. ∎

Lemma 2 *For each $k \geq 1$, the clique width of graphs in $Free_m(S_{k,k,k})$ is bounded by a constant.*

Proof. The statement of the lemma will be deduced from Theorem 2 and Lemma 1. Specifically, we shall show that every connected graph $G \in Free_m(S_{k,k,k})$ belongs to $[L_{2k+1}]_{2k-2}$. To this end, consider a path P of length $2k - 2$, and a cycle C of length at least $2k + 1$ in G. If G does not contain such P or C, the proposition is obvious. Assume P and C are vertex disjoint. Since G is connected, there must be a path P' whose endpoints belong to C and P, and the remaining vertices are outside C and P. Then the union of C, P and P' contains a subgraph isomorphic to $S_{k,k,k}$. This contradiction shows that P and C contain a vertex in common. Therefore, the graph obtained from G by deletion of the vertices of P belongs to L_{2k+1}, and the proposition follows. ∎

Theorem 3 *Let X be a monotone graph class. If $S \nsubseteq X$, then the clique-width of graphs in X is bounded by a constant.*

Proof. If $S \nsubseteq X$, then for any graph $H \in S - X$ we have $X \subseteq Free_m(H)$. Without loss of generality we may assume that every connected component of H is of the form $S_{k,k,k}$ for some $k \geq 1$ (obviously every graph in $S - X$ can be extended to such $H \in S - X$). Let $H_1, \ldots, H_s$ be the connected components of H. We will show that graphs in $Free_m(H)$ are of bounded clique-width by induction on s. If $s = 1$, the proposition follows from Lemma 2. If $s > 1$, then $Free_m(H) \subseteq Free_m(H_s) \cup [Free_m(H - V(H_s))]_{3k+1}$, where $3k+1$ is the number of vertices in $S_{k,k,k}$. Thus, the proposition follows by the inductive hypothesis and Theorem 2. ∎

4 Subclasses of Bipartite Graphs

It has been mentioned in Section 2 that the clique-width of bipartite graphs is unbounded since general bipartite graphs constitute a superfactorial class. In fact, several stronger propositions can be derived from Theorem 1 and some available results.

A bipartite graph is chordal bipartite if it does not contain induced cycles of length more than four. Spinrad has shown in [23] that the number of chordal bipartite graphs is $\Omega(2^{\Omega(n \log^2 n)})$. Thus, the chordal bipartite graphs form a superfactorial class and hence are not of bounded clique-width.

Another example comes from the well-known results on the maximum number of edges in bipartite graphs containing no $K_{p,p}$ as an induced subgraph. Denoting the class of $K_{p,p}$-free bipartite graphs by X^p, we have (see, e.g., [3, 10]):

$$c_1 n^{2-\frac{2}{p+1}} < \log_2 X_n^p < c_2 n^{2-\frac{1}{p}} \log_2 n \ .$$

In particular, the class of C_4-free bipartite graphs (i.e. X^2) is superfactorial and hence is not of bounded clique-width due to Theorem 1.

The result for C_4-free bipartite graphs has been improved in [15] in the following way. For each odd k, the authors present an infinite family of n-vertex bipartite graphs of girth (the length of a smallest cycle) at least $k + 5$ and of size (the number of edges) at least $2^{t-k-2} n^{1+\frac{1}{k-t+1}}$. Consequently, for each odd $k \geq 1$, $(C_4, \dots, C_{k+3})$-free bipartite graphs constitute a superfactorial class and hence are not of bounded clique-width.

In addition, the clique-width has been proven to be unbounded for several factorial classes of bipartite graphs such as square grids [6] or bipartite permutation graphs [4].

Several positive results on this topic refer to subclasses of bipartite graphs defined by forbidding the graph $S_{i,j,k}$ for some particular values of i, j and k. Obviously, the clique-width of claw-free bipartite graphs is bounded, since every connected graph in this class is either a path or a cycle. In [16], this simple result has been extended to the class of $S_{1,2,2}$-free bipartite and in [18], to $S_{1,2,3}$-free bipartite graphs. Notice that the latter class contains the bipartite analog of cographs [13] and some of their extensions [12,17]. In the class of $S_{2,2,2}$-free bipartite graphs the clique-width is unbounded, since it contains all bipartite permutation graphs. An interesting subclass of $S_{2,2,2}$-free bipartite graphs, defined by one additional forbidden bipartite subgraph, has been characterized recently in [2]. Combining that characterization with Theorem 2 and several more results we derive the following theorem.

Theorem 4 *The clique-width of $(S_{2,2,2}, A)$-free bipartite graphs is bounded, where A is the graph obtained from a P_6 by joining two vertices of degree two at distance three by an edge.*

Proof. We call a connected bipartite graph G *prime* if any two distinct vertices of G have different neighborhoods. The structure of $(S_{2,2,2}, A)$-free bipartite graphs has been characterized in [2] as follows: any prime graph in this class is either a caterpillar or a long circular caterpillar or an almost complete bipartite graph. A caterpillar is a tree that becomes a path by removing the pendant vertices. A circular caterpillar G is a graph that becomes a cycle C_k by removing the pendant vertices. We call G a long circular caterpillar if $k > 4$. A bipartite graph G is almost complete if for every vertex x in G, there is at most one unlike-colored vertex non-adjacent to x.

It has been proven in [8] that the clique-width of a graph G coincides with the clique-width of a maximal prime induced subgraph of G. Thus, to determine the clique-width in the class under consideration we restrict ourselves to caterpillars, long circular caterpillars and almost complete bipartite graphs. Caterpillars are trees and hence of bounded clique-width. Almost complete bipartite graphs are $S_{1,2,3}$-free and hence of bounded clique-width too [18]. For a circular caterpillar G, we apply Theorem 2: deleting a single vertex from G results in a forest. Therefore, the clique-width of circular caterpillars is bounded as well. ∎

5   Subclasses of Claw-Free Graphs

The clique-width of general claw-free graphs is unbounded, because they contain all co-bipartite graphs. In this section we analyze subclasses of claw-free graphs defined by a single additional forbidden subgraph, and provide a complete classification of such classes with respect to the clique-width.

We begin with an auxiliary construction that will permit us to establish unboundedness of the clique-width in some particular claw-free graphs.

Consider an $n \times n$ grid G, i.e. a graph with the vertex set $\{v_{i,j} : 1 \leq i, j \leq n\}$ and the edge set $\{(v_{i,j}, v_{k,l}) : (k = i \text{ and } l = j+1) \text{ or } (k = i+1 \text{ and } l = j)\}$. We shall say that vertex $v_{i,j}$ belongs to row i and column j. The vertices of the grid will be called *black*. Now let us construct a graph H_n from the grid G as follows. First, introduce on each edge of G a new vertex, called *white*. Now black vertices are pairwise nonadjacent, and each of them has at most 4 white neighbors: upper, lower, left and right. We shall say that upper and lower neighbors of a black vertex $v_{i,j}$ belong to column j, while its right and left neighbors are in row i. Finally, for each black vertex, we connect its upper neighbor to the right one, and the left neighbor to the lower one. An example of H_n with $n = 3$ is represented in Figure 2.

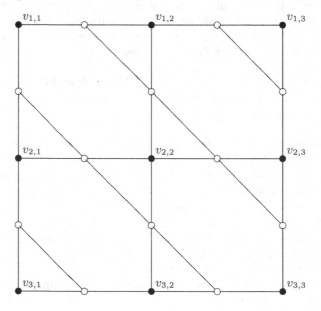

Fig. 2. Graph H_3

Lemma 3 $cwd(H_n) \geq n$.

Proof. Let T be a tree defining H_n. Given a node a in T, we denote by $T(a)$ the subtree of T rooted at a. The label of a vertex v of the graph H_n at the node a of T is defined as the label that v has immediately before the operation a is applied.

Let a be a lowest $\oplus$-node in T such that the graph defined by $T(a)$ contains a full row and a full column of H_n, and denote by b and c the two sons of a in T. Let us color the vertices of H_n in $T(b)$ and $T(c)$ by red and blue, respectively, and all the other vertices by gray. We denote the color of a vertex u by $color(u)$ and say that u is non-gray (respectively, non-red, non-blue) if $color(u)$ is different from gray (respectively, red, blue).

Let u be a red vertex and v a non-gray vertex. Assume there is a non-red vertex w which is adjacent to u but not to v. Since w is non-red, the edge (w, u) appears in H_n under some operation of type η located outside of $T(a)$. Therefore, to avoid introducing the edge (w, v) under the same operation, vertices u and v must have different labels at node a. The same can be proved if u is a blue vertex and w is non-blue. We summarize the above arguments as the following observation:

(1) If u is red (respectively, blue), v is non-gray and there is a non-red (respectively, non-blue) vertex w which is adjacent to u but not to v, then u and v must have different labels at node a.

Due to the choice of a, H_n contains a row and a column with no gray vertex. Let us denote a row without gray vertices by r, and assume first that there is neither blue nor red column in H_n. Then every column j must have a vertex with different color than that of $v_{r,j}$. We denote by w_j a nearest to $v_{r,j}$ vertex in the same column with $color(w_j) \neq color(v_{r,j})$. We then let u_j denote the vertex in the column j which is adjacent to w_j and located between w_j and $v_{r,j}$ (in particular, u_j may coincide with $v_{r,j}$). Denote $Q := \{u_1, u_2, \ldots, u_n\}$. Notice that every vertex $u_j \in Q$ is non-gray and w_j is adjacent to u_j but not to any other vertex in Q. Hence, due to the observation (1), all the vertices in Q have different labels at node a.

We have analyzed the case when there is neither blue nor red column in H_n. Now let H_n contain a blue column. Then obviously it cannot contain a red row. Moreover, it cannot contain a blue row due to the choice of a. We hence argue by symmetry. The case when H_n contains a red column is similar. ∎

It is a trivial observation that graph H_n is $(K_{1,3}, K_4, C_4, K_4 - e)$-free. Together with the preceding lemma this implies

Corollary 1 *The clique-width of* $(K_{1,3}, K_4, C_4, K_4 - e)$*-free graphs is unbounded.*

Lemma 4 *The class of* $(K_{1,3}, 2K_2)$*-free graphs is superfactorial.*

Proof. The class of graphs complement to those under the question contains all (K_3, C_4)-free graphs and hence all C_4-free bipartite graphs. From Section 4 we know that C_4-free bipartite graphs constitute a superfactorial class. ∎

Let D_k denote the class of graphs with vertex degree at most k, and $\overline{D_k}$ the class of complementary graphs.

Lemma 5 *Every $(K_{1,3}, \overline{K}_{1,3})$-free graph belongs to $D_2 \cup [\overline{D_2}]_6$.*

Proof. If a $(K_{1,3}, \overline{K}_{1,3})$-free graph G contains no triangle, then $G \in D_2$. Indeed, if a vertex v in G has at least three neighbors x, y, z, then either v, x, y, z induce a claw (if x, y, z are pairwise non-adjacent) or v, x, y is a triangle (if x is adjacent to y).

Assume now that G contains a triangle $T = \{x, y, z\}$. Let A be the set of vertices outside T adjacent to each vertex in the triangle, and B the set of remaining vertices. Obviously, no triple in A forms a co-triangle else a claw arises in G. Hence the subgraph of G induced by the set $V(G) - B$ belongs to $\overline{D_2}$. It remains to show that $|B| \leq 6$.

First, notice that every vertex v in B has at least one neighbor in T, otherwise v, x, y, z induce a $\overline{K}_{1,3}$ in G. Hence we may partition B into three subsets: B_1, the vertices adjacent to x but not to y; B_2, the vertices adjacent to y but not to z; B_3, the vertices adjacent to z but not to x. Every two vertices a, b in B_1 are adjacent else x, a, b, y would induce a claw. Any triangle in B_1 together with y would induce a $\overline{K}_{1,3}$. Thus, $|B_1| \leq 2$. Similarly, $|B_2| \leq 2$ and $|B_3| \leq 2$. Hence $|B| \leq 6$. ∎

Lemma 6 *Every connected (claw,paw)-free graph belongs to $D_2 \cup \overline{D_1}$.*

Proof. It has been proven in [20] that any connected paw-free graph is either triangle-free or $\overline{K}_{1,2}$-free, i.e. a complete multipartite graph. All triangle-free graphs without claw belong to D_2. Now let G be a connected claw- and $\overline{K}_{1,2}$-free graph. In other words, the complement to G is a $\overline{K}_{1,3}$-free graph whose every component is a clique. Obviously, $\overline{G}$ contains at least 2 connected components, otherwise G is not connected. But then every component of $\overline{G}$ consists of at most 2 vertices else a $\overline{K}_{1,3}$ arises. Hence $\overline{G} \in D_1$. ∎

Corollary 2 *The clique-width of $(K_{1,3}, \overline{K}_{1,3})$-free graphs and (claw,paw)-free graphs is bounded.*

Proof. Clearly, every connected graph in D_2 is either a cycle or a path, hence the clique-width of graphs in D_2 is bounded. It has been proven in [9] that $cwd(\overline{G}) \leq 2cwd(G)$ for any graph G. Consequently, the clique-width is bounded for graphs in $\overline{D_2} \supset \overline{D_1}$. This proposition in conjunction with Theorem 2 and Lemmas 5,6 produces the conclusion. ∎

We now are in a position to prove the main result of this section.

Theorem 5 *Let H be an arbitrary graph and X, the class of $(K_{1,3}, H)$-free graphs. Then the clique-width of graphs in X is bounded if and only if H is either a P_4 or a paw or the complement to a claw or one of their induced subgraphs.*

Proof. If H contains either a co-triangle or a C_5, then $(K_{1,3}, H)$-free graphs include all co-bipartite graphs. Consequently, the clique-width of $(K_{1,3}, H)$-free graphs is unbounded in that case.

If H contains a $2K_2$, then the class of $(K_{1,3}, H)$-free graphs is superfactorial (Lemma 4) and hence is not of bounded clique-width (Theorem 1).

If H contains a C_4 or a K_4 or $K_4 - e$, then the clique-width of $(K_{1,3}, H)$-free graphs is unbounded due to Corollary 1.

Thus, it remains to consider the case when H is a $(\overline{K_3}, 2K_2, C_4, C_5, K_4, K_4 - e)$-free graph. From $(2K_2, C_4, C_5)$-freeness we conclude that H is a split graph, i.e. a graph whose vertices can be partitioned into a clique C and an independent set I [11]. Clearly, C contains at most 3 vertices (otherwise a K_4 arises) and I contains at most 2 vertices (otherwise a $\overline{K_3}$ arises).

If $|C| = 3$, then each vertex in I is adjacent to at most one vertex in C else H contains either a K_4 or $K_4 - e$. Therefore, I may contain at most one vertex v else a $\overline{K_3}$ appears. If v has a neighbor in C, then H is a paw, otherwise it is the complement to a claw. In both cases, the clique-width of $(claw, H)$-free graphs is bounded due to Corollary 2.

If $|C| = 2$ and $|I| = 2$, then at least one vertex in I has less than two neighbors in C else H is a $K_4 - e$. Taking into account $\overline{K_3}$-freeness of H, we hence conclude that H is either a P_4 or a paw or the complement to a claw. In the remaining cases, H is an induced subgraph of one of these graphs. ∎

References

[1] Alekseev V.: On lower layers of the lattice of hereditary classes of graphs. Diskretn. Anal. Issled. Oper. Ser. 1 **4** (1) (1997) 3–12 (in Russian)

[2] Boliac R., Lozin V.V.: An attractive class of bipartite graphs. Discuss. Math. Graph Theory **21** (2) (2001) 293–301

[3] Bollobas B.: Extremal graph theory. Acad. Press, London (1978)

[4] Brandstädt A., Lozin V.V.: On the linear structure and clique-width of bipartite permutation graphs. Ars Combinatoria, to appear

[5] Chen H.-C., Wang Y.-L.: An efficient algorithm for generating Prüfer codes from labeled trees. Theory Comput. Systems **33** (2000) 97–105

[6] Courcelle B.: Graph grammars, monadic second-order logic and the theory of graph minors. Contemporary Mathematics **147** (1993) 565–590

[7] Courcelle B., Engelfriet J., Rozenberg G.: Handle-rewriting hypergraphs grammars. J. Comput. System Sci. **46** (2) (1993) 218–270

[8] Courcelle B., Makowsky J.A., Rotics U.: Linear time solvable optimization problems on graphs of bounded clique-width. Theory Comput. Syst. **33** (2) (2000) 125–150

[9] Courcelle B., Olariu S.: Upper bounds to the clique-width of a graph. Discrete Appl. Math. **101** (1-3) (2000) 77–114

[10] Erdös P., Spencer J.: Probabilistic methods in combinatorics. Probability and Mathematical Statistics, Vol. 17. Academic Press, New York-London (1974)

[11] Foldes S., Hammer P.L.: Split graphs. Congres. Numer. **19** (1977) 311–315

[12] Fouquet J.-L., Giakoumakis V., Vanherpe J.-M.: Bipartite graphs totally decomposable by canonical decomposition. Internat. J. Found. Comput. Sci. **10** (4) (1999) 513–533

[13] Giakoumakis V., Vanherpe J.-M.: Bi-complement reducible graphs. Adv. in Appl. Math. **18** (4) (1997) 389–402

[14] Golumbic M.C., Rotics U.: On the clique-width of some perfect graph classes. Internat. J. Found. Comput. Sci. **11** (3) (2000) 423–443

[15] Lazebnik F., Ustimenko V.A., Woldar A.J.: A new series of dense graphs of high girth. Bull. Amer. Math. Soc. (N.S.) **32** (1) (1995) 73–79

[16] Lozin V.V.: E-free bipartite graphs. Diskretn. Anal. Issled. Oper. Ser. 1, **7** (1) (2000) 49–66 (in Russian)

[17] Lozin V.V.: On a generalization of bi-complement reducible graphs. Lecture Notes in Comput. Sci. **1893** (2000) 528–538

[18] Lozin V.V.: Bipartite graphs without a skew star. Discrete Math. **256** (2002), to appear

[19] Makowsky J.A., Rotics U.: On the clique-width of graphs with few P_4's. Internat. J. Found. Comput. Sci. **10** (3) (1999) 329–348

[20] Olariu S.: Paw-free graphs. Inform. Process. Lett. **28** (1) (1988) 53–54

[21] Prüfer H.: Neuer Beweis eines satzes über Permutationen. Archiv der Mathematik und Physik **27** (1918) 742–744

[22] Scheinerman E.R., Zito J.: On the size of hereditary classes of graphs. J. Combin. Theory, Ser. B **61** (1) (1994) 16–39

[23] Spinrad J. P.: Nonredundant 1's in Γ-free matrices. SIAM J. Discrete Math. **8** (2) (1995) 251–257

The Probability of a Rendezvous Is Minimal in Complete Graphs

Martin Dietzfelbinger*

Technische Universität Ilmenau
Fakultät für Informatik und Automatisierung
98684 Ilmenau, Germany
martin.dietzfelbinger@tu-ilmenau.de

Abstract. In a connected simple graph G the following random experiment is carried out: each node chooses one of its neighbors uniformly at random. We say a rendezvous occurs if there are adjacent nodes u and v such that u chooses v and v chooses u. Métivier *et al.* (2000) asked whether it is true that the probability for a rendezvous to occur in G is at least as large as the probability of a rendezvous if the same experiment is carried out in the complete graph on the same number of nodes. In this paper we show that this is the case.

1 Introduction

The following random experiment was studied by Métivier, Saheb, and Zemmari [3]. Let $G = (V, E)$ be an undirected connected graph (representing a network) with $|V| = n \geq 2$ nodes. Each node $u \in V$ independently chooses one of its neighbors uniformly at random. We say *there is a rendezvous* (and consider this as a *"success"*) if there is an edge (u, v) in G such that u chooses v and v chooses u. Let $s(G) \in [0, 1]$ be the probability that there is a rendezvous if this experiment is carried out in G.

As usual, K_n denotes the complete graph with n nodes. The following is known. (For the proofs see [3]; J M. Robson had first proved $s(G) \geq 1 - 1/\sqrt{e}$.)

(i) $s(G) \geq 1 - e^{-n/2(n-1)}$ for every n-node graph G;
(ii) $s(K_n) \to 1 - 1/\sqrt{e}$ for $n \to \infty$ (convergence from above).

Since obviously $e^{-n/2(n-1)} \to 1/\sqrt{e}$ for $n \to \infty$, asymptotically the complete graphs achieve the minimum rendezvous probability, namely $1 - 1/\sqrt{e} \approx 0.39347$. It is natural to ask whether for each n the complete graph minimizes the rendezvous probability among all n-node graphs. Métivier *et al.* [3] stated the following assertion as an open problem.

Theorem 1. *If G is a connected graph with $n \geq 2$ nodes, then $s(G) \geq s(K_n)$.*

---

* Part of this work was done while the author was visiting the Max-Planck-Institut für Informatik, Saarbrücken, Germany.

P. Bose and P. Morin (Eds.): ISAAC 2002, LNCS 2518, pp. 55–66, 2002.

It is the purpose of this paper to prove this statement.

For a thorough discussion of the randomized rendezvous protocol, in particular the study of its behaviour in many example graph families like trees, rings, and graphs with bounded degree, and for applications in distributed computing, the reader is referred to [3].

A simple idea how the theorem might be proved is to consider the following step: to an incomplete graph G add some edge to obtain a graph G'. If this could always be done in such a way that $s(G) \geq s(G')$, then starting from an arbitrary graph G_0 on n nodes we could iterate this step, until we finally reach the complete graph K_n, never increasing the rendezvous probability. Unfortunately, it can be shown that the operation of adding an edge is not generally monotone: there are graphs G and G' such that G' results from G by adding one edge, but $s(G) < s(G')$. This was observed by Austinat in his Diplom thesis [1]. (His example is described in [3].) Even worse, there are graphs G so that for *each* graph G' obtained from G by adding one edge we have $s(G) < s(G')$. In [1], the following example is attributed to J. M. Robson (Fig. 1). Let G_n be the graph

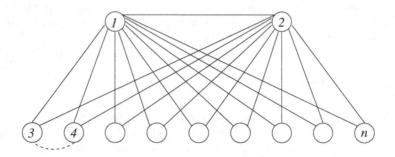

Fig. 1. Adding any edge increases the rendezvous probability

on node set $\{1, \ldots, n\}$ with edge set $\{(1,2)\} \cup \{(1,i),(2,i) \mid i = 3, \ldots, n\}$. It is not hard to see that $s(G_n) \to \frac{3}{4}$ for $n \to \infty$. On the other hand, adding one edge means adding (u, v) for some $u, v \in \{3, \ldots, n\}, u \neq v$. For symmetry reasons, we may assume that $u = 3, v = 4$. Call the resulting graph G'_n. Again, it is easy to see that $s(G'_n) \to \frac{7}{9}$ for $n \to \infty$. Since $\frac{7}{9} - \frac{3}{4} = \frac{1}{36}$, we have $s(G_n) < s(G'_n)$ for n large enough.

In our proof of the theorem, we modify the idea just sketched as follows. We generalize the random experiments by admitting certain *nonuniform* distributions at the nodes.

As before, let a connected graph $G = (V, E)$ with n nodes be given. Associated with each node u is a probability distribution on the set $N(u) = \{v \mid (u, v) \in E\}$ of its neighbors in G, given by numbers $p_{uv} \in [0, 1], v \in N(u)$, with

$$\sum_{v \in N(u)} p_{uv} = 1, \text{ for every } u \in V . \tag{1}$$

Node u chooses $v \in N(u)$ with probability p_{uv}, independently of the random choices of the other nodes. We assume

$$p_{uv} \geq \frac{1}{n-1}, \text{ for every } u \in V, v \in N(u). \tag{2}$$

This requirement has the effect that if u has maximal degree $n-1$ in G, then u chooses each of its neighbors with the same probability. In particular, if G is the complete graph K_n, all probabilities are equal, and we are back at the original experiment. — A combination of a graph and a probability distribution is denoted by $(G, (p_{uv})_{u \in V, v \in N(u)})$, or $(G, (p_{uv}))$ for short. If $(p_{uv})_{u \in V, v \in N(u)}$ is given by the context, we may also drop the (p_{uv}) part and simply write G. With $s(G, (p_{uv}))$ or $s(G)$ we denote the probability that the random experiment creates a rendezvous at at least one of the edges.

Given a graph $(G, (p_{uv}))$ with probabilities, we may add a new edge $(\hat{u}, \hat{v})$ to G as follows. We arrange it so that $\hat{u}$ chooses $\hat{v}$ with probability exactly $\frac{1}{n-1}$, and so that the probabilities that $\hat{u}$ chooses one of its other neighbors are decreased accordingly, but so that each one remains at least $\frac{1}{n-1}$. At the other node $\hat{v}$ probabilities are rearranged similarly. Later we show that adding a new edge to $(G, (p_{uv}))$ in this way does not increase the probability of a rendezvous if $s(G, (p_{uv}))$ is not larger than $\alpha = (\sqrt{5} - 1)/2 \approx 0.61803$. We will see below that $s(K_n) \leq 0.6$ for $n \geq 5$; this entails that it is easy to guarantee that $s(G, (p_{uv})) < \alpha$ in the relevant cases.

The overall argument now runs as follows. We start with some connected graph $G_0 = (V, E_0)$ on n nodes; we may assume $s(G_0) < \alpha$. The initial probabilities

$$p_{uv} = 1/|N(u)|, \text{ for } u \in V, (u, v) \in E_0 ,$$

are chosen so that the uniform distribution is represented. Adding edges one by one, in the manner just described, we run through a sequence of graphs with *nonuniform* distributions, never increasing the probability for a rendezvous, until finally we reach the complete graph K_n with all probabilities being equal to $\frac{1}{n-1}$. This implies $s(G_0) \geq s(K_n)$, as desired.

In the rest of the paper, the theorem is proved. In Section 2 some known facts and some preliminary arguments are provided. In Section 3 we analyze the step of adding one edge with nonuniform probabilities and prove the theorem. The proof of a central, but technical lemma is supplied in Section 4.

2   Basics

In this section, we list some facts about the randomized rendezvous experiment, as can be found in [3]. We also discuss the theorem for very small graphs. Unless stated otherwise, the following discussion applies to an arbitrary connected n-node graph $G = (V, E)$, $n \geq 2$.

The probability that there is a rendezvous at edge $e = (u, v)$ is

$$p_e = \frac{1}{|N(u)| \cdot |N(v)|}.$$

An important measure is

$$m(G) = \sum_{e \in E} p_e,$$

the expected number of edges at which a rendezvous occurs. Clearly, $s(G) \leq m(G)$. Further, it is not hard to show that

$$m(G) \geq \frac{n}{2(n-1)} \quad (= m(K_n)). \tag{3}$$

In order to obtain a lower bound for $s(G)$ we need the following basic fact, which is stated in [1] (with a proof by J. M. Robson) and also in [3, Proof of Prop. 15]. We give a simple proof without calculations.

Fact 2. *Let $B \subseteq E$ be an arbitrary set of edges, and let $e_0 \in E - B$. Let C_B be the event that no rendezvous occurs along any edge in B; let R_{e_0} be the event that there is a rendezvous on edge e_0. Then $\mathbf{P}(R_{e_0} \mid C_B) \geq \mathbf{P}(R_{e_0}) = p_{e_0}$.*

Proof. We must show that $\mathbf{P}(R_{e_0} \cap C_B) \geq \mathbf{P}(R_{e_0}) \cdot \mathbf{P}(C_B)$. — Let $e_0 = (u,v)$, and let C'_B denote the event that the choices of the nodes in $\{1, \ldots, n\} - \{u,v\}$ do not create a rendezvous on any edge in B, disregarding the edges incident with u or v. We observe that (i) $R_{e_0} \cap C_B = R_{e_0} \cap C'_B$ (if there is a rendezvous on edge e_0, there cannot be a rendezvous on any edge in B incident with u or v); (ii) the events C'_B and R_{e_0} are independent; (iii) $C_B \subseteq C'_B$, and estimate

$$\mathbf{P}(R_{e_0} \cap C_B) \overset{(i)}{=} \mathbf{P}(R_{e_0} \cap C'_B) \overset{(ii)}{=} \mathbf{P}(R_{e_0}) \cdot \mathbf{P}(C'_B) \overset{(iii)}{\geq} \mathbf{P}(R_{e_0}) \cdot \mathbf{P}(C_B),$$

as desired. $\square$

Using Fact 2, it is easy to see that

$$s(G) \geq 1 - \prod_{e \in E} (1 - p_e) \geq 1 - e^{-m(G)}, \tag{4}$$

which, together with (3), implies $s(G) \geq 1 - e^{-n/2(n-1)}$. Clearly, $1 - e^{-n/2(n-1)} \to 1 - 1/\sqrt{e} \approx 0.39347$ for $n \to \infty$.

Let $\mathcal{M}(G, j)$ denote the set of all matchings $M \subseteq E$ in G that contain exactly j edges. The inclusion-exclusion principle from combinatorics implies

$$s(G) = \sum_{j \geq 1} (-1)^{j-1} \sum_{M \in \mathcal{M}(G,j)} \prod_{e \in M} p_e. \tag{5}$$

From (5), one obtains the following "closed" formula for $s(K_n)$.

$$s(K_n) = \sum_{j \geq 1} (-1)^{j+1} \binom{n}{2j} \cdot \frac{(2j)!}{j! \cdot 2^j} \frac{1}{(n-1)^{2j}}. \tag{6}$$

The first summand in this sum is $m(K_n) = n/2(n-1)$. From (6), it is easily seen that

$$s(K_n) \to \sum_{j \geq 1} (-1)^{j+1} \frac{1}{j! \cdot 2^j} = 1 - e^{-1/2}, \text{ for } n \to \infty. \tag{7}$$

On the other hand, (6) can be used to calculate $s(K_n)$ for small n:

n	2	3	4	5	6	7	8
$s(K_n)$	1	$\frac{3}{4}$	$\frac{17}{27}$	$\frac{145}{256}$	$\frac{1653}{3125}$	$\frac{7847}{15552}$	$\frac{401491}{823543}$
decimal value	1.0	0.75	0.6296...	0.56640625	0.52896	0.5045...	0.4875...

Lemma 3. *If $n \geq 5$ then $s(K_n) < 0.6$.*

Proof. By the table, $s(K_5) = \frac{145}{256} < 0.6$. If $n \geq 6$, we use the simple fact that $s(K_n) < m(K_n) = \binom{n}{2} \cdot \frac{1}{(n-1)^2} = n/2(n-1) \leq 6/10$.   □

For graphs with very few nodes, Theorem 1 is proved by inspection. The cases $n = 2$ and $n = 3$ are trivial. For $n = 4$, we recall yet another fact from [3], namely that $s(G) = 1$ if G is a tree. There are only five different incomplete connected graphs on 4 nodes, which are depicted in Fig. 2 below.

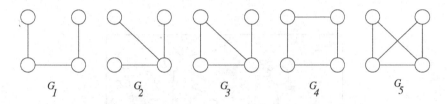

Fig. 2. The 5 incomplete connected graphs on 4 nodes

Graphs G_1 and G_2 are trees, hence $s(G_1) = s(G_2) = 1$. For the other graphs we use the inclusion-exclusion formula (5) to obtain $s(G_3) = \frac{5}{6}$, $s(G_4) = \frac{7}{8}$, and $s(G_5) = \frac{13}{18}$; all these values are larger than $s(K_4) = \frac{17}{27}$.

3   Adding an Edge in the Nonuniform Case

In view of the remarks made in the last section, we assume from here on that $n \geq 5$ and $s(K_n) < 0.6$.

Assume a connected graph $G = (V, E)$ on node set $V = \{1, \dots, n\}$ is given together with a family $(p_{uv})_{u \in V, v \in N(u)}$ of probability distributions, which are assumed to satisfy (1) and (2). We imagine that each node u, independently of the others, chooses one of its neighbors, governed by the distribution $(p_{uv})_{v \in N(u)}$. This induces a probability space. Probabilities in this probability space are denoted as $\mathbf{P}_{(G,(p_{uv}))}(\cdot)$ or as $\mathbf{P}_G(\cdot)$. We say a rendezvous occurs at edge (u, v) if u has chosen v and v has chosen u. This event is denoted by $R_{(u,v)}$. The event $\bigcup_{e \in E} R_e$ (there is some rendezvous) is denoted by R_G; the probability $\mathbf{P}_G(R_G)$ of this event is abbreviated as $s(G)$.

We describe in detail the operation of adding one edge to $(G, (p_{uv}))$, in case G is not the complete graph. By renaming we may assume that $(1, 2)$ is not in E.

We form $G' = (V, E')$ by $E' := E \cup \{(1,2)\}$ and fix new probabilities as follows, cf. Fig. 3. Let

$$p'_{12} = p'_{21} = \frac{1}{n-1} \, ; \tag{8}$$

for $u \in N(1)$, let $p'_{1u} = p_{1u} - \varepsilon_u$ for some $\varepsilon_u \geq 0$ such that

$$p'_{1u} \geq \frac{1}{n-1}, \text{ for } u \in N(1), \text{ and } \sum_{u \in N(1)} \varepsilon_u = \frac{1}{n-1}; \tag{9}$$

for $v \in N(2)$, let $p'_{2v} = p_{2v} - \delta_v$ for some $\delta_v \geq 0$ such that

$$p'_{2v} \geq \frac{1}{n-1}, \text{ for } v \in N(2), \text{ and } \sum_{v \in N(2)} \delta_v = \frac{1}{n-1}; \tag{10}$$

finally, let

$$p'_{uv} = p_{uv}, \text{ if } u, v \in V - \{1,2\}, v \in N(u). \tag{11}$$

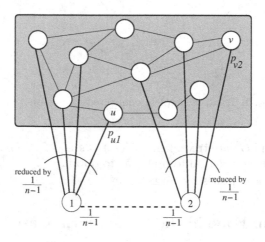

Fig. 3. Inserting an edge with minimal probabilities

The remainder of this section is devoted to the proof of the following central lemma concerning $s(G) = s(G, (p_{uv}))$ and $s(G') = s(G', (p'_{uv}))$. It uses the number $\alpha = \frac{1}{2}(\sqrt{5} - 1) \approx 0.61803$, which is the unique solution of the equation $x + x^2 = 1$ in $[0, 1]$.

Lemma 4 (Main Lemma). If $s(G) \leq \alpha$, then $s(G') \leq s(G)$.

Proof. In the following, we write $\mathbf{P}(\cdot)$ for $\mathbf{P}_{(G,(p_{uv}))}(\cdot)$, $\mathbf{P}'(\cdot)$ for $\mathbf{P}_{(G',(p'_{uv}))}(\cdot)$, R for R_G and R' for $R_{G'}$. Consider

$$C = \{\text{no rendezvous occurs among any two of the nodes } 3, \ldots, n\}. \tag{12}$$

The event C is not affected by the choices of nodes 1 and 2, hence:

$$\mathbf{P}(C) = \mathbf{P}'(C). \tag{13}$$

Now $\overline{C}$ implies R, hence

$$s(G) = \mathbf{P}(R) = \mathbf{P}(R \mid C) \cdot \mathbf{P}(C) + \mathbf{P}(\overline{C}) = \mathbf{P}(R \mid C) \cdot \mathbf{P}(C) + 1 - \mathbf{P}(C). \tag{14}$$

Similarly (using (13)),

$$s(G') = \mathbf{P}'(R') = \mathbf{P}'(R' \mid C) \cdot \mathbf{P}(C) + 1 - \mathbf{P}(C). \tag{15}$$

In case $\mathbf{P}(C) = 0$ the lemma is trivially true. Thus we may assume from now on that $\mathbf{P}(C) > 0$. By (14) and (15), to prove the lemma it is sufficient to show

$$\mathbf{P}'(R' \mid C) \le \mathbf{P}(R \mid C). \tag{16}$$

In the proof of (16), the following abbreviations are helpful:

$$p_u = p_{1u},$$
$$p_u^* = \mathbf{P}(u \text{ chooses } 1 \mid C), \quad \text{for } u \in N(1);$$
$$q_v = p_{2v},$$
$$q_v^* = \mathbf{P}(v \text{ chooses } 2 \mid C), \quad \text{for } v \in N(2);$$
$$\beta_{uv}^* = \mathbf{P}(u \text{ chooses } 1 \text{ and } v \text{ chooses } 2 \mid C), \quad \text{for } u \in N(1), v \in N(2).$$

Note that the probabilities p_u^*, q_v^*, and β_{uv}^* would not change if $\mathbf{P}'(\cdot)$ instead of $\mathbf{P}(\cdot)$ was used.

In a way similar to Fact 2 condition C increases the probability that u chooses 1 (resp. that v chooses 2):

Lemma 5. (a) $p_u^* \ge p_{u1}$ (hence $p_u^* \ge \frac{1}{n-1}$), for $u \in N(1)$.

(b) $q_v^* \ge q_{v2}$ (hence $q_v^* \ge \frac{1}{n-1}$), for $v \in N(2)$.

Proof. (Cf. Fact 2.) It is sufficient to prove (a). Fix $u \in N(1)$. Let C' be the event that the choices of the nodes in $\{3, \ldots, n\} - \{u\}$ do not create a rendezvous among themselves (not regarding what u does). We observe

(i) $\{u \text{ chooses } 1\} \cap C' = \{u \text{ chooses } 1\} \cap C$ (indeed, if u chooses 1, a rendezvous in $\{3, \ldots, n\}$ can only occur among nodes in $\{3, \ldots, n\} - \{u\}$);
(ii) the events C' and $\{u \text{ chooses } 1\}$ are independent;
(iii) $C \subseteq C'$,

to obtain:

$$p_u^* = \mathbf{P}(u \text{ chooses } 1 \mid C) = \mathbf{P}(\{u \text{ chooses } 1\} \cap C)/\mathbf{P}(C)$$
$$\overset{(i)}{=} \mathbf{P}(\{u \text{ chooses } 1\} \cap C')/\mathbf{P}(C) \overset{(ii)}{=} \mathbf{P}(u \text{ chooses } 1) \cdot \mathbf{P}(C')/\mathbf{P}(C)$$
$$\overset{(iii)}{\ge} \mathbf{P}(u \text{ chooses } 1) = p_{u1}.$$

$\square$

Define the events

$$R_1 = \bigcup_{u \in N(1)} R_{(1,u)} \quad \text{(there is a rendezvous involving node 1), and}$$

$$R_2 = \bigcup_{v \in N(2)} R_{(2,v)} \quad \text{(there is a rendezvous involving node 2).}$$

The corresponding events in the probability space for G' (not including the event that there is a rendezvous along the new edge $(1,2)$) are denoted by R_1' and R_2'. Now we write out the two probabilities in (16) in full. Using the obvious fact that for $u, \hat{u} \in N(1)$, $u \neq \hat{u}$ the events $\{1 \text{ chooses } u\}$ and $\{1 \text{ chooses } \hat{u}\}$ are disjoint, we obtain

$$\mathbf{P}(R \mid C) = \mathbf{P}(R_1 \mid C) + \mathbf{P}(R_2 \mid C) - \mathbf{P}(R_1 \cap R_2 \mid C)$$
$$= \sum_{u \in N(1)} p_u p_u^* + \sum_{v \in N(2)} q_v q_v^* - \sum_{u \in N(1), v \in N(2)} p_u q_v \beta_{uv}^*. \tag{17}$$

In $(G', (p'_{uv}))$, the choices made by nodes 1 and 2 are independent of C; further, if there is a rendezvous at $(1,2)$, there cannot be a rendezvous between 1 or 2 and any other node. Hence we obtain

$$\mathbf{P}'(R' \mid C) = \mathbf{P}'(R_{(1,2)} \mid C) + \mathbf{P}'(R_1' \mid C) + \mathbf{P}'(R_2' \mid C) - \mathbf{P}'(R_1' \cap R_2' \mid C)$$
$$= p'_{12} p'_{21} + \sum_{u \in N(1)} (p_u - \varepsilon_u) p_u^* + \sum_{v \in N(2)} (q_v - \delta_v) q_v^* - \sum_{u \in N(1), v \in N(2)} (p_u - \varepsilon_u)(q_v - \delta_v) \beta_{uv}^*$$
$$= \frac{1}{(n-1)^2} + \sum_{u \in N(1)} p_u p_u^* - \sum_{u \in N(1)} \varepsilon_u p_u^* + \sum_{v \in N(2)} q_v q_v^* - \sum_{v \in N(2)} \delta_v q_v^* -$$
$$- \sum_{u \in N(1), v \in N(2)} (p_u q_v - p_u \delta_v - \varepsilon_u q_v + \varepsilon_u \delta_v) \beta_{uv}^*. \tag{18}$$

Subtracting (17) from (18), and using the obvious fact that $\varepsilon_u \delta_v \geq 0$, we get

$$\mathbf{P}'(R' \mid C) - \mathbf{P}(R \mid C) \leq \tag{19}$$
$$\leq \frac{1}{(n-1)^2} - \sum_{u \in N(1)} \varepsilon_u p_u^* - \sum_{v \in N(2)} \delta_v q_v^* + \sum_{u \in N(1), v \in N(2)} (p_u \delta_v + \varepsilon_u q_v) \beta_{uv}^*.$$

At this point it becomes apparent that in order to proceed we must establish a relation between $\beta_{uv}^* = \mathbf{P}(u \text{ chooses } 1 \text{ and } v \text{ chooses } 2 \mid C)$ on the one hand and $p_u^* = \mathbf{P}(u \text{ chooses } 1 \mid C)$ and $q_v^* = \mathbf{P}(v \text{ chooses } 2 \mid C)$ on the other. It is clear that $\beta_{uv}^* = 0$ for $u = v$. If $u \neq v$, then in the unconditioned situation the events $\{u \text{ chooses } 1\}$ and $\{v \text{ chooses } 2\}$ are independent; in the probability space we obtained by conditioning on C this is no longer the case. The next lemma states that these two events are at least "negatively correlated" under the condition that C is true: if one of them happens, the other one becomes less likely. It is by no means obvious that this is so; the somewhat tricky proof will be supplied in the next section.

Lemma 6. $\beta_{uv}^* \leq p_u^* q_v^*$, *for* $u \in N(1), v \in N(2)$.

Using Lemma 6 we may continue from inequality (19) as follows:

$$\mathbf{P}'(R' \mid C) - \mathbf{P}(R \mid C) \leq \qquad\qquad\qquad\qquad\qquad (20)$$

$$\leq \frac{1}{(n-1)^2} - \sum_{u \in N(1)} \varepsilon_u p_u^* - \sum_{v \in N(2)} \delta_v q_v^* + \sum_{u \in N(1), v \in N(2)} (p_u \delta_v + \varepsilon_u q_v) p_u^* q_v^* =$$

$$= \frac{1}{(n-1)^2} - \sum_{u \in N(1)} \varepsilon_u p_u^* - \sum_{v \in N(2)} \delta_v q_v^* +$$

$$+ \Big(\sum_{u \in N(1)} p_u p_u^* \Big) \Big(\sum_{v \in N(2)} \delta_v q_v^* \Big) + \Big(\sum_{u \in N(1)} \varepsilon_u p_u^* \Big) \Big(\sum_{v \in N(2)} q_v q_v^* \Big)$$

$$= \frac{1}{(n-1)^2} - \Big(\sum_{u \in N(1)} \varepsilon_u p_u^* \Big) \Big(1 - \sum_{v \in N(2)} q_v q_v^* \Big) - \Big(1 - \sum_{u \in N(1)} p_u p_u^* \Big) \Big(\sum_{v \in N(2)} \delta_v q_v^* \Big)$$

$$= \frac{1}{(n-1)^2} - \Big(\sum_{u \in N(1)} \varepsilon_u p_u^* \Big) \mathbf{P}(\overline{R}_2 \mid C) - \mathbf{P}(\overline{R}_1 \mid C) \Big(\sum_{v \in N(2)} \delta_v q_v^* \Big).$$

By Lemma 5 we have $p_u^* \geq \frac{1}{n-1}$ and $q_v^* \geq \frac{1}{n-1}$, and by (9) and (11) we have $\sum_{u \in N(1)} \varepsilon_u = \frac{1}{n-1}$ and $\sum_{v \in N(2)} \delta_v = \frac{1}{n-1}$. Substituting this into (20) and simplifying we obtain

$$\mathbf{P}'(R' \mid C) - \mathbf{P}(R \mid C) \leq$$

$$\leq \frac{1}{(n-1)^2} - \frac{1}{(n-1)^2} \cdot \mathbf{P}(\overline{R}_2 \mid C) - \mathbf{P}(\overline{R}_1 \mid C) \cdot \frac{1}{(n-1)^2} =$$

$$= \frac{1}{(n-1)^2} \cdot (\mathbf{P}(R_1 \mid C) + \mathbf{P}(R_2 \mid C) - 1).$$

Thus, in order to prove (16) it is sufficient to show

$$\mathbf{P}(R_1 \mid C) + \mathbf{P}(R_2 \mid C) \leq 1,$$

or, equivalently,

$$\mathbf{P}(R_1 \cup R_2 \mid C) + \mathbf{P}(R_1 \cap R_2 \mid C) \leq 1. \qquad\qquad (21)$$

We know from (14) that $\mathbf{P}(R) = \mathbf{P}(R \mid C)\mathbf{P}(C) + 1 \cdot (1 - \mathbf{P}(C))$. This means that $\mathbf{P}(R)$ is a convex combination of $\mathbf{P}(R \mid C)$ and 1, hence

$$\mathbf{P}(R_1 \cup R_2 \mid C) = \mathbf{P}(R \mid C) \leq \mathbf{P}(R).$$

Further, using Lemma 6 again, we get

$$\mathbf{P}(R_1 \cap R_2 \mid C) = \sum_{u \in N(1), v \in N(2)} p_u q_v \beta_{uv}^*$$

$$\leq \sum_{u \in N(1), v \in N(2)} p_u q_v p_u^* q_v^* = \Big(\sum_{u \in N(1)} p_u p_u^* \Big) \Big(\sum_{v \in N(2)} q_v q_v^* \Big)$$

$$= \mathbf{P}(R_1 \mid C) \cdot \mathbf{P}(R_2 \mid C) \leq \mathbf{P}(R \mid C)^2$$

$$\leq \mathbf{P}(R)^2.$$

Adding the last two inequalities yields

$$\mathbf{P}(R_1 \cup R_2 \mid C) + \mathbf{P}(R_1 \cap R_2 \mid C) \le \mathbf{P}(R) + \mathbf{P}(R)^2 \le \alpha + \alpha^2 = 1,$$

since we have assumed that $s(G) = \mathbf{P}(R) \le \alpha = \frac{1}{2}(\sqrt{5} - 1)$, and $\alpha + \alpha^2 = 1$. Thus, (21) and hence (16) holds, and Lemma 4 is proved.   □

4   Proof of the Correlation Lemma

This section is devoted to the proof of Lemma 6. In order to carry out a proof by induction, we formulate a more general statement.

Lemma 7. *Fix two distinct nodes in G, called 1 and 2 for simplicity, and a nonempty set $U \subseteq V - \{1, 2\}$. Let $u, v \in U$ be arbitrary, and consider the events*

$$A = \{u \text{ chooses } 1\}, \qquad B = \{v \text{ chooses } 2\}, \qquad and$$
$$C = \{\text{there is no rendezvous among any two nodes in } U\}.$$

Then
$$\mathbf{P}(A \cap B \mid C) \le \mathbf{P}(A \mid C) \cdot \mathbf{P}(B \mid C).$$

(Lemma 6 follows by considering $U = V - \{1, 2\}$.)

Proof. If $\Pr(C) = 0$, there is nothing to prove. Thus assume $\Pr(C) > 0$ and note that then the statement of the Lemma is equivalent to

$$\mathbf{P}(A \cap B \cap C)\mathbf{P}(C) \le \mathbf{P}(A \cap C)\mathbf{P}(B \cap C). \tag{22}$$

We prove (22) by induction on $|U|$.

Base case $|U| = 1$: In this case u and v are identical, and hence A and B are disjoint events. Thus $\mathbf{P}(A \cap B \cap C) = 0$, and we are done.

Base case $|U| = 2$: In view of the argument in the previous case we may assume that $U = \{u, v\}$. If u and v are not adjacent, $\Pr(C) = 1$, and there is nothing to show. So assume edge (u, v) is present. The only place where a rendezvous can occur in $\{u, v\}$ is the edge (u, v). Hence $A = A \cap C$ and $B = B \cap C$, and we get

$$\mathbf{P}(A \cap B \cap C)\mathbf{P}(C) = \mathbf{P}(A \cap B)\mathbf{P}(C) \le p_{u1}p_{v2} = \mathbf{P}(A \cap C)\mathbf{P}(B \cap C).$$

Induction step: Let $|U| \ge 3$. In view of the argument in case $|U| = 1$ we may assume that u and v are different. Let

$$W = U - \{u, v\}, \text{ and}$$
$$D = \{\text{there is no rendezvous among any two nodes in } W\}.$$

Observe that $C \subseteq D$ and hence $\Pr(D) > 0$. We write out the probabilities that occur in (22) in more detail. It is obvious that the events A, B, and D are independent, and that $A \cap B \cap C = A \cap B \cap D$. Thus,

$$\mathbf{P}(A \cap B \cap C) = \mathbf{P}(A \cap B \cap D) = \mathbf{P}(A)\mathbf{P}(B)\mathbf{P}(D) = p_{u1}p_{v2}\mathbf{P}(D). \tag{23}$$

We consider events that describe that u resp. v is involved in a rendezvous:

$$S = \{\exists s \in N(u) \cap W : \text{ there is a rendezvous at } (u, s) \},$$
$$T = \{\exists t \in N(v) \cap W : \text{ there is a rendezvous at } (v, t) \}.$$

Now note that for $B \cap D$ to occur there are two possibilities: either v chooses 2 and there is no rendezvous in U at all (event $B \cap C$) *or* v chooses 2 and there is a rendezvous in U, but none in W — but then u must be involved in some rendezvous (event $B \cap D \cap S$). Thus,

$$\mathbf{P}(B \cap C) = \mathbf{P}(B \cap D) - \mathbf{P}(B \cap D \cap S) = \mathbf{P}(B)\mathbf{P}(D) - \mathbf{P}(B)\mathbf{P}(D \cap S)$$
$$= p_{v2}(\mathbf{P}(D) - \mathbf{P}(D \cap S)). \tag{24}$$

Similarly, we get

$$\mathbf{P}(A \cap C) = \mathbf{P}(A \cap D) - \mathbf{P}(A \cap D \cap T) = \mathbf{P}(A)\mathbf{P}(D) - \mathbf{P}(A)\mathbf{P}(D \cap T)$$
$$= p_{u1}(\mathbf{P}(D) - \mathbf{P}(D \cap T)). \tag{25}$$

Finally, we note that for D to occur there are three possibilities: either there is no rendezvous in U (event C) *or* node u or node v are involved in some rendezvous with nodes in W and there is no rendezvous in W (event $(S \cup T) \cap D$) *or* there is a rendezvous at edge (u, v) and none in W (event $D \cap \{u$ chooses v and v chooses $u\}$). Thus,

$$\mathbf{P}(C) = \mathbf{P}(D) - \mathbf{P}(D \cap S) - \mathbf{P}(D \cap T) + \mathbf{P}(D \cap S \cap T)$$
$$- \mathbf{P}(D \cap \{u \text{ chooses } v \text{ and } v \text{ chooses } u\}) \tag{26}$$
$$\leq \mathbf{P}(D) - \mathbf{P}(D \cap S) - \mathbf{P}(D \cap T) + \mathbf{P}(D \cap S \cap T).$$

Equations (23), (24), (25) and inequality (26) imply that in order to prove (22), it is sufficient to show that

$$p_{u1}p_{v2}\mathbf{P}(D)(\mathbf{P}(D) - \mathbf{P}(D \cap S) - \mathbf{P}(D \cap T) + \mathbf{P}(D \cap S \cap T)) \tag{27}$$
$$\leq p_{v2} \cdot (\mathbf{P}(D) - \mathbf{P}(D \cap S)) \cdot p_{u1} \cdot (\mathbf{P}(D) - \mathbf{P}(D \cap T)).$$

By multiplying out and cancelling we see that (27) is equivalent to $\mathbf{P}(D)\mathbf{P}(D \cap S \cap T) \leq \mathbf{P}(D \cap S)\mathbf{P}(D \cap T)$, or

$$\mathbf{P}(S \cap T \mid D) \leq \mathbf{P}(S \mid D)\mathbf{P}(T \mid D). \tag{28}$$

We can prove (28) by expanding the involved events and applying the induction hypothesis to distinguished nodes u and v and node set W. (For a similar calculation, cf. the proof of (22).)

$$\mathbf{P}(S \cap T \mid D) = \sum_{\substack{s \in N(u) \cap W \\ t \in N(v) \cap W}} p_{us}p_{vt}\,\mathbf{P}(s \text{ chooses } v \text{ and } t \text{ chooses } u \mid D)$$

$$\leq \sum_{s \in N(u) \cap W} \sum_{t \in N(v) \cap W} p_{us}p_{vt}\,\mathbf{P}(s \text{ chooses } v \mid D)\mathbf{P}(t \text{ chooses } u \mid D)$$

$$= \left(\sum_{s \in N(u) \cap W} p_{us}\,\mathbf{P}(s \text{ chooses } v \mid D) \right)\left(\sum_{t \in N(v) \cap W} p_{vt}\,\mathbf{P}(t \text{ chooses } u \mid D) \right)$$

$$= \mathbf{P}(S \mid D)\mathbf{P}(T \mid D).$$

Thus, (22) holds, and the induction step is complete. $\qquad\square$

5   Concluding Remarks

We have proved one basic relation in the framework of the probability spaces created by the rendezvous experiment. It turned out that a careful argumentation is required to deal with the "long-distance effects" of the condition "no rendezvous inside node set U". Overall, it seems that the subtleties of this probability space are not yet fully understood, neither for the case of uniform probabilities nor for the more general case of nonuniform ones. It would be nice to have more powerful tools that allow us to prove statements like Lemma 6 more easily.

The following observation points to one technique that may turn out to be helpful. We may generalize Fact 2, as follows. Let A and B be disjoint edge sets, and let C_A (C_B) be the event that there is no rendezvous along any edge in A (B). Then $\mathbf{P}(C_A \cap C_B) \leq \mathbf{P}(C_A) \cdot \mathbf{P}(C_B)$. While there is no obvious way to prove this directly, it can be shown to be a consequence of the powerful techniques from the theory of "negative association of random variables", as developed in detail in [2].

Finally we remark that one cannot hope to calculate $s(G)$ from G efficiently, as the mapping $G \mapsto s(G)$ is #P-complete (forthcoming joint work with H. Tamaki).

Acknowledgement. The author thanks Volker Diekert for introducing him to the problem, Holger Austinat for making his Diplom thesis accessible and for helpful remarks, as well as Y. Métivier, N. Saheb, and A. Zemmari for interesting electronic discussions on the rendezvous problem.

References

[1] H. Austinat, Verteilte Algorithmen zur Koordinatorwahl in Netzwerken, Diplomarbeit Nr. 1727, Universität Stuttgart, Fakultät Informatik, 1999, 66 pages.

[2] D. P. Dubhashi and D. Ranjan, Balls and bins: A study in negative dependence, Random Structures and Algorithms **13**(2), 99–124 (1998).

[3] Y. Métivier, N. Saheb, and A. Zemmari, Randomized rendezvous, in: G. Gardy and A. Mokkadem (Eds.), Algorithms, Trees, Combinatorics, and Probabilities, Trends in Mathematics, Birkhäuser, 2000, pp. 183–194. Also see the technical report version at `ftp://ftp.labri.u-bordeaux.fr/pub/Local/Info/ Publications/Rapports-internes/RR-122800.ps.gz`.

On the Minimum Volume of a Perturbed Unit Cube

Jin-Yi Cai

Computer Sciences Department
University of Wisconsin
Madison, WI 53706 USA
jyc@cs.wisc.edu
http://www.cs.wisc.edu/~jyc/

Abstract. We give exact bounds to the minimum volume of a parallelepiped whose spanning vectors are perturbations of the n unit vectors
by vectors of length at most ϵ. This extends Micciancio's recent sharp
bounds to all possible values of ϵ. We also completely determine all possible perturbations with length at most ϵ that achieve this minimum
volume.

1   Problem Statement

Suppose Q is the unit cube in $\mathbf{R}^n$ spanned by the n unit vectors $e_1 = (1, 0, \ldots, 0)$,
$e_2 = (0, 1, \ldots, 0), \ldots, e_n = (0, 0, \ldots, 1)$,

$$Q = \{\sum_{i=1}^{n} a_i e_i \mid 0 \le a_i \le 1, 1 \le i \le n\}. \tag{1}$$

Suppose a perturbation x_i is added to each e_i, where each x_i can be any small
vector in $\mathbf{R}^n$. Now consider the parallelepiped spanned by $u_1 = e_1 + x_1, u_2 = e_2 + x_2, \ldots, u_n = e_n + x_n$,

$$P = \{\sum_{i=1}^{n} a_i u_i \mid 0 \le a_i \le 1, 1 \le i \le n\}, \tag{2}$$

subject to the condition that all perturbation vectors x_i are of 2-norm at most
ϵ: $||x_i|| \le \epsilon$. We ask what is the minimum volume

$$f_n(\epsilon) = \min\{\mathrm{vol}(P)\}, \tag{3}$$

where the minimum is taken over all $||x_i|| \le \epsilon$. As the volume function is contin-
uous, and the minimization is over a compact set, clearly the minimum exists.
This problem came up in the work on the connection of worst-case/average-
case complexity for lattice problems. It was implicit in the work of Ajtai [1]. In
[3] a lower bound $f_n(\epsilon) \ge 1 - \epsilon n$ was proved. This bound was an improvement
and resulted in an improved worst-case/average-case connection in [3] over those

P. Bose and P. Morin (Eds.): ISAAC 2002, LNCS 2518, pp. 67–78, 2002.
© Springer-Verlag Berlin Heidelberg 2002

of [1]. The exact nature of the function $f_n(\epsilon)$ remained somewhat mysterious, and was explicitly asked to be determined in [2]. Recently Micciancio [4] gave a sharp bound that determines the function $f_n(\epsilon)$ for a subinterval of its domain of definition. In this paper we completely characterize this function $f_n(\epsilon)$.

While our determination does not currently improve the worst-case/average-case connection for lattice problems, the problem seems to be sufficiently natural and we hope it can be useful in the future.

2 Preliminaries

The volume of a parallelepiped spanned by $u_1, u_2, \ldots, u_n$ is given by the absolute value of the determinant of $I + X$. Thus

$$f_n(\epsilon) = \min\{|\det(I + X)|\}, \tag{4}$$

where the $n \times n$ matrix X has column vectors $x_i, 1 \le i \le n$, and I is the $n \times n$ unit matrix. Clearly f_n is monotonically non-increasing by definition, and starts off with $f_n(0) = 1$.

We first show that

Lemma 1. *Let $n \ge 1$. For any $\epsilon < \frac{1}{\sqrt{n}}$, $f_n(\epsilon) > 0$, and for any $\epsilon \ge \frac{1}{\sqrt{n}}$, $f_n(\epsilon) = 0$. Moreover, For any $\epsilon < \frac{1}{\sqrt{n}}$, the determinant $\det(I + X) > 0$, and thus the absolute value sign in the definition of f_n is unnecessary for all $\epsilon \le \frac{1}{\sqrt{n}}$.*

To show that $\det(I + X)$ is always positive for $||x_i|| < 1/\sqrt{n}$, we apply the Cauchy-Schwarz inequality to x_i and get the 1-norm $||x_i||_1 < 1$. Therefore the matrix $I + X$ has the property of strict central dominance by the column: $\forall j, 1 \le j \le n$,

$$1 + x_{jj} > \sum_{i \ne j} |x_{ij}|. \tag{5}$$

Such a matrix must have a positive determinant.

First, the determinant must be non-zero, or else, there exists a non-zero (row) vector v^T, such that

$$v^T(I + X) = 0. \tag{6}$$

We can assume the infinity norm of v is 1, and say $1 = |v_1| \ge |v_i|$. By replacing v with $-v$, we may assume $v_1 = 1$. Then

$$1 + x_{11} + \sum_{i \ge 2} x_{i1} v_i \ge 1 + x_{11} - \sum_{i \ge 2} |x_{i1}| > 0, \tag{7}$$

a contradiction.

Then a deformation argument shows that $\det(I + X) > 0$, for otherwise, for some $0 < t < 1$, $\det(I + tX) = 0$, contradicting to what has just been shown.

To see that for $\epsilon = 1/\sqrt{n}$, $f_n(\epsilon) = 0$, we take each $x_i = -\frac{1}{n}(1,1,\ldots,1)^T$, we observe that the matrix $I + X$ is singular, having all columns sum to 0.

So the only interesting values for ϵ are within $0 \le \epsilon \le 1/\sqrt{n}$. Suppose now the perturbation matrix X achieves the minimum volume, for a given ϵ in that range. Micciancio [4] gave a precise bound for f_n where $0 \le \epsilon \le \sqrt{\frac{1}{n} - \frac{1}{n^2}}$. His argument is short and pretty, and will be our starting point. For the sake of completeness we will present his argument first.

Micciancio observed the following necessary condition for the matrix X: For every dimension $1 \le i \le n$, the perturbation x_i must be perpendicular to the hyperplane spanned by $\{e_j + x_j \mid j \ne i, 1 \le j \le n\}$, and must be of maximum norm ϵ. This is clear in terms of the geometry and the fact that $\det(I + X)$ is always non-negative in that range. In matrix terms

$$X^T(I + X) = \epsilon^2 I + \mathrm{diag}(x_{11}, x_{22}, \ldots, x_{nn}). \tag{8}$$

Denote by T the diagonal matrix consisting of exactly the diagonal entries of X, then it follows that $X^T = \epsilon^2 I + T - X^T X$, which is symmetric, and hence so is X, and

$$X^2 + X = \epsilon^2 I + T. \tag{9}$$

With suitable renaming of the dimensions, and a suitable permutations of the columns of X we may assume X is grouped according to equal values on the diagonal, i.e.,

$$T = \begin{pmatrix} t_1 I_1 & 0 & \cdots & 0 \\ 0 & t_2 I_2 & \cdots & 0 \\ \vdots & \vdots & \ddots & \vdots \\ 0 & 0 & \cdots & t_k I_k \end{pmatrix} \tag{10}$$

where $1 \le k \le n$, each I_j has dimension m_j, $\sum_{j=1}^k m_j = n$, and if $i \ne j$ then $t_i \ne t_j$.

Since X is symmetric, there is an orthogonal matrix W such that $W^T X W$ is a diagonal matrix with eigenvalues of X on the diagonal. T being a polynomial of X, it follows that $W^T T W$ is also diagonal with the eigenvalues of T on the diagonal, thus it is a permutation of the diagonal entries of T. Thus there is a permutation matrix Π such that if we let $U = W\Pi$, then U is orthogonal, $U^T T U = T$, and $U^T X U$ is still a diagonal matrix with eigenvalues of X on the diagonal.

Partition the matrices $U = (U_{ij})$ and $X = (X_{ij})$ according to the dimensions $m_1, m_2, \ldots, m_k$, then by $TU = UT$ we have $t_i U_{ij} = t_j U_{ij}$ and thus $U_{ij} = 0$ if $i \ne j$. Therefore U is block diagonal, we name it $U = \mathrm{diag}(U_1, \ldots, U_k)$, each U_j is orthogonal. Since $U^T X U$ is diagonal, $U_i^T X_{ij} U_j = 0$ if $i \ne j$, and thus $X_{ij} = 0$ as well. So X is block diagonal as well, and we rename it $X = \mathrm{diag}(X_1, \ldots, X_k)$. We have arrived at the following necessary condition:

$$X_i^2 + X_i = (\epsilon^2 + t_i)I, \tag{11}$$

for each block of dimension m_i, $1 \le i \le k$.

Note that if the eigenvalues of X are λ_i, then the volume determinant is $\Pi_{i=1}^n(1+\lambda_i)$, and this product decomposes over the subspaces. It follows that the minimization problem is decomposed into the subproblems over all these blocks (except the perturbation vectors are still subject to the overall bound $1/\sqrt{n}$.) It is also clear that if any $m_i = 1$, the optimum is unique for that block with $X_i = -\epsilon$. Thus we consider in the following any $m_i \geq 2$. Rename, for some i, $m = m_i$, $Y = X_i$, $V = U_i$, $\tau = -t_i$, then we have

$$Y^2 + Y = (\epsilon^2 - \tau)I. \tag{12}$$

Micciancio's theorem [4] is the following

Theorem 1 (Micciancio). *For all $\epsilon < \sqrt{\frac{1}{n} - \frac{1}{n^2}}$, $f_n(\epsilon) = (1-\epsilon)^n$ and the minimum is uniquely achieved with $X = -\epsilon I$.*

He also noted the upper bound that $f_n(\epsilon) \leq 1 - \epsilon\sqrt{n}$ by taking all entries of X to $-\epsilon/\sqrt{n}$. Hence he has

$$0 < f_n(\epsilon) \leq \min\{(1-\epsilon)^n, 1 - \epsilon\sqrt{n}\} \tag{13}$$

for all $\epsilon < 1/\sqrt{n}$.

Theorem 1 will follow if $Y = -\epsilon I$ for each block, under the restriction on ϵ (and it follows that in this case there is in fact only one block).

Our main theorem is to prove the following refinement which completely characterizes the function $f_n(\epsilon)$.

Theorem 2. *Let $n \geq 1$. Let $A_n(\epsilon) = (1-\epsilon)^n$, for $0 \leq \epsilon \leq \frac{1}{\sqrt{n}}$. Let $B_n(\epsilon) = x(1-x)^{n-1}$, where*

$$x = \frac{1}{n} - \sqrt{\epsilon^2 - \frac{1}{n} + \frac{1}{n^2}}, \tag{14}$$

for $\sqrt{\frac{1}{n} - \frac{1}{n^2}} \leq \epsilon \leq \frac{1}{\sqrt{n}}$.

Then there is a unique cross over point c_n, satisfying $\sqrt{\frac{1}{n} - \frac{1}{n^2}} \leq c_n \leq \frac{1}{\sqrt{n}}$, such that

$$f_n(\epsilon) = \begin{cases} A_n(\epsilon) \text{ for } 0 \leq \epsilon \leq c_n \\ B_n(\epsilon) \text{ for } c_n \leq \epsilon \leq \frac{1}{\sqrt{n}} \end{cases} \tag{15}$$

Moreover, for $0 \leq \epsilon < c_n$ the minimum $A_n(\epsilon)$ is achieved uniquely by $X = -\epsilon I$, and for $c_n < \epsilon \leq \frac{1}{\sqrt{n}}$, the minimum $B_n(\epsilon)$ is achieved by exactly 2^{n-1} distinct perturbations, given by

$$X = \mu I + \xi \begin{pmatrix} \epsilon_1 \\ \epsilon_2 \\ \vdots \\ \epsilon_n \end{pmatrix} (\epsilon_1 \quad \epsilon_2 \quad \cdots \quad \epsilon_n), \tag{16}$$

where $\epsilon_1 = 1$ and $\epsilon_i = \pm 1$, for $i > 1$, and

$$\mu = -\frac{1}{n} + \sqrt{\epsilon^2 - \frac{1}{n} + \frac{1}{n^2}}. \tag{17}$$

and

$$\xi = -\frac{1}{n} + \frac{2}{n^2} - \frac{2}{n}\sqrt{\epsilon^2 - \frac{1}{n} + \frac{1}{n^2}}, \tag{18}$$

Finally c_n is asymptotically $\frac{1}{\sqrt{n}}(1 - e^{-\sqrt{n}})$.

The rest of this paper is to give a proof to this theorem.

Returning to (12), let $V^T Y V = \mathrm{diag}(\lambda_1, \dots, \lambda_m)$, then each eigenvalue of Y is real and satisfies

$$\lambda_i^2 + \lambda_i = \epsilon^2 - \tau. \tag{19}$$

Thus $\lambda_i = \mu_+$ or μ_-, where

$$\mu_\pm = -\frac{1}{2} \pm \sqrt{\frac{1}{4} + \epsilon^2 - \tau}. \tag{20}$$

Lemma 2. *For all $n \geq 1$, in order to achieve minimum volume for the $m \times m$ block Y, if $\epsilon < \sqrt{\frac{1}{n} - \frac{1}{n^2}}$, then all eigenvalues take the value μ_+. Moreover, for all $n > 4$ and $\epsilon \leq 1/\sqrt{n}$, to achieve minimum volume, there can be at most one eigenvalue $\lambda_i = \mu_-$, all others take the value μ_+.*

The first claim in Lemma 2 was from [4] and Theorem 1 was proved from that. Indeed, in that case Y is a scalar matrix, and to have $\det(I + Y)$ minimum obviously it must take the value $-\epsilon I_m$, and thus in fact there is only one block, $k = 1$, $m = n$, and $X = -\epsilon I_n$.

We prove Lemma 2.

The case $n = 1$ is vacuously true since in this case $\sqrt{\frac{1}{n} - \frac{1}{n^2}} = 0$. So we suppose $n \geq 2$.

We first claim that $\epsilon < 1/2$ if either $n \geq 2$ and $\epsilon < \sqrt{\frac{1}{n} - \frac{1}{n^2}}$, or $n > 4$ and $\epsilon \leq 1/\sqrt{n}$. This is clear since for $n \geq 2$, $\sqrt{\frac{1}{n}\left(1 - \frac{1}{n}\right)} \leq \sqrt{\frac{1}{2}\left(1 - \frac{1}{2}\right)} = 1/2$, and for $n > 4$, $1/\sqrt{n} < 1/2$ as well.

Since $-\tau$ is a diagonal entry, $|\tau| \leq \epsilon$. Thus $\frac{1}{4} + \epsilon^2 - \tau \geq \left(\frac{1}{2} - \epsilon\right)^2$, and $\sqrt{\frac{1}{4} + \epsilon^2 - \tau} \geq |\frac{1}{2} - \epsilon| = \frac{1}{2} - \epsilon$, for $\epsilon < 1/2$. So, $\mu_+ \geq -\epsilon$, and $\mu_- \leq -(1 - \epsilon)$. It follows that $\mu_+ > -1/2$ and $\mu_- < -1/2$. In particular $|\mu_-| > 1/2 > \epsilon$ and $\mu_+ \neq \mu_-$.

If $m = 1$, the block Y is trivially scalar. To achieve minimum volume, clearly $Y = -\epsilon I_1$. Since $|\mu_-| > \epsilon$, this $-\epsilon$, the sole eigenvalue of Y, must be μ_+.

So we may assume $m \geq 2$. Write Y by columns $Y = (y_1, \ldots, y_m)$, and consider the square of Frobenius norm $||Y||_F^2 = \sum_{i=1}^m ||y_i||^2 = m\epsilon^2$. This is invariant under the orthogonal transformation $V^T Y V$, and thus

$$m\epsilon^2 = \sum_{i=1}^m \lambda_i^2. \tag{21}$$

If there are exactly ℓ of them taking μ_-, then $m\epsilon^2 = \ell\mu_-^2 + (m - \ell)\mu_+^2$. If $\ell \geq 1$, then $m\epsilon^2 \geq \mu_-^2 + (m-1)\mu_+^2$, since $|\mu_-| \geq |\mu_+|$, which is obvious from (20). Let $g(\xi)$ be the quadratic function $g(\xi) = (1/2+\xi)^2 + (m-1)(1/2-\xi)^2$, for $\xi \geq 0$. Then it is easy to show by differentiation that g has a unique minimum at $\xi = 1/2 - 1/m$, and we get $m\epsilon^2 \geq g(1/2 - 1/m) = 1 - 1/m$. So

$$\epsilon^2 \geq \frac{1}{m}\left(1 - \frac{1}{m}\right), \tag{22}$$

which is $\geq \frac{1}{n}\left(1 - \frac{1}{n}\right)$, for $2 \leq m \leq n$. This proves the first part of Lemma 2.

Suppose now $n > 4$. Assume for a contradiction that $\ell \geq 2$. By the same argument that $|\mu_-| \geq |\mu_+|$, $m\epsilon^2 \geq 2\mu_-^2 + (m-2)\mu_+^2$. Let $h(\xi)$ be the following quadratic function $h(\xi) = 2(1/2 + \xi)^2 + (m-2)(1/2 - \xi)^2$, for $\xi \geq 0$. Again it is easy to show that $h(\xi)$ has a unique minimum at $\xi = 1/2 - 2/m$, and we get $m\epsilon^2 \geq 2 - 4/m$. Using $\epsilon^2 \leq 1/n$, we have $2 - 4/m \leq m/n \leq 1$ which implies that $m \leq 4$. Moreover, if $m = 4$ we get $2 - 1 \leq 4/n$ and if $m = 3$ we get $2 - 4/3 \leq 3/n$. Being an integer, in either cases, we can conclude that $n \leq 4$. So if $n > 4$ and $\ell \geq 2$, then $m \leq 2$. However by definition $\ell \leq m$, and so $\ell = m = 2$.

But if $m = 2$ and $\ell = 2$ we have a scalar matrix Y. In this case, the eigenvalue is the diagonal entry $-\tau$, and $|\tau| \leq \epsilon$ implies that it is actually μ_+ for both and not μ_-. So in fact $\ell = 0$, a contradiction.

Lemma 2 is proved.

Continuing the argument further, assuming $\ell \geq 1$, we had (22), from which and $\epsilon^2 \leq 1/n$, we get $m/n \geq 1 - 1/m \geq 1/2$. Thus $m \geq n/2$. Substituting back we get $m/n \geq 1 - 2/n$, from which we get $m \geq n - 2$, a happy situation where the estimate improves itself. Substituting back again we get $m/n \geq (n-3)/(n-2)$. m being an integer and $n > 4$, we finally derive that $m \geq n - 1$. In particular such a block Y with more than one eigenvalues, if it exists, is unique.

Lemma 3. *Let $n > 4$. In order to achieve minimum volume, the number of blocks Y with more than one eigenvalues is at most one, and if such a block exists, its dimension m is either n or $n - 1$. Furthermore one of its eigenvalues is taken with multiplicity $m - 1$.*

3   Rank 1 Perturbation of a Scalar Matrix

For any $m \geq 2$, let the symmetric matrix Y be an $m \times m$ block as in Section 2 having the property that it has two distinct eigenvalues, one of which μ is of multiplicity $m - 1$. Then $Y - \mu I$ is of rank 1. Thus, there exists non-zero vectors

y and z, such that $Y - \mu I = yz^T$. Say $y_i \neq 0$ and $z_j \neq 0$. By being symmetric, $y_i z_j = y_j z_i \neq 0$. Let $\xi = y_i z_i \neq 0$, take out ξ and rename y and z, we get

$$Y - \mu I = \xi yz^T, \tag{23}$$

where $y_i = z_i = 1$. Note that $\forall k, 1 \leq k \leq n$, by symmetry, $z_k = y_i z_k = y_k z_i = y_k$, and so $y = z$.

$$Y - \mu I = \xi yy^T. \tag{24}$$

Recall that all diagonal entries of Y are the same $-\tau$, take any diagonal entry we get $-\tau - \mu = \xi y_j^2$. As $\xi \neq 0$, all $|y_j|$ are equal and $|y_j| = y_i = 1$. Therefore Y takes the following form,

$$Y = \mu I + \xi \begin{pmatrix} \epsilon_1 \\ \epsilon_2 \\ \vdots \\ \epsilon_m \end{pmatrix} \begin{pmatrix} \epsilon_1 & \epsilon_2 & \cdots & \epsilon_m \end{pmatrix}, \tag{25}$$

where all $\epsilon_j = \pm 1$. We may further write $\epsilon_1 = 1$, which then uniquely specifies 2^{m-1} choices of ϵ_j.

It follows that $(Y - \mu I)^2 = m\xi(Y - \mu I)$. Expanding, we have $Y^2 = (2\mu + m\xi)Y - (\mu^2 + m\xi\mu)I$. Compare this to the matrix equation $Y^2 + Y = (\epsilon^2 - \tau)I$, and noting that $\{I, Y\}$ are linearly independent (i.e., Y is not a scalar matrix), we get

$$1 + 2\mu + m\xi = 0 \tag{26}$$
$$\epsilon^2 - \tau + \mu^2 + m\xi\mu = 0 \tag{27}$$

Also by taking trace in (25),

$$m\tau + m\mu + m\xi = 0. \tag{28}$$

So

$$\tau + \mu + \xi = 0. \tag{29}$$

Substituting τ from (29) to (27), and use (26), we get

$$0 = \epsilon^2 + \xi + \mu(1 + \mu + m\xi) = \epsilon^2 + \xi - \mu^2. \tag{30}$$

Finally from (30) substituting ξ in (26), we can solve for μ in

$$m\mu^2 + 2\mu + (1 - m\epsilon^2) = 0, \tag{31}$$

to get

$$\mu = -\frac{1}{m} \pm \sqrt{\epsilon^2 - \frac{1}{m} + \frac{1}{m^2}}. \tag{32}$$

As $(Y - \mu I)^2 = m\xi(Y - \mu I)$, the eigenvalues of $Y - \mu I$ are 0 with multiplicity $m-1$ and $m\xi$ with multiplicity one, thus Y has eigenvalues μ with multiplicity $m-1$ and $\mu + m\xi$ with multiplicity one. Hence the determinant $\det(I + Y)$ has the form $(1 + \mu)^{m-1}(1 + \mu + m\xi)$. Since $1 + \mu + m\xi = -\mu$ from (26), we get

$$\det(I + Y) = -\mu \cdot (1 + \mu)^{m-1}. \tag{33}$$

But which sign $\pm$ does μ take in equation (32)?

We claim that it must be the $+$ sign.

Denote by $z = \sqrt{\epsilon^2 - \frac{1}{m} + \frac{1}{m^2}}$, then

$$\det(I + Y) = \left(\frac{1}{m} + z\right) \cdot \left(1 - \frac{1}{m} - z\right)^{m-1} \tag{34}$$

if the minus sign $-$ is taken; and

$$\det(I + Y) = \left(\frac{1}{m} - z\right) \cdot \left(1 - \frac{1}{m} + z\right)^{m-1} \tag{35}$$

if the plus sign $+$ is taken. We want to show that the determinant is smaller with the $+$ sign. We omit the details here but it can be shown that the following polynomial in z has non-negative coefficients, and is positive

$$g(z) = \left(\frac{1}{m} + z\right) \cdot \left(1 - \frac{1}{m} - z\right)^{m-1} - \left(\frac{1}{m} - z\right) \cdot \left(1 - \frac{1}{m} + z\right)^{m-1} \tag{36}$$

We conclude that in order to have minimum determinant, the rank 1 perturbation matrix Y has

$$\mu = -\frac{1}{m} + \sqrt{\epsilon^2 - \frac{1}{m} + \frac{1}{m^2}}. \tag{37}$$

Also from (26) and (29)

$$\xi = -\frac{1}{m} + \frac{2}{m^2} - \frac{2}{m}\sqrt{\epsilon^2 - \frac{1}{m} + \frac{1}{m^2}}, \tag{38}$$

and

$$\tau = \frac{2}{m} - \frac{2}{m^2} - \left(1 - \frac{2}{m}\right)\sqrt{\epsilon^2 - \frac{1}{m} + \frac{1}{m^2}}. \tag{39}$$

The following Lemma can be shown

Lemma 4. *Let any $n \geq m \geq 2$. Suppose Y is an $m \times m$ block with two distinct eigenvalues, one of which μ is of multiplicity $m - 1$. In order to achieve minimum volume, Y takes the form (25), with μ given in (37) and ξ given in (38). Furthermore, all these 2^{m-1} distinct blocks Y do achieve the same determinant given in (33).*

4 Full Dimensional Block $m = n$

Suppose $n \geq 3$. In this section we suppose a full dimensional block Y exists with more than one distinct eigenvalues, as stipulated in Lemma 2. From Section 3 we know that, with $m = n$, Y is given in the form of Eqn (25), and the determinant is given in Eqn (33), where μ is given in Eqn (37).

As $\epsilon < \frac{1}{\sqrt{n}}$, we see that

$$\mu = -\frac{1}{n} + \sqrt{\epsilon^2 - \frac{1}{n} + \frac{1}{n^2}} < 0. \tag{40}$$

Consider the function

$$g(x) = x(1-x)^{n-1}. \tag{41}$$

It is the determinant function $\det(I + Y)$ if we let $x = -\mu > 0$ as a function of ϵ. As ϵ varies from $\sqrt{\frac{1}{n} - \frac{1}{n^2}}$ to $\frac{1}{\sqrt{n}}$, x strictly monotonically varies from $1/n$ down to 0.

Now we view $g(x)$ as a function of x, for $0 \leq x \leq 1/n$. By differentiation, $g'(x) = (1-x)^{n-2}(1 - nx)$, which is positive for the range $0 \leq x < 1/n$. Hence $g(x)$ is strictly monotonically increasing (as x increases from 0 to $1/n$), taking values from 0 to $g(1/n) = \frac{1}{n}\left(1 - \frac{1}{n}\right)^{n-1}$ each exactly once.

Our next task is to compare this with the "trivial" bound corresponding to the matrix $-\epsilon I$, i.e., when all the eigenvalues are equal. In this case, the determinant $\det(I + Y)$ is $(1 - \epsilon)^n$.

When ϵ increases from $\sqrt{\frac{1}{n} - \frac{1}{n^2}}$ to $\frac{1}{\sqrt{n}}$, both $g(x)$, now viewed as a function of ϵ, and the "trivial" bound $(1 - \epsilon)^n$, are strictly monotonically decreasing. We wish to show that there is a unique cross over.

Let $D(x) = g(x) - (1 - \epsilon)^n$, where

$$\epsilon = \sqrt{\left(x - \frac{1}{n}\right)^2 + \frac{1}{n} - \frac{1}{n^2}} \tag{42}$$

is now viewed as a function of x, in $0 \leq x \leq 1/n$.

We can show that $D\,|_{x=0} < 0$ and $D\,|_{x=1/n} > 0$, and $D'(x) > 0$ for $0 < x < 1/n$. This implies a unique cross over. The proof is a detailed analysis using both asymptotics for large n and numerical verifications for small n. We omit the proof here due to page limit.

Regarding the asymptotic location of this cross over in terms of ϵ, note that between $\sqrt{\frac{1}{n} - \frac{1}{n^2}}$ and $\frac{1}{\sqrt{n}}$, the bound $(1 - \epsilon)^n$ is asymptotically $e^{-\sqrt{n}}$. On the other hand, for $0 \leq x \leq 1/n$, $1 \geq (1 - x)^{n-1} \geq e^{-1}$ and thus $g(x) = \Theta(x)$. It follows that the cross over happens at around $x = \Theta(e^{-\sqrt{n}})$ asymptotically. In terms of ϵ, this happens at around $\epsilon \approx \frac{1}{\sqrt{n}}(1 - \Theta(e^{-\sqrt{n}}))$. One can get a bit more precise. Since from the above it is known that at cross over $x \approx \Theta(e^{-\sqrt{n}})$, then $(1 - x)^{n-1} \approx 1$, and $g(x) \approx x$, thus $\epsilon \approx \frac{1}{\sqrt{n}}(1 - e^{-\sqrt{n}})$ asymptotically.

Lemma 5. *For all $n \geq 3$, comparing the volume achieved by the perturbation matrix $-\epsilon I_n$ and that of a full dimensional block $(m = n)$ as given in Lemma 4, there is a unique cross over point c_n, satisfying $\sqrt{\frac{1}{n} - \frac{1}{n^2}} \leq c_n \leq \frac{1}{\sqrt{n}}$, such that*

$$(1 - \epsilon)^n < x(1 - x)^{n-1} \tag{43}$$

for $\sqrt{\frac{1}{n} - \frac{1}{n^2}} \leq \epsilon < c_n$, and

$$(1 - \epsilon)^n > x(1 - x)^{n-1} \tag{44}$$

for $c_n < \epsilon \leq \frac{1}{\sqrt{n}}$, where $x = \frac{1}{n} - \sqrt{\epsilon^2 - \frac{1}{n} + \frac{1}{n^2}}$. Moreover, c_n is asymptotically $\frac{1}{\sqrt{n}}(1 - e^{-\sqrt{n}})$.

5   Can Minimum Be Achieved with $m = n - 1$?

Let $n \geq 3$. Let us suppose there is a co-1 dimensional block Y, which has more than one distinct eigenvalues as stipulated in Lemma 2, that achieves minimum volume. Of course the left over 1-dimensional block must contribute $1 - \epsilon$ to the volume, and the total determinant is given as $(1 - \epsilon)y(1 - y)^{m-1}$, where $m = n - 1$, and y is given as follows

$$y = \frac{1}{m} - \sqrt{\epsilon^2 - \frac{1}{m} + \frac{1}{m^2}}. \tag{45}$$

ϵ satisfies the global constraint that $\epsilon \leq 1/\sqrt{n}$. Furthermore, in order that the eigenvalues of Y be real, we must further restrict $\epsilon \geq \sqrt{\frac{1}{m} - \frac{1}{m^2}}$. In the following, we will denote by ℓ and r the left and the right end points of this interval,

$$\ell = \sqrt{\frac{1}{n-1} - \frac{1}{(n-1)^2}}, \tag{46}$$

$$r = \frac{1}{\sqrt{n}}. \tag{47}$$

Note that $r > \ell$ since $r^2 - \ell^2 = \frac{1}{n(n-1)^2}$.

We want to compare $x(1-x)^{n-1}$ with $(1-\epsilon)y(1-y)^{n-2}$, the possible minimum volumes corresponding to a full dimensional block versus a co-1 dimensional block together with an extra one dimensional block. The common interval of definition is $[\ell, r]$, as the interval for the n-dimensional case is $[\sqrt{\frac{1}{n} - \frac{1}{n^2}}, \frac{1}{\sqrt{n}}]$ and its left most point $\sqrt{\frac{1}{n} - \frac{1}{n^2}} < \ell$, for all $n \geq 3$.

Define

$$F_n(\epsilon) = x(1 - x)^{n-1} \tag{48}$$

$$= \left(\frac{1}{n} - \sqrt{\epsilon^2 - \frac{1}{n} + \frac{1}{n^2}}\right)\left(1 - \frac{1}{n} + \sqrt{\epsilon^2 - \frac{1}{n} + \frac{1}{n^2}}\right)^{n-1} \tag{49}$$

and

$$\widehat{F}_n(\epsilon) = (1 - \epsilon)y(1 - y)^{n-2} \tag{50}$$

$$= (1 - \epsilon)\left(\frac{1}{m} - \sqrt{\epsilon^2 - \frac{1}{m} + \frac{1}{m^2}}\right)\left(1 - \frac{1}{m} + \sqrt{\epsilon^2 - \frac{1}{m} + \frac{1}{m^2}}\right)^{m-1} \tag{51}$$

where $m = n - 1$.

From the last section, both F_n and $\widehat{F}_n$ are strictly monotonically decreasing in ϵ. We want to show that for any $\epsilon \in [\ell, r]$,

$$F_n(\epsilon) < \widehat{F}_n(\epsilon). \tag{52}$$

Our strategy is to show that the end points of the interval satisfy

$$F_n(\ell) < \widehat{F}_n(r), \tag{53}$$

from which it follows that $\forall \epsilon \in [\ell, r]$,

$$F_n(\epsilon) \leq F_n(\ell) < \widehat{F}_n(r) \leq \widehat{F}_n(\epsilon). \tag{54}$$

It turns out that our strategy works for all $n \geq 4$, and the inequality (52) for the special case $n = 3$ can be proved separately. Both proofs of (53) for all $n \geq 4$ and (52) for $n = 3$ are careful analysis and are omitted here.

Lemma 6. *For all $n \geq 3$, the possibility of an $m = n - 1$ dimensional block with more than one eigenvalues together with an additional one dimensional block never produces a smaller volume than the corresponding n dimensional block with more than one eigenvalues, for the common interval of respective definitions.*

6 Putting It All Together

To put all this information together, we note that for $n > 4$ we already have enough to prove Theorem 2.

As noted already in Lemma 2 the case $\epsilon < \sqrt{\frac{1}{n} - \frac{1}{n^2}}$ has the unique minimizing $X = -\epsilon I$ and $f_n(\epsilon) = (1 - \epsilon)^n$. More generally a minimizing block Y is unique, $Y = -\epsilon I$, if it has only one eigenvalue (hence a scalar matrix).

Suppose $\sqrt{\frac{1}{n} - \frac{1}{n^2}} \leq \epsilon \leq \frac{1}{\sqrt{n}}$, and suppose $n > 4$. Then by Lemma 2 and Lemma 3 a minimizing X has at most one block which is non-scalar, and if it exists, it's unique and its dimension is either n or $n - 1$. Moreover, by Lemma 2 the characterization of such a block from Section 3 applies. Hence by Lemma 4 the only competing minimizing determinants $\det(I + Y)$ take the form $(1 - \epsilon)^n$, or $x(1 - x)^{n-1}$ or $(1 - \epsilon)y(1 - y)^{m-1}$, where $m = n - 1$, and x and y are as given in Section 4 and 5.

As shown in Lemma 6, $(1 - \epsilon)y(1 - y)^{m-1}$ in fact is always greater than $x(1 - x)^{n-1}$. Moreover by Lemma 5 there is a unique cross over between $(1 - \epsilon)^n$ and $x(1 - x)^{n-1}$. This completes the proof of Theorem 2 for all $n > 4$.

Finally we deal with all possible cases where the dimension $n \leq 4$.

The case $n = 1$ is trivial. By (14) $x(1 - x)^{n-1} = 1 - \epsilon$ in this case.

Let $n = 2$. Then the expression for $x(1 - x)^{n-1}$ evaluates to $1/2 - \epsilon^2$. Note that $(1 - \epsilon)^2 - (1/2 - \epsilon^2) = 2(1/2 - \epsilon)^2 \geq 0$, and strictly so for all $\epsilon \neq \frac{1}{2}$. Thus in this case, $x(1 - x)^{n-1}$ is the minimum through out the interval $[\sqrt{\frac{1}{n} - \frac{1}{n^2}}, \frac{1}{\sqrt{n}}] = [1/2, 1/\sqrt{2}]$. In other words, the cross over happened immediately at the left most point of the interval.

Let $n = 3$. If there is no block with more than one eigenvalues, then the minimum volume is $(1 - \epsilon)^3$. Suppose there is such a block, which of course must be of dimension either 2 or 3. The case of $m = 2 = n - 1$ with $(1 - \epsilon) \cdot (1/2 - \epsilon^2)$ can be dismissed by Lemma 6 of Section 5 as inferior to the $m = n = 3$ case with volume $x(1 - x)^{n-1} = (1/3 - \sqrt{\epsilon^2 - 2/9})(2/3 + \sqrt{\epsilon^2 - 2/9})^2$, (even though $(1 - \epsilon) \cdot (1/2 - \epsilon^2)$ is better than $(1 - \epsilon)^3$ throughout the common domain of definition $[1/2, 1/\sqrt{3}]$.)

Now Lemma 5 applies, and we conclude that Theorem 2 holds. The cross over happens in this case ($n = 3$) between $\sqrt{2}/3$ and $1/2$.

Let $n = 4$. If there are no block with more than one eigenvalues, then the minimum is $(1 - \epsilon)^4$. Suppose there is such a block. If this block size is 4, then we have the competing volume $x(1 - x)^3$ as before, which would have a unique cross over. From Lemma 6 we can dismiss the possibility of block size 3. This leaves the possibility of $m = 2$. Referring to $\mu_\pm$ in Eqn (20) this block must have minimum volume $(1/2 - \sqrt{1/4 + \epsilon^2 - \tau})(1/2 + \sqrt{1/4 + \epsilon^2 - \tau})$. Taking trace, $-2\tau = -1$ and $\tau = 1/2$. Substituting back we get $(1/2 - \sqrt{\epsilon^2 - 1/4})(1/2 + \sqrt{\epsilon^2 - 1/4}) = 1/2 - \epsilon^2$. This would have been smaller than $(1 - \epsilon)^2$ except $\mu_\pm$ in Eqn (20) are real numbers, which implies that $\epsilon \geq 1/2$. But here $n = 4$ and we are required to have $\epsilon \leq 1/\sqrt{n} = 1/2$, and at $\epsilon = 1/2$, $f_4(1/2) = 0$. Thus this possibility of $m = 2$ when $n = 4$ can be dismissed. This completes the proof of Theorem 2.

References

1. M. Ajtai. Generating hard instances of lattice problems. In *Proc. 28th Annual ACM Symposium on the Theory of Computing*, 1996. Full version available from ECCC, *Electronic Colloquium on Computational Complexity* TR96-007, at http://www.eccc.uni-trier.de/eccc/.
2. Jin-Yi Cai. Some Recent Progress on the Complexity of Lattice Problems. Plenary Talk. In the Proceedings of *The 14th Annual IEEE Conference on Computational Complexity*, 158–177, 1999.
3. J.-Y. Cai and A. Nerurkar. An Improved Worst-Case to Average-Case Connection for Lattice Problems. In *Proc. 38th IEEE Symposium on Foundations of Computer Science (FOCS)*, 1997, 468–477.
4. Daniele Micciancio. Minimal volume of almost cubic parallelepipeds. Manuscript. To appear.

Non-Delaunay-Based Curve Reconstruction

Sumanta Guha[12], Paula Josiah[2], Anoop Mittal[2], and Son Dinh Tran[2]

[1] Computer Science & Information Management
Asian Institute of Technology
P.O. Box 4 Klong Luang, Pathumthani 12120, THAILAND
guha@ait.ac.th
[2] Electrical Engineering & Computer Science
University of Wisconsin-Milwaukee
Milwaukee, WI 53211, USA
{guha, pwjosiah, amittal, dinhtran}@uwm.edu

Abstract. A new non-Delaunay-based approach is presented to reconstruct a curve, lying in 2- or 3-space, from a sampling of points. The underlying theory is based on bounding curvature to determine monotone pieces of the curve. Theoretical guarantees are established. The implemented algorithm, based heuristically on the theory, proceeds by iteratively partitioning the sample points using an octree data structure. The strengths of the approach are (a) simple implementation, (b) efficiency – experimental performance compares favorably with Delaunay-based algorithms, (c) robustness – curves with multiple components and sharp corners are reconstructed satisfactorily, and (d) potential extension to surface reconstruction.

1 Introduction

The twin problems of curve and surface reconstruction are roughly as follows: given a sample of points from an unknown curve or surface in an Euclidean space, reconstruct the curve or surface with some assurance of accuracy. Here we focus on reconstructing a curve in $\mathbb{R}^2$ or $\mathbb{R}^3$, though the approach proposed extends to curves in higher-dimensional spaces and, possibly, surfaces as well.

The specific problem is, given a sample of points W from a curve c, reconstruct c by forming a polygonal line l, called a polygonal reconstruction, that joins the points of W according to their adjacencies on c. Early approaches were made by several researchers [4,5,12,13,18,19,22]. Recently, beginning with Amenta, Bern and Eppstein [3], with subsequent refinements by others [7,8,15], a method has developed that is characterized by (a) the sampling density not being required to be uniform – but rather it is a function of the local feature size, a number that, roughly, captures the level of detail of a curve at a point, (b) reconstruction using a selection of edges from the Delaunay triangulation of the sample points, and (c) provable guarantees on the quality of the reconstruction. See Edelsbrunner [11] for a survey of Delaunay-based shape reconstruction techniques. Another recent approach of note is by Giesen [14], also [2], based on solving the TSP for the sample points.

P. Bose and P. Morin (Eds.): ISAAC 2002, LNCS 2518, pp. 79–90, 2002.

In this paper we give results in the spirit of the Delaunay-based method of Amenta, Bern, Eppstein and others, in that a theoretical guarantee related to the local feature size is proved. However, our approach is new and essentially different from existing ones. It is not Delaunay-based – the strategy is to bound the curvature and extract monotone pieces of a given curve. Methods are mostly differential geometric in nature. In our main theorems, for 2- and 3-space, the following is proved:

If the local feature size of a curve c in 2-space (3-space) is at least L everywhere and if a sample W of size n from c is such that any point of c is at most a straight-line distance of 0.25L (0.151L) from some point of W, then one can determine a polygonal reconstruction of c from W in time $O(n \log n)$.

The sampling requirement is weaker if the curve is known a priori to be planar. Compared with some recent Delaunay-based algorithms our requirements are more stringent in that uniform bounds are required on local feature size and sampling density. However, our algorithm does not require Delaunay computation which can be expensive. In fact, the complexity of a Delaunay triangulation of n points in 3-space may be $\Omega(n^2)$, whereas we have a time guarantee of $O(n \log n)$. Further, our implementation [21], based heuristically on the theory, is extremely simple and proceeds by iteratively refining a 3-dimensional grid of boxes using an octree data structure. Experiments show it is efficient and gives output of at least the quality of existing methods.

2   Theory

2.1   Plane Curves

Our plan is simple:

If, somehow, one can determine that a plane curve c is monotone w.r.t. a given line l the problem becomes trivial: the order along c of points of the sample set W is identical to the order of their projections along l.

Now, for c to be *non-monotone* w.r.t. a line l_1, it must be intersected at two points by a line perpendicular to l_1, which implies that there is a tangent at some point on c perpendicular to l_1. It follows that, for c to be non-monotone w.r.t. to two lines l_1 and l_2, the tangent along c must turn at least as much as the angle between l_1 and l_2.

However, the rate at which the tangent turns w.r.t. the length of the curve is measured by the curvature of c. Therefore, if we can bound its curvature, a sufficiently short piece of c will be monotone w.r.t. at least one of two lines that have a sufficiently large angle between them.

Through this subsection, *unless otherwise stated*, by a curve we mean a *simple regular connected curve* $c : [a, b] \to \mathbb{R}^2$, *of class at least* C^2.

Often the same symbol is used to denote a curve as a function as well as the image of that function in $\mathbb{R}^2$. The reader is referred to any standard text in differential geometry, eg., [10], for the definition of the *curvature* $\kappa(p)$ at a point

p on a curve c and related concepts. To save space proofs of results already in accessible literature are omitted.

Lemma 1 (Pestov-Ionin). *Let c be a simple closed curve that is C^1 and piecewise C^2, and suppose that the magnitude of the curvature of c anywhere is at most K. Then a disc of radius $1/K$ lies in (the closure of) the interior of c.*

Proof. The reader is referred to the paper by Pestov and Ionin [20] or the more accessible one by Ahn et al [1] that contains a proof (actually of the case when $K = 1$, but generalized easily), as well as another application of the lemma in computational geometry. □

Lemma 2. *Let c be a curve such that the magnitude of the curvature of c anywhere is at most $K (> 0)$; i.e., the radius of curvature at any point of c is at least $R = 1/K$. Suppose further that there exist two points on c such that the tangents at these points are perpendicular.*
Then the diameter of c is at least $\sqrt{2}R$.

Proof. Consider a closest pair, distance measured by *arc length*, amongst pairs of points on c with perpendicular tangents. W.l.o.g. assume the points in such a pair are the origin $(0,0)$ and (A, B), and that tangents there are parallel to the x- and y-axis, respectively. See Figure 1(a). Between $(0,0)$ and (A, B), c is constrained to lie in the axes-parallel rectangle with corners at $(0,0)$ and (A, B). For, otherwise, between $(0,0)$ and (A, B) the curve intersects twice at least one of the 4 lines on which a side of this rectangle lies, implying a tangent parallel to one of the axes between two of the intersection points. This in turn implies the existence of a pair of points with tangents parallel to the axes that are closer than $(0,0)$ and (A, B), contradicting the hypothesis on $(0,0)$ and (A, B).

By reflecting c about the x-axis, the y-axis, and the origin, one gets three more curves c_1, c_2 and c_3 so that, after suitable translations, $\bar{c} = c \cup c_1 \cup c_2 \cup c_3$ is a simple closed curve satisfying the hypotheses of Lemma 1. Therefore, $\bar{c}$ contains a disc of radius R in its interior. However, $\bar{c}$, by its construction, lies in a rectangle with sides of length $2A$ and $2B$. It follows that $(0,0)$ and (A, B) are at least $\sqrt{2}R$ apart. □

Definitions. The *medial axis* of a curve c in $\mathbb{R}^2$ is the closure of the set of points in $\mathbb{R}^2$ that have more than one closest point on c. The *local feature size* $f(p)$ at $p \in c$ is the least distance of p to the medial axis.

A curve c is said to be *x-monotone* (*y-monotone*) if any line parallel to the y-axis (x-axis) intersects at most one point of c.

The following is well-known but a proof is given as we are not aware of a reference containing one.

Lemma 3. *For any point p on a curve c in $\mathbb{R}^2$ the local feature size $f(p)$ is at most the radius of curvature at p.*

Proof. Assume a finite radius of curvature at p (i.e., curvature $\kappa(p) > 0$), for if not there is nothing to prove.

The center of curvature $C(p)$ (i.e., the center of the osculating circle) at p is the limit of the center of circles through three non-collinear points p, p', and p'' on c as $p' \to p$ and $p'' \to p$. Take p' and p'' to be points on either side of p on c that are equidistant (using straight-line distance) from p. $C(p)$ is then the limit of the intersection of the perpendicular bisectors of the chords pp' and pp'', as $p' \to p$ and $p'' \to p$. See Figure 1(b). The intersection of these two perpendicular bisectors is always equidistant from both chords pp' and pp'', so taking limits it is seen that $C(p)$ is, in fact, itself a point of the medial axis. The result follows. See Figure 1(c) for a situation where the local feature size is strictly smaller than the radius of curvature. $\square$

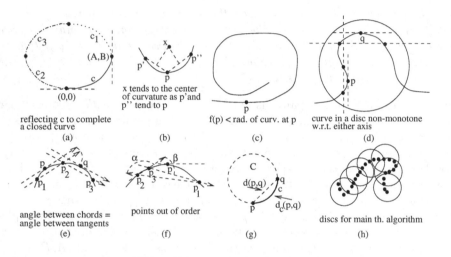

Fig. 1. Illustrations for the theory.

Lemma 4 ([3]). *Let D be a disc containing a point p of a curve c such that the diameter of D is at most $f(p)$.*

Then $c \cap D$ is connected and, therefore, itself a curve.

Proof. (Recall that we require a curve to be connected.) Refer to Corollary 3 of Section 5 of the paper by Amenta et al [3]. $\square$

Lemma 5. *Let c be a curve such that the local feature size at any point of c is at least L. Suppose now that c intersects a disc D of diameter at most L.*

Then $c \cap D$ is either x-monotone or y-monotone (or both).

Proof. Suppose, if possible, that c is neither x-monotone nor y-monotone in D. Let $p \in c \cap D$. The diameter of D is at most $L \leq f(p)$. Therefore, by Lemma 4, $c_1 = c \cap D$ is a curve lying in D.

Since c_1 is not x-monotone there is a line parallel to the y-axis that intersects c_1 in at least two points. It follows that there exists a point, say p, on c_1 whose tangent is parallel to the y-axis. Similarly, there exists a point, say q, on c_1 whose tangent is parallel to the x-axis. See Figure 1(d).

By Lemma 3, L bounds below the radii of curvature at points of c_1. Since the tangent directions at p and q are perpendicular the diameter of c_1 is at least $\sqrt{2}L$ by Lemma 2. But this is impossible as c_1 lies in D whose diameter is at most L. The result follows.   □

Definitions. A finite set W of points, all lying on a curve c, is a *sample* of c. A curve c is said to be ω-*sampled* by W (or W is an ω-sample of c) if, for each point $p \in c$, the *length of the arc* along c from p to some point of W is at most ω. *Remark*: This definition is different from that prevalent in current literature but is convenient for our purpose for the moment. Later, more will be said about the relation between our definition and others.

If W is a sample of a *non-closed* curve c an *ordering* of W by c is the linear ordering of the points of W imposed by a traversal of c from one endpoint to the other. Note that this ordering is unique up to reversal.

Henceforth, in this subsection, to avoid technicalities with the ordering of a sample *we shall always assume a curve to be non-closed* – this requirement is not essential to the theory and can be removed.

If W is a sample of c a *polygonal reconstruction* of c from W is the polygonal line obtained by joining points of W, adjacent in an ordering of W by c, with straight-line segments.

Lemma 6. *Let a curve c be x-monotone (y-monotone) and suppose that W is a sample of c.*

Then the ordering of the points of W by x-coordinate (y-coordinate) is induced by c.

Proof. Straight-forward.   □

Even if it is known that a piece of a curve c is either x- or y-monotone, say by an application of Lemma 5, how does one identify which of the orderings by x- or y-coordinate is indeed induced by c? As will be seen the solution is to use a sufficiently dense sample of c.

Lemma 7. *Let c be a curve such that the magnitude of the curvature at any point of c is at most $K (> 0)$ and let W be an ω-sample of c, for some constant ω. Suppose p_1, p_2 and p_3 are any three consecutive points of W in an ordering by c.*

Then the angle between the directed chords p_1p_2 and p_2p_3 is at most $4\omega K$, where the angle between two directed lines l_1 and l_2 is measured by the smaller of the angles that l_1 has to turn to match directions with l_2.

Proof. There is a point p between p_1 and p_2 whose tangent is parallel to the chord p_1p_2. Similarly, there is a point q between p_2 and p_3 whose tangent is parallel to the chord p_2p_3. See Figure 1(e). By the constraint on sampling density the arc length between p and q is at most 4ω.

If a curve c is parameterized by arc length s (so $c = c(s)$) and $\theta(s)$ is the angle made by the tangent at a point $p = c(s)$ with some fixed line, it is known that $|\kappa(p)| = |\frac{d\theta}{ds}(p)|$.

Since $|\kappa(p)|$ and, therefore, $|\frac{d\theta}{ds}(p)|$ is bounded above by K, it follows by elementary calculus that the angle between the tangents at p and q is at most $4\omega K$. $\square$

In the opposite direction is the following:

Lemma 8. *Let c be a curve such that the magnitude of the curvature at any point of c is at most $K (> 0)$ and let W be an ω-sample of c, for some constant ω. Suppose the ordering of the points of W by y-coordinate (x-coordinate) is not induced by c.*

Then there exist three points p_1, p_2 and p_3 in W, consecutive in the ordering by y-coordinate (x-coordinate), such that the angle between the directed chords $p_1 p_2$ and $p_2 p_3$ is at least $\pi - 4\omega K$.

Proof. Suppose the ordering of the points of W by y-coordinate is not induced by c. One may then assume w.l.o.g. that there exist three points p_1, p_2 and p_3, such that (a) $y(p_1) < y(p_2) < y(p_3)$ (where $y(p_i)$ denotes the y-coordinate of p_i), and (b) p_2, p_3 and p_1 occur consecutively in an ordering by c. See Figure 1(f).

It follows that there exists a point $p \in c$, lying between p_1 and p_2, whose tangent is parallel to the x-axis. By choosing points of W closer to p if necessary one can assume that p_1, p_2 and p_3 are, in fact, consecutive in the ordering by y-coordinate.

By the previous lemma the angle β between the directed chords $p_2 p_3$ and $p_3 p_1$ is at most $4\omega K$. The angle α between the directed chords $p_1 p_2$ and $p_2 p_3$ is then seen to be at least $\pi - \beta \geq \pi - 4\omega K$.

The cases when the p_i are arranged differently along c may be handled similarly. $\square$

Definitions. We return to a discussion of sampling density. First, another definition is given that is similar to ω-sampling but removes the dependence of the distance measure on arc length along the given curve: a curve c is said to be *Euclidean ω-sampled* by W if, for each point $p \in c$, the *Euclidean distance* from p to some point of W is at most ω. The definition most common in recent literature, introduced by Amenta et al [3], is essentially a *non-uniform* version of Euclidean ω-sampling, that we call *LFS-sampling*: a curve c is said to be LFS ϵ-*sampled* by W if, for each point $p \in c$, the Euclidean distance from p to some point of W is at most $\epsilon f(p)$ (thus sampling density may vary with local feature size).

To mimic the framework of [3] we wish to state the main theorem in terms of Euclidean ω-sampling. This first requires:

Lemma 9. *Given two points p and q on a curve c let $d_c(p, q)$ and $d(p, q)$ denote the distance between the two points measured by arc length and Euclidean distance, respectively. Suppose further that the local feature size at any point of c is at least L.*

If $d(p,q) \leq \sqrt{2}L$ then $d_c(p,q) \leq \frac{\pi}{2\sqrt{2}}d(p,q)$. It follows that if W is an Euclidean ω-sample of c, where $\omega \leq \sqrt{2}L$, then W is a $\frac{\pi}{2\sqrt{2}}\omega$-sample of c.

Proof. The proof is similar to that of Lemma 2. As, by Lemma 3, L bounds below the radii of curvature at points of c, one can apply Lemma 1 to place c outside the interior of a circle C of radius L. With the hypotheses of the lemma it is easy to see that ratio $d_c(p,q)/d(p,q)$ can reach no higher value than when p and q are, in fact, on such a circle C of radius L at an angle of $\frac{\pi}{2}$ apart, with c coinciding with C between them. See Figure 1(g). In this case $d_c(p,q)/d(p,q) = \frac{\pi}{2\sqrt{2}}$ and the result follows.   □

Theorem 1 (Main 2-Space). *Let c be a curve such that the local feature size at any point of c is at least L and let W be an Euclidean ω-sample of c of size n, where $\omega \leq L/4$.*

Then one can determine a polygonal reconstruction of c from W in time $O(n \log n)$.

Proof. Henceforth, for this proof, by a disc is always meant one of diameter L, and $D(p)$ denotes the disc centered at p. We make the following two observations.

1. For a disc D an ordering of the points of $W \cap D$ by c may be determined as follows:
   Sort the points in $W \cap D$ by both x- and y-coordinates. By Lemma 5, c is either x- or y-monotone in D. By Lemma 9, W is an ω'-sample with $\omega' \leq \frac{\pi}{8\sqrt{2}}L$. Using Lemmas 6, 7 and 8, and keeping in mind that curvature along c is bounded below by $1/L$, we conclude both
   (a) if a sorted order of $W \cap D$ by one of the coordinates *is* induced by c, then successive chords in that order make an angle of at most $\frac{\pi}{2\sqrt{2}}$, and,
   (b) if a sorted order of $W \cap D$ by one of the coordinates is *not* induced by c, then there exists a pair of successive chords in that order making an angle of at least $\pi - \frac{\pi}{2\sqrt{2}}$.
   Consequently, an examination of the angles made by successive chords in either sorted orders of the points of $W \cap D$ determines an ordering of the points of $W \cap D$ by c.
2. As W is an Euclidean ω-sample of c, with $\omega \leq L/4$, two points of W successive in an ordering by c are at most $L/2$ apart. Therefore, the radius of $D(p)$ being $L/2$, $D(p)$ contains both the points preceding and succeeding p in an ordering of W by c (if either such exists).

We now outline an algorithm that proves the theorem. Let $\mathcal{D}$ be an initially empty set of discs. The invariant that is maintained through the algorithm is an ordering of the points of $W \cap \cup\{D : D \in \mathcal{D}\}$ by c.

For some point $p \in W$ use Observation 1 to find an ordering of the points of $W \cap D(p)$ by c. Insert $D(p)$ into $\mathcal{D}$. As long as there are points of W outside $\cup\{D : D \in \mathcal{D}\}$ do the following:

Use Observation 1 to find an ordering of the points of $W \cap D(q_i)$, where q_i, $1 \leq i \leq 2$, are the endpoints of $W \cap \cup\{D : D \in \mathcal{D}\}$ in the ordering by c;

use Observation 2 to merge the orderings of $W \cap \cup \{D : D \in \mathcal{D}\}$ and those of $W \cap D(q_i)$, $1 \leq i \leq 2$, by c; insert $D(q_i)$, $1 \leq i \leq 2$, into $\mathcal{D}$.

See Figure 1(h). The time complexity of $O(n \log n)$ is dominated by point location and sorting operations.   □

2.2   Curves in 3-Space

Though there are significant technical differences in implementation our basic strategy for 3D curve reconstruction is similar to that for plane curves:

If, somehow, one can determine that a curve c is monotone w.r.t. one of the three coordinate axes the problem becomes trivial.

Now, for a curve c to be *non-monotone* w.r.t. *all three* axes, its tangent must turn along c some minimum angle (which we shall prove – see Lemma 10). Therefore, if we can bound its curvature, a sufficiently short piece of c will be monotone w.r.t. at least one of the three coordinate axes.

Definitions. Through this subsection by a curve we mean a *simple regular connected curve* $c : [a, b] \rightarrow \mathbb{R}^3$, *of class at least* C^2.

A curve c is said to be *x-monotone*, or monotone w.r.t. the x-axis, if any plane parallel to the yz-plane intersects c in at most one point. Similarly define c to be *y-monotone* or *z-monotone*.

The definitions of *medial axis* and *local feature size* for curves in 3-space are similar to those for plane curves.

Through the rest of this subsection we rapidly develop the theory for 3D-space, highlighting analogies with the planar case and omitting most details.

Lemma 10. *Suppose a curve c is not monotone w.r.t. any of the three coordinate axes. Then there exists a pair of points on c such that the angle between the tangents at these points is at least $\pi/3$ (tangents are taken to be vectors directed by an orientation of the curve).*

Proof. Clearly, if c is not x-monotone then it has a tangent parallel to the yz-plane. It follows that if c is not monotone w.r.t. any of the coordinate axes then it has tangents parallel to each of the coordinate planes.

The lemma now follows from the following:

Lemma 11. *If three non-null vectors α, β and γ are parallel to the three coordinate planes, respectively, then the angle between at least two of them is at least $\pi/3$.*

Proof. Omitted.   □ □

Lemma 12. *Let c be a curve such that the magnitude of the curvature of c anywhere is at most $K (> 0)$; i.e., the radius of curvature at any point of c is at least $R = 1/K$. Suppose further that c is not monotone w.r.t. any of the coordinate axes.*

Then the diameter of c is at least $\frac{\pi}{3\sqrt{3}} R$.

Proof. This is the analogue in 3-space of Lemma 2. However, Pestov-Ionin can no longer be applied and the proof uses another differential geometric method. Details are omitted.   □

The 3D analogues of Lemmas 3 and 4 are straightforward and lead to the following analogue of Lemma 5.

Lemma 13. *Let c be a curve such that the local feature size at any point of c is at least L. Suppose now that c intersects a ball B of diameter less than $\frac{\pi}{3\sqrt{3}}L$.*

Then $c \cap B$ is monotone w.r.t. at least one coordinate axis.   □

In 3-space only weaker analogues of Lemmas 7 and 8 hold – proofs of both are omitted:

Lemma 14. *Let c be a curve such that the magnitude of the curvature at any point of c is at most K (> 0) and let W be an ω-sample of c, for some constant ω. Suppose p_1, p_2 and p_3 are any three consecutive points of W in an ordering by c.*

Then the angle between the directed chords p_1p_2 and p_2p_3 is at most $6\omega K$, where the angle between two directed lines l_1 and l_2 is measured by the smaller of the angles that l_1 has to turn to match directions with l_2.   □

Lemma 15. *Let c be a curve such that the magnitude of the curvature at any point of c is at most K (> 0) and let W be an ω-sample of c, for some constant ω. Suppose the ordering of the points of W by z-coordinate (x-coordinate, y-coordinate) is not induced by c.*

Then there exist three points p_1, p_2 and p_3 in W, consecutive in the ordering by z-coordinate (x-coordinate, y-coordinate), such that the angle between the directed chords p_1p_2 and p_2p_3 is at least $\pi - 6\omega K$.   □

The definitions of *ω-sampling* and *Euclidean ω-sampling*, however, remain the same in 3-space as in 2-space, as does the relation between the two – refer Lemma 9 – though the proof, omitted here, is more technical. As in the planar situation we make the convenient assumption that *curves are non-closed*.

Finally, we have:

Theorem 2 (Main 3-Space). *Let c be a curve such that the local feature size at any point of c is at least L and let W be an Euclidean ω-sample of c of size n, where $\omega \leq 0.151L$.*

Then one can determine a polygonal reconstruction of c from W in time $O(n \log n)$.

Proof. Similar to the main theorem for 2-space using balls instead of disks – details are omitted.   □

Remark: It is interesting that our claims for curves in 3-space can be proved solely requiring bounds on curvature and none on torsion.

3 Implementation

A direct implementation of the algorithm described in either main theorem is typically not possible as, in practice, c is an unknown curve whose bounds for its radius of curvature and local feature size can at best be guessed. Therefore, a priori we do not know a "right" disk or ball size. Instead, our software, which is written for 3-space but reconstructs plane curves as well, starts with a large *3D-box* containing the region of interest, then iteratively refines it using an *octree* data structure while simultaneously partitioning sample points into successively smaller boxes.

The practical algorithm that is implemented in a software package called PointsNBoxes (freely available over the internet [21]) is the following:

Algorithm
  1. Determine an equal-sided bounding box (i.e., cube) B. This is the initial node of the octree structure O.
  2. Sort the points by x-, y-, and z-coordinates and store in vectors V^x, V^y, and V^z, respectively. The points themselves are not stored in the vectors, but rather pointers into the vector of input points. However, for the sake of simplicity we continue to refer to the pointers as points.
  3. Compute angles between successive chords in the orders of points of all three vectors V^x, V^y, and V^z.
     If all angles, for V^x (or V^y, or V^z), are at most a fixed threshold value Δ then accept and output the points in the order of V^x (or V^y, or V^z) (we make a heuristic choice of $\Delta = 30°$ in most of our experiments that seems to work well).
     If not, split B into eight equal boxes B_i, $1 \le i \le 8$, and, correspondingly, assign 8 children to B in O. Linearly scan V^x, V^y, and V^z and, according to which B_i each point lies in, assign it to V_i^x, V_i^y, and V_i^z, respectively (this step, not accidentally, is similar to the construction phase for orthogonal range search structures [6]).
  4. Repeat step 3, with B_i in place of B, V_i^x in place of V^x, V_i^y in place of V^y, and V_i^z in place of V^z, for $1 \le i \le 8$.
  5. ("stitching" together pieces in each box)
     Proceed sequentially from some starting box and grow the connected curve box by box. For a current endpoint p of the curve look for a "best partner" endpoint in a neighboring box. O is enhanced so that, for each box, neighboring boxes can be listed. The implementation uses a heuristic to determine best partner: for each endpoint $\bar{p}$ in a neighboring box compute the two angles between the chord $p\bar{p}$ and the chords joining p and $\bar{p}$ to their adjacent points in their respective boxes, and choose as best partner the one making "smallest" such angles. We do not discuss details or handling of technicalities, eg., a box with only one point. Further, before accepting a partner $\bar{p}$ for p on the basis of such angle comparisons, our implementation requires the distance between $\bar{p}$ and p to be at most a heuristically determined value depending on the local box sizes (details are omitted). For, otherwise, we tend

to find a (bad) partner for p in situations where it should have no partner eg., near endpoints of the original curve.

Multiple components are detected when a first sequential pass does not cover all sample points; in which case another pass is begun.

The implementation ensures integrity of the output in that the reconstruction is always a union of polygonal curves – all points are of degree at most two.

Experiments with various curves have shown efficiency and output quality that compares favorably with Delaunay-based methods. The former was expected as sorting n points, followed by linearly many angle computations, is almost certain to be cheaper than global Delaunay computation. In practice, sampling density is determined by the scanning technology employed and likely to be fairly uniform through the curve. In which case Delaunay-based methods are even more costly if there is a large sample from a "less detailed" piece of the curve, as they perform a Delaunay computation regardless, while such pieces are rapidly reconstructed by our method. See our website [21] for figures of octrees not subdividing in regions of low detail.

In fact, because the implementation splits an octree box only if it is unable to find a polygonal reconstruction of the curve in that box, in practice it (a) does respond to level of detail – in that it rapidly reconstructs pieces of the curve with little detail, and (b) does not require uniform sampling – as computations in individual boxes are independent (these two items being the claimed strengths of current Delaunay-based methods).

The implementation does well even with multiple components and sharp corners (see the website [21] for an example figure), the latter causing particular difficulty with Delaunay-based methods [9].

4   Future Work

1. Improve theoretical guarantees: it would be particularly interesting to prove a claim regarding reconstruction by our method using an LFS ϵ-sample *a la* the scheme of [3].
2. Remove the disconnect between the theoretical algorithm and the octree-based implementation.
3. Reconstruct surfaces – this is an intriguing project. The paradigm is similar: project sample points on to the three coordinate planes and lift back a Delaunay triangulation of a "good" projection if one is found; if not, partition the points into smaller boxes.

   We recently learned of the work of Gopi et al [16,17] where similar ideas have already been implemented with good resutls – Gopi et al first estimate normals to the surface at all sample points and, subsequently, project likely neighborhood sets of points on to corresponding tangent planes.

Acknowledgements. We thank Herbert Edelsbrunner for pointing out an error in an earlier version and helpful comments. We thank an anonymous referee for pointing out references [16,17] and numerous useful suggestions.

References

1. H-K. Ahn, O. Cheong, J. Matoušek, A. Vigneron, Reachability by paths of bounded curvature in convex polygons, *Proc. ACM Symp. on Comp. Geom.* (2001) 251–259.
2. E. Althaus, K. Mehlhorn, Polynomial time TSP-based curve reconstruction, *Proc. ACM-SIAM Symp. on Disc. Alg.* (2000) 686–695.
3. N. Amenta, M. Bern, D. Eppstein, The crust and the β-skeleton: combinatorial curve reconstruction, *Graphical Models and Image Processing* **60/2** (1998) 125–135.
4. D. Attali, r-regular shape reconstruction from unorganized points, *Proc. ACM Symp. on Comp. Geom.* (1997) 248–253.
5. F. Bernardini, C.L. Bajaj, Sampling and reconstructing manifolds using α-shapes, *Proc. 9th Canadian Conf. on Comp. Geom.* (1997) 193–198.
6. M. de Berg, M. van Kreveld, M. Overmars, O. Schwarzkopf, *Computational Geometry: Algorithms and Applications*, 2nd edn., Springer-Verlag (2000).
7. T. K. Dey, P. Kumar, A simple provable algorithm for curve reconstruction, *Proc. ACM-SIAM Symp. Disc. Alg.* (1999) 893–894.
8. T.K. Dey, K. Mehlhorn, E. Ramos, Curve reconstruction: connecting dots with good reason, *Comput. Geom. Theory Appl.* **15** (2000) 229–244.
9. T.K. Dey, R. Wenger, Reconstructing curves with sharp corners, *Proc. ACM Symp. on Comp. Geom.* (2000) 233–241.
10. M. Do Carmo, *Differential Geometry of Curves and Surfaces*, Prentice Hall (1976).
11. H. Edelsbrunner, Shape reconstruction with Delaunay complex, *LNCS 1380, LATIN'98: Theoretical Informatics* (1998) 119–132.
12. H. Edelsbrunner, D.G. Kirkpatrick, R. Seidel, On the shape of sets of points on the plane, *IEEE Trans. on Info. Theory* **29** (1983) 71–78.
13. L.H. de Figueiredo, J. de Miranda Gomes, Computational morphology of curves, *The Visual Comp.* **11** (1995) 105–112.
14. J. Giesen, Curve reconstruction, the traveling salesman problem and Menger's theorem on length, *Proc. ACM Symp. on Comp. Geom.* (1999) 207–216.
15. C.M. Gold, J. Snoeyink, A one-step crust and skeleton extraction algorithm, *Algorithmica* **30** (2001), 144–163.
16. M. Gopi, *Theory and Practice of Sampling and Reconstruction for Manifolds with Boundaries*, Ph.D. Dissertation (2001) http://www.cs.unc.edu/~gopi/dis.pdf.
17. M. Gopi, S. Krishnan, C.T. Silva, Surface reconstruction based on lower dimensional localized Delaunay triangulation, *Proc. EUROGRAPHICS 2000* (2000) C467–C478.
18. M. Melkemi, $\mathcal{A}$-shapes of a finite point set, *Proc. ACM Symp. on Comp. Geom.* (1997) 367–372.
19. J. O'Rourke, H. Booth, R. Washington, Connect-the-dots: a new heuristic, *Computer Vision, Graphics, and Image Proc.* **39** (1987) 258–266.
20. G. Pestov, V. Ionin, On the largest possible circle imbedded in a given closed curve, *Dok. Akad. Nauk SSSR* **127** (1959) 1170–1172.
21. PointsNBoxes, freely downloadable curve reconstruction software, http://www.cs.uwm.edu/~pnb.
22. R.C. Veltkamp, The γ-neighborhood graph, *Comput. Geom. Theory Appl.* **1** (1992) 227–246.

Cutting a Country for Smallest Square Fit

Marc van Kreveld[1] and Bettina Speckmann[2]*

[1] Institute for Information and Computing Sciences, Utrecht University,
marc@cs.uu.nl
[2] Institute for Theoretical Computer Science, ETH Zürich,
speckman@inf.ethz.ch

Abstract. We study the problem of cutting a simple polygon with n vertices into two pieces such that – if we reposition one piece disjoint of the other, without rotation – they have the minimum possible bounding square. If we cut with a single horizontal or vertical segment, then we can compute an optimal solution for a convex polygon with n vertices in $O(n)$ time. For simple polygons we give an $O(n^4\alpha(n)\log n)$ time algorithm.

1 Introduction

When browsing through the Rand McNally's Road Atlas of the U.S.A., it appears that not every state is shown on one single or double page. Sometimes the northern and southern halves of a state are shown on two consecutive double pages. Since a state can be modeled by a simple polygon, the aspect ratio of a double page is fixed, and the map scale is supposed to be the same on both double pages, the problem can be modeled as covering a simple polygon by two equal-size axis-aligned squares of smallest size. The squares represent the pages, and by scaling any fixed aspect ratio can be handled. Algorithmically, the problem of covering a simple polygon with two pages is not very difficult and can be solved in linear time. In the case of three map pages, a more involved algorithm can still achieve linear running time, which was shown by Hoffmann [5].

Fig. 1. Fitting a country into a square.

* Supported by the Berlin-Zürich Graduate Program "Combinatorics, Geometry, and Computation", financed by the German Science Foundation (DFG) and ETH Zürich.

Another method the Rand McNally's Road Atlas uses to make states fit better on pages is to cut them into two pieces, rearrange these, and fit them on a single or a double page. For example, the western part of Florida can be cut off and placed at a different position to allow a larger map scale. In some edition of the road atlas this technique was applied to seven of the states.

This paper discusses an abstracted version of the problem of cutting a country or state optimally and fitting the pieces on a page. A country is represented by a simple polygon and we cut this polygon with either a horizontal or vertical line segment into two simply-connected pieces. We then rearrange the pieces by translation only and do not allow them to intersect. An optimal cut and an optimal placement are the ones that result in the smallest enclosing square. For a convex polygon with n vertices we compute an optimal solution, i.e., an optimal cut and an optimal placement, in $O(n)$ time. For a simple polygon we give an $O(n^4 \alpha(n) \log n)$ time algorithm.

In various papers, Daniels and Milenkovic report results on containment, motivated by marker making in the textile industry. Generally, multiple-part, translational containment problems are NP-hard [2] and known algorithms contain the number of parts in the exponent. Simplified to be comparable to our case, fitting two convex polygons with n vertices inside a constant size polygon takes $O(n)$ time [7], if the polygons are non-convex it requires $O(n^4)$ time [2]. If minimization of the enclosing polygon is of interest, then Milenkovic shows a lower bound of $\Omega(n^4)$ for a related problem [6].

Alt and Hurtado [1] discuss packing convex polygons into a rectangle of smallest size, measured either by area or by perimeter. They study both packing with overlap and without overlap. Of the latter type, they show that a smallest rectangle can be found in linear time for two convex polygons when no rotation is allowed.

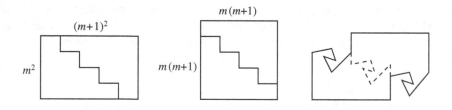

Fig. 2. Dissecting a rectangle and a simple polygon to form a square.

Dissections have been studied at length in Frederickson's book [3]. It contains an interesting example of a rectangle that requires a dissection consisting of many segments so that after repositioning one piece a square is formed, see Figure 2. To the right, there is another example to show that optimal cuts may have a complex shape.

The algorithmic problems and results mentioned above are different from ours because we compute both a cut and a packing into a smallest square. Previous papers compute only a packing (and papers on cutting are less related to ours). Combinatorially, there are a linear number of possible places for a horizontal or vertical cut through a simple polygon. Still our bound in the convex case remains linear and in the non-convex case it is worse only by a factor of $O(\alpha(n)\log n)$ compared with algorithms that do packing only. In our solution to the non-convex case most of the complications do not arise from the linearly many combinatorially distinct cuts to consider, but from discretizing the position of the cut while maintaining the optimality and efficiency.

In Section 2 we present the algorithm for convex polygons and in Section 3 we describe the algorithm for non-convex polygons. We conclude with some open problems.

2   Cutting a Convex Polygon

Assume we are given a convex polygon P with n vertices. Our goal is to find a vertical line segment C that cuts P into two parts A and B and then to translate A to a position in which A and B do not properly intersect, such that the smallest enclosing axis-parallel square S for A and B is minimized. If we also allow horizontal cuts, then we repeat the algorithm with x- and y-coordinates exchanged.

In order to efficiently find an optimal solution, i.e. a configuration that is determined by an optimal cut and an optimal placement, we identify certain properties of a subset of the optimal solutions. In particular, we show that there is always an optimal solution such that (i) either A or B determine the x or the y-span of S, (ii) A and B are in contact, and (iii) both A and B are in contact with either the left or the right side of S (see Lemma 1). Restricting ourselves to search for optimal solutions that have the properties just described and incorporating the following two observations yields a surprisingly simple algorithm that computes an optimal solution in linear time.

First, we define the NW chain of a convex polygon Q to be the polygonal chain that is the part of the boundary of Q from the leftmost vertex to the topmost vertex, clockwise. The SW chain, NE chain, and SE chain are defined similarly. Horizontal and vertical edges are included in the chain their leftmost endpoint – clockwise – is adjacent to.

Observation 1 *Given two convex polygons Q and R, and a square S, such that $s = \max\{x\text{-}span(R), y\text{-}span(R)\}$, then if Q and R both fit in S, there is a placement where Q (i) touches the left side of S, and R in the NW chain, or (ii) touches the left side of S, and R in the SW chain, or (iii) touches the right side of S, and R in the NE chain, or (iv) touches the right side of S, and R in the SE chain.*

Applied to A and B this observation translates to: If B is the polygon that determines the span of S then there are only four canonical placements for A.

Second, we can observe that if we consider an optimal solution which has the properties described above, then while shifting the cut to shrink one polygon and grow the other, always keeping them in contact, the contact moves only in one direction along the boundary of each polygon, passing over each vertex at most once. For a specific constellation this observation translates to:

Observation 2 *If A and B fit in S, and the x-span of B is s, and A is in contact with the left side of S and the NW chain of B, then when the cut is shifted to shrink B and grow A, the contact between A and B can only go rightward on B (clockwise) and rightward on A (counterclockwise), assuming A keeps touching the left side of S and the NW chain of B.*

Similar observations hold if A is in contact with any other chain of B or if B determines the y-span of S.

Our algorithm now actually consists of eight incremental algorithms: The cut starts four times at the left side of P, implying that B is the larger polygon that determines the size of S. The other four times the cut starts at the right side of P and A is the larger polygon. If we start at the left side we initialize with $A = \emptyset$ and $B = P$ and grow A and shrink B (therefore also S) while keeping A to the NW (resp. SW, NE, or SE) of B. We then start with $A = P$ and $B = \emptyset$ and grow B and shrink A (therefore also S) while keeping B to the NW (resp. SW, NE, or SE) of A. During the run of each algorithm we maintain the contact points on A and B and a bounding square S whose size corresponds to the x or y-span of either A or B depending on the constellation we are currently processing. The algorithms terminate whenever A and B do not fit into S anymore.

There are two types of events our algorithm has to process: (i) the cutting line passes over a vertex of A or B and (ii) the contact point between A and B proceeds to a next vertex or edge on A or B. Based on the observations above it is straightforward to see that each algorithm only needs to process $O(n)$ events, each at constant cost.

Whenever an algorithm terminates we use the current information on the contacts between A, B, and S to compute the optimum in-between the last two events. Finally, the minimum of the minimal square sizes found by the eight algorithms is the true minimum square size we set out to find.

Now all that remains is to prove the following lemma:

Lemma 1. *For a given convex polygon P there exists an optimal vertical cut C and an optimal translation of A and B that put A and B without intersection into a square S with side length s and one of the following holds: (a) the x-span or y-span of B is equal to s and if A is not empty, then it is in contact with B and with the left or right side of S or (b) the x-span or y-span of A is equal to s and if B is not empty, then it is in contact with A and with the left or right side of S.*

Proof. (sketch) Assume that an optimal solution is given, where S is the enclosing square with side length s, the pieces are polygons A and B, and the cut is C. We will transform this solution into one that satisfies the statement in the lemma, without increasing the side length s of S.

Assume first that the angles of the cut edge corners α_t and α_b of A sum up to at most π. Then we will show how to transform to case (a) of the lemma. Otherwise, the angles of the cut edge corners β_t and β_b of B sum up to at most π because P is convex. Then it follows by symmetry that we can transform to case (b). Hence we only need to show the first part, and we assume that $\alpha_t + \alpha_b \leq \pi$.

Let ℓ be a line that separates the interiors of A and B. First assume that ℓ can be vertical. We shift the cut edge to make B grow and A shrink simultaneously. This cannot increase the x-span of A and B, so we can continue until either A becomes empty and $B = P$, or the y-span of B becomes s. In both cases we are done.

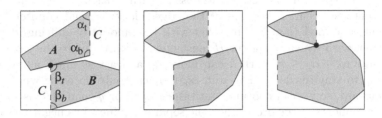

Fig. 3. The three cases of the proof.

Now assume that ℓ cannot be vertical. By symmetry we may assume that A lies above ℓ and B lies below ℓ, see Figure 3. Again we need to consider two cases, one where ℓ has positive slope and one where ℓ has negative slope. These cases are not symmetric. However, in this sketch we will discuss only one. Furthermore, the case of a horizontal line ℓ follows in exactly the same way, but here it is not treated explicitly because it increases the complexity of the formulations.

Since B is below and right of ℓ, and A above and left of ℓ, we can assume that S and B are in contact at both the bottom and right side of S, otherwise we can move B to make this true. Similarly, we can assume that A is in contact with the top and left side of S, or we can move A inside S to make this happen. If A and B can be separated by a vertical line in this new configuration, then we are done as just shown, so we assume that this is not the case. If A and B are in contact, then the tangent point can only be the lower endpoint of the cut edge of A or the upper endpoint of the cut edge of B (see the left two pictures in Fig. 3), otherwise we obtain a contradiction with the convexity of P. Furthermore, because of the positive slope of ℓ, the angle assumption on A, and the fact that A and B come from one convex polygon, we can show that the highest point of A is the upper endpoint of the cut edge.

The main observation is that we can shift the cut to grow B and shrink A simultaneously, until the either the x-span or the y-span of B is equal to s. During this shift, we will not move A, B, or S in horizontal direction, only vertically. We must show that A and B still fit in S vertically, until the x-span or y-span of B is s. Then we can reposition A to satisfy the other criteria of the lemma.

We next show that the cut can be shifted without forcing the square S to increase in size. When shifting the cut leftward, A loses a trapezoidal region bounded from the left and right by vertical sides. Because $\alpha_b + \alpha_t \leq \pi$, the left side of the trapezoid is at most as long as the right side. Because of the shift, piece A can move upward in S with an amount depending on the slope of the edge of A counterclockwise from the cut edge, and the amount of shifting. This follows from the observation that the highest point of A must be the upper endpoint of the cut edge of A. At the same time, B grows, and pushes itself and/or A upward with respect to the bottom side of S. If A and B were not in contact yet, they may come in contact. Three possible situations are shown in Figure 3. Which contacts occur, the slopes of the edges clockwise and counterclockwise of the cut edge in A (not B!), together with the amount of shifting, determine the amount with which A and/or B must move upward due to the size increase of B. Since $\alpha_b + \alpha_t \leq \pi$, part A can move up at least as much as B need be moved up to stay inside S. The sum $\alpha_b + \alpha_t$ can only decrease when the cut shifts leftward, so we can continue until the x-span or y-span of B is s.

The arguments of the proof when the separating line ℓ has negative slope are similar and we omit them here. After arriving in the situation where the x-span or y-span of B is s, either A is empty, or we can move A until it is in contact with B.   □

Theorem 1. *Given a convex polygon P with n vertices, we can determine a vertical cut C of P into subpolygons A and B, and a non-intersecting placement of A with respect to B, such that the smallest enclosing square of A and B is minimized in $O(n)$ time.*

3   Cutting a Simple Polygon

Assume we are given a not necessarily convex simple polygon P with n vertices. Our goal is to find a vertical line segment C that cuts P into two parts A and B and then to translate A to a position in which A and B do not properly intersect, such that the smallest enclosing (axis-parallel) square S for A and B is minimized.

In an optimal solution A and B will be in contact (more precisely, there is an optimal solution with A and B in contact). Furthermore, we can assume that a vertex v of A lies on an edge e of B. We first choose v and e in P and then consider all cuts C that partition P such that v and e are part of different subpolygons. Consider a vertical decomposition of P: There are a number of *separating trapezoids* such that a cut through any of these trapezoids separates v and e. The separating trapezoids can be linearly ordered such that a cut through the first one results in the smallest A and largest B and cuts through the following trapezoids grow A while shrinking B. In Figure 4 the separating trapezoids are shaded grey and ordered from left to right. For convenience we add the vertices of the trapezoidal decomposition as vertices to (the boundary of) P. This does not restrict the problem and the number of vertices of P is still in $O(n)$. The edge e is now an edge of the polygon including the vertices induced by the trapezoidation, so it can be a subedge of an edge of the original input polygon (see Fig. 4).

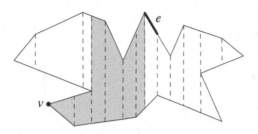

Fig. 4. A polygon P with v and e chosen; the separating trapezoids are shaded grey.

We first describe only the cases where the edge e of B lies on the boundary of P and is not the one cut edge of B. Similarly, we assume first that v is not an endpoint of the cut edge of A and that e is not adjacent to the cut edge of B. We make these assumptions to explain the general idea and we will show later how to handle all cases.

All positions with $v \in A$ in contact with $e \in B$ can be represented by an interval I of the real line. At each end of I, vertex v coincides with one of the endpoints of e. For ease of description, we assume B to be stationary and A translating with respect to B. Since v must be in contact with e, A will translate along the vector between the two endpoints of the edge e. We are interested in those positions where A and B are non-intersecting. A and B intersect if and only if they each have a line segment that properly intersect. In a degenerate case this may not be true, but we will not consider such cases here. If we consider one line segment e_A of A in translation, then in the general case, e_A intersects a (steady) line segment e_B of B along some stretch, but possibly not before and/or not after it. The positions on interval I where e_A and e_B intersect are a subinterval of I. Since this holds for all edges of A and B, there are $O(n^2)$ intervals on I that define in what position of A with respect to B they intersect. We will store these intervals in a segment tree, and augment every internal node μ with a boolean FREE that indicates whether in the interval I_μ of I, represented by μ, there is at least one position that is not covered by any of the $O(n^2)$ intervals in the subtree rooted at μ. Since this boolean is only valid for the intervals stored in the subtree rooted at μ, there may be an interval stored at an ancestor of μ which makes no position for μ non-covered. The boolean helps to answer queries with a query interval I_q, to locate the leftmost or rightmost non-covered position in I_q, in $O(\log n)$ time. We will use the term "free" throughout the description for a placement of A with respect to B so that they do not properly intersect.

We do not have to store the intervals explicitly at the nodes of the segment tree. Instead, we store a counter at each node that represents how many intervals are stored at that node as in a segment tree for stabbing counting queries. The counts and augmentation by booleans can be maintained under insertion and deletion of any interval in $O(\log n)$ time. The segment tree needs $O(n^2)$ storage because it represents $O(n^2)$ intervals.

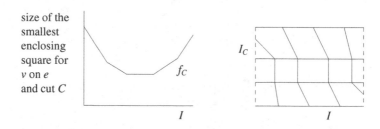

Fig. 5. The function f_C for a given cut C (left); the patches for all positions on I and cuts inside a trapezoid (right).

Assuming that the position of the cut C is fixed, let f_C be the function that maps the position of v on e (or, equivalently, a point on I) to the side length of the smallest enclosing square of A and B in the specified position. The function f_C is defined on every point of I, regardless whether this position is free. Over the interval I f_C is a piecewise linear continuous function consisting of at most five pieces (see Fig. 5). If there are five pieces, then the first two pieces have negative slope, the middle piece is horizontal and is the minimum of the function, and the last two pieces have positive slope. The function f_C is easy to compute in $O(n)$ time. The optimal free placement for A and B with v on e is either some position on I that realizes the minimum of f_C, or the rightmost position on I left of the minimum, or the leftmost position right of the minimum. Note that with the help of the segment tree we can determine the optimum placement inside any query interval I_q in $O(\log n)$ time with at most three queries. We query with the subinterval of I_q that has negative slope to find the rightmost free position, we query with the subinterval of I_q with constant slope for any free position, and we query with the subinterval of I_q with positive slope for the leftmost free position.

Assume next that the cut C jumps from trapezoid boundary to trapezoid boundary. Assume first that this makes A larger by exactly one trapezoid and B smaller by that same trapezoid. More generally, it can happen that a large part of B suddenly goes to A, like the upper left part of P in Figure 4. We can treat this as a sequence of single trapezoids going from B to A. This may make A temporarily disconnected, but this will not affect the algorithm. We will not query the segment tree for a free position until we reach the next possible cut. When we jump to a next trapezoid boundary A loses its former cut edge, it gains a new cut edge, and it gains two edges from B. Similarly, B loses its former cut edge and two more edges and gains a new cut edge. In total, four edges change for each of A and B. These eight edges were involved in $O(n)$ of the $O(n^2)$ intervals on I. To update the segment tree, we perform $O(n)$ insertions and deletions of intervals. Since the FREE-information and the counts in the segment tree can be maintained in $O(1)$ time per node visited, one update takes only $O(\log n)$ time. Then we have the segment tree for the next cut. We can determine the new function f_C and query for the new optimum with three queries. In total, taking one whole trapezoid from B and adding it onto A takes $O(n \log n)$ time.

In general it is the case that the optimum cut and position of v on e does not occur at a trapezoid boundary, but somewhere in the middle between two boundaries. We handle this as follows: We consider simultaneously the position of v on e and the exact position of the cut between two consecutive trapezoid boundaries. When the cut progresses from one trapezoid boundary to the next, we could maintain all changes that occur to the $O(n^2)$ intervals on I. However, there can be a cubic number of events (endpoint swaps) between intervals that change and intervals that do not change – this is too costly to maintain. Instead we keep changing and non-changing intervals separate in our solution. We represent the exact position of the cut as a second dimension added to I, which yields another interval I_C, and we consider the rectangular region $R = I \times I_C$ (see Fig. 5). Any point in this region represents a position of v on e and a position of the cut in the trapezoid under consideration. The lower side of R corresponds to the situation where the cut trapezoid is completely part of B and the upper side corresponds to the situation where the cut trapezoid is completely part of A. The $O(n^2)$ intervals resulting from one edge of A and one edge of B now become regions of one of two types, see Figure 6. These regions show where

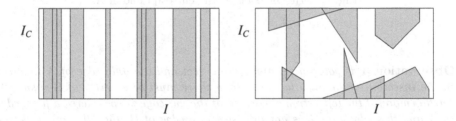

Fig. 6. The rectangle $R = I \times I_C$: the rectangular forbidden regions (left) and all other forbidden regions (right).

the edges intersect, so they are forbidden regions for the free placement of A with respect to B. If edges $e_A \in A$ and $e_B \in B$ do not change when the cut is moved in the trapezoid, then we get a rectangular region $i \times I_C$ in R, where i was the original interval we got for e_A and e_B. There are $O(n^2)$ such rectangles. All pairs of edges that involve the cut or one of the edges that go from B to

A when the cut progresses, define $O(n)$ differently shaped regions. The regions involving an edge adjacent to the cut edge of A are triangular or quadrilateral, and more specifically, are the region vertically above some line segment inside R. This line segment may have one endpoint on the upper side of R, which determines whether it is triangular or quadrilateral of shape. Similarly, any region involving an edge adjacent to the cut edge of B is the region vertically below a line segment.

The regions involving the cut edge of A or B are also simple polygonal shapes. However, their shape is such that a vertical line could intersect them twice inside R and therefore the complexity of the boundary of the forbidden regions inside R may be quadratic, even if we do not consider the rectangles. Due to the complexity of these regions the solution we are about to describe would not be efficient, so we need to represent them differently. We make use of the following simple geometric observation (see Fig. 7):

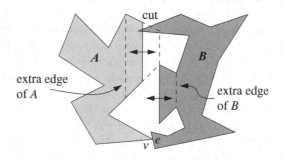

Fig. 7. Vertex of B inside the cut trapezoid of A.

Observation 3 *If polygons A and B intersect and the only edge of A involved in the insection is the cut edge of A, then there must be a vertex of B inside A. Furthermore, if the left extreme position of the cut edge is considered a fixed edge of A and this edge also does not intersect any edge of B, then the cut trapezoid of A must contain a vertex of B.*

This observation shows that the only other forbidden regions we must add are the ones where the cut trapezoid of A contains a vertex of B or the cut trapezoid of B contains a vertex of A. This is true if we consider the left side of the cut trapezoid of A to be a fixed edge of A and the right edge of the cut trapezoid to be a fixed edge of B. The extra edge of A was actually used when we considered the situation at the trapezoid boundary just before. So the corresponding intervals are already in the segment tree. For the extra edge of B we insert $O(n)$ intervals in the segment tree.

The new forbidden regions that spring from this observation are the pentagonal regions vertically above or below two adjacent line segments. Summarizing, we have three types of forbidden regions inside R:

1. $O(n^2)$ vertical rectangles inside R, that cross R completely in the vertical direction. These are represented in the segment tree.
2. $O(n)$ regions vertically above a line segment inside R.
3. $O(n)$ regions vertically below a line segment inside R.

We compute the lower envelope of the regions of the second type and the upper envelope of the regions of the third type. Only the regions in-between can contain placements for A and B and a choice of the cut, such that A and B do not intersect. We combine the envelopes by a left-to-right scan to compute the regions in-between. These regions are bounded by at most $O(n\alpha(n))$ line segments and can be computed in $O(n \log n)$ time [4].

The function f that gives the side length of the smallest enclosing square as a function of a point in R (a position of v on e and position of the cut) is now a piecewise linear bivariate function (see Fig. 5). If the cut is fixed, then the function f_C has the shape we noted before. This holds for every cut. The function f may have up to fifteen patches. Two horizontal lines partition the rectangle R into three slabs, the middle slab is partitioned by four vertical lines, and the other two slabs are partitioned by four diagonal lines. The twelve lines partitioning the three slabs connect to each other. The function f gives the side length of the enclosing square regardless of whether A and B intersect.

We now overlay the $O(1)$ patch boundaries of f with the $O(n\alpha(n))$ line segments of the region in-between the envelopes. We take the collection S of all $O(n\alpha(n))$ edges we have in this arrangement, take their x-extents, and query in the segment tree to find both the leftmost and the rightmost free position in the query interval. We use the function f to determine the size of the smallest enclosing square for each answer in $O(1)$ time. In total, the queries to find candidate solutions take $O(n\alpha(n) \log n)$ time. We select the smallest answer, which is the best solution for a given vertex v in contact with a given edge e and a given cut trapezoid.

Lemma 2. *The optimal placement and cut for a given v, e, and cut trapezoid is a leftmost or rightmost free placement on a segment of S.*

Proof. By construction, any segment of S lies inside of only one patch of f or is part of a patch boundary. Over each patch a linear function defines the side length of the smallest enclosing square. Obviously, we need a free placement so that A and B do not intersect. The lemma follows.   □

We already described how to update the segment tree so that it becomes valid for the next separating trapezoid. The update requires $O(n \log n)$ time in total and since we go through at most $O(n)$ trapezoids, the total update time becomes $O(n^2 \log n)$. The initial costs for constructing the segment tree are also $O(n^2 \log n)$, which brings the total cost for one edge e, one vertex v, and all cuts that separate them to $O(n^2\alpha(n) \log n)$. Concluding, if the optimal cut of polygon P and position of the parts A and B has some vertex-edge contact not at the cut edge, then we can determine it in $O(n^4\alpha(n)\log n)$ time.

The remaining issue is a situation where the optimal cut and position has a contact between A and B that involves an edge or vertex of the cut edge of A or B. A careful case analysis shows that this situation can be handled without affecting the running time of our algorithm asymptotically. Due to the limited amount of space available for this abstract we omit the details of this analysis. However, they can be found in the full version of the paper.

Theorem 2. *Given a simple polygon P with n vertices, we can determine a vertical cut C of P into subpolygons A and B, and a non-intersecting placement of A with respect to B, such that the smallest enclosing square of A and B is minimized in $O(n^4\alpha(n)\log n)$ time and $O(n^2)$ space.*

4 Open Problems

The problem of cutting a polygon into two and rearranging the parts gives rise to various interesting and very difficult problems. Our algorithm breaks down if we allow straight line cuts that need not to be horizontal or vertical. An even more general problem arises if we allow several line segments or even any curve as the cut and still wish to rearrange the parts optimally to fit inside a square. It might also be of interest to study the problem if we do not restrict ourselves to translations but also allow rotations while rearranging the parts.

References

1. H. Alt and F. Hurtado. Packing convex polygons into rectangular boxes. In *Discrete and Computational Geometry – Japanese Conference, JCDCG 2000*, number 2098 in Lect. Notes in Comp. Science, pages 67–80. Springer, 2001.
2. K. Daniels and V. J. Milenkovic. Multiple translational containment, part i: An approximate algorithm. *Algorithmica*, 19(1–2):148–182, September 1997.
3. Greg Frederickson. *Dissections: Plane and Fancy*. Cambridge University Press, 1997.
4. J. Hershberger. Finding the upper envelope of n line segments in $O(n\log n)$ time. *Inform. Process. Lett.*, 33:169–174, 1989.
5. Michael Hoffmann. Covering polygons with few rectangles. In *Abstracts 17th European Workshop Comput. Geom.*, pages 39–42. Freie Universität Berlin, 2001.
6. V. Milenkovic. Translational polygon containment and minimal enclosure using linear programming based restriction. In *Proc. 28th Annu. ACM Sympos. Theory Comput.*, pages 109–118, 1996.
7. Victor J. Milenkovic. Multiple translational containment, part ii: Exact algorithm. *Algorithmica*, 19(1–2):183–218, September 1997.

On the Emptiness Problem for Two-Way NFA with One Reversal-Bounded Counter *

Zhe Dang[1]**, Oscar H. Ibarra[2], and Zhi-Wei Sun[3]

[1] School of Electrical Engineering and Computer Science
Washington State University
Pullman, WA 99164, USA
zdang@eecs.wsu.edu
[2] Department of Computer Science
University of California
Santa Barbara, CA 93106, USA
[3] Department of Mathematics
Nanjing University
Nanjing 210093, China

Abstract. We show that the emptiness problem for two-way nondeterministic finite automata augmented with one reversal-bounded counter (i.e., the counter alternates between nondecreasing and nonincreasing modes a fixed number of times) operating on bounded languages (i.e., subsets of $w_1^* \ldots w_k^*$ for some nonnull words $w_1, \ldots, w_k$) is decidable, settling an open problem in [11,12]. The proof is a rather involved reduction to the solution of a special class of Diophantine systems of degree 2 via a class of programs called two-phase programs. The result has applications to verification of infinite state systems.

1 Introduction

Automata theory tries to answer questions concerning the relationship between formal languages and automata that recognize the languages. A fundamental decision question concerning any class of language recognizers is whether the emptiness problem (for the class) is decidable, i.e., whether there exists an algorithm to decide the following question: given an arbitrary machine M in the class, is the language accepted by M empty? Decidability of emptiness can lead to the decidability of other questions such as containment, equivalence, etc.

The simplest recognizers are the finite automata. It is well-known that all the different varieties of finite automata (one-way, two-way, etc.) are effectively equivalent, and the class has a decidable emptiness problem. When the two-way finite automaton is augmented with a storage device, such as a counter, a pushdown stack or a Turing machine tape, emptiness becomes undecidable (no algorithms exist). In fact, it follows from a result in [19] that the emptiness problem is undecidable for two-way finite automata augmented with one counter (even on a unary input alphabet). If one restricts the

---

* The work by Oscar H. Ibarra has been supported in part by NSF Grant IIS-0101134.
** Corresponding author.

P. Bose and P. Morin (Eds.): ISAAC 2002, LNCS 2518, pp. 103–114, 2002.
© Springer-Verlag Berlin Heidelberg 2002

machines to make only a finite number of turns on the input tape, the emptiness problem is still undecidable, even for the case when the input head makes only one turn [11]. However, for such machines with one-way input, the emptiness problem is decidable, since they are simply pushdown automata with a unary stack alphabet.

Restricting the operation of the counter in a two-way one-counter machine makes the emptiness problem decidable for some classes. For example, it has been shown that emptiness is decidable for two-way counter machines whose input head is finite-crossing (i.e., for all inputs, the number of times the input head crosses the boundary between any two adjacent cells is bounded by a fixed number) and whose counter is reversal-bounded (i.e., the number of alternations between nondecreasing mode and nonincreasing mode is bounded by a fixed number, independent of the input) [11]. Interestingly, when the two-way input is unrestricted but the counter is reversal-bounded, emptiness is decidable when the machine is deterministic and accepts a bounded language (i.e., a subset of $w_1^* \ldots w_k^*$ for some nonnull words $w_1, \ldots, w_k$) [10]. This result was later shown to hold for the general case when the the input is not over a bounded language [12]. These machines are quite powerful. They can accept fairly complex languages. For example, such a machine can recognize the language consisting of strings of the form $0^i 1^j$ where i divides j. A question left open in [11,12] is whether the aforementioned decidability of emptiness holds for *nondeterministic* machines (over bounded or unbounded languages). Our main result settles this question for the bounded case. More precisely, we show that the emptiness problem for two-way nondeterministic finite automata augmented with a reversal-bounded counter over bounded languages is decidable. At present, we are not able to generalize this result to the case when the input to the machine does not come from a bounded language. We note that when the machines are augmented with two reversal-bounded counters, emptiness is undecidable, even when the machines are deterministic and accept only bounded languages [11].

We believe that our main theorem will find applications in the area of verification. We note that the recent interest in counter machine models [4,5,7,13,6] is not motivated by investigations of their formal language properties but by their applications to model checking of infinite-state systems, motivated by the recent successes of model-checking techniques for finite-state systems [2,3,21,15,22]. The main result in this paper would be useful in establishing a number of new decidability results concerning various verification problems for infinite-state systems containing integer counters and parameterized constants.

The rest of this paper is organized as follows. In Section 2, we introduce some known results on reversal-bounded counters and number theory. These results are used in the proof of our main theorem. In Section 3, we show a decidable class of Diophantine systems of degree 2. The Diophantine systems are used in Section 4 to establish that a class of simple programs has a decidable emptiness problem. The main theorem follows in Section 5 by reducing it to the simple programs. We conclude in Section 6 with a verification example. Due to space limitation, most of the proofs are not included in the paper.

2   Preliminaries

Let c be a nonnegative integer. A c-counter machine is a two-way nondeterministic finite automaton with input endmarkers (two-way NFA) augmented with c counters, each of which can be incremented by 1, decremented by 1, and tested for zero. We assume, w.l.o.g., that each counter can only store a nonnegative integer, since the sign can be stored in the states. If r is a nonnegative integer, let 2NCM(c,r) denote the class of c-counter machines where each counter is r *reversal-bounded*; i.e., it makes at most r alternations between nondecreasing and nonincreasing modes in any computation; e.g., a counter whose values change according to the pattern 0 1 1 2 3 4 $\underline{4}$ $\underline{3}$ 2 1 $\underline{0}$ $\underline{1}$ 1 $\underline{0}$ is 3-reversal, where the reversals are underlined. For convenience, we sometimes refer to a machine in the class as a 2NCM(c,r). A 2NCM(c,r) is *finite-crossing* if there is a positive integer d such that in any computation, the input head crosses the boundary between any two adjacent cells of the input no more than d times. Note that a 1-crossing 2NCM(c,r) is a one-way nondeterministic finite automaton augmented with c r-reversal counters. 2NCM(c) will denote the class of c-counter machines whose counters are r-reversal bounded for some given r. For deterministic machines, we use 'D' in place of 'N'. If M is a machine, $L(M)$ denotes the language that M accepts.

A language is *strictly bounded* over k letters $a_1, a_2, \ldots, a_k$ if it is a subset of $a_1^* a_2^* \ldots a_k^*$. A language is *bounded* over k nonnull words $w_1, w_2, \ldots, w_k$ if it is a subset of $w_1^* w_2^* \ldots w_k^*$. A straightforward argument shows that a machine of any type studied in this paper accepts a nonempty bounded language if and only if there is another machine of the same type that accepts a nonempty strictly bounded language. So when dealing with the emptiness question for machines over bounded languages, we need only handle the case when the machines accept strictly bounded languages. There are other equivalent definitions of "boundedness" that we will use in the paper.

We will also need the following results.

Theorem 1. *The emptiness problem is decidable for the following classes:*
(a) 2DCM(1) [12].
(b) 2NCM(c) over a unary alphabet (i.e., over a bounded language on 1 letter) [12].
(c) finite-crossing 2NCM(c) for every c [11,9].

Let **N** be the set of nonnegative integers. Let X be a finite set of nonnegative integer variables. An *atomic Presburger relation* on X is either an atomic linear relation

$$\Sigma_{x \in X} a_x x < b,$$

or a mod constraint $x \equiv_d c$, where a_x, b, c and d are integers with $0 \le c < d$. A Presburger formula can always be constructed from atomic Presburger relations using $\neg$ and $\wedge$. Presburger formulas are closed under quantification. Let Q be a set of n-tuples $(j_1, \ldots, j_n)$ in $\mathbf{N}^n$. Q is *Presburger-definable* if there is a Presburger formula $p(x_1, \ldots, x_n)$ such that the set of nonnegative integer solutions of p is exactly Q. It is known that Q is a semilinear set iff Q is Presburger-definable [8]. One may already notice that, for the purpose of this paper, we define a Presburger formula only over nonnegative integer variables (instead of integer variables).

Let $T(B, x)$ be a Presburger formula in two nonnegative integer variables B and x. T is *unitary* if T is a conjunction of atomic Presburger relations and each atomic linear relation in T is in the form of $a_1 B + a_2 x < b$ with $a_2 \in \{-1, 0, 1\}$. We say T is *1-mod-free* (resp. *2-mod-free*) if T does not contain any mod constraints in the form of $B \equiv_d b$ (resp. $x \equiv_d b$) for any b, d. We say T is *mod-free* if T is 1-mod-free and 2-mod-free. T is a *point* if T is $x = a \wedge B = b$ for some $a, b \in \mathbf{N}$. T is a *line* if T is $x = aB + b$, or T is $B = b$ (called a vertical line), for some $a, b \in \mathbf{N}$. T is a *sector* if T is $x \geq aB + b$, or T is $aB + b \leq x \leq a'B + b'$, for some $a < a', b, b' \in \mathbf{N}$. Observe that if T is mod-free and unitary, then T can be written into a (finite) disjunction of points, lines, and sectors. T is *single* if T is a point, a line, or a sector.

An *atomic $\mathcal{D}$-formula* over nonnegative integer variables $x_1, \ldots, x_n$ is either

$$f(x_1, \ldots, x_n) = 0$$

or a divisibility

$$f(x_1, \ldots, x_n) | g(x_1, \ldots, x_n)$$

where f and g are linear polynomials with integer coefficients. A *$\mathcal{D}$-formula* can be built from atomic $\mathcal{D}$-formulas using $\wedge$, $\vee$, and $\exists$. Notice that a Presburger formula is also a $\mathcal{D}$-formula. If a $\mathcal{D}$-formula does not contain $\exists$-quantifiers, the formula is called a *ground formula*. A set Q of n-tuples $(j_1, \ldots, j_n)$ in $\mathbf{N}^n$ is *$\mathcal{D}$-definable* if there is a $\mathcal{D}$-formula $p(x_1, \ldots, x_n)$ such that the set of nonnegative integer solutions of p is exactly Q. The following is Lipshitz's Theorem [16].

Theorem 2. *The satisfiability of $\mathcal{D}$-definable formulas is decidable.*

We will also need two basic results in number theory.

Theorem 3. *Let m_1, m_2 be positive integers and r_1, r_2 be nonnegative integers. The following two items are equivalent:*

(1) There is a nonnegative integer solution of n to $m_1 | (n - r_1) \wedge m_2 | (n - r_2)$,
(2) $\gcd(m_1, m_2) | r_1 - r_2$. [17]

The following is a well-known theorem of Frobenius (cf. [1,14,20]).

Theorem 4. *Let $a_1, \ldots, a_n$ be positive integers. Then there exists a positive integer b_0 such that, for each integer $b \geq b_0$ with $\gcd(a_1, \ldots, a_n) | b$, the linear equation*

$$a_1 x_1 + \ldots + a_n x_n = b$$

has nonnegative integer solutions.

The main theorem of the paper is that the emptiness problem for 2NCM(1) over bounded languages is decidable. The next three sections constitute the entire proof. We first investigate a class of decidable Diophantine systems of degree 2 in Section 3. Then, we show that the emptiness problem for so-called "two-phase programs" is decidable in Section 4. The main theorem follows in Section 5 by reducing the emptiness problem for 2NCM(1) over bounded languages to the emptiness problem for two-phase programs.

3   A Decidable Class of Diophantine Systems of Degree 2

It is well-known that, in general, it is undecidable to determine if a Diophantine system of degree 2 (i.e., a finite set of Diophantine equations of degree 2) has a nonnegative integral solution, cf. [18]. In this section, we find a nontrivial decidable class of Diophantine systems of degree 2. This result will be used in our later proof.

Let $u, s_1, \ldots, s_m, t_1, \ldots, t_n, B_1, \ldots, B_k$ be nonnegative integer variables. A *positive linear polynomial* over $B_1, \ldots, B_k$ is in the form of $a_0 + a_1 B_1 + \ldots + a_k B_k$ where each a_i, $0 \leq i \leq k$, is a nonnegative integer. In this section, $U, U_i, \Phi, \Phi_i, V, V_j, \Gamma, \Gamma_j$ $(1 \leq i \leq m, 1 \leq j \leq n)$ are positive linear polynomials over $B_1, \ldots, B_k$. Consider the following inequalities

$$\sum_{1 \leq i \leq m} U_i s_i + U \leq u \leq \sum_{1 \leq i \leq m} U_i s_i + U + \sum_{1 \leq i \leq m} \Phi_i s_i + \Phi \tag{1}$$

and

$$\sum_{1 \leq j \leq n} V_j t_j + V \leq u \leq \sum_{1 \leq j \leq n} V_j t_j + V + \sum_{1 \leq j \leq n} \Gamma_j t_j + \Gamma. \tag{2}$$

Θ is a predicate on nonnegative integer k-tuples such that, for all nonnegative integers $B_1, \ldots, B_k$, $\Theta(B_1, \ldots, B_k)$ is true iff the conjunction of (1) and (2) has a nonnegative integer solution for $u, s_1, \ldots, s_m, t_1, \ldots, t_n$. The following lemma states that Θ is effectively $\mathcal{D}$-definable; i.e., a $\mathcal{D}$-formula defining Θ can be computed from the description of (1) and (2). The proof uses Theorem 4 and Theorem 3.

Lemma 1. *The predicate $\Theta(B_1, \ldots, B_k)$ defined above is effectively $\mathcal{D}$-definable.*

4   Two-Phase Programs

In this section, we introduce an intermediate machine model called simple programs. A simple program is intended to model a class of nondeterministic programs with a single nondecreasing counter and a number of parameterized constants. For instance, consider the following simple program

    Input (B_1, B_2, B_3);
    1: $x := 0$;
    2: Increment x by any amount (nondeterministically chosen)
       between B_1 and $2B_1$;
    3: Nondeterministically goto 4, 5, or 7;
    4: Increment x by B_2;
    5: Increment x by B_3;
    6: Goto 2;
    7: Halt.

In the program, the input nonnegative integer variables remain unchanged during computation; i.e., they are parameterized constants. Each increment made on the counter satisfies some Presburger constraint in two variables; e.g., $B_1 \leq \delta \leq 2B_1$ holds for the increment δ made in step 2 above. A two-phase program is simply a pair of simple programs G_1 and G_2 that share the same array of input variables $B_1, \ldots, B_k$. We are interested in the following question: is there an assignment for $B_1, \ldots, B_k$ such that the counter in G_1 and the counter in G_2 have the same value when both G_1 and G_2 halt? A decidable answer to this question will be given in this section. The reader might have noticed that there is some inherent relationship between two-phase programs and 2NCM(1) over bounded languages. Indeed, this intermediate result will be used in the next section to prove our main theorem. Before we proceed further, we need a formal definition.

A *simple program* G is a tuple

$$\langle S, B_1, \ldots, B_k, x, \mathbf{T}, E \rangle$$

where

- S is a finite set of control states, with two special states designated as the initial state and the final state.
- $B_1, \ldots, B_k$ are k input (nonnegative integer) variables,
- x is the nonnegative integer counter which is always nondecreasing,
- $\mathbf{T}$ is a finite set of Presburger formulas on two nonnegative integer variables,
- $E \subseteq S \times \{1, \ldots, k\} \times \mathbf{T} \times S$ is a finite set of *edges*. Each edge $\langle s, i, T, s' \rangle$ in E denotes a transition from state s to state s' while incrementing the counter x according to the *evolution pair* (i, T).

The semantics of G is defined as follows. A configuration $(s, v_1, \ldots, v_k, u)$ in $S \times \mathbf{N}^k \times \mathbf{N}$ is a tuple of a control state s, values $v_1, \ldots, v_k$ for the input variables $B_1, \ldots, B_k$, and value u for the counter x.

$$(s, v_1, \ldots, v_k, u) \to^G (s', v'_1, \ldots, v'_k, u')$$

denotes a *one-step transition* satisfying the following conditions:

- There is an edge $\langle s, i, T, s' \rangle$ in G connecting state s to state s',
- The value of each input variable does not change; i.e., $(v_1, \ldots, v_k) = (v'_1, \ldots, v'_k)$,
- The evolution pair (i, T) is satisfied; i.e., $T(v_i, u' - u)$ is true (hence, $u \leq u'$ since T is defined on nonnegative integers).

A *path* is a finite sequence

$$(s_0, v_1, \ldots, v_k, u_0) \ldots (s_i, v_1, \ldots, v_k, u_i) \ldots (s_m, v_1, \ldots, v_k, u_m)$$

for some $m \geq 1$ such that $(s_i, v_1, \ldots, v_k, u_i) \to^G (s_{i+1}, v_1, \ldots, v_k, u_{i+1})$ for each $0 \leq i \leq m - 1$. In particular, if $u_0 = 0$ (the counter starts from 0), s_0 is the initial state and s_m is the final state, then G *accepts* $(v_1, \ldots, v_k, u_m)$.

A *two-phase program* G_{+-} consists of two simple programs G_+ and G_- that share the same S, input variables $B_1, \ldots, B_k$ and $\mathbf{T}$. We shall use x_+ (resp. x_-) to denote the counter in the *positive* (resp. *negative*) program G_+ (resp. G_-). A k-tuple of nonnegative integer values $v_1, \ldots, v_k$ is *accepted by the two-phase program* G_{+-} if there is a counter values u such that $(v_1, \ldots, v_k, u)$ is accepted by both G_+ and G_-. We shall use $L(G_{+-})$ to denote all the k-tuples accepted by G_{+-}. $L(G_{+-})$ is called the tuple language accepted by G_{+-}. A two-phase program models some one counter system where the counter starts from 0 and, after a number of increments followed by a number of decrements, moves back to 0. In G_{+-}, the positive program models the increasing phase and the negative program models the decreasing phase (but the counter in the negative program is always increasing). Therefore, we need further argue whether the total increments made by the positive program equals the total increments made by the negative program. The main result of this section is that the tuple language accepted by a two-phase program G_{+-} is $\mathcal{D}$-definable. The proof first shows that it suffices to consider a special form of a two-phase program G_{+-}: each $T \in \mathbf{T}$ is a point, a line, or a sector. Then, the result follows by making use of Lemma 1.

Theorem 5. *The tuple language accepted by a two-phase program is $\mathcal{D}$-definable.*

Consider a finite set of two-phase programs $\mathcal{G}$, each of which has k-ary input $B_1, ...,$ B_k. The Presburger emptiness problem for $\mathcal{G}$ is to decide, given a Presburger formula $R(B_1, \ldots, B_k)$, whether there is some input $B_1, \ldots, B_k$ accepted by each program in $\mathcal{G}$. Since $R(B_1, \ldots, B_k)$ is $\mathcal{D}$-definable and $\mathcal{D}$-definability is closed under intersection, we have

Theorem 6. *The Presburger emptiness problem for a finite set of two-phase programs is decidable.*

5 2NCM(1) over Bounded Languages

Before we discuss 2NCM(1,r), we first look at a property of a 2NCM(1,0) M over a unary input (i.e., a two-way NFA with a unary input tape augmented with a nondecreasing (i.e., monotonic) counter). The input is in the form of

$$\not{c}\, \underbrace{a \ldots a}_{B} \$$$

of size B for some B, where $\not{c}$ and $\$$ are the left and right endmarkers. M works exactly as a two-way NFA except that, at some move (i.e., a left move, a right move, or a stationary move), M can increment the counter by 1. Suppose the counter initially starts from 0. When the input head is initially at the left endmarker, we use M_{LL} (resp. M_{LR}) to denote the restricted version of M that M returns to the left (resp. right) endmarker upon acceptance (during which M does not read the endmarkers). When the input head is initially at the right endmarker, M_{RR} and M_{RL} are defined similarly. We use $T_{LL}(B, x)$ (resp. $T_{LR}(B, x), T_{RR}(B, x), T_{RL}(B, x)$) to stand for the fact that M_{LL} (resp. M_{LR}, M_{RR}, M_{RL}) accepts the input of size B and upon acceptance, the counter has value x.

If we allow the input head to return to the endmarkers for multiple times, $T(B, x)$ can not be characterized by a Presburger formula. For instance, let M be such as machine. M keeps scanning the input (of size $B \geq 1$) from $\not{c}$ to $\$$ and back, while incrementing the counter. M nondeterministically accepts when $\$$ is reached. Obviously, $T_{LR}(B, x)$ now is exactly $\exists n(2nB + B|x)$ that is not Presburger. However, with the restrictions of M_{LR}, $T(B, x)$ is Presburger. The proof uses a complex loop analysis technique.

Lemma 2. $T_{LL}(B, x)$, $T_{LR}(B, x)$, $T_{RR}(B, x)$, *and* $T_{RL}(B, x)$ *are Presburger for any M specified above.*

Lemma 2 also works for a stronger version of M. We assume the counter in M is r-reversal-bounded for some r (instead of increasing only). Notice that the counter when decreasing can have negative values. We may similarly define restricted machines $M_{LL}, M_{LR}, M_{RL}, M_{RR}$. We shall use $T_{LL}^+(B, x)$ (resp. $T_{LL}^-(B, x)$) to denote that $x \geq 0$ and x (resp. $-x$) is the final value of the reversal-bounded counter in M_{LL} on input of size B. Similarly, we may define $T_{LR}^+(B, x), T_{LR}^-(B, x)$ etc.

Lemma 3. $T_{LL}^+(B, x)$, $T_{LL}^-(B, x)$, $T_{LR}^+(B, x)$, $T_{LR}^-(B, x)$, $T_{RR}^+(B, x)$, $T_{RR}^-(B, x)$, $T_{RL}^+(B, x)$, and $T_{RL}^-(B, x)$ are Presburger for any M specified above and the counter in M is reversal-bounded.

In the rest of this section, we focus on the emptiness problem for 2NCM(1,r) on bounded languages. A slightly different definition of boundedness, but equivalent to the one we gave in Section 2 with respect to decidability of emptiness is the following. A *k-bounded language* is a subset of

$$b_1 a_1^* b_2 a_2^* \ldots b_k a_k^* b_{k+1}$$

where b_i, $1 \leq i \leq k + 1$, is the i-th delimiter, and each block a_i^* between the two delimiters b_i and b_{i+1} is the i-th block. A bounded language is a k-bounded language for some k. Recall that a 2NCM(1,r) is a two-way NFA augmented with an r-reversal-bounded counter. When the input language of a 2NCM(1,r) M is restricted to a bounded language, M is called a 2NCM(1,r) over a bounded language.

Let M be a 2NCM(1,1) working on a k-bounded language. Let

$$w = b_1 1^{B_1} \ldots b_k 1^{B_k} b_{k+1}$$

be an input word where 1^{B_i} is the i-th block of symbol 1's with length B_i. Sometimes, we simply call the input as $(B_1, \ldots, B_k)$. Without loss of generality, we assume that the counter x in M, when M accepts the input, returns to 0 and the input head is on delimiter b_{k+1} with M being at the final state. An accepting computation C of M can be divided into a number of *segments*. Each segment is associated with a state pair (s, s') and a block 1^{B_i}. In the sequel, we shall use $\alpha, \beta, \ldots$ to denote a segment. We have the following four cases:

   (1). (a LL-segment) M, at state s, reads the $i + 1$-th delimiter and M returns to the $i + 1$-th delimiter with state s', during which M only reads symbols in 1^{B_i}.

   (2). (a LR-segment) M, at state s, reads the i-th delimiter and M returns to the $i + 1$-th delimiter with state s', during which M only reads symbols in 1^{B_i}.

   (3). (a RR-segment) M, at state s, reads the i-th delimiter and M returns to the i-th delimiter with state s', during which M only reads symbols in 1^{B_i}.

   (4). (a RL-segment) M, at state s, reads the $i + 1$-th delimiter and M returns to the i-th delimiter with state s', during which M only reads symbols in 1^{B_i}.

   A segment is *positive* (resp. *negative*) if the net counter change is ≥ 0 (resp. < 0) on the segment. Therefore, since the counter is one reversal-bounded, C can be treated as a sequence C_+ of positive segments followed by a sequence C_- of negative segments. Obviously, since C is accepting, the total increments $\Delta(C_+)$ of the counter on C_+ equals the total decrements $\Delta(C_-)$ of the counter on C_-.

We use a *segment symbol* $+_{s,s',d,i}$ (resp. $-_{s,s',d,i}$) to abstract a positive (resp. negative) segment associated with state pair (s, s'), $d \in \{LL,LR,RR,RL\}$, and i-th block 1^{B_i}. According to Lemma 2 and Lemma 3, on a segment, the relationship between the absolute values of counter changes and the length of the block associated with the segment can be characterized by a Presburger formula (i.e., *the formula of the segment symbol*). Now, a two-phase program G_{+-} can be constructed such that each segment symbol $+_{s,s',d,i}$ corresponds to a transition in G_+ as follows (in below, T is the formula of the segment symbol):

- If $d = LL$, then the transition is $((s, i + 1), i, T, (s', i + 1))$.
- If $d = LR$, then the transition is $((s, i), i, T, (s', i + 1))$.
- If $d = RR$, then the transition is $((s, i), i, T, (s', i))$.
- If $d = RL$, then the transition is $((s, i + 1), i, T, (s', i + 1))$.

Similarly, transitions in G_- can be constructed from symbols $-_{s,s',d,i}$. Let G_{+-}^{s,i_0} be a two-phase program consisting of G_+ and G_- such that

- the initial state of G_+ is $(s_0, i = 1)$ where s_0 is the initial state of M,
- the final state of $G_{-}+$ is (s, i_0),
- the initial state of G_- is (s, i_0),
- the final state of G_- is $(s_f, i = k + 1)$ where s_f is the final state of M.

It is noticed that the final state of G_+ equals the initial state of G_-. It is observed that $(B_1, \ldots, B_k)$ is accepted by M iff there are some state s and some $1 \le i_0 \le k + 1$ such that,

$(B_1, \ldots, B_k)$ is accepted by G_{+-}^{s,i_0}.

Since there are only finitely many choices of s and i_0, from Theorem 5, we obtain that the bounded language accepted by 2NCM(1,1) is effectively $\mathcal{D}$-definable.

Lemma 4. *The bounded language accepted by a 2NCM(1,1) is effectively $\mathcal{D}$-definable.*

Next, we show that the bounded language accepted by 2NCM(1,r) for any r is $\mathcal{D}$-definable. 2NCM(1,r), when $r > 1$, is more complex than 2NCM(1,1). However, we will show that we can effectively reduce the emptiness of 2NCM(1,r) M_r into the emptiness of the "intersection" of finitely many 2NCM(1,1)'s. We may assume w.l.o.g that, on a k-bounded input word

$$b_1 1^{B_1} \ldots b_k 1^{B_k} b_{k+1},$$

M_r makes a counter reversal only when it is reading one of the delimiters $b_1, \ldots, b_{k+1}$. (Otherwise, we may insert up to r many new delimiters $c_1, \ldots, c_r$ to an input word of M_r and construct a new 2NCM(1,r) M_r' working on the new $k + r$-bounded word. M_r' simulates M_r properly and makes sure that, whenever M_r makes the i-th reversal, M_r' is reading the delimiter c_i. It is not difficult to show that, if the the bounded language accepted by M_r' is $\mathcal{D}$-definable, then so is the bounded language accepted by M_r.)

The r-reversal-bounded counter x behaves like this:

$$\nearrow, \searrow, \ldots, \nearrow, \searrow$$

(each $\nearrow$ stands for a nondecreasing phase; each $\searrow$ stands for a nonincreasing phase). Two consecutive phases of $\nearrow$ and $\searrow$ are called a *round*. Without loss of generality, we assume that r is odd and x makes exactly r reversals, so x has precisely $m = 1 + \frac{r-1}{2}$ rounds. We also assume that the machine starts with zero counter and accepts with zero counter. If $w = (B_1, \ldots, B_k)$ is an input to M_r, $PAD(w)$ is a string $(B_1, \ldots, B_k, E_1, \ldots, E_{m-1})$. That is, $PAD(w)$ is $(B_1, \ldots, B_k)$ padded with some $(m-1)$-bounded word $(E_1, \ldots, E_{m-1})$. Note that a given k-bounded word w has many $PAD(w)$'s.

A "trace" of the computation of M_r can be represented by a $2(m-1)$-tuple $\alpha = \langle d_1, q_1, d_2, q_2, \ldots, d_{m-1}, q_{m-1} \rangle$, where at the end of round $i = 1, \ldots, m-1$, M_r is at delimiter d_i (since M_r is about to reverse) in state q_i. Clearly, there are only a finite number of such α's. We will construct m 2NCM(1,1)'s $\hat{M}_1, \ldots, \hat{M}_m$ such that:

(*)   a k-bounded word w is in $L(M_r)$ iff $\alpha PAD(w)$ is in $L(\hat{M}_1) \cap \ldots \cap L(\hat{M}_m)$ for some α and $PAD(w)$.

If w is an input to M_r, the input to each $\hat{M}_i$ is a string of the form $\alpha PAD(w)$. For $i = 2, \ldots, m-1$, $\hat{M}_i$ carries out the following two phases:

1. Restores the value of the counter to E_{i-1}, then moves it's input head to delimiter d_{i-1}, and then simulates M_r starting in state q_{i-1}. In the simulation, $\hat{M}_i$ ignores α and the paddings.
2. When M_r completes a round and starts to reverse (i.e., increments) the counter, $\hat{M}_i$ "remembers" the delimiter e_i and state s_i (when the reversal occurs), and goes to block E_i and verifies that the current value of the counter is E_i (note that if such is the case, the counter would be zero after checking). Then $\hat{M}_i$ moves its input head to the leftmost symbol and accepts if $d_i = e_i$ and $q_i = s_i$.

For $i = 1$, $\hat{M}_1$ does not need the restoration phase, but simulates M_r starting in state q_0 (the initial state of M_r). It also executes phase 2. For $i = m$, $\hat{M}_m$ executes the restoration phase only and accepts if M_r, after completing a round, accepts. Notice that, in the above construction, each E_i is used to denote the counter value of M_r at the end of each round. It is easy to verify that (*) above holds and each $\hat{M}_i$ is indeed a 2NCM(1,1). Hence, from Lemma 4 noticing that $\mathcal{D}$-definability is closed under intersection, union (over the α's) and $\exists$-quantification (for eliminating the padding E_i's), we have finally proved the main theorem of the paper that settles the open problem in [11,12].

Theorem 7. *The bounded language accepted by 2NCM(1,r) is effectively $\mathcal{D}$-definable. Therefore, the emptiness problem for 2NCM(1,r) over bounded languages is decidable.*

6   Conclusion

We showed that the emptiness problem for two-way nondeterministic finite automata augmented with one reversal-bounded counter operating on bounded languages is decidable, settling an open problem in [11,12]. The proof was a rather involved reduction to the solution of a special class of Diophantine systems of degree 2 via a class of programs called two-phase programs. The result has applications to verification of infinite state systems.

For instance, consider a nondeterministic transition system M containing nonnegative integer parameterized constants $A, B_1, \ldots, B_k$ and a nonnegative integer counter x, which starts at 0. M's transition involves nondeterministically changing the state (from among a finite number of control states) and updating the counter by performing one of the instructions $x := x + B_i$, $1 \leq i \leq k$. M *terminates* if, M reaches some control state s with $x \leq A$, and any further execution of an updating instruction from s will make $x > A$. In practice, M can be used to model a buffer controller design that handles k types $b_1, \ldots, b_k$ of blocks. Every block of type b_i is with size B_i. The use of parameterized constants is common at the design stage; the constants are concretized in specific implementations. An instruction of the form $x := x + B_i$ on the edge from s to s' means that a block of type b_i is put into the buffer. Notice that, the graph of M makes the controller select blocks of various types according to some regular ordering. x and A in the controller represent the "used" and maximal capacity of the buffer, respectively. Hence, M terminates at the moment when the buffer does not overflow ($x < A$) and putting any additional block according to the ordering into the buffer will make the buffer overflow. Consider the following "efficiency problem" for M: for any $A, B_1, \ldots, B_k$ that satisfy a Presburger formula $P(A, B_1, \ldots, B_k)$ (e.g., a design constraint like $A > B_1 > \ldots > B_k$), when M terminates, the unused buffer size is less than each B_i (i.e., the buffer is maximally used). From the main result in this paper, the efficiency problem is decidable. To see this, we formulate the negation of the problem as follows: Are there values for $A, B_1, \ldots, B_k$ satisfying P such that there is a value x and M terminates with $A - x > B_i$ for some i? Let L be the bounded language representing the nonnegative integer tuples of $(A, B_1, \ldots, B_k)$ that satisfy the negation. It is not hard to construct a two-way nondeterministic finite automaton augmented with one reversal-bounded counter to accept bounded language L. We leave the details to the reader.

Thanks go to WSU PhD student Gaoyan Xie for discussions on the above example.

References

1. A. Brauer. On a problem of partitions. *Amer. J. Math.*, 64:299–312, 1942.
2. E. M. Clarke and E. A. Emerson. Design and synthesis of synchronization skeletons using branching time temporal logic. In *Workshop of Logic of Programs*, volume 131 of *Lecture Notes in Computer Science*. Springer, 1981.
3. E. M. Clarke, E. A. Emerson, and A. P. Sistla. Automatic verification of finite-state concurrent systems using temporal logic specifications. *ACM Transactions on Programming Languages and Systems*, 8(2):244–263, April 1986.
4. H. Comon and Y. Jurski. Multiple counters automata, safety analysis and Presburger arithmetic. In *Proc. 10th Int. Conf. Computer Aided Verification (CAV'98)*, volume 1427 of *Lecture Notes in Computer Science*, pages 268–279. Springer, 1998.
5. H. Comon and Y. Jurski. Timed automata and the theory of real numbers. In *Proc. 10th Int. Conf. Concurrency Theory (CONCUR'99)*, volume 1664 of *Lecture Notes in Computer Science*, pages 242–257. Springer, 1999.
6. Zhe Dang, O. H. Ibarra, and R. A. Kemmerer. Decidable Approximations on Generalized and Parameterized Discrete Timed Automata. In *Proceedings of the 7th Annual International Computing and Combinatorics Conference (COCOON'01)*, volume 2108 of *Lecture Notes in Computer Science*, pages 529–539. Springer, 2001.

7. A. Finkel and G. Sutre. Decidability of reachability problems for classes of two counters automata. In *Proc. 17th Ann. Symp. Theoretical Aspects of Computer Science (STACS'2000), Lille, France, Feb. 2000*, volume 1770 of *Lecture Notes in Computer Science*, pages 346–357. Springer, 2000.
8. S. Ginsburg and E. Spanier. Semigroups, Presburger formulas, and languages. *Pacific J. of Mathematics*, 16:285–296, 1966.
9. E. M. Gurari and O. H. Ibarra. The complexity of decision problems for finite-turn multi-counter machines. *Journal of Computer and System Sciences*, 22:220–229, 1981.
10. E. M. Gurari and O. H. Ibarra. Two-way counter machines and Diophantine equations. *Journal of the ACM*, 29(3):863–873, 1982.
11. O. H. Ibarra. Reversal-bounded multicounter machines and their decision problems. *Journal of the ACM*, 25(1):116–133, January 1978.
12. O. H. Ibarra, T. Jiang, N. Tran, and H. Wang. New decidability results concerning two-way counter machines. *SIAM J. Comput.*, 24:123–137, 1995.
13. O. H. Ibarra, J. Su, Zhe Dang, T. Bultan, and R. A. Kemmerer. Counter machines: decidable properties and applications to verification problems. In *Proceedings of the 25th International Symposium on Mathematical Foundations of Computer Science (MFCS 2000)*, volume 1893 of *Lecture Notes in Computer Science*, pages 426–435. Springer-Verlag, 2000.
14. R. Kannan. Lattice translates of a polytope and the Frobenius problem. *Combinatorica*, 12:161–177, 1992.
15. K.L. McMillan. *Symbolic Model Checking*. Kluwer Academic Publishers, Norwell Massachusetts, 1993.
16. L. Lipshitz. The Diophantine problem for addition and divisibility. *Transactions of AMS*, 235:271–283, 1978.
17. K. Mahler. On the Chinese remainder theorem. *Math. Nachr.*, 18:120–122, 1958.
18. Y. V. Matiyasevich. *Hilbert's Tenth Problem*. MIT Press, 1993.
19. M. Minsky. Recursive unsolvability of Post's problem of Tag and other topics in the theory of Turing machines. *Ann. of Math.*, 74:437–455, 1961.
20. J. L. Ramirez-Alfonsin. Complexity of the Frobenius problem. *Combinatorica*, 16:143–147, 1996.
21. A. P. Sistla and E. M. Clarke. Complexity of propositional temporal logics. *Journal of ACM*, 32(3):733–749, 1983.
22. M. Y. Vardi and P. Wolper. An automata-theoretic approach to automatic program verification (preliminary report). In *Proceedings 1st Annual IEEE Symp. on Logic in Computer Science, LICS'86, Cambridge, MA, USA, 16–18 June 1986*, pages 332–344, Washington, DC, 1986. IEEE Computer Society Press.

Quantum Multi-prover Interactive Proof Systems with Limited Prior Entanglement

Hirotada Kobayashi[1][2] and Keiji Matsumoto[1]

[1] Quantum Computation and Information Project,
Exploratory Research for Advanced Technology,
Japan Science and Technology Corporation,
5-28-3 Hongo, Bunkyo-ku, Tokyo 113-0033, Japan
{hirotada, keiji}@qci.jst.go.jp
[2] Department of Information Science, The University of Tokyo,
7-3-1 Hongo, Bunkyo-ku, Tokyo 113-0033, Japan

Abstract. This paper gives the first formal treatment of a quantum analogue of multi-prover interactive proof systems. It is proved that the class of languages having quantum multi-prover interactive proof systems is necessarily contained in NEXP, under the assumption that provers are allowed to share at most polynomially many prior-entangled qubits. This implies that, in particular, without any prior entanglement among provers, the class of languages having quantum multi-prover interactive proof systems is equal to NEXP. Related to these, it is shown that, if a prover does not have his private qubits, the class of languages having quantum single-prover interactive proof systems is also equal to NEXP.

1 Introduction

After Deutsch [10] gave the first formal treatment of quantum computation, a number of papers have provided evidence that quantum computation has much more power than classical computation for solving certain computational tasks, including notable Shor's integer factoring algorithm [23]. Watrous [25] showed that it might be also the case for single-prover interactive proof systems, by constructing a constant-round quantum interactive protocol for a PSPACE-complete language, which is impossible for classical interactive proof systems unless the polynomial-time hierarchy collapses to AM [2,16]. A natural question to ask is how strong a quantum analogue of multi-prover interactive proof systems is. This paper gives the first step for this question, by proving that the class of languages having quantum multi-prover interactive proof systems is necessarily contained in NEXP, under the assumption that provers are allowed to share at most polynomially many prior-entangled qubits. This might even suggest the possibility that, under such an assumption, quantum multi-prover interactive proof systems are weaker than classical ones, since Cleve [9] reported that some classical two-prover interactive proofs do not work with a pair of cheating provers sharing polynomially many EPR pairs.

P. Bose and P. Morin (Eds.): ISAAC 2002, LNCS 2518, pp. 115–127, 2002.

Interactive proof systems were introduced by Babai [2] and Goldwasser, Micali, and Rackoff [15]. An interactive proof system consists of an interaction between a computationally unbounded prover and a polynomial-time probabilistic verifier. The prover attempts to convince the verifier that a given input string satisfies some property, while the verifier must check the validity of the assertion of the prover. It is well-known that the class IP of languages having interactive proof systems is equal to PSPACE, shown by Shamir [22] based on the work of Lund, Fortnow, Karloff, and Nisan [20], and on the result of Papadimitriou [21].

Quantum interactive proof systems were introduced by Watrous [25] in terms of quantum circuits. He showed that every language in PSPACE has a three-message quantum interactive proof with exponentially small one-sided error. A consecutive work of Kitaev and Watrous [18] showed that any quantum interactive protocol, even with two-sided bounded error, can be parallelized to a three-message quantum protocol with exponentially small one-sided error. They also showed that the class QIP of languages having quantum interactive proof systems is necessarily contained in EXP.

A multi-prover interactive proof system, introduced by Ben-Or, Goldwasser, Kilian, and Wigderson [5], is an extension of a (single-prover) interactive proof system in which a verifier communicates with not only one but multiple provers, while provers cannot communicate with each other and cannot know messages exchanged between the verifier and other provers. A language L has a multi-prover interactive proof system if, for some k denoting the number of provers, there exists a verifier V such that (i) for every input in L, there exist provers $P_1, \ldots, P_k$ that can convince V with certainty, and (ii) for every input not in L, no set of provers $P'_1, \ldots, P'_k$ can convince V with probability more than $1/2$. Babai, Fortnow, and Lund [3], combining the result by Fortnow, Rompel, and Sipser [14], showed that the class MIP of languages having multi-prover interactive proof systems is equal to NEXP. A sequence of papers [7,8,11,19] led to a result of Feige and Lovász [12] that every language in NEXP has a two-prover interactive proof system with just one round (i.e. two messages) of communication (meaning that the verifier sends one question to each of the provers in parallel, then receives their responses), with exponentially small one-sided error.

In this paper we first define quantum multi-prover interactive proof systems by naturally extending the quantum single-prover model. Perhaps the most important and interesting difference between quantum and classical multi-prover interactive proofs is that provers may share entanglement *a priori*. Particular cases are protocols with two provers initially sharing lots of EPR pairs. For the sake of generality, we may allow protocols with any number of provers and with any kind of prior entanglement, not limited to EPR-type ones. Although sharing classical randomness among provers does not change the power of classical multi-prover interactive proofs (unless zero-knowledge properties are taken into account [4]), sharing prior entanglement does have a possibility both to strengthen and to weaken the power of quantum multi-prover interactive proofs. In fact, while sharing prior entanglement may increase the power of cheating provers as shown by Cleve [9], it may be possible for a quantum verifier to turn the prior entanglement among provers to his advantage.

The main result of this paper is to show the NEXP upper bound for quantum multi-prover interactive proof systems under the assumption that provers are allowed to share at most polynomially many prior-entangled qubits. That is, polynomially many prior-entangled qubits among provers cannot be advantageous to a quantum verifier. As a special case of this result, it is proved that, if provers do not share any prior entanglement with each other, the class of languages having quantum multi-prover interactive proof systems is equal to NEXP. Another result related to these is that, in the case the prover does not have his private qubits, the class of languages having quantum single-prover interactive proof systems is also equal to NEXP. This special model of quantum single-prover interactive proofs can be regarded as a quantum counterpart of a probabilistic oracle machine [14,13,3] in the sense that there is no private space for the prover during the protocol, and thus we call this model a *quantum oracle circuit*. Our result shows that quantumization of probabilistic oracle machines does not change the power of the model.

To prove the NEXP upper bound of quantum multi-prover interactive proof systems, a key idea is to bound the number of private qubits of provers without diminishing the computational power of them. Suppose that each prover has only polynomially many private qubits during the protocol. Then the total number of qubits of the quantum multi-prover interactive proof system is polynomial-bounded, and it can be simulated classically in non-deterministic exponential time. Now the point is whether space-bounded quantum provers (i.e. provers can apply any unitary transformations on their spaces, but the number of qubits in their spaces is bounded polynomial with respect to the input length) are as powerful as space-unbounded quantum provers or not. Under the assumption that provers are allowed to share at most polynomially many prior-entangled qubits, we show that, even with only polynomially many private qubits, each prover can do everything that he could with as many qubits as he likes, in the sense that the verifier cannot distinguish the difference at all. For this, we also prove one fundamental property on quantum information theory, which itself is also of interest and worth while stating.

2   Definitions

2.1   Quantum Multi-prover Interactive Proof Systems

Here we give a formal definition of quantum multi-prover interactive proof systems by extending quantum single-prover model defined by Watrous [25].

Let k be the number of provers, and let $\Sigma = \{0,1\}$ be the alphabet set. For every input $x \in \Sigma^*$ of length $n = |x|$, the entire system of quantum k-prover interactive proof system consists of $q(n) = q_V(n) + \sum_{i=1}^{k}(q_{\mathcal{M}_i}(n) + q_{\mathcal{P}_i}(n))$ qubits, where $q_V(n)$ is the number of qubits that are private to a verifier V, each $q_{\mathcal{P}_i}(n)$ is the number of qubits that are private to a prover P_i, and each $q_{\mathcal{M}_i}(n)$ is the number of message qubits used for communication between V and P_i. Note that no communication is allowed between different provers P_i and P_j. It is assumed that q_V and each $q_{\mathcal{M}_i}$ are polynomial-bounded functions. Moreover, without loss of generality, we may assume that

$q_{\mathcal{M}_1} = \cdots = q_{\mathcal{M}_k} = q_{\mathcal{M}}$ and $q_{\mathcal{P}_1} = \cdots = q_{\mathcal{P}_k} = q_{\mathcal{P}}$. Accordingly, the entire system consists of $q(n) = q_{\mathcal{V}}(n) + k(q_{\mathcal{M}}(n) + q_{\mathcal{P}}(n))$ qubits.

Given polynomial-bounded functions $m, q_{\mathcal{V}}, q_{\mathcal{M}} \colon \mathbb{Z}^+ \to \mathbb{N}$, an m-*message* $(q_{\mathcal{V}}, q_{\mathcal{M}})$-*restricted quantum verifier* V for a quantum k-prover interactive proof system is a polynomial-time computable mapping of the form $V \colon \Sigma^* \to \Sigma^*$. For every input $x \in \Sigma^*$ of length n, V uses at most $q_{\mathcal{V}}(n)$ qubits for his private space and at most $q_{\mathcal{M}}(n)$ qubits for communication with each prover. The string $V(x)$ is interpreted as a $\lfloor m(n)/2 + 1 \rfloor$-tuple $(V(x)_1, \ldots, V(x)_{\lfloor m(n)/2+1 \rfloor})$, with each $V(x)_j$ a description of a polynomial-time uniformly generated quantum circuit acting on $q_{\mathcal{V}}(n) + kq_{\mathcal{M}}(n)$ qubits. One of the private qubits of the verifier is designated as the output qubit. The notion of polynomial-time uniformly generated families of quantum circuits we use here is along the lines in [25,18]. Furthermore, it suffices to cosider only unitary quantum circuits. See [1] for a detailed description of the equivalence of the unitary and non-unitary quantum circuit models in view of time complexity.

Given polynomial-bounded functions $m, q_{\mathcal{M}} \colon \mathbb{Z}^+ \to \mathbb{N}$ and a function $q_{\mathcal{P}} \colon \mathbb{Z}^+ \to \mathbb{N}$, an m-*message* $(q_{\mathcal{M}}, q_{\mathcal{P}})$-*restricted quantum prover* P_i for each $i = 1, \ldots, k$ is a mapping of the form $P_i \colon \Sigma^* \to \Sigma^*$. For every input $x \in \Sigma^*$ of length n, each P_i uses at most $q_{\mathcal{P}}(n)$ qubits for his private space and at most $q_{\mathcal{M}}(n)$ qubits for communication with the verifier. The string $P_i(x)$ is interpreted as a $\lfloor m(n)/2 + 1/2 \rfloor$-tuple $(P_i(x)_1, \ldots, P_i(x)_{\lfloor m(n)/2+1/2 \rfloor})$, with each $P_i(x)_j$ a description of a quantum circuit acting on $q_{\mathcal{M}}(n) + q_{\mathcal{P}}(n)$ qubits. No restrictions are placed on the complexity of the mapping P_i (i.e., each $P_i(x)_j$ can be an arbitrary unitary transformation). Furthermore, for some function $q_{\text{ent}} \colon \mathbb{Z}^+ \to \mathbb{N}$ satisfying $q_{\text{ent}} \leq q_{\mathcal{P}}$, each P_i may have at most $q_{\text{ent}}(n)$ qubits among his private qubits that are prior-entangled with some private qubits of other provers. Such a prover P_i is called q_{ent}-*prior-entangled*. For the sake of generality, we allow any kind of prior entanglement, not limited to EPR-type ones.

An m-*message* $(q_{\mathcal{V}}, q_{\mathcal{M}}, q_{\mathcal{P}})$-*restricted quantum k-prover interactive proof system* consists of an m-message $(q_{\mathcal{V}}, q_{\mathcal{M}})$-restricted quantum verifier V and m-message $(q_{\mathcal{M}}, q_{\mathcal{P}})$-restricted quantum provers $P_1, \ldots, P_k$. If $P_1, \ldots, P_k$ are q_{ent}-prior-entangled, such a quantum k-prover interactive proof system is called q_{ent}-*prior-entangled*. Let $\mathcal{V} = l_2(\Sigma^{q_{\mathcal{V}}})$, each $\mathcal{M}_i = l_2(\Sigma^{q_{\mathcal{M}}})$, and each $\mathcal{P}_i = l_2(\Sigma^{q_{\mathcal{P}}})$ denote the Hilbert spaces corresponding to the private qubits of the verifier, the message qubits between the verifier and the ith prover, and the private qubits of the ith prover, respectively. Given a verifier V, provers $P_1, \ldots, P_k$, and an input x of length n, we define a circuit $(P_1(x), \ldots, P_k(x), V(x))$ acting on $q(n)$ qubits as follows. If $m(n)$ is odd, circuits $P_1(x)_1, \ldots, P_k(x)_1$, $V(x)_1$, $\ldots$, $P_1(x)_{(m(n)+1)/2}, \ldots, P_k(x)_{(m(n)+1)/2}$, $V(x)_{(m(n)+1)/2}$ are applied in sequence, each $P_i(x)_j$ to $\mathcal{M}_i \otimes \mathcal{P}_i$, and each $V(x)_j$ to $\mathcal{V} \otimes \mathcal{M}_1 \otimes \cdots \otimes \mathcal{M}_k$. If $m(n)$ is even, circuits $V(x)_1, P_1(x)_1, \ldots, P_k(x)_1, \ldots$, $V(x)_{m(n)/2}, P_1(x)_{m(n)/2}, \ldots, P_k(x)_{m(n)/2}, V(x)_{m(n)/2+1}$ are applied in sequence. Note that the order of applications of the circuits of the provers at each round has actually no sense since the space $\mathcal{M}_i \otimes \mathcal{P}_i$ on which the circuits of the ith prover act is separated from each other prover.

At any given instant, the state of the entire system is a unit vector in the space $\mathcal{V} \otimes \mathcal{M}_1 \otimes \cdots \otimes \mathcal{M}_k \otimes \mathcal{P}_1 \otimes \cdots \otimes \mathcal{P}_k$. For instance, in the case $m(n) = 3$, given an input x of length n, the state of the system after all of the circuits of the provers and the verifier have been applied is

$$V_2 P_{k,2} \cdots P_{1,2} V_1 P_{k,1} \cdots P_{1,1} |\psi_{\text{init}}\rangle,$$

where each V_j and $P_{i,j}$ denotes the extension of $V(x)_j$ and $P_i(x)_j$, respectively, to the space $\mathcal{V} \otimes \mathcal{M}_1 \otimes \cdots \otimes \mathcal{M}_k \otimes \mathcal{P}_1 \otimes \cdots \otimes \mathcal{P}_k$ by tensoring with the identity, and $|\psi_{\text{init}}\rangle \in \mathcal{V} \otimes \mathcal{M}_1 \otimes \cdots \otimes \mathcal{M}_k \otimes \mathcal{P}_1 \otimes \cdots \otimes \mathcal{P}_k$ denotes the initial state. In the initial state $|\psi_{\text{init}}\rangle$ for q_{ent}-prior-entangled proof systems, only the first $q_{\text{ent}}(n)$ qubits in each $\mathcal{P}_i$ may be entangled with other qubits in $\mathcal{P}_1 \otimes \cdots \otimes \mathcal{P}_k$. All the qubits other than these prior-entangled ones are initially in the $|0\rangle$-state.

For every input x, the probability that the $(k+1)$-tuple $(P_1, \ldots, P_k, V)$ accepts x is defined to be the probability that an observation of the output qubit in the basis of $\{|0\rangle, |1\rangle\}$ yields $|1\rangle$, after the circuit $(P_1(x), \ldots, P_k(x), V(x))$ is applied to the initial state $|\psi_{\text{init}}\rangle$.

Although k, the number of provers, has been treated to be constant so far, the above definition can be naturally extended to the case that $k: \mathbb{Z}^+ \to \mathbb{N}$ is a function of the input length n. In what follows, we treat k as a function. Note that the number of provers possible to communicate with the verifier must be bounded polynomial in n.

Definition 1. *Given   polynomial-bounded   functions $k, m: \mathbb{Z}^+ \to \mathbb{N}$,   a function $q_{\text{ent}}: \mathbb{Z}^+ \to \mathbb{N}$,   and   functions $a, b: \mathbb{Z}^+ \to [0,1]$,   a   language L is in $\mathrm{QMIP}(k, m, q_{\text{ent}}, a, b)$ iff there exist polynomial-bounded functions $q_\mathcal{V}, q_\mathcal{M}: \mathbb{Z}^+ \to \mathbb{N}$ and an m-message $(q_\mathcal{V}, q_\mathcal{M})$-restricted quantum verifier V for a quantum k-prover interactive proof system such that, for every input x of length n,*

(i) *if $x \in L$, there exist a function $q_\mathcal{P}: \mathbb{Z}^+ \to \mathbb{N}$ satisfying $q_\mathcal{P} \geq q_{\text{ent}}$ and a set of k quantum provers $P_1, \ldots, P_k$ of m-message $(q_\mathcal{M}, q_\mathcal{P})$-restricted q_{ent}-prior-entangled such that $(P_1, \ldots, P_k, V)$ accepts x with probability at least $a(n)$,*

(ii) *if $x \notin L$, for all functions $q'_\mathcal{P}: \mathbb{Z}^+ \to \mathbb{N}$ satisfying $q'_\mathcal{P} \geq q_{\text{ent}}$ and all sets of k quantum provers $P'_1, \ldots, P'_k$ of m-message $(q_\mathcal{M}, q'_\mathcal{P})$-restricted q_{ent}-prior-entangled, $(P'_1, \ldots, P'_k, V)$ accepts x with probability at most $b(n)$.*

Let $\mathrm{QMIP}(poly, poly, q_{\text{ent}}, a, b)$ denote   the   union   of   the   classes $\mathrm{QMIP}(k, m, q_{\text{ent}}, a, b)$ over all polynomial-bounded functions k and m. The class QMIP of languages having quantum multi-prover interactive proof systems is defined as follows.

Definition 2. *A language L is in QMIP iff there exists a function $q_{\text{ent}}: \mathbb{Z}^+ \to \mathbb{N}$ such that, for any function $q'_{\text{ent}}: \mathbb{Z}^+ \to \mathbb{N}$ satisfying $q'_{\text{ent}} \geq q_{\text{ent}}$, L is in $\mathrm{QMIP}(poly, poly, q'_{\text{ent}}, 1, 1/2)$.*

Next we define the class $\mathrm{QMIP}^{(\text{l.e.})}$ of languages having quantum multi-prover interactive proof systems with at most polynomially many prior-entangled qubits.

Definition 3. *A language L is in* $\mathrm{QMIP}^{(\mathrm{l.e.})}$ *iff there exists a polynomial-bounded function* $q_{\mathrm{ent}} \colon \mathbb{Z}^+ \to \mathbb{N}$ *such that, for any polynomial-bounded function* $q'_{\mathrm{ent}} \colon \mathbb{Z}^+ \to \mathbb{N}$ *satisfying* $q'_{\mathrm{ent}} \geq q_{\mathrm{ent}}$, *$L$ is in* $\mathrm{QMIP}(poly, poly, q'_{\mathrm{ent}}, 1, 1/2)$.

Finally we define the class $\mathrm{QMIP}^{(\mathrm{n.e.})}$ of languages having quantum multi-prover interactive proof systems without any prior entanglement.

Definition 4. *A language L is in* $\mathrm{QMIP}^{(\mathrm{n.e.})}$ *iff L is in* $\mathrm{QMIP}(poly, poly, 0, 1, 1/2)$.

2.2 Quantum Oracle Circuits

Consider a situation in which a verifier can communicate with only one prover, but the prover does not have his private qubits. We call this model a *quantum oracle circuit*, since it can be regarded as a quantum counterpart of a probabilistic oracle machine [14,13,3] in the sense that there is no private space for the prover during the protocol. For the definition of quantum oracle circuits, we use slightly different terminologies from those in the previous subsection so that they are fitted to the term 'oracle' rather than 'prover'.

Given polynomial-bounded functions $m, q_V, q_O \colon \mathbb{Z}^+ \to \mathbb{N}$, an *$m$-oracle-call (q_V, q_O)-restricted quantum verifier* V for a quantum oracle circuit is a $2m$-message (q_V, q_O)-restricted quantum verifier for a quantum single-prover interactive proof system. A *q_O-restricted quantum oracle* O for an m-oracle-call (q_V, q_O)-restricted quantum verifier is a $2m$-message $(q_O, 0)$-restricted quantum prover. Note that our definition of a quantum oracle completely differs from the commonly used one by Bennett, Bernstein, Brassard, and Vazirani [6].

Definition 5. *Given a polynomial-bounded function* $m \colon \mathbb{Z}^+ \to \mathbb{N}$ *and functions* $a, b \colon \mathbb{Z}^+ \to [0, 1]$, *a language L is in* $\mathrm{QOC}(m, a, b)$ *iff there exist polynomial-bounded functions* $q_V, q_O \colon \mathbb{Z}^+ \to \mathbb{N}$ *and an m-oracle-call (q_V, q_O)-restricted quantum verifier V for a quantum oracle circuit such that, for every input x of length n,*

(i) if $x \in L$, there exists a q_O-restricted quantum oracle O for V such that V with access to O accepts x with probability at least $a(n)$,

(ii) if $x \notin L$, for all q_O-restricted quantum oracles O' for V, V with access to O' accepts x with probability at most $b(n)$.

Let $\mathrm{QOC}(poly, a, b)$ denote the union of the classes $\mathrm{QOC}(m, a, b)$ over all polynomial-bounded functions m. The class QOC of languages accepted by quantum oracle circuits is defined as follows.

Definition 6. *A language L is in* QOC *iff L is in* $\mathrm{QOC}(poly, 1, 1/2)$.

3 $\mathrm{QMIP}^{(\mathrm{l.e.})} \subseteq \mathrm{NEXP}$

Now we show that every language having a quantum multi-prover interactive proof system is necessarily in NEXP under the assumption that provers are allowed to share at most polynomially many prior-entangled qubits.

A key idea is to bound the number of private qubits of provers without diminishing the computational power of them. More precisely, we prove Lemma 9 in Subsection 3.2, which claims that, for any protocol of quantum multi-prover interactive proofs with at most polynomially many prior-entangled qubits, there exists a quantum multi-prover interactive protocol with the same number of provers and with the same number of messages, in which each prover uses only polynomially many qubits for his private space with respect to the input length, and the probability of acceptance is exactly same as that of the original one.

For simplicity, in this section and after, we often drop the argument x and n in the various functions defined in the previous section. We also assume that operators acting on subsystems of a given system are extended to the entire system by tensoring with the identity, when it is not confusing.

3.1 Useful Properties on Quantum Fundamentals

First we state two useful properties on quantum information theory, which play key roles in the proof of Lemma 9. Indeed, an interesting and important point in this paper is how to combine and apply these two to the theory of quantum multi-prover interactive proof systems.

The first property we state is a well-known property below.

Theorem 7 ([24,17]). *Let* $|\phi\rangle, |\psi\rangle \in \mathcal{H}_1 \otimes \mathcal{H}_2$ *satisfy* $\mathrm{tr}_{\mathcal{H}_2} |\phi\rangle\langle\phi| = \mathrm{tr}_{\mathcal{H}_2} |\psi\rangle\langle\psi|$. *Then there is a unitary transformation* U *over* $\mathcal{H}_2$ *such that* $(I_{\mathcal{H}_1} \otimes U)|\phi\rangle = |\psi\rangle$, *where* $I_{\mathcal{H}_1}$ *is the identity operator over* $\mathcal{H}_1$.

The second property, Theorem 8 below, is a key property first shown in this paper. The authors believe that this property itself is worth while stating in quantum information theory. The proof is omitted due to limitations of space.

Theorem 8. *Fix a state* $|\phi\rangle$ *in* $\mathcal{H}_1 \otimes \mathcal{H}_2 \otimes \mathcal{H}_3$ *and a unitary transformation* U *over* $\mathcal{H}_2 \otimes \mathcal{H}_3$ *arbitrarily, and let* $|\psi\rangle$ *denote* $(I_{\mathcal{H}_1} \otimes U)|\phi\rangle$. *Then, for any Hilbert space* $\mathcal{H}'_3$ *of* $\dim(\mathcal{H}'_3) \leq \dim(\mathcal{H}_3)$ *such that there exists a state* $|\phi'\rangle$ *in* $\mathcal{H}_1 \otimes \mathcal{H}_2 \otimes \mathcal{H}'_3$ *satisfying* $\mathrm{tr}_{\mathcal{H}'_3} |\phi'\rangle\langle\phi'| = \mathrm{tr}_{\mathcal{H}_3} |\phi\rangle\langle\phi|$, *there exist a Hilbert space* $\mathcal{H}''_3$ *of* $\dim(\mathcal{H}''_3) = (\dim(\mathcal{H}_2))^2 \cdot \dim(\mathcal{H}'_3)$ *and a state* $|\psi'\rangle$ *in* $\mathcal{H}_1 \otimes \mathcal{H}_2 \otimes \mathcal{H}''_3$ *such that* $\mathrm{tr}_{\mathcal{H}''_3} |\psi'\rangle\langle\psi'| = \mathrm{tr}_{\mathcal{H}_3} |\psi\rangle\langle\psi|$.

3.2 QMIP$^{(\mathrm{l.e.})}$ $\subseteq$ NEXP

Now we are ready to show the following lemma.

Lemma 9. *Let* $k, m, q_{\mathcal{V}}, q_{\mathcal{M}}, q_{\mathrm{ent}} \colon \mathbb{Z}^+ \to \mathbb{N}$ *be polynomial-bounded functions and* V *be an* m-message $(q_{\mathcal{V}}, q_{\mathcal{M}})$-restricted quantum verifier for a quantum k-prover interactive proof system. Then, for any function $q_{\mathcal{P}} \colon \mathbb{Z}^+ \to \mathbb{N}$ sat-isfying $q_{\mathcal{P}} \geq q_{\mathrm{ent}}$ *and any set of* m-message $(q_{\mathcal{M}}, q_{\mathcal{P}})$-restricted q_{ent}-prior-entangled quantum provers $P_1, \ldots, P_k$, *there exists a set of* m-message $(q_{\mathcal{M}}, q_{\mathrm{ent}} + 2\lfloor m/2 + 1/2 \rfloor q_{\mathcal{M}})$-restricted q_{ent}-prior-entangled quantum provers $P'_1, \ldots, P'_k$ *such that, for every input* x, *the probability of accepting* x *by* $(P'_1, \ldots, P'_k, V)$ *is exactly equal to the one by* $(P_1, \ldots, P_k, V)$.

Proof. It is assumed that $q_{\mathcal{P}} \geq q_{\text{ent}} + 2\lfloor m/2 + 1/2 \rfloor q_{\mathcal{M}}$, since there is nothing to show in the case $q_{\mathcal{P}} < q_{\text{ent}} + 2\lfloor m/2 + 1/2 \rfloor q_{\mathcal{M}}$. It is also assumed that the values of m are even, and thus $2\lfloor m/2 + 1/2 \rfloor q_{\mathcal{M}} = m q_{\mathcal{M}}$ (odd cases can be dealt with a similar argument).

Given a protocol $(P_1, \ldots, P_k, V)$ of an m-message $(q_{\mathcal{V}}, q_{\mathcal{M}}, q_{\mathcal{P}})$-restricted q_{ent}-prior-entangled quantum k-prover interactive proof system, we first show that P_1 can be replaced by an m-message $(q_{\mathcal{M}}, q_{\text{ent}} + m q_{\mathcal{M}})$-restricted q_{ent}-prior-entangled quantum prover P_1' such that the probability of acceptance by $(P_1', P_2, \ldots, P_k, V)$ is exactly equal to the one by $(P_1, \ldots, P_k, V)$ on every input. Having shown this, we repeat the same process for each of provers to construct a protocol $(P_1', P_2', P_3, \ldots, P_k, V)$ from $(P_1', P_2, \ldots, P_k, V)$ and so on, and finally we obtain a protocol $(P_1', \ldots, P_k', V)$ in which all of $P_1', \ldots, P_k'$ are m-message $(q_{\mathcal{M}}, q_{\text{ent}} + m q_{\mathcal{M}})$-restricted q_{ent}-prior-entangled. We construct P_1' by showing, for every input x, how to construct each $P_1'(x)_j$ from the original $P_1(x)_j$. In what follows, each $P_i(x)_j$ and $P_i'(x)_j$ will be abbreviated as $P_{i,j}$ and $P_{i,j}'$, respectively.

Let each $|\psi_j\rangle, |\phi_j\rangle \in \mathcal{V} \otimes \mathcal{M}_1 \otimes \cdots \otimes \mathcal{M}_k \otimes \mathcal{P}_1 \otimes \cdots \otimes \mathcal{P}_k$, for $1 \leq j \leq m/2$, denote a state of the original m-message $(q_{\mathcal{V}}, q_{\mathcal{M}}, q_{\mathcal{P}})$-restricted q_{ent}-prior-entangled quantum k-prover interactive proof system defined in an inductive manner by

$$
\begin{aligned}
|\phi_1\rangle &= V_1 |\psi_{\text{init}}\rangle, \\
|\phi_j\rangle &= V_j P_{k,j-1} \cdots P_{1,j-1} |\phi_{j-1}\rangle, \quad 2 \leq j \leq m/2, \\
|\psi_j\rangle &= P_{1,j} |\phi_j\rangle, \quad\quad\quad\quad\quad\quad 1 \leq j \leq m/2.
\end{aligned}
$$

Here $|\psi_{\text{init}}\rangle \in \mathcal{V} \otimes \mathcal{M}_1 \otimes \cdots \otimes \mathcal{M}_k \otimes \mathcal{P}_1 \otimes \cdots \otimes \mathcal{P}_k$ is the initial state in which the first $q_{\text{ent}}(n)$ qubits in each $\mathcal{P}_j$ may be entangled with private qubits of other provers than P_j. All the qubits other than these prior-entangled qubits are the $|0\rangle$-states in the state $|\psi_{\text{init}}\rangle$. Note that $\text{tr}_{\mathcal{M}_1 \otimes \mathcal{P}_1} |\psi_j\rangle\langle\psi_j| = \text{tr}_{\mathcal{M}_1 \otimes \mathcal{P}_1} |\phi_j\rangle\langle\phi_j|$, for each $1 \leq j \leq m/2$.

We define each $P_{1,j}'$ inductively. To define $P_{1,1}'$, consider the states $|\phi_1\rangle$ and $|\psi_1\rangle$. Let $|\phi_1'\rangle = |\phi_1\rangle$. Since all of the last $q_{\mathcal{P}} - q_{\text{ent}}$ qubits in $\mathcal{P}_1$ in the state $|\phi_1\rangle$ are the $|0\rangle$-states and $|\psi_1\rangle = P_{1,1}|\phi_1\rangle$, by Theorem 8, there exists a state $|\psi_1'\rangle$ in $\mathcal{V} \otimes \mathcal{M}_1 \otimes \cdots \otimes \mathcal{M}_k \otimes \mathcal{P}_1 \otimes \cdots \otimes \mathcal{P}_k$ such that $\text{tr}_{\mathcal{P}_1} |\psi_1'\rangle\langle\psi_1'| = \text{tr}_{\mathcal{P}_1} |\psi_1\rangle\langle\psi_1|$ and all but the first $q_{\text{ent}} + 2 q_{\mathcal{M}}$ qubits in $\mathcal{P}_1$ are the $|0\rangle$-states in the state $|\psi_1'\rangle$. Furthermore we have

$$
\text{tr}_{\mathcal{M}_1 \otimes \mathcal{P}_1} |\psi_1'\rangle\langle\psi_1'| = \text{tr}_{\mathcal{M}_1 \otimes \mathcal{P}_1} |\psi_1\rangle\langle\psi_1| = \text{tr}_{\mathcal{M}_1 \otimes \mathcal{P}_1} |\phi_1\rangle\langle\phi_1| = \text{tr}_{\mathcal{M}_1 \otimes \mathcal{P}_1} |\phi_1'\rangle\langle\phi_1'|.
$$

Therefore, by Theorem 7, there exists a unitary transformation $Q_{1,1}$ acting on $\mathcal{M}_1 \otimes \mathcal{P}_1$ such that $Q_{1,1}|\phi_1'\rangle = |\psi_1'\rangle$ and $Q_{1,1}$ is of the form $P_{1,1}' \otimes I_{q_{\mathcal{P}} - q_{\text{ent}} - m q_{\mathcal{M}}}$, where $P_{1,1}'$ is a unitary transformation acting on qubits in $\mathcal{M}_1$ and the first $q_{\text{ent}} + m q_{\mathcal{M}}$ qubits of $\mathcal{P}_1$, and $I_{q_{\mathcal{P}} - q_{\text{ent}} - m q_{\mathcal{M}}}$ is the $(q_{\mathcal{P}} - q_{\text{ent}} - m q_{\mathcal{M}})$-dimensional identity matrix.

Assume that $Q_{1,j}$, $|\phi_j'\rangle$, and $|\psi_j'\rangle$ have been defined for each j, $1 \leq j \leq \xi \leq m/2 - 1$, to satisfy

$$
\begin{aligned}
- \ |\phi_1'\rangle &= V_1 |\psi_{\text{init}}\rangle, \\
|\phi_j'\rangle &= V_j P_{k,j-1} \cdots P_{2,j-1} Q_{1,j-1} |\phi_{j-1}'\rangle, \quad 2 \leq j \leq \xi, \\
|\psi_j'\rangle &= Q_{1,j} |\phi_j'\rangle, \quad\quad\quad\quad\quad\quad\quad\quad\quad 1 \leq j \leq \xi.
\end{aligned}
$$

- $\mathrm{tr}_{\mathcal{P}_1}|\psi_j\rangle\langle\psi_j| = \mathrm{tr}_{\mathcal{P}_1}|\psi_j'\rangle\langle\psi_j'|$, $1 \le j \le \xi$.
- All but the first $q_{\mathrm{ent}} + 2(j-1)q_{\mathcal{M}}$ qubits in $\mathcal{P}_1$ are the $|0\rangle$-states in $|\phi_j'\rangle$.
- All but the first $q_{\mathrm{ent}} + 2jq_{\mathcal{M}}$ qubits in $\mathcal{P}_1$ are the $|0\rangle$-states in $|\psi_j'\rangle$.

Notice that $Q_{1,1}$, $|\phi_1'\rangle$, and $|\psi_1'\rangle$ defined above satisfy such conditions. Define $Q_{1,\xi+1}$, $|\phi_{\xi+1}'\rangle$, and $|\psi_{\xi+1}'\rangle$ in the following way to satisfy the above four conditions for $j = \xi + 1$.

Let $U_\xi = V_{\xi+1}P_{k,\xi}\cdots P_{2,\xi}$ and define $|\phi_{\xi+1}'\rangle = U_\xi|\psi_\xi'\rangle$. Then all but the first $q_{\mathrm{ent}} + 2\xi q_{\mathcal{M}}$ qubits in $\mathcal{P}_1$ are the $|0\rangle$-states in $|\phi_{\xi+1}'\rangle$, since none of $P_{2,\xi},\ldots,P_{k,\xi},V_{\xi+1}$ acts on the space $\mathcal{P}_1$ and $|\psi_\xi'\rangle$ satisfies the fourth condition. Since $\mathrm{tr}_{\mathcal{P}_1}|\psi_\xi\rangle\langle\psi_\xi| = \mathrm{tr}_{\mathcal{P}_1}|\psi_\xi'\rangle\langle\psi_\xi'|$, by Theorem 7, there exists a unitary transformation A_ξ acting on $\mathcal{P}_1$ such that $A_\xi|\psi_\xi'\rangle = |\psi_\xi\rangle$. Thus we have

$$|\psi_{\xi+1}\rangle = P_{1,\xi+1}U_\xi A_\xi|\psi_\xi'\rangle = P_{1,\xi+1}A_\xi U_\xi|\psi_\xi'\rangle = P_{1,\xi+1}A_\xi|\phi_{\xi+1}'\rangle.$$

Hence, by Theorem 8, there exists a state $|\psi_{\xi+1}'\rangle$ such that $\mathrm{tr}_{\mathcal{P}_1}|\psi_{\xi+1}'\rangle\langle\psi_{\xi+1}'| = \mathrm{tr}_{\mathcal{P}_1}|\psi_{\xi+1}\rangle\langle\psi_{\xi+1}|$ and all but the first $q_{\mathrm{ent}} + 2(\xi+1)q_{\mathcal{M}}$ qubits in $\mathcal{P}_1$ are the $|0\rangle$-states in the state $|\psi_{\xi+1}'\rangle$. Since $P_{1,\xi+1}$ and A_ξ act only on $\mathcal{M}_1 \otimes \mathcal{P}_1$, we have

$$\mathrm{tr}_{\mathcal{M}_1\otimes\mathcal{P}_1}|\psi_{\xi+1}'\rangle\langle\psi_{\xi+1}'| = \mathrm{tr}_{\mathcal{M}_1\otimes\mathcal{P}_1}|\psi_{\xi+1}\rangle\langle\psi_{\xi+1}| = \mathrm{tr}_{\mathcal{M}_1\otimes\mathcal{P}_1}|\phi_{\xi+1}'\rangle\langle\phi_{\xi+1}'|.$$

Therefore, by Theorem 7, there exists a unitary transformation $Q_{1,\xi+1}$ acting on $\mathcal{M}_1 \otimes \mathcal{P}_1$ such that $Q_{1,\xi+1}|\phi_{\xi+1}'\rangle = |\psi_{\xi+1}'\rangle$. It follows that $Q_{1,\xi+1}$ is of the form $P_{1,\xi+1}' \otimes I_{q_{\mathcal{P}}-q_{\mathrm{ent}}-mq_{\mathcal{M}}}$, where $P_{1,\xi+1}'$ is a unitary transformation acting on qubits in $\mathcal{M}_1$ and the first $q_{\mathrm{ent}} + mq_{\mathcal{M}}$ qubits of $\mathcal{P}_1$, because all of the last $q_{\mathcal{P}} - q_{\mathrm{ent}} - mq_{\mathcal{M}}$ qubits in $\mathcal{P}_1$ are the $|0\rangle$-states in both of the states $|\phi_{\xi+1}'\rangle$ and $|\psi_{\xi+1}'\rangle$. One can see that $Q_{1,\xi+1}$, $|\phi_{\xi+1}'\rangle$, and $|\psi_{\xi+1}'\rangle$ satisfy the four conditions above by their construction.

Having defined $Q_{1,j}, |\phi_j'\rangle, |\psi_j'\rangle$ for each $1 \le j \le m/2$, compare the state just before the final measurement is performed in the original protocol and that in the modified protocol applying $Q_{1,j}$'s instead of $P_{1,j}$'s. For $U_{m/2} = V_{m/2+1}P_{k,m/2}\cdots P_{2,m/2}$, let $|\phi_{m/2+1}\rangle = U_{m/2}|\psi_{m/2}\rangle$ and $|\phi_{m/2+1}'\rangle = U_{m/2}|\psi_{m/2}'\rangle$. These $|\phi_{m/2+1}\rangle$ and $|\phi_{m/2+1}'\rangle$ are exactly the states we want to compare. Noticing that $\mathrm{tr}_{\mathcal{P}_1}|\psi_{m/2}\rangle\langle\psi_{m/2}| = \mathrm{tr}_{\mathcal{P}_1}|\psi_{m/2}'\rangle\langle\psi_{m/2}'|$, we have $\mathrm{tr}_{\mathcal{P}_1}|\phi_{m/2+1}\rangle\langle\phi_{m/2+1}| = \mathrm{tr}_{\mathcal{P}_1}|\phi_{m/2+1}'\rangle\langle\phi_{m/2+1}'|$, since none of $P_{2,m/2},\ldots,P_{k,m/2},V_{m/2+1}$ acts on $\mathcal{P}_1$. Thus we have

$$\mathrm{tr}_{\mathcal{P}_1\otimes\cdots\otimes\mathcal{P}_k}|\phi_{m/2+1}\rangle\langle\phi_{m/2+1}| = \mathrm{tr}_{\mathcal{P}_1\otimes\cdots\otimes\mathcal{P}_k}|\phi_{m/2+1}'\rangle\langle\phi_{m/2+1}'|,$$

which implies that the verifier V cannot distinguish $|\phi_{m/2+1}'\rangle$ from $|\phi_{m/2+1}\rangle$ at all. Hence, for every input x, the probability of accepting x in the protocol $(Q_1, P_2, \ldots, P_k, V)$ is exactly equal to the one in the original protocol $(P_1, \ldots, P_k, V)$, and Q_1 uses only $q_{\mathrm{ent}} + mq_{\mathcal{M}}$ qubits in his private space. In the protocol $(Q_1, P_2, \ldots, P_k, V)$, each $Q_{1,j}$ is described as $Q_{1,j} = P_{1,j}' \otimes I_{q_{\mathcal{P}}-q_{\mathrm{ent}}-mq_{\mathcal{M}}}$, where $P_{1,\xi+1}'$ is a unitary transformation acting on qubits in $\mathcal{M}_1$ and the first $q_{\mathrm{ent}} + mq_{\mathcal{M}}$ qubits of $\mathcal{P}_1$. Consequently, these $P_{1,j}'$'s form the desired m-message $(q_{\mathcal{M}}, q_{\mathrm{ent}} + mq_{\mathcal{M}})$-restricted quantum prover P_1'.

Now we repeat the above process for each of provers, and finally we obtain a protocol $(P_1', \ldots, P_k', V)$ in which all k provers are m-message $(q_\mathcal{M}, q_\text{ent} + mq_\mathcal{M})$-restricted quantum provers. It is obvious that, for every input x, the probability of accepting x in the protocol $(P_1', \ldots, P_k', V)$ is exactly equal to the one in the original protocol $(P_1, \ldots, P_k, V)$, and we have the assertion.   □

From Lemma 9, it is straightforward to show the following lemma.

Lemma 10. *For any polynomial-bounded functions $k, m, q_\text{ent} : \mathbb{Z}^+ \to \mathbb{N}$, $\mathrm{QMIP}(k, m, q_\text{ent}, 1, 1/2) \subseteq \mathrm{NEXP}$.*

Hence we have the following theorem.

Theorem 11. $\mathrm{QMIP}^{(\mathrm{l.e.})} \subseteq \mathrm{NEXP}$.

Note that our upper bound of NEXP holds even if we allow protocols with two-sided bounded error, since the proof of Lemma 9 does not depend on the accepting probabilities a, b, and Lemma 10 can be easily modified to two-sided bounded error cases.

4 $\mathrm{QMIP}^{(\mathrm{n.e.})} = \mathrm{QOC} = \mathrm{NEXP}$

In the previous section, we proved that the class of languages having quantum multi-prover interactive proof systems is necessarily contained in NEXP under the assumption that provers are allowed to share at most polynomially many prior-entangled qubits. As a special case of this, it is proved in this section that, if provers do not share any prior entanglement with each other, the class of languages having quantum multi-prover interactive proof systems is equal to NEXP. Another result related to this is that QOC is also equal to NEXP, or in other words, the class of languages having quantum single-prover interactive proof systems is also equal to NEXP if a prover does not have his private qubits.

The inclusions $\mathrm{QMIP}^{(\mathrm{n.e.})} \subseteq \mathrm{NEXP}$ and $\mathrm{QOC} \subseteq \mathrm{NEXP}$ directly come from Lemma 10. Thus it is sufficient for our claim to show $\mathrm{NEXP} \subseteq \mathrm{QMIP}^{(\mathrm{n.e.})} \subseteq \mathrm{QOC}$. Fortunately, in the cases without prior entanglement, it is easy to show that a quantum verifier can successfully simulate any classical multi-prover protocol, in particular, a one-round two-prover classical interactive protocol that can verify a language in NEXP with exponentially small one-sided error [12]. Thus, we have the following theorem and corollary.

Theorem 12. $\mathrm{NEXP} \subseteq \mathrm{QMIP}^{(\mathrm{n.e.})}$.

Corollary 13. *For prior-unentangled cases, if a language L has a quantum multi-prover interactive proof system with two-sided bounded error, then L has a two-message quantum two-prover interactive proof system with exponentially small one-sided error.*

The remainder of this section is devoted to the proof of $\mathrm{QMIP}^{(\mathrm{n.e.})} \subseteq \mathrm{QOC}$.

Lemma 14. *Let* $k, m\colon \mathbb{Z}^+ \to \mathbb{N}$ *be polynomial-bounded functions, and* $a, b\colon \mathbb{Z}^+ \to [0,1]$ *be functions satisfying* $a \geq b$. *Then* $\mathrm{QMIP}(k, m, 0, a, b) \subseteq \mathrm{QOC}(k\lfloor (m+1)/2 \rfloor, a, b)$.

Proof. For simplicity, we assume that the values of m are even, and thus $k\lfloor (m+1)/2 \rfloor = km/2$ (odd cases can be proved with a similar argument).

Let L be a language in $\mathrm{QMIP}(k, m, 0, a, b)$. Then, from Definition 1 together with Lemma 9, there exist polynomial-bounded functions $q_V, q_M \colon \mathbb{Z}^+ \to \mathbb{N}$ and an m-message (q_V, q_M)-restricted quantum verifier V for a quantum k-prover interactive proof system such that, for every input x of length n, (i) if x is in L, there exists a set of m-message (q_M, mq_M)-restricted quantum provers $P_1, \dots, P_k$ without prior entanglement such that $(P_1, \dots, P_k, V)$ accepts x with probability at least $a(n)$, and (ii) if x is not in L, for all sets of m-message (q_M, mq_M)-restricted quantum provers $P'_1, \dots, P'_k$ without prior entanglement, $(P'_1, \dots, P'_k, V)$ accepts x with probability at most $b(n)$.

We construct a $km/2$-oracle-call verifier V^{QOC} of a quantum oracle circuit as follows. Let us consider that quantum registers (collections of qubits upon which various transformations are performed) $\mathbf{W}$, $\mathbf{M}_i$, and $\mathbf{P}_i$, for $1 \leq i \leq k$, are prepared among the private qubits of the verifier V^{QOC}, and quantum registers $\mathbf{M}$ and $\mathbf{P}$ are prepared among the qubits for oracle calls. $\mathbf{W}$ consists of q_V qubits, each $\mathbf{M}_i$ and $\mathbf{M}$ consist of q_M qubits, and each $\mathbf{P}_i$ and $\mathbf{P}$ consist of $q_P = mq_M$ qubits. Let $\mathcal{W}^{\mathrm{QOC}}$, each $\mathcal{M}_i^{\mathrm{QOC}}$, and each $\mathcal{P}_i^{\mathrm{QOC}}$ denote the Hilbert spaces corresponding to the registers $\mathbf{W}$, $\mathbf{M}_i$, and $\mathbf{P}_i$, respectively. Take the Hilbert space $\mathcal{V}^{\mathrm{QOC}}$ corresponding to the qubits private to the verifier V^{QOC} as $\mathcal{V}^{\mathrm{QOC}} = \mathcal{W}^{\mathrm{QOC}} \otimes \mathcal{M}_1^{\mathrm{QOC}} \otimes \cdots \otimes \mathcal{M}_k^{\mathrm{QOC}} \otimes \mathcal{P}_1^{\mathrm{QOC}} \otimes \cdots \otimes \mathcal{P}_k^{\mathrm{QOC}}$. Accordingly, the number of private qubits of V^{QOC} is $q_{\mathcal{V}}^{\mathrm{QOC}} = q_V + k(q_M + q_P) = q_V + k(m+1)q_M$. Let $\mathcal{M}^{\mathrm{QOC}}$ and $\mathcal{P}^{\mathrm{QOC}}$ denote the Hilbert spaces corresponding to the registers $\mathbf{M}$ and $\mathbf{P}$, respectively. Take the Hilbert space $\mathcal{O}^{\mathrm{QOC}}$ corresponding to the qubits for oracle calls as $\mathcal{O}^{\mathrm{QOC}} = \mathcal{M}^{\mathrm{QOC}} \otimes \mathcal{P}^{\mathrm{QOC}}$. Accordingly, the number of qubits for oracle calls is $q_{\mathcal{O}}^{\mathrm{QOC}} = q_M + q_P = (m+1)q_M$.

Consider each V_j, the jth quantum circuit of the verifier V of the original quantum k-prover interactive proof system, which acts on $\mathcal{V} \otimes \mathcal{M}_1 \otimes \cdots \otimes \mathcal{M}_k$. For each j, let U_j^{QOC} be just the same unitary transformation as V_j and U_j^{QOC} acts on $\mathcal{W}^{\mathrm{QOC}} \otimes \mathcal{M}_1^{\mathrm{QOC}} \otimes \cdots \otimes \mathcal{M}_k^{\mathrm{QOC}}$, corresponding to that V_j acts on $\mathcal{V} \otimes \mathcal{M}_1 \otimes \cdots \otimes \mathcal{M}_k$. Define the verifier V^{QOC} of the corresponding quantum oracle circuit in the following way:

- At the first transformation of V^{QOC}, V^{QOC} first applies U_1^{QOC}, and then swaps the contents of $\mathbf{M}_1$ for those of $\mathbf{M}$.
- At the $((j-1)k+1)$-th transformation of V^{QOC} for each $2 \leq j \leq m/2$, V^{QOC} first swaps the contents of $\mathbf{M}$ and $\mathbf{P}$ for those of $\mathbf{M}_k$ and $\mathbf{P}_k$, respectively, then applies U_j^{QOC}, and finally swaps the contents of $\mathbf{M}_1$ and $\mathbf{P}_1$ for those of $\mathbf{M}$ and $\mathbf{P}$.
- At the $((j-1)k+i)$-th transformation of V^{QOC} for each $2 \leq i \leq k$, $1 \leq j \leq m/2$, V^{QOC} first swaps the contents of $\mathbf{M}$ and $\mathbf{P}$ for those of $\mathbf{M}_{i-1}$

and $\mathbf{P}_{i-1}$, respectively, then swaps the contents of $\mathbf{M}_i$ and $\mathbf{P}_i$ for those of $\mathbf{M}$ and $\mathbf{P}$.

It is easy to check that this V^{QOC} derives the desired quantum oracle circuit. $\square$

The inclusion $\mathrm{QMIP}^{(\mathrm{n.e.})} \subseteq \mathrm{QOC}$ immediately follows from Lemma 14. Thus we have the following theorem.

Theorem 15. $\mathrm{QMIP}^{(\mathrm{n.e.})} = \mathrm{QOC} = \mathrm{NEXP}$.

Acknowledgement. The authors are grateful to Richard E. Cleve for explaining how an entangled pair of provers can cheat a classical verifier in some cases. The authors would also like to thank Lance J. Fortnow and Hiroshi Imai for their valuable comments on an earlier version of this paper.

References

1. D. Aharonov, A. Yu. Kitaev, and N. Nisan. Quantum circuits with mixed states. In *Proceedings of the Thirtieth Annual ACM Symposium on Theory of Computing*, pages 20–30, 1998.
2. L. Babai. Trading group theory for randomness. In *Proceedings of the Seventeenth Annual ACM Symposium on Theory of Computing*, pages 421–429, 1985.
3. L. Babai, L. J. Fortnow, and C. Lund. Non-deterministic exponential time has two-prover interactive protocols. *Computational Complexity*, 1(1):3–40, 1991.
4. M. Bellare, U. Feige, and J. Kilian. On the role of shared randomness in two prover proof systems. In *Proceedings of the 3rd Israel Symposium on the Theory of Computing and Systems*, pages 199–208, 1995.
5. M. Ben-Or, S. Goldwasser, J. Kilian, and A. Wigderson. Multi-prover interactive proofs: how to remove the intractability assumptions. In *Proceedings of the Twentieth Annual ACM Symposium on Theory of Computing*, pages 113–131, 1988.
6. C. H. Bennett, E. Bernstein, G. Brassard, and U. V. Vazirani. Strengths and weaknesses of quantum computing. *SIAM Journal on Computing*, 26(5):1510–1523, 1997.
7. J.-Y. Cai, A. Condon, and R. J. Lipton. On bounded round multi-prover interactive proof systems. In *Proceedings of the 5th Annual Structure in Complexity Theory Conference*, pages 45–54, 1990.
8. J.-Y. Cai, A. Condon, and R. J. Lipton. PSPACE is provable by two provers in one round. *Journal of Computer and System Sciences*, 48(1):183–193, 1994.
9. R. E. Cleve. An entangled pair of provers can cheat. Talk at the Workshop on Quantum Computation and Information, California Institute of Technology, November 2000.
10. D. Deutsch. Quantum theory, the Church-Turing principle and the universal quantum computer. *Proceedings of the Royal Society of London A*, 400:97–117, 1985.
11. U. Feige. On the success probability of two provers in one-round proof systems. In *Proceedings of the 6th Annual Structure in Complexity Theory Conference*, pages 116–123, 1991.
12. U. Feige and L. Lovász. Two-prover one-round proof systems: their power and their problems (extended abstract). In *Proceedings of the Twenty-Fourth Annual ACM Symposium on Theory of Computing*, pages 733–744, 1992.

13. L. J. Fortnow. *Complexity-Theoretic Aspects of Interactive Proof Systems*. PhD thesis, Department of Mathematics, Massachusetts Institute of Technology, May 1989.
14. L. J. Fortnow, J. Rompel, and M. Sipser. On the power of multi-prover interactive protocols. *Theoretical Computer Science*, 134(2):545–557, 1994.
15. S. Goldwasser, S. Micali, and C. Rackoff. The knowledge complexity of interactive proof systems. *SIAM Journal on Computing*, 18(1):186–208, 1989.
16. S. Goldwasser and M. Sipser. Private coins versus public coins in interactive proof systems. In S. Micali, editor, *Randomness and Computation*, volume 5 of *Advances in Computing Research*, pages 73–90. JAI Press, 1989.
17. L. P. Hughston, R. Jozsa, and W. K. Wootters. A complete classification of quantum ensembles having a given density matrix. *Physics Letters A*, 183:14–18, 1993.
18. A. Yu. Kitaev and J. H. Watrous. Parallelization, amplification, and exponential time simulation of quantum interactive proof systems. In *Proceedings of the Thirty-Second Annual ACM Symposium on Theory of Computing*, pages 608–617, 2000.
19. D. Lapidot and A. Shamir. Fully parallelized multi prover protocols for NEXP-time. *Journal of Computer and System Sciences*, 54(2):215–220, 1997.
20. C. Lund, L. J. Fortnow, H. Karloff, and N. Nisan. Algebraic methods for interactive proof systems. *Journal of the ACM*, 39(4):859–868, 1992.
21. C. H. Papadimitriou. Games against nature. *Journal of Computer and System Sciences*, 31(2):288–301, 1985.
22. A. Shamir. IP = PSPACE. *Journal of the ACM*, 39(4):869–877, 1992.
23. P. W. Shor. Polynomial-time algorithms for prime factorization and discrete logarithms on a quantum computer. *SIAM Journal on Computing*, 26(5):1484–1509, 1997.
24. A. Uhlmann. Parallel transport and "quantum holonomy" along density operators. *Reports on Mathematical Physics*, 24:229–240, 1986.
25. J. H. Watrous. PSPACE has constant-round quantum interactive proof systems. In *Proceedings of the 40th Annual Symposium on Foundations of Computer Science*, pages 112–119, 1999.

Some Remarks on the L-Conjecture

Qi Cheng

School of Computer Science, the University of Oklahoma,
Norman, OK 73072, USA
qcheng@cs.ou.edu

Abstract. In this paper, we show several connections between the L-conjecture, proposed by Burgisser [3], and the boundedness theorem for the torsion of elliptic curves. Assuming the L-conjecture, a sharper bound is obtained for the number of torsions over extensions of k on an elliptic curve over a number field k, which improves Masser's result [6]. It is also shown that the Torsion Theorem for elliptic curves [10] follows directly from the WL-conjecture, which is a much weaker version of the L-conjecture. Since the WL-conjecture differs from the trivial lower bound only at the coefficient, this result provides an interesting example where increasing the coefficient in a trivial lower bound of straight-line complexity is difficult and important.

1 Introduction

The relation between the number of distinct rational roots of a polynomial and the straight-line complexity of the polynomial has been studied by several researchers. Lipton [5] proved that polynomials with many distinct rational roots can not be evaluated by a short straight-line program, unless integer factorization is easy. This observation was explicitly formulated by Blum, Cucker, Shub and Smale [1] in the so called τ-conjecture. Let f be any univariate integral polynomial in x. The conjecture claims that

$$z(f) \le (\tau(f) + 1)^c,$$

where $z(f)$ is the number of distinct rational roots of f, $\tau(f)$ is the length of the shortest straight-line program computing f from 1 and x, and c is an absolute constant. It is rather surprising that they showed that this conjecture implies that $NP_{\mathbf{C}} \ne P_{\mathbf{C}}$. Proving τ-conjecture (or disproving it in case that it is false) is also known as Smale's 4th problem "integer zeros of a polynomial of one variable". It is believed that solving this problem is very hard, and even partial solutions will have great impacts on researches of algebraic geometry and computational complexity. For Smale's list of the most important problems for the mathematicians of 21st century, see [12].

In [3], a question was raised whether a similar conjecture is true for polynomials over any number field k when $\tau(f)$ is replaced by $L(f)$, which is the length of the shortest straight-line program computing $f(x)$ from x and any constants. More precisely, it is conjectured that

P. Bose and P. Morin (Eds.): ISAAC 2002, LNCS 2518, pp. 128–136, 2002.
© Springer-Verlag Berlin Heidelberg 2002

Conjecture 1. (L-conjecture) Given a number field k, there exists a constant c depending only on k, such that for any $f \in k[x]$,

$$N_d(f) \leq (L(f) + d)^c,$$

where $N_d(f)$ is the number of distinct irreducible factors of f over k with degree at most d.

It is easy to see that the L-conjecture over $\mathbf{Q}$ implies the τ-conjecture. In this paper, we examine the implications of L-conjecture in algebraic geometry and number theory. In particular, we derive some results from this conjecture. These results cover some fundamental problems in number theory and algebraic geometry. Some of them are still open. Those which have been settled require the most advanced mathematical tools.

1.1   Summary of Results

First we show that if the L-conjecture is true, then Masser's result [6] on the number of torsions on an elliptic curve can be improved. His results are summarized as follows.

Proposition 1. *Let k be a number field and $E : Y^2 = 4x^3 - g_2 x - g_3$ be an elliptic curve over k. There is a positive effective constant c, depending only on the degree of k, such that the torsion subgroup of $E(K)$ with $[K : k] = D$ has cardinality at most $c\sqrt{w}D(w + \log D)$, where w is the absolute logarithmic height of $(1 : g_2 : g_3)$ in $\mathbf{P}_k^2$.*

We prove the following theorem:

Theorem 1. *Use the notations in the above proposition. The cardinality of the torsion subgroup of $E(K)$ is at most $c_1 D^{c_2}$ where c_1 and c_2 are constants depending only on k, if the L-conjecture is true in the number field k.*

Note that in Theorem 1, the constants are only dependent on the number field k, and are independent of the curve. The bound in Theorem 1 is lower than Masser's bound when k is fixed and w is large. For example, if $w > 2^D$, Masser's bound is exponential in D but our bound is polynomial in D.

We then study how hard it is to prove the L-conjecture. To this end, we consider a much weaker statement. We call it WL-conjecture, standing for *weaker L-conjecture*.

Conjecture 2. (WL-conjecture) Let k be a number field. For any univariate polynomial $f \in k[x]$, there exists two constants $c_1 > 0$ and $0 \leq c_2 < 1/72$ such that

$$z_k(f) \leq c_1 2^{c_2 L(f)},$$

where $z_k(f)$ is the number of distinct roots over k of f.

Not only the WL-conjecture is a special case of L-conjecture, but also it is much weaker than the L-conjecture, as in the WL-conjecture the number of zeros is bounded from above by an exponential function in $L(f)$ while in L-conjecture the number of zeros is bounded from above by a polynomial function in $L(f)$. The other way to view the WL-conjecture is that it states a lower bound of straight-line complexity:

$$L(f) \geq c_3 \log z_k(f) + c_4$$

where $c_3 > 72$ and c_4 are two constants depending only on k. Note that $L(f) \geq \log z_k(f)$ is obviously true over any field k because the degree of f is at most $2^{L(f)}$. Unlike the L-conjecture, the WL-conjecture differs from the trivial lower bound only at the coefficient. Nevertheless, we show that the WL-conjecture is hard to prove in the sense that the following famous Torsion Theorem is a direct consequence of the WL-conjecture.

Theorem 2. *(Torsion Theorem for Elliptic Curves) Let E be an elliptic curve defined over a number field k. Then the number of torsions in $E(k)$ is bounded from above by a constant depending only on k.*

This theorem is a part of the following result, also known as the Uniformly Boundedness Theorem (UBT).

Theorem 3. *(Strong Torsion Theorem for Elliptic Curves) Let E be an elliptic curve defined over a number field k, then the number of torsions in $E(k)$ is bounded from above by a constant depending only on $m = [k : \mathbf{Q}]$.*

We can also prove that if c_1 in the WL-conjecture depends only on $[k : \mathbf{Q}]$, the UBT follows from the WL-conjecture. All the results in this paper are obtained by studying the division polynomial $P_n(x)$. We showed that if there is one point on an elliptic curve $E(k)$ with order n, then the division polynomial $P_n(x)$ must have at least $(n-1)/2$ distinct solutions in k. On the other hand, $P_n(x)$ can be computed by a straight-line program of length at most $72 \log n + 60$. The WL-conjecture is violated if n is bigger than a certain constant.

1.2   Related Work

Boneh [2] also observed the connections between the straight-line complexity of polynomials and the bound of torsions over a number field on an abelian variety. In his report, no bound better than currently known was obtained for the number of torsions on elliptic curves. And he assumed the hardness of integer factorization, which essentially means that for any number field k and any constant c,

$$L(f) \geq (\log z_k(f))^c,$$

if $L(f)$ is sufficiently large. The condition that the integer factorization is hard is much stronger than the WL-conjectre. In our paper, we obtain a better bound for the number of torsion points over extension fields on an elliptic curve and improve

Masser's results, which is the best estimation currently known. We also assume the WL-conjecture to study the property of the torsions. Our results indicate that proving a better coefficient in the trivial straight-line complexity lower bound is very hard, hence illustrate the difficulties of proving a superpolynomial lower bound and proving the hardness of factoring.

2 Motivations

The Torsion Theorem and the Strong Torsion Theorem for elliptic curves are very important in number theory and algebraic geometry. The Strong Torsion Theorem has been settled recently. The proof of these theorems, even in the case of the Torsion Theorem, is by no means easy. Having said this, we agree that there is no quantitive measurement on how difficult a mathematical statement can be proved. But certainly if a statement was open for years and was tried by a lot of researchers, even though it was solved eventually, we can safely say that this theorem is hard to prove. In case of the Torsion Theorem for elliptic curves, it is impossible to prove it without deploying the most advanced tools in modern number theory and algebraic geometry.

In this paper, we find surprising connections between the torsion theorem and the WL-conjecture. We show that the number of torsions over number fields will be severely limited if the WL-conjecture is true, hence the Torsion Theorem follows directly from the WL-conjecture. Our argument is simple and elementary. It is quite astonishing, considering how weak the WL-conjecture is and how much effort people have put into proving the Torsion Theorem.

Although the UBT has been proved, our result is still interesting. First it indicates what a future proof of the WL-conjecture looks like if it exists. The proofs of mathematical statements are usually categorized into two kinds: elementary and analytical. An elementary one uses direct reasoning and relys on the combinatorial arguments. An analytical proof, on the other hand, uses the tools from complex analysis. The modular forms are sometimes involved in analytical proofs. The famous examples of analytical proofs include the proofs of the Fermat Last Theorem and the UBT. These theorems are believed not to have elementary proofs. To the contrary, the results in theoretical computer science are usually obtained by elementary methods.

Our result clearly indicates that in order to prove the WL-conjecture, we have to look at the modular form and some highly advanced tools in algebraic geometry, which are not apparently related to the theory of computational complexity. There should be no elementary proof of the WL-conjecture, unless the Torsion Theorem can be proved in an elementary way. Since it is unlikely that the Torsion Theorem for elliptic curves has an elementary proof, our result implies that even improving the coefficient in a trivial straight-line complexity lower bound requires the advanced analytical tools.

Secondly, the $(W)L$-conjecture claims that there is no short straight-line program to compute a polynomial which has many distinct roots in a fixed number field. This is a typical lower bound result in computational complexity. It is a common belief that we lack techniques to obtain lower bounds in computational complexity. Our results add one more example to this phenomenon and shed light on its reason. The Torsion Theorem may be viewed as a lower bound result in computational complexity. Although we don't believe that the Torsion Theorem is equivalent to the WL-conjecture, we think that we can learn a lot from the proof of UBT to construct the proofs the WL-conjecture and even the L-conjecture.

3   Straight-Line Programs of Polynomials

A straight-line program of a polynomial over a field k is a sequence of ring operations, which outputs the polynomial in the last operation. Formally,

Definition 1. *A straight-line program of a polynomial $f(x)$ is a sequence of instructions. The i-th instruction is*

$$v_i \leftarrow v_m \circ v_n \quad or \quad v_i \leftarrow c_i$$

where $\circ \in \{+, -, *\}$, $i > m$, $i > n$, c_i is x or any constant in k, and the last variable v_n equals to $f(x)$. The length of the program is the number of the instructions. The length of the shortest straight-line program of $f(x)$ is called the straight-line complexity of $f(x)$ and is denote by $L(f)$.

The polynomial x^n has a straight-line complexity at most $2 \log n$. In some cases, a straight-line program is a very compact description of a polynomial. It can represent a polynomial with a huge number of terms in a small length. For example, the polynomial $(x + 1)^n$ can be computed using the repeated squaring technique and hence has a straight-line complexity at most $2 \log n$, while it has $n + 1$ terms.

The number of distinct roots over a number field k of a polynomial with small straight-line complexity seems limited. For example, the equation $(x + 1)^n = 0$ has only one distinct root. The equation $x^n - 1 = 0$ has n distinct roots, but if we fix a number field k, the number of distinct roots in k will not increase even if we increase n. The relation between the number of distinct roots of a polynomial f over a number field and $L(f)$ is not well understood.

4   Division Polynomials

An elliptic curve E over a field k is a smooth cubic curve. If the characteristic of k is neither 2 nor 3, we may assume that the elliptic curve is given by an equation of the form

$$y^2 = x^3 + ax + b, \qquad a, b \in k.$$

Let K be an extension field of k. The solution set of the equation in K, plus the infinity point, forms an abelian group. We use $E(K)$ to denote the group.

We call a point *torsion* if it has a finite order in the group. The x-coordinates of the torsions of order $n > 3$ are the solutions of $P_n(x)$, the n-th division polynomial of E. The polynomial $P_n(x)$ can be computed recursively. The recursion formula can be found in a lot of literatures. For completeness we list them below.

$$P_1 = 1$$
$$P_2 = 1$$
$$P_3 = 3x^4 + 6ax^2 + 12bx - a^2$$
$$P_4 = 2(x^6 + 5ax^4 + 20bx^3 - 5a^2x^2 - 4abx - 8b^2 - a^3)$$
$$P_{4n+1} = 16(x^3 + ax + b)P_{2n+2}P_{2n}^3 - P_{2n-1}P_{2n+1}^3$$
$$P_{4n+2} = P_{2n+1}(P_{2n+3}P_{2n}^2 - P_{2n-1}P_{2n+2}^2)$$
$$P_{4n+3} = P_{2n+3}P_{2n+1}^3 - 16(x^3 + ax + b)P_{2n}P_{2n+2}^3$$
$$P_{4n+4} = P_{2n+2}(P_{2n+4}P_{2n+1}^2 - P_{2n}P_{2n+3}^2)$$

Note that our division polynomials are a little different from the division polynomial $\psi_n(x, y)$ in some literatures, for example the Silverman's book [11]. When n is even, $P_n = \psi_n$. When n is odd, $P_n = (2y)^{-1}\psi_n$. We use P_n because it is a univariate polynomial and the computation of P_n does not involve division.

Lemma 1. $L(P_n) \leq 72 \log n + 60$.

Proof. We deploy the dynamical programming technique to construct the straight-line program. Before evaluating $P_n(x)$, we need to evaluate up to 5 division polynomials with indices around $n/2$, according to the recursion. For the same reason, in order to compute these 5 or less division polynomials, we need to compute up to 8 division polynomials with indices about $n/4$. However, this does not mean that the number of division polynomials we need to evaluate in each recursion level grows unlimitedly as the level increases. In fact, in order to evaluate the list of division polynomials $P_i(x), P_{i+1}(x), \cdots, P_{i+j}(x)$, we only need to evaluate $P_{\lceil i/2 \rceil - 2}, P_{\lceil i/2 \rceil - 1}, \cdots, P_{\lfloor (i+j)/2 \rfloor + 1}, P_{\lfloor (i+j)/2 \rfloor + 2}$. If $j > 7$, the latter list is shorter than the former one. On the other hand, if $j \leq 7$, then the latter list contains at most 8 polynomials. Hence if we want to evaluate $P_n(x)$, we only go through $\log n$ recursion levels and evaluate at most $8 \log n$ number of $P_i(x)$'s. Evaluating any $P_i(x)$ requires at most 9 more ring operations from the division polynomials in the previous level. The overhead of computing P_1, P_2, P_3, P_4 and $16(x^3 + Ax + B)$ is less than 60 steps. The total number of arithmetic operations is thus less than $72 \log n + 60$.

Lemma 2. *If $P_n(x)$ has one solution in K which is a x-coordinate of a point P in $E(K)$ of order n, then it must have at least $(n-1)/2$ distinct solutions in K.*

Proof. If there exists a point P in $E(K)$ with order n, then the points $P, 2P, 3P, \cdots, (n-1)P$ are distinct, none of them is the infinity point 0 and all of them have orders dividing n. The x-coordinates of these points are in K and they are the roots of $P_n(x)$. Two points have different x-coordinates, unless the sum of these two points are 0. If n is odd, we have exactly $(n-1)/2$ distinct x-coordinates. If n is even, we have $n/2$ distinct x-coordinates.

5   Outlines of Proofs

5.1   Improving Masser's Bound

Now we are ready to prove Theorem 1.

Proof. Suppose there is a point $P \in E(K)$ of order n. Denote the x-coordinates of $P, 2P, 3P, \cdots, (n-1)P$ by $x_1, x_2, \cdots, x_{n-1}$ respectively. According to Lemma 2 there are at least $(n-1)/2$ different numbers in $x_1, x_2, \cdots, x_{n-1}$. All of them are the roots of the n-th division polynomial $P_n(x)$ of the curve E. We also know that the minimal polynomial over k of $x_i \in K$, $1 \leq i \leq n-1$, has degree at most $[K : k] = D$. Hence there are at least $(n-1)/(2D)$ factors of $P_n(x)$ which have degrees less than or equal to D. According to the L-conjecture, we have

$$(n-1)/(2D) \leq N_D(P_n) \leq (L(P_n) + D)^c \leq (72 \log n + 60 + D)^c.$$

This gives us $n \leq c_1 D^{c+2}$ for a constant c_1 independent of the curve E.

5.2   Proving the Torsion Theorem from the WL-Conjecture

The rank and torsion of the Mordell-Weil group of an abelian variety are two central topics in the study of algebraic geometry. Although the research on the rank shows only slow progress, remarkable achievements have been made in the research on torsion, culminating in the recent proof of the Uniform Boundedness Theorem(UBT).

First we briefly review the history. The Torsion Theorem in some special cases was conjectured by Beppo Levi as early as the beginning of the 20th century. Mazur proved the case of $k = \mathbf{Q}$ in his landmark paper [7] in 1977. He also gave the bound 16 explicitly. Mazur's result requires a deep research on modular forms. Little progress was made on the torsion theorem until in 1992 Kamienny announced his ground-breaking result [4]. He settled the cases when k is any quadratic number field, and suggested the techniques to attack the whole conjecture. His method led to the proofs of the Strong Torsion Theorem for $d \leq 14$. It was Merel [8] who finally managed to prove the Strong Torsion Theorem for all the positive integers d in 1996. The following effective version of the UBT was proved by Parent [9].

Proposition 2. *Let $\mathcal{E}$ be an elliptic curve over a number field K. Denote the order of the torsion subgroup of $E(K)$ by N. Let $d = [K : \mathbf{Q}]$. Suppose that p is a prime divisor of N, and p^n is the largest power of p dividing N. We have*

$$p \leq (1 + 3^{d/2})^2,$$
$$p^n \leq 65(3^d - 1)(2d)^6.$$

Now we start to prove the Torsion Theorem from the WL-conjecture. Suppose an elliptic curve E/K has a point with order n in K. The division polynomial $P_n(x)$ has at least $(n-1)/2$ distinct solutions over K, i.e. $z(P_n) \geq (n-1)/2$.

But $P_n(x)$ can be computed by a straight-line program of length at most $72 \log n + 60$, i.e. $L(f) \le 72 \log n + 60$. If the WL-conjecture is true, we have

$$(n-1)/2 \le z(P_n) \le c_1 2^{c_2 L(P_n)} \le c_1 2^{c_2(72 \log n + 60)},$$

which is possible only if

$$n \le (3c_1 2^{60c_2})^{\frac{1}{1-72c_2}}.$$

The c_1 and c_2 in the right hand side depend only on k. This argument shows that the Torsion Theorem is the direct consequence of the WL-conjecture. Similar arguments show that if in the WL-conjecture c_1 depends only on $[k : \mathbf{Q}]$, then the Strong Torsion Theorem follows from the WL-conjecture as well.

6 Conclusion

In this paper, we improved Masser's upper bound of the number of torsion points over extension number fields on an elliptic curve assuming the WL-conjecture. We showed that the Torsion Theorem is a direct consequence of the WL-conjecture, hence there is no elementary proof of the WL-conjecture, unless there is an elementary proof of the Torsion Theorem.

Although the Torsion Theorem has been proved, we believe that our second result is still interesting, because it shows that an elementary proof of the WL-conjecture unlikely exists by the reduction technique, which was widely used in computer science to show a problem is easier than the other problem. Our results strongly suggest that in order to construct a proof of the $(W)L$-conjecture, we should learn from the proof of the Torsion Theorem.

References

1. Lenore Blum, Felipe Cucker, Michael Shub, and Steve Smale. *Complexity and Real Computation.* Springer-Verlag, 1997.
2. Dan Boneh. Studies in computational number theory with applications to cryptography. Technical report, Princeton University, 1996.
3. Peter Burgisser. On implications between P-NP-Hypotheses: Decision versus computation in algebraic complexity. In *Proceedings of MFCS*, volume 2136 of *Lecture Notes in Computer Science*, 2001.
4. S. Kamienny. Torsion points on elliptic curves and q-coefficients of modular forms. *Inventiones Mathematicae*, 109:221–229, 1992.
5. Richard J. Lipton. Straight-line complexity and integer factorization. In *Algorithmic number theory (Ithaca, NY, 1994)*, pages 71–79, Berlin, 1994. Springer.
6. D. W. Masser. Counting points of small height on elliptic curves. *Bull. Soc. Math. France*, 117(2):247–265, 1989.
7. B. Mazur. Rational isogenies of prime degree. *Invent. Math.*, 44, 1978.
8. L. Merel. Bounds for the torsion of elliptic curves over number fields. *Invent. Math.*, 124(1-3):437–449, 1996.
9. P. Parent. Effective bounds for the torsion of elliptic curves over number fields. *J. Reine Angew. Math*, 506:85–116, 1999.

10. Alice Silverberg. Open questions in arithmetic algebraic geometry. In *Arithmetic Algebraic Geometry(Park City, UT, 1999)*, volume 9 of *Institute for Advanced Study/Park City Mathematics Series*, pages 83–142. American Mathematical Society, 2001.
11. J.H. Silverman. *The arithmetic of elliptic curves*. Springer-Verlag, 1986.
12. S. Smale. Mathematical problems for the next century. In V. Arnold, M. Atiyah, P. Lax, and B. Mazur, editors, *Mathematics: Frontiers and Perspectives, 2000*. AMS, 2000.

A Framework for Network Reliability Problems on Graphs of Bounded Treewidth[*]

Thomas Wolle

Institute of Information and Computing Sciences, Utrecht University
P.O.Box 80.089, 3508 TB Utrecht, The Netherlands
thomasw@cs.uu.nl

Abstract. In this paper, we consider problems related to the network reliability problem, restricted to graphs of bounded treewidth. We look at undirected simple graphs with each vertex and edge a number in $[0, 1]$ associated. These graphs model networks in which sites and links can fail, with a given probability, independently of whether other sites or links fail or not. The number in $[0, 1]$ associated to each element is the probability that this element does not fail. In addition, there are distinguished sets of vertices: a set S of servers, and a set L of clients. This paper presents a dynamic programming framework for graphs of bounded treewidth for computing for a large number of different properties Y whether Y holds for the graph formed by the nodes and edges that did not fail. For instance, it is shown that one can compute in linear time the probability that all clients are connected to at least one server, assuming the treewidth of the input graph is bounded. The classical S-terminal reliability problem can be solved in linear time as well using this framework. The method is applicable to a large number of related questions. Depending on the particular problem, the algorithm obtained by the method uses linear, polynomial, or exponential time.

1 Introduction

In this paper, we investigate the problem to compute the probability that a given network has a certain property. We use the following model: To each site (vertex) of the network, a value in $[0, 1]$ is assigned, which expresses the probability that this site is working. Failures of links (edges) between sites can be simulated by introducing further vertices, see Section 5.1 for details. We assume that the occurrence of failures of elements are (statistically) independent. Given a network $G = (V, E)$, with associated probabilities of non-failure, and a set of vertices $S \subseteq V$, the S-terminal reliability problem asks for the probability, that there is a connection between every pair of vertices of S. For general graphs or networks this problem is #P-complete [8]. When restricting the set of input networks to graphs of bounded treewidth, linear time algorithms are possible.

[*] This research was partially supported by EC contract IST-1999-14186: Project ALCOM-FT (Algorithms and Complexity - Future Technologies).

P. Bose and P. Morin (Eds.): ISAAC 2002, LNCS 2518, pp. 137–149, 2002.
© Springer-Verlag Berlin Heidelberg 2002

Much work has been done on algorithms for graphs of bounded treewidth. A general description of a common 'bottom up' method for such graphs can be found in [3] or [4]. The (special case of the) S-terminal reliability problem where only edges can fail and nodes are perfectly reliable was considered by Arnborg and Proskurowski [1]. They show a linear time algorithm for this problem on partial k-trees, i.e., graphs of treewidth at most k. Mata-Montero generalized this to partial k-trees in which edges as well as nodes can fail [7]. Rosenthal introduced in [9] a decomposition method, which was applied by Carlier and Lucet to solve network reliability problems with edge and node failures on general graphs [5]. Carlier, Manouvrier and Lucet solved with this approach the 2-edge connected reliability problem for graphs of bounded pathwidth [6]. We will generalize both the type of the problems which can be solved, and the method used in [5] and [6].

For graphs of bounded treewidth with associated non-failure probabilities, we present a generic framework which is much more general than previously presented work. A large number of questions can be answered, which can be more complicated than the classical S-terminal reliability problem. These questions ask for the probability that the graph of the 'up' elements has a certain property. We allow that there are two types of special vertices: a set S of servers and a set L of clients (as well as other vertices that serve simply to build connections). With this technique, we can answer questions such as: 'What is the probability that all clients are connected to at least one server?' and 'What is the expected number of components of the graph of the non-failed elements that contain a vertex of S or L?' Many of these questions are shown to be solvable in linear or polynomial time when G has bounded treewidth.

The properties one can ask for, cover the connectivity of vertices of the set S among each other, and also the connectivity between vertices of S and vertices of L. Furthermore, the number of connected components which contain a vertex of S or L can be dealt with. The more information needed to answer a question, the higher the running time of our algorithm. This starts with linear time and can increase to exponential time (for solving a problem for which all possible information is necessarily preserved, and hence no diminishing of information and running time is possible).

2   Definitions

Definition 1. *A* tree decomposition *of a graph $G = (V, E)$ is a pair (T, X) with $T = (I, F)$ a tree, and $X = \{X_i \mid i \in I\}$ a family of subsets of V, one for each node of T, such that*

- $\bigcup_{i \in I} X_i = V$.
- *for all edges $\{v, w\} \in E$ there exists an $i \in I$ with $\{v, w\} \subseteq X_i$.*
- *for all $i, j, k \in I$: if j is on the path in T from i to k, then $X_i \cap X_k \subseteq X_j$.*

The width *of a tree decomposition $((I, F), \{X_i \mid i \in I\})$ is $\max_{i \in I} |X_i| - 1$. The* treewidth *of a graph G is the minimum width over all tree decompositions of G.*

A tree decomposition (T, X) *is* nice, *if* T *is rooted and binary, and the nodes are of four types:*

- Leaf nodes i *are leaves of* T *and have* $|X_i| = 1$.
- Introduce nodes i *have exactly one child* j *with* $X_i = X_j \cup \{v\}$ *for some vertex* $v \in V$.
- Forget nodes i *have exactly one child* j *with* $X_i = X_j \setminus \{v\}$ *for some vertex* $v \in V$.
- Join nodes i *have two children* j_1, j_2 *with* $X_i = X_{j_1} = X_{j_2}$

A tree decomposition can be converted into a nice tree decomposition of the same width in linear time (see [4]). A subgraph of $G = (V, E)$ induced by $V' \subseteq V$ is a graph $G[V'] = (V', E')$ with $E' = E \cap (V' \times V')$. A graph G can consist of connected components $O_1, ..., O_q$. In that case, we write $G = O_1 \cup ... \cup O_q$, where $\cup$ denotes the disjoint union of graphs, and each O_i $(1 \leq i \leq q)$ is connected. Throughout this paper, $G = (V, E)$ denotes a fixed, undirected, simple graph, with V the set of vertices, E the set of edges and k its treewidth. $T = (I, F)$ is a nice tree decomposition with width k of the graph G. The term *vertex* refers to a vertex of the graph and *node* refers to a vertex of the tree (decomposition).

A vertex v of the network or graph is either *up* or *down*. That means the network site represented by v is either functioning or it is in a failure state. The probability that v is up is denoted by $p(v)$. A *scenario* f assigns to each vertex its state of operation: v is up ($f(v) = 1$) or down ($f(v) = 0$). Hence, a scenario describes which vertices are up and which are down in the network. For the scenario f, $W^{f=1}$ is the set of all vertices of W which are up and $W^{f=0} = W \setminus W^{f=1}$, for $W \subseteq V$. The probability of a scenario f is:

$$\Pr(f) = \prod_{v \in V^{f=1}} p(v) \cdot \prod_{v \in V^{f=0}} 1 - p(v) \, .$$

Clearly, there are $2^{|V|}$ scenarios. The elements in $L \subseteq V$ are representing special objects, the *clients*, and the elements of $S \subseteq V$ are the *servers*, where $n_S = |S|$ and $n_L = |L|$. We say $W \subseteq V$ is connected in G, if for any two vertices of W there is a path joining them. One can easily model edge failures, by introducing a new vertex on each edge, and then using the methods for networks with only node failures. Thus, to ease presentation, we will assume from now on that edges do not fail. The following definition summarises this paragraph.

Definition 2. *Let* $G = (V, E)$ *be a graph.*

- $p : V \to [0, 1]$ *assigns to each vertex its probability to be up*
- $f : V \to \{0, 1\}$ *assigns to each vertex its state for this scenario* f
- $W^{f=q} = \{v \in W \mid f(v) = q\}$, *for* $q \in \{0, 1\}$, $W \subseteq V$, $f : V \to \{0, 1\}$
- $L \subseteq V$ *is the set of clients of* G, $n_L = |L|$
- $S \subseteq V$ *is the set of servers of* G, $n_S = |S|$

3 The Technique

3.1 More Special Definitions

We assume to have a nice tree decomposition $(T, X) = ((I, F), \{X_i \mid i \in I\})$ of $G = (V, E)$. As described in many articles, e.g. [3] and [4], we will use a bottom up approach with this tree decomposition. We will compute partial solutions for the subgraphs corresponding to the subtrees of T step by step. These partial solutions are contained in the roots of the subtrees. To compute this information, we only need the information stored in the children of a node, which is typical for dynamic programming. The following definition provides us the terminology for the subgraphs.

Definition 3. *Let* $G = (V, E)$ *be a graph and* $(T = (I, F), X = \{X_i \mid i \in I\})$ *a nice tree decomposition of* G.

- $V_i = \{v \in X_j \mid j = i \text{ or } j \text{ is an descendant of } i\}$
- $G_i = G[V_i]$
- $L_i = L \cap V_i$
- $S_i = S \cap V_i$

Blocks. For a certain subgraph G_i, we consider the scenarios of G restricted to V_i. Every scenario f causes G_i to be decomposed into one or more connected components $O_1, ..., O_q$, i.e. $G[V_i^{f=1}] = O_1 \cup ... \cup O_q$. For these components, we consider the intersection with X_i These intersection sets $B_1 = O_1 \cap X_i, ..., B_r = O_q \cap X_i$ are called *blocks*. Thus, each scenario f specifies a multiset of blocks. (It is a multiset since more than one component may have an empty intersection with X_i.) Since we are looking at reliability problems with a set L of clients and a set S of servers, it is important to store the information whether $O_j \cap L^{f=1} \neq \emptyset$ and/or $O_j \cap S^{f=1} \neq \emptyset$. If this is the case we add $|O_j \cap L^{f=1}|$ L-flags and/or $|O_j \cap S^{f=1}|$ S-flags to the block B_j. Hence, the blocks for a scenario f reflect the connections between the vertices of X_i in $G[V_i^{f=1}]$ as well as the number of servers and clients of the components of $G[V_i^{f=1}]$. We use the notation '$\#L\text{flags}(B)$' to refer to the number of L-flags of block B (for S-flags, we use $\#S\text{flags}(B)$).

Definition 4. *A block* B *of the graph* G_i *is a (perhaps empty) subset of* X_i *possibly extended by* L- *and* S-*flags. For* $\{v_1, ..., v_t\} \subseteq X_i$, *we write* $B = \{v_1, ..., v_t\}$ *or* $B = \{v_1, ..., v_t\}_{S...S}^{L...L}$, *respectively.*

Classes. For each scenario f, we associate the multiset of its blocks to node i. We call this representation of f its *class* C^f. To G_i we associate a multiset of classes, one for each possible scenario. Note that two different scenarios can have identical classes, but these will be considered as different objects in the multiset of the classes of G_i. Later in this paper, equivalence relations on scenarios (and hence on classes) are defined, and thus equivalence classes will be formed that can correspond to more than one scenario. The most obvious of these is to assume that scenarios are equivalent when their classes are identical, but also less obvious equivalence relations will be considered. It will be seen that the algorithmic techniques for computing the collection of classes of the graphs G_i and their corresponding probabilities will carry over with little modification to algorithms for computing the collections of equivalence classes for many equivalence relations.

Let $G[V_i^{f=1}] = O_1 \cup ... \cup O_q$. Then we have the following blocks $B_1 = O_1 \cap X_i, ..., B_q = O_q \cap X_i$, which simply make up the class $C^f = (B_1, ..., B_q)$. A block B of a class C^f has x L-flags, if and only if the component O corresponding to B contains exactly x vertices of L. The same is also true for the S-flags. The next definition summarise this.

Definition 5. *A class* $C^f = (B_1, ..., B_q)$ *of* G_i *is a multiset of blocks of* G_i, *such that scenario* f *specifies* $B_1 = O_1 \cap X_i, ..., B_q = O_q \cap X_i$. *At this,* $\#Lflags(B_j) = |O_j \cap L|$ *and* $\#Sflags(B_j) = |O_j \cap S|$, *for* $j = 1, ..., q$, *and* $G[V_i^{f=1}] = O_1, ..., O_q$.

We denote by n_b the maximum number of blocks of a class over all classes, and by n_c the maximum number of classes of a node over all nodes. The numbers n_c and n_b are crucial points to the algorithm running time as we will see. Thus, in the subsequent Section 4.1 we look at equivalence relations between classes to reduce their number.

Example classes. Let $X_i = \{u, v, w\}$. Some of the possible classes for i are:

- $C_1 = (\{uv\}_S^{LL}\{w\}\{ \}^L\{ \}_S)$; all vertices are up, u, v are in one component containing two clients and one server, w is in another component without clients and servers, there are two components with empty intersection with X_i, one of them contains a client the other a server
- $C_2 = (\{vw\}\{ \}\{ \}\{ \}_S)$; v and w are up and connected, there are three components with empty intersection with X_i, one of them contains a server
- $C_3 = (\{vw\}\{ \}^S)$; v and w are up and connected, there is one components with empty intersection with X_i and this component contains a vertex of S

Here, the classes C_2 and C_3 are somehow 'similar'. They only differ in the number of empty blocks without flags. Such blocks neither give information about clients or servers nor can be used to create additional connections. Hence, it can be sensible to define C_2 and C_3 to be equivalent (see Section 4.1).

3.2   The Method for Network Reliability Problems

As already mentioned, we compute the classes for every node of the nice tree decomposition using a bottom up approach. However, we will not only compute the classes C but their probabilities $\Pr(C)$ as well. For a class C of node i, $\Pr(C)$ is the probability that G_i is in a scenario f which is represented by C. Lemma 1 and similar statements for leaf, forget and join nodes, show us that therefore we only need the (already computed) classes of the children of the node, insofar they exist. Once we have the probabilities of all classes of all nodes, and especially of the root, we easily can solve problems by summing up the requested probabilities.

Starting with the leaves of the nice tree decomposition, we compute for each node all classes that are possible for this node, and their corresponding probabilities. In this computation, the classes of its child(ren) play at this a decisive role. When 'walking up' the tree decomposition to a node i the classes will be refined and split into other classes representing the states of the vertices of X_i and their connections. Once more, we refer to Section 4.1 in which equivalence relations for reducing the number of classes are described. Then we will see that classes are not only refined, while walking up the tree, but at the same time, some classes will 'collapse' into one (equivalence) class, since they become equivalent.

3.3   How to Compute the Classes and Their Probabilities?

Due to space constraints, we only discuss introduce nodes here.

Introduce Nodes. Let i be an introduce node with child j, such that $X_i = X_j \cup \{v\}$. We use the classes and their probabilities of node j to compute the classes and their probabilities of node i. For any state of v (up or down, in L, in S, or not) we compute and add the classes to those of i. This is done by considering the following two cases for each class C_j of j, and modifying it to get a class C_i of i:

- v is up; We add v to each block $B \in C_j$, which contains a neighbour of v, since there is a connection between v and the component corresponding to B. If v is a client or a server, then we add an L- or S-flag to block B. Now, we merge the blocks of this class as described below. The probability of the resulting new class C_i is: $\Pr(C_i) = p(v) \cdot \Pr(C_j)$.
- v is down; We will not make any changes to the class C_j to get C_i, since no new connections can be made and no flags were added. However, we compute the probability: $\Pr(C_i) = (1 - p(v)) \cdot \Pr(C_j)$.

Lemma 1. *There is an algorithm that computes the classes and their probabilities of an introduce node i correctly, given all classes and their probabilities of the child j of node i.*

Such an algorithm performing the description above, can be implemented to run in time $O(n_c \cdot n_b^2 \cdot k^2)$.

Merging of Class Blocks. We describe a procedure that is a subroutine of the introduce- and join-node-algorithm. It is used to maintain the structure of classes (and has nothing to do with equivalence classes). Due to an introduction of a new vertex (see above) or joining two classes a class can contain two blocks with nonempty intersection. Between two components O_a and O_b corresponding to such blocks B_a and B_b exists a connection via a vertex of $B_a \cap B_b$. This hurts the proper structure of a class. Hence, we *merge* such non-disjoint blocks B_a and B_b into one block B. We have to add to B the number of L flags of both blocks B_a and B_b. On the other hand, we must subtract the number of clients which belong to both blocks B_a and B_b, because they are counted twice. The same applies to servers. An algorithm for merging can be implemented to run in $O(k^2 \cdot n_b^2)$ time.

4   Results

Our results can be divided into two groups. The first group does not use an equivalence relation between classes. Its consideration is a kind of warm up for the second group, which uses equivalence relations between classes (see Section 4.1).

So far, a class C of a node i represents only one scenario f of G_i, and $\Pr(C)$ is the probability that this scenario occurs for G_i. The blocks of C reflect the components and the flags the number of vertices of S and L in each component of the graph $G[V_i^{f=1}]$. Hence, any question which asks for the probability of a certain property concerning the components and connectivity of vertices of S and L can be answered. Clearly, only graphs can have such properties, but we will generalise this to scenarios and classes.

A scenario f for node i *has* property Y, if $G[V_i^{f=1}]$ has property Y. A class C of node i *has* property Y, if the scenario represented by C has property Y.

If we want to use classes for solving problems, it is also very important to be able to make the decision whether a class has a certain property just by considering the information given by the representation of the class (namely the blocks of the class). A class C of node i *shows* property Y, if the information given by the representation of C is sufficient to determine that C has Y.

The following definition describes all the properties that can be *checked* with this approach. Therefore an equivalence between showing and having is necessary.

Definition 6. *A property Y of G_i can be* checked *by using classes of i, if for all classes C of i hold:*

$$C \text{ has property } Y \iff C \text{ shows property } Y$$

After computing all classes for node i in a bottom up manner, we can use these classes to easily solve several problems for G_i.

Lemma 2. *Let Y be a property that can be checked by the classes of i. The probability that the subgraph of G_i induced by the up vertices has property Y equals the sum over the probabilities of the classes of node i which show this property Y.*

From the definitions and lemma above, we easily conclude that each question that asks for the probability of a property Y of G_i can be answered, if Y can be checked using the classes of i.

Since there are $O(2^{|V_i|})$ scenarios, this results in a method with exponential running time, because the number of classes is a crucial point for the running time of the algorithm. Thus, in the next Section 4.1, we consider how to reduce the number of classes. However, the more classes per node we 'allow' the more information we conserve during the bottom up process and hence the more problems can be solved.

4.1   Reducing the Number of Classes

In this section we will have a global look at the strategy. The correctness and restrictions follow in subsequent sections. We will use the terms class and equivalence class, which can cause confusion. Classes refer to 'objects' containing only one scenario, while equivalence classes may represent several classes or scenarios, respectively. Indeed, it is the case that equivalence classes can be used in the algorithms instead of classes.

By using an equivalence relation R between classes it is possible to reduce the number of objects, i.e. equivalence classes, handled by the algorithms. Another interpretation is that we define an equivalence relation between scenarios, since so far each class represents only one scenario. It is sensible to define classes to be equivalent, only if they belong to the same node. Equivalent classes must be 'similar' with regards to the way they are processed by the algorithms. We have to show that R is 'preserved' by the algorithms, i.e. if we have two equivalent classes of a node as input to an algorithm, then the resulting classes are also equivalent. (This is described in detail in Section 4.3.) Additionally, we have to take care that the equivalence relation R we have chosen, is suitable to solve our problem.

Then we are able to use the algorithms with the equivalence classes (or better: with their representatives) instead of classes. We have to modify the algorithms slightly, since we have to check after processing each node, if two classes or two equivalence classes became equivalent. If this is the case, we join them into one equivalence class with new probability the sum of the joined (equivalence) classes. The following list of steps summarises what we have to do.

- Step 1: We define an equivalence relation R which is fine enough to solve our problem (see Section 4.2). On the other hand, R should be 'coarse' to reduce the number of equivalence classes and hence the running time.
- Step 2: We must modify our algorithms to maintain equivalence classes, i.e. it is necessary to check for equivalent (equivalence) classes after processing each node. If we find such two (equivalence) classes, we keep only one of them with accumulated probability. For doing this, we add some code to the algorithms, which can be implemented to need $O(n_c^2 \cdot n_b^2)$ steps.
- Step 3: We have to show the correctness of using the representatives of R instead of classes, i.e. that the algorithms preserve R (see Section 4.3).

- Step 4: We analyse the running time by estimating the maximum number of equivalence classes and the maximum number of blocks.

4.2   Information Content of Equivalence Classes

As at the beginning of Section 4, we introduce some notation for equivalence classes. Therewith, we want to express the information 'contained' in equivalence classes. We clarify what it means for a relation to be fine enough, and we give a lemma that tells us what we can do with relations that are fine enough.

When using an equivalence relation, we define classes to be equivalent which may not be equal. By doing this, we lose the information that these classes were different and also why they were different. However, we reduce the number of classes, as intended. Clearly, using no equivalence relation or one which defines every class (scenario) to be in an extra equivalence class provides as much information as possible. Such a relation would be the finest we can have. For solving the S-terminal reliability problem a rather coarse equivalence relation is sufficient. Hence, we are faced with a trade-off between information content and efficiency. In analogy to the definitions at the beginning of Section 4, we give the following definition for equivalence classes.

An equivalence class C^* of classes for node i *has* property Y, if each class belonging to C^* has property Y. An equivalence class C^* of node i *shows* property Y, if its representative shows Y. Despite the abstract level, we formalise when equivalence classes provide enough information for solving problems.

Definition 7. *An equivalence relation R is fine enough for property Y, if for all nodes i of the tree decomposition holds: For all equivalence classes C^* of node i must hold:*

$$(\forall f \text{ represented by } C^* : f \text{ has property } Y)$$

$$\vee$$

$$(\forall f \text{ represented by } C^* : f \text{ has not property } Y)$$

That means R is fine enough for Y, if there is no equivalence class which contains a scenario that has property Y and another one which does not have property Y. In the same flavour as Lemma 2, the next one tells us that under certain conditions, we can use equivalence classes to solve problems for G after computing all equivalence classes of all nodes until we reached the root r.

Lemma 3. *Let R be an equivalence relation which is fine enough for property Y that can be checked using classes. The probability that G_i has property Y equals the sum over the probabilities of the equivalence classes of node i which show this property Y.*

4.3   When and Why Is Using Equivalence Relations Correct?

We already know, when an equivalence relation is fine enough. In this section we will see under which conditions we can use (the representatives of) the equivalence classes during the execution of the algorithm. The choice of the equivalence relation R cannot be made arbitrarily. As mentioned in Step 1, R has to be fine enough to provide enough information to solve the problem. On the other hand R should be coarse. Furthermore, R must have some structural properties, i.e. R must be 'respected or preserved' by the algorithms. This targets to the property that if two classes are equivalent before processed by an algorithm, they are also equivalent after that. To capture exactly the notion of 'preserving an equivalence relation', some definitions are required. However, in this extended abstract, we only look at the algorithm for introduce-nodes. The upper index x in C_j^x in this definition represents not a scenario, but an index to distinguish different classes of node j.

Definition 8. *Let i be an introduce node with child j with $X_i = X_j \cup \{v\}$. C_j^x is a class of node j which results in the classes $C_i^{x,up}$ and $C_i^{x,down}$ of node i, for v is up or down, respectively. An introduce-node-algorithm* preserves *the equivalence relation R if it holds:*

$$\forall C_j^1, C_j^2 : (C_j^1, C_j^2) \in R \Longrightarrow (C_i^{1,up}, C_i^{2,up}) \in R \wedge (C_i^{1,down}, C_i^{2,down}) \in R$$

If we use only (the representatives of) the equivalence classes in the algorithms, we can process all classes (scenarios) represented by this equivalence class by only one 'central execution' of an algorithm. The result will be, again, equivalence classes. When and why is this correct? The next lemma, which is easy to see, tells us that indeed all classes (scenarios) represented by one equivalence class of R can be handled by just one process of an algorithm if R is preserved by this algorithms.

Lemma 4. *When using equivalence classes of R as input for the forget-, introduce- and join-node-algorithm, then the result will be equivalence classes of R as well, if R is preserved by the forget-, introduce- and join-node-algorithm.*

That means, we do not have to be careful whether an algorithm 'hurts' R by creating a result that actually is not an equivalence class of R. We are now ready to bring everything together.

Theorem 1. *Let R be an equivalence relation of classes that is fine enough for property Y which can be checked using classes. Furthermore, let R be preserved by the forget-, introduce- and join-node-algorithm. Then we can use our framework to compute the probability that the subgraph of G_i induced by the up vertices has property Y.*

The running time depends on the relation R. If R has a finite number of equivalence classes, then the algorithm is linear time. If the number of equivalence classes is bounded by a polynomial in the number of vertices of G, then the algorithm uses polynomial time.

4.4   Solvable Problems

To give an exhaustive list of problems that can be handled with this approach, is beyond the scope of this paper, because many equivalence relations can be used, and with each such relation very often many questions can be answered. Fortunately, the last theorem tells us something about solvable problems. We simply can say that every problem that asks for the probability of a property Y of G can be solved, if Y can be checked using classes. Therefore we should find an equivalence relation R as coarse as possible, but still fine enough for Y. Furthermore, to apply the framework described in this paper, R must be preserved by the algorithms. Nevertheless, we list some example properties.

To get relations that allow linear running time, we can restrict the maximum number of blocks and the number of flags per block to be constant. With such relations we can answer questions like: 'What is the probability that all clients are connected to at least one server?' or 'What is the probability that all servers are useful, i.e. have a client connected to them?' We can also use only one kind of special vertices, e.g. only servers. With such a relation we are able to give an answer to 'What is the probability that all servers are connected?', which is the classical S-terminal reliability problem. With additional ideas and modifications, it is also possible to answer the following question in linear time: 'What is the expected number of components that contain at least one vertex of S (of L; of S and L)?'   .

A reasonable assumption could be to consider the number of servers n_S to be small. We can use a different S-flag for each server. In this case, a relation which does not conflate empty blocks with S-flags, but only with L-flags, can be utilised. Further, each block can have multiple flags of each type. Unfortunately, this relation leads to a maximum number of classes bounded exponentially in n_S. With this relation we can answer the question with which probability certain servers are connected, and with what probability server x has at least y clients connected to it (x, y are integers). We can also determine the 'most useless' server, i.e. the server with fewest average number of clients connected to it. Of course, this relation enables us to compute the expected number of components with at least one server.

Polynomial running time in n_L and n_S is needed by relations which bound n_b by a constant, however the number of flags per block is only bounded by n_L and n_S. Such relations enable properties like: at least x clients are not connected to a server, or at most y server are not connected to a client, as well as at least x clients are connected to at least one server while at least y servers are connected to at least one client.

5   Discussion

5.1   Edge Failures

As mentioned earlier edge failures can be simulated by vertex failures. Therefore, we place a new vertex on each edge. Its reliability will be the one of the corresponding edge. All edges of the resulting graph will be defined to be perfectly reliable. We clearly increase the number of vertices of the graph by doing this. The resulting graph has then $|V| + |E|$ vertices. For arbitrary graphs it holds $|E| \leq \frac{|V| \cdot (|V|-1)}{2}$ which would mean a potential quadratic increment. When restricted to graphs of treewidth at most k, we have $|E| \leq k \cdot |V| - \frac{k \cdot (k+1)}{2}$ (see [2]). Hence, only an increment linear in $|V|$, if k considered a constant. The effect on the treewidth of a graph itself is also rather secondary. We consider an edge $\{u, v\}$. In our tree decomposition, we have one node i with $\{u, v\} \subseteq X_i$. When placing vertex w on this edge, we can simply attach a new node $\{u, v, w\}$ to the node i. It is easy to see that this results in a proper tree decomposition and does not increase its width, if the width is at least 2.

5.2   Treewidth vs. Pathwidth

In Section 4.3, we required a relation R to be preserved by all three algorithms. We can refrain from this demand, if we look at graphs of bounded pathwidth. A path decomposition of G is a tree decomposition (T, X) of G whereas T is a path. Hence, join-nodes do not appear and thus R has not to be preserved by the join-node-algorithm. This may enable more relations, but at the other hand we have to use a path decomposition.

5.3   Running Times

The running time for our algorithm heavily depends on the chosen relation R. We can see it is very important to choose a relation 'as coarse as possible'. Furthermore, the framework provides enough possibilities for using additional tricks, ideas and modifications. It may be necessary to use them to get the best running time. One general possibility to decrease the running time would be to delete classes that can never have the required property as soon as possible. However, the performance gain is not easy to analyse.

Acknowledgements. Many thanks to our colleagues Hans Bodlaender and Frank van den Eijkhof for their very useful discussions and comments, and also thanks to Peter Lennartz and Ian Sumner.

References

1. S. Arnborg, A. Proskurowski: *Linear time algorithms for NP-hard problems restricted to partial k-trees.* Discrete Appl. Math. 23, (1989), 11-24
2. H. L. Bodlaender: *A linear-time algorithm for finding tree-decompositions of small treewidth.* SIAM J. Comput. 25/6 (1996), 1305-1317
3. H. L. Bodlaender: *A tourist guide through treewidth.* Acta Cybernet. 11 (1993), 1-23
4. H. L. Bodlaender: *Treewidth: Algorithmic techniques and results.* In I. Privara and P. Ruzicka, editors, Proceedings of the 22nd International Symposium on Mathematical Foundations of Computer Science, MFCS '97, LNCS 1295, (1997), 19-36

5. J. Carlier, C. Lucet: *A decomposition algorithm for network reliability evaluation.* Discrete Appl. Math. 65, (1996), 141-156

6. C. Lucet, J.-F. Manouvrier, J. Carlier: *Evaluating Network Reliability and 2-Edge-Connected Reliability in Linear Time for Bounded Pathwidth Graphs.* Algorithmica 27, (2000), 316-336

7. E. Mata-Montero: *Reliability of Partial k-tree Networks.* Ph.D. Thesis, Technical report: CIS-TR-90-14, University of Oregon, (1990)

8. J. S. Provan, M. O. Ball: *The complexity of counting cuts and of computing the probability that a graph is connected.* SIAM J. Comput. 12/4, (1983), 777-788

9. A. Rosenthal: *Computing the reliability of complex networks.* SIAM J. Appl. Math. 32, (1977), 384-393

A Faster Approximation Algorithm for 2-Edge-Connectivity Augmentation*

Anna Galluccio[1] and Guido Proietti[2,1]

[1] Istituto di Analisi dei Sistemi ed Informatica "Antonio Ruberti",
CNR, Viale Manzoni 30, 00185 Roma, Italy.
galluccio@iasi.rm.cnr.it.
[2] Dipartimento di Informatica, Università di L'Aquila,
Via Vetoio, 67010 L'Aquila, Italy.
proietti@di.univaq.it.

Abstract. Given a weighted graph G with n vertices and m edges, the *2-edge-connectivity augmentation* problem is that of finding a minimum weight set of edges of G to be added to a spanning subgraph H of G to make it 2-edge-connected. Such a problem is well-known to be NP-hard, but it becomes solvable in polynomial time if H is a depth-first search tree of G, and the fastest algorithm for this special case runs in $\mathcal{O}(m + n \log n)$ time. In this paper, we sensibly improve such a bound, by providing an efficient algorithm running in $\mathcal{O}(M \cdot \alpha(M,n))$ time, where α is the classic inverse of the Ackermann's function and $M = m \cdot \alpha(m,n)$. This algorithm has two main consequences: First, it provides a faster 2-approximation algorithm for the general 2-edge-connectivity augmentation problem; second, it solves in $\mathcal{O}(m \cdot \alpha(m,n))$ time the problem of maintaining, by means of a minimum weight set of edges, the 2-edge-connectivity of a 2-edge-connected communication network undergoing an edge failure, thus improving the previous $\mathcal{O}(m + n \log n)$ time bound.

Keywords: Edge-Connectivity Augmentation, Network Survivability, Approximation Algorithms.

1 Introduction

The increasing attention towards *reliability* aspects of communication often asks for strengthening the degree of (either edge- or vertex-) connectivity of an already existing network.

For the edge-connectivity case, which is of interest for this paper, such a task can be formulated as a classical *augmentation problem* on graphs: Given a k-edge-connected, real weighted, undirected graph $G = (V, E)$, with $|V| = n$

---

* This work has been partially supported by the CNR-Agenzia 2000 Program, under Grants No. CNRC00CAB8 and CNRG003EF8, and by the Research Project REAL-WINE, partially funded by the Italian Ministry of Education, University and Research.

and $|E| = m$, given an h-edge connected spanning subgraph $H = (V, E')$ of G, with $h \geq 1$, and given an integer $h < \lambda \leq k$, find a minimum weight set of edges $\text{AUG}_\lambda(H, G)$ in $E \setminus E'$ whose addition to H increases its edge-connectivity to λ.

Such a problem turns out to be NP-hard [2], and thus most of the research in the past focused on the design of approximation algorithms for solving it. In particular, for the special case $\lambda = 2$, which is of interest for this paper, efficient approximation algorithms for finding $\text{AUG}_2(H, G)$ are known. More precisely, for the weighted case, the best performance ratio is 2 [4,9], while for the unweighted case, Nagamochi and Ibaraki developed a $(51/26 + \epsilon)$-approximation algorithm, for any constant $\epsilon > 0$ [11]. Analogous versions of augmentation problems for vertex-connectivity and for directed graphs have been widely studied, and we refer the interested reader to the following comprehensive papers [3,10].

Apart from designing approximated solutions, researchers have also investigated the problem of characterizing polynomially solvable cases of the problem. This can be done by giving additional constraints on the structure of the pair (H, G). First, Eswaran and Tarjan[2] proved that $\text{AUG}_2(H, G)$ can be found in polynomial time if G is complete and all edges have weight 1, i.e., all potential links between sites may be activated at the same cost. Afterwards, Watanabe and Nakamura extended this result to any desired edge-connectivity value [14], and faster algorithms in this scenario have been proposed in [5].

An interesting connection between designing approximation algorithms and characterizing polynomially solvable cases arises from [9], where, when developing a 2-approximation algorithm for finding $\text{AUG}_2(H, G)$, the authors implicitly prove that if H is a *depth-first search (DFS) tree* of G, then it can be augmented in polynomial time, in the following way: First of all, transform G into a directed graph G', by directing all the tree-edges towards the root r of H and all the non-tree edges towards the leaves; then, set the weight of all the tree-edges to 0; finally, compute an *optimum branching* in G' with respect to r, that is a minimum spanning tree containing a directed path from r to every vertex in G'. This can be done in $\mathcal{O}(m + n \log n)$ time, by making use of the algorithm proposed in [6]. It is worth noting that this operation represents the bottle-neck of the 2-approximation algorithm developed in [9]. Hence, finding a faster algorithm for augmenting a DFS-tree would result in a faster approximation algorithm for the general 2-edge-connectivity augmentation problem. This is exactly what this paper is concerned about. More specifically, we present an algorithm for augmenting a DFS-tree which runs in $\mathcal{O}(m \cdot \alpha(m, n) \cdot \alpha(m \cdot \alpha(m, n), n))$ time and $\mathcal{O}(m \cdot \alpha(m, n))$ space, where α is the classic inverse of the Ackermann's function defined in [12]. Notice that $\mathcal{O}(\alpha(m, n) \cdot \alpha(m \cdot \alpha(m, n), n)) = \mathcal{O}(\alpha^2(m, n))$, and therefore, for all practical purposes, our algorithm is linear.

We also show that our algorithm can be used for maintaining the 2-edge-connectivity in a 2-edge-connected communication network undergoing an edge failure. Indeed, as soon as an edge fails, one can be interested in finding a minimum weight set of *replacement edges* whose addition to the network allows to maintain the 2-edge-connectivity (without removing all the original, still working, edges). As we will see later, this problem reduces to a 2-edge-connectivity augmentation problem on a Hamiltonian path, and the running time of our algorithm for this special case decreases to $\mathcal{O}(m \cdot \alpha(m, n))$. In this way, we improve the previous known $\mathcal{O}(m + n \log n)$ time bound [7].

The paper is organized as follows: in Section 2 we recall some basic graph terminology; in Section 3 we show how to augment a DFS-tree, while in Section 4, we present the technique to efficiently implement such an algorithm, and we give a detailed space and time complexity analysis; afterwards, in Section 5, we apply these results to improve the running time of the 2-approximation algorithm for the general 2-edge-connectivity augmentation problem and to handle edge failures in 2-edge-connected networks; finally, in Section 6, we list some possible extensions of our work.

2   Basic Definitions

Let $G = (V, E)$ be an undirected graph, where V is the set of vertices and $E \subseteq V \times V$ is the set of edges. G is said to be *weighted* if there exists a real function $w : E \mapsto \mathbb{R}$, otherwise G is *unweighted*. A graph $H = (V(H), E(H))$ is called a *subgraph* of G if $V(H) \subseteq V$ and $E(H) \subseteq E$. If $V(H) = V$, then H is called a *spanning subgraph* of G. The weight of H is defined as $w(H) = \sum_{e \in E(H)} w(e)$.

A *simple path* (or a *path* for short) in G is a subgraph H of G with $V(H) = \{v_0, \dots, v_k | v_i \neq v_j$ for $i \neq j\}$ and $E(H) = \{(v_i, v_{i+1}) | 0 \leq i < k\}$. A *cycle* is a path whose endvertices v_0 and v_k coincide. A spanning path of G is called a *Hamiltonian path* of G. A graph G is *connected* if, for any $u, v \in V$, there exists a path $P(u, v)$ in G.

A *rooted tree* is a connected acyclic graph with a privileged vertex distinguished from the others. Let T denote a spanning tree of G rooted at $r \in V$. Edges in T are called *tree edges*, while the remaining edges of G are called *non-tree edges*. A non-tree edge (u, v) *covers* all the tree edges along the (unique) path from u to v in T. Let $P(r, x)$ denote the unique path in T between r and $x \in V(T)$. Any vertex y in $P(r, x)$ is called an *ancestor* of x in T. Symmetrically, x is called a *descendant* of any vertex y in $P(r, x)$. A *depth-first search* tree of G, *DFS-tree* for short, is a rooted spanning tree T of G such that, for any non-tree edge (u, v), either u is an ancestor of v in T, or v is an ancestor of u in T.

A graph G is said to be *k-edge-connected*, where k is a positive integer, if the removal of any $k - 1$ distinct edges from G leaves G connected. Given an h-edge-connected spanning subgraph H of a k-edge-connected graph G, and a positive integer $h < \lambda \leq k$, finding a λ-*augmentation* of H in G means to select a minimum weight set of edges in $E \setminus E(H)$, denoted as $\text{AUG}_\lambda(H, G)$, such that the spanning subgraph $H' = (V, E(H) \cup \text{AUG}_\lambda(H, G))$ of G is λ-edge-connected.

3   Augmenting a DFS-Tree

Let $T = (V, E')$ be a DFS-tree of G rooted at $r \in V$. Let $v_1, v_2, \dots, v_n = r$ be a numbering of the vertices of G as obtained by any fixed postorder visit of T. In the remainder of the paper, edges will be specified by letting the second endvertex (named the *tail*) be the farthest one from r.

First of all, we introduce the notation we will use in the remainder of the paper (see also Figure 1). Let T_k denote the subgraph of T induced by the edge set $\{e_1, \ldots, e_k\}$. A *covering* of T_k is a set of edges in $E \setminus E'$ which cover all the edges of T_k. Let $V(v_i \rightsquigarrow v_j)$ and $E'(v_i \rightsquigarrow v_j)$ denote the set of vertices and the set of edges on the (unique) path in T between v_i and v_j, respectively, and let $F(e_k)$ denote the set of non-tree edges covering the tree edge e_k. Let $E'(v_i)$ denote the (possibly empty) set of tree edges joining v_i with its children in T. Finally, given a tree edge $e_k = (v_h, v_k)$ and a non-tree edge $f = (v_s, v_t) \in F(e_k)$, we define

$$E'(e_k, f) = \bigcup_{v_i \in V(v_k \rightsquigarrow v_t)} E'(v_i) \setminus E'(v_k \rightsquigarrow v_t). \tag{1}$$

Notice that if $v_k = v_t$, then $E'(e_k, f) = E'(v_k)$.

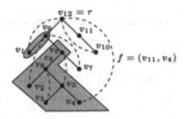

$f = (v_{11}, v_4)$

Fig. 1. A DFS-tree (solid edges) in a graph G, where edge weights are omitted for the sake of readability. Then, if we focus on edge $e_6 = (v_8, v_6)$, we have that T_6 is shaded, $F(e_6) = \{(v_9, v_2), (v_8, v_5), f = (v_{12}, v_4)\}$, $E'(e_6, f) = \{(v_6, v_2), (v_5, v_3)\}$.

The algorithm is based on dynamic programming and consists of $n - 1$ iterations. During the k-th iteration, the algorithm associates with the tree edge e_k a *selected edge* $\epsilon(e_k) \in F(e_k)$, defined as follows:

Definition 1. *Let $e_k = (v_h, v_k)$ be a tree edge, and let $f \in F(e_k)$; let ϕ be defined recursively as*

$$\phi(e_k, f) = w(f) + \sum_{e_j \in E'(e_k, f)} \min_{f_i \in F(e_j)} \{\phi(e_j, f_i)\}. \tag{2}$$

Then, a selected edge $\epsilon(e_k)$ is such that

$$\phi(e_k, \epsilon(e_k)) = \min_{f \in F(e_k)} \{\phi(e_k, f)\}. \tag{3}$$

Notice that by plugging (3) into (2), we have that

$$\phi(e_k, f) = w(f) + \sum_{e_j \in E'(e_k,f)} \phi(e_j, \epsilon(e_j)), \tag{4}$$

and if v_k coincides with the tail of f, then from (1) we have that (4) reduces to

$$\phi(e_k, f) = w(f) + \sum_{e_j \in E'(v_k)} \phi(e_j, \epsilon(e_j)). \tag{5}$$

We call this value the *basic weight* of f, and we indicate it by $\phi_0(f)$.

Let S_k denote the subtree of T rooted at v_k plus the edge e_k. It is easy to see that the following set of selected edges defines a covering of S_k

$$X(e_k) = \begin{cases} \epsilon(e_k) & \text{if } v_k \text{ is a leaf,} \\ \epsilon(e_k) \cup \bigcup_{e_j \in E'(e_k,\epsilon(e_k))} X(e_j) & \text{otherwise.} \end{cases} \tag{6}$$

Then, after the $n-1$ iterations have been completed, the algorithm returns the set

$$\text{SOL} = \bigcup_{e_j \in E'(r)} X(e_j). \tag{7}$$

Although it is not hard to realize that SOL contains a minimum weight set of edges of G covering T, for a rigorous proof of correctness we refer the reader to the full version of the paper (see the Appendix).

4   Implementing Efficiently the Algorithm

In this section, we describe an efficient implementation of our algorithm, and we give a detailed space and time complexity analysis. It is worth noting that, to the best of our knowledge, there are no previous algorithms in the literature which are based on our technique.

4.1   Finding a Selected Edge

To compute efficiently the selected edges, we use an auxiliary graph called *transmuter* [13], representing the set of *fundamental cycles* of G with respect to T, i.e., the cycles of G containing only one non-tree edge. Such a transmuter, say $D_G(T)$, is a directed graph containing one *source* vertex $s(e_k)$ for each tree edge e_k in T, one *sink* vertex $t(f)$ for each non-tree edge f in G, plus a number of additional vertices of indegree 2. The main property of a transmuter is that there is a directed path from a given source vertex $s(e_k)$ to a given sink vertex $t(f)$ if and only if edges e_k and f belong to the same fundamental cycle in G with respect to T.

Let us sketch how the transmuter can be used to compute the selected edges. Basically, the correctness of the procedure will rely on the fact that, given any two non-tree edges $f, f' \in F(e_k)$ such that $\phi(e_k, f) \leq \phi(e_k, f')$, then, for any tree edge e_h with $h > k$ and such that $f, f' \in F(e_h)$, it follows that $\phi(e_h, f) \leq \phi(e_h, f')$.

With each vertex in the transmuter, we associate a *label*, consisting of a couple $\langle f, \phi(e_k, f) \rangle$, where f denotes a non-tree edge, while $\phi(e_k, f)$ is the *key* of the label and is obtained from (4); furthermore, a vertex can be marked as *unreached, active* or *reached*. At the beginning, all the vertices in the transmuter are marked *unreached*, and their labels are undefined. During the k-th iteration, the algorithm considers the edge e_k, and $D_G(T)$ is processed as follows:

1. Visit all the vertices of $D_G(T)$ which are reachable from $s(e_k)$, by stopping the descent along a path as soon as either a vertex marked *reached*, or a sink vertex is encountered. Mark all the reached vertices as *active*.
2. If an active sink vertex $t(f)$ has an undefined label, then label it with the corresponding couple $\langle f, \phi(e_k, f) = \phi_0(f) \rangle$, where $\phi_0(f)$ is computed from (5), by possibly making use of the labels associated with the edges in $E'(v_k)$ (which have already been processed); then, mark $t(f)$ as *reached*.
3. If an active (either sink or internal) vertex has a defined label (namely, it was labelled $\langle f, \phi(e_h, f) \rangle$ and marked *reached* at a previous iteration $h < k$), then update its label to $\langle f, \phi(e_k, f) \rangle$ (where $\phi(e_k, f)$ is defined by (4) and is computed by using the efficient technique specified in the next section); then, mark $t(f)$ as *reached*.
4. Process the active vertices of $D_G(T)$ in reverse topological order, by backtracking over the edges used during the descent from the source vertex; during this process, label a given active vertex with a minimum-key label among those of its reached successors, and then mark it as *reached*.

As we will prove in the full paper (see the Appendix), at the end of Step 4, vertex $s(e_k)$ will receive a label $\langle \epsilon(e_k), \phi(e_k, \epsilon(e_k)) \rangle$, where $\epsilon(e_k)$ coincides with an edge in $F(e_k)$ satisfying (3).

4.2 Updating the Labels in the Transmuter

To perform the label updating in $D_G(T)$, we use an adaptation of the technique proposed in [13] to compute functions defined on paths in trees.

We start by creating a *forest* $\mathcal{F}(G)$ of trees. Initially, $\mathcal{F}(G)$ is composed of n singletons $\nu_1, \ldots, \nu_n$, where vertex ν_j is associated with vertex $v_j \in V$. With each vertex ν_j in $\mathcal{F}(G)$, we associate a key $\kappa(\nu_j)$, initially set equal to 0. The following instructions manipulate $\mathcal{F}(G)$:

- *Link*(ν_i, ν_j): combine the trees with roots ν_i and ν_j into a single tree rooted in ν_i, adding the edge $e = (\nu_i, \nu_j)$;
- *Update*(ν_j, x): if ν_j is the root of a tree, then set $\kappa(\nu_j) := \kappa(\nu_j) + x$;
- *Eval*(ν_j): find the root of the tree currently containing ν_j, say ν_i, and return the *sum* of all the keys on the path from ν_j to ν_i.

Note that $Eval(\nu_j)$ assumes that a pointer to element ν_j is obtained in constant time.

The sequence of operations in $\mathcal{F}(G)$ goes hand in hand with the postorder visit of T and with the processing of $D_G(T)$. More precisely, immediately before edge e_k is considered by the augmentation algorithm, we perform the following steps:

1. Compute $\Delta(v_k) = \displaystyle\sum_{e_j \in E'(v_k)} \phi(e_j, \epsilon(e_j))$, where these ϕ values are stored in the labels of the source vertices of the transmuter which have already been processed;
2. For each j such that $e_j \in E'(v_k)$, perform $Update(\nu_j, \Delta(v_k) - \phi(e_j, \epsilon(e_j)))$;
3. For each j such that $e_j \in E'(v_k)$, perform $Link(\nu_k, \nu_j)$.

Furthermore, when edge e_k is considered, a label updating in $D_G(T)$ is computed as follows: Let $f = (v_s, v_t)$ be a non-tree edge covering e_k, and let $V(v_k \rightsquigarrow v_t) = \langle v_{i_1} = v_k, v_{i_2}, \ldots, v_{i_q} = v_t \rangle$; then, we have that an $Eval(\nu_t)$ operation returns

$$\sum_{j=1}^{q} \kappa(\nu_{i_j}) = \kappa(\nu_{i_1}) + \sum_{j=2}^{q} \kappa(\nu_{i_j})$$

and being $\kappa(\nu_{i_1}) = \kappa(\nu_k) = 0$, this can be rewritten as

$$\sum_{j=2}^{q} \left(\Delta(v_{i_{j-1}}) - \phi\left(e_{i_j}, \epsilon(e_{i_j})\right) \right) = \sum_{e_j \in E'(e_k, f)} \phi(e_j, \epsilon(e_j)) - \sum_{e_j \in E'(v_t)} \phi(e_j, \epsilon(e_j)). \tag{8}$$

Hence, from (4) and (5), we have that

$$\phi\left(e_k, f\right) = \phi_0(f) + Eval(\nu_t), \tag{9}$$

and therefore, as soon as we associate with each non-tree edge f its basic weight, we have that a label updating can be performed through an $Eval$ operation.

4.3   Analysis of the Algorithm

We can finally prove the following:

Theorem 1. *Let $G = (V, E)$ be a 2-edge-connected weighted graph with n vertices and m edges, and let T be a DFS-tree of G. Then, $\mathrm{Aug}_2(T, G)$ can be computed in $\mathcal{O}\big(m \cdot \alpha(m, n) \cdot \alpha(m \cdot \alpha(m, n), n)\big)$ time and $\mathcal{O}(m \cdot \alpha(m, n))$ space.*

Proof. The time complexity follows from the use of the transmuter and the corresponding label updating. Concerning the label updating, this is performed through the manipulation of $\mathcal{F}(G)$. First of all, notice that the computation of $\Delta(v_i)$, for all $v_i \in V$, costs $\mathcal{O}(n)$ time. Moreover, since the *Eval* instruction requires the sum of the keys associated with the vertices, which is an *associative* operation over the *group* of real numbers, we can apply the *path compression with balancing* technique described in [13] to modify the execution of the various operations occurring in $\mathcal{F}(G)$. Hence, a sequence of p *Eval* and n *Link* and *Update* operations can be performed in $\mathcal{O}((p+n)\cdot\alpha(p+n,n))$ time [13]. Since each edge of $D_G(T)$ corresponds at most to a single *Eval* instruction, and given that $D_G(T)$ has size $\mathcal{O}(m\cdot\alpha(m,n))$ [13], it follows that $p = \mathcal{O}(m\cdot\alpha(m,n))$, and therefore the total time needed to handle the label updating is $\mathcal{O}\big(m\cdot\alpha(m,n)\cdot\alpha(m\cdot\alpha(m,n),n)\big)$.

Since $D_G(T)$ can be built in $\mathcal{O}(m \cdot \alpha(m,n))$ time and each edge in $D_G(T)$ is visited a constant number of times by the algorithm, and given that all the remaining operations (more precisely those corresponding to (5) and (7)) can be performed by using linear time and space, the claim follows.   □

5   Applications

5.1   A Faster 2-Approximation for the General 2-Augmentation Problem

The main consequence of our algorithm is a faster 2-approximation algorithm for the general 2-edge-connectivity augmentation problem, obtained by improving the running time of the core operation of the algorithm by Khuller and Thurimella [9]. More specifically, the following can be proved:

Theorem 2. *Let $G = (V, E)$ be a 2-edge-connected weighted graph with n vertices and m edges, and let $H = (V, E')$ be a connected spanning subgraph of G. Then, there exists an $\mathcal{O}\big(m \cdot \alpha(m,n) \cdot \alpha(m \cdot \alpha(m,n), n)\big)$ time and $\mathcal{O}(m \cdot \alpha(m,n))$ space algorithm to increase the edge-connectivity of H to 2 with weight less than twice of the optimum.*

Proof. Let us briefly recall the algorithm developed in [9]. First of all, find in $\mathcal{O}(m)$ time the 2-edge-connected components of H [1], and shrink each one of them to a single vertex; in this way, H will result in a tree $H' = (V', E_1)$, whose edges are the bridges of H. For any pair of vertices $u, v \in V'$, select an edge of minimum weight in $E \setminus E'$ joining u and v (if any), and let E_2 be the union of all these edges. It is not hard to see that E_2 can be found in $\mathcal{O}(m)$ time and $\text{AUG}_2(H, G) \subseteq E_2$. Afterwards, root H' at an arbitrary node, say r. Then, for each edge $e = (u, v) \in E_2$, compute in $\mathcal{O}(m \cdot \alpha(m, n))$ time [8] the *least common ancestor* of u and v in H', say z. If $z \neq u, v$, then set $E_3 = E_2 \setminus \{e\} \cup \{e' = (z, u), e'' = (z, v)\}$, with $w(e') = w(e'') = w(e)$. Finally, direct all the edges in E_1 towards the root setting their weight to 0, and direct all the edges in E_3 towards the leaves, and compute on this set of directed edges in $\mathcal{O}(m + n \log n)$ time [6] an optimum branching with respect to r. The subset of edges in E_3 in this branching, once considered as undirected, augment to 2 the edge-connectivity of H, and moreover has weight less than twice of $\text{AUG}_2(H, G)$. From this the authors get an $\mathcal{O}(m + n \log n)$ time 2-approximation algorithm for the general 2-edge-connectivity augmentation problem. However, it is easy to see that H' is a DFS-tree with respect to $G' = (V', E_3)$, and therefore, we can directly find $\text{AUG}_2(H', G')$ by means of our algorithm, without calling the optimum branching procedure, and thus sensibly improving the overall running time, with just a small overhead in the used space. □

5.2   Maintaining 2-Edge-Connectivity through Augmentation

The above result has an interesting application for solving a survivability problem on networks, that is the problem of adding to a given 2-edge-connected network undergoing a transient edge failure, the minimum weight set of edges needed to reestablish the 2-edge-connectivity. In this way, extensive (in terms of both computational efforts and set-up costs) network restructuring is avoided. Such a problem can be solved in $\mathcal{O}(m + n \log n)$ time and linear space [7]. We now show how to improve this result to $\mathcal{O}(m \cdot \alpha(m, n))$ time and space.

First, we prove that the time complexity of the algorithm presented in the previous sections can be further decreased if the DFS-tree is actually a Hamiltonian path. In fact, the following holds:

Corollary 1. *Let $G = (V, E)$ be a 2-edge-connected weighted graph with n vertices and m edges, and let $\Pi = (V, E')$ be a Hamiltonian path of G. Then, $\text{AUG}_2(\Pi, G)$ can be computed in $\mathcal{O}(m \cdot \alpha(m, n))$ time and space.*

Proof. In fact, for any path edge e_k and any non-path edge $f = (v_s, v_t) \in F(e_k)$, we have that $E'(e_k, f) = (v_t, v_{t-1})$, and therefore, from (4), it follows that $\phi(e_k, f)$ is constantly equal to its basic weight (5). In other words, label updating in the transmuter is not required, and thus the time overhead induced by the use of $\mathcal{F}(G)$ is avoided. From this and from the analysis performed in Theorem 1, the thesis follows. □

Now, let $H = (V, E')$ be a 2-edge-connected spanning subgraph of a weighed graph $G = (V, E)$. Let $G - e$ denote the graph obtained from G by removing an edge $e \in E$. In the following, we assume that G is 3-edge-connected, so that $G - e$ is (at least) 2-edge-connected. Given an edge $e \in E'$, if $H - e$ is not 2-edge-connected, then we say that e is *vital* for H. In the sequel, an edge e removed from H will always be considered as vital for H.

Let $\text{AUG}_2(H - e, G - e)$ be a minimum weight set of edges in $E \setminus E'$ such that the spanning subgraph $H' = (V, E \setminus \{e\} \cup \text{AUG}_2(H - e, G - e))$ of $G - e$ is 2-edge-connected. Using the results of the previous sections, we prove that $\text{AUG}_2(H - e, G - e)$ can be computed efficiently. More precisely:

Theorem 3. *Let $G = (V, E)$ be a 3-edge-connected, real weighted graph with n vertices and m edges. Let $H = (V, E')$ be a 2-edge-connected spanning subgraph of G. Then, for any vital edge $e \in E'$, we have that $\text{AUG}_2(H - e, G - e)$ can be computed in $\mathcal{O}(m \cdot \alpha(m, n))$ time and space.*

Proof. After the removal of e from H, every 2-edge-connected component in $H - e$ can be contracted into a single vertex in $\mathcal{O}(m)$ time [1], resulting in a path $\Pi = (V', E_1)$ whose edges are the bridges of $H - e$. Consider now the minimum weight set of edges $E_2 \subseteq E \setminus E'$ which connect vertices in V', and let $G' = (V', E_1 \cup E_2)$. It is not hard to see that E_2 can be found in $\mathcal{O}(m)$ time and that $\textsc{Aug}_2(H - e, G - e) = \textsc{Aug}_2(\Pi, G')$. Therefore, since Π is a Hamiltonian path of G', from Corollary 1 the thesis follows. $\square$

6   Conclusions

The technique we presented in this paper to solve efficiently the $\textsc{Aug}_2(H, G)$ problem when H is a DFS-tree of G is, from our point of view, of independent interest and can be applied to a larger class of similar graph problems. In particular, the monotone increasing of ϕ when moving towards the root, as pointed out in Section 4.1, seems to be the crucial property that has to be satisfied in order to apply our technique.

Our algorithm is efficient, but it is hard to believe that its running time is optimal. Apart from trying to improve it, many interesting problems remain open: (1) the extension to vertex-connectivity augmentation problems; (2) the extension of the results to the case in which all the possible edge failures in H are considered, aiming at providing a faster solution than that obtained by repeatedly applying our algorithms, once for the failure of each vital edge in H.

References

1. A.V. Aho, J.E. Hopcroft and J.D. Ullman, *The design and analysis of computer algorithms*, Addison Wesley, (1974).
2. K.P. Eswaran and R.E. Tarjan, Augmentation problems, *SIAM Journal on Computing*, **5** (1976) 653–665.
3. A. Frank, Augmenting graphs to meet edge-connectivity requirements, *SIAM Journal on Discrete Mathematics*, **5** (1992) 25–53.
4. G.N. Frederickson and J. Jájá, Approximation algorithms for several graph augmentation problems, *SIAM Journal on Computing*, **10**(2) (1981) 270–283.
5. H.N. Gabow, Application of a poset representation to edge-connectivity and graph rigidity, *Proc. 32nd Ann. IEEE Symp. on Foundations of Computer Science (FOCS'91)*, IEEE Computer Society, 812–821.
6. H.N. Gabow, Z. Galil, T. Spencer and R.E. Tarjan, Efficient algorithms for finding minimum spanning trees in undirected and directed graphs, *Combinatorica*, **6** (1986) 109–122.
7. A. Galluccio and G. Proietti, Polynomial time algorithms for edge-connectivity augmentation of Hamiltonian paths, *Proc. 12th Ann. Int. Symp. on Algorithms and Computation (ISAAC'01)*, December 19-21, 2001, Christchurch, New Zealand, Vol. 2223 of Lecture Notes in Computer Science, Springer, 345–354.
8. D. Harel and R.E. Tarjan, Fast algorithms for finding nearest common ancestors, *SIAM Journal on Computing*, **13**(2) (1984) 338–355.
9. S. Khuller and R. Thurimella, Approximation algorithms for graph augmentation, *Journal of Algorithms*, **14**(2) (1993) 214–225.

10. S. Khuller, Approximation algorithms for finding highly connected subgraphs, in *Approximation Algorithms for NP-Hard Problems*, Dorit S. Hochbaum Eds., PWS Publishing Company, Boston, MA, 1996.

11. H. Nagamochi and T. Ibaraki, An approximation for finding a smallest 2-edge-connected subgraph containing a specified spanning tree, *Proc. 5th Annual International Computing and Combinatorics Conference (COCOON'99)*, Vol. 1627 of Lecture Notes in Computer Science, Springer, 31–40.

12. R.E. Tarjan, Efficiency of a good but not linear set union algorithm, *Journal of the ACM*, **22** (1975) 215–225.

13. R.E. Tarjan, Applications of path compression on balanced trees, *Journal of the ACM*, **26**(4) (1979) 690–715.

14. T. Watanabe and A. Nakamura, Edge-connectivity augmentation problems, *Journal of Computer and System Sciences*, **35**(1) (1987) 96–144.

Appendix: Correctness of the Algorithm

In this appendix we prove the correctness of the algorithm presented in Section 3.

Theorem 4. *The algorithm is correct.*

Proof. The algorithm clearly returns a set of edges covering T. Hence, in order to prove that the algorithm is correct, it remains to show that the weight of the solution found by the algorithm equals $w(\text{AUG}_2(T, G))$.

Let $e_k = (v_h, v_k)$ and let $E'(k) = \{e_j \in E'(v_i) \mid v_i \in V(v_n \rightsquigarrow v_h) \wedge j \leq k\}$. The solution provided by the algorithm at iteration k, say $\text{SOL}(k)$, is defined as follows

$$\text{SOL}(k) = \bigcup_{e_j \in E'(k)} X(e_j). \tag{10}$$

We have to prove that $w(\text{SOL}) = w(\text{AUG}_2(T, G))$. Let $\text{OPT}(k)$ denote a minimum weight set of edges of G covering $T_k, 1 \leq k \leq n - 1$. Notice that $w(\text{OPT}(n-1)) = w(\text{AUG}_2(T, G))$. We shall prove that $w(\text{SOL}(k)) = w(\text{OPT}(k))$, for $k = 1, \ldots, n-1$, from which, being $\text{SOL} = \text{SOL}(n-1)$, the thesis will follow.

The proof is by induction on k. For $k = 1$, the thesis follows trivially. Assume the thesis is true up to $k - 1 < n - 1$, i.e., $w(\text{SOL}(i)) = w(\text{OPT}(i))$ for $i = 1, \ldots, k - 1$. We shall prove that $w(\text{OPT}(k)) = w(\text{SOL}(k))$.

Clearly, $\text{OPT}(k)$ has to contain a non-tree edge covering e_k, say f. Since T is a DFS-tree, the insertion of f in $\text{OPT}(k)$ can only affect the edges of $\text{OPT}(k-1)$ covering S_k. Thus, if $X^*(e_j)$ is the set of edges of $\text{OPT}(k)$ covering the subtree S_j, then

$$\text{OPT}(k) = X^*(e_k) \cup \bigcup_{e_j \in E'(k) \setminus \{e_k\}} X^*(e_j).$$

Since $\bigcup_{e_j \in E'(k) \backslash \{e_k\}} X^*(e_j) = \text{OPT}(i)$, for some $i < k$, we have, by induction, that

$$w(\text{OPT}(i)) = w(\text{SOL}(i)) = \sum_{e_j \in E'(k) \backslash \{e_k\}} w(X(e_j)).$$

Hence, it suffices to show that $w(X^*(e_k)) = w(X(e_k))$. From (6) and (4), it is easy to see that, if the transmuter correctly computes the selected edges, then $w(X(e_k)) = \phi(e_k, \epsilon(e_k))$.

In the following, we prove that by processing the transmuter as described, at the end of the k-th iteration, for each tree edge $e_j, j \leq k$, a corresponding selected edge $\epsilon(e_j)$ can be retrieved from the label associated with the source vertex $s(e_j)$ of $D_G(T)$. More precisely, the following holds:

Lemma 1. *At the end of the k-th iteration, the label associated with the source vertex $s(e_j), j \leq k$, contains a selected edge for e_j.*

Proof. The proof is by induction on k. For $k = 1$, the thesis follows trivially. Assume the thesis is true up to $k - 1 < n - 1$, and let $\langle f, \phi(e_k, f) \rangle$ be the label associated with $s(e_k)$ at the end of the k-th iteration. Then, it suffices to prove that f is a selected edge for e_k; in fact, since the source vertices have zero indegree, their labels, once assigned, cannot be modified.

For the sake of contradiction, let us assume that there exists a non-tree edge $f' \in F(e_k)$ such that $\phi(e_k, f') < \phi(e_k, f)$. First of all, observe that edges in $F(e_k)$ can be partitioned into two sets: $F_{\text{new}}(e_k)$, containing all the non-tree edges whose tail coincides with v_k, and $F_{\text{old}}(e_k) = F(e_k) \backslash F_{\text{new}}(e_k)$. Clearly, f' cannot belong to $F_{\text{new}}(e_k)$, since in such a case it would be associated with a sink vertex $t(f')$ marked as *active* at iteration k; therefore, its label receives the key $\phi(e_k, f') = \phi_0(f') < \phi(e_k, f)$, and will eventually reach $s(e_k)$, contradicting the assumptions. Hence, it must be $f' \in F_{\text{old}}(e_k)$. In this case, f' must cover an edge in $E'(v_k)$, say $e_i = (v_k, v_i)$. This means, the paths from $s(e_i)$ to $t(f')$ and from $s(e_k)$ to $t(f')$ in $D_G(T)$ share one or more vertices. Among these vertices, let v be the one closest to $s(e_k)$ along the latter path. Two cases are possible:

(1) $v = t(f')$: in this case, the label of $t(f')$ is updated to $\langle f', \phi(e_k, f') \rangle$, with $\phi(e_k, f') < \phi(e_k, f)$, and therefore it will eventually reach $s(e_k)$, contradicting the assumptions.

(2) v is an internal vertex: in this case, from the inductive step and from the fact that the indegree of the internal vertices of $D_G(T)$ is 2, we have that before the label updating at iteration k, v was labelled with an edge $f'' \in F(e_i) \cap F(e_k)$ such that $\phi(e_i, f'') \leq \phi(e_i, f')$. Hence, from (4) and from the fact that $E'(e_k, f) = E'(e_i, f) \cup \{E'(v_k) \backslash \{e_i\}\}$, it follows that

$$\phi(e_k, f'') = \phi(e_i, f'') + \sum_{e_j \in E'(v_k) \backslash \{e_i\}} \phi(e_j, \epsilon(e_j)) \leq$$

$$\phi\left(e_i, f'\right) + \sum_{e_j \in E'(v_k)\setminus\{e_i\}} \phi(e_j, \epsilon(e_j)) = \phi\left(e_k, f'\right) < \phi(e_k, f),$$

and therefore the label $\langle f'', \phi(e_k, f'') \rangle$ will eventually reach $s(e_k)$, contradicting the assumptions.   □

From the above lemma and from the definition of $\phi(e_k, \epsilon(e_k))$, it follows that for any covering $Y(e_k)$ of S_k, we have that $w(X(e_k)) \leq w(Y(e_k))$. In particular, $w(X(e_k)) \leq w(X^*(e_k))$, from which, being $w(X^*(e_k))$ minimal, the thesis follows.   □

Tree Spanners on Chordal Graphs: Complexity, Algorithms, Open Problems

A. Brandstädt[1], F.F. Dragan[2], H.-O. Le[1]*, and V.B. Le[1]

[1] Institut für Theoretische Informatik, Fachbereich Informatik,
Universität Rostock, 18051 Rostock, Germany.
{ab,hoang-oanh.le,le}@informatik.uni-rostock.de
[2] Dept. of Computer Science, Kent State University, Ohio, USA.
dragan@cs.kent.edu

Abstract. A *tree t-spanner* T in a graph G is a spanning tree of G such that the distance in T between every pair of vertices is at most t times their distance in G. The TREE t-SPANNER problem asks whether a graph admits a tree t-spanner, given t. We substantially strengthen the hardness result of Cai and Corneil [*SIAM J. Discrete Math.* 8 (1995) 359–387] by showing that, for any $t \geq 4$, TREE t-SPANNER is **NP**-complete even on chordal graphs of diameter at most $t+1$ (if t is even), respectively, at most $t + 2$ (if t is odd). Then we point out that every chordal graph of diameter at most $t - 1$ (respectively, $t - 2$) admits a tree t-spanner whenever $t \geq 2$ is even (respectively, $t \geq 3$ is odd), and such a tree spanner can be constructed in linear time.

The complexity status of TREE 3-SPANNER still remains open for chordal graphs, even on the subclass of undirected path graphs that are strongly chordal as well. For other important subclasses of chordal graphs, such as very strongly chordal graphs (containing all interval graphs), 1-split graphs (containing all split graphs) and chordal graphs of diameter at most 2, we are able to decide TREE 3-SPANNER efficiently.

1 Introduction and Results

All graphs considered are connected. For two vertices in a graph G, $d_G(x, y)$ denotes the distance between x and y; that is, the number of edges in a shortest path in G joining x and y. The value $\mathrm{diam}(G) := \max d_G(x, y)$ is the *diameter* of the graph G.

Let $t \geq 2$ be a fixed integer. A spanning tree T of a graph G is a *tree t-spanner* of G if, for every pair of vertices x, y of G, $d_T(x, y) \leq t \cdot d_G(x, y)$. TREE t-SPANNER is the following problem: Given a graph G, does G admit a tree t-spanner?

There are many applications of tree spanners in different areas; especially in distributed systems and communication networks. In [1], for example, it was shown that tree spanners can be used as models for broadcast operations; see also [21]. Moreover, tree spanners also appear in biology [2], and in [25], tree

---

* Research of this author was supported by DFG, Project no. Br1446-4/1

P. Bose and P. Morin (Eds.): ISAAC 2002, LNCS 2518, pp. 163–174, 2002.

spanners were used in approximating the bandwidth of graphs. We refer to [7,6, 22,24] for more background information on tree spanners.

In [6] Cai and Corneil gave a linear time algorithm solving TREE 2-SPANNER and proved that TREE t-SPANNER is **NP**-complete for any $t \geq 4$. A graph is *chordal* if it does not contain any chordless cycle of length at least four. For a popular subclass of chordal graph, the strongly chordal graphs, Brandstädt *et al.* [3] proved that, for every $t \geq 4$, TREE t-SPANNER is solvable in linear time. Indeed, they show that every strongly chordal graph admits a tree 4-spanner. In contrast, one of our results is

Theorem 1 *For any $t \geq 4$, TREE t-SPANNER is **NP**-complete on chordal graphs of diameter at most $t + 1$ (if t is even), respectively, of at most $t + 2$ (if t is odd).*

Comparing with a recent result due to Papoutsakis [20], it is interesting to note that the union of two tree t-spanners, $t \geq 4$, may contain chordless cycles of any length. This perhaps indicates the difficulty in proving Theorem 1. Indeed, our reduction from 3SAT to TREE t-SPANNER given in Section 2 is quite involved.

Moreover, to the best of our knowledge, Theorem 1 is the first hardness result for TREE t-SPANNER on a restricted, well-understood graph class. Notice that in [12] it is shown that TREE t-SPANNER, $t \geq 4$, is **NP**-complete on planar graphs *if* the integer t is part of the input.

In view of the diameter constraints in Theorem 1, we note that TREE t-SPANNER remains open on chordal graphs of diameter t (t is even) and of diameter $t - 1$, t or $t + 1$ (if t is odd). For "smaller" diameter we have

Theorem 2 *For any even integer t, every chordal graph of diameter at most $t - 1$ admits a tree t-spanner, and such a tree spanner can be constructed in linear time. For any odd integer t, every chordal graph of diameter at most $t - 2$ admits a tree t-spanner, and such a tree spanner can be constructed in linear time.*

We were able also to show that chordal graphs of diameter at most $t - 1$ (t is odd) admit tree t-spanners if and only if chordal graphs of diameter 2 admit tree 3-spanners. This result is used to show that every chordal graph of diameter at most $t - 1$ (t is odd), if it is planar or a k-tree, for $k \leq 3$, has a tree t-spanner and such a tree spanner can be constructed in polynomial time. Note that, for any fixed t, there is a 2-tree without a tree t-spanner [16]. So, even those kind of results are of interest. Unfortunately, the reduction above (from arbitrary odd t to $t = 3$) is of no direct use for general chordal graphs because not every chordal graph of diameter at most 2 admits a tree 3-spanner. One of our theorems characterizes those chordal graphs of diameter at most 2 that admit such spanners.

We now discuss TREE t-SPANNER on important subclasses of chordal graphs. It is well-known that chordal graphs are exactly the intersection graphs of subtrees in a tree [13]. Thus, intersection graphs of paths in a tree, called *path graphs*, form a natural subclass of chordal graphs. TREE t-SPANNER remains unresolved even on this natural subclass of chordal graphs.

The complexity status of TREE 3-SPANNER remains a long standing open problem. However, it can be solved efficiently for many particular graph classes, such as *cographs* and *complements of bipartite graphs* [5], *directed path graphs* [17] (hence for all interval graphs [16,18,23]), *split graphs* [5,16,25], *permutation graphs* and *regular bipartite graphs* [18], *convex bipartite graphs* [25], and recently for *planar graphs* [12]. In [5,19,20], some properties of graphs admitting a tree 3-spanner are discussed.

On chordal graphs, however, TREE 3-SPANNER remains even open on path graphs which are strongly chordal as well. For some important subclasses of chordal graphs we can decide TREE 3-SPANNER efficiently. Graphs considered in the theorem below are defined in Section 5.

Theorem 3 *All very strongly chordal graphs and all 1-split graphs admit a tree 3-spanner, and such a tree 3-spanner can be constructed in linear time.*

Theorem 4 *For a given chordal graph $G = (V, E)$ of diameter at most 2, TREE 3-SPANNER can be decided in $O(|V||E|)$ time. Moreover, a tree 3-spanner of G, if it exists, can be constructed within the same time bound.*

Theorem 3 improves previous results on tree 3-spanners in interval graphs [16,18, 23] and on split graphs [5,16,25]. The complexity status of TREE t-SPANNER on chordal graphs considered in this paper is summarized in Figure 1 and Table 1.

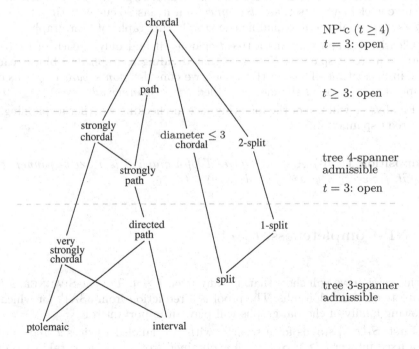

Fig. 1. The complexity status of TREE t-SPANNER on chordal graphs and important subclasses

Table 1. The complexity status of TREE t-SPANNER on chordal graphs under diameter constraints

Diameter at most	Complexity
$t + 2$, $t \geq 5$ odd	**NP**-complete
$t + 1$, $t \geq 4$ even	**NP**-complete
$t + 1$, $t \geq 3$ odd	?
t, $t \geq 3$	?
$t - 1$, $t \geq 5$ odd	?
$t - 1$, $t = 3$	polynomial time
$t - 1$, $t \geq 2$ even	linear time
$t - 2$, $t \geq 3$ odd	linear time

Notations and definitions not given here may be found in any standard textbook on graphs and algorithms. We write xy for the edge joining vertices x, y; x and y are also called *endvertices* of xy. For a set C of vertices, $N(C)$ denotes the set of all vertices outside C adjacent to a vertex in C; $N(x)$ stands for $N(\{x\})$ and $deg(x)$ stands for $|N(x)|$. We set $d(v, C) := \min\{d(v, x) \; : \; x \in C\}$. The *eccentricity* of a vertex v in G is the maximum distance from v to other vertices in G. The *radius* $r(G)$ of G is the minimum of all eccentricities and the *diameter* $\operatorname{diam}(G)$ of G is the maximum of all eccentricities. A *cutset* of a graph is a set of vertices whose deletion disconnects the graph. A graph is *non-separable* if it has no one-element cutset, and *triconnected* if it has no cutset with ≤ 2 vertices. *Blocks* in a graph are maximal non-separable subgraphs of that graph.

Clearly, a graph contains a tree t-spanner if and only if each of its blocks contains a tree t-spanner. Note also that dividing a graph into blocks can be done in linear time. Thus, in this paper, we consider *non-separable graphs* only. Graphs having a tree t-spanner are called *tree t-spanner admissible*.

Finally, we will use of the following fact in checking whether a spanning tree is a tree t-spanner.

Proposition 1 ([6]) *A spanning tree T of a graph G is a tree t-spanner if and only if, for every edge xy of G, $d_T(x, y) \leq t$.*

2   NP-Completeness, $t \geq 4$

In this section we will show that, for any fixed $t \geq 4$, TREE t-SPANNER is **NP**-complete on chordal graphs. The proof is a reduction from 3SAT, for which the following family of chordal graphs will play an important role.

First, $S_1[x, y]$ stands for a triangle with two labeled vertices x and y. Next, for a fixed integer $k \geq 1$, $S_{k+1}[x, y]$ is obtained from $S_k[x, y]$ by taking to every edge $e \neq xy$ in $S_k[x, y]$ that belongs to exactly one triangle a new vertex v_e and joining v_e to exactly the two endvertices of e. We write also S_k for $S_k[x, y]$ for some suitable labeled vertices x, y. See Figure 2.

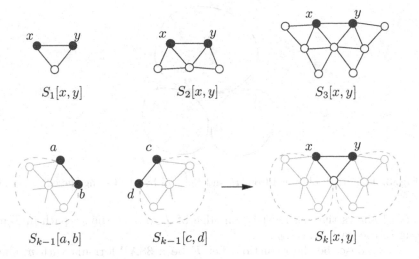

Fig. 2. The graph $S_k[x,y]$ obtained from $S_{k-1}[a,b]$ and $S_{k-1}[c,d]$ by identifying $b=d$ and joining $x=a$ with $y=c$

Equivalently, $S_{k+1}[x,y]$ is obtained from two disjoint $S_k[a,b]$ and $S_k[c,d]$ by identifying the two vertices c,d to a vertex z and joining the vertices $x:=a$ and $y:=b$ by an edge. With this notation, z is the common neighbor of x and y in $S_{k+1}[x,y]$, and we call $S_k[x,z]$ and $S_k[y,z]$ the two *corresponding* S_k in $S_{k+1}[x,y]$.

We denote by $S_k[x,y)$ the graph $S_k[x,y]-y$, that is, the graph obtained from $S_k[x,y]$ by deleting the vertex y. The following observations collect basic facts on S_k used in the reduction later.

Observation 1

(1) *For every* $v \in S_k[x,y]$, $d_{S_k[x,y]}(v,\{x,y\}) \le \lceil \frac{k}{2} \rceil$,
(2) $S_k[x,y]$ *has a tree* $(k+1)$-*spanner containing the edge* xy,
(3) $S_k[x,y)$ *has a tree* k-*spanner* T *such that, for each neighbor* y' *of* y *in* $S_k[x,y)$, $d_T(x,y') \le k$.

Proofs of these and all other results are omitted in this extended abstract. They will be given in the journal version.

Observation 2 *Let H be an arbitrary graph and let e be an arbitrary edge of H. Let K be an $S_k[x,y]$ disjoint from H. Let G be the graph obtained from H and K by identifying the edges e and xy; see Figure 3. Suppose that T is a tree t-spanner in G, $t > k$, such that the xy-path in T belongs to H. Then*

(1) $d_T(x,y) \le t-k$, *and*
(2) *there exists an edge* $uv \in K$ *with* $d_T(u,v) \ge d_T(x,y)+k$.

Part (1) of Observation 2 indicates a way to force an edge xy to be a tree edge, or to force a path of the tree to belong to certain part of the graph: Choosing

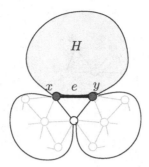

Fig. 3. The graph obtained from H and $S_k[x,y]$ by identifying the edge $e = xy$

$k = t - 1$ shows that xy must be an edge of T, or else the xy-path in T must belong to the part $S_{t-1}[x,y]$.

We now describe the reduction. Let F be a 3SAT formula with m clauses $C_j = (c_{j1}, c_{j2}, c_{j3})$ over n variables v_i. Set $\ell := \lfloor \frac{t}{2} \rfloor - 2$ and $\lambda := \lceil \frac{\ell}{2} \rceil$. Since $t \geq 4$, $\ell \geq 0$ and $\lambda \geq 0$.

For each variable v_i create the graph $G(v_i)$ as follows.

- Set $q_i^0 := v_i$, $q_i^{\ell+1} := \overline{v_i}$. We will use $u \in \{q_i^0, q_i^{\ell+1}\}$ as a vertex in our graph as well as a literal in the given 3SAT formula F.
- Take a clique Q_i on $\ell + 2$ vertices $q_i^0, \dots, q_i^\ell, q_i^{\ell+1}$.
- For each edge $xy \in \{q_i^k q_i^{k+1} : 0 \leq k \leq \ell\}$ create an $S_{t-1}[x,y]$. No two of the S_{t-1} have a vertex in common unless those in $\{x,y\}$.
- Take a chordless path on vertices $s_i^0, s_i^1, \dots, s_i^\lambda$ and edges $s_i^k s_i^{k+1}$, $0 \leq k < \lambda$.
- Connect each s_i^k, $0 \leq k \leq \lambda$, to exactly q_i^0 and $q_i^{\ell+1}$.
- For each edges $xy \in \{s_i^k s_i^{k+1} : 0 \leq k < \lambda\}$ create an $S_{t-2}[x,y]$.
- For each edges $xy \in \{s_i^0 q_i^0, s_i^0 q_i^{\ell+1}, s_i^\lambda q_i^0, s_i^\lambda q_i^{\ell+1}\}$ create an $S_{t-(\ell+2)}[x,y]$.

Note that the clique Q_i is a cutset of $G(v_i)$ and the components of $G(v_i) - Q_i$ are chordal. Thus, $G(v_i)$ is a chordal graph. See also Figure 4.

For each clause C_j create the graph $G(C_j)$ as follows. If t is even, $G(C_j)$ is simply a single vertex a_j. If t is odd, $G(C_j)$ is the graph $S_{t-1}[a_j^1, a_j^2]$. In any case, $G(C_j)$ is a chordal graph.

Finally, the graph $G = G(F)$ is obtained from all $G(v_i)$ and $G(C_j)$ by identifying all vertices s_i^0 to a single vertex s, and adding the following additional edges:

- connect every vertex in Q_i with every vertex in $Q_{i'}$, $i \neq i'$. Thus, the cliques Q_i, $1 \leq i \leq n$, form together a clique Q in G,
- for each literal $u_i \in \{q_i^0, q_i^{\ell+1}\}$, if $u_i \in C_j$ then connect u_i with a_j, respectively, with a_j^1 and a_j^2, according to the parity of t.

The description of the graph $G = G(F)$ is complete. Clearly, G can be constructed in polynomial time.

Lemma 1 G is chordal, and $\operatorname{diam}(G) \leq \begin{cases} t+1 \text{ if } t \text{ is even,} \\ t+2 \text{ if } t \text{ is odd.} \end{cases}$

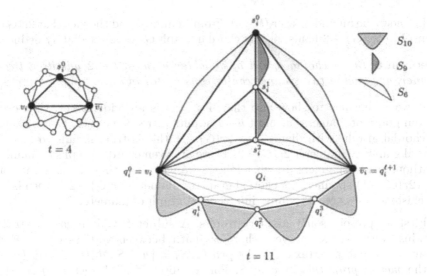

Fig. 4. The graph $G(v_i)$

Lemma 2 *Suppose G admits a tree t-spanner. Then F is satisfiable.*

Lemma 3 *Suppose F is satisfiable. Then G admits a tree t-spanner.*

Theorem 1 follows from Lemmas 1-3. We remark that the chordal graph G constructed above always admits a tree $(t+1)$-spanner.

3   Tree Spanners in Chordal Graphs of "Smaller" Diameter

It is known [8,9] that for a chordal graph G, $\mathrm{diam}(G) \geq 2r(G) - 2$ holds. This already yields the following.

Theorem 5 *Let $t \geq 2$ be an even integer. Every chordal graph of diameter at most $t-1$ admits a tree t-spanner, and such a tree spanner can be constructed in linear time.*

We remark that there are chordal graphs of diameter t without tree t-spanner. Thus, Theorem 5 is best possible under diameter constraints.

Corollary 1 *Every chordal graph of diameter at most 3 has a tree 4-spanner, and such a tree spanner can be constructed in linear time.*

It remains an interesting open question whether existence of a tree 3-spanner in a given chordal graph of diameter at most 3 can be tested in polynomial time.

Lemma 4 *Every chordal graph G admits a tree $(2r(G))$-spanner, and such a tree spanner can be constructed in linear time.*

Let now t be an odd integer ($t \geq 3$). From Lemma 4 and the fact that $2r(G) \geq \text{diam}(G) \geq 2r(G) - 2$ holds for any chordal graph G, we immediately deduce.

Theorem 6 *Every chordal graph of diameter at most $t - 2$ admits a tree t-spanner, and such a tree spanner can be constructed in linear time.*

It would be nice to show also that, if $t \geq 3$ is an odd integer, then every chordal graph of diameter at most $t - 1$ admits a tree t-spanner. But, although for chordal graphs with $\text{diam}(G) \geq 2r(G) - 1$ this is true, it fails to hold for chordal graphs of diameter $2r(G) - 2$. There are even chordal graphs of diameter 2 without tree 3-spanners. In what follows we will show that the existence of a tree $(2r(G) - 1)$-spanner in a chordal graph of diameter $2r(G) - 2$ "depends" on the existence of a tree 3-spanner in a chordal graph of diameter 2.

First we present some auxiliary results. A subset $S \subseteq V$ is *m–convex* if S contains every vertex on every chordless path between vertices of S. For a subset $S \subseteq V$ and a vertex $v \in V$, let $proj(v, S) = \{u \in S : d_G(v, u) = d_G(v, S)\}$ be the *metric projection* of v to S. For a subset $X \subseteq V$, let $proj(X, S) = \bigcup_{v \in X} proj(v, S)$. A subset $A \subseteq V$ is a *two-set* in G if $d_G(v, u) \leq 2$ holds for every $v, u \in A$.

Lemma 5 *Let G be a (not necessarily chordal) graph. The metric projection $proj(A, S)$ of any two-set A to an m–convex set S of G is a two-set.*

Lemma 6 *In every chordal graph $G = (V, E)$ of diameter $2r(G) - 2$ there exists a two-set S such that $d_G(v, S) \leq r(G) - 2$ for every $v \in V$. Moreover such a two-set can be determined within time $O(|V|^3)$.*

Lemma 7 *Every maximal by inclusion two-set of a chordal graph is m-convex.*

Theorem 7 *Chordal graphs of diameter $2r(G) - 2$ admit tree $(2r(G) - 1)$-spanners if and only if chordal graphs of diameter 2 admit tree 3-spanners. Moreover, if a tree 3-spanner of any chordal graph of diameter 2 can be found in polynomial time, then a tree $(2r(G) - 1)$-spanner of a chordal graph of diameter $2r(G) - 2$ can be found in polynomial time, too.*

We do not know how to use this theorem for general chordal graphs (since not all chordal graphs of diameter 2 have tree 3-spanners), but this theorem could be very useful for those hereditary subclasses of chordal graphs where each graph of diameter 2 is tree 3-spanner admissible. Then, for every graph of diameter at most $t - 1$ from those classes, a tree t-spanner will exist and it could be found in polynomial time if corresponding tree 3-spanner is constructable in polynomial time. For an arbitrary chordal graph G with $\text{diam}(G) = 2r(G) - 2$, it can happen that a chordal graph of diameter at most 2, generated by a two-set of G (found as described in Lemma 6 and Theorem 7), does not have a tree 3-spanner, but yet G itself admits a $(2r(G) - 1)$-spanner. We are still working on TREE $(2r(G) - 1)$-SPANNER problem in chordal graphs of diameter $2r(G) - 2$. It is natural to ask whether a combination of Theorem 7 and Theorem 9 works.

4   Tree 3-Spanners in Chordal Graphs of Diameter 2

In this section, we give an application of Theorem 7 as well as a criterion for the tree 3-spanner admissibility of chordal graphs of diameter at most 2.

A graph G is *non-trivial* if it has at least one edge.

Lemma 8 *Let G be a non-trivial chordal graph of diameter at most 2. If G does not contain a clique K_5 on five vertices, then G has a dominating edge, i.e., an edge $e \in E$ such that $d_G(v, e) \leq 1$ for any $v \in V$.*

Since neither planar graphs nor 3-trees have cliques on 5 vertices and any graph with a dominating edge is trivially tree 3-spanner admissible, we conclude.

Corollary 2 *Let G be a non-trivial graph of diameter at most 2. If G is a planar chordal graph or a k-tree for $k \leq 3$, then G has a dominating edge and hence a tree 3-spanner.*

As we mentioned in introduction, there is no constant t such that planar chordal graphs or k-trees ($k \geq 2$) are tree t-spanner admissible. So, it is interesting to mention the following result.

Theorem 8 *Every chordal graph of diameter at most $t - 1$, if it is planar or a k-tree ($k \leq 3$), has a tree t-spanner and such a tree spanner can be constructed in polynomial time.*

In what follows we will assume that G is an arbitrary chordal graph which admits a tree 3-spanner T. Note that any tree of diameter at most 2 is a star and any tree of diameter 3 has a dominating edge (in this case T is called a *bistar*).

Lemma 9 *For any maximal (by inclusion) clique C of G one of the following conditions holds.*

a) *C induces a star in T,*
b) *either C induces a bistar in T or there is a vertex $v \notin C$ such that $C \cup \{v\}$ induces a bistar in T.*

Clearly, T is a star only if G has an universal vertex, and the diameter of T is 3 only if G has a dominating edge. The following theorem handles the case of all chordal graphs of diameter at most 2. Unfortunately, not every such graph has a dominating edge. There are chordal graphs of diameter 2 which do not have any tree 3-spanners, and there are chordal graphs of diameter 2 that have a tree 3-spanner but all those spanners are of diameter 4. Theorem 4 will follow from this theorem.

Theorem 9 *A chordal graph G of diameter at most 2 admits a tree 3-spanner if and only if there is a vertex v in G such that any connected component of the second neighborhood of v has a dominating vertex in $N(v)$.*

5   Tree Spanners in Strongly Chordal Graphs and k-Split Graphs

For an integer $k \geq 3$, a k-*sun* consists of a k-clique $\{v_1, \ldots, v_k\}$ and a k-vertex stable set $\{u_1, \ldots, u_k\}$, and edges $u_i v_i, u_i v_{i+1}$, $1 \leq i < k$, and $u_k v_k, u_k v_1$. A chordal graph is *strongly chordal* [11] if it does not contain a k-sun as an induced subgraph. In [3], it is proved that every strongly chordal graph admits a tree 4-spanner and such a tree spanner can be constructed in linear time. Not every strongly chordal graph has a tree 3-spanner. Actually, TREE 3-SPANNER remains open on strongly chordal graphs.

A k-*planet* is obtained from a k-path $v_1 v_2 v_3 \cdots v_k$ and a triangle abc by adding edges av_i, $1 \leq i \leq k-1$ and bv_i, $2 \leq i \leq k$; see Figure 5.

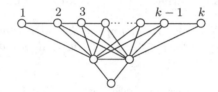

Fig. 5. A k-planet

Definition 1 *A chordal graph is called* very strongly chordal *if it does not contain a k-planet as an induced subgraph.*

As a 3-sun is a 3-planet and every k-sun ($k \geq 4$) contains an induced 4-planet, the class of very strongly chordal graphs is properly contained in the class of strongly chordal graphs. Moreover, the class of very strongly chordal graphs contains all interval graphs and all distance hereditary chordal graphs, called *ptolemaic graphs* [15]. The nice feature of this subclass of strongly chordal graphs is

Theorem 10 *Every very strongly chordal graph admits a tree 3-spanner and such a tree spanner can be constructed in linear time.*

Another well-known subclass of strongly chordal graphs consists of the intersection graphs of directed paths in a rooted directed tree, called *directed path graphs*. The class of directed path graphs generalizes interval graphs naturally, and contains all ptolemaic graphs, and is tree 3-spanner admissible [17].

The intersection graphs of paths in a tree are called (undirected) *path graph*. We call shortly a graph *strongly path graph* if it is strongly chordal as well as a path graph. Clearly, every directed path graph is a strongly path graph, but not vice versa. Indeed, there are many strongly path graphs having no tree 3-spanner (while every directed path graph does [17]). Moreover, in contrast to strongly chordal graphs, for every t, there is a path graph having no tree t-spanner [16].

A *split graph* is one whose vertex set can be partitioned into a clique and a stable set. Split graphs are exactly those chordal graphs whose complements are chordal as well. It is known (and easy to see; cf. [5,16,25]) that every split graph admits a tree 3-spanner. We are going to describe a new subclass of chordal graphs containing all split graphs and still are tree 3-spanner admissible.

First, for an arbitrary graph G let $S(G)$ be the set of all simplicial vertices of G. We also use $S(G)$ for the subgraph of G induced by $S(G)$.

Lemma 10 *If $G \setminus S(G)$ has a tree $(t-1)$-spanner then G has a tree t-spanner.*

Definition 2 *For an arbitrary graph G and an integer $k \geq 0$ let $G_k := G_{k-1} \setminus S(G_{k-1})$; $G_0 := G$. A graph G is called k-split if G_k is a clique.*

Clearly, 0-split graphs are exactly the cliques, and all split graphs are 1-split but not vice versa. The following fact is probably known.

Proposition 2 *A graph G is chordal if and only if G is k-split for some k.*

Theorem 11 *Every k-split graph admits a tree $(k+2)$-spanner.*

Corollary 3 *All 1-split graphs, hence all split graphs, admit a tree 3-spanner, and a such a tree 3-spanner can be constructed in linear time, given the set of all simplicial vertices.*

Note that the existence of a tree $(k+2)$-spanner in k-split graphs is best possible: there are many k-split graphs without tree $(k+1)$-spanner; for example, the 3-sun is 1-split (even split) and has no tree 2-spanner.

6 Conclusion

In this paper we have proved that, for any $t \geq 4$, TREE t-SPANNER is **NP**-complete on chordal graphs of diameter at most $t+1$ (if t is even), respectively, at most $t+2$ (if t is odd), improving the hardness result in [6] on a restricted well-understood graph class. We have shown that every chordal graph G of diameter at most $t-1$ is tree t-spanner admissible if $\operatorname{diam}(G) \neq 2r(G) - 2$.

The complexity of TREE t-SPANNER remains unresolved on chordal graphs of diameter t (if t is even) and of diameter t or $t+1$ (if t is odd). TREE t-SPANNER remains also open on path graphs and the case $t = 3$ remains even open on path graphs that are strongly chordal graphs as well. However, we have shown that all very strongly chordal graphs, a subclass of strongly chordal graphs that contains all interval graphs and all ptolemaic graphs, are tree 3-spanner admissible, and a tree 3-spanner for a given very strongly chordal graph can be constructed in linear time. This improves known results on tree 3-spanners in interval graphs [16,18,23]. We have also improved known results on tree 3-spanners in split graphs [5,16,25] by showing that all 1-split graphs, a subclass of chordal graphs containing all split graphs, are tree 3-spanner admissible, and a tree 3-spanner for a 1-split graph can be constructed in linear time, given the set of its simplicial vertices. We presented a polynomial time algorithm for the TREE 3-SPANNER problem on chordal graphs of diameter at most 2.

Many questions remain still open. Among them:

(1) Can TREE 3-SPANNER be decided efficiently on (strongly) chordal graphs?
(2) Can TREE $(2r(G)-1)$-SPANNER be decided efficiently on chordal graphs of diameter $2r(G) - 2$?
(3) What is the complexity of TREE t-SPANNER for chordal graphs of diameter at most t?

References

1. B. Awerbuch, A. Baratz, D. Peleg, Efficient broadcast and light-weighted spanners, *manuscript*, 1992
2. H.-J. Bandelt, A. Dress, Reconstructing the shape of a tree from observed dissimilarity data, *Adv. Appl. Math.* 7 (1986) 309-343
3. A. Brandstädt, V. Chepoi, F. Dragan, Distance approximating trees for chordal and dually chordal graphs, *J. Algorithms* 30 (1999) 166-184
4. A. Brandstädt, V.B. Le, J. Spinrad, Graph Classes: A Survey, *SIAM Monographs on Discrete Math. Appl.*, (SIAM, Philadelphia, 1999)
5. L. Cai, Tree spanners: Spanning trees that approximate the distances, *Ph.D. thesis*, University of Toronto, 1992
6. L. Cai, D.G. Corneil, Tree spanners, *SIAM J. Discrete. Math.* 8 (1995) 359-387
7. L. Cai, D.G. Corneil, Tree spanners: An overview, *Congressus Numer.* 88 (1992) 65-76
8. G.J. Chang, G.L. Nemhauser, The k-domination and k-stability problems on sunfree chordal graphs, *SIAM. J. Alg. Disc. Meth.* 5 (1984) 332-345
9. V.D. Chepoi, Centers of triangulated graphs, *Math. Notes* 43 (1988) 82-86
10. V.D. Chepoi, F.F. Dragan, Linear-time algorithm for finding a center vertex of a chordal graph, *Lecture Notes in Computer Science* 855 (1994) 159-170
11. M. Farber, Characterizations of strongly chordal graphs, *Discrete Math.* 43 (1983) 173-189
12. S.P. Fekete, J. Kremer, Tree spanners in planar graphs, *Discrete Appl. Math.* 108 (2001) 85-103
13. F. Gavril, The intersection graphs of subtrees in trees are exatly the chordal graphs, *J. Combin. Theory* (B) 16 (1974) 47-56
14. M.C. Golumbic, Algorithmic Graph Theory and Perfect Graphs (Academic Press, New York, 1980)
15. E. Howorka, A characterization of ptolemaic graphs, *J. Graph Theory* 5 (1981) 323-331
16. Hoàng-Oanh Le, Effiziente Algorithmen für Baumspanner in chordalen Graphen, *Diploma thesis*, Dept. of mathematics, technical university of Berlin, 1994
17. H.-O Le, V.B. Le, Optimal tree 3-spanners in directed path graphs, *Networks* 34 (1999) 81-87
18. M.S. Madanlal, G. Venkatesan, C. Pandu Rangan, Tree 3-spanners on interval, permutation and regular bipartite graphs, *Inform. Process. Lett.* 59 (1996) 97-102
19. I.E. Papoutsakis, Two structure theorems on tree spanners, *M.Sc. thesis*, Dept. of Computer Science, University of Toronto, 1999
20. I.E. Papoutsakis, On the union of two tree spanners of a graph, *Preprint*, 2001
21. D. Peleg, Distributed Computing: A Locality-Sensitive Approach, *SIAM Monographs on Discrete Math. Appl.*, (SIAM, Philadelphia, 2000)
22. D. Peleg, A. Schaeffer, Graph spanners, J. Graph Theory 13 (1989) 99-116
23. E. Prisner, Distance approximating spanning trees, in: Proc. STACS'97, *Lecture Notes in Computer Science, Vol.* 1200 (Springer, Berlin, 1997) 499-510
24. J. Soares, Graph spanners: A survey, *Congressus Numer.* 89 (1992) 225-238
25. G. Venkatesan, U. Rotics, M.S. Madanlal, J.A. Makowsky, C. Pandu Ragan, Restrictions of minimum spanner problems, *Information and Computation* 136 (1997) 143-164

An Asymptotic Fully Polynomial Time Approximation Scheme for Bin Covering*

Klaus Jansen[1] and Roberto Solis-Oba[2]

[1] Institut für Informatik und Praktische Mathematik,
Universität zu Kiel,
Christian Albrechts Platz 4, 24098 Kiel, Germany,
kj@informatik.uni-kiel.de
[2] Department of Computer Science,
University of Western Ontario,
London, Ontario, N6A 5B7, Canada,
solis@cs.uwo.ca

Abstract. In the bin covering problem, given a list $L = (a_1, \ldots, a_n)$ of items with sizes $s_L(a_i) \in (0, 1)$, the goal is to find a packing of the items into bins such that the number of bins that receive items of total size at least 1 is maximized. This is a dual problem to the classical bin packing problem. In this paper we present the first asymptotic fully polynomial-time approximation scheme (AFPTAS) for the bin covering problem.

1 Introduction

Given a list $L = (a_1, \ldots, a_n)$ of items with sizes $s_L(a_i) \in (0, 1)$, the bin covering problem is to find a packing of the items into bins such that the number of bins that receive items of total size at least 1 is maximized. This problem is considered to be a kind of dual to the classical bin packing problem, and sometimes it is called the dual bin packing problem.

The bin covering problem is NP-hard [2], and, furthermore, it is hard to approximate within a factor $\alpha > \frac{1}{2}$ of the optimum. To see this, observe that an α-approximation algorithm, $\alpha > \frac{1}{2}$, for the problem applied to instances in which the total size of the items in L is 2, would solve the partition problem. On the positive side the next-fit algorithm, that places elements in a bin until their size is at least 1 before proceeding to fill the next bin, achieves a performance ratio of $\frac{1}{2}$ [1]. A similar situation holds for the on-line version of the bin covering problem since it is known [5] that every on-line algorithm A can fill at most half of the optimum number of bins, and the next-fit heuristic fills at least half of the optimum number of bins [1].

* Supported in part by EU Project APPOL I + II, Approximation and Online Algorithms, IST-1999-14084, and IST-2001-30012, by the EU Research Training Network ARACNE, Approximation and Randomized Algorithms in Communication Networks, HPRN-CT-1999-00112, and by the Natural Sciences and Engineering Research Council of Canada grant R3050A01.

P. Bose and P. Morin (Eds.): ISAAC 2002, LNCS 2518, pp. 175–186, 2002.
© Springer-Verlag Berlin Heidelberg 2002

Interestingly, if the number of bins filled by an optimum solution is large, then algorithms with performance ratio better than $1/2$ can be designed. Given a list L of items and an algorithm A for the bin covering problem, let $A(L)$ be the number of bins that A can fill with items of total size at least 1. Let $OPT(L)$ be the numbers of bins covered by an optimal algorithm. In this paper we are interested in studying the asymptotic worst case ratio of algorithm A, defined as

$$R_\infty^A = \liminf_{OPT(L)\to\infty} \frac{A(L)}{OPT(L)}.$$

Algorithms with asymptotic worst case ratios $\frac{2}{3}$ and $\frac{3}{4}$ are given in [1,2,3], Recently, an algorithm with asymptotic ratio $1 - \varepsilon$, for any constant value $\varepsilon > 0$ was designed by Csirik, Johnson, and Kenyon [4]. This algorithm runs in time polynomial in n but exponential in $1/\varepsilon$.

In this paper we present an asymptotic fully polynomial time approximation scheme (AFPTAS) for the bin covering problem, i.e., an algorithm with asymptotic worst case ratio $1 - \varepsilon$, for any fixed $\varepsilon > 0$, and running time polynomial in both n and $1/\varepsilon$.

Theorem 1. *There is an algorithm A that, given a list L of n items with sizes $s(a_i) \in (0,1)$ and a positive number $\epsilon > 0$, produces a bin covering of L such that*

$$A(L) \geq (1 - \epsilon)OPT(L) - O(1/\varepsilon^3).$$

The time complexity of A is polynomial in n and $1/\epsilon$.

This algorithm narrows the gap between the approximability of the bin packing problem and that of the bin covering problem [4,11], and it solves the question first posed in [1] of whether an AFPTAS for the bin covering problem exists.

Our algorithm is a fast implementation of the algorithm in [4] that does not have the exponential dependency of the time complexity on $1/\varepsilon$. The algorithm in [4] solves approximately the bin covering problem by first formulating it as an integer program, and then by rounding a basic optimum solution of a linear program relaxation of the integer program.

We get a faster algorithm by using the potential price directive decomposition method [7,8,10,14] to find a near optimum solution for the above linear program. The application of this method to our problem requires the efficient solution of two variants of the knapsack problem. Most versions of the knapsack problem studied in the literature [9,12,13] can be approximated arbitrarily close to the optimum in polynomial time. Interestingly, one of our variants is hard to approximate within a factor $1 + \varepsilon$ of the optimum for small values $\varepsilon > 0$. Despite this, we can still find an approximate solution for the linear program in time polynomial in n and $1/\varepsilon$, since we prove that the optimum solution of the combined knapsack problems can be efficiently approximated. We present an efficient algorithm for transforming the near optimum solution for the linear program into a basic feasible solution, which is then rounded to yield an asymptotic fully polynomial time approximation scheme for the bin covering problem.

2 An Asymptotic Polynomial Time Approximation Scheme

In this section we briefly review the asymptotic polynomial time approximation scheme for the bin covering problem described in [4]. In the next section we show how to design an asymptotic fully polynomial time approximation scheme for the problem. For any set $T \subseteq L$, let $s_L(T) = \sum_{a_i \in T} s_L(a_i)$ be the sum of item sizes in list T. Clearly, $s_L(L) \geq OPT(L)$. Without loss of generality we assume that $s_L(L) \geq 2$, $\varepsilon < 1/2$, and $1/\varepsilon$ is integer.

Partition the set L of items into 3 classes, $\mathcal{L}, \mathcal{M}$, and $\mathcal{S}$, respectively formed by *large, medium,* and *small* items. These sets are defined as follows: (a) if $n < \lfloor s_L(L) \rfloor (1 + 1/\varepsilon)$ then $\mathcal{L} = L$, $\mathcal{M} = \mathcal{S} = \emptyset$; (b) otherwise, $\mathcal{L}$ consists of the largest $\lfloor s_L(L) \rfloor / \varepsilon$ items in L, $\mathcal{M}$ contains the next largest $\lfloor s_L(L) \rfloor$ items, and $\mathcal{S}$ includes the remaining elements.

We need to consider two cases: $s_L(L) > 13/\varepsilon^3$ and $s_L(L) \leq 13/\varepsilon^3$.

1. **Case 1:** $s_L(L) > 13/\varepsilon^3$. In this case the sizes of large items are rounded as follows. First, sort the large items non-increasingly by size. Then, partition the items into $1/\varepsilon^2$ groups $G_1, G_2, \ldots, G_{1/\varepsilon^2}$ such that each group G_i, $1 \leq i \leq |\mathcal{L}| \bmod (1/\varepsilon^2)$ has $\lceil |\mathcal{L}|\varepsilon^2 \rceil$ items, and each one of the remaining groups has $\lfloor |\mathcal{L}|\varepsilon^2 \rfloor$ items. These groups are formed by placing the largest size items in G_1, the next largest size items in G_2, and so on. For each group G_i, round the sizes of the items down to the size of the smallest item in the group.

   Let H be the set of different sizes after rounding. Note that $|H| \leq 1/\varepsilon^2$. For an item size $u \in H$, let $n(u)$ be the number of large items of rounded size u. Let $s(a_i)$ be the rounded size of item a_i. For small and medium items a_i, let $s(a_i) = s_L(a_i)$. Given a set of items $T \subseteq L$, let $s(T)$ be the sum of the rounded sizes of the items in T.

   We define a *bin configuration* C as a set of large items of total rounded size smaller than 2. For a bin configuration C and size $u \in H$, let $n(u, C)$ be the number of large items of size u in C. Let ζ be the set of all possible bin configurations. The bin covering problem can be expressed as an integer program by using a variable x_C for each bin configuration C that indicates the number of bins in the optimum solution that contain items according to configuration C. If in this integer program we relax the integrality constraint on the variables x_C and we assume that each small item can be split into smaller pieces, then we get the following linear program.

$$\text{Max} \sum_{C \in \zeta} x_C \tag{1}$$

$$\text{s.t.} \sum_{C \in \zeta} n(v, C)x_C \leq n(v), \quad \forall v \in H,$$

$$\sum_{\substack{C \in \zeta \\ s(C) < 1}} (1 - s(C))x_C \leq s(\mathcal{S}),$$

$$x_C \geq 0, \quad \forall C \in \zeta.$$

We note that the first constraint implies that not all large items might be placed in the bins. Solve this linear program, and let $(x_C^*)_{C \in \zeta}$ be a basic optimum solution. Round each value x_C^* down to the nearest integer value $y_C^* = \lfloor x_C^* \rfloor$. The large items are placed in a set D of bins according to the rounded solution $(y_C^*)_{C \in \zeta}$.

After the large items are placed in a set D of bins of size $|D| = \sum_{C \in \zeta} y_C^*$, the medium and small items are greedily placed into those bins in D that contain items with total size smaller than 1: medium and small items are placed into an under-filled bin until the total size of the items in the bin is at least 1.

Note that this process must fill all bins in D with items of total size at least 1. This is because $|D| \leq \lfloor s(L) \rfloor$ and so the total excess that the greedy placement of medium and small items assigns to (previously) under-filled bins is at most $s(\mathcal{M})$. This fact and the second condition of linear program (1) ensure that all bins in D are filled to size at least 1.

The above algorithm needs to find a basic optimum solution for linear program (1), and this requires time exponential in $1/\varepsilon$ since the linear program has $O(n^{1/\varepsilon^2})$ variables. In [4] it is proved that the algorithm fills at least $OPT(L)(1 - 5\varepsilon) - 4$ bins, and thus, it is an asymptotic polynomial time approximation scheme for this case.

2. **Case 2:** $s_L(L) \leq 13/\varepsilon^3$. In this case, $\mathcal{L}$ contains at most $13(1 + 1/\varepsilon)/\varepsilon^3 < 20/\varepsilon^4$ items. Since there is a constant number of large items, exhaustive search can be used to place them into the maximum number of bins such that the total unused space in the under-filled bins is at most $s_L(\mathcal{S})$. Then, the medium and small items are placed as before. This process also yields a solution of value at least $OPT(L)(1 - 5\varepsilon) - 4$.

3   Approximate Solution of the Linear Program

In this section we describe an algorithm for approximately solving linear program (1) in time polynomial in n and $1/\varepsilon$. As we show, this is enough to derive an asymptotic fully polynomial time approximation scheme for the bin covering problem. First, we modify the above definition of a bin configuration so as to simplify the description of our algorithm. If the set of small items is empty, then we define a bin configurations as a set of large items of total rounded size smaller than 2, but at least 1. Hence, if $\mathcal{S} = \emptyset$ the second constraint of linear program (1) disappears. If $\mathcal{S} \neq \emptyset$, then we use the same definition of bin configuration as above.

As before, let us assume that $\varepsilon < 1/2$. Consider first the case when $s_L(L) \leq 13/\epsilon^3$. Since $OPT(L) \leq 13/\varepsilon^3$ is a constant, this case does not affect the asymptotic performance of an algorithm for the bin covering problem. We use the next-fit algorithm for this case, which fills at least $OPT(L) - O(1/\epsilon^3)$ bins. For the rest of the paper we concentrate on the case when $s_L(L) > 13/\epsilon^3$ is not constant.

Let X^* be the value of an optimum solution $(x_C^*)_{C \in \zeta}$ for linear program (1). Since $s_L(L) > 13/\epsilon^3$ and $\varepsilon < 1/2$ then $X^* > 1$. Also, every item has size at most 1, and so at most n bins can be filled by the optimum solution $(x_C^*)_{C \in \zeta}$; therefore, $1 < X^* \le s(L) \le n$.

We can use the potential price directive decomposition method [8,10] to find a near optimal solution for linear program (1), but first, we need to re-write the linear program in a special form. This special form requires that we know a $(1 - \varepsilon)$-approximation of the value X^*. To find such an approximation, we first divide the interval $[1, n]$ into sub-intervals of size ε. We, then, use binary search over these intervals and the procedure described below to find the subinterval $(k^*\varepsilon, (k^* + 1)\varepsilon]$ that contains X^*. This value $k^*\varepsilon$ is the desired approximation to X^* since $(1 - \varepsilon)X^* \le k^*\varepsilon \le X^*$. This approach for finding the approximate value of X^* requires that we test at most $O(\ln(n/\varepsilon))$ possible values of the form $k\varepsilon$.

Given a value $k\varepsilon$, as described above, we re-write the linear program as follows.

$$\lambda^* = \min\{\lambda \mid \sum_{C \in \zeta} \frac{n(v,C)}{n(v)} x_C \le \lambda, \quad \forall v \in H, (x_C)_{C \in \zeta} \in B_k \text{ and}$$
$$\sum_{C \in \zeta \text{ s.t. } s(C)<1} \frac{1-s(C)}{s(S)} x_C \le \lambda, \quad \forall (x_C)_{C \in \zeta} \in B_k\}, \tag{2}$$

where

$$B_k = \{(x_C)_{C \in \zeta} \mid \sum_{C \in \zeta} x_C = k\varepsilon, \text{ and } x_C \ge 0 \text{ for all } C \in \zeta\}.$$

Remember that if $S = \emptyset$ then the second constraint of linear program (2) disappears; we should keep this in mind for the sequel. A feasible solution of value $\lambda^* = 1$ for this new linear program (2) is a solution for the original linear program (1) of value $k\varepsilon$.

Linear program (2) is a *convex block-angular resource sharing problem* [8], and the price directive decomposition method [8,10] can be used to solve it to any given precision $\rho > 0$. In the next section we show how to use this algorithm to either (a) find a solution $(x_C)_{C \in \zeta} \in B_k$ with value $\sum_{C \in \zeta} \frac{n(v,C)}{n(v)} x_C \le (1 + \rho)$ for every $v \in H$ and such that $\sum_{C \in \zeta \text{ s.t. } s(C)<1} \frac{(1-s(C))}{s(S)} x_C \le (1 + \rho)$, or (b) to conclude that there is no solution with value $\sum_{C \in \zeta} \frac{n(v,C)}{n(v)} x_C \le 1$ for all $v \in H$ and such that $\sum_{C \in \zeta \text{ s.t. } s(C)<1} \frac{(1-s(C))}{s(S)} x_C \le 1$. In the second case we know that we have selected a value $k\varepsilon > k^*\varepsilon$ and so in the binary search we choose next a smaller value for k. In the first case we select a larger value for k. At the end we find a solution $(x'_C)_{C \in \zeta}$ with value $\sum_{C \in \zeta} x'_C = k^*\varepsilon \ge (1 - \epsilon)X^*$, and such that

$$\sum_{C \in \zeta} \frac{n(v,C)}{n(v)} x'_C \le (1 + \rho), \quad \forall v \in H, \text{ and}$$
$$\sum_{C \in \zeta \text{ s.t. } s(C)<1} \frac{1-s(C)}{s(S)} x_C \le (1 + \rho).$$

To transform $(x'_C)_{C \in \zeta}$ into a feasible solution $(\tilde{x}_C)_{C \in \zeta}$ for linear program (1) we set $(\tilde{x}_C)_{C \in \zeta} = (x'_C(1 - \rho))_{C \in \zeta}$, so that

$$\sum_{C \in \zeta} \frac{n(v,C)}{n(v)} \tilde{x}_C \le (1 + \rho)(1 - \rho) = 1 - \rho^2 \le 1, \forall v \in H \text{ and}$$
$$\sum_{C \in \zeta \text{ s.t. } s(C)<1} \frac{1-s(C)}{s(S)} \tilde{x}_C \le (1 + \rho)(1 - \rho) = 1 - \rho^2 \le 1.$$

Furthermore, the value of the new solution $(\tilde{x}_C)_{C \in \zeta}$ is

$$\sum_{C \in \zeta} \tilde{x}_C \geq (1 - \rho)(1 - \epsilon)X^* \geq (1 - 2\epsilon)X^*, \qquad (3)$$

if we choose for ρ any value no larger than $\varepsilon/(1 - \varepsilon)$. Selecting $\rho = \varepsilon$ satisfies the condition and implies that $\sum_{C \in \zeta} \tilde{x}_C \geq (1 - 2\epsilon)X^*$.

3.1   Solving the Convex Block-Angular Problem

For convenience we re-write problem (2) in compact matrix form as

$$\lambda^* = \min\{\lambda \mid A(x_C)_{C \in \zeta} \leq \lambda e, (x_C)_{C \in \zeta} \in B_k\}, \qquad (4)$$

where e is the $|H| + 1$ dimensional unit vector and A is the constraint matrix corresponding to linear program (2). The price directive decomposition method [8,10] finds an approximate solution of problem (4) by starting with an arbitrary vector $(x'_C)_{C \in \zeta} \in B_k$ and iteratively adjusting this vector until a solution with the desired precision is found. To get a solution of value at most $(1 + \varepsilon)\lambda^*$, this algorithm needs to perform $O(|H|(\varepsilon^{-2} + \ln |H|))$ iterations [10]. In each iteration the algorithm must solve the following *block problem*:

$$\min \{y^T A(x_C)_{C \in \zeta} | (x_C)_{C \in \zeta} \in B_k\}, \text{ where} \qquad (5)$$

$y = (y_1, y_2, \ldots, y_{|H|+1})$ is a non-negative *price vector*.

In the next section we show how to solve this block problem efficiently. The non-zero components of the solution $(x_C)_{C \in \zeta}$ of problem (5) determines a set of configurations. A linear combination of these configurations and the configurations in the current solution $(x'_C)_{C \in \zeta}$ define the new value for the vector $(x'_C)_{C \in \zeta}$ (for complete details see [8]).

We notice that the final solution $(x'_C)_{C \in \zeta}$ might be formed by up to

$$O(|H|(\frac{1}{\epsilon^2} + \ln |H|)) = O(\frac{1}{\epsilon^4})$$

configurations, since in each iteration we might add one configuration to the solution. But, observe that there are at most $1 + 1/\epsilon^2$ linear constraints in the linear program and, therefore, a basic feasible solution for it consists of at most $O(1/\varepsilon^2)$ configurations.

We can transform $(x'_C)_{C \in \zeta}$ into another feasible solution for (4) with the same value, but at most $1 + 1/\varepsilon^2$ positive components x'_C. To see this consider a set Q of $1/\epsilon^2 + 2$ configurations C with positive variables $x'_C > 0$ in the solution $(x'_C)_{C \in \zeta}$. Let M be the submatrix of A formed by the $1/\epsilon^2 + 2$ columns corresponding to those variables. Solve the system $M(w_C)_{C \in Q} = 0$, where $(w_C)_{C \in Q}$ is a non-zero $(2 + 1/\varepsilon^2)$-dimensional vector. There exists always a non-zero solution of this system since matrix M is singular.

Then modify the solution $(x'_C)_{C \in \zeta}$ by making $x'_C = x'_C + \delta w_C$ for all configurations $C \in Q$, where $|\delta|$ is the smallest value such that at least one of the new components $x'_C + \delta w_C$ is zero. Note that the new solution $(x'_C)_{C \in \zeta}$ has reduced the number of non-zero components by at least one, and furthermore, it is a feasible solution for (5) of the same value $(1 + \varepsilon)\lambda^*$ as before. We repeat this process until we find a solution $(x'_C)_{C \in \zeta}$ with at most $1 + 1/\varepsilon^2$ positive components.

3.2   The Block Problem

Consider block problem (5). Since B_k is a simplex, an optimum solution for (5) is attained at a vertex $(\bar{x}_C)_{C \in \zeta}$ of B_k corresponding to a single configuration C^*, i.e. $\bar{x}_{C^*} = k\varepsilon$ and $\bar{x}_C = 0$ for all other configurations $C \in \zeta$, $C \neq C^*$. Thus, to solve (5) it is sufficient to find a bin configuration C^* with smallest associated *price value* $y^T A[C^*]$, where $A[C^*]$ is the column of A corresponding to configuration C^*.

Depending on the total size of the items in a configuration C, the configuration's price $y^T A[C]$ can be either

- $\sum_{u \in C} n(s(u), C) y_{s(u)} / n(s(u))$ if $s(C) \geq 1$, or
- $\sum_{u \in C} n(s(u), C) y_{s(u)} / n(s(u)) + (1 - s(C)) y_{|H|+1} / s(S)$ if $s(C) < 1$.

Therefore, configuration C^* can be found by solving the following two integer programs. Variable z_v denotes the number of items of size v chosen for the solution. In the first integer program we optimize over all configurations with size smaller than 1.

$$\tau_1 = \min \sum_{v \in H} \frac{y_v}{n(v)} z_v + \frac{y_{|H|+1}}{s(S)} \left(1 - \sum_{v \in H} s(v) z_v\right)$$
$$\text{s.t.} \qquad \sum_{v \in H} s(v) z_v \leq 1, \tag{6}$$
$$z_v \in \{0, 1, \dots, n(v)\}.$$

In the second integer program, we optimize over all configurations $C \in \zeta$ with size $s(C) \geq 1$.

$$\tau_2 = \min \sum_{v \in H} \frac{y_v}{n(v)} z_v$$
$$\text{s.t.} \qquad \sum_{v \in H} s(v) z_v \geq 1, \tag{7}$$
$$z_v \in \{0, 1, \dots, n(v)\}.$$

Configuration C^* has price value $y^T A[C^*] = \min\{\tau_1, \tau_2\}$. The above integer programs (6) and (7) are variations of the knapsack problem with bounded variables. Problem (7) is the minimization version of the knapsack problem, and it is not difficult to show that a simple dynamic programming algorithm solves it in polynomial time within a factor $(1 + \varepsilon)$ of the optimum.

Problem (6), however, is more complicated to solve approximately. To see this, assume that the coefficients $\frac{y_v}{n(v)}$ are all equal to 0, and $y_{|H|+1} = 1$. Moreover, assume that the optimum solution for the problem consists of a set U of elements of total size exactly equal to 1. Then, the value of the optimum solution is 0. A $(1 + \varepsilon)$-approximation algorithm $\mathcal{A}$ for the problem has to find in polynomial time a solution of value 0, and therefore it must solve the subset sum problem exactly.

Note, however, that we do not need to solve approximately both problems (6) and (7), but rather we must find the approximate solution of problem

$$\tau^* = \min\{\tau_1, \tau_2\}. \tag{8}$$

Bounding the Value of the Optimum Solution for the Block Problem.
To solve problem (8), let us first find lower and upper bounds for τ^*. Every item
$u \in L$ has price $y_{s(u)}/n(s(u))$, and if the total size S' of the items placed in a bin
is smaller than 1, then we might think that the empty part of the bin is filled
with a dummy element of size exactly $1 - S'$ and price $(1 - S')y_{|H|+1}/s(S)$. It
is convenient to think then, that besides the items in L there is an additional
dummy item a_{n+1} of variable size (and price) that we can always use to completely fill a bin. Let L' be a list containing all elements from L plus element a_{n+1}.
Then, problem (8) can be thought as that of selecting the minimum price set of
elements from L' of total size at least 1.

Let $r(u) = \frac{y_{s(u)}/n(s(u))}{s(u)}$ for all items $u \in L$, and $r(a_{n+1}) = \frac{y_{|H|+1}}{s(S)}$. Let us
order the items $u \in L'$ in non-decreasing order of price-to-size ratio $r(u)$. Note
that we do not need to consider elements u with ratio $r(u)$ larger than $r(a_{n+1})$
because such elements cannot belong to an optimum solution for problem (8).

Take the items $u \in L'$ in non-decreasing order of ratio $r(u)$ and place them
in a bin of size 1 until we find the first item w_1 that overflows the bin. (We
consider that item a_{n+1} always overflows the bin.) Item w_1 is not placed in the
bin. Let R_1 be the set of items that get placed into the bin. Let $s(R_1) \leq 1$ be
the total size of the elements in R_1, let $L^<(R_1)$ be the set of items not in R_1,
each of size smaller than $1 - s(R_1)$. Similarly, let $L^>(R_1)$ be the set of items not
in R_1 of size larger than $1 - s(R_1)$. Find the smallest price item p_1 in $L^>(R_1)$.
Note that R_1 plus item p_1 forms a feasible solution for problem (8).

Take the elements u in $L^<(R_1)$ in non-decreasing order of ratio $r(u)$ and
add them to the bin containing items R_1 until we find the first item w_2 that
would overflow the bin. Let R_2 $(w_2 \notin R_2)$ be the new set of items in the bin
(including those from R_1), and let p_2 be the smallest price item in $L^>(R_2)$. We
have built a second feasible solution $R_2 \cup \{p_2\}$. We repeat the above process
with the elements in $L^<(R_2)$ to get a new feasible solution, and so on. Let
$R_1 \cup \{p_1\}, R_2 \cup \{p_2\}, \dots, R_k \cup \{p_k\}$ be the feasible solutions computed as described
above. Note that $p_k = a_{n+1}$. Let $R \cup \{p\}$ be the solution with minimum price in
this group, and let

$$P = \frac{1}{2} \left[\sum_{u \in R} y_{s(u)}/n(s(u)) + y_{s(p)}/n(s(p)) \right].$$

Lemma 1. $P \leq \tau^* \leq 2P$.

Proof. Let z^* be an optimum solution for problem (8). If z^* does not contain
any items from sets $L^>(R_i)$, $i = 1, \dots, k - 1$, then z^* consists of a subset of
the items in $R_k \cup \{p_k\}$. Hence, because of the order in which the items are
considered, the total price of $R_k \cup \{p_k\}$ is equal to the price of z^*. Therefore,
$P \leq \sum_{u \in R_k} y_{s(u)}/n(s(u)) + y_{s(p_k)}/n(s(p_k)) = \tau^* \leq 2P$.

The case when z^* contains at least one item $u \in L^>(R_i)$ is more complicated,
and so we omit the proof here. $\qquad\square$

Solving the Block Problem. We round up each one of the coefficients $\frac{y_v}{n(v)}$ in (6) and (7) to the nearest value of the form $i\frac{\varepsilon}{n}P$, where $i \in \{0, 1, \ldots, \lceil\frac{2n}{\varepsilon}\rceil\}$. By Lemma 1 performing this rounding the value of an optimum solution z^* for problem (8) increases by at most $\varepsilon P \leq \varepsilon\tau^*$. Therefore, an optimum solution for problem (8) with rounded coefficients $\frac{y_v}{n(v)}$ has value at most $(1 + \varepsilon)$ times the optimum solution for the problem with the original coefficients.

We can find an optimum solution for problem (8) with rounded coefficients by using dynamic programming as follows. Let the elements in list L be $a_1, a_2, \ldots,$ a_n. Let $M(i, \ell)$ be the maximum size $s(T)$ of a subset $T \subseteq \{a_1, \ldots, a_i\}$ of price value $\sum_{a_i \in T} y_{s(a_i)}/n(s(a_i)) = \ell\frac{\varepsilon}{n}P$ for $i = 0, 1, \ldots, n$, $\ell = 0, 1, \ldots, \lceil\frac{2n}{\varepsilon}\rceil$. The values $M(i, \ell)$ can be computed by solving the following recurrence equation.

$$M(0, \ell) = 0 \quad \text{for all } \ell = 0, 1, \ldots, \lceil 2n/\varepsilon \rceil$$
$$M(i, \ell) = \max \{M(i - 1, \ell), M(i - 1, \ell - r_i) + s(a_i)\},$$
$$\text{where } y_{s(a_i)}/n(s(a_i)) = r_i\varepsilon P/n, \text{ for all } i = 1, \ldots, n,$$
$$\text{and } \ell = 1, \ldots, \lceil 2n/\varepsilon \rceil.$$

A simple algorithm for solving this recurrence uses $O(n^2/\varepsilon)$ time. Once we have computed the values $M(i, \ell)$, we find the smallest price value $\ell'\frac{\varepsilon}{n}P$ for which there is a solution of size $M(n, \ell') \geq 1$. The following theorem shows how to use these values $M(i, \ell)$ to find a $(1 + \varepsilon)$-approximation for problem (8).

Theorem 2. *An optimum solution for problem (8) has value at most*

$$(1 + \varepsilon) \min \left\{ i\frac{\varepsilon}{n}P + \frac{y_{|H|+1}}{s(\mathcal{S})}\sigma(1 - M(n, i)) \mid i = 0, \ldots, \ell' \right\},$$

where $\sigma(x) = \max\{x, 0\}$. Hence, a $(1 + \varepsilon)$-approximation for problem (8) can be computed in $O(n^2/\varepsilon)$ time.

Proof. Consider problem (8) with coefficients $\frac{y_v}{n(v)}$ rounded as described above. By the above discussion, the optimum solution for this problem is at most $(1 + \varepsilon)$ times larger than the optimum solution for the problem with the original coefficient values. Hence, to prove the lemma we just need to show that an optimum solution for the rounded problem has value

$$\min \left\{ i\frac{\varepsilon}{n}P + \frac{y_{|H|+1}}{s(\mathcal{S})}\sigma(1 - M(n, i)) \mid i = 0, \ldots, \ell' \right\}.$$

Let $\hat{\tau}_1$ and $\hat{\tau}_2$ be optimum solutions for the rounded problems (6) and (7), respectively. Let $\hat{\tau} = \min\{\hat{\tau}_1, \hat{\tau}_2\}$, and the $\hat{z}$ be a solution of value $\hat{\tau}$. Note that if $\hat{\tau} = \hat{\tau}_2$, then the set $T \subseteq L$ used to compute the value $M(n, \ell')$ is the minimum price set of size at least 1, and so $\sum_{a_i \in T} y_{s(a_i)}/n(s(a_i)) = \ell'\frac{\varepsilon}{n}P = \hat{\tau}_2$.

The proof for the case when $\hat{\tau} = \hat{\tau}_1$ is slightly more complicated, and so we omit it here. $\qquad\square$

3.3   Time Complexity and Performance Guarantee

The running time of the algorithm for approximately solving linear program (1) is

$$O(|H|(\frac{1}{\varepsilon^2} + \ln |H|) \ln(\frac{n}{\varepsilon}) \max\{n^2/\varepsilon, |H| \ln \ln(|H|/\varepsilon)\}).$$

The term $O(|H| \ln \ln(|H|/\varepsilon))$ is the overhead per iteration of the potential price directive decomposition method [8,10]. Since $|H| = O(1/\varepsilon^2)$, then this time complexity is

$$O\left(\frac{1}{\varepsilon^5} \ln(\frac{n}{\varepsilon}) \max\{n^2, \frac{1}{\varepsilon} \ln \ln(\frac{1}{\varepsilon^3})\}\right).$$

This solution is modified as described in Section 3.1 so that the number of fractional values in the resulting vector x_C is at most $O(1/\varepsilon^2)$. This post-processing can be performed in $O(\frac{1}{\varepsilon^4}\mathcal{M}(\frac{1}{\varepsilon^2}))$ time, where $\mathcal{M}(n)$ is the time to invert an $(n \times n)$ matrix.

We now need to show that the solution computed by the algorithm is asymptotically very close to the optimum. Consider the partition of the set of large items into groups G_i as described in Section 2. Let $\mathcal{L}_{inf}$ ($\mathcal{L}_{sup}$) denote the set of large items obtained by rounding the size of each item of group G_i down to the size of the smallest item of group G_i (up to the size of the smallest item of group G_{i-1} for $i > 1$ and to 1 for $i = 1$). Then, $\mathcal{L}_{inf} < \mathcal{L}_{sup}$ in the sense that there is a bijection between the items such that each item of $\mathcal{L}_{inf}$ is no larger than the corresponding item of $\mathcal{L}_{sup}$. It follows that the maximum objective values, $X^*(\mathcal{L}_{inf})$ and $X^*(\mathcal{L}_{sup})$, of the linear program (1) corresponding to the instances $\mathcal{L}_{inf}$ and $\mathcal{L}_{sup}$ satisfy:

$$X^*(\mathcal{L}_{inf}) \leq X^*(\mathcal{L}_{sup}).$$

Furthermore, since each group G_i has at most $\lfloor |\mathcal{L}|\varepsilon^2 \rfloor + 1$ items, then by removing the $\lfloor |\mathcal{L}|\varepsilon^2 \rfloor$ smallest items from $\mathcal{L}_{inf}$ and adding $\lfloor |\mathcal{L}|\varepsilon^2 \rfloor$ items of size 1 we get a new set $\mathcal{L}'$ such that $\mathcal{L}_{inf} < \mathcal{L}' < \mathcal{L}_{sup}$. Hence,

$$X^*(\mathcal{L}_{sup}) \leq X^*(\mathcal{L}_{inf}) + |\mathcal{L}|\varepsilon^2 + 1.$$

By using the following two inequalities (that can be found in [4]),

$$\begin{aligned} X^*(\mathcal{L}_{sup}) &\geq OPT(\mathcal{L} \cup \mathcal{S}), \\ OPT(\mathcal{L} \cup \mathcal{S}) &\geq OPT(L) - \lfloor s_L(L) \rfloor \varepsilon - (1 - \varepsilon), \end{aligned}$$

where $OPT(\mathcal{L} \cup \mathcal{S})$ is the maximum number of bins that can be covered with items from $\mathcal{L} \cup \mathcal{S}$, we can show that:

$$\sum_{C \in \zeta} \tilde{x}_C \geq (1 - 2\varepsilon)X^*(\mathcal{L}_{inf}) \geq \sum_{C \in \zeta} \tilde{x}_C \geq (1 - 2\varepsilon)OPT(L) - (5/2)\varepsilon s_L(L) - 1.$$

We transform this solution $(\tilde{x}_C)_{C \in \zeta}$ into a feasible solution for the bin covering problem by rounding down each variable $\tilde{x}_C$ to the nearest integer, i.e., we get an integral solution $(y_C)_{C \in \zeta}$ by making $y_C = \lfloor \tilde{x}_C \rfloor$, for all $C \in \zeta$. This solution has value $\sum_C y_C \geq \sum_C \tilde{x}_C - (1 + 1/\varepsilon^2)$ since, as we showed at the end of Section 3.1, our solution $(\tilde{x}_C)_{C \in \zeta}$ has at most $\frac{1}{\varepsilon^2} + 1$ non-zero components. Now we are ready to prove Theorem 1.

Proof. **(Theorem 1)** By the above inequalities,

$$\sum_{C \in \zeta} y_C \geq (1 - 2\varepsilon)OPT(L) - (5/2)\varepsilon s_L(L) - 2 - \frac{1}{\varepsilon^2}.$$

Since $OPT(L) > (s_L(L) - 1)/2$, then $-(5/2)\varepsilon s_L(L) > -5\varepsilon OPT(L) - (5/2)\varepsilon$. For $\varepsilon \leq 1/2$, this gives a solution with value $\sum_{C \in \zeta} y_C \geq (1 - 7\varepsilon)OPT(L) - \frac{2}{\varepsilon^2}$. Now, if $s_L(L) > \frac{13}{\varepsilon^3}$ we get $-\frac{2}{\varepsilon^2} > -\frac{2\varepsilon}{13}s_L(L) > -\frac{2\varepsilon}{13}(2OPT(L) + 1) = -\frac{4\varepsilon}{13}OPT(L) - \frac{2\varepsilon}{13} > -\varepsilon OPT(L) - 1$. Therefore, the total number of bins covered by our solution is at least $(1 - 8\varepsilon)OPT(L) - 1$.

If $s_L(L) \leq \frac{13}{\varepsilon^3}$, then our algorithm covers $OPT(L) - O(1/\varepsilon^3)$ bins.   □

4   Conclusions

We have presented an asymptotic fully polynomial time approximation scheme for the bin covering problem. This algorithm produces a solution of value $(1 - \varepsilon)OPT(L) - O(1/\varepsilon^3)$, for any value $\varepsilon > 0$. By using a more complicated algorithm we are able decrease the length of the solution to $(1 - \varepsilon)OPT(L) - o(1/\varepsilon^3)$. We can also slightly improve the time complexity of the algorithm presented here. We give the details in the full version of the paper.

It still remains as an open question [4] whether it is possible to design an asymptotic fully polynomial time approximation scheme for the problem that produces a solution of length $OPT(L) - o(OPT(L))$.

References

1. S.B. Assman, D.S. Johnson, D.J. Kleitman and J.Y-T. Leung, On the dual of the one-dimensional bin packing problem, *Journal on Algorithms* 5 (1984), 502–525.
2. S.B. Assman, Problems in Discrete Applied Mathematics, *PhD thesis, Department of Mathematics, MIT, Cambridge, MA*, 1983.
3. J. Csirik, J.B.G. Frenk, M. Labbe and S. Zhang, Two simple algorithms for bin covering, *Acta Cybernetica* 14 (1999), 13–25.
4. J. Csirik, D.S. Johnson and C. Kenyon, Better approximation algorithms for bin covering, *Proceedings of SIAM Conference on Discrete Algorithms* (2001).
5. J. Csirik and V. Totik, On-line algorithms for a dual version of bin packing, *Discrete Applied Mathematics*, 21 (1988), 163–167.

6. W.F. de la Vega and C.S. Lueker, Bin packing can be solved within $1 + \epsilon$ in linear time, *Combinatorica*, 1 (1981), 349–355.
7. M.D. Grigoriadis and L.G. Khachiyan, Fast approximation schemes for convex programs with many blocks and coupling constraints, *SIAM Journal on Optimization*, 4 (1994), 86–107.
8. M.D. Grigoriadis and L.G. Khachiyan, Coordination complexity of parallel price-directive decomposition, *Mathematics of Operations Research*, 21 (1996), pp. 321–340.
9. O.H. Ibarra and C.E. Kim, Fast approximation algorithms for the knapsack and sum of subset problem, *Journal of the ACM*, 22 (1975), 463–468.
10. K. Jansen and H. Zhang, Approximate algorithms for general packing problems with modified logarithmic potential function, to appear: *Proceedings 2nd International Conference on Theoretical Computer Science*, Montreal, 2002.
11. N. Karmarkar and R.M. Karp, An efficient approximation scheme for the one-dimensional bin packing problem, *Proceedings 23rd Annual Symposium on Foundations of Computer Science* (1982), 312–320.
12. H. Kellerer and U. Pferschy, A new fully polynomial approximation scheme for the knapsack problem, *Proceedings 1st International Workshop on Approximation Algorithms for Combinatorial Optimization* (1998), 123–134.
13. E. Lawler, Fast approximation algorithms for knapsack problems, *Mathematics of Operations Research*, 4 (1979), 339–356.
14. S.A. Plotkin, D.B. Shmoys, and E. Tardos, Fast approximation algorithms for fractional packing and covering problems, *Mathematics of Operations Research*, 20 (1995), 257–301.

Improved Approximation Algorithms for Max-2SAT with Cardinality Constraint

Markus Bläser and Bodo Manthey[*]

Institut für Theoretische Informatik
Universität zu Lübeck
Wallstraße 40, 23560 Lübeck, Germany
{blaeser,manthey}@tcs.mu-luebeck.de

Abstract. The optimization problem Max-2SAT-CC is Max-2SAT with the additional cardinality constraint that the value one may be assigned to at most K variables. We present an approximation algorithm with polynomial running time for Max-2SAT-CC. This algorithm achieves, for any $\epsilon > 0$, approximation ratio $\frac{6+3\cdot e}{16+2\cdot e} - \epsilon \approx 0.6603$. Furthermore, we present a greedy algorithm with running time $O(N \log N)$ and approximation ratio $\frac{1}{2}$. The latter algorithm even works for clauses of arbitrary length.

1 Introduction

The maximum satisfiability problem (Max-SAT) is a central problem in combinatorial optimization. An instance of Max-SAT is a set of Boolean clauses over variables $x_1, \ldots, x_n$. Our goal is to find an assignment for the variables $x_1, \ldots, x_n$ that satisfies the maximum number of clauses. We may also associate a nonnegative weight with each clause. In this case, we are looking for an assignment that maximizes the sum of the weights of the satisfied clauses. If every clause has length at most ℓ, we obtain the problem Max-ℓSAT. The currently best approximation algorithm for Max-SAT is due to Asano and Williamson [3] and achieves approximation ratio 0.7846. In the case of Max-ℓSAT, we are particulary interested in the value $\ell = 2$. The best positive and negative results currently known for Max-2SAT are 0.931 by Feige and Goemans [7] and $\frac{21}{22} \approx 0.954$ by Håstad [9], respectively.

In this work, we consider Max-SAT and Max-ℓSAT with cardinality constraint. In addition to C, we get an integer K as input. The goal is to find an assignment that maximizes the number (or sum of weights) of satisfied clauses among all assignments that give the value one to at most K variables. We call the resulting problems Max-SAT-CC and Max-ℓSAT-CC. Note that lower bounds for the approximability of Max-SAT and Max-ℓSAT are also lower bounds for Max-SAT-CC and Max-ℓSAT-CC, respectively. The corresponding decision problems are natural complete problems in parameterized complexity, see e.g. Downey and Fellows [5].

---

[*] Birth name: Bodo Siebert. Supported by DFG research grant Re 672/3.

P. Bose and P. Morin (Eds.): ISAAC 2002, LNCS 2518, pp. 187–198, 2002.
© Springer-Verlag Berlin Heidelberg 2002

An important special case of Max-SAT-CC is the maximum coverage problem (MCP). An instance of MCP is a collection of subsets $S_1, \ldots, S_n$ of some universe $U = \{u_1, \ldots, u_m\}$ and an integer K. We are asked to cover as many elements as possible from U with (at most) K of the given subsets. MCP can be considered as a natural dual to the maximum set cover problem. In the weighted version, each u_j also has a nonnegative weight w_j. We can convert an instance of MCP into a corresponding instance of Max-SAT-CC with only positive literals: for every $u_j \in U$ we have a clause c_j containing all literals x_i with $u_j \in S_i$.

A simple greedy algorithm for MCP achieves approximation ratio $1 - e^{-1}$ (see Cornuéjols et al. [4]). On the other hand, Feige [6] showed that no polynomial time algorithm can have a better approximation ratio, unless $\mathsf{NP} \subseteq \mathsf{DTime}(n^{\log \log n})$. For a restricted version of MCP, where each element occurs in at most ℓ sets, Ageev and Sviridenko [1] presented a $\left(1 - \left(1 - \frac{1}{\ell}\right)^{\ell}\right)$-approximation algorithm with polynomial running time. This algorithm yields approximation ratio $\frac{3}{4}$ for $\ell = 2$. By the above reduction, this restricted version of MCP is equivalent to Max-ℓSAT-CC with only positive literals.

Sviridenko [10] designed a $(1 - e^{-1})$-approximation algorithm for general Max-SAT-CC with polynomial running time. This is again tight, since even the version with only positive literals allows no better approximation, unless $\mathsf{NP} \subseteq \mathsf{DTime}(n^{\log \log n})$. Sviridenko raised the question whether it is possible to obtain better approximation algorithms for Max-ℓSAT for small values of ℓ.

New Results. As our first result, we present an approximation algorithm for Max-2SAT-CC with polynomial running time. This approximation algorithm achieves an approximation performance of $\frac{6+3 \cdot e}{16+2 \cdot e} - \epsilon \approx 0.6603$, for any $\epsilon > 0$. Thus, we give a positive answer to Sviridenko's question for $\ell = 2$. (Note that $1 - e^{-1} \approx 0.6321$.)

Second, we present a simple greedy algorithm for Max-SAT-CC with running time $O(N \log N)$ and prove that its approximation performance is $\frac{1}{2}$. We give an example to show that this approximation ratio is tight. Thus, in contrast to MCP, this greedy approach is not optimal for Max-SAT-CC.

2   A 0.6603-Approximation Algorithm for Max-2SAT-CC

Consider a set $C = \{c_1, \ldots, c_m\}$ of clauses of length at most two over the variables $x_1, \ldots, x_n$ and assign to each clause c_j a nonnegative weight w_j. A clause is called *pure*, if either all literals in it are positive or all of them are negative. Let $J_=$ be the set of indices of the pure clauses. (Note that clauses of length one are pure.) $J_{\neq} = \{1, 2, \ldots, m\} \setminus J_=$ denotes the set of all indices corresponding to *mixed* clauses. For each clause c_j, we define sets of indices $I_j^+, I_j^- \subseteq \{1, \ldots, n\}$ as follows: $i \in I_j^+$ iff x_i occurs positive in c_j and $i \in I_j^-$ iff x_i occurs negative in c_j.

Our 0.6603-approximation algorithm (see Figure 1) works as follows. As input, it gets a clause set C and an integer K. We first solve the following relaxed linear program LP_k for each $0 \le k \le K$:

APROX (C, K, ϵ)

Input: Clause set C over variables $x_1, \ldots, x_n$,
          nonnegative integer $K \leq n$,
          an $\epsilon > 0$.
Output: Assignment with at most K ones.

1: **for** $0 \leq k \leq K$ **do**
2:    Solve the linear program LP_k.
      Let M_k be the value of an optimum solution.
3: Choose $k_{\max}$ such that $M_{k_{\max}}$ is maximized.
   Let $(y^\star, z^\star)$ be the corresponding optimum solution of $\mathrm{LP}_{k_{\max}}$.
4: $A_1 :=$ ROUNDING PROCEDURE 1 $(C, k_{\max}, y^\star, z^\star)$.
5: $A_2 :=$ ROUNDING PROCEDURE 2 $(C, k_{\max}, y^\star, z^\star, \epsilon)$.
6: Return the assignment A_i $(i = 1, 2)$ that satisfies the maximum
   number of clauses.

Fig. 1. The approximation algorithm

$$\text{maximize} \quad \sum_{j=1}^{m} w_j \cdot z_j$$

$$\text{subject to} \quad \sum_{i \in I_j^+} y_i + \sum_{i \in I_j^-} (1 - y_i) \geq z_j \qquad (j = 1, \ldots, m),$$

$$\sum_{i=1}^{n} y_i = k,$$

$$0 \leq z_j \leq 1 \qquad (j = 1, \ldots, m),$$

$$0 \leq y_i \leq 1 \qquad (i = 1, \ldots, n).$$

Variable y_i corresponds to Boolean variable x_i and variable z_j to clause c_j. This is essentially the same relaxed linear program as used by Goemans and Williamson [8] for Max-SAT. We have just added the cardinality constraint $\sum_{i=1}^{n} y_i = k$.

For $0 \leq k \leq K$, let M_k be the value of an optimum solution of LP_k and choose $k_{\max}$ such that $M_{k_{\max}}$ is maximal. Let $(y^\star, z^\star)$ be an optimum solution of the relaxed linear program $\mathrm{LP}_{k_{\max}}$. We round $y^\star$ in two different ways to obtain two assignments each with at most K ones. The assignment satisfying the larger number of clauses is a 0.6603-approximation to an optimum assignment. We solve LP_k for each k separately and do not replace the cardinality constraint by $\sum_{i=1}^{n} y_i \leq K$, since the first rounding procedure can only be applied if $\sum_{i=1}^{n} y_i$ is integral.

The quality of each of the two rounding procedures depends on the distribution of the values of $z_j^\star$ among pure and mixed clauses. In the remainder of the analysis, δ is chosen such that $\sum_{j \in J_=} z_j^\star = \delta \cdot \sum_{j=1}^{m} z_j^\star$. Rounding Procedure 1 is favorable, if δ is large, whereas Rounding Procedure 2 is advantageous, if δ is small.

For the sake of simplicity, we only consider the unweighted case in the following analysis (i.e., all w_j equal one). However, it is possible to transfer the analysis for the unweighted case to the weighted case with only marginal extra effort.

2.1 Rounding Procedure 1

In this section, we present a simple deterministic rounding procedure, which is based on Ageev and Sviridenko's "pipage rounding" [1]. The solution obtained by this rounding procedure has weight at least $(\frac{3}{8} + \frac{3}{8}\delta) \cdot \sum_{j=1}^{m} z_j^\star$.

ROUNDING PROCEDURE 1 $(C, k, y^\star, z^\star)$

Input: Clause set C over variables $x_1, \ldots, x_n$,
               nonnegative integer $k \leq n$,
               optimum solution $(y^\star, z^\star)$ of LP_k.
Output: Assignment with exactly k ones.

1: Let $a^\star = y^\star$.
2: **while** $a^\star$ has two noninteger coefficients $a_{i_1}^\star$ and $a_{i_2}^\star$ **do**
3:     Apply a "pipage rounding" step to $a^\star$ as described in the text.
4: Return the assignment $a^\star$.

Fig. 2. Rounding Procedure 1

First, we modify the set C of clauses to obtain a new set $\hat{C}$ of clauses. For each $j \in J_=$, we add c_j to $\hat{C}$. For each $j \in J_{\neq}$, we do the following: assume that $c_j = x_{i_1} \vee \overline{x_{i_2}}$. If $y_{i_1}^\star \geq 1 - y_{i_2}^\star$, we add the clause x_{i_1} to $\hat{C}$. Otherwise, we add $\overline{x_{i_2}}$ to $\hat{C}$. Formally, we treat $\hat{C}$ as a multiset, since two different mixed clauses may be transformed into the same clause.

For further analysis, we consider the relaxed linear program $\hat{LP}_k$ corresponding to the set $\hat{C}$ of clauses. (Note that we do not need $\hat{LP}_k$ to apply the rounding procedure but only to analyze the approximation ratio of the rounded solution.) For given $\eta \in [0,1]^n$, let Π_η denote the linear program obtained from $\hat{LP}_k$ by substituting every variable y_i by the corresponding η_i. Let $\hat{z}^\star$ be the optimum solution of $\Pi_{y^\star}$. (Note that due to the structure of Π_η, the optimum solution is unique.) By the construction of $\hat{C}$, $(y^\star, \hat{z}^\star)$ is a solution of $\hat{LP}_k$ fulfilling

$$\sum_{j=1}^{m} \hat{z}_j^\star \geq \delta \cdot \sum_{j=1}^{m} z_j^\star + \tfrac{1}{2}(1 - \delta) \cdot \sum_{j=1}^{m} z_j^\star = (\tfrac{1}{2} + \tfrac{1}{2}\delta) \cdot \sum_{j=1}^{m} z_j^\star. \qquad (1)$$

(Note that $\hat{z}_j^\star = z_j^\star$ if $j \in J_=$, otherwise $\hat{z}_j^\star$ is either $y_{i_1}^\star$ or $1 - y_{i_2}^\star$.) All clauses in $\hat{C}$ are pure, in other words, $\hat{J}_{\neq} = \emptyset$. (We use the same naming conventions for $\hat{C}$ and $\hat{LP}_k$ as for C and LP_k, we just add a "$\hat{\ }$".) The fact that $\hat{J}_{\neq}$ is empty allows us to apply "pipage rounding". To this aim, let

$$\hat{F}(y) = \sum_{j=1}^m \left(1 - \prod_{i \in \hat{I}_j^+} (1 - y_i) \cdot \prod_{i \in \hat{I}_j^-} y_i \right).$$

Note that for each j, either $\hat{I}_j^+$ or $\hat{I}_j^-$ is empty. If η is a $\{0,1\}$-valued vector of length n, then $\hat{F}(\eta)$ is exactly the number of satisfied clauses when we assign to each Boolean variable x_i the value η_i. Furthermore, if ζ denotes the optimum solution of $\hat{\Pi}_\eta$, then $\sum_{j=1}^m \zeta_j \geq \hat{F}(\eta)$.

Goemans and Williamson [8] proved that for any $\eta \in [0,1]^n$ and any ζ that is an optimal solution of $\hat{\Pi}_\eta$, we have

$$\hat{F}(\eta) \geq \tfrac{3}{4} \sum_{j=1}^m \zeta_j. \tag{2}$$

(In general, the factor $\tfrac{3}{4}$ has to be replaced by $1 - (1 - \tfrac{1}{\ell})^\ell$ where ℓ is the maximum clause length.)

Every clause in $\hat{C}$ contains either only positive or only negative literals. Thus, the univariate quadratic polynomial $\Phi_{i_1,i_2,\eta}$ defined by

$$\Phi_{i_1,i_2,\eta}(\epsilon) = \hat{F}(\eta_1, \ldots, \eta_{i_1-1}, \eta_{i_1} - \epsilon, \eta_{i_1+1}, \ldots, \eta_{i_2} + \epsilon, \ldots)$$

is convex for all choices of indices i_1, i_2, since the coefficient of ϵ^2 is nonnegative by the fact that for all j either $\hat{I}_j^+$ or $\hat{I}_j^-$ is empty.

Now consider a vector $\eta \in [0,1]^n$ with $\sum_{i=1}^n \eta_i = k$ and assume that η is not a $\{0,1\}$-vector. Then there are two indices i_1 and i_2 such that $\eta_{i_1}, \eta_{i_2} \in (0,1)$. Let $\epsilon_1 = \min\{\eta_{i_1}, 1 - \eta_{i_2}\}$ and $\epsilon_2 = -\min\{\eta_{i_2}, 1 - \eta_{i_1}\}$. By the convexity of $\Phi_{i_1,i_2,\eta}$, we have either

$$\Phi_{i_1,i_2,\eta}(\epsilon_1) \geq \hat{F}(\eta) \qquad \text{or} \qquad \Phi_{i_1,i_2,\eta}(\epsilon_2) \geq \hat{F}(\eta).$$

Let η' be the vector obtained from η as follows. If the first of the inequalities above is fulfilled, we replace η_{i_1} by $\eta_{i_1} - \epsilon_1$ and η_{i_2} by $\eta_{i_2} + \epsilon_1$. Otherwise, if the second one is fulfilled, we replace η_{i_1} by $\eta_{i_1} - \epsilon_2$ and η_{i_2} by $\eta_{i_2} + \epsilon_2$. The vector η' has at least one more $\{0,1\}$-entry than η by the choice of ϵ_1 and ϵ_2. By the construction of η', we have

$$\hat{F}(\eta') \geq \hat{F}(\eta). \tag{3}$$

Now we start with the initial optimum solution $(y^\star, z^\star)$ of LP_k and treat it as a solution of $\hat{LP}_k$. Then we repeatedly apply a "pipage rounding" step to $y^\star$ as described above. After at most n such steps, we have a $\{0,1\}$-vector $a^\star$. Since a "pipage rounding" step never changes the sum of the vector elements, $a^\star$ has exactly k ones. We have

$$\hat{F}(a^\star) \geq \hat{F}(y^\star) \geq \tfrac{3}{4} \sum_{j=1}^m \hat{z}_j^\star,$$

where the first inequality follows from repeated application of Inequality 3 and the second is simply Inequality 2. Thus by Inequality 1, we obtain

$$\hat{F}(a^\star) \geq \left(\tfrac{3}{8} + \tfrac{3}{8}\delta\right) \cdot \sum_{j=1}^{m} z_j^\star \,,$$

which proves the next lemma.

Lemma 1. *Let $a^\star \in \{0,1\}^n$ be the assignment with exactly k ones obtained by applying Rounding Procedure 1 to the solution $(y^\star, z^\star)$. Then $a^\star$ satisfies at least $\left(\tfrac{3}{8} + \tfrac{3}{8}\delta\right) \cdot N$ many clauses, where N is the number of clauses that are satisfied by an optimum assignment with k ones.*

2.2   Rounding Procedure 2

The rounding procedure presented in the previous section yields a good approximation ratio if δ is large. In this section we focus our attention on mixed clauses. We present a rounding procedure which works well especially if δ is small.

ROUNDING PROCEDURE 2 $(C, k, y^\star, z^\star, \epsilon)$

Input: Clause set C over variables $x_1, \ldots, x_n$,
                nonnegative integer k with $0 \leq k \leq n$,
                optimum solution $(y^\star, z^\star)$ of LP_k,
                $\epsilon > 0$.

Output: Assignment with at most k ones.

1: **if** k is not sufficiently large **then**
2:     Try all assignments with at most k ones.
         Choose the one that satisfies the maximum number of clauses.
3: **else**
4:     **do** k times
5:         Draw and replace one index from the set $\{1, 2, \ldots, n\}$ at random.
             The probability of choosing i is $\frac{y_i^\star}{k}$.
6:     Set $x_i = 1$ iff i was drawn at least once in the last step.
7: Return the assignment computed.

Fig. 3. Rounding Procedure 2

The rounding procedure is described in Figure 3. It works as follows. We draw and replace k times an index out of the set $\{1, 2, \ldots, n\}$, where i is drawn with probability $\frac{y_i^\star}{k}$. (Note that $\sum_{i=1}^{n} y_i^\star = k$.) Let S be the set of indices drawn. Then we set $x_i = 1$ iff $i \in S$. The assignment obtained assigns the value one to at most k variables.

Now we have to estimate the probability that a clause is satisfied by the assignment obtained. For pure clauses we use the estimate given by Sviridenko [10, Theorem 1, Cases 1 and 3].

Lemma 2 (Sviridenko [10]). *Assume that c_j is a pure clause. Then the probability that c_j is satisfied by the random assignment is*

$$\Pr(c_j \text{ is satisfied}) \geq \left(1 - e^{-1}\right) \cdot z_j^* .$$

For mixed clauses, the estimate given by Sviridenko [10, Theorem 1, Case 2] can be improved.

Lemma 3. *Assume that c_j is a mixed clause. Then for every $\epsilon > 0$ there is a $k_0 \in \mathbb{N}$ such that for all $k \geq k_0$ the probability that c_j is satisfied by the random assignment is*

$$\Pr(c_j \text{ is satisfied}) \geq \left(\tfrac{3}{4} - \epsilon\right) \cdot z_j^* .$$

To prove Lemma 3, we need the following lemma.

Lemma 4. *For every $\alpha, \beta \in [0,1]$ and $k \geq \frac{4}{\ln(4 \cdot \epsilon + 1)}$ we have*

$$1 - e^{-\beta} \cdot \left(1 - e^{-\frac{4}{k} - \alpha}\right) \geq \left(\tfrac{3}{4} - \epsilon\right) \cdot \min\left\{1, \beta + 1 - \alpha\right\} .$$

Proof. Let $\epsilon > 0$ be some arbitrary fixed constant. Throughout this proof, we substitute $\mu = \alpha - \beta$, $\nu = \alpha + \beta$, and $\xi = \frac{4}{k}$. We consider the function

$$f(\mu, \nu) := \frac{1 - e^{\frac{\mu - \nu}{2}} + e^{-\xi - \nu}}{\min\{1, 1 - \mu\}} = \frac{1 - e^{-\beta} \cdot \left(1 - e^{-\frac{4}{k} - \alpha}\right)}{\min\{1, \beta + 1 - \alpha\}} .$$

Our aim is to find the minimum of the function for $0 \leq \alpha \leq 1$ and $0 \leq \beta \leq 1$. We restrict ourselves to $\mu < 1$, since $\lim_{\mu \to 1, \mu < 1} f(\mu, \nu) \to +\infty > \tfrac{3}{4}$.

Consider the partial derivative

$$f_\nu(\mu, \nu) = \frac{1}{\min\{1, 1 - \mu\}} \cdot \left(\tfrac{1}{2} \cdot e^{\frac{\mu - \nu}{2}} - e^{-\xi - \nu}\right) .$$

We have $f_\nu(\mu, \nu) = 0$ iff $\nu = \ln 4 - \mu - 2 \cdot \xi$. Furthermore, we have

$$f_{\nu\nu}(\mu, \nu) = \frac{1}{\min\{1, 1 - \mu\}} \cdot \left(-\tfrac{1}{4} \cdot e^{\frac{\mu - \nu}{2}} + e^{-\xi - \nu}\right) > 0$$

for $\nu = \ln 4 - \mu - 2 \cdot \xi$ and $-1 \leq \mu < 1$. Thus, the only local minima of f are obtained for $\nu = \ln 4 - \mu - 2 \cdot \xi$. Since f has no local maximum, these are the only values to be considered in the sequel and we can restrict our attention to the function

$$g(\mu) = f(\mu, \ln 4 - \mu - 2 \cdot \xi) = \frac{1}{\min\{1, 1 - \mu\}} \cdot \left(1 - \tfrac{1}{4} \cdot e^{\xi + \mu}\right) .$$

For $\mu \leq 0$, $g(\mu) = 1 - \tfrac{1}{4} \cdot e^{\xi + \mu}$ is monotonically decreasing. For $\mu \geq 0$, $g(\mu) = \frac{1}{1 - \mu} \cdot \left(1 - \tfrac{1}{4} \cdot e^{\xi + \mu}\right)$ is monotonically increasing. Thus, g reaches its minimum for $\mu = 0$ and we have

$$f(\mu, \nu) \geq f(0, \ln 4 - 2 \cdot \xi) = 1 - \tfrac{1}{4} \cdot e^\xi .$$

We choose $k = \frac{4}{\xi} \geq \frac{4}{\ln(4 \cdot \epsilon + 1)}$ and obtain $f(\mu, \nu) \geq \tfrac{3}{4} - \epsilon$. $\square$

Proof (of Lemma 3). Assume that $c_j = x_{i_1} \vee \overline{x_{i_2}}$. Then we have

$$\Pr(c_j \text{ is satisfied}) = \Pr(i_1 \in S \vee i_2 \notin S)$$
$$\geq 1 - e^{-y_{i_1}^\star}\left(1 - e^{-\frac{4}{k}-y_{i_2}^\star}\right)$$
$$\geq \left(\tfrac{3}{4} - \epsilon\right) \cdot \min\left\{1, \left(y_{i_1} + (1 - y_{i_2})\right)\right\}$$
$$\geq \left(\tfrac{3}{4} - \epsilon\right) \cdot z_j^\star .$$

The first inequality follows from Sviridenko's results, which hold for all k above some constant k_1 (independent of $y_{i_1}^\star$ and $y_{i_2}^\star$). The second one follows from Lemma 4. The last inequality follows from $z_j^\star \leq y_{i_1} + (1 - y_{i_2})$ and $z_j^\star \leq 1$. □

If $k < k_0 := \max\left\{\frac{4}{\ln(4 \cdot \epsilon + 1)}, k_1\right\}$, we can try all assignments with at most k ones in polynomial time. Thus, our algorithm solves the problem exactly in this case.

By Lemma 2, we have an expected weight of at least $\delta \cdot \left(1 - e^{-1}\right) \cdot \sum_{j=1}^{m} z_j^\star$ for pure clauses. For mixed clauses we have an expected weight of at least $(1 - \delta) \cdot \left(\tfrac{3}{4} - \epsilon\right) \cdot \sum_{j=1}^{m} z_j^\star$ by Lemma 3.

The randomized rounding procedure presented in this section can be derandomized using the method of conditional expectation (see e.g. Alon et al. [2]).

Overall, we obtain the following lemma.

Lemma 5. *For any $\epsilon > 0$, if we apply Rounding Procedure 2 to the optimal solution $(y^\star, z^\star)$, we obtain an assignment with at most k ones that satisfies at least*

$$\left(\left(\tfrac{1}{4} - \tfrac{1}{e}\right) \cdot \delta + \tfrac{3}{4} - \epsilon\right) \cdot N$$

clauses, where N is the maximum number of clauses that can be satisfied by an assignment with exactly k ones. □

2.3   Analysis of the Approximation Ratio

Our approximation algorithm (see Figure 1) solves the linear program LP_k for every $0 \leq k \leq K$. It chooses $k_{\max}$ such that $M_{k_{\max}}$ is maximized. Then it applies both Rounding Procedure 1 and 2 to this optimum solution and obtains two assignments. Finally, it returns the assignment satisfying the larger number of clauses. The approximation ratio of the algorithm is, for an arbitrary $\epsilon > 0$,

$$\min_{0 \leq \delta \leq 1} \max\left\{\left(\tfrac{3}{8} \cdot \delta + \tfrac{3}{8}\right), \left(\tfrac{1}{4} - \tfrac{1}{e}\right) \cdot \delta + \tfrac{3}{4} - \epsilon\right\}.$$

By some simple calculations, we obtain the following theorem.

Theorem 1. *For every $\epsilon > 0$, there is a polynomial time approximation algorithm for* Max-2SAT-CC *with approximation ratio $\frac{6 + 3 \cdot e}{16 + 2 \cdot e} - \epsilon$.* □

Note that $\frac{6 + 3 \cdot e}{16 + 2 \cdot e} > 0.66031$.

3   A Fast Greedy Algorithm for Max-SAT-CC

The approximation algorithm presented in Section 2 surely has polynomial running time. However, it involves solving a linear program K times. The same is true for Sviridenko's $(1 - e^{-1})$-approximation algorithm for arbitrary clause lengths. Thus, a faster algorithm might be desirable for practical applications. Figure 4 shows a simple greedy algorithm working for arbitrary clause lengths. As the main result of the present section, we prove that it has approximation performance $\frac{1}{2}$. Again for the sake of simplicity, we present the algorithm only for the case of unweighted clauses. It can be extended to handle weighted clauses in a straight forward manner.

The algorithm can easily be transformed into a $\frac{1}{2}$-approximation algorithm for the problem where we are asked to find an optimum asignment with *exactly* K ones. For this purpose, we just have to add a statement similar to the one in lines 1–2 that does the following: if $K = n$, then it returns the assignment giving all variables the value one.

GREEDY (C, K)

Input: Clause set C over variables $x_1, \ldots, x_n$,
          nonnegative integer K with $0 \leq K \leq n$.
Output: Assignment G_K with at most K ones.

1: **if** $K = 0$ **then**
2:    Let G_K be the assignment that assigns zero to all variables and return.
3: **else**
4:    Let $p = \max\{p_1, \ldots, p_n\}$ and $q = \max\{q_1, \ldots, q_n\}$.
     Set $\xi = 1$, if $p \geq q$. Otherwise, set $\xi = 0$.
     Choose an index i_0 such that $p_{i_0} = p$, if $\xi = 1$, and $q_{i_0} = q$, otherwise.
5:    Substitute $x_{i_0} \mapsto \xi$ and remove all trivial clauses.
     Let C' be the clause set obtained.
6:    $G' := \text{GREEDY}(C', K - \xi)$.
7:    Let G_K be the assignment, that behaves on $\{x_1, \ldots, x_n\} \setminus \{x_{i_0}\}$ like G'
     and assigns x_{i_0} the value ξ.

Fig. 4. The greedy algorithm

3.1   Analysis of the Approximation Ratio

For each variable x_i, let p_i be the number of clauses in which x_i appears positive and q_i be the number of clauses in which x_i appears negative. Let $p = \max\{p_1, \ldots, p_n\}$ and $q = \max\{q_1, \ldots, q_n\}$. Basically, the algorithm chooses an index i_0 such that by specializing x_{i_0}, we satisfy the largest possible number of clauses that can be satisfied by substituting only one variable. Then we proceed recursively.

For the analysis, let A_k, $0 \leq k \leq n$, be an optimum assignment with at most k ones for C and let Opt_k be the number of clauses satisfied by A_k. In the same way we define A'_k and Opt'_k for C'. The next two lemmata are crucial for analyzing the approximation performance of the greedy algorithm.

Lemma 6. *For all $1 \leq k \leq n$, we have $\mathrm{Opt}_k + q \geq \mathrm{Opt}_{k-1} \geq \mathrm{Opt}_k - p$.*

Proof. We start with the first inequality: if A_{k-1} is also an optimum assignment with at most k ones, then we are done. Otherwise, if we change a zero of A_{k-1} into a one, then we get an assignment with at most k ones. By the definition of q, this assignment satisfies at least $\mathrm{Opt}_{k-1} - q$ clauses. Consequently, $\mathrm{Opt}_k + q \geq \mathrm{Opt}_{k-1}$.

The second inequality follows in a similar fashion: if A_k has at most $k-1$ ones, then we are done. Otherwise, if we change a one in A_k into a zero, we get an assignment with at most $k-1$ ones. By the definition of p, this assignment satisfies at least $\mathrm{Opt}_k - p$ clauses. $\qquad\square$

Corollary 1. *For all $1 \leq k \leq n$, we have $\mathrm{Opt}'_k + q \geq \mathrm{Opt}'_{k-1} \geq \mathrm{Opt}'_k - p$.*

Proof. The proof of Lemma 6 surely works for C' if we define p' and q' accordingly. By the maximality of p and q, we may replace p' and q' by p and q. $\quad\square$

Lemma 7. *For all $1 \leq k \leq n$, we have*

$$A_k(x_{i_0}) = 1 \ \Rightarrow \ \mathrm{Opt}'_{k-1} \geq \mathrm{Opt}_k - p \ \ and$$
$$A_k(x_{i_0}) = 0 \ \Rightarrow \ \mathrm{Opt}'_k \geq \mathrm{Opt}_k - q \,.$$

Proof. In the first case, if we restrict A_k to $\{x_1, \ldots, x_n\} \setminus \{x_{i_0}\}$, we get an assignment with at most $k-1$ ones. It satisfies at least $\mathrm{Opt}_k - p$ clauses of C'.

The second case follows in the same way: if we restrict A_k to $\{x_1, \ldots, x_n\} \setminus \{x_{i_0}\}$, we get an assignment with at most k ones that satisfies at least $\mathrm{Opt}_k - q$ clauses of C'. $\qquad\square$

Theorem 2. *Algorithm* GREEDY *returns an assignment G_K with at most K ones that satisfies at least $\frac{1}{2} \mathrm{Opt}_K$ clauses.*

Proof. The proof is by induction on the recursion depth. If the depth is zero (i.e., $K = 0$), then GREEDY obviously returns the optimum assignment.

Now assume that GREEDY has approximation performance $\frac{1}{2}$ on all instances that can be solved with recursion depth at most d and assume that C is an instance requiring recursion depth $d+1$. We distinguish two cases, namely $\xi = 1$ and $\xi = 0$. Each case has two subcases, namely $A_K(x_{i_0}) = 1$ and $A_K(x_{i_0}) = 0$.

We start with $\xi = 1$. Let N_K denote the number of clauses satisfied by G_K. If $A_K(x_{i_0}) = 1$, then

$$
\begin{aligned}
N_K &\geq \tfrac{1}{2} \mathrm{Opt}'_{K-1} + p && \text{(by the induction hypothesis)} \\
&\geq \tfrac{1}{2} \mathrm{Opt}_K + \tfrac{1}{2}p && \text{(by Lemma 7)} \\
&\geq \tfrac{1}{2} \mathrm{Opt}_K \,.
\end{aligned}
$$

If $A_K(x_{i_0}) = 0$, then

$$N_K \geq \tfrac{1}{2}\text{Opt}'_{K-1} + p \qquad \text{(by the induction hypothesis)}$$
$$\geq \tfrac{1}{2}(\text{Opt}'_K - p) + p \qquad \text{(by Corollary 1)}$$
$$\geq \tfrac{1}{2}(\text{Opt}'_K + q) \qquad \text{(since } p \geq q)$$
$$\geq \tfrac{1}{2}\text{Opt}_K \qquad \text{(by Lemma 7).}$$

This completes the case $\xi = 1$.

The case $\xi = 0$ is handled as follows: if $A_K(x_{i_0}) = 1$, then we have

$$N_K \geq \tfrac{1}{2}\text{Opt}'_K + q \qquad \text{(by the induction hypothesis)}$$
$$\geq \tfrac{1}{2}(\text{Opt}'_{K-1} - q) + q \qquad \text{(by Corollary 1)}$$
$$\geq \tfrac{1}{2}(\text{Opt}'_{K-1} + p) \qquad \text{(since } q \geq p)$$
$$\geq \tfrac{1}{2}\text{Opt}_K \qquad \text{(by Lemma 7).}$$

If $A_K(x_{i_0}) = 0$, then

$$N_K \geq \tfrac{1}{2}\text{Opt}'_K + q \qquad \text{(by the induction hypothesis)}$$
$$\geq \tfrac{1}{2}\text{Opt}_K + \tfrac{1}{2}q \qquad \text{(by Lemma 7)}$$
$$\geq \tfrac{1}{2}\text{Opt}_K .$$

This completes the proof. $\qquad\qquad\qquad\qquad\qquad\qquad\qquad\qquad\qquad\qquad$ $\square$

The approximation factor proved in the previous theorem is tight, a worst case example is the following: We have two clauses $x_1 \vee x_2$ and $\overline{x_1}$, each with weight 1, a clause x_1 with weight ϵ, and $K = 1$. (If we allow multisets as clause sets, then this can be transformed into an unweighted instance.) The optimum assignment gives x_1 the value zero and x_2 the value one. This satisfies all clauses but x_1 and we get weight 2. GREEDY however gives x_1 the value one and thus only achieves weight $1 + \epsilon$. Thus, in contrast to the maximum coverage problem (i.e., clause sets with only positive literals), this greedy approach does not achieve the optimum approximation factor of $1 - e^{-1}$.

3.2   Estimating the Running Time

The greedy algorithm presented above can be implemented such that its running time is $O(N \log N)$, where N is the length of the input. For the analysis, let $r_i = \max\{p_i, q_i\}$ be the maximum number of clauses that can be satisfied by setting x_i appropriately.

We start with building a heap containing the values r_i ($1 \leq i \leq n$). Then we extract the variable x_i with maximum r_i from the heap and set it to an appropriate value. After that we have to update the $r_{i'}$ values of some variables $x_{i'}$ and maintain the heap. Finally, we continue the recursion.

Let us estimate the running time. Let n_j be the number of variables of clause c_j and m_i be the number of occurences of variable x_i.

In recursion depth t, we extract variable x_{i_t} from the heap and set it to either zero or one. Together with maintaining the heap, this requires a running time of $O(\log n + m_{i_t})$.

Let C_t be the set of clauses that will be satisfied by setting x_{i_t} in depth t. (Note that the sets C_t are pairwise disjoint.) We have to update the value $r_{i'}$ of all variables $x_{i'}$ that occur in clauses of C_t. These are at most $\sum_{c_j \in C_t} n_j$. Together with maintaining the heap, this requires a running time of $O\left(\sum_{c_j \in C_t} n_j \cdot \log n\right)$.

Thus the overall running time is

$$O\left(\sum_{t=1}^{n}\left(\log n \cdot \left(1 + \sum_{c_j \in C_t} n_j\right) + m_{i_t}\right)\right) \subseteq O\left(N \cdot \log N\right).$$

4   Conclusions

We have presented an approximation algorithm with approximation performance $\frac{6+3\cdot e}{16+2\cdot e} - \epsilon$ (for an arbitrary $\epsilon > 0$) for Max-2SAT-CC, the Max-2SAT problem with the additional constraint that the value one may be assigned to at most K variables. Thus, we are able to give a positive answer to Sviridenko's question [10] whether Max-SAT-CC can be approximated better than $1 - e^{-1}$ if the clause length is bounded. Our approach can be extended to handle larger values of ℓ. Since there are more types of mixed clauses, the analysis becomes more complicated.

Furthermore, we have presented a greedy algorithm for Max-SAT-CC with running time $O(N \log N)$, which achieves a tight approximation ratio of $\frac{1}{2}$.

References

1. A. A. Ageev and M. I. Sviridenko. Approximation algorithms for maximum coverage and max cut with given sizes of parts. In *Proc. of the 7th Int. Conf. on Integer Programming and Combinatorial Optimization (IPCO)*, volume 1620 of *Lecture Notes in Comput. Sci.*, pages 17–30. Springer, 1999.
2. N. Alon, J. H. Spencer, and P. Erdös. *The Probabilistic Method*. John Wiley and Sons, 1992.
3. T. Asano and D. P. Williamson. Improved approximation algorithms for MAX SAT. *J. Algorithms*, 42(1):173–202, 2002.
4. G. P. Cornuéjols, M. L. Fisher, and G. L. Nemhauser. Location of bank accounts to optimize float: An analytic study of exact and approximate algorithms. *Management Science*, 23:789–810, 1977.
5. R. G. Downey and M. R. Fellows. *Parameterized Complexity*. Springer, 1999.
6. U. Feige. A threshold of $\ln n$ for approximating set cover. *J. ACM*, 45(4):634–652, 1998.
7. U. Feige and M. X. Goemans. Approximating the value of two prover proof systems, with applications to MAX 2SAT and MAX DICUT. In *Proc. of the 3rd Israel Symp. on the Theory of Comput. and Systems (ISTCS)*, pages 182–189, 1995.
8. M. X. Goemans and D. P. Williamson. New $\frac{3}{4}$-approximation algorithms for the maximum satisfiability problem. *SIAM J. Discrete Math.*, 7(4):656–666, 1994.
9. J. Håstad. Some optimal inapproximability results. *J. ACM*, 48(4):798–859, 2001.
10. M. I. Sviridenko. Best possible approximation algorithm for MAX SAT with cardinality constraint. *Algorithmica*, 30(3):398–405, 2001.

A Better Approximation for the Two-Stage Assembly Scheduling Problem with Two Machines at the First Stage

Yoshiyuki Karuno[1] and Hiroshi Nagamochi[2]

[1] Kyoto Institute of Technology, Sakyo, Kyoto 606-8585, Japan.
karuno@ipc.kit.ac.jp
[2] Toyohashi University of Technology, Toyohashi, Aichi 441-8580, Japan.
naga@ics.tut.ac.jp

Abstract. In this paper, we consider the two-stage assembly scheduling problem with two machines at the first stage. The problem has been known to be strongly NP-hard and a 1.5-approximation algorithm has been obtained. We give a 1.37781-approximation algorithm that runs in $O(n^2 \log n)$ time for an instance with n jobs.

1 Introduction

The two-stage assembly scheduling problem introduced by Potts et al. [4] can be described as follows. The two-stage assembly system consists of m dedicated machines at the first stage and a single assembly machine at the second stage. There are n jobs (or products) to be processed in the assembly system. A job consists of m components, and each component is manufactured on a specific machine at the first stage. It is ready to start assembling the m components on the assembly machine at the second stage when all of the components have been manufactured at the first stage. Each machine at the first stage can manufacture at most one component at a time, and also the assembly machine can assemble at most one job at a time. By saying that a machine processes a *task*, we mean that a machine at the first stage manufactures a component of a job or the assembly machine at the second stage assembles the prepared m components to complete a job. No interruption is allowed for processing tasks. The objective is to find an optimal schedule that minimizes the maximum completion time, i.e., the makespan.

A schedule is a *permutation* one if the processing orderings of jobs are identical over all machines. For the special case of $m = 1$ (i.e., the traditional two-machine flowshop scheduling problem), it is well known that an optimal schedule can be obtained by Johnson's algorithm [2] in $O(n \log n)$ time, and it is a permutation schedule. Potts et al. [4] proved that the problem is strongly NP-hard even when $m = 2$, while an optimal schedule is still an *optimal permutation schedule* (i.e., the one with the minimum makespan among all permutation schedules) for arbitrary m. They also showed that any permutation schedule has a makespan at most twice the optimal, and moreover, presented an $O(mn + n \log n)$

P. Bose and P. Morin (Eds.): ISAAC 2002, LNCS 2518, pp. 199–210, 2002.

time algorithm which delivers a permutation schedule with a makespan at most $(2 - \frac{1}{m})$ times the optimal. Independently of the work by Potts et al. [4], Lee et al. [3] considered the problem with $m = 2$. They also proved that the problem is strongly NP-hard, and designed a 1.5-approximation algorithm. We remark that the $(2 - \frac{1}{m})$-approximation of Potts et al. [4] becomes identical to the 1.5-approximation algorithm when $m = 2$. Hariri and Potts [1] gave lower bounds and some dominance rules, and developed a branch-and-bound algorithm for the problem. In their paper, computational results were reported for problem instances tested with up to 8000 jobs and 10 first-stage machines.

The assembly scheduling problem is important in practice, since it appears in various applications where a set of modules are produced in individual sectors, followed by assembling or packaging them. In fact, a fire engine assembly plant motivated the work by Lee et al. [3]. Each fire engine consists of three main components, the body, the chassis and the engine. In the plant, the body and chassis of a fire engine are produced (in parallel) at two different department in-house, while the engine is purchased from external. Since the engines are delivered directly from the supplier, the assembly of a fire engine can start when both the body and chassis have been produced at the individual departments. Potts et al. [4] gave a different example, which appears in the production of customized personal computers. A customer orders a specific set of parts such as a central processing unit, a hard disk, a display monitor, a keyboard, etc. In this application, these parts are produced at the first stage, and packaging them is performed at the second stage.

In this paper, we consider the case of $m = 2$. Let $\tilde{M}$ denote the machine at the second stage, and $\mathcal{L}$ be the set of indices for the machines at the first stage. We have n jobs, and denote by $\mathcal{P}$ the set of jobs. Each job $p \in \mathcal{P}$ consists of tasks $J_\ell(p)$ on machines M_ℓ at the first stage for $\ell \in \{0, 1\}$ and task $\tilde{J}(p)$ on $\tilde{M}$ at the second stage. The problem instance is denoted by $I = (\mathcal{P}, \mathcal{L}, J)$. For an instance I, we denote by $opt(I)$ the minimum makespan for processing all jobs in I. We may also use $J_\ell(p)$ (or $\tilde{J}(p)$) to denote the processing time of the task.

In this paper, we prove the next result.

Theorem 1. *For the two-stage assembly scheduling problem with two machines at the first stage, a 1.37781-approximation solution can be obtained in $O(n^2 \log n)$ time.*

2   Basic Properties

2.1   Lower Bounds for General m

Let $B = \tilde{J}(1) + \tilde{J}(2) + \cdots + \tilde{J}(n)$, $A_\ell = J_\ell(1) + J_\ell(2) + \cdots + J_\ell(n)$, $\ell \in \mathcal{L}$ and $A = \max_{\ell \in \mathcal{L}} A_\ell$.

We here observe some lower bounds on the minimum makespan $opt(I)$.

Lemma 1. *Let $I = (\mathcal{P}, \mathcal{L}, J)$ be an instance.*

(i) $opt(I) \geq \max\{A, B\}$.

(ii) *For any $p \in \mathcal{P}$, $opt(I) \geq \tilde{J}(p) + \max_{\ell \in \mathcal{L}} J_\ell(p)$.*
(iii) *For any $p_1, p_2 \in \mathcal{P}$, $opt(I) \geq \tilde{J}(p_1) + \tilde{J}(p_2) + \min_{i=1,2}\{\max_{\ell \in \mathcal{L}} J_\ell(p_i)\}$.*
(iv) *For any subinstance $I' = (\mathcal{P}', \mathcal{L}', J)$ of $(\mathcal{P}, \mathcal{L}, J)$ such that $\mathcal{P}' \subseteq \mathcal{P}$ and $\mathcal{L}' \subseteq \mathcal{L}$, $opt(I) \geq opt(I')$.*

It is known [4] that there always exists a permutation schedule among optimal schedules. Thus, in the sequel, we concentrate on only permutation schedules. In other words, we consider an ordering π of all jobs in $\mathcal{P}$.

For a given instance $I = (\mathcal{P}, \mathcal{L}, J)$ and a schedule π of $\mathcal{P}$, we denote by $C_I(\pi)$ the makespan required to process jobs in I according to the schedule π.

Lemma 2. *Let $I = (\mathcal{P}, \mathcal{L}, J)$ be an instance. For a constant $c \in (0,1)$, if there is a subset $\mathcal{P}' \subseteq \mathcal{P}$ such that $\sum_{p \in \mathcal{P}'} \tilde{J}(p) \geq (1-c) \cdot B$, then we can obtain a $(1+c)$-approximation solution in $O(|\mathcal{L}|n \cdot (|\mathcal{P}'|)!)$ time.*

Proof. Let $B' = \sum_{p \in \mathcal{P}'} \tilde{J}(p)$. We first consider the instance $I' = (\mathcal{P}, \mathcal{L}, J')$ such that $\tilde{J}(p)$ is set to be 0 for each $p \in \mathcal{P} - \mathcal{P}'$ (where other processing times at the first and second stages remain unchanged). Then it is not difficult to see that there is an optimal schedule π' to I' such that all the jobs in $\mathcal{P} - \mathcal{P}'$ are processed after jobs in $\mathcal{P}'$ are all finished. Moreover, an ordering of $\mathcal{P} - \mathcal{P}'$ can be chosen arbitrarily once the ordering of $\mathcal{P}'$ is optimized. Therefore, an optimal schedule π' to I' can be obtained in $O(|\mathcal{L}|n \cdot (|\mathcal{P}'|)!)$ time by inspecting all orderings over $\mathcal{P}'$.

Notice that $C_{I'}(\pi')$ is a lower bound on $opt(I)$. Thus, $opt(I) \geq \max\{C_{I'}(\pi'), B\}$. On the other hand, no more $B - B'$ time is required than $C_{I'}(\pi')$ when jobs in I are processed according to π'. That is, $opt(I) \leq C_I(\pi') \leq C_{I'}(\pi') + (B - B')$. Hence we have $\frac{C_I(\pi')}{opt(I)} \leq \frac{C_{I'}(\pi')}{\max\{C_{I'}(\pi'), B\}} \leq \frac{C_{I'}(\pi') + (B-B')}{\max\{C_{I'}(\pi'), B\}} \leq 1 + c$. □

2.2 Johnson's Rule for $m = 1$

When $m = 1$ ($\mathcal{L} = \{0\}$), the problem is known to be polynomially solvable by Johnson's algorithm [2]. For an instance $I' = (\mathcal{P}, \{0\}, J)$, we denote $\mathcal{P}_I^+ = \{p \in \mathcal{P} \mid J_0(p) < \tilde{J}(p)\}$, $\mathcal{P}_I^0 = \{p \in \mathcal{P} \mid J_0(p) = \tilde{J}(p)\}$ and $\mathcal{P}_I^- = \{p \in \mathcal{P} \mid J_0(p) > \tilde{J}(p)\}$. We partition the set $\mathcal{P}$ of jobs into two disjoint classes $\mathcal{P}_I^1$ and $\mathcal{P}_I^2$ so that $\mathcal{P}_I^1 \subseteq \mathcal{P}_I^+ \cup \mathcal{P}_I^0$ and $\mathcal{P}_I^2 \subseteq \mathcal{P}_I^- \cup \mathcal{P}_I^0$ hold. Notice that such a partition may not be unique if $\mathcal{P}_I^0 \neq \emptyset$. Let π^1 be an ordering of $\mathcal{P}_I^1$ arranged in the nondecreasing order with respect to the processing times at the first stage, and π^2 be an ordering of $\mathcal{P}_I^2$ arranged in the nonincreasing order with respect to the processing times at the second stage. Then an optimal schedule is obtained by $\pi = [\pi^1, \pi^2]$.

2.3 Properties for $m = 2$

We show some properties of the problem with $m = 2$. Let $\mathcal{L} = \{0, 1\}$, and let M_0 and M_1 denote two machines at the first stage.

For a real $\lambda \in [0, 1]$, we denote $J_\lambda(p) = (1 - \lambda)J_0(p) + \lambda J_1(p)$ and by $I(\lambda)$ the instance $(\mathcal{P}, \{0\}, J_\lambda)$ with a single machine at the first stage.

Lemma 3. [1] *Let $I = (\mathcal{P}, \{0, 1\}, J)$ be an instance. For any real $\lambda \in [0, 1]$, $opt(I) \geq opt(I(\lambda))$.*

We easily observe the following property.

Lemma 4. (i) *Let π_0^* be an optimal schedule to the instance $I(0)$. If $C_{I(1)}(\pi_0^*) \leq C_{I(0)}(\pi_0^*)$, then π_0^* is optimal to the instance I.*
(ii) *Let π_1^* be an optimal schedule to the instance $I(1)$. If $C_{I(0)}(\pi_1^*) \leq C_{I(1)}(\pi_1^*)$, then π_1^* is optimal to the instance I.*

When jobs in $\mathcal{P}$ are indexed as $1, 2, \ldots, n$, we may denote $\sum_{i \leq p \leq j} J_\lambda(p)$ for $\lambda \in [0, 1]$ by $J_\lambda[i, j]$.

Lemma 5. *For any $\lambda \in [0, 1]$ and ordering π of $\mathcal{P}$, $\min\{C_{I(0)}(\pi), C_{I(1)}(\pi)\} \leq (1 + \frac{1}{4}) \max\{A, C_{I(\lambda)}(\pi)\}$ holds.*

Proof. Let $\mu = \max\{A, C_{I(\lambda)}(\pi)\}$. We derive a contradiction by assuming that for some $\lambda \in (0, 1)$ and ordering π, $C_{I(1)}(\pi) > (1 + \frac{1}{4})\mu$ and $C_{I(0)}(\pi) > (1 + \frac{1}{4})\mu$ hold. W.l.o.g. assume $\pi = [1, 2, \ldots, n]$. Let $h \in \mathcal{P}$ be the earliest job such that when all jobs in $\mathcal{P}$ are processed according to π no idle time occurs to process all jobs $h, h + 1, \ldots, n$ at the second stage of $I(1)$. Let $h' \in \mathcal{P}$ be defined similarly for $I(0)$.

Since the makespan $C_{I(1)}(\pi)$ is longer than $C_{I(\lambda)}(\pi)$ by $\frac{1}{4}\mu$, π must have finished processing job h at the first stage in $I(\lambda)$ at least by $\frac{1}{4}\mu$ time unit earlier than $I(1)$. This implies that

$$J_1[1, h] > J_\lambda[1, h] + \frac{1}{4}\mu.$$

Hence $\lambda \neq 1$. By substituting $J_\lambda[1, h] = (1 - \lambda)J_0[1, h] + \lambda J_1[1, h]$, we have $J_1[1, h] > (1 - \lambda)J_0[1, h] + \lambda J_1[1, h] + \frac{1}{4}\mu$. Thus,

$$J_1[1, h] > J_0[1, h] + \frac{1}{(1 - \lambda)} \cdot \frac{1}{4}\mu.$$

By symmetry, we have

$$J_0[1, h'] > J_1[1, h'] + \frac{1}{\lambda} \cdot \frac{1}{4}\mu \text{ and } \lambda \neq 0.$$

W.l.o.g. assume $h \leq h'$. Hence $J_1[1, h'] \geq J_1[1, h]$. From the above two inequalities, we have

$$\begin{aligned} J_1[1, h] - J_1[1, h'] &> J_0[1, h] - J_0[1, h'] + (\tfrac{1}{1-\lambda} + \tfrac{1}{\lambda}) \cdot \tfrac{1}{4}\mu \\ &\geq J_0[1, h] - J_0[1, h'] + \mu \\ &\geq 0, \end{aligned}$$

where we use the fact that $\frac{1}{\lambda} + \frac{1}{1-\lambda} \geq 4$ ($\lambda \in (0, 1)$) and $J_0[1, h'] \leq A \leq \mu$. This, however, contradicts $J_1[1, h'] \geq J_1[1, h]$. $\square$

3 Sequence of Schedules

Potts et al. [4] proved that any optimal schedule to the instance $I(\lambda)$ for $\lambda = 1/2$ is a 1.5-approximation solution to the original instance I. In this paper, we prove that by an adequate choice of $\lambda \in (0,1)$ an optimal schedule to the instance $I(\lambda)$ is a 1.37781-approximation solution to I except for some tractable cases. To figure out such a λ, we solve the instance $I(\lambda)$ for all λ running from $\lambda = 0$ to $\lambda = 1$. To facilitate this, we introduce a sequence of schedules which starts with an optimal schedule to $I(0)$ and ends up with an optimal one to $I(1)$.

In what follows, for an instance $I(\lambda)$ ($\lambda \in [0,1]$), we define partition $\mathcal{P}$ into two subsets $\mathcal{P}^1_{I(\lambda)} = \{p \in \mathcal{P} \mid J_\lambda(p) \leq \tilde{J}(p)\}$ and $\mathcal{P}^2_{I(\lambda)} = \{p \in \mathcal{P} \mid J_\lambda(p) > \tilde{J}(p)\}$, where we call a job in $\mathcal{P}^1_{I(\lambda)}$ (resp., $\mathcal{P}^2_{I(\lambda)}$) type 1 (resp., type 2) with respect to $I(\lambda)$.

For a schedule (i.e., an ordering) π of $\mathcal{P}$, two jobs p and p' are called consecutive if they appear consecutively (as p, p' or p', p) in π. For two schedules π and π', we say that a schedule π' is obtained from π by switching a consecutive pair if π contains consecutive jobs p and p' and π' is obtained just by exchanging p and p' at their positions. We also say that a schedule π' is obtained from a schedule π by shifting a job if π' is constructed by removing a job p from π and inserting the job p into some other position of π. For a sequence $[\pi_1, \pi_2, \ldots, \pi_K]$ of schedules of $\mathcal{P}$ is called successive if for any $i = 1, 2, \ldots, K - 1$, π_{i+1} is obtained from π_i by switching a consecutive pair or by shifting a job.

In what follows, we consider a successive sequence of schedules with the following property.

Lemma 6. *For a given instance $I = (\mathcal{P}, \{0,1\}, J)$, we can construct in $O(n^2 \log n)$ time a successive sequence $\Pi = [\pi_1, \pi_2, \ldots, \pi_K]$ ($K \leq 4n^2$) of schedules of $\mathcal{P}$ such that*

(i) *for each $i = 1, 2, \ldots, K$, π_i is an optimal schedule to the instance $I(\lambda)$ with some real $\lambda \in [0,1]$. In particular, π_1 is an optimal schedule to the instance $I(0)$ and π_K is an optimal schedule to the instance $I(1)$.*

(ii) *for any two consecutive schedules π_i and π_{i+1}, if π_{i+1} is obtained from π_i by switching a consecutive pair p and p', then there is a $\lambda \in [0,1]$ such that $opt(I(\lambda)) = C_{I(\lambda)}(\pi_i) = C_{I(\lambda)}(\pi_{i+1})$, $J_\lambda(p) \leq \tilde{J}(p)$ and $J_\lambda(p') \leq \tilde{J}(p')$.*

(iii) *for any two consecutive schedules π_i and π_{i+1}, if π_{i+1} is obtained from π_i by shifting a job q, then there is a $\lambda \in [0,1]$ such that $opt(I(\lambda)) = C_{I(\lambda)}(\pi_i) = C_{I(\lambda)}(\pi_{i+1})$ and $J_\lambda(q) = \tilde{J}(q)$.*

Proof. See Appendix 1. □

4 Algorithm

Based on Lemma 6, our 1.37781-approximation algorithm finds a schedule π^{apx} of $\mathcal{P}$ as follows. We denote $\alpha = 0.37781$. We first assume that

$$\mathcal{P} \text{ contains no more than two jobs } p \text{ such that } \tilde{J}(p) \geq \frac{(1-\alpha)B}{3} \qquad (1)$$

(since otherwise by Lemma 2 we can construct a $(1+\alpha)$-approximation solution to I).

We compute a sequence Π in Lemma 6. We then assume that

$$C_{I(0)}(\pi_1) < C_{I(1)}(\pi_1) \text{ and } C_{I(0)}(\pi_K) > C_{I(1)}(\pi_K) \tag{2}$$

(otherwise we can obtain an optimal solution to I by Lemma 4). Thus, by Lemma 6, there exist a consecutive pair of schedules π_k, π_{k+1} and a real $\lambda^* \in [0,1]$ such that

$$\begin{aligned} C_{I(0)}(\pi_k) &\leq C_{I(1)}(\pi_k), \quad C_{I(0)}(\pi_{k+1}) \geq C_{I(1)}(\pi_{k+1}), \\ opt(I(\lambda^*)) &= C_{I(\lambda^*)}(\pi_k) = C_{I(\lambda^*)}(\pi_{k+1}). \end{aligned} \tag{3}$$

From a sequence Π, such a pair π_k and π_{k+1} can be found in $O(n \log n)$ time by conducting a binary search on a sequence of $K = O(n^2)$ schedules.

We here use the following property, where the proof will be given in the next section.

Lemma 7. *Under the assumption of (1), one of the above two schedules π_k and π_{k+1} is a $(1+\alpha)$-approximation solution to I.*

Thus, by choosing one of π_k and π_{k+1}, we obtain a $(1+\alpha)$-approximation solution π^{apx} to I. The entire algorithm is described as follows.

Algorithm SEQ
Input: An instance $I = (\mathcal{P}, \mathcal{L}, J)$.
Output: A $(1+\alpha)$-approximation solution π^{apx} to I.
begin
1   **if** $\mathcal{P}$ contains some three jobs p with $\tilde{J}(p) \geq \frac{(1-\alpha)B}{3}$
2   **then** Output a $(1+\alpha)$-approximation solution π^{apx}
        constructed by Lemma 2, and halt
3   **else** Compute a successive sequence $\Pi = [\pi_1, \pi_2, \ldots, \pi_K]$
        of schedules of $\mathcal{P}$ in Lemma 6;
4   **if** $C_{I(1)}(\pi_1) \leq C_{I(0)}(\pi_1)$ **then** Output $\pi^{apx} := \pi_1$, and halt
5   **else if** $C_{I(0)}(\pi_K) \leq C_{I(1)}(\pi_K)$ **then** Output $\pi^{apx} := \pi_K$, and halt
6   **else** Find k and $\lambda^* \in [0,1]$ that satisfy (3);
7   Output a better schedule from π_k and π_{k+1} as π^{apx}, and halt
           end /* if */
        end /* if */
     end /* if */
end. /* SEQ /*

5   Correctness

To show the correctness of algorithm SEQ, it suffices to prove Lemma 7. Let

$$\gamma = \max \left\{ \max_{\lambda \in [0,1]} opt(I(\lambda)), \max_{p_1, p_2 \in \mathcal{P}} (\tilde{J}(p_1) + \tilde{J}(p_2) + \min_{i=1,2} \{\max_{\ell=0,1} J_\ell(p_i)\}) \right\}$$
$$(\geq \max\{A, B\}),$$

which is a lower bound on $opt(I)$ by Lemma 1(iii) and Lemma 3. By Lemma 5 with $\lambda = \lambda^*$ and $\pi = \pi_k$, we have

$$\min\{C_{I(0)}(\pi_k), C_{I(1)}(\pi_k)\} \le (1 + \frac{1}{4}) \max\{A, opt(I(\lambda^*))\} < (1+\alpha)\gamma.$$

Similarly we obtain $\min\{C_{I(0)}(\pi_{k+1}), C_{I(1)}(\pi_{k+1})\} < (1+\alpha)\gamma$.

If one of $C_{I(1)}(\pi_k)$ and $C_{I(0)}(\pi_{k+1})$ is less than or equal to $(1+\alpha)\gamma$, then by (3) π_k or π_{k+1} is a $(1+\alpha)$-approximation solution to I. To prove Lemma 7, we assume that both $C_{I(1)}(\pi_k)$ and $C_{I(0)}(\pi_{k+1})$ are greater than $(1+\alpha)\gamma$ and derive a contradiction. From the assumption and the above inequalities, we have

$$C_{I(1)}(\pi_k) > (1+\alpha)\gamma \ge C_{I(0)}(\pi_k) \text{ and } C_{I(0)}(\pi_{k+1}) > (1+\alpha)\gamma \ge C_{I(1)}(\pi_{k+1}).$$

This implies that $\pi_k \ne \pi_{k+1}$ and $\lambda^* \ne 0,1$ (otherwise we would have $\gamma \ge opt(I(\lambda^*)) = C_{I(0)}(\pi_{k+1}) > (1+\alpha)\gamma$ or $\gamma \ge opt(I(\lambda^*)) = C_{I(1)}(\pi_k) > (1+\alpha)\gamma$). From these inequalities, we have

$$C_{I(1)}(\pi_k) > (1+\alpha)\gamma \ge C_{I(1)}(\pi_{k+1}) \text{ and } C_{I(0)}(\pi_{k+1}) > (1+\alpha)\gamma \ge C_{I(0)}(\pi_k).$$

There are two cases to be distinguished:
Case-1: π_{k+1} is obtained from π_k by switching a consecutive pair.
Case-2: π_{k+1} is obtained from π_k by shifting a job.

5.1   Analysis of Case-1

W.l.o.g. denote $\pi_k = [1, 2, \ldots, n]$ and let h and $h'(= h+1)$ be the consecutive jobs to be switched to obtain π_{k+1}. Thus, $\pi_{k+1} = [1, 2, \ldots, h-1, h' = h+1, h, h+2, \ldots, n]$. By noting that the difference of π_k and π_{k+1} is the position of h and h' and that $C_{I(1)}(\pi_k) > C_{I(1)}(\pi_{k+1})$, we can conclude that h is the earliest job such that no idle time occurs to process all jobs $h, h+1(= h'), h+2, \ldots, n$ at the second stage of $I(1)$ when all jobs in $\mathcal{P}$ are processed according to π_k. Analogously by $C_{I(0)}(\pi_{k+1}) > C_{I(0)}(\pi_k)$, we see that h' is the earliest job such that π_{k+1} requires no idle time to process all jobs $h'(= h+1), h, h+2, h+3, \ldots, n$ at the second stage of $I(0)$.

We use the following notations:

$$z = J_1[1, h-1] + J_1(h), \quad x = J_0[1, h-1] + J_0(h), \quad y = \lambda^* z + (1-\lambda^*)x,$$

$$z' = J_0[1, h-1] + J_0(h'), \quad x' = J_1[1, h-1] + J_1(h'), \quad y' = \lambda^* x' + (1-\lambda^*)z'.$$

Since the makespan $C_{I(1)}(\pi_k)$ is longer than $C_{I(\lambda^*)}(\pi_k)$ at least by $\alpha\gamma$, π_k must have finished processing job h at the first stage in $I(\lambda^*)$ at least by $\alpha\gamma$ time unit earlier than in $I(1)$. This implies that

$$z > y + \alpha\gamma. \tag{4}$$

Similarly we have

$$z' > y' + \alpha\gamma. \tag{5}$$

By definition of y and (4), we have $z > \lambda^* z + (1-\lambda^*)x + \alpha\gamma$. Hence

$$z > \frac{1}{1-\lambda^*}\alpha\gamma + x. \tag{6}$$

By definition of y' and (5), we have $z' - \alpha\gamma > y' = \lambda^* x' + (1-\lambda^*)z'$, from which

$$x' < z' - \frac{1}{\lambda^*}\alpha\gamma.$$

With this and (6), we have

$$J_1(h) \geq z - x' > \frac{1}{1-\lambda^*}\alpha\gamma + x - z' + \frac{1}{\lambda^*}\alpha\gamma \geq 4\alpha\gamma - A,$$

where we use the fact that $\frac{1}{1-\lambda} + \frac{1}{\lambda} \geq 4$ and $z' \leq A$. By symmetry, we have

$$J_0(h') \geq z' - x > 4\alpha\gamma - A.$$

Since $J_{\lambda^*}(h) \leq \tilde{J}(h)$ and $J_{\lambda^*}(h') \leq \tilde{J}(h')$ hold by Lemma 6(ii), we have

$$\tilde{J}(h) + \tilde{J}(h') \geq J_{\lambda^*}(h) + J_{\lambda^*}(h') \geq \lambda^* J_1(h) + (1-\lambda^*)J_0(h') > 4\alpha\gamma - A.$$

Thus, by Lemma 1(iii), $\tilde{J}(h) + \tilde{J}(h') + \min\{J_1(h), J_0(h')\} > 8\alpha\gamma - 2A \geq (8\alpha - 2)\gamma$ is a lower bound on $opt(I)$. Hence we have

$$\gamma > (8\alpha - 2)\gamma,$$

which is a contradiction to $\alpha > 1/4$.

5.2   Analysis of Case-2

Let λ^* be the real in (3). W.l.o.g. denote $\pi_k = [1, 2, \ldots, n]$ and assume that π_{k+1} is obtained from π_k by shifting job q. By Lemma 6(iii), $J_{\lambda^*}(q) = \tilde{J}(q)$ holds. Assume that q is of type 2 with respect to $I(\lambda^*)$ (the other case where q is of type 1 can be treated symmetrically by exchanging the rolls of π_k and π_{k+1}). Let p be the job of type 1 such that the job q is inserted between p and $p+1$. Thus $\pi_{k+1} = [1, 2, \ldots, p, q, p+1, \ldots, q-1, q+1, \ldots, n]$. By $\pi_k \neq \pi_{k+1}$, we note that

$$p + 1 < q.$$

Note that each job $j \in \{p+1, p+2, \ldots, q\}$ satisfies

$$\tilde{J}(j) \geq J_{\lambda^*}(q) = \tilde{J}(q) \tag{7}$$

since Johnson's rule implies that $\tilde{J}(j) \geq J_{\lambda^*}(j) \geq J_{\lambda^*}(q)$ if job j is of type 1 in $I(\lambda^*)$ and $\tilde{J}(j) \geq \tilde{J}(q) = J_{\lambda^*}(q)$ otherwise (if j is of type 2).

Let h be the earliest job such that π_k requires no idle time to process all jobs $h, h+1, h+2, \ldots, n$ at the second stage of $I(1)$. By $C_{I(1)}(\pi_k) > C_{I(1)}(\pi_{k+1})$, we can conclude that

$$p + 1 \leq h \leq q - 1$$

(where we see that $h \neq q$ since otherwise $J_{\lambda^*}[1,q] + \tilde{J}[q,n] = C_{I(1)}(\pi_k) > C_{I(1)}(\pi_{k+1}) \geq J_{\lambda^*}[1,q] + +\tilde{J}(q-1) + \tilde{J}[q+1,n]$ would imply $\tilde{J}(q) > \tilde{J}(q-1)$, contradicting (7)).

Let h' the earliest job such that π_{k+1} requires no idle time to process h' and all jobs after h' at the second stage of $I(0)$. By $C_{I(0)}(\pi_{k+1}) > C_{I(0)}(\pi_k)$, we see that

$$h' = q \quad \text{or} \quad p+1 \leq h' \leq q-1.$$

We use the following notations:

$$z = J_1[1,h], \quad x = J_0[1,h], \quad y = \lambda^* z + (1-\lambda^*)x,$$

$$z' = J_0[1,p] + J_0(q) + J_0[p+1,h'],$$

$$x' = J_1[1,p] + J_1(q) + J_1[p+1,h'], \quad y' = \lambda^* x' + (1-\lambda^*)z'.$$

As in Case-1, we have

$$z > y + \alpha\gamma, \quad z' > y' + \alpha\gamma,$$

and thereby

$$z - x' > (\frac{1}{1-\lambda^*} + \frac{1}{\lambda^*})\alpha\gamma - A, \quad z' - x > (\frac{1}{1-\lambda^*} + \frac{1}{\lambda^*})\alpha\gamma - A.$$

Hence if $h' \neq q$ then it holds $h' < h$ since otherwise $x' \geq J_1[1,h'] \geq J_1[1,h] = z$ would contradict $z - x' > 4\alpha\gamma - A \geq (4\alpha - 1)\gamma \geq 0$.

We here consider two subcases (a) $q \geq p+3$ and (b) $q = p+2$.

(a) $q \geq p+3$. We see that $J_0(q) \geq z' - x$ since $h' = q$ implies $J_0(q) \geq J_0[1,p] + J_0(q) - J_0[1,h]$ while $h' \neq q$ means that $h' < h$ and $J_0(q) \geq J_0[1,h'] + J_0(q) - J_0[1,h]$. Hence

$$J_0(q) \geq z' - x > (\frac{1}{1-\lambda^*} + \frac{1}{\lambda^*})\alpha\gamma - A \geq ((\frac{1}{1-\lambda^*} + \frac{1}{\lambda^*})\alpha - 1)\gamma.$$

By (1), $\frac{(1-\alpha)B}{3} > \tilde{J}(j)$ must hold for one of these three jobs $j = p+1, p+2, q$. Hence

$$\frac{(1-\alpha)B}{3} > \tilde{J}(j) \geq J_{\lambda^*}(q) \quad \text{(by (7))}$$
$$\geq (1-\lambda^*)J_0(q) > (1-\lambda^*)((\frac{1}{1-\lambda^*} + \frac{1}{\lambda^*})\alpha - 1)\gamma$$
$$\geq (2\sqrt{\alpha} - 1)\gamma,$$

where we use the fact that $f(\lambda) = (1-\lambda)((\frac{1}{1-\lambda} + \frac{1}{\lambda})\alpha - 1)$ attains the minimum at $\lambda = \sqrt{\alpha}$ over the range $\lambda \in (0,1)$. However, $(1-\alpha)/3 > (2\sqrt{\alpha} - 1)$ is a contradiction.

(b) $q = p+2$. In this case, $h' = q$ and $h = p+1$ are only possible choices. Since $h = p+1$ and $h' = q$ are consecutive in π_k, we can carry over the argument of Case-1 and obtain

$$J_1(h) > (\frac{1}{1-\lambda^*} + \frac{1}{\lambda^*})\alpha\gamma - A, \quad J_0(h') > (\frac{1}{1-\lambda^*} + \frac{1}{\lambda^*})\alpha\gamma - A,$$

$$J_{\lambda^*}(h) + J_{\lambda^*}(h') > 4\alpha\gamma - A \geq (4\alpha - 1)\gamma.$$

By Lemma 6(iii), $q = h'$ satisfies

$$\tilde{J}(h') = J_{\lambda^*}(h').$$

If job h is of type 1 in $I(\lambda^*)$, then $\tilde{J}(h) \geq J_{\lambda^*}(h)$ holds and we are done in the similar manner of Case-1. Assume that h is of type 2 in $I(\lambda^*)$, indicating that $\tilde{J}(h) \geq \tilde{J}(h')$ by Johnson's rule. By the definition of γ

$$
\begin{aligned}
\gamma &\geq \min\{J_1(h), J_0(h')\} + \tilde{J}(h) + \tilde{J}(h') \\
&\geq \min\{J_1(h), J_0(h')\} + 2J_{\lambda^*}(h') \\
&\geq \min\{J_1(h), J_0(h')\} + 2(1 - \lambda^*)J_0(h') \\
&\geq (3 - 2\lambda^*)((\tfrac{1}{1-\lambda^*} + \tfrac{1}{\lambda^*})\alpha - 1)\gamma. \\
&> 1.00003\gamma,
\end{aligned}
$$

where we use the fact that for $\alpha = 0.37781$, $f(x) = (3 - 2x)((\frac{1}{1-x} + \frac{1}{x})\alpha - 1)$ takes the minimum over $x \in (0, 1)$ at some value $x^* \in (0.544591, 0.544592)$. This, however, is a contradiction.

References

1. Hariri, A. M. A. and Potts, C. N.: A branch and bound algorithm for the two-stage assembly scheduling problem, European Journal of Operational Research **103** (1997) 547–556
2. Johnson, S. M.: Optimal two- and three-stage production schedules with setup times included, Naval Research Logistics Quarterly **1** (1954) 61–68
3. Lee, C.-Y., Cheng, T. C. E. and Lin, B. M. T.: Minimizing the makespan in the 3-machine assembly-type flowshop scheduling problem, Management Science **39** (1993) 616–625
4. Potts, C. N., Sevast'janov, S. V., Strusevich, V. A., Van Wassenhove, L. N. and Zwaneveld, C. M.: The two-stage assembly scheduling problem: Complexity and approximation, Operations Research **43** (1995) 346–355

Appendix 1: Proof of Lemma 6

To obtain such a sequence in the lemma, we consider how the Johnson's optimal schedule $\pi(\lambda)$ for an instance $I(\lambda)$ changes when λ runs over $[0, 1]$. An optimal schedule due to Johnson's rule is given by $\pi = [\pi^1, \pi^2]$ such that π^1 is an nondecreasing order of jobs $p \in \mathcal{P}^1_{I(\lambda)}$ with respect to $J_\lambda(p)$ and π^2 is an nonincreasing order of jobs $p \in \mathcal{P}^2_{I(\lambda)}$ with respect to $\tilde{J}(p)$. Such a schedule may not be unique if there is a pair of jobs $p, p' \in \mathcal{P}^1_{I(\lambda)}$ with $J_\lambda(p) = J_\lambda(p')$ or a pair of jobs $p, p' \in \mathcal{P}^2_{I(\lambda)}$ with $\tilde{J}(p) = \tilde{J}(p')$. Thus there may be $\Omega(n!)$ such schedules $\pi = [\pi^1, \pi^2]$ for the same λ. As to the ordering of $\mathcal{P}^2_{I(\lambda)}$, we break ties for such a pair p and p' by their job indices and assume that π^2 of $\mathcal{P}^2_{I(\lambda)}$ is always uniquely determined.

For a technical reason, we prepare a dummy job $\tilde{p}$ associated with each job $p \in \mathcal{P}$. Let $\tilde{\mathcal{P}} = \{\tilde{p} \mid p \in \mathcal{P}\}$, and define $J_0(\tilde{p}) = \tilde{J}(p)$, $J_1(\tilde{p}) = \tilde{J}(p)$ for $\tilde{p} \in \tilde{\mathcal{P}}$.

For each job $\hat{p} \in \mathcal{P} \cup \tilde{\mathcal{P}}$, let $L(\hat{p})$ be the linear function with a variable $\lambda \in [0,1]$ such that $g(\hat{p})\lambda + J_0(\hat{p})$, where the gradient $g(\hat{p})$ of $\hat{p}$ is given by $J_1(\hat{p}) - J_0(\hat{p})$. (note that for $\tilde{p} \in \tilde{\mathcal{P}}$, $L(\tilde{p}) = \tilde{J}(p)$, i.e., a constant over $\lambda \in [0,1]$).

Let $LINE$ denote the set of all linear functions $L(\hat{p})$, $\hat{p} \in \mathcal{P} \cup \tilde{\mathcal{P}}$. A pair of two functions $L, L' \in LINE$ with $g(L) \neq g(L')$ is called an $intersection$, and its value $\theta(L, L')$ is defined to be the λ at which L and L' take the same value. Let Λ be the set of all intersections (L, L') with $\theta(L, L') \in [0, 1]$. Observe that Johnson's schedule to instance $I(\lambda)$ changes at a λ that is given by $\theta(L, L')$ with some intersection $(L, L') \in \Lambda$.

In general, some two intersections may have the same θ value. To avoid this technically, we perturb each of the input data $J_1(p)$, $J_0(p)$, $\tilde{J}(p)$ for $p \in \mathcal{P}$ by a sufficiently small amount. It is not difficult to see that by an adequate perturbation no two intersections have the same (perturbed) θ values and all intersections can be arranged in a total order with respect to the perturbed θ values. We will show later that such a total order $\sigma(\Lambda)$ of Λ can be obtained in $O(n^2 \log n)$ time. We here how to construct a successive sequence Π of schedules in Lemma 6 based on such an order $\sigma(\Lambda)$.

Let $\sigma(\Lambda) = [\rho_1, \rho_2, \ldots, \rho_r]$ ($r \leq |\Lambda| \leq (2n)^2$), where each ρ_i denotes an intersection $(L, L') \in \Lambda$. For an intersection $\rho = (L(\hat{p}), L(\hat{p}'))$, $\hat{p}, \hat{p}' \in \mathcal{P} \cup \tilde{\mathcal{P}}$, we write

$$\hat{p} <_\rho \hat{p}'$$

if $L(\hat{p})$ is below $L(\hat{p}')$ when λ is smaller than $\theta(L(\hat{p}), L(\hat{p}'))$ (so either $\hat{p} <_\rho \hat{p}'$ or $\hat{p}' <_\rho \hat{p}$ holds).

We first construct an ordering π'_1 of $\mathcal{P} \cup \tilde{\mathcal{P}}$ in the nondecreasing order with respect to J_0 values such that for any two $\hat{p}, \hat{p}' \in \mathcal{P} \cup \tilde{\mathcal{P}}$ with $J_0(\hat{p}) = J_0(\hat{p}')$, $\hat{p}$ appears before $\hat{p}'$ in π'_1 if $\hat{p} <_{\rho_i} \hat{p}'$ holds for some ρ_i.

Starting with π'_1, we construct orderings $\pi'_2, \pi'_3, \ldots, \pi'_{r+1}$ in such a way that π'_{i+1} is obtained from π'_i by flipping the positions of $\hat{p}$ and $\hat{p}'$ such that $\rho_i = (L(\hat{p}), L(\hat{p}'))$ (notice that such $\hat{p}$ and $\hat{p}'$ appear consecutively in π'_i).

For each ordering π'_i, we regard a job $p \in \mathcal{P}$ as a type 1 job if and only if p appears before the corresponding dummy job $\tilde{p}$ in π'_i. Then we obtain a unique ordering of these type 1 jobs in each π'_i, and denote by π''_i this ordering of the type 1 jobs. Recall that an ordering of type 2 jobs is uniquely determined (by an appropriate tie-breaking). Thus we discard duplications (if any) from the sequence $\pi''_1, \pi''_2, \ldots, \pi''_{r+1}$. For the resulting sequence $\{\pi''_{i_1}, \pi''_{i_2}, \ldots, \pi''_{i_K}\}$ ($K \leq 4n^2$), we obtain the desired sequence Π satisfying Lemma 6 by attaching the ordering of type 2 jobs to each π''_{i_j}.

Finally we prove that an order $\sigma(\Lambda)$ of Λ can be obtained in $O(n^2 \log n)$ time. W.l.o.g. assume that jobs in $\hat{p} \in \mathcal{P} \cup \tilde{\mathcal{P}}$ are indexed as

$$\hat{p}_1, \hat{p}_2, \ldots, \hat{p}_{2n} \tag{8}$$

so that it holds

$$g(\hat{p}_1) \leq g(\hat{p}_2) \leq \cdots \leq g(\hat{p}_{2n}). \tag{9}$$

Let ϵ be a positive real. Then for each $p \in \mathcal{P}$, we consider perturbed values $J_1'(p) := J_1(p) - \epsilon^{n-i+1}$, $J_0'(p) := J_0(p) - \epsilon^{n-i+1}$, and $\tilde{J}'(p) := \tilde{J}(p) - \epsilon^{n-j+1}$, where i is the index with $p = \hat{p}_i$ and j is the index with $\tilde{p} = \hat{p}_j$. Note that such a perturbation does not change the gradient of any linear function $L(\hat{p})$. The next claim shows that a desired ordering $\sigma(\Lambda)$ can be easily obtained without explicitly recomputing the perturbed θ values.

Claim 1. (i) *any two intersections in Λ no longer take the same perturbed θ value for a sufficiently small $\epsilon > 0$ and an nondecreasing ordering of Λ with respect to the θ values perturbed by an ϵ becomes uniquely determined and remains unchanged as ϵ approaches to 0.*
(ii) *For each intersection $\rho = (L, L') \in \Lambda$, let $key(\rho)$ consist of four entries $\theta(L, L'), j, -g(\hat{p}_j) + g(\hat{p}_i), -i$, where i and j be the indices of jobs $\hat{p}_i$ and $\hat{p}_j$ such that $L = L(\hat{p}_i)$, $L' = L(\hat{p}_j)$ and $i < j$ hold. Then the ordering $\sigma(\Lambda)$ in (i) is obtained by arranging intersections ρ in Λ in the lexicographically increasing order with respect to $key(\rho)$.*

Proof. Assume that two intersections $(L(\hat{p}_i), L(\hat{p}_j))$ $(i < j)$ and $(L(\hat{p}_h), L(\hat{p}_k))$ $(h < k)$ take the same θ value. The θ value of the former intersection is given by

$$\frac{J_0(\hat{p}_i) - J_0(\hat{p}_j)}{g(\hat{p}_j) - g(\hat{p}_i)},$$

and the latter by

$$\frac{J_0(\hat{p}_h) - J_0(\hat{p}_k)}{g(\hat{p}_k) - g(\hat{p}_h)}.$$

After the perturbation, these values become respectively

$$\frac{J_0(\hat{p}_i) - J_0(\hat{p}_j)}{g(\hat{p}_j) - g(\hat{p}_i)} + \frac{\epsilon^{n-j+1} - \epsilon^{n-i+1}}{g(\hat{p}_j) - g(\hat{p}_i)}$$

and

$$\frac{J_0(\hat{p}_h) - J_0(\hat{p}_k)}{g(\hat{p}_k) - g(\hat{p}_h)} + \frac{\epsilon^{n-k+1} - \epsilon^{n-h+1}}{g(\hat{p}_k) - g(\hat{p}_h)}.$$

By the indexing rule of $\hat{p}_1, \hat{p}_2, \dots, \hat{p}_{2n}$, we see that if $j > k$ then $\frac{\epsilon^{n-j+1} - \epsilon^{n-i+1}}{g(\hat{p}_j) - g(\hat{p}_i)} > \frac{\epsilon^{n-k+1} - \epsilon^{n-h+1}}{g(\hat{p}_k) - g(\hat{p}_h)} > 0$ (since ϵ is much smaller than $g(\hat{p}_k) - g(\hat{p}_h)$).

Similarly $\frac{\epsilon^{n-j+1} - \epsilon^{n-i+1}}{g(\hat{p}_j) - g(\hat{p}_i)} > \frac{\epsilon^{n-k+1} - \epsilon^{n-h+1}}{g(\hat{p}_k) - g(\hat{p}_h)} > 0$ holds if $j = k$ and $g(\hat{p}_j) - g(\hat{p}_i) < g(\hat{p}_k) - g(\hat{p}_h)$ hold or if $j = k$, $g(\hat{p}_j) - g(\hat{p}_i) = g(\hat{p}_k) - g(\hat{p}_h)$ and $i < h$ hold. This implies that no two intersections have the same θ value any more after the perturbation. We see that (i) holds. Moreover, intersections that coincide before the perturbation are now arranged in the lexicographical order with respect to key, proving (ii). $\square$

The desired order $\sigma(\Lambda)$ can be obtained in $O(n^2 \log n)$ time by sorting elements in Λ with respect to key in Claim 1(ii).

Queaps

John Iacono[1] and Stefan Langerman[2]*

[1] Dept. of Computer and Information Science, Polytechnic University
jiacono@poly.edu
[2] School of Computer Science, McGill University
sl@cgm.cs.mcgill.ca

Abstract. We present a new priority queue data structure, the *queap*, that executes insertion in $O(1)$ amortized time and Extract-min in $O(\log(k + 2))$ amortized time if there are k items that have been in the heap longer than the item to be extracted. Thus if the operations on the queap are first-in first-out, as on a queue, each operation will execute in constant time. This idea of trying to make operations on the least recently accessed items fast, which we call the queueish property, is a natural complement to the working set property of certain data structures, such as splay trees and pairing heaps, where operations on the most recently accessed data execute quickly. However, we show that the queueish property is in some sense more difficult than the working set property by demonstrating that it is impossible to create a queueish binary search tree, but that many search data structures can be made almost queueish with a $O(\log \log n)$ amortized extra cost per operation.

1 Introduction

Self adjusting data structures are those structures that use a simple restructuring heuristic to implement their operations, and forgo any use of explicit balance information. Examples of self adjusting structures include the splay trees[14], pairing heaps[6] and skew heaps[15]. These structures share many of the per operation runtimes as the corresponding best comparison based structures in the amortized sense, while using less memory and being significantly easier to code. For example, splay trees execute k insert, delete, and search operations on a tree of size at most n in time $O(k \log n)$, which is the same as for AVL trees[1], or any of the other balanced binary search trees.

However, self adjusting data structures do more than just match the existing performance of their non-self adjusting counterparts; they are able to execute many seemingly natural classes of searches in time much faster than in $O(\log n)$ time. Currently, there exists no complete analysis of splay trees, instead their runtime is characterized by several different theorems, none of which imply all of the others: The static optimality theorem, the static finger theorem, the working set theorem (all from [14]), the sequential access theorem [16,17] and the

---

* Research is supported by grants from MITACS, FCAR and CRM.

P. Bose and P. Morin (Eds.): ISAAC 2002, LNCS 2518, pp. 211–218, 2002.

dynamic finger theorem [2,3]. Each of these theorems targets a seemingly natural class of sequences of searches, and shows how these desirable classes of sequences run in $o(\log n)$ amortized time per operation. These theorems have been joined by a number of conjectures about the runtime of splay trees: most notably the dynamic optimality conjecture [14], the unified conjecture [10] and the key-independent optimality conjecture [11].

The splay tree theorems and conjectures, although intended to merely bound the runtime of splay trees, can be viewed as characterizations of distribution sensitive behavior that can be used in a much broader context. These properties thus deserve study in their own right, as studying these properties gives us insight into the possibilities and limitations of classic comparison based data structures. Recent work has established the relationships amongst the known properties, thus establishing a hierarchy of properties [10]. New data structures that are, in some ways, advantageous to splay trees have been engineered from several of the properties. Still other work has applied the properties, which were originally formulated for use in analyzing dictionaries, to heaps [9], planar point location [7] and point searching [4].

The properties that splay trees have, and are conjectured to have, show that splay trees execute many "natural sequences" quickly. In this paper we ask the intentionally vague question "Do splay trees execute all natural sequences quickly?" The answer is no, as we show that there is a class of sequences of operations that can be executed on a queue in constant time per operation that requires $\Omega(\log n)$ to execute on splay trees, and, in fact any binary search tree.

We introduce a new property, which we call the *queueish* property. As this property is a variant of the working set property of splay trees, we present the working set property first: The working set property of splay trees states that the time to search item x is $O(\lg w(x))$[1], where $w(x)$ is the number of distinct items that have been accessed (an access is a search or an insert) since x's last access. Informally, in a data structure with the working set property, accesses to items recently accessed are faster than accesses to items that have not been accessed in a while. The queueish property states the complementary idea that in any structure with this property, an access to x is fast if x is one of the least recently accessed items. Formally, a data structure is said to be queueish if the time to search item x is $O(\lg q(x))$, where $q(x) = n - 1 - w(x)$ is the number of distinct items that have *not* been accessed since x's last access, and n is the number of elements in the data structure. Any structure with the queueish property can execute any repeated sequence of n distinct search operations in constant time per operation. The queueish property is seemingly as natural as the working set property, yet we prove in section 5 that splay trees, and in fact any binary search tree structure cannot have the queueish property. More precisely, we show that an amortized lower bound of $\Omega(\log n)$ per access on a binary search tree still holds even when $q(x) = 0$ for every element accessed. We do however provide a general transformation in Section 4 that turns many data structures, including any standard dynamic dictionary and a point location data structure, into one with the queueish property plus an additional $O(\lg \lg n)$ time per operation. This

---

[1] To ease the notation, we denote $\lg x = \log_2 x$ for $x \geq 2$, and $\lg 0 = \lg 1 = 1$.

transformation produces a data structure which is not a binary search tree, and so does not contradict our previous statement.

We also present a positive result for priority queues. We say that a priority queue has the queueish property if the amortized time to insert is $O(1)$ and the amortized time to extract-min x is $O(\lg q(x))$ where $q(x)$ is the number of items that have been in the priority queue longer than x. A similar definition for the working set property can be made. It has been shown that pairing heaps, a form of self-adjusting heap, have the working set property [9]. In section 2, we describe a priority queue data structure with the queueish property, which we call the queap. Queaps might prove useful in some situations. For example if a queap is used for sorting an array, the running time will be $O(n \lg(I/n))$ where I is the number of inversions in the array to sort. More generally, queaps could be advantageous for implementing discrete event simulations such as *future event set* algorithms. In such algorithms, one wishes to simulate some complex system, such as the replacement of broken light bulbs in a building. The algorithm maintains a list of future events (e.g. the breaking time of every currently used light bulb), and at every step extracts the next occurring event in the list (e.g. the first light bulb to break), processes the event (replace the light bulb) and inserts some eventual new events (the time at which the newly replaced light bulb breaks), according to some probability distribution. If the distribution is such that the new events are inserted within the k last occurring events in the list (deterministically or on average), then finding the next event will only require $O(\lg k)$ amortized time (deterministically or on average) for a queap, while insertions will take worst case and amortized constant time. For a survey on discrete event simulation, see e.g. [5, Chapter XIV.5].

The paper proceeds as follows: Section 2 provides a description of queaps, and Section 3 provides the analysis of queaps. A general transformation that turns many data structures into a queueish one with an additional cost of $O(\lg \lg n)$ per operation is found in Section 4. Section 5 provides the proof that no standard binary search tree structure can be queueish.

2   Queaps

Our data structure supports the operations:

$New(Q)$: Creates a new empty Queap.
$Insert(Q, x)$: inserts element x into Q.
$Minimum(Q)$: returns a pointer to the element in Q whose key is minimum.
$Delete(Q, x)$: deletes element x from Q.
$Delete\_Min(Q)$: returns a pointer to the element in Q whose key is minimum, and removes that element from Q.

With both delete operations running in amortized $O(\lg q(x))$ time for an element x and all other operations running in worst case (and amortized) constant time.

Consider the list $x_1, \ldots, x_n$ of elements currently in the data structure, in the order in which they were inserted. These elements will be split into two groups: The *old* set $x_1, \ldots, x_k$ and the *new* set $x_{k+1}, \ldots, x_n$ for some k, $1 \le k \le n$.

The old set will be stored in the leaves of a $(2, 4)$-tree T, from left to right in the order $x_1, \ldots, x_k$, preceded by a dummy leaf x_0 with infinite key value. A $(2, 4)$-tree is a balanced tree structure in which all internal nodes have degree between 2 and 4, and all leaves have same depth [8,13]. In such trees, insertions and deletions of leaves can be performed in amortized constant time per operation. Note that we are using the structural properties of $(2, 4)$-trees, and we are not using them as binary search trees. Each internal node v of T contains h_v, a pointer to the leaf with the minimum key value in its subtree (denoted T_v). Furthermore, with each node v on the path from the root to the leftmost leaf will be stored the pointer c_v to the leaf with the smallest key in $T - T_v$. For these nodes, the value h_v will not be maintained. We also keep a pointer to the leftmost leaf x_0 in T. Note that c_{x_0} contains a pointer to the leaf with the minimum key value in T.

The new keys $x_{k+1}, \ldots, x_n$ will be stored in a linked list L in that order, and a pointer to the element with minimum key value in L will be kept in a variable $minL$. Here is a description of the operations:

$New(Q)$: Create a new empty tree T and list L. $k \leftarrow 0$, $n \leftarrow 0$.
$Insert(Q, x)$: Insert x to the right of L, update $minL$. $n \leftarrow n + 1$.
$Minimum(Q)$: If $key(minL) < key(c_{x_0})$, return $minL$ else return c_{x_0}.
$Delete(Q, x)$: If $x \in L$, insert all the elements of L in T and L becomes empty, update h_v for all nodes v whose children are new or have been modified, up to the root of T. Walk down from the root to x_0, updating the c_v values. $k \leftarrow n$.
Delete x from T, walk in T on the path from x to x_0, correcting h_v values, and then c_v values. $n \leftarrow n - 1$, $k \leftarrow k - 1$.
$Delete\_Min(Q)$: $min \leftarrow Minimum(Q)$. $Delete(Q, min)$. Return min.

3   Running Time Analysis

We will analyze the amortized cost of the operations using the potential method. The potential function for a queap $Q = (T, L)$ will be $\phi(Q) = c|L|$ for some constant c to be specified later. If the actual cost of an operation is t, the amortized cost for that operation is defined to be $\tilde{t} = t + \phi(Q') - \phi(Q)$ where Q and Q' correspond to the data structure before and after the operation, respectively. Since $\phi(Q) \geq 0$ for any Q, and is zero for an empty data structure Q, we know that the total amortized cost of any sequence of operations is an upper bound on the actual cost of that sequence.

Note that we are using $(2, 4)$-trees, a data structure whose cost is already amortized, as a substructure. In the following analysis, we will use the amortized costs for insertions and deletions on those trees. This would be equivalent to using the actual costs of these operations and adding the potential function of the $(2, 4)$-trees to our $\phi(Q)$. We now describe the amortized cost for each operation.

$Insert(Q, x)$: The actual cost of this operation is $O(1)$. Since the size of the list L grows by one, the potential increases by c, which is a constant, and so the amortized cost of this operation is $O(1) + c = O(1)$.

$Minimum(Q)$: This operation doesn't modify the data structure, so its amortized cost is equal to its actual cost, $O(1)$.

$Delete(Q, x)$: If the deleted leaf is in T, then the cost of the delete operation in the tree is $O(1)$. We then still have to correct the h_v and c_v pointers on the path from x to x_0. Let r be the highest node on that path. If r is k levels above the leaves, then the subtree of the left child of r contains at least 2^{k-1} leaves, all of which have been inserted before x. Thus $q(x) > 2^{k-1}$ and $k = O(\lg q(x))$ and so is the cost of this operation. Since this doesn't modify the potential function, the amortized cost is also $O(\lg q(x))$.

If $x \in L$, we first need to insert all the elements of L in the tree T. This has a cost of $a|L|$ for some constant a (amortized over $(2, 4)$-tree operations). Once these insertions are done, the update of the h_v and c_v pointers for all nodes that are new or have been structurally modified is certainly no more than the very cost of the insertions, and so the total time spent so far is bounded by $2a|L|$.

At this point, if the root of T has not been modified, we still need to continue correcting h_v values for nodes v whose children's h values have been modified, up to the root of T. This costs at most $O(\lg |T|)$. We now walk from the root of T down to x_0, correcting c_v values, for a cost of $O(\lg |T|)$. The total spent so far is now no more than $2a|L| + O(\lg |T|)$, but note that if x was in L, then all elements in T have been inserted before x and $q(x) > |T|$. So we have spent $2a|L| + O(\lg q(x))$ and have caused a drop in potential of $c|L|$ since all elements of the list have been removed from it. So if $c > 2a$, the amortized cost for merging T and L is $O(\lg q(x))$. We still have to delete x from the updated tree, but this can be done in $O(\lg q(x))$ as seen above.

$Delete\_Min(Q)$: The amortized cost of this operation is just the sum of the costs of the operations it uses, i.e. $O(\lg q(x))$.

4 Weakly Queueish Data Structures

We here present a simple scheme to transform data structures for searching problems into an amortized queuish equivalent with an additive $O(\log \log n)$ access cost. A *searching problem* consists of maintaining a set A of objects with respect to some binary relation $\square[x, a]$. A query $Q(x, A)$ will return some object $a \in A$ for which $\square[x, a]$ is true, if such an object exists. For ease of notation we will write $q(x)$ for $q(a)$ if a is the object returned by $Q(x, A)$. It is assumed that the $\square$ operator is computable in $O(1)$ time.

Theorem 1. *Let $D(A)$ be a data structure for solving a searching problem with query operation Q on a set A with n elements. If $D(B')$ for any $B' \subset B$, $|B| = O(|B'|^2)$ can be constructed in $O(|B'|)$ time given $D(B)$, and supports $O(\lg |B'|)$ query times, then $D(A)$ can be transformed into a data structure supporting $O(\lg q(x) + \lg \lg n)$ amortized query times.*

Proof. The data structure consists of a series of data structures $D_1, D_2, ..., D_k$ and a series of queues $Q_1, Q_2, ..., Q_k$. The concatenation of all these queues contains all the elements in increasing order of the last access times on these elements, i.e. Q_1 contains the least recently accessed items and Q_k contains the most recently accessed items. The queues are implemented as doubly linked lists, and we will refer to the most recently accessed items in Q_i as being in the *head* of Q_i, and to its least recently accessed items as being in the *tail* of Q_i.

The elements of Q_i are all present in D_i but D_i might contain some elements not in Q_i. Pointers are maintained between each element in D_i and its corresponding element in one of the queues even if that queue is not Q_i. Every element in Q_i contains the number i of the queue it is in. For $i < k$, the size of D_i is $2^{(2^i)}$ and the size of Q_i is greater than $2^{(2^i-1)}$. D_k will contain all the elements present in the whole data structure, and $k = \lfloor \lg \lg n \rfloor$. This will ensure that $|Q_k| > 2^{(2^k-1)}$. Let $m_i = |D_i| - |Q_i|$, the number of elements in D_i not present in Q_i. For this data structure we will use the potential function

$$\phi = c \sum_{i=1}^{k-1} m_i (\lfloor \lg \lg n \rfloor - i)$$

At creation, $|D_i| = |Q_i|$ for all $i < k$, and so $\phi = 0$. Every time an element x is looked up, we query it sequentially in $D_1, D_2, \ldots$ until it is found in D_i. We then know the queue Q_j that contains x and $j \geq i$. Remove x from Q_j and insert it in the head of Q_k. The actual cost of this operation is $O(2^i) = O(\lg q(x))$, and the increase in potential is $c(\lfloor \lg \lg n \rfloor - j)$, so the amortized cost so far is $O(\lg q(x) + \lg \lg n)$, but $|Q_j|$ might have just become too small. If that happens, we remove $r = 2^{(2^j)} - |Q_j|$ elements from the tail of Q_{j+1} to insert them at the head of Q_j, and we reconstruct D_j from Q_j. The actual cost of this operation is no more than ar for some constant a. But the potential has now decreased by $cr(\lfloor \lg \lg n \rfloor - j)$ for structure j and increased by $cr(\lfloor \lg \lg n \rfloor - (j+1))$ for structure $j + 1$, and so altogether, it has decreased by cr. So if $c > a$, this reconstructing operation has a null amortized cost. We now might still have to reconstruct the structure $j + 1$ and so on, but all this is also done for free, and so the amortized query time for x is $O(\lg q(x) + \lg \lg n)$. □

Note that if the original data structure allows $O(1)$ time deletions when supplied with a pointer to the element (e.g. marking the deleted elements), then the transformed data structure could support amortized $O(\lg \lg n)$ time deletions by simultaneously deleting elements from D_i whenever they are removed from Q_i, and reconstructing the structure whenever some constant fraction of the elements have been deleted. If insertions are supported in the original data structure, they can be supported in the transformed structure with the same amortized times by always inserting in D_k and rebuilding the structure whenever n grows by some constant factor.

Corollary 1. *Given a totally ordered set A of n elements, it is possible to maintain them in a dictionnary data structure supporting $O(\lg q(x) + \lg \lg n)$ query times to access element $x \in A$.*

Proof. Let $D(A)$ be an array keeping the elements of A in sorted order. Given $D(B)$ and a set $B' \subset B$, the elements of B' given as unordered indices in $D(B)$ and $|B| = O(|B'|^2)$, it is possible to construct $D(B')$ in $O(|B'|)$ time using radix sorting. □

Corollary 2. *Given a set of n disjoint triangles in R^2, it is possible to maintain them in a point location data structure so that given a point $p \in R$, the triangle Δ containing p, if it exists, can be found in $O(\lg q(\Delta) + \lg \lg n)$.*

Proof. The data structure $D(A)$ for a set A of triangles will be the combination of the point location data structure of Kirkpatrick [12], and an array containing the vertices of the triangles in A in sorted order of their x coordinate. Given $D(B)$ and a set $B' \subset B$, and $|B| = O(|B'|^2)$, it is possible as in the previous corollary to reconstruct the sorted array of vertices for $D(B')$. Then, knowing the sorted order of the vertices for B', the point-location data structure of Kirkpatrick can be constructed in $O(|B'|)$ time. □

It is interesting to note that no data structure is known that approaches the working set running times for a point location data structure.

5   No Queueish Binary Search Trees

In this section we show that no dictionary implemented as a standard binary search tree structure can be queueish. By a standard binary search tree, we mean one that can traverse one edge in the tree or perform one rotation in unit cost. More precisely, we will consider an initial binary search tree T_0 containing n elements and a sequence s of nodes of T_0, and denote by $\chi(s, T_0)$ the minimum cost of accessing the nodes in the sequence s, starting from the tree T_0, where the cost for traversing an edge and the cost for performing a rotation in the tree is 1. We can assume without loss of generality that the nodes in T_0 are indexed in symmetric order by integers from 1 to n. Wilber [18] proves:

Theorem 2 (Wilber [18]). *There exists a sequence s^* of length n such that for any initial search tree T_0 containing n elements, $\chi(s^*, T_0) = \Omega(n \lg n)$.*

Or in other words, the amortized cost per operation for sequence s^* on any standard binary search tree structure is $\Omega(\lg n)$. To strengthen the statement of our theorem, we introduce a weaker version of the queueish property: a dictionnary data structure is said to be *weakly queueish* if a search to item x runs in amortized time $O(\lg q(x)) + o(\lg n)$. We can now prove our theorem:

Theorem 3. *No standard binary search tree structure can be queueish or even weakly queueish.*

Proof. Let $\hat{s}$ be the sequence formed by repeating the sequence s^* $\lg n$ times. By Wilber's theorem, $\chi(\hat{s}, T_0) = \Omega(n(\lg n)^2)$ for any T_0. On the other hand, any queueish or weakly queueish dictionnary data structure will run the first occurence of s^* in $O(n \lg n)$ since $q(x) \leq n$ by definition, and for the remainder of the sequence, $q(x) = 0$ and so the total cost for accessing the sequence $\hat{s}$ on a queueish data structure would be $O(n \lg n)$, and $o(n(\lg n)^2)$ on a weakly queueish data structure. This implies that a queueish or even a weakly queueish dictionnary data structure cannot be implemented as a binary search tree. □

For example, this theorem implies that the data structure presented in the previous section, which is weakly queueish, could not be implemented as a binary search tree and keep the same running times.

References

1. G. M. Adel'son-Vel'skii and E. M. Landis. An algorithm for the organization of information. *Soviet. Math.*, 3:1259–1262, 1962.
2. R. Cole. On the dynamic finger conjecture for splay trees. Part II: The proof. Technical Report Computer Science TR1995-701, New York Univerity, 1995.
3. R. Cole, B. Mishra, J. Schmidt, and A. Siegel. On the dynamic finger conjecture for splay trees. Part I: Splay sorting log n-block sequences. Technical Report Computer Science TR1995-700, New York Univerity, 1995.
4. E. Demaine, J. Iacono, and S. Langerman. Proximate point searching. In *Proc. 14th Canad. Conf. on Computational Geometry*, pages 1–4, 2002.
5. L. Devroye. *Nonuniform random variate generation*. Springer-Verlag, New York, 1986.
6. M. L. Fredman, R. Sedgewick, D. D. Sleator, and R. E. Tarjan. The pairing heap: A new form of self-adjusting heap. *Algorithmica*, 1:111–129, 1986.
7. M. T. Goodrich, M. Orletsky, and K. Ramaiyer. Methods for achieving fast query times in point location data structures. In *Proc. 8th ACM-SIAM Sympos. Discrete Algorithms*, pages 757–766, 1997.
8. S. Huddleston and K. Mehlhorn. A new data structure for representing sorted lists. *Acta Inform.*, 17:157–184, 1982.
9. J. Iacono. Improved upper bounds for pairing heaps. In *7th Scandinavian Workshop on Algorithm Theory*, pages 32–45, 2000.
10. J. Iacono. Alternatives to splay trees with $O(\log n)$ worst-case access times. In *Proc. 12th ACM-SIAM Sympos. Discrete Algorithms*, pages 516–522, 2001.
11. J. Iacono. *Distribution Sensitive Data Structures*. PhD thesis, Rutgers University – New Brunswick, 2001.
12. D. G. Kirkpatrick. Optimum search in planar subdivisions. *SIAM J. Comput.*, 12(1):28–35, 1983.
13. D. Maier and S. C. Salveter. Hysterical B-trees. *Inform. Process. Lett.*, 12:199–202, 1981.
14. D. D. Sleator and R. E. Tarjan. Self-adjusting binary trees. *JACM*, 32:652–686, 1985.
15. D. D. Sleator and R. E. Tarjan. Self-adjusting heaps. *SIAM Journal of Computing*, 15:52–69, 1986.
16. R. Sundar. *Amoritzed Complexity of Data Structures*. PhD thesis, New York University, 1991.
17. R. E. Tarjan. Sequential access in splay trees takes linear time. *Combinatorica*, 5:367–378, 1985.
18. R. Wilbur. Lower bounds for accessing binary search trees with rotations. In *Proc. 27th Symp. on Foundations of Computer Science*, pages 61–69, 1986.

Funnel Heap – A Cache Oblivious Priority Queue

Gerth Stølting Brodal[1],[*],[**] and Rolf Fagerberg[1],[*]

BRICS[***], Department of Computer Science,
University of Aarhus, Ny Munkegade, DK-8000 Århus C, Denmark.
{gerth,rolf}@brics.dk

Abstract. The cache oblivious model of computation is a two-level memory model with the assumption that the parameters of the model are unknown to the algorithms. A consequence of this assumption is that an algorithm efficient in the cache oblivious model is automatically efficient in a multi-level memory model. Arge et al. recently presented the first optimal cache oblivious priority queue, and demonstrated the importance of this result by providing the first cache oblivious algorithms for graph problems. Their structure uses cache oblivious sorting and selection as subroutines. In this paper, we devise an alternative optimal cache oblivious priority queue based only on binary merging. We also show that our structure can be made adaptive to different usage profiles.

1 Introduction

External memory models are formal models for analyzing the memory access patterns of algorithms on modern computer architectures with several levels of memory and caches. The cache oblivious model, recently introduced by Frigo et al. [13], is based on the I/O model of Aggarwal and Vitter [1], which has been the most widely used external memory model—see the surveys by Arge [2] and Vitter [14]. Both models assume a two-level memory hierarchy where the lower level has size M and data is transfered between the two levels in blocks of B elements. The difference is that in the I/O model the algorithms are aware of B and M, whereas in the cache oblivious model these parameters are unknown to the algorithms and I/Os are handled automatically by an optimal off-line cache replacement strategy.

Frigo et al. [13] showed that an efficient algorithm in the cache oblivious model is automatically efficient on each level of a multi-level memory model. They also presented optimal cache oblivious algorithms for matrix transposition, FFT, and sorting. Cache oblivious search trees which match the search cost of the standard (cache aware) B-trees [4] were presented in [6,8,9,11]. Cache oblivious

* Partially supported by the Future and Emerging Technologies programme of the EU under contract number IST-1999-14186 (ALCOM-FT).
** Supported by the Carlsberg Foundation (contract number ANS-0257/20).
*** Basic Research in Computer Science, www.brics.dk, funded by the Danish National Research Foundation.

algorithms have also been given for problems in computational geometry [6,10], for scanning dynamic sets [5], and for layout of static trees [7]. Recently, the first cache oblivious priority queue was developed by Arge et al. [3], who also showed how this result leads to several cache oblivious graph algorithms. The structure of Arge et al. uses existing cache oblivious sorting and selection algorithms as subroutines.

In this paper, we present an alternative cache oblivious priority queue, Funnel Heap, based only on binary merging. Essentially, our structure is a single heap-ordered tree with binary mergers in the nodes and buffers on the edges. It was inspired by the cache oblivious merge sort algorithm Funnelsort presented in [13] and simplified in [10]. Like the priority queue of Arge et al., our data structure supports the operations INSERT and DELETEMIN using amortized $O(\frac{1}{B} \log_{M/B} \frac{N}{B})$ I/Os per operation, under the so-called tall cache assumption $M \geq B^2$. Here, N is the total number of elements inserted.

For a slightly different algorithm we give a refined analysis, showing that the priority queue adapts to different profiles of usage. More precisely, we show that the ith insertion uses amortized $O(\frac{1}{B} \log_{M/B} \frac{N_i}{B})$ I/Os, where N_i can be defined in any of the following three ways: (a) N_i is the number of elements present in the priority queue when performing the ith insert operation, (b) if the ith inserted element is removed by a DELETEMIN operation prior to the jth insertion then $N_i = j - i$, or (c) N_i is the maximum rank that the ith inserted element has during its lifetime in the priority queue, where rank denotes the number of smaller elements present in the priority queue. DELETEMIN is amortized for free since the work is charged to the insertions. These results extend the line of research taken in [12], where (a) and (c) are called size profile and max depth profile, respectively.

We note that as in [10], we can relax the tall cache assumption by changing parameters in the construction. More precisely, for any $\varepsilon > 0$ a data structure only assuming $M \geq B^{1+\varepsilon}$ can be made, at the expense of $\log_{M/B}(x)$ becoming $\frac{1}{\varepsilon} \log_M(x)$ in the expressions above for the running times. We leave the details to the full paper.

This paper is organized as follows. In Section 2 we introduce the concept of mergers and in Section 3 we describe our priority queue. Section 4 gives the analysis of the presented data structure. Finally, Section 5 gives the analysis based on different profiles of usage.

2   Mergers

Our data structure is based on *binary mergers*. A binary merger takes as input two sorted streams of elements and delivers as output the sorted stream formed by merging of these. A *merge step* moves one element from the head of an input stream to the tail of the output stream. The heads of the input streams and the tail of the output stream reside in *buffers* holding a limited number of elements. A buffer is simply an array of elements, plus fields storing its capacity and pointers to its first and last elements. Figure 1 shows a binary merger.

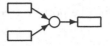

Fig. 1. A binary merger.

Binary mergers may be combined to *binary merge trees* by letting the output buffer of one merger be an input buffer of another. In other words, a binary merge tree is a binary tree with mergers at the internal nodes and buffers at the edges. The leaves of the tree are buffers containing the streams to be merged. See Figure 3 for example of a merge tree. Note that we describe a merger and its output buffer as separate entities mainly in order to visualize the binary merge process. In an actual implementation, the two will probably be identified, and merge trees will simply be binary trees of buffers.

Invoking a binary merger in a merge tree means performing merge steps until its output buffer is full (or both input streams are exhausted). If an input buffer gets empty during the process (but the corresponding stream is not exhausted), it is filled by invoking the merger having this buffer as output buffer. If both input streams of a merger get exhausted, its output stream is marked as exhausted. The resulting recursive procedure, except for the issue of exhaustion, is shown in Figure 2 as FILL(v). An invocation FILL(r) of the root r of the merge tree produces the next part of a stream which is the merge of the streams at the leaves of the tree.

Procedure FILL(v)
    while v's output buffer is not full
        if left input buffer empty
            FILL(left child of v)
        if right input buffer empty
            FILL(right child of v)
        perform one merge step

Fig. 2. Invoking a merger.

One particular merge tree is the *k-merger*. In this paper, we only consider $k = 2^i$ for i a positive integer. A k-merger is a perfectly balanced binary merge tree with $k - 1$ binary mergers, k input streams, and buffers of specific sizes. The size of the output buffer of the root is k^3. The sizes of the remaining buffers are defined recursively: Let the *top tree* be the subtree consisting of all nodes of depth at most $\lceil i/2 \rceil$, and let the subtrees rooted by nodes at depth $\lceil i/2 \rceil + 1$ be the *bottom trees*. The buffers at edges between nodes at depth $\lceil i/2 \rceil$ and depth $\lceil i/2 \rceil + 1$ all have size $\lceil k^{3/2} \rceil$, and the sizes of the remaining buffers are defined by recursion on the top tree and the bottom trees. A 16-merger is illustrated in Figure 3.

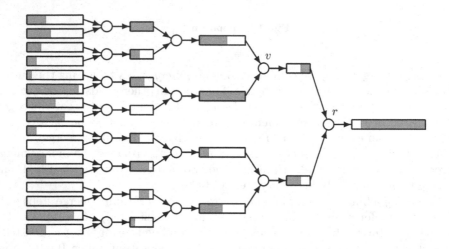

Fig. 3. A 16-merger consisting of 15 binary mergers. Shaded regions are the occupied parts of the buffers. The procedure FILL(r) has been called on the root r, and is currently performing merge steps at its left child v.

To achieve I/O efficiency in the cache oblivious model, the memory layout of a k-merger is also defined recursively. The entire k-merger is laid out in contiguous memory locations, first the top tree, then the middle buffers, and finally the bottom trees, and this layout is applied recursively within the top tree and each of the bottom trees.

The k-merger structure was defined by Frigo et al. [13] for use in their cache oblivious mergesort algorithm Funnelsort. The algorithm described above for invoking a k-merger appeared in [10], and is a simplification of the original one. For both algorithms, the following lemma holds [10,13].

Lemma 1. *The invocation of the root of a k-merger uses $O(k + \frac{k^3}{B} \log_{M/B} k^3)$ I/Os, if $M \geq B^2$. The space required for a k-merger is $O(k^2)$, not counting the space for the input and output streams.*

3   The Priority Queue

Our data structure consists of a sequence of k-mergers with double-exponentially increasing k, linked together in a list as depicted in Figure 4, where circles are binary mergers, rectangles are buffers, and triangles are k-mergers. The entire structure constitutes a single binary merge tree. Roughly, the growth of k is given by $k_{i+1} = k_i^{4/3}$.

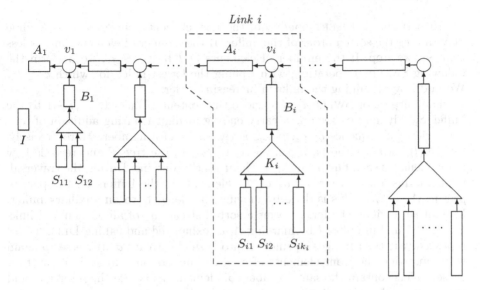

Fig. 4. The priority queue based on binary mergers.

More precisely, let k_i and s_i be values defined inductively as follows,

$$
\begin{aligned}
(k_1, s_1) &= (2, 8)\,, \\
s_{i+1} &= s_i(k_i + 1)\,, \\
k_{i+1} &= \lceil\lceil s_{i+1}^{1/3}\rceil\rceil\,,
\end{aligned}
\tag{1}
$$

where $\lceil\lceil x\rceil\rceil$ denotes the smallest power of two above x, i.e. $\lceil\lceil x\rceil\rceil = 2^{\lceil \log x\rceil}$. Link i in the linked list consists of a binary merger v_i, two buffers A_i and B_i, and a k_i-merger K_i with k_i input buffers $S_{i1}, \ldots, S_{ik_i}$. We refer to B_i, K_i, and $S_{i1}, \ldots, S_{ik_i}$ as the *lower part* of the link. The size of both A_i and B_i is k_i^3, and the size of each S_{ij} is s_i. Link i has an associated counter c_i for which $1 \le c_i \le k_i + 1$. Its initial value is one. It will be an invariant that $S_{ic_i}, \ldots, S_{ik_i}$ are empty.

Additionally, the structure contains one insertion buffer I of size s_1. All buffers contain a (possibly empty) sorted sequence of elements. The structure is laid out in memory in the order I, link 1, link 2, $\ldots$, and within link i the layout order is c_i, A_i, v_i, B_i, K_i, S_{i1}, S_{i2}, $\ldots$, S_{ik_i}.

The linked list of buffers and mergers constitute one binary tree T with root v_1 and with sorted sequences of elements on the edges. We maintain the invariant that this tree is heap-ordered, i.e. when traversing any path towards the root, elements will be passed in decreasing order. Note that the invocation of a binary merger maintains this invariant. The invariant implies that if buffer A_1 is non-empty, the minimum element in the queue will be in A_1 or in I.

To perform a DELETEMIN operation, we first call FILL(v_1) if buffer A_1 is empty. We then remove the smallest of the elements in A_1 and I from its buffer, and return it.

To perform an INSERT operation, the new element is inserted into I while maintaining the sorted order of the buffer. If the number of elements in I is less than s_1, we stop. If the number of elements in I becomes s_1, we perform the following SWEEP(i) operation, with i being the lowest index for which $c_i \leq k_i$. We find i by examining the links in increasing order.

The purpose of SWEEP(i) is to move the content of links $1, \ldots, i-1$ to the buffer S_{ic_i}. It may be seen as a carry ending at digit i during addition of one, if we view the sequence $c_1, c_2, c_3, \ldots$ as the digits of a number. More precisely, SWEEP(i) traverses the path p from A_1 to S_{ic_i} in the tree T and records how many elements each buffer on this path currently contains. During the traversal, it also forms a sorted stream σ_1 of the elements in the buffers on the part of p from A_i to S_{ic_i}. This is done by moving the elements to an auxiliary buffer. In another auxiliary buffer, it forms a sorted stream σ_2 of all elements in links $1, \ldots, i-1$ and in buffer I by marking A_i as exhausted and calling DELETEMIN repeatedly. It then merges σ_1 and σ_2 into a single stream σ, traverses p again while inserting the front elements of σ in the buffers on p in such a way that these buffers contain the same numbers of elements as before the insertion, and then inserts the remaining part of σ in S_{ic_i}. Finally, it resets c_ℓ to one for $\ell = 1, 2, \ldots, i-1$ and increments c_i by one.

4   Analysis

4.1   Correctness

By the discussion above, correctness of DELETEMIN is immediate. For INSERT, we must show that the two invariants are maintained and that S_{ic_i} does not overflow when calling SWEEP(i).

After an INSERT, the new contents in the buffers on the path p are the smallest elements in σ, distributed exactly as the old contents. Hence, an element on this path can only be smaller than the element occupying the same location before the operation. It follows that the heap-order invariant is maintained.

The lower part of link i is emptied each time c_i is reset to one. This implies that the invariant requiring $S_{ic_i}, \ldots, S_{ik_i}$ to be empty is maintained. It also implies that the lower part of link i never contains more than the number of elements inserted into $S_{i1}, S_{i2}, \ldots, S_{ik_i}$ by the k_i SWEEP(i) operations occurring since last time c_i was reset. From the definition (1) we by induction on i get $s_i = s_1 + \sum_{j=1}^{i-1} k_j s_j$ for all i. If follows by induction on time that the number of elements inserted into S_{ic_i} during SWEEP(i) is at most s_i.

4.2   Complexity

Most of the work performed is the movement of elements upwards in the tree T during invocations of binary mergers in T. We account for the I/Os incurred during the filling of a buffer by charging them evenly to the elements filled into the buffer, except when an A_i or B_i buffer is not filled completely due to exhaustion, where we account for the I/Os by other means.

We claim that the number of I/Os charged to an element during its ascent in T from an input stream of K_i to the buffer A_1 is $O(\frac{1}{B}\log_{M/B} s_i)$, if we identify elements residing in buffers on the path p at the beginning of SWEEP(i) with those residing at the same positions in these buffers at the end of SWEEP(i).

To prove the claim, we assume that the maximal number of small links are kept in cache always—the optimal cache replacement strategy of the cache oblivious model can only incur fewer I/Os. More precisely, let Δ_i be the space occupied by links 1 to i. From (1) we have $s_i^{1/3} \le k_i < 2s_i^{1/3}$, so the $\Theta(s_i k_i)$ space usage of $S_{i1},\ldots,S_{ik_i}$ is $\Theta(k_i^4)$, which by Lemma 1 dominates the space usage of link i. Also from (1) we have $s_i^{4/3} < s_{i+1} < 3s_i^{4/3}$, so s_i and k_i grows doubly-exponentially with i. Hence, Δ_i is dominated by the space usage of link i, implying $\Delta_i = \Theta(k_i^4)$. We let i_M be the largest i for which $\Delta_i \le M$ and assume that links 1 to i_M are kept in cache always.

Consider the ascent of an element from K_i to B_i for $i > i_M$. By Lemma 1, each invocation of the root of K_i incurs $O(k_i + \frac{k_i^3}{B}\log_{M/B} k_i^3)$ I/Os. From $M < \Delta_{i_M+1}$ and the above discussion, we have $M = O(k_i^4)$. The tall cache assumption $B^2 \le M$ gives $B = O(k_i^2)$, which implies $k_i = O(k_i^3/B)$. As we are not counting invocations of the root of K_i where B_i is not filled completely, i.e. where the root is exhausted, it follows that each element is charged $O(\frac{1}{B}\log_{M/B} k_i^3) = O(\frac{1}{B}\log_{M/B} s_i)$ I/Os to ascend through K_i and into B_i.

The element can also be charged during insertion into A_j for $j = i_M,\ldots,i$. The filling of A_j incurs $O(1 + |A_j|/B)$ I/Os. From $B = O(k_{i_M+1}^2) = O(k_{i_M}^{8/3})$ and $|A_j| = k_j^3$, we see that the last term dominates. Therefore an element is charged $O(1/B)$ per buffer A_j, as we only charge when the buffer is filled completely. From $M = O(k_{i_M+1}^4) = O(s_{i_M}^{16/9}) = O(s_{i_M})$, we by the doubly-exponentially growth of s_j get that $i - i_M = O(\log\log_M s_i) = O(\log_M s_i) = O(\log_{M/B} s_i)$. Hence, the ascent through K_i dominates over insertions into A_j for $j = i_M,\ldots,i$, and the claim is proved.

To prove the I/O complexity of our structure stated in the introduction, we note that by induction on i, at least s_i insertions take place between each call to SWEEP(i). A call to SWEEP(i) inserts at most s_i elements in $S_{i c_i}$. We let the last s_i insertions preceding the call to SWEEP(i) pay for the I/Os charged to these elements during their later ascent through T. By the claim above, this cost is $O(\frac{1}{B}\log_{M/B} s_i)$ I/Os per insertion. We also let these insertions pay for the I/Os incurred by SWEEP(i) during the formation and placement of streams σ_1, σ_2, and σ, and for I/Os incurred by filling buffers which become exhausted. We claim that these can be covered without altering the $O(\frac{1}{B}\log_{M/B} s_i)$ cost per insertion.

The claim is proved as follows. The formation of σ_1 is done by a traversal of the path p. By the specified layout of the data structure (including the layout of k-mergers), this traversal is part of a linear scan of the part of memory between A_1 and the end of K_i. Such a scan takes $O((\Delta_{i-1} + |A_i| + |B_i| + |K_i|)/B) = O(k_i^3/B) = O(s_i/B)$ I/Os. The formation of σ_2 has already been accounted for by charging ascending elements. The merge of σ_1 and σ_2 into σ and the placement of σ are not more costly than a traversal of p and $S_{i c_i}$, and hence also incur $O(s_i/B)$ I/Os. To account for the I/Os incurred when filling buffers which become exhausted, we note that B_i, and therefore also A_i, can only become exhausted once between each call to SWEEP(i). From $|A_i| = |B_i| = k_i{}^3 = \Theta(s_i)$ it follows that charging each call to SWEEP(i) an additional cost of $O(\frac{s_i}{B} \log_{M/B} s_i)$ I/Os will cover all such fillings, and the claim is proved.

In summary, charging the last s_i insertions preceding a call to SWEEP(i) a cost of $O(\frac{1}{B} \log_{M/B} s_i)$ I/Os each will cover all I/Os incurred by the data structure. Given a sequence of operation on an initial empty priority queue, let $i_{\max}$ be the largest i for which SWEEP(i) takes place. We have $s_{i_{\max}} \leq N$, where N is the number of insertions in the sequence. An insertion can be charged by at most one call to SWEEP(i) for $i = 1, \ldots, i_{\max}$, so by the doubly-exponentially growth of s_i, the number of I/Os charged to an insertion is

$$O \left(\sum_{k=0}^{\infty} \frac{1}{B} \log_{M/B} N^{(3/4)^k} \right) = O \left(\frac{1}{B} \log_{M/B} N \right).$$

The amortized number of I/Os for a DELETEMIN is actually zero, as all occurring I/Os have been charged to insertions.

5   Profile Adaptive Performance

To make the complexity bound depend on N_ℓ, we make the following changes to our priority queue. Let r_i denote the number of elements residing in the lower part of link i. The value of r_i is stored at v_i and will only need to be updated when removing an element from B_i and when a call to SWEEP(i) creates a new S_{ij} list (in the later case $r_1, \ldots, r_{i-1}$ are reset to zero).

The only other modification is the following change of the call to SWEEP(i). Instead of finding the lowest index i where $c_i \leq k_i$, we find the lowest index i where either $c_i \leq k_i$ or $r_i \leq k_i s_i/2$. If $c_i \leq k_i$, SWEEP(i) proceeds as described Section 3, and c_i is incremented by one. Otherwise $c_i = k_i + 1$ and $r_i \leq k_i s_i/2$, in which case we will recycle one of the S_{ij} buffers. If there exists an input buffer S_{ij} which is empty, we use S_{ij} as the destination buffer for SWEEP(i). If all S_{ij} are nonempty, the two input buffers S_{ij_1} and S_{ij_2} with the smallest number of elements contain at most s_i elements in total. Assume without loss of generality $\min S_{ij_1} \geq \min S_{ij_2}$, where $\min S$ denotes the smallest element in stream S. We merge the content of S_{ij_1} and S_{ij_2} into S_{ij_2}. Since $\min S_{ij_1} \geq \min S_{ij_2}$ the heap order remains satisfied. Finally we apply SWEEP(i) with S_{ij_1} as the destination buffer.

5.1   Analysis

The correctness follows as in Section 4.1, except that the last induction on time is slightly extended. We must now use that $k_i \geq 2$ implies $k_i s_i/2 + s_i \leq k_i s_i$ to argue that SWEEP(i) will not make the lower part of link i contain more than $k_i s_i$ elements in the case where $c_i = k_i + 1$ and $r_i \leq k_i s_i/2$.

For the complexity, we as in Section 4.2 only have to consider the case where $i > i_M$. We note that in the modified algorithm, the additional number of I/Os required by SWEEP(i) for locating and merging S_{ij_1} and S_{ij_2} is $O(k_i + s_i/B)$ I/Os. As seen in Section 4.2, this is dominated by $O(\frac{s_i}{B}\log_{M/B} s_i)$, which is the number of I/Os already charged to SWEEP(i) in the analysis.

We will argue that SWEEP(i) collects $\Omega(s_i)$ elements from links $1,\ldots,i-1$ that have been inserted since the last call to SWEEP(j) with $j \geq i$, and that for half of these elements the value N_ℓ is $\Omega(s_i)$. The claimed amortized complexity $O(\frac{1}{B}\log_{M/B} N_\ell)$ then follows as in Section 4.2, except that we now charge the cost of SWEEP(i) to these $\Omega(s_i)$ elements.

The main property of the modified algorithm is captured by the following invariant:

> For each i, the links $1,\ldots,i$ contain in total at most $\sum_{j=1}^{i}|A_j| = \sum_{j=1}^{i} k_j^3$ elements which have been removed from A_{i+1} by the binary merger v_i since the last call to SWEEP(j) with $j \geq i+1$.

Here, we after a call to SWEEP($i+1$) define all elements in A_j to have been removed from A_ℓ for $1 \leq j < \ell \leq i+1$. When an element e is removed from A_{i+1} by v_i and is output to A_i, then all elements in the lower part of link i must be larger than e. All elements removed from A_{i+1} since the last call to SWEEP(j) with $j \geq i+1$ were smaller than e. These elements must either be stored in A_i or have been removed from A_i by the merger in v_{i-1}. It follows that at most $\sum_{j=1}^{i-1}|A_j| + |A_i| - 1$ elements removed from A_{i+1} are present in links $1,\ldots,i$. Hence, the invariant remains valid after moving e from A_{i+1} to A_i. By definition, the invariant remains valid after a call to SWEEP(i).

A call to SWEEP(i) will create a stream with at least $s_1 + \sum_{j=1}^{i-1} k_j s_j/2 \geq s_i/2$ elements. By the above invariant, at least $t = s_i/2 - \sum_{j=1}^{i-1}|A_j| = s_i/2 - \sum_{j=1}^{i-1} k_j^3$ $= \Omega(s_i)$ elements must have been inserted since the last call to SWEEP(j) with $j \geq i$. Finally, for each of the three definitions of N_ℓ in Section 1 we for at least $t/2$ of the t elements have $N_\ell \geq t/2$, because:

(a) For each of the $t/2$ most recently inserted elements, at least $t/2$ elements were already inserted when these elements where inserted.

(b) For each of the $t/2$ earliest inserted elements, at least $t/2$ other elements have been inserted before they themselves get deleted.

(c) The $t/2$ largest elements each have (maximum) rank at least $t/2$.

This proves the complexity stated in Section 1.

References

1. A. Aggarwal and J. S. Vitter. The input/output complexity of sorting and related problems. *Communications of the ACM*, 31(9):1116–1127, Sept. 1988.
2. L. Arge. External memory data structures. In *Proc. 9th Annual European Symposium on Algorithms (ESA)*, volume 2161 of *LNCS*, pages 1–29. Springer, 2001.
3. L. Arge, M. A. Bender, E. D. Demaine, B. Holland-Minkley, and J. I. Munro. Cache-oblivious priority queue and graph algorithm applications. In *Proc. 34th Ann. ACM Symp. on Theory of Computing*, pages 268–276. ACM Press, 2002.

4. R. Bayer and E. McCreight. Organization and maintenance of large ordered indexes. *Acta Informatica*, 1:173–189, 1972.
5. M. Bender, R. Cole, E. Demaine, and M. Farach-Colton. Scanning and traversing: Maintaining data for traversals in a memory hierarchy. In *Proc. 10th Annual European Symposium on Algorithms (ESA)*, 2002. To appear.
6. M. Bender, R. Cole, and R. Raman. Exponential structures for cache-oblivious algorithms. In *Proc. 29th International Colloquium on Automata, Languages, and Programming (ICALP)*, volume 2380 of *LNCS*, pages 195–207. Springer, 2002.
7. M. Bender, E. Demaine, and M. Farach-Colton. Efficient tree layout in a multi-level memory hierarchy. In *Proc. 10th Annual European Symposium on Algorithms (ESA)*, 2002. To appear.
8. M. A. Bender, E. Demaine, and M. Farach-Colton. Cache-oblivious B-trees. In *Proc. 41st Ann. Symp. on Foundations of Computer Science*, pages 399–409. IEEE Computer Society Press, 2000.
9. M. A. Bender, Z. Duan, J. Iacono, and J. Wu. A locality-preserving cache-oblivious dynamic dictionary. In *Proc. 13th Ann. ACM-SIAM Symp. on Discrete Algorithms*, pages 29–39, 2002.
10. G. S. Brodal and R. Fagerberg. Cache oblivious distribution sweeping. In *Proc. 29th International Colloquium on Automata, Languages, and Programming (ICALP)*, volume 2380 of *LNCS*, pages 426–438. Springer, 2002.
11. G. S. Brodal, R. Fagerberg, and R. Jacob. Cache oblivious search trees via binary trees of small height. In *Proc. 13th Ann. ACM-SIAM Symp. on Discrete Algorithms*, pages 39–48, 2002.
12. M. J. Fischer and M. S. Paterson. Fishspear: A priority queue algorithm. *Journal of the ACM*, 41(1):3–30, 1994.
13. M. Frigo, C. E. Leiserson, H. Prokop, and S. Ramachandran. Cache-oblivious algorithms. In *40th Annual Symposium on Foundations of Computer Science*, pages 285–297. IEEE Computer Society Press, 1999.
14. J. S. Vitter. External memory algorithms and data structures: Dealing with massive data. *ACM Computing Surveys*, 33(2):209–271, June 2001.

Characterizing History Independent Data Structures*

Jason D. Hartline, Edwin S. Hong, Alexander E. Mohr,
William R. Pentney, and Emily C. Rocke

Department of Computer Science,
University of Washington, Seattle, WA 98195.
{hartline,edhong,amohr,bill,ecrocke}@cs.washington.edu

Abstract. We consider history independent data structures as proposed for study by Teague and Naor [3]. In a history independent data structure, nothing can be learned from the representation of the data structure except for what is available from the abstract data structure. We show that for the most part, strong history independent data structures have canonical representations. We also provide a natural less restrictive definition of strong history independence and characterize how it restricts allowable representations. We also give a general formula for creating dynamically resizing history independent data structures and give a related impossibility result.

1 Introduction

On April 16, 2000, the New York Times published an article regarding the CIA's role in the overthrow of the Iranian government in 1953. In addition to the article, the Times' website posted a CIA file from 1954 which detailed the actions of various revolutionaries involved in the plot. The Times opted to black out many of the names mentioned in the document; some of the people referred to were still alive and residing in Iran and could have been put at risk for retribution. The file was published as an Adobe PDF file that contained the original document in its entirety and an overlay covering up parts of the document. Shortly after releasing the document, some Internet users reverse engineered the overlay and made the original document available on the Web. In an environment where information is valuable, private, incriminating, etc., data structures that retain information about previous operations performed upon them can cause considerable problems; the Times' blunder represents a particularly grievous instance of this. History independent data structures are designed not to reveal any information beyond that necessarily provided by the contents of the data structure.

The idea of maintaining a data structure so that no extraneous information is available was first explicitly studied by Micciano [2]. This work studied "oblivious

---

* The full version of this extended abstract is available from:
  http://www.cs.washington.edu/research/computation/theory-night
  /papers/hist-indep.ps

P. Bose and P. Morin (Eds.): ISAAC 2002, LNCS 2518, pp. 229–240, 2002.

trees" where no information about past operations could be deduced from the pointer structure of the nodes in the search tree. In [4], Snyder studied bounds on the performance of search, insert and delete functions on uniquely represented data structures, which employ a canonical representation for each possible state of the data structure. More stringent history independence requirements were studied by Naor and Teague in [3]. In their model, the entire memory representation of a history independent data structure, not just the pointer structure, must not divulge information about previous states of the data structure. Following [3] we consider two types of history independence: *weak history independence*, in which we assume that a data structure will only be observed once; and *strong history independence*, in which case the data structure may be observed multiple times. A data structure is history independent if nothing can be learned from the data structure's memory representation during these observations except for the current abstract state of the data structure.

In Section 3 we give a simple definition of strong history independence. In the full version of this paper we show that this definition equivalent to that of [3]. Under this definition, any strong history independent implementation of a data structure must satisfy a natural canonicality criterion (Section 4). For example, a strongly history independent implementation of a hash table has the property that up to randomness in the initialization of the hash table, e.g., the choice of hash functions, the hash table's representation in memory is deterministically given by its contents. This answers an open question posed in [3] about the necessity of canonical representations.

In Section 5 we consider a natural relaxation of strong history independence, where non-canonical representations and randomness can be used. However, we show that even under this less restrictive definition, there are still very stringent limitations on using non-canonical representations.

Finally, in Section 6 we discuss the issue of creating dynamically resizing history independent data structures. We give a general technique for dynamically resizing weak history independent data structures in amortized constant time against a non-oblivious adversary. We prove that no such technique exists for strongly history independent dynamically resizing data structures. This result provides insight into the open problem of whether there is a complexity separation between weak and strong history independence [3].

2   Preliminaries

The results presented in this paper apply to history independent data structures in general. To this end we must have a general understanding of data structures. An *abstract data structure* defines the set of operations for a data structure and its semantics.

A data structure's *state* is the current value or contents of the data structure as specified by its abstract data structure. A data structure's *representation* in memory for any given state is the physical contents of memory that represent that state. An *implementation* of a data structure gives a map from any valid representation/operation pair to a new representation (and possible output).

We assume that our abstract data structure is deterministic.[1] That is, each operation takes one state deterministically to another state. Define the *state transition graph* to be the directed graph induced on states (as vertices) of the data structure by the operations (directed edges). It is useful to consider the following trichotomy of state transition graphs according to standard graph properties:

- The graph may be acyclic (a DAG). This is the case if it is not possible to return to a previously visited state. Examples are the union-find data structure [5] and hash tables that do not support the delete operation.
- The graph may be strongly connected, i.e., all states are mutually reachable. Hash tables (with delete operation), queues, stacks, etc. are all examples of such data structures. We define these (below) as *reversible* data structures because they do not support irreversible operations.
- Otherwise the graph is a combination of the above consisting of strongly connected components that are interconnected acyclically.

Definition 1. *A data structure is* reversible *if its state transition graph is strongly connected.*

Let A, B, and C represent states of the data structure (i.e., vertices in the graph). Let X and Y represent sequences of operations on the data structure (i.e., directed paths in the graph) with $[X^k]$ being the sequence of operations of X repeated k times and $[X, Y]$ the sequence of operations consisting of the operations of X followed by the operations in Y. We say that B is *reachable* from A, notated $A \to B$, if some non-empty sequence of operations takes state A to state B. If X is such a sequence of operations, we say $A \xrightarrow{X} B$. If A is reachable from B and B is reachable from A then A and B are *mutually reachable*, notated $A \rightleftharpoons B$. This is synonymous with saying that states A and B are in the same strongly connected component of the state transition graph. We say $\oslash \xrightarrow{X} A$ if, on data structure initialization, the sequence of operations X produces state A.

Note though that if $A \rightleftharpoons B$ for some B, then $A \rightleftharpoons A$; though in general, a state A is not necessarily mutually reachable with itself (E.g., if the state transition graph is acyclic). We call a state not mutually reachable with itself a *transient* state.

The data structures that we consider may use randomization in choosing which representation to use for any given state. In the context of history independent data structures, randomization over representations is useful both for efficiency and for maintaining history independence.

Let **a** and **b** denote representations of states A and B respectively. It will be convenient to view a state as the set of representations that represent that state, thus $\mathbf{a} \in A$. Let X be such that $A \xrightarrow{X} B$. Let $\mathbf{Pr}\left[\mathbf{a} \xrightarrow{X} \mathbf{b}\right]$ denote the probability that, starting from representation **a** of A, the sequence X of operations on the data structure yields representation **b** of B. We say **b** is *reachable* from **a**, denoted

---

[1] Our results can be extended to randomized abstract data structures by noting that they are equivalent to deterministic abstract data structures when the user picks which operation to apply randomly.

$\mathbf{a} \to \mathbf{b}$, if there is some sequence of operations X such that $\mathbf{Pr}\left[\mathbf{a} \xrightarrow{X} \mathbf{b}\right] > 0$. We say $\mathbf{a} \rightleftharpoons \mathbf{b}$ if $(\mathbf{a} \to \mathbf{b}) \land (\mathbf{b} \to \mathbf{a})$. We say representation $\mathbf{a}$ of state A is *reachable* if it is reachable from data structure startup, i.e., $\oslash \to \mathbf{a}$. We henceforth only consider representations that are reachable. We say $\oslash \to \mathbf{a}$ if there exists X such that $(\oslash \xrightarrow{X} A) \land \mathbf{Pr}\left[\oslash \xrightarrow{X} \mathbf{a}\right] > 0$.

The representation of the data structure encapsulates everything known about it. Therefore $\mathbf{Pr}\left[\mathbf{a} \xrightarrow{X} \mathbf{b}\right]$ must be independent of any states or representations of the data structure prior to entering representation $\mathbf{a}$ of state A. Thus, a data structure behaves like a Markov chain on its representations except that the transition probabilities are based on which operation is performed. This is formalized in Note 1.

Note 1. If $A \xrightarrow{X} B$ and $B \xrightarrow{Y} C$ then,

$$\mathbf{Pr}\left[\mathbf{a} \xrightarrow{X} \mathbf{b} \xrightarrow{Y} \mathbf{c}\right] = \mathbf{Pr}\left[\mathbf{a} \xrightarrow{X} \mathbf{b}\right] \cdot \mathbf{Pr}\left[\mathbf{b} \xrightarrow{Y} \mathbf{c}\right].$$

3   History Independence

The goal of history independence is to prevent information from being leaked though the representation of a data structure in the case that it is observed by an outside party. As such, the requirements for history independence depend on the nature of the potential observations. Following [3] we define weak history independence for the case where the data structure is only observed once, e.g., when a laptop is lost. Alternatively, a strong history independent data structure allows for multiple observations of the data structure without giving any information about the operations between the observations beyond that implied by the states observed.

Definition 2 (Weak History Independence). *A data structure implementation is* weakly history independent *if, for any two sequences of operations X and Y that take the data structure from initialization to state A, the distribution over memory after X is performed is identical to the distribution after Y. That is:*

$$(\oslash \xrightarrow{X} A) \land (\oslash \xrightarrow{Y} A) \implies \forall \mathbf{a} \in A, \ \mathbf{Pr}\left[\oslash \xrightarrow{X} \mathbf{a}\right] = \mathbf{Pr}\left[\oslash \xrightarrow{Y} \mathbf{a}\right].$$

Definition 3 (Strong History Independence (SHI)). *A data structure implementation is* strongly history independent *if, for any two (possibly empty) sequences of operations X and Y that take a structure in state A to state B, the distribution over representations of B after X is performed on a representation $\mathbf{a}$ is identical to the distribution after Y is performed on $\mathbf{a}$. That is:*

$$(A \xrightarrow{X} B) \land (A \xrightarrow{Y} B) \implies \forall \mathbf{a} \in A, \ \forall \mathbf{b} \in B, \ \mathbf{Pr}\left[\mathbf{a} \xrightarrow{X} \mathbf{b}\right] = \mathbf{Pr}\left[\mathbf{a} \xrightarrow{Y} \mathbf{b}\right].$$

Here, A may be the null (pre-initialization) state, $\oslash$, in which case $\mathbf{a}$ is the empty representation, $\oslash$. Thus, strongly history independent data structures are a subset of weakly history independent data structures. Although this definition differs from the arguably more complex one given by Naor and Teague [3], we show in the full paper that they are in fact equivalent.

For strong history independent data structures, because the $\mathbf{Pr}\left[\mathbf{a} \xrightarrow{X} \mathbf{b}\right]$ does not depend on the path X, we can introduce the notation $\mathbf{Pr}[\mathbf{a} \to \mathbf{b}]$ to mean $\mathbf{Pr}\left[\mathbf{a} \xrightarrow{X} \mathbf{b}\right]$ for any X taking $A \xrightarrow{X} B$. We now discuss some useful properties of "$\rightleftharpoons$" and "$\to$" on history independent data structures.

Transitivity of "$\to$" and "$\rightleftharpoons$": *Under* SHI, $(\mathbf{a} \to \mathbf{b}) \wedge (\mathbf{b} \to \mathbf{c}) \implies (\mathbf{a} \to \mathbf{c})$, *and similarly,* $(\mathbf{a} \rightleftharpoons \mathbf{b}) \wedge (\mathbf{b} \rightleftharpoons \mathbf{c}) \implies (\mathbf{a} \rightleftharpoons \mathbf{c})$.
Reflexivity: $\mathbf{a} \rightleftharpoons \mathbf{a}$ *if and only if* $A \rightleftharpoons A$.
Symmetry of "$\to$": *Under* SHI, *for states* A *and* B *with* $A \rightleftharpoons B$, $\mathbf{a} \to \mathbf{b} \implies \mathbf{b} \to \mathbf{a}$.

Transitivity follows immediately from the definition of reachable. See the full version of the paper for the proof of reflexivity and symmetry.

4   Canonical Representations and *SHI*

Note that if a data structure has canonical representations for each state, it is necessarily *SHI*. An open question from [3] is whether the converse is true: does strong history independence necessarily imply that the data structure has canonical representations? In this section we answer the question by showing that up to randomization on transition between strongly connected components of the state transition graph the representations are canonical. For example, any strongly history independent implementation of a reversible data structure, e.g., a hash table, has canonical representations up to initial randomness, e.g., choice of hash functions.

Lemma 1. *Under* SHI, *if* $\mathbf{a}$ *and* $\mathbf{a}'$ *are both representations of state* A *with* $\mathbf{a} \to \mathbf{a}'$, *then* $\mathbf{a} = \mathbf{a}'$.

Proof. We show the contrapositive. Let E be the empty sequence of operations. For $\mathbf{a} \neq \mathbf{a}'$, we have $\mathbf{Pr}\left[\mathbf{a} \xrightarrow{E} \mathbf{a}'\right] = 0$ because the data structure is not allowed to change unless an operation is performed. Thus, by *SHI*, $\mathbf{Pr}[\mathbf{a} \to \mathbf{a}'] = 0$ so $\mathbf{a} \not\to \mathbf{a}'$.   □

Intuitively, this requires A to have a canonical representation after it is visited for the first time. Before the first visit it may have a distribution of possible representations. Now we extend the requirement for a canonical representation of A to the point where the data structure first hits a state reachable from A. From Lemma 1 by transitivity of "$\to$" we have:

Corollary 1. *Under* SHI, *let* $\mathbf{a}$ *and* $\mathbf{b}$ *be representations of states* A *and* B *such that* $\mathbf{a} \rightleftharpoons \mathbf{b}$. *For all representations* $\mathbf{b}'$ *of* B, $\mathbf{a} \rightleftharpoons \mathbf{b}'$ *iff* $\mathbf{b}' = \mathbf{b}$.

Corollary 1 shows that after reaching the first state of a strongly connected component of the state transition graph, there is a chosen canonical representation for every other state in the component. This is a complete answer to the question of whether canonical representations are necessary for strong history independence. In particular we have the following specification of the corollary:

Theorem 1. *For a reversible data structure to be* SHI, *a canonical representation for each state must be determined during the data structure's initialization.*

Again, examples of reversible data structures - ones in which all states are mutually reachable - are hash tables (that include the delete operation), queues, stacks, etc.

5   Less Restricted *SHI*

The definition of strong history independence above is highly restrictive, requiring, as just seen, canonical representations of states (after some initial random choices) to qualify. As evident by the proof of Lemma 1, a key contributor to its restrictiveness is that we require that the observer not be able to distinguish a nonempty sequence of operations from an empty one. A natural question would be whether anything can be gained by relaxing this requirement. In the this section, we discuss a relaxation of strong history independence which allows the empty sequence to be distinguished from any other sequence of operations. We show that while this does allow for randomness over representations, the randomness allowed is still very restricted.

5.1   Definition

The following definition introduces a form of strong history independence more permissive than the *SHI* definition.

Definition 4 (*SHI**). *A data structure implementation is strongly history independent if, for any two* nonempty *sequences of operations X and Y,*

$$(A \xrightarrow{X} B) \wedge (A \xrightarrow{Y} B) \implies \forall\, \mathbf{a} \in A,\ \forall\, \mathbf{b} \in B,\ \mathbf{Pr}\left[\mathbf{a} \xrightarrow{X} \mathbf{b}\right] = \mathbf{Pr}\left[\mathbf{a} \xrightarrow{Y} \mathbf{b}\right].$$

SHI* permits an observer to distinguish an empty operation sequence from a nonempty one without considering it to be a violation of history independence. However, given that the operation sequence is nonempty, the observer may still not glean information about how many operations were performed, or what the operations were, besides what is inherent in the observed state of the abstract data structure.

Arguably, *SHI** is more practically useful than *SHI*. In essence, it is nearly as strong as *SHI* if operations on the data structure are expected to be considerably more frequent than observations of it. If the opposite were true, that observations are more frequent than operations, then the situation approaches constant surveillance, in which case the observer can simply record each state as it occurs, making history independence useless.

In the remainder of this section we show that, despite this added flexibility, there are still very strict requirements to which a *SHI** data structure must adhere. For example, each state in a reversible *SHI** data structure must have a "canonical distribution" over representations that is chosen during the initialization process. Moreover, each operation on the data structure must result in the representation being explicitly resampled from the state's canonical distribution. As an example, this precludes a hash table implementation from using a randomized method for resolving conflicts, as all conflicts would have to be remembered and re-resolved after every operation.

5.2   Canonical Representation Distributions

Unlike in the case of *SHI*, under *SHI** a single state A may have distinct representations $\mathbf{a}$ and $\mathbf{a}'$ with $\mathbf{a} \rightleftharpoons \mathbf{a}'$ (contrast with Lemma 1).

Lemma 2. *Under* SHI*, *if* $\mathbf{a}$ *and* $\mathbf{a}'$ *are both representations of state* A *with* $\mathbf{a} \rightleftharpoons \mathbf{a}'$, *then for any representation* $\mathbf{b}$ *of state* B *with* $A \to B$, $\mathbf{Pr}[\mathbf{a} \to \mathbf{b}] = \mathbf{Pr}[\mathbf{a}' \to \mathbf{b}]$.

Proof. For the case that $\mathbf{Pr}[\mathbf{a} \to \mathbf{b}] = \mathbf{Pr}[\mathbf{a}' \to \mathbf{b}] = 0$ the lemma is true. Thus assume without loss of generality that $\mathbf{Pr}[\mathbf{a} \to \mathbf{b}] > 0$.

Since $A \rightleftharpoons A$, let W be any nonempty series of operations taking $A \xrightarrow{W} A$. Let X be any nonempty series of operations taking $A \xrightarrow{X} B$. Consider the series of operations $Q_N = [(W)^N, X]$ as performed on representation $\mathbf{a}$. Let a_i be a random variable for the representation of state A after i performances of W and let $\mathcal{E}_N$ be the random event that $a_i = \mathbf{a}'$ for some $i \in \{1, \dots, N\}$. First, by Note 1 and *SHI**, we have:

$$\mathbf{Pr}\left[\mathbf{a} \xrightarrow{Q_N} \mathbf{b} \mid \mathcal{E}_N\right] = \mathbf{Pr}[\mathbf{a}' \to \mathbf{b}].$$

Now we show that $\lim_{N \to \infty} \mathbf{Pr}[\mathcal{E}_N] = 1$. As observed in Note 1, once conditioned on the representation at a_{i-1} the value of a_i can not depend on anything but a_{i-1}. The series of representations $\{a_1, a_2, \dots\}$ can thus be viewed as a first-order Markov chain, where the states of the Markov chain are the representations of state A reachable from $\mathbf{a}$. Since all such representations are mutually reachable, the Markov chain is irreducible.

We now employ the following basic property of irreducible Markov chains: *In the limit, if the expected number of visits to a state is unbounded then the probability that the state is visited approaches one.* Let $\alpha = \mathbf{Pr}[a_i = \mathbf{a}'] = \mathbf{Pr}[\mathbf{a} \to \mathbf{a}']$. The expected number of occurrences of $\mathbf{a}'$ after N steps is $\sum_{i=1}^{N} \mathbf{Pr}[a_i = \mathbf{a}'] = \alpha N$. Since α is constant, this is unbounded as N increases. Thus, $\lim_{N \to \infty} \mathbf{Pr}[\mathcal{E}_N] = 1$.

To complete the proof, by *SHI**,

$$\mathbf{Pr}[a \to b] = \mathbf{Pr}\left[a \xrightarrow{Q_N} b\right]$$

$$= \mathbf{Pr}\left[a \xrightarrow{Q_N} b \mid \mathcal{E}_N\right]\mathbf{Pr}[\mathcal{E}_N] + \mathbf{Pr}\left[a \xrightarrow{Q_N} b \mid \neg\mathcal{E}_N\right](1 - \mathbf{Pr}[\mathcal{E}_N])$$

$$= \mathbf{Pr}[a' \to b]\,\mathbf{Pr}[\mathcal{E}_N] + \mathbf{Pr}\left[a \xrightarrow{Q_N} b \mid \neg\mathcal{E}_N\right](1 - \mathbf{Pr}[\mathcal{E}_N]).$$

In the limit as N increases, this quantity approaches $\mathbf{Pr}[a' \to b]$. However, since *SHI** guarantees that it is constant as N changes, it must be that $\mathbf{Pr}[a \to b] = \mathbf{Pr}[a' \to b]$. □

We now give the main theorem of this section which shows that, among other things, reversible data structures have canonical distributions.

Theorem 2 (Canonical Distributions). *Under strong history independence, for any* a, b, c *with* $a \rightleftharpoons b$ *and* $a \to c$, $\mathbf{Pr}[a \to c] = \mathbf{Pr}[b \to c]$.

Proof. First note that $b \rightleftharpoons a \to c$ gives $b \to c$. By definition of "$\to$", there exists at least one sequence of operations X with $A \xrightarrow{X} B$, and at least one Y with $B \xrightarrow{Y} C$. For any such X and Y, a $\xrightarrow{[X,Y]} c$ must go through some representation b' of B. By symmetry and transitivity of "$\rightleftharpoons$", $b' \rightleftharpoons b$. Let $\langle b \rangle$ be the set of all such b'. Thus,

$$\mathbf{Pr}[a \to c] = \mathbf{Pr}\left[a \xrightarrow{[X,Y]} c\right] = \sum_{b' \in \langle b \rangle} \mathbf{Pr}\left[a \xrightarrow{X} b' \xrightarrow{Y} c\right] \quad \text{(By } SHI^*\text{)}$$

$$= \sum_{b' \in \langle b \rangle} \mathbf{Pr}\left[a \xrightarrow{X} b'\right] \cdot \mathbf{Pr}\left[b' \xrightarrow{Y} c\right] \quad \text{(Note 1)}$$

$$= \sum_{b' \in \langle b \rangle} \mathbf{Pr}\left[a \xrightarrow{X} b'\right] \cdot \mathbf{Pr}\left[b \xrightarrow{Y} c\right] \quad \text{(Lemma 2)}$$

$$= \mathbf{Pr}\left[b \xrightarrow{Y} c\right] \cdot \sum_{b' \in \langle b \rangle} \mathbf{Pr}\left[a \xrightarrow{X} b'\right]$$

$$= \mathbf{Pr}\left[b \xrightarrow{Y} c\right] \cdot 1 = \mathbf{Pr}[b \to c].$$

□

5.3   Order of Observations

In the full paper we continue these arguments by showing that the probability distribution over representations for a set of observed states is independent of the order in which the states are observed.[2] Note, however, that the states must be observed in an order consistent with the state transition graph. This is only interesting for states that can be observed in a different order, i.e., states that are mutually reachable (in the same connected component of the state transition graph). As an example, for a and b such that $A \rightleftharpoons B$, the result implies that $\mathbf{Pr}[\varnothing \to a \to b] = \mathbf{Pr}[\varnothing \to b \to a]$. This order independence implies that our definition of strong history independence is in fact equivalent to that of [3]:

---

[2] Since *SHI* is more restrictive than *SHI** it suffices to show this result for *SHI**.

Definition 5 (*NT-SHI* and *NT-SHI**). *Let S_1 and S_2 be sequences of operations and let $P_1 = \{i_1^1, \ldots, i_\ell^1\}$ and $P_2 = \{i_1^2, \ldots, i_\ell^2\}$ be two lists of observation points such that for all $b \in \{1, 2\}$ and $1 \le j \le \ell$ we have $1 \le i_j^b \le |S_b|$ and the state following the i_j^1 prefix of S_1 and the i_j^2 prefix of S_2 are identical (for NT-SHI* we further require that for $j \ne k$ that $i_j^b \ne i_k^b$). A data structure implementation is strongly history independent if for any such sequences the joint distribution over representations at the observation points of P_1 and the corresponding points of P_2 are identical.*[3]

Note that this definition allows the observations of the operations to be made out of order, i.e., $i_j^b > i_{j+1}^b$. It is for this reason that the order invariance described above is required to show that *SHI* and *NT-SHI* are equivalent.

6   Dynamic Resizing Data Structures

Many open addressing, i.e., array based, data structures use dynamic resizing techniques of doubling or halving in size as they grow or shrink. The open addressing hash table or the array based queue are classic examples. This technique combined with an amortized analysis yields data structures with amortized constant time operations. A weakly history-independent hash table that dynamically resizes is presented in [3]. The resizing scheme for the hash table generalizes to convert any weakly history independent array-based data structure with constant time operations and linear time resize into a dynamically resizing data structure with amortized constant time per operation against an oblivious adversary. Dynamic resizing for oblivious adversaries are discussed in more detail in the full paper; in this section we will focus on the non-oblivious case.

A non-oblivious adversary is allowed access to the random bits flipped during the running of the algorithm. Such an adversary can adapt a sequence of operations to the data structure as it is running in order to make it perform poorly. For example, a hash table with randomized choice of hash functions does not perform well against a non-oblivious adversary because the adversary, knowing the choice of hash function, can choose to only insert elements that hash to the same location.

We provide the following results on dynamically resizing history independent data structures for non-oblivious adversaries:

- There is a general method for making any history independent data structure with constant time operations and linear time resize into a weakly history independent dynamically resizing data structure with amortized constant time operations against a non-oblivious adversary.
- In contrast, there is no general method for making a strongly history independent array based data structure with a linear time resize operation into a strongly history independent dynamically resizing data structure with amortized constant time operations against a non-oblivious adversary.

---

[3] [3] does not define *NT-SHI**. We include it here as the natural relaxation to nonempty sequences for comparison to *SHI**.

We will tailor our description of these techniques to data structures with explicitly defined size, capacity, and unit insertion and unit deletion operations such as a hash table or an array based queue. Note that if there is no deletion operation then the standard approach of doubling the capacity when the data structure is full, i.e. when the size is equal to the capacity, is efficient for non-oblivious adversaries. If there is no insertion operation then the data structure cannot grow and resizing is irrelevant. Let n denote the size of the data structure and N the capacity.

Non-oblivious Adversary and Weak History Independence

We show how to make any fixed capacity history independent (weakly or strongly) array-based data structure with a linear time resize operation into a constant time amortized dynamically resizing weakly history independent data structure. The principle behind this method is to maintain the invariant that N is random variable that is uniform on $\{n, \ldots, 2n-1\}$ while each insert or delete operation only causes the data structure to resize with probability $O(1/n)$.

We show below how to modify the insert function of any data structure to maintain our invariant on N. The delete function is similar. See the full version of the paper for details. Although N is a random variable the actual value of N is known when working with the data structure.

Insert:

1. if $N = n$
   - Resize data structure to size uniform on $\{n+1, \ldots, 2(n+1) - 1\}$.
2. Otherwise (i.e., $N > n$)
   - With probability $2/(n+1)$ resize to $N = 2n$ or $N = 2n + 1$, that is:
   $$N \leftarrow \begin{cases} 2n & \text{with probability } 1/(n+1) \\ 2n + 1 & \text{with probability } 1/(n+1) \\ \text{no change} & \text{otherwise} \end{cases}$$
3. Insert new item.

To show correctness we must show that given a call to insert with N uniform on $\{n, \ldots, 2n-1\}$, after insert the new capacity is uniform on $\{n+1, \ldots, 2(n+1) - 1\}$. Clearly if step 1 occurs, the new capacity is uniform as desired. On the other hand, if step 2 occurs the probability that N is unchanged is $1 - 2/(n+1) = (n-1)/(n+1)$. Since each of these $n-1$ possible values for the old N is equally likely, the probability that N is any one of them is $1/(n+1)$. Clearly $\Pr[N = 2n] = \Pr[N = 2n + 1] = 1/(n+1)$. Thus, the new N is uniformly distributed over $\{n+1, \ldots, 2n+1\}$.

We now show that the runtime of our insert (due to resizing) is expected constant time. Assuming that a resize takes linear time, we need the probability of resize of be $O(1/n)$. Step 1, which always incurs a resize, occurs with probability $1/n$. Step 2 incurs a resize with probability $2/(n+1)$. Thus the total probability of resize is less than $3/n$; since the resize operation is linear in n, the expected time spent in resizing is $O(1)$.

Non-oblivious Adversary and Strong History Independence

The technique employed in the previous section for making amortized dynamically resizing weak history independent data structures fails when strong history independence is required. The technique described maintains a canonical distribution of possible capacities for each n such that the probability is $O(1/n)$ that the capacity needs to be changed on an insert or delete (to maintain the correct distribution on capacities). However, strongly history independent data structures cannot use such a technique because in states that are mutually reachable, randomness over choice of representation must be completely regenerated during each operation (Theorem 2).

We will show any strongly history independent data structure that has

- non-constant unamortized time resizes, and
- insert operations that can be undone in constant number of operations (i.e., delete),

has amortized non-constant time operations against a non-oblivious adversary. Thus, there is no general technique for taking a constant time strongly history data structure with inserts and deletes that has a linear time resize operation for changing the capacity and making it resizable.

Consider, for example, the *deque* (double-ended queue) that supports operations *inject* and *eject* from the front of the deque and *push* and *pop* from the rear of the deque. The insert operations *inject* and *push* have corresponding delete operations *eject* and *pop*. A weakly history independent deque can be implemented in an array in the same way a *queue* is implemented in an array. The implementation of a strongly history independent deque is an open question, as is the implementation of a queue. However, the result we now prove tells us that either there is a strongly history independent deque with constant time resize or there is no non-oblivious amortized constant time resizing strongly history independent deque. This would provide a separation result between strong history independence and weak history independence because weak history independent amortized constant time resizing deques exist.

Theorem 3. *Against a non-oblivious adversary, any strongly history independent data structure that dynamically resizes (in non-constant unamortized time) and has a sequence of undoable insert operations has non-constant amortized resizes.*

Proof. Let N' be a large value (we will take it in the limit for this result). Consider any sequence of insert operations $X_1, \ldots, X_{N'}$ that we will perform in order on the data structure, taking it from state S_0 to state $S_{N'}$. Let $\bar{X}_i$ be the operation(s) that undo(es) X_i.

Note that any data structure in which operations are undoable is reversible. Thus, all states are mutually reachable, and once we initialize the data structure and end up in some representation s_0 of state S_0, the reachable representations of state S_i are limited to those that are mutually reachable from s_0, i.e., s_i such that $s_0 \rightleftharpoons s_i$. Furthermore, the probability that s_i is the representation of state S_i is exactly $\mathbf{Pr}[s_0 \to s_i]$ for either the case that we arrive in S_i from applying operation X_i from state S_{i-1} or from applying $\bar{X}_{i+1}$ from state S_{i+1}. Conditioned on $\mathbf{s}_0$ being the initial representation, let $p_i^{N'}$ be the probability that the representation of state S_i has capacity at least N' when performing X_i from S_{i-1} or $\bar{X}_{i+1}$ from S_{i+1}. By SHI^*, $\mathbf{Pr}[\mathbf{s}_{i-1} \to \mathbf{s}_i] = \mathbf{Pr}[\mathbf{s}_0 \to \mathbf{s}_i]$. Thus,

$$p_i^{N'} = \sum_{\mathbf{s}' \in \langle \mathbf{s}_i \rangle} \{ \mathbf{Pr}[\mathbf{s}_0 \to \mathbf{s}'] \; : \; \mathbf{s}' \text{ has capacity at least } N' \}.$$

All representations of state $S_{N'}$ have capacity at least N' because they have size N'. Thus, $p_{N'}^{N'} = 1$. Also, for N' suitably large, $p_0^{N'} = 0$. As such, there must be an k such that $p_k^{N'} < 1/2$ and $p_{k+1}^{N'} \geq 1/2$. The probability of resize on transition from S_k to S_{k-1}, given initial representation $\mathbf{s}_0$, is thus at least $(1 - p_k^{N'}) \times p_{k+1}^{N'} \geq 1/4$. As this resize takes (non-amortized) linear time, the sequence of operations given by $Y = [X_1, \dots, X_k, [X_{k+1}, \bar{X}_{k+1}]^k]$ takes at least $O(k)$ times the unamortized cost of resize for $3k$ operations, which yields an amortized cost per operation on the same order as the cost of resize (which we have assumed to be non-constant).   □

7   Conclusions

We have shown the relationship between strong history independence and canonical representations. In doing so we have proposed a new definition, SHI^*, of strong history independence that allows non-empty sequences of operations to be distinguished from empty ones. We leave as an open question whether there is a complexity separation between the two, i.e., does there exist any interesting data structure that has an efficient SHI^* implementation but not SHI?

We have also given a general technique for dynamically resizing weak history independent data structures. We have shown that for a standard efficiency metric this is not possible for strong history independent data structures. We show that efficient dynamic resizing under this metric is not possible for strong history independent data structures, but leave as an open question whether there exist strong history independent data structures that would benefit from such a resizing.

References

1. T. H. Cormen, C. E. Leiserson, and R. L. Rivest. *Introduction to Algorithms*. MIT Press, Cambridge, MA, 1990.
2. D. Micciancio. Oblivious data structures: Applications to cryptography. In *Proc. of 29th ACM Symposium on Theory of Computing*, pages 456–464, 1997.
3. M. Naor. and V. Teague. Anti-persistence: History Independent Data Structures. In *Proc. of 33nd Symposium Theory of Computing*, May 2001.
4. L. Synder. On Uniquely Represented Data Structures. In *Proc. of 28th Symposium on Foundations of Computer Science*, 1977.
5. Robert E. Tarjan. Efficiency of a good but not linear set union algorithm. *Journal of the ACM*, 22:215–225, 1975.

Faster Fixed Parameter Tractable Algorithms for Undirected Feedback Vertex Set

Venkatesh Raman[1], Saket Saurabh[2], and C. R. Subramanian[1]

[1] The Institute of Mathematical Sciences, Chennai 600 113, India.
{vraman,crs}@imsc.ernet.in
[2] Chennai Mathematical Institute, Chennai 600 017, India
saurabh@cmi.ac.in

Abstract. We give a $O(\max\{12^k, (4\lg k)^k\} \cdot n^\omega)$ algorithm for testing whether an undirected graph on n vertices has a feedback vertex set of size at most k where $O(n^\omega)$ is the complexity of the best matrix multiplication algorithm. The previous best fixed parameter tractable algorithm for the problem took $O((2k+1)^k n^2)$ time. The main technical lemma we prove and use to develop our algorithm is that that there exists a constant c such that, if an undirected graph on n vertices with minimum degree 3 has a feedback vertex set of size at most $c\sqrt{n}$, then the graph will have a cycle of length at most 12. This lemma may be of independent interest.

We also show that the feedback vertex set problem can be solved in $O(d^k kn)$ for some constant d in regular graphs, almost regular graphs and (fixed) bounded degree graphs.

1 Introduction

We explore efficient fixed parameter algorithms for the undirected feedback vertex set problem under the framework introduced by Downey and Fellows[8]. Given an undirected graph on n vertices and an integer parameter k, the feedback vertex set problem asks whether the given graph has a set of k vertices whose removal results in an acyclic graph. Throughout this paper, all our graphs are undirected and unweighted.

In the framework of parameterized complexity, a problem with input size n and parameter k is fixed parameter tractable (FPT) if there exists an algorithm to solve the problem in $O(f(k)n^{O(1)})$ time where f is any function of k. Such an algorithm is quite useful in practice for small ranges of k (against a naive $n^{k+O(1)}$ algorithm). Some of the well known fixed parameter tractable problems include parameterized versions of VERTEX COVER, MAXSAT and MAX CUT (see [8]). On the contrary, for the parameterized versions of problems like CLIQUE and DOMINATING SET the best known algorithms have only $n^{k+O(1)}$ running time (see [8] for details). The central aim of the study of the parameterized complexity is to identify problems exhibiting this contrasting behaviour.

The feedback vertex set problem is known to be fixed parameter tractable through an algorithm taking $O((2k+1)^k n^2)$ time [8]. See [5] for a recent random-

P. Bose and P. Morin (Eds.): ISAAC 2002, LNCS 2518, pp. 241–248, 2002.
© Springer-Verlag Berlin Heidelberg 2002

ized $O(4^k kn)$ algorithm that finds a minimum feedback vertex set with probability at least $1 - (1 - 4^{-k})^{c4^k}$ where k is the size of the minimum feedback vertex set and $c > 0$ is an arbitrary constant.

One practical theme in the area of parameterized complexity is to optimize on the function $f(k)$ for problems which are known to be fixed parameter tractable. Some of the problems that have attracted wide attention in this direction are the parameterized versions of Vertex Cover[11], [6], 3-dimensional matching[8], [7], planar dominating set[1], [2] and MAXSAT[12], [3]. In this paper we make the first serious attempt to reduce the function $f(k)$ for the feedback vertex set problem.

We substantially improve the known $f(k)$ function for the problem by developing an algorithm whose running time is $O((12 + 4 \lg k)^k n^\omega)$ where ω is the exponent in the runtime for the best Matrix Multiplication algorithm. En route to the algorithm, we also prove an interesting structural result that any graph on n vertices with a feedback vertex set size at most $c\sqrt{n}$ for some fixed constant c, will have a cycle of length at most 12. This result may be of independent interest. We further improve the function to d^k for some constant d for bounded degree graphs and regular graphs.

The next section mentions some preliminary results relating to the feedback vertex set problem. Section 3 proves the main structural theorem outlined above. Section 4 gives the new algorithm based on the structural theorem proved in Section 3 and using the results described in Section 2. Section 5 describes the improved algorithm for bounded degree and regular graphs. Finally Section 6 concludes with some remarks and open problems.

2   Preliminaries

The following results are well known in the literature on Feedback set problems. See, for example, [4] for proofs. For the sake of ease of description, we deal with multigraphs (where self-loops and parallel edges are allowed).

Lemma 1. *Let G be an undirected, connected multigraph. Perform the following steps until it is not possible, to get a multigraph G'. Then G has a feedback vertex set of size at most k if and only G' has a feedback vertex set of size at most k.*

1. *If G has a vertex of degree 1, remove it (along with the edge incident on it).*
2. *If G has a vertex x of degree 2 (which is not a self-loop) adjacent to vertices y and z, short circuit by removing x and joining and y and z by a new edge (even if y and z were adjacent earlier).*

Clearly the graph G' obtained is either a graph on one vertex with a self loop (in which case G has a feedback vertex set of size 1) or G' is such that each vertex has degree at least three. G' can be constructed in $O(m)$ steps where m is the number of edges in G. Thus without loss of generality, we can assume G to have minimum degree 3.

It is well-known that shortest cycles can be found quickly (see [10]).

Lemma 2. *Given an undirected graph G, a shortest cycle (if there is any) in G can be found in $O(\min\{mn, n^\omega\})$ time where n^ω is the running time of the best-known algorithm for multiplying two n by n matrices.*

The following is a classical result due to Erdös and Posa[9] from which the first approximation algorithm with a ratio of $O(\lg n)$ was proved for this problem.

Lemma 3. *Any graph G with minimum degree at least 3 has a cycle of length at most $2\lg n + 1$.*

Now consider the following algorithm (first observed in [13]) to determine whether G has a feedback vertex set of size at most k. Perform the preprocessing steps above to make the minimum degree 3. Find a shortest cycle C in the resulting graph, which is of length at most $2\lg n + 1$. Since any feedback vertex set must contain at least one vertex from every cycle, G has a feedback vertex set of size at most k if and only if $G - v$ has a feedback vertex set of size at most $k - 1$ for some $v \in C$. Thus we can determine whether or not G has a feedback vertex set of size k by checking recursively whether $G - v$ has a feedback vertex set of size at most $k - 1$ for every $v \in C$. Thus we have

Theorem 1. *Given a graph G on n vertices, and an integer parameter k, we can determine whether or not G has a feedback vertex set of size at most k in $O((2\lg n + 1)^k n^\omega)$ time where ω is the exponent on n in the algorithm to multiply two n by n matrices.*

Note that since $(\lg n)^k \leq (3k\lg k)^k + n$ for all n and k, this gives another fixed paramater tractable algorithm for the feedback vertex set problem.

In Section 4, using a structural result, we employ a different technique for small values of k, and use this algorithm only for large enough k and improve the overall complexity of the algorithm.

3   A Structural Theorem for Graphs Having Small Feedback Vertex Set

Theorem 2. *If a graph G on n vertices with minimum degree 3 has a feedback vertex set of size k, then there exists a cycle in G of length atmost 12 as long as $(n - k) > 18 \cdot \binom{k}{2}$.*

Proof. Let G be a graph on n vertices with minimum degree 3 having a feedback vertex set F of size k. Then, the induced graph $T = G - F$ has $n - k$ vertices and T is a forest.

Let $L = \{x \in T : deg_T(x) \leq 2\}$.

Here $deg_T(x)$ denotes the degree of the vertex x inside the forest T. Since the minimum degree of the graph is three, every vertex in L will have some neighbour in F. Let a pair of vertices (a, b) where $a, b \in G - F$ and possibly $a = b$, be *close* if **either** $a = b$ and $deg_T(a) \leq 1$ **or** $a, b \in L$, $a \neq b$, and the distance between a and b in T (denoted $d_T(a, b)$) is at most 4.

We claim that there are at least $|V(G - F)|/18$ close pairs.

Before we prove the claim, we argue that the claim implies the theorem. Suppose there is a close pair (a, b) with $a \neq b$ and a and b having a common neighbor, say c, in F. Then the path in T between a and b together with the edges (a, c) and (b, c) will form a cycle of length at most 6. So assume that there is no *such* close pair.

Let a and b be an arbitrary close pair. Then, either $a = b$ and $deg_T(a) \leq 1$ or $a \neq b$, $d_T(a, b) \leq 4$ and a and b do not share any common neighbor in F. In the first case, pick two distinct neighbors c, d (the choices are arbitrary) of a in F. In the second case, pick a neighbor $c \in F$ of a and a neighbor $d \in F$ of b (again, the choices are arbitrary) with $c \neq d$. Associate with the close pair (a, b) the pair (c, d) in F. Thus associate with every close pair, a pair of vertices in F. Since there are $(n - k)/18 > \binom{k}{2}$ close pairs, there will be atleast two close pairs (a_1, b_1) and (a_2, b_2) which are associated with (hence adjacent to) a pair (f, g) in F. The paths in T between a_1 and b_1 and a_2 and b_2 together with the edges $(a_1, f), (a_2, f), (b_1, g), (b_2, g)$ will give rise to a cycle of length atmost 12.

So it suffices to prove the claim. It is also clear that it suffices to prove that for every connected component T' (a tree) of $T = G - F$ on $v = |V(T')|$ vertices, there are atleast $v/18$ close pairs in T'. Clearly, this is true if T' is an isolated vertex a since (a, a) is a close pair. If T' is a tree on more than one vertex, without loss of generality, we can assume that $T' = T = G - F$. Since in a tree on v vertices at most $2v/3$ vertices can have degree 3 or more, it follows that $|L| \geq (n - k)/3$.

Let $L' = \{x \in L : d_T(x, y) > 4 \ \forall y \in L\}$

I.e. L' consists of those vertices of L which are at a distance of at least 5 from every other vertex in L. Let $N_2(x)$ be the set of all vertices in T, other than x, which are at a distance of at most 2 from x in T. The following observations are easy to verify.

1. $x, y \in L' \Rightarrow N_2(x) \cap N_2(y) = \phi$.
2. $x \in L' \Rightarrow N_2(x)$ consists of only vertices with degree 3 or more in T, I.e. $N_2(x) \cap L = \emptyset$.
3. $x \in L' \Rightarrow |N_2(x)| \geq 3$.

The last claim follows since x will have at least one neighbour in T and that will have degree at least 3 by the second observation.

Thus $3|L'| \leq |V(T) - L| \leq 2(n - k)/3$ from which it follows that $|L'| \leq 2(n - k)/9$ and $|L - L'| \geq (n - k)/9$.

Thus at least $1/9$ fraction of the vertices of $G - F$ have some close neighbour. Thus, there are at least $(n - k)/18$ close pairs, which is what we wanted to show. $\square$

Corollary 1. *There exists a constant c such that, for sufficiently large n, if a graph on n vertices with minimum degree 3 has a feedback vertex set of size at most $c\sqrt{n}$, then there exists a cycle in the graph of length atmost 12.*

4   New Algorithm

Based on the structure thoerem described in the last section, we describe a new algorithm in a recursive fashion as follows. This algorithm uses (for large values of k) the algorithm described in Section 2.

Algorithm FBVS(G, k) (* G is a multigraph, $k \geq 0$ *)
(returns a feedback vertex set of size k in G if there is one and returns NO otherwise).

- *Step 0*: If G is acyclic, answer YES and return $\emptyset$.
- *Step 0'*: If $k = 0$ and G contains a cycle, answer NO and EXIT.
- *Step 1*: Preprocess G (by deleting pendant vertices and by short circuiting degree two vertices as described in Section 2) to get a graph G' whose minimum degree is at least 3. G' may be now a multigraph.
- *Step 2*: Find a shortest cycle C in G'. Let $l \geq 1$ be its length. If $k \leq c\sqrt{n}$ and $l > 12$, then answer NO. Here c is as in Corollary 1.
- *Step 3*: If for some vertex $v \in C$, FBVS$(G' - v, k - 1)$ is true then answer YES and return $\{v\} \cup FBVS(G' - v, k - 1)$, else answer NO.

Correctness :   The proof uses induction on k. For $k = 0$, the answer should be YES if and only if G is acyclic and this is happening. For $k > 0$, assume the algorithm is correct for all $k' < k$. We need to show that whenever the algorithm answers YES (or NO), the answer is indeed YES (or NO). From the arguments of Section 2, it follows that G has a feedback vertex set of size k if and only if G' has a feedback vertex set of size k.

Case 1: Algorithm exits at Step 3. Since some vertex of C has to be in any Feedback vertex set of size k, it follows from the inductive hypothesis that this step correctly answers.

Case 2: Algorithm exits at Step 2. In this case, $k \leq c\sqrt{n}$ and length of C is more than 12. Thus, by Corollary 1, G does not have a feedback vertex set of size k and the algorithm correctly answers NO.

Running Time :   Let n denote the number of vertices and m the number of edges (self-loops and parallel edges included) in G.

Step 1 takes $O(m)$ time where m is the number of edges in the graph, and Step 2 takes $O(\min\{mn, n^\omega\})$ time by Lemma 2. Note that G' has atmost m edges (all included). Also, G' always has a cycle of length atmost $2(\log_2 n + 1)$.

Assume that algorithm executes Step 3. Then, either $k > c\sqrt{n}$ or $l \leq 12$. In any case, $l \leq \max\{12, 4 \lg k\}$. If $T(n, m, k)$ denotes the running time, then

$$T(n, m, k) \leq O(\ \min(nm, n^\omega)\ ) + \max(12, 4\log_2 k) \cdot T(n, m, k - 1)$$
$$= O(\ (\max(12, 4\log_2 k))^k \cdot \min(nm, n^\omega)\ )$$

Thus we have the following theorem.

Theorem 3. *Given an undirected graph G on n vertices, and an integer parameter k, we can determine whether or not G has a feedback vertex set of size at most k in time $O(\max\{12^k, (4 \lg k)^k\} n^\omega)$ time.*

5 Improved Algorithm for Bounded Degree Graphs and Regular Graphs

In this section, we give improved algorithms for bounded degree graphs and regular graphs. These algorithms are based on lower bounds of the size of the minimum feedback vertex set in these classes of graphs.

Lemma 4. *Let $G(V, E)$ be a graph on n vertices with minimum degree $\delta \geq 3$ and maximum degree Δ. Then the size of the minimum feedback vertex set for G is more than $n(\delta - 2)/2(\Delta - 1)$.*

Proof. Let F be a minimum feedback vertex set for G, and let E_F be the set of edges with at least one end point in F. Since $G - F$ is a forest, there are at most $n - |F| - 1$ edges in $G - F$. Thus

$$\Delta|F| \geq |E_F| \geq |E| - n + |F| + 1 > n\delta/2 - n + |F|$$

which implies

$$(\Delta - 1)|F| > n(\delta - 2)/2$$

or $|F| > n(\delta - 2)/2(\Delta - 1)$. $\square$

Using this lemma, we give improved algorithms for regular, almost regular (defined later) and bounded degree graphs.

5.1 Regular Graphs

Setting $\delta = \Delta \geq 3$ in the lemma 4, we get

Corollary 2. *Let $G(V, E)$ be an r-regular graph on n vertices with $r \geq 3$. Then the size of the minimum feedback vertex set for G is more than $n/4$.*

Using Corollary 2, we can obtain the following fixed parameter tractable algorithm for regular graphs, based on the technique of reduction to kernel [8].

> If G is 1-regular, then G is acyclic. If G is 2-regular, G is a vertex disjoint collection of cycles. If the number of cycles is atmost k, answer YES, output a solution obtained by picking one (arbitrary) vertex from each cycle and answer NO otherwise.
> Otherwise, G is r-regular for some $r \geq 3$. If $k \leq n/4$, answer NO, otherwise $n < 4k$ and try all k element subsets S of V to check whether $G - S$ is acyclic. If the answer is yes for any subset S then answer YES (S is a feedback vertex set of size k) and answer NO otherwise (the graph has no feedback vertex set of size at most k otherwise).

The correctness of the algorithm is based on the claim above. Also the running time of the algorithm is $\binom{4k}{k}kn$ which is $O((4e)^k kn)$. Thus we have

Theorem 4. *There is an there is an $O((4e)^k kn)$ algorithm to determine whether a given r-regular graph on n vertices has a feedback vertex set of size at most k.*

5.2 Almost Regular Graphs

Let us call a graph G *almost regular* if $\Delta(G) - \delta(G) \leq c$ for some constant c. Then we have,

Corollary 3. *Let $G(V, E)$ be an almost regular graph on n vertices where $3 \leq \delta$ and $\Delta - \delta \leq c$. Then the size of the minimum feedback vertex set for G is more than $n(\Delta - c - 2)/2(\Delta - 1)$.*

Using Corollary 3, we can design a fixed parameter tractable algorithm for almost regular graphs (as we did for regular graphs) to prove the following.

Theorem 5. *For some constant $d > 1$, there is an $O(d^k kn)$ algorithm to determine whether an almost regular G (with $\delta \geq 3$) has a feedback vertex set of size at most k.*

5.3 Bounded Degree Graphs

Here, we assume that G is an arbitrary graph for which Δ is bounded above by an arbitrary but fixed positive integer. We can preprocess (as explained in Section 2) G so as to get a G' with $\delta(G') \geq 3$ and $\Delta(G') \leq \Delta(G)$ such that G has a feedback vertex set of size k if and only if G' has one such set. Hence, using Lemma 4, we can design a fixed parameter algorithm (as done in the case of regular graphs) and obtain the following theorem.

Theorem 6. *For each fixed positive integer D, there is some positive constant $d > 1$ depending only on D such that there is an $O(d^k kn)$ algorithm to determine if a given G on n vertices with $\Delta(G) \leq D$ has a feedback vertex set of size at most k.*

6 Conclusions and Open Problems

In this paper, we obtained faster algorithms for parameterized feedback vertex set problem on undirected graphs. Our result achieves a significant improvement in the dependence on k (the parameter) of the running time. Our result implies that the problem of checking if an undirected G has a feedback vertex set of size $O(\lg n / \lg \lg n)$ is solvable in polynomial time. It is worth exploring whether the $f(k)$ for the parameterized feedback vertex set problem in general graphs can be made to d^k for some constant d.

References

1. J. Alber, H. L. Bodlaender, H. Fernau and R. Niedermeier, 'Fixed Parameter Algorithms for Dominating Set and Related Problems on Planar Graphs', *Lecture Notes in Computer Science* **1851** (2000) 93–110; to appear in *Algorithmica*.
2. J. Alber, H. Fan, M. R. Fellows, H. Fernau, R. Niedermeier, F. Rosamand and U. Stege, 'Refined Search Tree Techniques for Dominating Set on Planar Graphs', *Lecture Notes in Computer Science* **2136** (2001) 111–122.

3. N. Bansal and V. Raman, 'Upper Bounds for MAX-SAT further improved, *Lecture Notes in Computer Science* **1741** (1999) 247–258.
4. R. Bar-Yehuda, D. Geiger, J. Naor, R. M. Roth, 'Approximation Algorithms for the Feedback Vertex Set Problem with Applications to Constraint Satisfaction and Bayesian Inference', *Proceedings of the 5th Annual ACM-SIAM Symposium on Discrete Algorithms*, Arlington, Virginia, (1994) 344–354.
5. A Becker, R. Bar-Yehuda and D. Geiger, 'Random Algorithms for the Loop Cutset Problem', *Journal of Artificial Intelligence Research* **12** (2000) 219–234.
6. J. Chen, I. A. Kanj and W. Jia, 'Vertex Cover, Further Observations and Further Improvements', *Lecture Notes in Computer Science* **1665** (1999) 313–324.
7. J. Chen, D. K. Friesen, W. Jia and I. A. Kanj, 'Using Nondeterminism to Design Efficient Deterministic Algorithms' in *Proceedings of 21st Foundations of Software Technology and Theoretical Computer Science (FST TCS) conference*, Lecture Notes in Computer Science, Springer Verlag **2245** (2001) 120–131.
8. R. Downey and M. R. Fellows, 'Parameterized Complexity', Springer Verlag 1998.
9. P. Erdos and L. Posa, 'On the maximal number of disjoint circuits of a graph', *Publ Math. Debrecen* **9** (1962) 3–12.
10. A. Itai and M. Rodeh, "Finding a minimum circuit in a graph", *SIAM Journal on Computing,* **7** (1978) 413–423.
11. R. Niedermeier and P. Rossmanith, 'Upper Bounds for Vertex Cover: Further Improved', *in Proceedings of the Symposium on Theoretical Aspects of Computer Science (STACS)*, Lecture Notes in Computer Science **1563** (1999) 561–570.
12. R. Niedermeier and P. Rossmanith, 'New Upper Bounds for Maximum Satisfiability', *Journal of Algorithms* **36** (2000) 63-68.
13. V. Raman, 'Parameterized Complexity,' in *Proceedings of the 7th National Seminar on Theoretical Computer Science*, Chennai, India (1997), 1–18.

An $O(pn + 1.151^p)$-Algorithm for p-Profit Cover and Its Practical Implications for Vertex Cover*

Ulrike Stege[1], Iris van Rooij[2], Alex Hertel[1], and Philipp Hertel[1]

[1] Dept. of Comp. Sc., Univ. of Victoria, Victoria B.C.
stege@cs.uvic.ca,awkkh@shaw.ca
[2] Dept. of Psych., Univ. of Victoria, Victoria B.C.,
irisvr@uvic.ca

Abstract. We introduce the problem PROFIT COVER which finds application in, among other areas, psychology of decision-making. A common assumption is that *net value* is a major determinant of human choice. PROFIT COVER incorporates the notion of net value in its definition. For a given graph $G = (V, E)$ and an integer $p > 0$, the goal is to determine $PC \subseteq V$ such that the profit, $|E'| - |PC|$, is at least p, where E' are the by PC covered edges. We show that p-PROFIT COVER is a parameterization of VERTEX COVER. We present a fixed-parameter-tractable (fpt) algorithm for p-PROFIT COVER that runs in $O(p|V| + 1.150964^p)$. The algorithm generalizes to an fpt-algorithm of the same time complexity solving the problem p-EDGE WEIGHTED PROFIT COVER, where each edge $e \in E$ has an integer weight $\text{w}(e) > 0$, and the profit is determined by $\sum_{e \in E'} \text{w}(e) - |PC|$. We combine our algorithm for p-PROFIT COVER with an fpt-algorithm for k-VERTEX COVER. We show that this results in a more efficient implementation to solve MINIMUM VERTEX COVER than each of the algorithms independently.

1 Introduction

We introduce the *profit problem* PROFIT COVER and study its classical and parameterized complexity. PROFIT COVER is an adaptation of the graph problem VERTEX COVER. We define the optimization version of this well known graph problem. For a given graph[1] $G = (V, E)$, a subset $V' \subseteq V$ is called a *vertex cover* for G if for each edge $(u, v) \in E$, $u \in V'$ or $v \in V'$.

MINIMUM VERTEX COVER (MVC)
Input: A graph $G = (V, E)$.
Output: A *minimum* vertex cover $V' \subseteq V$ for G (i.e., a vertex cover V' for G where $|V'|$ is minimized over all possible vertex covers for G).

---

* This research is supported by a UVic research grant and by NSERC grant 54194.
[1] All graphs considered in this article are simple and undirected.

P. Bose and P. Morin (Eds.): ISAAC 2002, LNCS 2518, pp. 249–261, 2002.

VERTEX COVER (VC) is known to be NP-complete [9]. Also its natural parameterization k-VERTEX COVER (k-VC) is known to be fixed-parameter tractable
[4,7] (here k denotes the size of the vertex cover to be determined). The fastest
known fixed-parameter-tractable algorithm that solves k-VC has a running time
of $O(|V|k + 1.285^k)$ [3].

To illustrate the characteristic nature of a profit problem, we consider a
problem with two goals: (1) Find a set of vertices that covers as many edges as
possible and (2) find a set of vertices that contains as few vertices as possible.
Satisfying either goal is trivial. However, if we want to *satisfy both goals at the
same time* we are confronted with the trade-off between the number of edges
to cover and the number of vertices needed to do that. MVC is a special case
of the problem described above. In MVC, goal (1) is given priority over goal
(2), such that the final goal is to cover *all* edges with as *few* vertices as possible.
When we consider MAXIMUM PROFIT COVER then the same two goals are given,
but neither takes precedence over the other. The goal is to determine a subset
$V' \subseteq V$ such that $|E'| - |V'|$ is maximized, where $E' \subseteq E$ are the edges covered
by the vertices in V'. $|E'| - |V'|$ is called the *profit* of V' for G.

MAXIMUM PROFIT COVER (MPC)

Input: A graph $G = (V, E)$.

Question: A *maximum profit cover* $V' \subseteq V$ (i.e., a subset $V' \subseteq V$ where
   $\text{profit}_{PC,G}(V') = |E_{PC,G}(V')| - |V'|$ is maximized over all possible subsets
   of $V' \subseteq V$. Here, $(u, v) \in E_{PC,G}(V')$ if $(u, v) \in E$ and $(u \in V'$ or $v \in V'))$.

The decision version of this problem is as follows.

PROFIT COVER (PC)

Input: A graph $G = (V, E)$, and an integer $p > 0$.

Question: Does there exist a subset $V' \subseteq V$ with $\text{profit}_{PC,G}(V') \geq p$?

Even though the questions asked in VC and PC are different, the problems are
closely related and thus the complexity of PC is of interest to fields in which VC
finds its application (e.g., data cleaning for multiple sequence alignments [4,21]
and phylogeny compatibility in computational biology [19]). We also describe a
new field of research in which VC, and PC in particular, find a natural application, viz. the psychology of decision-making (see [12] for a review). In this field,
EDGE WEIGHTED PROFIT COVER instantiates a useful generalization of PC.

EDGE WEIGHTED PROFIT COVER (EWPC)

Input: A graph $G = (V, E)$, where each edge $e \in E$ is associated a integer
   weight $w(e) > 0$, integer $p > 0$.

Question: Does there exist a subset $V' \subseteq V$ with $\text{profit}_{EWPC,G}(V') \geq p$?
   Here, $\text{profit}_{EWPC,G}(V') = \sum\limits_{e \in E_{PC,G}(V')} w(e) - |V'|$?[2]

1.1   Graph Problems in Human Decision Making

VC, PC and EWPC model choice situations that humans may encounter in their
everyday live. We discuss a scheduling problem as an illustration. Suppose a stu-

---

[2] As in PC, in EWPC $(u, v) \in E_{PC,G}(V')$ if (1) $(u, v) \in E$ and (2) $u \in V'$ or $v \in V'$.
For simplicity, for the weight of an edge (u, v) we write $w(u, v)$ instead of $w((u, v))$.

dent (DM) is planning a schedule for the upcoming year. There are a number of
activities that DM wants to undertake (e.g., attend a conference, take a course),
but for some pairs of these activities there exists a conflict in DM's world (e.g.,
a conference overlaps with a course). This situation can be modeled as follows.
Let $G = (V, E)$ be a graph such that the vertices represent the activities and
each edge (u, v) represents a conflict between u and v. Assume that DM wants
to exclude as few activities as possible from his or her schedule. Also, DM wants
to reduce the amount of conflict as much as possible. Since both goals cannot
be satisfied at the same time, DM has to formulate a goal that can be satisfied.
Let DM set as a goal to find *at most k* activities to exclude, such that there is
no conflict in the schedule. Then DM's choice situation is modeled by VC. Alternatively, DM can be interested in excluding a set of activities such that DM
makes at least p *profit*, where the profit is considered to be the number of conflicts resolved *minus* the number of activities excluded. This choice situation is
modeled by PC. In many real-world situations the value of activities and the
degree of conflict between pairs of activities may vary. Consider the situation
where the degree of conflict differs for different pairs of activities, but the value
of each activity is the same. If the goal is to establish a satisfactory profit by
excluding activities, then EWPC models the choice situation.

We have introduced and motivated the problems PC and EWPC. Complexity
analyses of such everyday human decision problems are relevant for the important topic of human rationality [13,14]. Specifically, complexity theory can aid
psychology in understanding the limitations of human rationality [18,20].

1.2   Overview

We show NP-completeness of the problems PC and EWPC (Section 3). We
further show that p-PROFIT COVER is a parameterization for VC. In Section 4,
we show both PC and EWPC have linear problem kernels for parameter p (of
size $2p$ for connected graphs and of size $3p - 3$ for general graphs) and thus both
problems are in FPT. We then extend the algorithm by a bounded search-tree
technique and re-kernelization. The resulting running time is $O(p|V|+1.150964^p)$
(Section 4.2). Section 5 discusses how to combine fpt-algorithms for different
parameterizations of a problem. As an example we discuss how the combination
of an algorithm for k-VC and our p-PC algorithm results in a practical and
efficient implementation solving the problems MVC and MPC.

2   Notation and Terminology

We assume basics in graph theory, algorithms, and complexity theory [9,11].
Let $G = (V, E)$ be a graph with vertex set V and edge set E. We define the
neighborhood $N_G(v)$ of a vertex $v \in V$ as the set of all vertices $u \in V$ with $(u, v) \in$
E. The *degree* of a vertex $v \in V$ is denoted by $\deg_G(v)$, where $\deg_G(v) = |N_G(v)|$.
The *set difference* of two sets V and W is denoted by $V \backslash W = \{v \in V | v \notin W\}$.
For a graph $G = (V, E)$ and a set V', $G - V'$ denotes the graph $G^* = (V^*, E^*)$,
such that $V^* = V \backslash V'$ and $E^* = E \backslash E'$ with $E' = \{(u, v) \in E | u \in V' \text{ or } v \in V'\}$.

If $V' = \{v\}$ we simply write $G - v$ instead of $G - V'$. A graph $S = (V_S, E_S)$ is a *subgraph* of a graph $G = (V, E)$ if $V_S \subseteq V$ and $E_S \subseteq E$. Let $S = (V_S, E_S)$ be a subgraph of $G = (V, E)$. Then $G - S$ denotes the graph $G - V_S$. For a rooted tree $T = (V_T, E_T)$ that consists of more than 2 vertices we assume that root r is of

Table 1. Branching vectors and their corresponding $|r|$.

Branching vector	Estimation for $|r|$
[3,8]	1.146139
[4,6]	1.150964
[4,7]	1.138182
[5,6]	1.134724

degree $\deg_T(r) \geq 2$. Let $v, w \in V_T$ such that w is parent of v. Then T_v denotes the by v *induced subtree* of T if T_v is the connected component of $T - (v, w)$ that contains vertex v.

Parameterized complexity was introduced by Downey and Fellows [4]. For introductory surveys see [1,5,6,8]. We denote instances of *parameterized decision problems* with (I, k) where I is the non-parameterized input and k denotes the problem parameter. Furthermore, (I, k) is called a *yes-instance* if the answer to the question asked in the problem is *yes*. If the answer is *no*, we call (I, k) a *no-instance*. A parameterized decision problem is called *fixed-parameter tractable* (in short *fpt*), if for every instance (I, k) it can be solved in time $\text{pol}(|I|) + f(k)$ where pol is a polynomial function and f is a function depending on k only. The complexity class containing all the fpt decision problems is denoted by FPT.[3]

A *problem kernel* for an instance (I, k) is $|I'|$ where (I', k') is the result of applying a polynomial-time algorithm on (I, k) such that $|I'| = f(k)$ for a function f [4]. Such a technique is called *kernelization*. It is known that a parameterized decision problem is in FPT if and only if it is kernelizable [5].

In this paper we also make use of the *bounded search-tree technique*. The goal is to maintain a bounded search tree for the different possible solutions of a given problem instance (I, k). A bounded search tree is a rooted tree bounded in the size by a function $f(k)$. We call the vertices in a search tree *nodes*. The nodes of the search tree are labeled by k-solution candidate sets. Since the size of the search tree depends only on the parameter, the search tree becomes constant size for fixed k. To estimate the size of such a search-tree we use linear recurrence relations with constant coefficients. For an fpt-algorithm and a search tree, we call $[b_1, b_2, \ldots, b_s]$ a *branching vector*[4] of *length* s, if the algorithm recursively makes calls for parameters of sizes $k - b_1$, $k - b_2$, ..., and $k - b_s$. This corresponds to the recurrence $t_k = t_{k-b_1} + t_{k-b_2} + \ldots + t_{k-b_s}$ with the characteristic polynomial $x^b = x^{b-b_1} + x^{b-b_2} + \ldots + x^{b-b_s}$ (here b is the maximum value of all b_i, $i \in$

---

[3] Parameterized decision problems that are not in FPT (unless W[1] = FPT) are called W[1]-hard, where W[1] $\supseteq$ FPT [4].

[4] The term branching vector was first introduced in [16]. For a first fpt-algorithm using this technique to estimate search trees see [2].

$\{1, \ldots, s\}$). If the characteristic polynomial has a single root r, then we can estimate the search-tree size with $O(|r|^k)$ [10]. For a list of branching vectors and their corresponding estimated search-tree size used in this article see Table 1.

The combination of kernelization and bounded search-tree techniques leads to a natural running time of $O(\mathrm{pol}(|I|) + f(k))$. In the case of an $O(|r|^k)$ search-tree size, a problem kernel of size $O(k^\alpha)$ and a kernelization time of $O(kn)$, a typical running time might be $O(kn + |r|^k k^\alpha)$. The technique *re-kernelization*, described first in [22], repeats the kernelization step after each branching in the search tree. Niedermeier and Rossmanith presented a better time-complexity analysis for a search-tree algorithm using re-kernelization [17]. Using their analysis the running time improves to $O(kn + |r|^k)$.

3   NP-Completeness of PC and EWPC

We show NP-completeness for PC via a reduction from VC. As a consequence, the more general problem EWPC is also NP-complete.[5] Before proving NP-completeness of PROFIT COVER we observe that for a given graph $G = (V, E)$ for each subset $PC \subseteq V$ with profit p, there exists a vertex cover $V' \supseteq PC$, $V' \subseteq V$, with profit at least p.

Observation 1. *Let* $G = (V, E)$ *be a graph and* $V' \subseteq V$ *with* $\mathrm{profit}_{\mathrm{EWPC},G}(V') = p$. *If there exists* $(u, v) \in E$ *with* $u, v \notin V'$, *then* $\mathrm{profit}_{\mathrm{EWPC},G}(V' \cup \{u\}) \geq p$.

Theorem 1. PROFIT COVER *is NP-complete.*

Proof. We reduce from VC. An instance for VC is given by a graph $G = (V, E)$ and an integer $k > 0$. We show, G has a vertex cover $V' \subseteq V$ of size k for G if and only if there is a subset $PC \subseteq V$ with $\mathrm{profit}_{\mathrm{PC},G}(PC) = |E| - k$. Let $V' \subseteq V$ be a vertex cover for G of size k. Then $\mathrm{E}_{\mathrm{PC},G}(V') = E$ and therefore $\mathrm{profit}_{\mathrm{PC},G}(V') = |\mathrm{E}_{\mathrm{PC},G}(V')| - |V'| = |E| - k$. Conversely, let $PC \subseteq V$ with $\mathrm{profit}_{\mathrm{PC},G}(PC) = p$. We distinguish two cases: (1) Let $\mathrm{E}_{\mathrm{PC},G}(PC) = E$. We rewrite $p = |E| - k$. We can conclude that $|PC| = k$ and thus PC is a vertex cover of size k. (2) Let $\mathrm{E}_{\mathrm{PC},G}(PC) \neq E$. We extend PC to a vertex cover V' by applying Observation 1 as long as there exist uncovered edges in G. Thus, $\mathrm{profit}_{\mathrm{PC},G}(V') \geq \mathrm{profit}_{\mathrm{PC},G}(PC) \geq p$. Then $\mathrm{E}_{\mathrm{PC},G}(V') = E$ and the vertex-cover size is $k = |V'| \leq |E| - \mathrm{profit}_{\mathrm{PC},G}(V') \leq |E| - p$.

Corollary 1. EDGE WEIGHTED PROFIT COVER *is NP-complete.*

The algorithm described in the proof above computes, in linear time, a vertex cover of size $k = |E| - p$ for a given set with profit p in G; especially we receive a minimum vertex cover for a given maximum profit cover.

---

[5] It is easily verified that both PC and EWPC are members of the class NP.

4   Fast fpt-Algorithms for p-PC and p-EWPC

In this section we present an fpt-algorithm for p-EWPC, the naturally parameterized version of EWPC. Note that PC is a special case of EWPC. Thus the algorithm applies also for p-PROFIT COVER (p-PC).

p-EDGE WEIGHTED PROFIT COVER (p-EWPC)
Input: A graph $G = (V, E)$, each edge $e \in E$ has a positive weight w(e) $\in$ $\mathbb{N} \backslash \{0\}$, integer $p > 0$.
Parameter: p
Question: Does there exist a subset $V' \subseteq V$ with profit$_{\text{EWPC},G}(V') \geq p$?

We first show that p-EWPC has a linear problem kernel and then present a bounded search-tree algorithm for p-EWPC which results in a time complexity of $O(p|V| + 1.150964^p)$.

4.1   A Linear Problem Kernel for p-EWPC

We present a linear-sized problem kernel for p-EWPC that we can construct in linear time.[6] We start by stating the Subgraph Lemma, the key property used to show the existence of a problem kernel ($|V| \leq 2p$) for connected graphs. We then present the six reduction rules that, together with the Component Lemma (cf. page 256), allow us to conclude a problem kernel for general graphs ($|V| \leq 3p-3$).

Lemma 1. (Subgraph Lemma) *Let $G = (V, E)$ be an edge-weighted graph and $S = (V_S, E_S)$ be a subgraph of G. If there exists a set V', $V' \subseteq V_S$, with profit$_{\text{PC},S}(V') = p$ then profit$_{\text{EWPC},G}(V') \geq p$.*

(K 1) If $v \in V$ with $\deg_G(v) = 0$ in an edge-weighted graph $G = (V, E)$ then (G, p) is a yes-instance for p-EWPC if and only if $(G - v, p)$ is a yes-instance for p-EWPC.

(K 2) Let $G = (V, E)$ be an edge-weighted graph and let $v \in V$ with $\deg_G(v) = 1$ and $N_G(v) = \{w\}$. Then (G, p) is a yes-instance for p-EWPC if and only if $(G - \{v, w\}, p - \text{profit}_{\text{EWPC},G}(\{w\}))$ is a yes-instance instance for p-EWPC.

(K 3) Let $G = (V, E)$ be an edge-weighted graph where neither (K 1) nor (K 2) apply. Let $v \in V$ with $N_G(v) = \{u, w\}$ and $(u, w) \in E$. Then (G, p) is a yes-instance for p-EWPC if and only if $(G - \{u, v, w\}, p - \text{profit}_{\text{EWPC},G}(\{u, w\}))$ is a yes-instance for p-EWPC.

(K 4) Let $G = (V, E)$ be an edge-weighted graph where the rules (K 1) to (K 3) do not apply. Let $v \in V$ with $N_G(v) = \{u, w\}$ and $(u, w) \notin E$. We define an edge-weighted graph $G^* = (V^*, E^*)$ with weight function w*(.) as follows. Let F_1 and F_2 be edge sets with

---

[6] We remark that the kernelization algorithm described here is much simpler than the algorithm that creates a linear kernel for k-VC by Chen *et al* [3]. Their algorithm requires after a $O(k|V|)$ preprocessing step another $O(k^3)$ procedure applying a theorem by Nemhauser and Trotter [15].

- $F_1 = \{(x, u)|(x, u) \in E \text{ and } (x, w) \in E\} \setminus \{(u, v)\}$ (i.e., $F_1 \subseteq E$) and
- F_2 consists of *new* edges, namely $F_2 = \{(x, u)|(x, u) \notin E \text{ and } (x, w) \in E\}$ (i.e., $F_2 \cap E = \emptyset$).

Let $V^* = V \setminus \{v, w\}$ and let $E^* = (E \setminus (\{(x, w)|x \in N_G(w)\} \cup \{(u, v)\})) \cup F_2$. We define $w^*(.)$ for G^* such that

- for all edges $e \in E^* \setminus (F_1 \cup F_2)$ $w^*(e) = w(e)$.
- for every $e \in F_1$, $e = (u, x)$. Then $w^*(u, x) = w(u, x) + w(w, x)$.
- for every $e \in F_2$, $e = (u, x)$. Then $w^*(u, x) = w(w, x)$.

Then (G, p) is a yes-instance for p-EWPC if and only if $(G^*, p - w(v, u) - w(v, w) + 1)$ is a yes-instance for p-EWPC.[7]

Proof. (*Sketch*) Let $V' \subseteq V$ with $\text{profit}_{\text{EWPC}, G}(V') = p$. We show there exists $V'' \subseteq V^*$ with $\text{profit}_{\text{EWPC}, G}(V'') \geq p - w(v, u) - w(v, w) + 1$. We distinguish the cases (1) $v \in V'$ and $u, w \notin V'$ and (2) $u, w \in V'$ and $v \notin V'$. (To see that these are the only cases, assume w.l.o.g. $v, u \in V'$ and $w \notin V'$. But then $\text{profit}_{\text{EWPC}, G}((V' \setminus \{v\}) \cup \{w\}) \geq p$ and we can consider case (2) instead.) Because of Observation 1, we can assume if $v \in V'$ and $u, w \notin V'$ that $N_G(u) \subseteq V'$ and $N_G(w) \subseteq V'$. Then $\text{profit}_{\text{EWPC}, G^*}(V' \setminus \{v\}) = p - \text{profit}_{\text{EWPC}, G}(\{v\}) = p - w(v, u) - w(v, w) + 1$. If $v \notin V'$ and $u, w \in V'$, then $\text{profit}_{\text{EWPC}, G^*}(V' \setminus \{w\}) = p - w(v, u) - w(v, w) + 1$.

Conversely, assume there exists $V' \subseteq V^*$ with $\text{profit}_{\text{EWPC}, G^*}(V') \geq p^*$. Because of Observation 1, we can assume that every edge in G^* is covered by V'. Therefore either $u \in V'$ or $N_{G^*}(u) \subseteq V'$. If $u \in V'$ then $\text{profit}_{\text{EWPC}, G}(V' \cup \{w\}) \geq p^* + w(v, u) + w(v, w) - 1$. If $u \notin V'$ then $\text{profit}_{\text{EWPC}, G}(V' \cup \{v\}) \geq p^* + w(v, u) + w(v, w) - 1$.

(K 5) Let $G = (V, E)$ be an edge-weighted graph and let $x \in V$ with $N_G(x) = \{u, v, w\}$. Assume that the rules (K 1) to (K 4) do not apply. If the subgraph of G induced by the vertices $x, u, v,$ and w is a clique, then (G, p) is a yes-instance for p-EWPC if and only if $(G - \{x, u, v, w\}, p - \text{profit}_{\text{EWPC}, G}(\{u, v, w\}))$ is a yes-instance for p-EWPC.

(K 6) Let $G = (V, E)$ be an edge-weighted graph and let $x \in V$ with $N_G(x) = \{u, v, w\}$. Assume that the rules (K 1) to (K 5) do not apply to G. Assume further that the subgraph of G induced by the vertices $x, u, v,$ and w contains exactly 5 edges, say $(u, w) \notin E$. Then (G, p) is a yes-instance for p-EWPC if and only $(G - v, p - \text{profit}_{\text{EWPC}, G}(\{v\}))$ is a yes-instance for p-EWPC.

Proof. (*Sketch*) Let $V' \subseteq V$ with $\text{profit}_{\text{EWPC}, G}(V') \geq p$. Because of Observation 1 we can assume that V' covers every edge in G. If $x \notin V'$ then $u, v,$ and w are included in V'. If $x \in V'$ then either v or 2 vertices of $\{u, v, w\}$ are included in V'. Say $u, w \in V'$. But then $\text{profit}_{\text{EWPC}, G}((V' \setminus \{x\}) \cup \{v\}) \geq p$. Therefore, in all cases we can assume that $v \in V'$. Thus, $\text{profit}_{\text{EWPC}, G-v}(V' \setminus$

---

[7] A similar reduction for vertex cover was first described in [22]. Chen *et al.* call this reduction *vertex folding* [3].

$\{v\}) \geq p - \text{profit}_{\text{EWPC},G}(\{v\})$. On the other hand, assume there exists $V^* \subseteq V \setminus \{v\}$ with $\text{profit}_{\text{EWPC},G-v}(V^*) = p^*$ then $\text{profit}_{\text{EWPC},G}(V^* \cup \{v\}) \geq p^* + \text{profit}_{\text{EWPC},G}(\{v\})$.

Note that after each application of (K 6) to G (on a vertex x) we can apply rule (K 4) (to vertex x).

Definition 1. *We call an edge-weighted graph $G = (V,E)$ reduced if (K 1) to (K 6) do not apply to G.*

Observation 2. *Let $G = (V,E)$ be a reduced and connected edge-weighted graph. Then $|V| \geq 5$.*

Theorem 2 shows the existence of a problem kernel for connected graphs. Note that this theorem also applies for non-reduced graphs with $|V| \geq 3$.

Theorem 2. *Let $G = (V,E)$ be a connected edge-weighted graph. If $|V| \geq 2p+1$ then (G,p) is a yes-instance for p-EWPC.*

To prove Theorem 2, we remark that each connected graph has a spanning subtree. In Lemma 2 we show that each tree consisting of at least $2p+1$ vertices has profit p. From the Subgraph Lemma (Lemma 1) it then follows that every connected graph consisting of at least $2p+1$ vertices has profit p.

Lemma 2. *Let $T = (V_T, E_T)$ be a tree with $|V_T| \geq 2p+1$. Then (T,p) is a yes-instance for p-EWPC.*

Proof. Let $T = (V_T, E_T)$ be a rooted tree with $|V_T| \geq 2p+1$. Since $|V_T| \geq 3$, we can pick a vertex $w \in V_T$ with $\deg_T(w) \geq 2$ and w is parent of leaves only. Let $T_w = (V_w, E_w)$ be the by w induced subtree of T. Then $|V_w| = \deg_T(w)$ and $\text{profit}_{\text{EWPC},T}(\{w\}) = \deg_T(w) - 1$. Consider $T - T_w$. Then $|V_T \setminus V_w| \geq 2p+1 - \deg_T(w)$ and $T - T_w$ has profit $p' = p - (\deg_T(w) - 1)$. To prove the claim it is enough to show that $|V_T \setminus V_w| \geq 2p'+1$. Since $p = p' + (\deg_T(w) - 1)$ it follows that $|V_T \setminus V_w| \geq 2p+1 - \deg_T(w) = 2(p' + \deg_T(w) - 1) + 1 - \deg_T(w) = 2p' + \deg_T(w) - 1$. We know that $\deg_T(w) \geq 2$ and thus $|V_T \setminus V_w| \geq 2p' + 1$.

With Theorem 2 above we determined a lower bound for the maximum profit in connected graphs, i.e. for each graph $G = (V,E)$ there is a subset $V' \subseteq V$ with profit at least $\lfloor \frac{|V|-1}{2} \rfloor$. It follows from (K 2) that the linear-time algorithm implied by the proof above solves MAXIMUM EWPC for trees. We generalize the kernelization result to disconnected graphs using the following lemma.

Lemma 3. *(Component Lemma) Let $G = (V,E)$ be a reduced edge-weighted graph consisting of at least $\frac{p}{2}$ connected components. Then (G,p) is a yes-instance for p-EWPC.*

Theorem 3. *Let $G = (V,E)$ be a reduced edge-weighted graph. Assume further that the Component Lemma does not apply. If $|V| \geq 3p - 2$ then (G,p) is a yes-instance for p-EWPC.*

Proof. Let $G_i = (V_i, E_i)$, $i = 1, \ldots, q$ $(q < \frac{p}{2})$, be the connected components of G. Let $|V| \geq 3p - 2$. Then $p \leq \frac{1}{2}|V| - \frac{p}{2} + 1$. (G, p) is a yes-instance because G has profit $\sum_{i=1}^{q} \lfloor \frac{|V_i|-1}{2} \rfloor \geq \sum_{i=1}^{q} \frac{|V_i|-1}{2} - \frac{q}{2} = \frac{1}{2} \sum_{i=1}^{q} |V_i| - q = \frac{1}{2}|V| - q \geq \frac{1}{2}|V| - \frac{p}{2} + 1$.

Definition 2. *For an instance (G, p) for p-EWPC, we say $G = (V, E)$ is kernelized if (K 1) to (K 6) and the Component Lemma do not apply, and $|V| \leq 3p - 3$.*

Corollary 2. p-PC *and* p-EWPC *are in FPT.*

4.2   A Bounded Search Tree

In the remainder of this section, let (G, p) be a kernelized instance for p-EWPC. From Observation 1 we can conclude the following branching step.

Basic branching step. Let (G, p) be an instance for p-EWPC. Then for a vertex v of G we can branch into instances (G_1, p_1) and (G_2, p_2) with $G_1 = G - v$, $p_1 = p - \text{profit}_{\text{EWPC}, G}(\{v\})$, $G_2 = G - (N_G(v) \cup \{v\})$, and $p_2 = p - \text{profit}_{\text{EWPC}, G}(N_G(v))$.

This branching step is the only branching step applied. To branch efficiently we branch on a highest degree vertex. After each branching step we re-kernelize before branching is repeated.

Theorem 4. *Let $G = (V, E)$ be a kernelized edge-weighted graph and let (G, p) be an instance for p-EWPC. Then the branching vector at the node labeled by (G, p) is $[a, b]$ where either $a \geq 4$, $b \geq 6$ or $a \geq 3$, $b \geq 8$.*

Proof. (*Sketch*) Let $G = (V, E)$ be a kernelized edge-weighted graph. We know that for all $x \in V$, $\deg_G(x) \geq 3$. We distinguish five cases. For every case we assume that the preceding cases do not apply.

Case 1. Let $x \in V$ with $\deg_G(x) \geq 6$. Branching on x results in the instances (G_1, p_1) and (G_2, p_2), $G_i = (V_i, E_i)$ $(i = 1, 2)$, with $G_1 = G - x$ and $G_2 = G - (N_G(x) \cup \{x\})$. Then we ask if (G_1, p_1) or (G_2, p_2) is a yes-instance for p-EWPC with $p_1 = p - \text{profit}_{\text{EWPC}, G}(\{x\})$ and $p_2 = p - \text{profit}_{\text{EWPC}, G}(N_G(x))$. We know that $\text{profit}_{\text{EWPC}, G}(\{x\}) \geq 5$. Because each vertex in $N_G(x)$ has a degree of at least 3 in G, there are at least 12 edges incident to the vertices in $N_G(x)$. Then $\text{profit}_{\text{EWPC}, G}(N_G(x)) \geq 12 - 6 = 6$. Thus, the worst branching vector resulting out of this case is $[5, 6]$.

Case 2. Let $x \in V$ with $\deg_G(x) = 5$. Because each vertex in $N_G(x)$ has a degree of at least 3 and the graph is reduced, there are at least 11 edges incident to $N_G(x)$. We know that $\text{profit}_{\text{EWPC}, G}(\{x\}) = 4$ and $\text{profit}_{\text{EWPC}, G}(N_G(x)) \geq 6$. Thus, the worst branching vector is $[4, 6]$.

Case 3. Let $\deg_G(x) = 4$. Because each vertex in $v \in V$ has $3 \leq \deg_G(v) \leq 4$ and G is kernelized, there are at least 10 edges incident to $N_G(x)$. We distinguish three cases. *Case 3a.* There are at least 12 edges incident to the vertices in $N_G(x)$. Since $\text{profit}_{\text{EWPC}, G}(\{x\}) = 3$ and $\text{profit}_{\text{EWPC}, G}(N_G(x)) \geq 8$, the worst branching vector is [3,8]. *Case 3b.* There are exactly 10 edges incident to the vertices in $N_G(x)$. Let $N(x) = \{a, b, c, d\}$. Then w.l.o.g. the edges incident to $N_G(x)$ are $(a, b), (c, d), (a, v_a), (b, v_b), (c, v_c)$, and (d, v_d), with $\{v_a, v_b, v_c, v_d\} \subseteq V$ but $v_a, v_b, v_c, v_d \notin N_G(x)$ (here v_a, v_b, v_c, v_d are not necessarily distinct). Branching on x results in (G_1, p_1) and (G_2, p_2) with $G_1 = G - x$ and $G_2 = G - (N_G(x) \cup \{x\})$. Since $\text{profit}_{\text{EWPC},G}(\{x\}) = 3$ and $\text{profit}_{\text{EWPC},G}(N_G(x)) \geq 6$, the worst branching vector is [3,6]. Since $\deg_{G_1}(a) = 2$ we can always reduce (G_1, p_1) further by applying (K 3) or (K 4) and improve the profit by at least 1. Thus, the worst case branching vector improves to [4,6]. *Case 3c.* Assume there are exactly 11 edges incident to $N_G(x)$. Branching on x results in (G_1, p_1) and (G_2, p_2) with $G_1 = G - x$ and $G_2 = G - (N_G(x) \cup \{x\})$. Since $\text{profit}_{\text{EWPC},G}(\{x\}) = 3$ and $\text{profit}_{\text{EWPC},G}(N_G(v)) \geq 7$, the worst branching vector is [3,7]. Because there exists $v \in N_{G_1}(x)$ with $\deg_{G_1}(v) = 2$ we can always reduce (G_1, p_1) further by applying (K 3) or (K 4) and improve the profit by at least 1. Thus the worst case branching vector improves to [4,7].

   Case 4. Let $\deg_G(x) = 3$, $N_G(x) = \{a, b, c\}$. Since G is reduced, we only have to consider two cases. *Case 4a.* Assume $(b, c), (a, c) \notin E$ but $(a, b) \in E$. Since G is kernelized and 3-regular, a and b have x as the only common neighbor. Branching on x results in (G_1, p_1) and (G_2, p_2) with $G_1 = G - x$ and $G_2 = G - (N_G(x) \cup \{x\})$. Since $\text{profit}_{\text{EWPC},G}(\{x\}) = 2$ and $\text{profit}_{\text{EWPC},G}(N_G(x)) \geq 5$, the worst branching vector so far is [2,5]. Because $\deg_{G_1}(a) = 2$, we can apply (K 4) to a (note that, since a and c are not adjacent, this does not affect the degree of c) and apply (K 3) or (K 4) to c. Thus the branching vector improves to [4,5]. Consider G_2. Since G_2 has at least one degree-2 vertex we can apply (K 3) or (K 4). The final worst case branching vector is [4,6]. *Case 4b.* Assume $(a, b), (a, c), (b, c) \notin E$. We know that G is 3-regular. Branching on x results in (G_1, p_1) and (G_2, p_2) with $G_1 = G - x$ and $G_2 = G - (N_G(x) \cup \{x\})$. Since $\text{profit}_{\text{EWPC},G}(\{x\}) = 2$ and $\text{profit}_{\text{EWPC},G}(N_G(x)) \geq 6$, the worst branching vector is [2,6]. We know $\deg_{G_1}(a) = \deg_{G_1}(b) = \deg_{G_1}(c) = 2$. Since a and b have no neighbor in common we can apply (K 3) or (K 4) to a and b and thus the branching vector improves to [4,6].

From Theorem 4 and Table 1 on page 252 we receive the estimated search-tree size for the algorithm described above.

Corollary 3. *The bounded search tree has a worst-case size of $O(1.150964^p)$.*

The kernelization step of the described fpt-algorithm can be realized in time $O(p|V|)$. Then the bounded-search-tree algorithm with the integrated re-kernelization after each branching step leads to the running time of $O(p|V| + 1.150964^p)$ (see Section 2 for a more detailed explanation of this complexity analysis).

Corollary 4. *The fpt-algorithm above has a running time of $O(p|V| + 1.150964^p)$.*

5   Efficient Implementations via Combining Tractable Parameterizations

We conclude from the NP-completeness proof for PC that there is a 1:1 correspondence between the yes-instances (G, p) for p-PC and the yes-instances $(G, |E| - p)$ for k-VC (Theorem 1). Thus we can consider p-PC and k-VC as two different parameterizations of VC.[8] With our fpt-algorithm for p-PC we presented an alternative strategy to attack VC. In this section we demonstrate how different fpt-parameterizations of a given decision problem can be combined to an algorithm that incorporates the advantages of all these algorithms. Therefore, the combined algorithm yields a more efficient implementation than any of the algorithms if implemented independently. As an example, we sketch how we can combine the presented fpt-algorithm for p-PC with any fpt-algorithm for k-VC.

The general flavor of common fpt-algorithms for k-VC [3,5,16] is analogous to our fpt-algorithm for p-PC (i.e., the algorithms consist of a combination of kernelization, bounded-search-tree technique, and often re-kernelization, as described in the last paragraph in Section 2). The last step in a kernelization for an instance (G, k) for k-VC is a check of the number of vertices in the reduced graph. If the number exceeds the problem-kernel size, then (G, k) is a no-instance. At each node in the search tree at least one vertex is picked and thus k is reduced by at least 1. Therefore, the bounded search tree has branches of length at most k. The computation of a branch in the search tree is finished as soon as k vertices are picked or a vertex cover is found. On the other hand, our p-PC algorithm returns as an answer yes for an instance (G, p) if, after reducing, the graph is larger than the problem kernel. The branch length in the search tree is bounded by p; the computation of a branch can be ended as soon as enough profit is allocated and therefore its length does not exceed $\frac{p}{3}$ (Theorem 4).

We construct a combined algorithm by (1) keeping track of both parameters during every step of the algorithm and (2) checking the possible decisions for both parameters at each stage of the implementation. First, we take advantage of the problem kernels for the different parameterizations. The problem kernel for k-VC allows no-instances to be identified early whereas the problem kernel for p-PC allows yes-instances to be identified early. If, for a given positive integer p, the problem kernel for p-PC is smaller than the problem kernel for k-VC (with $k = |E| - p$) then there are more instances for which an early decision can be made (thanks to the p-PC kernel) than if we solely use the problem kernel for k-VC. Conversely, if the kernel for p-PC is larger than the problem kernel for k-VC (with $k = |E| - p$) then there are more instances for which an early decision can be made (thanks to the k-VC kernel) than if we solely use the problem kernel for p-PC. Secondly, in the search tree, whenever a decision in terms of k or p $(= |E| - k)$ can be made after a branching step, the computation of this particular branch terminates. Note that a yes-decision in terms of p might be

---

[8] For another parameterization for VC consider the problem INDEPENDENT SET. For a given graph $G = (V, E)$ there exists a vertex cover of size k if and only if there exists an independent set of size $|V| - k$ [9]. INDEPENDENT SET, parameterized by the size of the independent set to be determined, is known to be W[1]-complete [4].

possible even before a vertex cover is found. Conversely, since a no-decision in terms of k can be made even before a vertex cover is found, we can decide early that it is impossible to allocate $|E| - k$ profit.

To solve for example MVC an fpt-algorithm has to be repeated for a given graph until the optimum value for the fixed parameter is found. Using a combined implementation as described above will speed up the process of early decisions and thus avoid a big part of the otherwise necessary exponential search via branching. We implemented the algorithm suggested above and the results reflect the expected speed-up when compared with an implemenation of a k-VC algorithm as published in [5].

6   Conclusions

We presented a new fpt-algorithm for VERTEX COVER. We first introduced a parameterization, p-PROFIT COVER, and showed that it is solvable in time $O(p|V| + 1.150964^p)$. We then showed how our fpt-algorithm for p-PROFIT COVER can be used to speed up existing algorithms for k-VERTEX COVER. This new approach of combining fpt-algorithms for different parameterizations of a problem provides an additional tool for applying parameterized complexity theory in practice.

Besides its useful relationship to VC, the problem PC is also of theoretical and practical interest in itself. We have demonstrated that in some contexts PC is a better model than VC (Section 1.1). For example, in certain problems in human decision making VC may pose unnecessary constraints on the set of acceptable solutions. Further, in many real-world situations conflicts are a matter of degree. Here EDGE WEIGHTED PROFIT COVER provides a natural and useful adaptation of VC to model those problem situations.

We have shown that both p-PC and p-EWPC have linear problem kernels of size $2p$ for connected graphs and size $3p - 3$ for disconnected graphs. The fast fpt-algorithm we presented for p-PC also solves the more general problem p-EWPC. Despite the small constant $c = 1.150964$ in the running time of $O(p|V| + c^p)$, this fpt-algorithm is much simpler to implement than the sophisticated k-VC algorithm presented in [3].

Acknowledgements. We thank Hausi A. Müller and Frank Ruskey for their feedback.

References

[1]   J. Alber, J. Gramm, R. Niedermeier, "Faster exact algorithms for hard problems: A parameterized point of view," *Discr. Mathematics* (2001) 229, 3–27.

[2]   R. Balasubramanian, M.R. Fellows, and V. Raman, "An improved fixed-parameter algorithm for Vertex Cover". *Inform. Proc. Letters* (1998) 65, 163–168.

[3]   J. Chen, I.A. Kanj, W. Jia, "Vertex Cover: Further observations and further improvements," *J. Algorithms* (2001), 41, 280–301.

[4]   R.G. Downey and M.R. Fellows, *Parameterized Complexity* (1999), Springer.

[5] R.G. Downey, M.R. Fellows, and U. Stege, "Parameterized Complexity: A Framework for Systematically Confronting Computational Intractability," *AMS-DIMACS Proc. Series* (1999) 49, 49–99.

[6] R.G. Downey, M.R. Fellows, and U. Stege, "Computational Tractability: The View From Mars," *Bulletin of the EATCS* (1999).

[7] M.R. Fellows, "On the complexity of vertex set problems," *Tech. Rep.* (1988), Computer Science Department, University of New Mexico.

[8] M.R. Fellows, "Parameterized Complexity: The Main Ideas and Connections To Practical Computing," *1st Dagstuhl Workshop on Exp. Algorithms* (2001).

[9] M.R. Garey and D.S. Johnson, *Computers and Intractability: A Guide to the Theory of NP-Completeness* (1979), Freeman.

[10] R.L. Graham, D.E. Knuth, and O. Patashnik, *Concrete Mathematics*, Addison-Wesley (1994).

[11] J. Gross and J. Yellen, *Graph theory and its applications* (1999), CRC Press.

[12] R. Hastie, "Problems for judgment and decision making," *Ann. Review of Psychology* (2001) 52, 653–683.

[13] K.I. Manktelow and D.E. Over (Eds.), *Rationality: Psychological and philosophical perspectives* (1993), London, Routledge.

[14] P.K. Moser (Ed.), *Rationality in action: Contemporary approaches* (1990), Cambridge University Press.

[15] G.L. Nemhauser and L.E. Trotter, "Vertex packing: structural properties and algorithms," *Mathematical Programming* (1975), 8, 232–248.

[16] R. Niedermeier and P. Rossmanith, "Upper Bounds for Vertex Cover Further Improved," In *Proc. of 16th STACS* (1999). LNCS 1563 , 561–570.

[17] R. Niedermeier, P. Rossmanith, "A general method to speed up fixed-parameter-tractable algorithms," *Inf. Proc. Letters* (2000) 73, 125–129.

[18] M. Oaksford and N. Chater, "Reasoning theories and bounded rationality," In K.I. Manktelow & D.E. Over (Eds.) *Rationality: Psychological and philosophical perspectives* (1993), 31–60, Routledge.

[19] J. Setubal and J. Meidanis. *Introduction to Computational Molecular Biology* (1997), PWS Publ. Comp.

[20] H.A. Simon, "Invariants of human behavior". *Ann. Rev. Psych.* (1990) 41(1), 1–19.

[21] U. Stege, *Resolving Conflicts from Problems in Computational Biology* (2000) Ph.D. thesis, No.13364, ETH Zürich.

[22] U. Stege and M.R. Fellows, "An Improved Fixed-Parameter-Tractable Algorithm for Vertex Cover," (1999) Tech. Rep. 318, Dept. of Comp. Science, ETH Zürich.

Exponential Speedup of Fixed-Parameter Algorithms on $K_{3,3}$-Minor-Free or K_5-Minor-Free Graphs*

Erik D. Demaine[1], Mohammad Taghi Hajiaghayi[1], and Dimitrios M. Thilikos[2]

[1] Laboratory for Computer Science, Massachusetts Institute of Technology,
200 Technology Square, Cambridge, MA 02139, U.S.A.
{hajiagha,edemaine}@theory.lcs.mit.edu,
[2] Departament de Llenguatges i Sistemes Informàtics, Universitat Politècnica de
Catalunya, Campus Nord – Mòdul C5, Desp. 211b, c/Jordi Girona Salgado, 1-3.
E-08034, Barcelona, Spain
sedthilk@lsi.upc.es

Abstract. We present a fixed-parameter algorithm that constructively solves the k-dominating set problem on graphs excluding one of K_5 or $K_{3,3}$ as a minor in time $O(4^{16.5\sqrt{k}}n^{O(1)})$, which is an exponential factor faster than the previous $O(2^{O(k)}n^{O(1)})$. In fact, we present our algorithm for any H-minor-free graph where H is a single-crossing graph (can be drawn in the plane with at most one crossing) and obtain the algorithm for $K_{3,3}(K_5)$-minor-free graphs as a special case. As a consequence, we extend our results to several other problems such as vertex cover, edge dominating set, independent set, clique-transversal set, kernels in digraphs, feedback vertex set and a series of vertex removal problems. Our work generalizes and extends the recent result of exponential speedup in designing fixed-parameter algorithms on planar graphs by Alber et al. to other (nonplanar) classes of graphs.

1 Introduction

According to a 1998 survey book [19], there are more than 200 published research papers on solving domination-like problems on graphs. Since this problem is very hard and NP-complete even for special kinds of graphs such as planar graphs, much attention has focused on solving this problem on a more restricted class of graphs. It is well known that this problem can be solved on trees [10] or even the generalization of trees, graphs of bounded treewidth [26]. The approximability of the dominating set problem has received considerable attention, but it is not known and it is not believed that this problem has constant factor approximation algorithms on general graphs [5].

* The work of the third author was supported by the IST Programme of the EU under contract number IST-1999-14186 (ALCOM-FT), the Spanish CICYT project TIC2000-1970-CE, and the Ministry of Education and Culture of Spain (Resolución 31/7/00 – BOE 16/8/00).

P. Bose and P. Morin (Eds.): ISAAC 2002, LNCS 2518, pp. 262–273, 2002.
© Springer-Verlag Berlin Heidelberg 2002

Downey and Fellows [14] introduced a new concept to handle NP-hardness called *fixed-parameter tractability*. Unfortunately, according to this theory, it is very unlikely that the k-dominating set problem has an efficient fixed-parameter algorithm for general graphs. In contrast, this problem is fixed-parameter tractable on planar graphs. The first algorithm for planar k-dominating set was developed in the book of Downey and Fellows [14]. Recently, Alber et al. [1] demonstrated a solution to the planar k-dominating set in time $O(4^{6\sqrt{34k}}n)$ (for an improvement of this result, proposed by Kanj and Perković, see [20]). Indeed, this result was the first nontrivial result for the parameterized version of an NP-hard problem where the exponent of the exponential term grows sublinearly in the parameter. One of the aims of this paper is to generalize this result to nonplanar classes of graphs.

A graph G is *H-minor-free* if H cannot be obtained from any subgraph of G by contracting edges. A graph is called a *single-crossing graph* if it can be drawn in the plane with at most one crossing. Similar to the approach of Alber et al., we prove that for a single-crossing graph H, the treewidth of any H-minor-free graph G having a k-dominating set is bounded by $O(\sqrt{k})$. As a result, we generalize current exponential speedup in fixed-parameter algorithms on planar graphs to other kinds of graphs and show how we can solve the k-dominating set problem on $K_{3,3}$-minor-free or K_5-minor-free graphs in time $O(4^{16.5\sqrt{k}}n^{O(1)})$. The genesis of our results lies in a result of Hajiaghayi et al. [18] on obtaining the local treewidth of the aforementioned class of graphs. The classes of $K_{3,3}$-minor-free graphs and K_5-minor-free graphs have been considered before, e.g. in [22,27].

Using the solution for the k-dominating set problem on planar graphs, Kloks et al. [9,17,23,24] and Alber et al. [1,2] obtained exponential speedup in solving other problems such as vertex cover, independent set, clique-transversal set, kernels in digraph and feedback vertex set on planar graphs. In this paper we also show how our results can be extended to these problems and many other problems such as variants of dominating set, edge dominating set and a series of vertex removal problems. The reader is referred to [12] for the full proofs of theorems in this paper.

2 Background

We assume the reader is familiar with general concepts of graph theory such as (un)directed graphs, trees and planar graphs. The reader is referred to standard references for appropriate background [8]. In addition, for exact definitions of various NP-hard graph-theoretic problems in this paper, the reader is referred to Garey and Johnson's book on computers and intractability [16].

Our graph terminology is as follows. All graphs are finite, simple and undirected, unless indicated otherwise. A graph G is represented by $G = (V, E)$, where V (or $V(G)$) is the set of vertices and E (or $E(G)$) is the set of edges. We denote an edge e in a graph G between u and v by $\{u, v\}$. We define n to be the number of vertices of a graph when it is clear from context. We define the *r-neighborhood* of a set $S \subseteq V(G)$, denoted by $N_G^r(S)$, to be the set of vertices at distance at most r from at least one vertex of $S \subseteq V(G)$; if $S = \{v\}$ we simply use the notation $N_G^r(v)$. The *union* of two disjoint graphs G_1 and G_2, $G_1 \cup G_2$, is a graph G such that $V(G) = V(G_1) \cup V(G_2)$ and $E(G) = E(G_1) \cup E(G_2)$.

For generalizations of algorithms on undirected graphs to directed graphs, we consider underlying graphs of directed graphs. The *underlying graph* of a directed graph H is the undirected graph G in which $V(G) = V(H)$ and $\{u, v\} \in E(G)$ if and only if $(u, v) \in E(H)$ or $(v, u) \in E(H)$.

Contracting an edge $e = \{u, v\}$ is the operation of replacing both u and v by a single vertex w whose neighbors are all vertices that were neighbors of u or v, except u and v themselves. A graph G is a *minor* of a graph H if H can be obtained from a subgraph of G by contracting edges. A graph class $\mathcal{C}$ is a *minor-closed* class if any minor of any graph in $\mathcal{C}$ is also a member of $\mathcal{C}$. A minor-closed graph class $\mathcal{C}$ is *H-minor-free* if $H \notin \mathcal{C}$. For example, a planar graph is a graph excluding both $K_{3,3}$ and K_5 as minors.

A *tree decomposition* of a graph $G = (V, E)$, denoted by $TD(G)$, is a pair (χ, T) in which $T = (I, F)$ is a tree and $\chi = \{\chi_i | i \in I\}$ is a family of subsets of $V(G)$ such that:

1. $\bigcup_{i \in I} \chi_i = V$;
2. $\forall_{e = \{u,v\} \in E}$ there exists an $i \in I$ such that both u and v belong to χ_i; and
3. $\forall_{v \in V}$, the set of nodes $\{i \in I | v \in \chi_i\}$ forms a connected subtree of T.

3   General Results on Clique-Sum Graphs

Suppose G_1 and G_2 are graphs with disjoint vertex-sets and $k \geq 0$ is an integer. For $i = 1, 2$, let $W_i \subseteq V(G_i)$ form a clique of size k and let G_i' $(i = 1, 2)$ be obtained from G_i by deleting some (possibly no) edges from $G_i[W_i]$ with both endpoints in W_i. Consider a bijection $h : W_1 \to W_2$. We define a *k-sum* G of G_1 and G_2, denoted by $G = G_1 \oplus_k G_2$ or simply by $G = G_1 \oplus G_2$, to be the graph obtained from the union of G_1' and G_2' by identifying w with $h(w)$ for all $w \in W_1$. The images of the vertices of W_1 and W_2 in $G_1 \oplus_k G_2$ form the *join set*. In the rest of this section, when we refer to a vertex v of G in G_1 or G_2, we mean the corresponding vertex of v in G_1 or G_2 (or both). It is worth mentioning that $\oplus$ is not a well-defined operator and it can have a set of possible results.

Let s be an integer where $0 \leq s \leq 3$ and $\mathcal{C}$ be a finite set of graphs. We say that a graph class $\mathcal{G}$ is a *clique-sum class* if any of its graphs can be constructed by a sequence of i-sums $(i \leq s)$ applied to planar graphs and graphs in $\mathcal{C}$. We call a graph *clique-sum* if it is a member of a clique-sum class. We call the pair $(\mathcal{C}, s)$ the *defining pair* of $\mathcal{G}$ and we call the maximum treewidth of graphs in $\mathcal{C}$ the *base* of $\mathcal{G}$ and the *base* of graphs in $\mathcal{G}$. A series of k-sums (not necessarily unique) which generate a clique-sum graph G are called *a decomposition of G into clique-sum operations*.

According to the result of [25], if $\mathcal{G}$ is the class of graphs excluding a single crossing graph (can be drawn in the plane with at most one crossing) H then $\mathcal{G}$ is a clique-sum class with defining pair $(\mathcal{C}, s)$ where the base of $\mathcal{G}$ is bounded by a constant c_H depending only on H. In particular, if $H = K_{3,3}$, the defining pair is $(\{K_5\}, 2)$ and $c_H = 4$ [28] and if $H = K_5$ then the defining pair is $(\{V_8\}, 3)$ and $c_H = 4$ [28]. Here by V_8 we mean the graph obtained from a cycle of length eight by joining each pair of diagonally opposite vertices by an edge. For more results on clique-sum classes see [13].

From the definition of clique-sum graphs, one can observe that, for any clique-sum graph G which excludes a single crossing graph H as a minor, any minor G' of G is also a clique-sum graph which excludes the same graph H as a minor.

We call a clique-sum graph class $\mathcal{G}$ α-recognizable if there exists an algorithm that for any graph $G \in \mathcal{G}$ outputs in $O(n^\alpha)$ time a sequence of clique sums of graphs of total size $O(|V(G)|)$ that constructs G. We call a graph α-recognizable if it belongs in some α-recognizable clique-sum graph class. Using the results in [21] and [4] one can verify the following.

Theorem 1 ([21,4]). *The class of K_5-minor-free ($K_{3,3}$-minor-free) graphs is a 2-recognizable (1-recognizable) clique-sum class.*

A *parameterized graph class* (or just *graph parameter*) is a family $\mathcal{F}$ of classes $\{\mathcal{F}_i, i \geq 0\}$ where $\bigcup_{i \geq 0} \mathcal{F}_i$ is the set of all the graphs and for any $i \geq 0$, $\mathcal{F}_i \subseteq \mathcal{F}_{i+1}$. Given two parameterized graph classes $\mathcal{F}^1$ and $\mathcal{F}^2$ and a natural number $\gamma \geq 1$ we say that $\mathcal{F}^1 \preccurlyeq_\gamma \mathcal{F}^2$ if for any $i \geq 0$, $\mathcal{F}_i^1 \subseteq \mathcal{F}_{\gamma \cdot i}^2$.

In the rest of this paper, we will identify a parameterized problem with the *parameterized graph class* corresponding to its "yes" instances.

Theorem 2. *Let $\mathcal{G}$ be an α-recognizable clique-sum graph class with base c and let $\mathcal{F}$ be a parameterized graph class. In addition, we assume that each graph in $\mathcal{G}$ can be constructed using i-sums where $i \leq s \leq 3$. Suppose also that there exist two positive real numbers β_1, β_2 such that:*

(1) For any $k \gtrsim 0$, planar graphs in $\mathcal{F}_k$ have treewidth at most $\beta_1 \sqrt{k} + \beta_2$ and such a tree decomposition can be found in linear time.

(2) For any $k \geq 0$ and any $i \leq s$, if $G_1 \oplus_i G_2 \in \mathcal{F}_k$ then $G_1, G_2 \in \mathcal{F}_k$

Then, for any $k \geq 0$, the graphs in $\mathcal{G} \cap \mathcal{F}_k$ all have treewidth $\leq \max\{\beta_1 \sqrt{k} + \beta_2, c\}$ and such a tree decomposition can be constructed in $O(n^\alpha + (\sqrt{k})^s \cdot n)$ time.

Theorem 3. *Let $\mathcal{G}$ be a graph class and let $\mathcal{F}$ be some parameterized graph class. Suppose also for some positive real numbers $\alpha, \beta_1, \beta_2, \delta$ the following hold:*

(1) For any $k \geq 0$, the graphs in $\mathcal{G} \cap \mathcal{F}_k$ all have treewidth $\leq \beta_1 \sqrt{k} + \beta_2$ and such a tree decomposition can be decided and constructed (if it exists) in $O(n^\alpha)$ time. We also assume testing membership in $\mathcal{G}$ takes $O(n^\alpha)$ time.

(2) Given a tree decomposition of width at most w of a graph, there exists an algorithm deciding whether the graph belongs in $\mathcal{F}_k$ in $O(\delta^w n)$ time.

Then there exists an algorithm deciding in $O(\delta^{\beta_1 \sqrt{k} + \beta_2} n + n^\alpha)$ time whether an input graph G belongs in $\mathcal{G} \cap \mathcal{F}_k$.

It is worth mentioning that Demaine et al. [11] very recently designed a polynomial-time algorithm to decompose any H-minor-free graph, where H is a single-crossing graph, into clique-sum operations. Thus $O(n^\alpha)$ is polynomial for these H-minor-free graphs.

4   Fixed-Parameter Algorithms for Dominating Set

In this section, we will describe some of the consequences of Theorems 2 and 3 on the design of efficient fixed-parameter algorithms for a series of parameterized problems where their inputs are clique-sum graphs.

A *dominating set* of a graph G is a set of vertices of G such that each of the rest of vertices has at least one neighbor in the set. We represent the k-*dominating set* problem with the parameterized graph class $\mathcal{DS}$ where $\mathcal{DS}_k$ contains graphs which have a dominating set of size $\leq k$. Our target is to show how we can solve the k-dominating set problem on clique-sum graphs, where H is a single-crossing graph, in time $O(c^{\sqrt{k}} n^{O(1)})$ instead of the current algorithms which run in time $O(c^k n^{O(1)})$ for some constant c. By this result, we extend the current exponential speedup in designing algorithms for planar graphs [2] to a more generalized class of graphs. In fact, planar graphs are both $K_{3,3}$-minor-free and K_5-minor-free graphs, where both $K_{3,3}$ and K_5 are single-crossing graphs.

According to the results in [20], condition (1) of Theorem 2 is satisfied for $\mathcal{DS}$ for $\beta_1 = 16.5$ and $\beta_2 = 50$. Also, the next lemma shows that condition (2) of Theorem 2 is also correct.

Lemma 1. *If $G = G_1 \oplus_m G_2$ has a k-dominating set, then both G_1 and G_2 have dominating sets of size at most k.*

We can now apply Theorem 2 for $\beta_1 = 16.5$ and $\beta_2 = \max\{50, c\}$.

Theorem 4. *If $\mathcal{G}$ is an α-recognizable clique-sum class of base c, then any member G of $\mathcal{G}$ where its dominating set has size at most k has treewidth at most $16.5\sqrt{k} + \max\{50, c\}$ and the corresponding tree decomposition of G can be constructed in $O(n^\alpha)$ time.*

Theorem 4 tells us that condition (1) of Theorem 3 is satisfied. Moreover, according to the results in [1,3] condition (2) of Theorem 3, is satisfied for the graph parameter $\mathcal{DS}$ when $\delta = 4$. Applying now theorems 3 and 4 we have the following.

Theorem 5. *There is an algorithm that in $O(4^{16.5\sqrt{k}} n + n^\alpha)$ time solves the k-dominating set problem for any α-recognizable clique-sum graph. Consequently, there is an algorithm that in $O(4^{16.5\sqrt{k}} n)$ $(O(4^{16.5\sqrt{k}} n + n^2))$ time solves the k-dominating set problem for $K_{3,3}$ (K_5)-minor-free graphs.*

5   Algorithms for Parameters Bounded by Dominating Set Number

We provide a general methodology for deriving fast fixed-parameter algorithms in this section. First, we consider the following theorem which is an immediate consequence of Theorem 3.

Theorem 6. *Let $\mathcal{G}$ be a graph class and let $\mathcal{F}^1, \mathcal{F}^2$ be two parameterized graph classes where $\mathcal{F}^1 \preccurlyeq_\gamma \mathcal{F}^2$ for some natural number $\gamma \geq 1$. Suppose also that there exist positive real numbers $\alpha, \beta_1, \beta_2, \delta$ such that:*

(1) For any $k \geq 0$, the graphs in $\mathcal{G} \cap \mathcal{F}_k^2$ all have treewidth $\leq \beta_1 \sqrt{k} + \beta_2$ and such a tree decomposition can be decided and constructed (if it exists) in $O(n^\alpha)$ time. We also assume testing membership in $\mathcal{G}$ takes $O(n^\alpha)$ time.
(2) There exists an algorithm deciding whether a graph of treewidth $\leq w$ belongs in $\mathcal{F}_k^1$ in $O(\delta^w n)$ time.
    Then:
(1) For any $k \geq 0$, the graphs in $\mathcal{G} \cap \mathcal{F}_k^1$ all have treewidth at most $\beta_1 \sqrt{\gamma k} + \beta_2$ and such a tree decomposition can be constructed in $O(n^\alpha)$ time.
(2) There exists an algorithm deciding in $O(\delta^{\beta_1 \sqrt{\gamma k} + \beta_2} n + n^\alpha)$ time whether an input graph G belongs in $\mathcal{G} \cap \mathcal{F}_k^1$.

The idea of our general technique is given by the following theorem that is a direct consequence of Theorems 4 and 6.

Theorem 7. *Let $\mathcal{F}$ be a parameterized graph class satisfying the following two properties:*

(1) It is possible to check membership in $\mathcal{F}_k$ of a graph G of treewidth at most w in $O(\delta^w n)$ time for some positive real number δ.
(2) $\mathcal{F} \preccurlyeq_\gamma \mathcal{DS}$.
    Then:
(1) Any clique sum graph G of base c in $\mathcal{F}_k$ has treewidth at most $\max\{16.5\sqrt{\gamma k} + 50, c\}$.
(2) We can check whether an input graph G is in $\mathcal{F}_k$ in $O(\delta^{16.5\sqrt{\gamma k}} n + n^\alpha)$ [1] on an α-recognizable clique-sum graph of base c.

Theorem 7 applies for several graph parameters. In particular it can be applied for the k-weighted dominated set problem, the k-dominating set problem with property Π, the Y-domination problem, the k-vertex cover problem, the k-edge dominating set problem, the k-edge trasversal set problem, the minimum maximum matching problem, the k-kernel problem in digraphs and the k-independent set problem. For more details on the analysis of each of these problems, see [12].

---

[1] In the rest of this paper, we assume that constants, e.g. c, are small and they do not appear in the powers, since they are absorbed into the O notation.

6   Fixed-Parameter Algorithms for Vertex Removal Problems

For any graph class $\mathcal{G}$ and any nonnegative integer k the graph class k-almost$(\mathcal{G})$ contains any graph $G = (V, E)$ where there exists a subset $S \subseteq V(G)$ of size at most k such that $G[V - S] \in \mathcal{G}$. We note that using this notation if $\mathcal{G}$ contains all the edgeless graphs or forests then k-almost$(\mathcal{G})$ is the class of graphs with vertex cover $\leq k$ or feedback vertex set $\leq k$.

A graph $G = (V, E)$ has a k-cut $S \subseteq V$ when $G[V - S]$ is disconnected and $|S| = k$. Let G_1, G_2 be two of the connected components of $G[V - S]$. Given a component $G_1 = (V_1, E_1)$ of $G[V - S]$ we define its *augmentation* as the graph $G[V_1 \cup S]$ in which we add all edges among vertices of S. We say a k-cut S *minimally separates* G_1 and G_2 if each vertex of S has a neighbor in G_1 and G_2. A graph $G = (V, E)$ has a *strong k-cut* $S \subseteq V$ if $|S| = k$ and $G[V - S]$ has at least k connected components and each pair of them is minimally separated by S. We say that G is the result of the multiple k-clique sum of $G_1, \dots, G_r$ with respect to some join set W if $G = G_1 \oplus_k \cdots \oplus_k G_r$ where the join set is always W and such that W is a strong k-cut of G.

Lemma 2. *Let k be a positive integer and let G be a graph with a strong k-cut S where $1 \leq k$. Then the treewidth of G is bounded above by the maximum of the treewidth of each of the augmented components of G after removing S.*

Lemma 3. *Let $G = (V, E)$ be a graph with a strong k-cut S where $1 \leq k \leq 3$. Then if G belongs to some minor-closed graph class $\mathcal{G}$ then any of the augmented components of G after removing S is also k-connected and belongs to $\mathcal{G}$.*

We now need the following adaptation of the results of [21] and [4] (Theorem 1).

Lemma 4. *Let G be a connected $K_{3,3}$-free graph and let $\mathcal{S}$ be the set of its strong i-cuts, $1 \leq i \leq 2$. Then G can be constructed after a sequence of multiple i-clique sums, $1 \leq i \leq 2$, applied to planar graphs or K_5's where each of these multiple sums has a member of $\mathcal{S}$ as join set. Moreover this sequence can be constructed by an algorithm in $O(n)$ time.*

Lemma 5. *Let G be a connected K_5-free graph and let $\mathcal{S}$ be the set of its strong i-cuts, $1 \leq i \leq 3$. Then G can be constructed after a sequence of multiple i-clique sums, $1 \leq i \leq 3$, applied to planar graphs or V_8's where each of these multiple sums has a member of $\mathcal{S}$ as join set. Moreover this sequence can be constructed by an algorithm in $O(n^2)$ time.*

Theorem 8. *Let $\mathcal{G}$ be a $K_{3,3}(K_5)$-minor-free graph class and let $\mathcal{F}$ be any minor-closed parameterized graph class. Suppose that there exist real numbers $\beta_0 \geq 4, \beta_1$ such that any planar graph in $\mathcal{F}_k$ has treewidth at most $\beta_1\sqrt{k} + \beta_0$ and such a tree decomposition can be found in linear time. Then graphs in $\mathcal{G} \cap \mathcal{F}_k$ all have treewidth $\leq \beta_1\sqrt{k} + \beta_0$ and such a tree decomposition can be constructed in $O(n)$ $(O(n^2))$ time.*

We define $\mathcal{T}_r$ to be the class of graphs with treewidth $\leq r$. It is known that for $1 \leq i \leq 2$, $\mathcal{T}_i$ is exactly the class of K_{i+2}-minor-free graphs.

Lemma 6. *Planar graphs in k-almost $(\mathcal{T}_2)$ have treewidth $\leq 16.5\sqrt{k}+50$. Moreover, such a tree decomposition can be found in linear time.*

We conclude the following general result:

Theorem 9. *Let $\mathcal{G}$ be any class of graphs with treewidth ≤ 2. Then any $K_{3,3}(K_5)$-minor-free graph in k-almost $(\mathcal{G})$ has treewidth $\leq 16.5\sqrt{k} + 50$. Moreover, such a tree decomposition can be found in $O(n)$ $(O(n^2))$ time.*

Combining Theorems 3 and 9 we conclude the following.

Theorem 10. *Let $\mathcal{G}$ be any class of graphs with treewidth ≤ 2. Suppose also that there exists an $O(\delta^w n)$ algorithm that decides whether a given graph belongs in k-almost $(\mathcal{G})$ for graphs of treewidth at most w. Then, one can decide whether a $K_{3,3}(K_5)$-minor-free graph belongs in k-almost $(\mathcal{G})$ in time $O(\delta^{16.5\sqrt{k}}n + n^\alpha)$.*

If $\{O_1, \ldots, O_r\}$ is a finite set of graphs, we denote as minor-excl$(O_1, \ldots, O_r)$ the class of graphs that are O_i-minor-free for $i = 1, \ldots, r$.

As examples of problems for which Theorems 9 and 10 can be applied, we mention the problems of checking whether a graph, after removing k vertices, is *edgeless* ($\mathcal{G} = \mathcal{T}_0$), or has *maximum degree* ≤ 2 ($\mathcal{G} = $ minor-excl$(K_{1,3})$), or becomes a *a star forest* ($\mathcal{G} = $ minor-excl(K_3, P_3)), or a *caterpillar* ($\mathcal{G} = $ minor-excl$(K_3$, subdivision of $K_{1,3})$), or a *forest* ($\mathcal{G} = \mathcal{T}_1$), or *outerplanar* ($\mathcal{G} = $ minor-excl$(K_4, K_{2,3})$), or *series-parallel*, or has treewidth $\leq k$ ($\mathcal{G} = \mathcal{T}_2$).

We consider the cases where $\mathcal{G} = \mathcal{T}_0$ and $\mathcal{G} = \mathcal{T}_1$. In particular we prove the following (for details, see [12]).

Theorem 11. *For any $K_{3,3}(K_5)$-minor-free graph G the following hold.*
(1) If G has a feedback vertex set of size at most k then G has treewidth at most $16.5\sqrt{k} + 50$.
(2) We can check whether G has a feedback vertex set of size $\leq k$ in $O(c_{\text{fvs}}^{16.5\sqrt{k}}n + n)$ $(O(c_{\text{fvs}}^{16.5\sqrt{k}}n + n^2))$ time, for some small constant c_{fvs}.
(3) If G has a vertex cover of size at most k then G has treewidth at most $4\sqrt{3}\sqrt{k} + 5$.
(4) We can check whether G has a vertex cover of size $\leq k$ in $O(2^{4\sqrt{3}\sqrt{k}}n + n)$ $(O(2^{4\sqrt{3}\sqrt{k}}n + n^2))$ time.
(5) We can check whether G has a vertex cover of size $\leq k$ in $O(2^{4\sqrt{3}\sqrt{k}}k+kn+n)$ $(O(2^{4\sqrt{3}\sqrt{k}}k + kn + n^2))$ time.

7   Further Extensions

In this section, we obtain fixed-parameter algorithms with exponential speedup for k-vertex cover and k-edge dominating set on graphs more general than

$K_{3,3}(K_5)$-minor-free graphs. Our approach, similar to the Alber et al.'s approach [2], is a general one that can be applied to other problems.

Baker [6] developed several approximation algorithms to solve NP-complete problems for planar graphs. To extend these algorithms to other graph families, Eppstein [15] introduced the notion of bounded local treewidth, defined formally below, which is a generalization of the notion of treewidth. Intuitively, a graph has bounded local treewidth (or locally bounded treewidth) if the treewidth of an r-neighborhood of each vertex $v \in V(G)$ is a function of r, $r \in \mathbb{N}$, and not $|V(G)|$.

The *local treewidth* of a graph G is the function $\mathrm{ltw}^G : \mathbb{N} \to \mathbb{N}$ that associates with every $r \in \mathbb{N}$ the maximum treewidth of an r-neighborhood in G. We set $\mathrm{ltw}^G(r) = \max_{v \in V(G)}\{\mathrm{tw}(G[N_G^r(v)])\}$, and we say that a graph class $\mathcal{C}$ has *bounded local treewidth (or locally bounded treewidth)* when there is a function $f : \mathbb{N} \to \mathbb{N}$ such that for all $G \in \mathcal{C}$ and $r \in \mathbb{N}$, $\mathrm{ltw}^G(r) \leq f(r)$.

A graph is called an *apex graph* if deleting one vertex produces a planar graph. Eppstein [15] showed that a minor-closed graph class $\mathcal{E}$ has bounded local treewidth if and only if $\mathcal{E}$ is H-minor-free for some apex graph H.

So far, the only graph classes studied with small local treewidth are the class of planar graphs [15] and the class of clique-sum graphs, which includes minor-free graphs like $K_{3,3}$-minor-free or K_5-minor-free graphs [18]. It has been proved that for any planar graph G, $\mathrm{ltw}^G(k) \leq 3k - 1$ [18], and for any $K_{3,3}$-minor-free or K_5-minor-free graph G, $\mathrm{ltw}^G(k) \leq 3k + 4$ [15]. For these classes of graphs, there are efficient algorithms for constructing tree decompositions.

Eppstein [15] showed how the concept of the kth outer face in planar graphs can be replaced by the concept of the kth layer (or level) in graphs of locally bounded treewidth. The kth layer (L_k) of a graph G consists of all vertices at distance k from an arbitrary fixed vertex v of $V(G)$. We denote *consecutive layers from i to j* by $L[i,j] = \cup_{i \leq k \leq j} L_k$.

Here we generalize the concept of *layerwise separation*, introduced in Alber et al.'s work [2] for planar graphs, to general graphs.

Let G be a graph layered from a vertex v, and r be the number of layers. A *layerwise separation of width w and size s* for G is a sequence $(S_1, S_2, \cdots, S_r)$ of subsets of V, with property that $S_i \subseteq \bigcup_{j=i}^{i+(w-1)} L_j$; S_i separates layers L_{i-1} and L_{i+w}; and $\sum_{j=1}^{r} |S_j| \leq s$.

A parameterized problem P has *Layerwise Separation Property (LSP)* of width w and size-factor d, if for each instance (G, k) of the problem P, graph G admits a layerwise separation of width w and size dk.

For example, we can obtain constants $w = 2$ and $d = 2$ for the vertex cover problem. In fact, consider a k-vertex cover C on a graph G and set $S_i = (L_i \cup L_{i+1}) \cap C$. The S_i's form a layerwise separation. Similarly, we can get constants $w = 2$ and $d = 2$ for the edge dominating set problem.

Lemma 7. *Let P be a parameterized problem on instance (G, k) that admits a problem kernel of size dk. Then the parameterized problem P on the problem kernel has LSP of width 1 and size-factor d.*

In fact, using Lemma 7 and the problem kernel of size $2k$ for the vertex cover problem, this problem has the LSP of width 1 and size-factor 2.

The proof of the following theorem is very similar to the proof of Theorem 14 of Alber et al.'s work [2] and hence omitted.

Theorem 12. *Suppose for a graph G, $ltw^G(r) \leq cr + d$ and a tree decomposition of width $ch + d$ can be constructed in $O(n^\alpha)$ for any h consecutive layers (h is a constant). Also assume G admits a layerwise separation of width w and size dk. Then we have $tw(G) \leq 2\sqrt{6dk} + cw + d$. Such a tree decomposition can be computed in time $O(n^\alpha)$.*

Now, since for any H-minor-free graph G, where H is a single-crossing graph, $ltw^G(r) \leq 3r + c_H$ and $tw(L[i,j]) \leq 3(j-i+1) + c_H$ [18], we have the following.

Corollary 1. *For any H-minor-free graph G, where H is a single-crossing graph, that admits a layerwise separation of width w and size dk, we have $tw(G) \leq 2\sqrt{6dk} + 3w + c_H$.*

Since we can construct the aforementioned kind of tree decompositions for $K_{3,3}(K_5)$-minor-free graphs in $O(n)(O(n^2))$ and their local treewidth is $3r + 4$ [18], the following result follows immediately.

Corollary 2. *For any $K_{3,3}(K_5)$-minor-free graph G, that admits a layerwise separation of width w and size dk, we have $tw(G) \leq 2\sqrt{6dk} + 3w + 4$. Such a tree decomposition can be computed in time $O(n)$ $(O(n^2))$.*

In fact, we have this general theorem.

Theorem 13. *Suppose for a graph G, $ltw^G(r) \leq cr + d$ and a tree decomposition of width $ch + d$ can be constructed in time $O(n^\alpha)$ for any h consecutive layers. Let P be a parameterized problem on G such that P has the LSP of width w and size-factor d and there exists an $O(\delta^w n)$-time algorithm, given a tree decomposition of width w for G, decides whether problem P has a solution of size k on G.*

Then there exists an algorithm which decides whether P has a solution of size k on G in time $O(\delta^{2\sqrt{6dk}+cw+d} n + n^\alpha)$.

8   Conclusions and Future Work

In this paper, we considered H-minor-free graphs, where H is a single-crossing graph, and proved that if these graphs have a k-dominating set then their treewidth is at most $c\sqrt{k}$ for a small constant c. As a consequence, we obtained exponential speedup in designing FPT algorithms for several NP-hard problems on these graphs, especially $K_{3,3}$-minor-free or K_5-minor-free graphs. In fact, our approach is a general one that can be applied to several problems which can be reduced to the dominating set problem as discussed in Section 5 or to problems that themselves can be solved exponentially faster on planar graphs [2]. Here, we present several open problems that are possible extensions of this paper.

One topic of interest is finding other problems to which the technique of this paper can be applied. Moreover, it would be interested to find other classes of graphs than H-minor-free graphs, where H is a single-crossing graph, on which the problems can be solved exponentially faster for parameter k.

For several problems in this paper, Kloks et al. [9,24,17,23] introduced a reduction to the problem kernel on planar graphs. Since $K_{3,3}$-minor-free graphs and K_5-minor-free graphs are very similar to planar graphs in the sense of having a linear number of edges and not having a clique of size six, we believe that one might obtain similar results for these graphs. Working in this area was beyond the scope of this paper, but still it would be instructive.

As mentioned before, Theorem 9 holds for any class of graphs with treewidth ≤ 2. It is an open problem whether it is possible to generalize it to apply to any class of graphs of treewidth $\leq h$ for arbitrary fixed h. Moreover, there exists no general method for designing $O(\delta^w n)$-time algorithms for vertex removal problems in graphs with treewidth $\leq w$. If this becomes possible, then Theorem 10 will have considerable algorithmic applications.

Finally, as a matter of practical importance, it would be interesting to obtain a constant coefficient better than 16.5 for the treewidth of planar graphs having a k-dominating set (or better than $4\sqrt{3}$ for the case of a k-vertex cover). Such a result would imply a direct improvement to our results and to all the results in [1,2,23,9,24].

Acknowledgments. We thank Prabhakar Ragde and Naomi Nishimura for their encouragement and help on this paper.

References

1. Jochen Alber, Hans L. Bodlaender, Henning Fernau, Ton Kloks, and Rolf Niedermeier. Fixed parameter algorithms for planar dominating set and related problems. *Algorithmica*, 33(4):461–493, 2002.
2. Jochen Alber, Henning Fernau, and Rolf Niedermeier. Parameterized complexity: Exponential speed-up for planar graph problems. In *Electronic Colloquium on Computational Complexity (ECCC)*. Germany, 2001.
3. Jochen Alber and Rolf Niedermeier. Improved tree decomposition algorithms for domination like problems. In *Latin American Theoretical Informatics*, pages 613–627. Lecture Notes in Computer Science, volume 2286, 2002.
4. Takao Asano. An approach to the subgraph homeomorphism problem. *Theoret. Comput. Sci.*, 38(2-3):249–267, 1985.
5. G. Ausiello, P. Crescenzi, G. Gambosi, V. Kann, A. Marchetti-Spaccamela, and M. Protasi. *Complexity and approximation*. Springer-Verlag, Berlin, 1999.
6. Brenda S. Baker. Approximation algorithms for NP-complete problems on planar graphs. *J. Assoc. Comput. Mach.*, 41(1):153–180, 1994.
7. Hans L. Bodlaender. A partial k-arboretum of graphs with bounded treewidth. *Theoret. Comput. Sci.*, 209(1-2):1–45, 1998.
8. John A. Bondy and U. S. R. Murty. *Graph Theory with Applications*. American Elsevier Publishing Co., Inc., New York, 1976.

9. Maw-Shang Chang, Ton Kloks, and Chuan-Min Lee. Maximum clique transversals. In *Proceedings of the 27th International Workshop on Graph-Theoretic Concepts in Computer Science*, pages 300–310. Boltenhagen, Germany, 2001.
10. Ernest J. Cockayne, Sue Goodman, and Stephen Hedetniemi. A linear time algorithm for the domination number of a tree. *Inform. Process. Lett.*, 4(1):41–44, 1975.
11. Erik D. Demaine, Mohammadtaghi Hajiaghayi, and Dimitrios M. Thilikos. 1.5-Approximation for Treewidth of Graphs Excluding a Graph with One Crossing as a Minor. In *Proceedings of the 5th International Workshop on Approximation Algorithms for Combinatorial Optimization*, 2002, to appear.
12. Erik D. Demaine, Mohammadtaghi Hajiaghayi, and Dimitrios M. Thilikos. Exponential speedup of fixed parameter algorithms on $K_{3,3}$-minor-free or K_5-minor-free graphs. Technical Report MIT-LCS-TR-838, Mass. Inst. Technology, March 2002.
13. Reinhard Diestel. Simplicial decompositions of graphs: a survey of applications. *Discrete Math.*, 75(1-3):121–144, 1989.
14. Rodney G. Downey and Michael R. Fellows. *Parameterized Complexity*. Springer-Verlag, New York, 1999.
15. David Eppstein. Diameter and treewidth in minor-closed graph families. *Algorithmica*, 27(3-4):275–291, 2000.
16. Michael R. Garey and David S. Johnson. *Computers and Intractability: A Guide to the Theory of NP-completeness*. W. H. Freeman and Co., 1979.
17. Gregory Gutin, Ton Kloks, and C. M. Lee. Kernels in planar digraphs. In *Optimization Online*. Mathematical Programming Society, Philadelphia, 2001.
18. Mohammadtaghi Hajiaghayi, Naomi Nishimura, Prabhakar Ragde, and Dimitrios M. Thilikos. Fast approximation schemes for $K_{3,3}$-minor-free or K_5-minor-free graphs. In *Euroconf. on Combinatorics, Graph Theory and Applications*, 2001.
19. Teresa W. Haynes, Stephen T. Hedetniemi, and Peter J. Slater. *Fundamentals of domination in graphs*. Marcel Dekker Inc., New York, 1998.
20. Iyad A. Kanj and Ljubomir Perković. Improved parameterized algorithms for planar dominating set. In *27th International Symposium on Mathematical Foundations of Computer Science*. Lecture Notes in Computer Science 2002.
21. André Kézdy and Patrick McGuinness. Sequential and parallel algorithms to find a K_5 minor. In *Proceedings of the Third Annual ACM-SIAM Symposium on Discrete Algorithms (Orlando, FL, 1992)*, pages 345–356, 1992.
22. Samir Khuller. Extending planar graph algorithms to $K_{3,3}$-free graphs. *Inform. and Comput.*, 84(1):13–25, 1990.
23. Ton Kloks and Leizhen Cai. Parameterized tractability of some (efficient) Y-domination variants for planar graphs and t-degenerate graphs. In *International Computer Symposium (ICS)*. Taiwan, 2000.
24. Ton Kloks, C.M. Lee, and Jim Liu. Feedback vertex sets and disjoint cycles in planar (di)graphs. In *Optimization Online*. Math. Prog. Soc., Philadelphia, 2001.
25. Neil Robertson and Paul Seymour. Excluding a graph with one crossing. In *Graph structure theory (Seattle)*, pages 669–675. Amer. Math. Soc., 1993.
26. Jan A. Telle and Andrzej Proskurowski. Practical algorithms on partial k-trees with an application to domination-like problems. In *Proceedings of Third Workshop on Algorithms and Data Structures (Montréal, 1993)*, pages 610–621.
27. Vijay V. Vazirani. NC algorithms for computing the number of perfect matchings in $K_{3,3}$-free graphs and related problems. *Inform. and Comput.*, 80(2):152–164, 1989.
28. Kehrer Wagner. Über eine Eigenschaft der eben Komplexe. *Deutsche Math.*, 2:280–285, 1937.

Casting a Polyhedron with Directional Uncertainty*

Hee-Kap Ahn[1], Otfried Cheong[2], and René van Oostrum[2]

[1] Imaging Media Research Center, Korea Institute of Science & Technology,
P.O. Box 131, CheongRyang, Seoul, South Korea.
[2] Institute of Information & Computing Sciences, Utrecht University,
P.O. Box 80.089, 3508TB Utrecht, The Netherlands.

Abstract. Casting is a manufacturing process in which molten material is poured into a cast (mould), which is opened after the material has solidified. As in all applications of robotics, we have to deal with imperfect control of the casting machinery. In this paper, we consider directional uncertainty: given a 3-dimensional polyhedral object, is there a polyhedral cast such that its two parts can be removed in opposite directions *with uncertainty* α without inflicting damage to the object or the cast parts? We give a necessary and sufficient condition for castability, and a randomized algorithm that verifies castability and produces two polyhedral cast parts for a polyhedral object of arbitrary genus. Its expected running time is $O(n \log n)$. The resulting cast parts have $O(n)$ vertices in total. We also consider the case where the removal direction is not specified in advance, and give an algorithm that finds all feasible removal directions with uncertainty α in expected time $O(n^2 \log n/\alpha^2)$.

1 Introduction

Casting is a manufacturing process in which molten material is poured into a cavity inside a *mould* (cast). After the liquid material has hardened, the mould is opened, and we are left with an object [6,12], which has the shape of the cavity.

An industrial CAD/CAM system can aid a part designer in verifying already *during the design* of an object whether the object in question can actually be manufactured using a casting process. At the basis of this verification is a geometric decision: is it possible to enclose the object in a mould that can be split into two parts, such that these two *cast parts* can be removed from the object without colliding with the object or each other. (We are not interested in casting processes where the mould has to be destroyed to remove the object.) Note that this is a preliminary decision meant to aid in part design—to physically create the mould for a part one needs to take into account other factors such as heat flow and how air can evade from the cavity.

This problem has been studied by Bose, Bremner, and van Kreveld [5], who considered the *sand casting model* relevant in iron casting, where the two cast

* This research was partially supported by the Hong Kong Research Grants Council.

P. Bose and P. Morin (Eds.): ISAAC 2002, LNCS 2518, pp. 274–285, 2002.
© Springer-Verlag Berlin Heidelberg 2002

parts have to be separated by a plane. Ahn et al. [1] gave, to our knowledge, the first complete algorithm to determine the castability of polyhedral parts for cast removal as we described above, under the assumption that the two cast parts have to be removed in opposite directions. This restriction is true for current casting machinery, and we will therefore assume it in this paper as well. Nevertheless, Ahn, Cheng, and Cheong [2] considered the castability of polyhedral parts in a relaxed model that may become relevant in the future.

The casting algorithms mentioned above assume perfect control of the casting machinery. When a cast part is removed, it is required that the part moves exactly in the specified direction. In practice, however, this will rarely be the case. As in all applications of robotics, we have to deal with imperfect control of the machinery, and a certain level of uncertainty in its movements [6]. When a facet of the object or of a cast part is almost parallel to the direction in which the cast parts are being moved, the two touching surfaces may damage each other when the mould is being opened. This can make the resulting object worthless, or it may wear away the surface of the mould so that it cannot be reused as often as desirable.

In Figure 1 (a), the mould can be opened by moving the two parts in direction $\vec{d}$ and $-\vec{d}$. If, however, due to imperfect control, the upper part is translated in direction $\vec{d'}$, it will destroy the object. The cast parts in (b) are redesigned so that both cast parts can be translated without damage in the presence of some uncertainty.

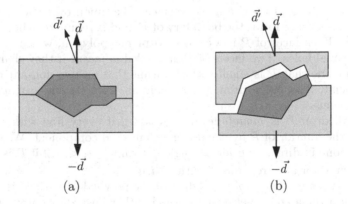

Fig. 1. (a) The upper part of the cast is stuck in direction $\vec{d'}$, (b) a redesigned cast.

In this paper, we consider directional uncertainty in the casting process: given a 3-dimensional polyhedral object, is there a polyhedral cast such that its two parts can be removed in opposite directions *with uncertainty* α without damage to the object or the cast parts? We call such an object *castable with uncertainty* α.

Directional uncertainty has been considered by researchers in motion planning, and robotics in general. A motion planning model with directional uncertainty was perhaps first proposed by Lozano-Pérez, Mason and Taylor [10]. An extensive treatment of motion planning with directional uncertainty is given in the book by Latombe [9].

We generalize the characterization of castable polyhedra by Ahn et al. [1] to incorporate uncertainty in the directions in which the cast parts are removed. A formal definition of our model is given in Section 2. It turns out that one of the main difficulties is to guarantee that the two cast parts are *polyhedral*—while this is trivial in the exact case, it requires approximation of a curved surface in our model with uncertainty. We give a randomized algorithm that verifies whether a polyhedral object of arbitrary genus is castable for a given direction of cast part removal and given uncertainty $\alpha > 0$. The expected running time of the algorithm is $O(n \log n)$, where n is the number of vertices of the input polyhedron. If the object is castable, the algorithm also computes two polyhedral cast parts with $O(n)$ vertices in total.

We then consider the case where the direction of cast part removal is not specified in advance. We give an algorithm that finds all possible removal directions in which the polyhedral object is castable with uncertainty $\alpha > 0$ in expected time $O(n^2 \log n / \alpha^2)$.

2 Preliminaries

Throughout this paper, $\mathcal{P}$ denotes a polyhedron, that is, a not necessarily convex solid bounded by a piece-wise linear surface. The union of vertices, edges, and facets on this surface forms the boundary of $\mathcal{P}$, and is required to be a connected 2-manifold. Each facet of $\mathcal{P}$ is a connected planar polygon, which is allowed to have polygonal holes. Two facets of $\mathcal{P}$ are called *adjacent* if they share an edge. We also assume that $\mathcal{P}$ is *simple*, which means that no two non-adjacent facets share a point. The polyhedron $\mathcal{P}$ may contain tunnels, and can indeed have arbitrary genus.

A polyhedron $\mathcal{P}$ is *monotone* in direction $\vec{d}$ if every line with direction $\vec{d}$ intersects the interior of $\mathcal{P}$ in at most one connected component. We say that $\mathcal{P}$ is α-monotone in direction $\vec{d}$ for an angle α with $0 \leq \alpha < \pi/2$ if $\mathcal{P}$ is monotone in direction $\vec{d'}$ for all directions $\vec{d'}$ with $\angle(\vec{d}, \vec{d'}) \leq \alpha$,

We say that a facet f of a polyhedron or polyhedral surface is α-*steep* in direction $\vec{d}$ if the angle β between a normal, either inward or outward, of f and $\vec{d}$ lies in the range $\pi/2 - \alpha < \beta < \pi/2 + \alpha$. A polyhedron or polyhedral surface is called α-*safe* in direction $\vec{d}$ if none of its facets is α-steep for that direction. Note that an α-monotone polyhedron in direction $\vec{d}$ is not necessarily α-safe in $\vec{d}$. For example, a convex polyhedron is α-monotone in any direction, but there always exists a direction in which some of its facets is α-steep. Conversely, a polyhedron can be α-safe without being α-monotone.

A *terrain* is the graph of a (possibly partially defined) continuous, piece-wise differentiable function with domain $\mathbb{R}^2$ and range $\mathbb{R}$. This means that a terrain is a surface with the property that every vertical line intersects it in at most one point. Hence, it is monotone in direction $\vec{z}$. We call a terrain α-*safe* if the normal vector of the surface makes an angle of at most $\pi/2 - \alpha$ with the vertical direction wherever it is defined. A terrain is *polyhedral* if the surface is piece-wise linear.

A *mould* $\mathcal{M}$ for a polyhedron $\mathcal{P}$ is a pair $(\mathcal{C}_r, \mathcal{C}_b)$ of two polyhedra $\mathcal{C}_r$ and $\mathcal{C}_b$, such that the interiors of $\mathcal{C}_r$, $\mathcal{C}_b$, and $\mathcal{P}$ are pairwise disjoint and the union $B := \mathcal{C}_r \cup \mathcal{P} \cup \mathcal{C}_b$ is a rectangular box that completely contains $\mathcal{P}$ in its interior. We call $\mathcal{C}_r$ and $\mathcal{C}_b$ the *red cast part* and the *blue cast part* of $\mathcal{M}$.

A mould $\mathcal{M}$ with opening direction $\vec{d}$ is α-feasible, if for each pair of directions $(\vec{d}_r, \vec{d}_b)$ with $\angle(\vec{d}, \vec{d}_r) \le \alpha$ and $\angle(-\vec{d}, \vec{d}_b) \le \alpha$, the red cast part $\mathcal{C}_r$ can be translated to infinity in direction $\vec{d}_r$ without colliding with $\mathcal{P}$ or $\mathcal{C}_b$, and the blue cast part $\mathcal{C}_b$ can be translated to infinity in direction $\vec{d}_b$ without colliding with $\mathcal{P}$. Note that the order of removing the cast parts is actually irrelevant.

A polyhedron $\mathcal{P}$ is α-*castable in direction* $\vec{d}$ if an α-feasible mould with opening direction $\vec{d}$ exists. For the special case $\alpha = 0$, we say that $\mathcal{P}$ is *castable* in direction $\vec{d}$.

The following simple lemma characterizes polyhedra castable in direction $\vec{d}$ [1].

Lemma 1. *A polyhedron $\mathcal{P}$ is castable in direction $\vec{d}$ if and only if it is monotone in direction $\vec{d}$.*

The main result of the present paper is a generalization of this result to α-castability. We state the result here—it will take us a few more pages to prove it.

Theorem 2. *A polyhedron $\mathcal{P}$ is α-castable in direction $\vec{d}$ if and only if $\mathcal{P}$ is α-monotone and α-safe in direction $\vec{d}$.*

The following lemma proves the necessity of the condition.

Lemma 3. *If a polyhedron $\mathcal{P}$ is α-castable in direction $\vec{d}$, then $\mathcal{P}$ is α-monotone and α-safe in direction $\vec{d}$.*

Proof. Assume that $\mathcal{P}$ is not α-safe, so a facet f is α-steep with respect to $\vec{d}$. A point p in the interior of f can be neither on the boundary of $\mathcal{C}_r$ nor on the boundary of $\mathcal{C}_b$, and so $\mathcal{P}$ is not α-castable in direction $\vec{d}$.

On the other hand, if $\mathcal{P}$ is α-castable in direction $\vec{d}$, it is castable in any direction $\vec{d}'$ with $\angle(\vec{d}, \vec{d}') \le \alpha$. By Lemma 1, it follows that $\mathcal{P}$ is monotone in direction $\vec{d}'$. It follows that $\mathcal{P}$ is α-monotone. $\qquad\square$

3   Finding a Mould

It remains to prove the sufficiency of the condition in Theorem 2. We do so by showing how to construct an α-feasible mould for any α-monotone and α-safe polyhedron. To simplify the presentation, we will assume, without loss of generality, that $\vec{d}$ is the upward vertical direction (the positive z-direction). We say that $\mathcal{P}$ is α-castable if it is α-castable in the vertical direction.

A facet of $\mathcal{P}$ is called an *up-facet* if its outward normal points upwards, and a *down-facet* if its outward normal points downwards. Assuming $\mathcal{P}$ is α-safe, there are no vertical facets, and so each facet is either an up-facet or a down-facet. Clearly an up-facet of $\mathcal{P}$ must be a facet of the red cast part $\mathcal{C}_r$, while a down-facet of $\mathcal{P}$ must be a facet of the blue cast part $\mathcal{C}_b$. The difficulty is finding the separating surface between $\mathcal{C}_r$ and $\mathcal{C}_b$ "elsewhere."

Assume that $\mathcal{P}$ is α-castable and that $(\mathcal{C}_r, \mathcal{C}_b)$ is an α-feasible mould for $\mathcal{P}$. Again we denote by B the axis-parallel box that forms the outside of the mould. We define the *blue parting surface* $\mathcal{S}_b$ as the common boundary of $\mathcal{C}_b$ and $\mathcal{C}_r \cup \mathcal{P}$, and the *red parting surface* $\mathcal{S}_r$ as the common boundary of $\mathcal{C}_r$ and $\mathcal{C}_b \cup \mathcal{P}$. Any upwards directed vertical line ℓ must intersect $\mathcal{C}_b$, $\mathcal{P}$, and $\mathcal{C}_r$ in this order, each in a single connected component that can be empty. It follows that both $\mathcal{S}_b$ and $\mathcal{S}_r$ are polyhedral terrains. The two terrains coincide except where they bound the polyhedron $\mathcal{P}$. If we let $\mathcal{S} := \mathcal{S}_b \cap \mathcal{S}_r$, define $\mathcal{S}_u$ to be the union of all up-facets, and $\mathcal{S}_d$ to be the union of all down-facets, we can write $\mathcal{S}_r = \mathcal{S} \cup \mathcal{S}_u$ and $\mathcal{S}_b = \mathcal{S} \cup \mathcal{S}_d$. The boundary of $\mathcal{S}$ is the set of silhouette edges of $\mathcal{P}$ (an edge is a silhouette edge if it separates an up-facet from a down-facet).

Constructing a mould therefore reduces to the construction of the terrain $\mathcal{S}$. For the special case $\alpha = 0$, Ahn et al. [1] gave a simple triangulation method for constructing $\mathcal{S}$. Unfortunately, for $\alpha > 0$ this construction does not necessarily produce an α-feasible mould, even when the polyhedron is α-castable. The problem is that even if the polyhedron is α-monotone and α-safe, the constructed terrain $\mathcal{S}$ may not be so. We now prove that it suffices to make sure this does not happen.

Lemma 4. *Let B be an axis-parallel box, and let S be an α-safe terrain separating the top and bottom facets of B. Let C be the part of B above S, and let $C' := B \setminus C$. Let $\vec{d}$ be the upward vertical direction, and let $\vec{d'}$ be such that $\angle(\vec{d}, \vec{d'}) \leq \alpha$. Then C can be translated to infinity in direction $\vec{d'}$ without colliding with C'.*

Proof. Assume the claim was false, and consider a point $p \in C$ that when translated in direction $\vec{d'}$ collides with a point $q \in C'$. The line segment pq lies completely inside B, and so its vertical projection onto S is a path π. Since p lies above one end-point of π, q lies below the other end-point, and the angle between pq and the xy-plane is greater than $\pi/2 - \alpha$, there must be a segment on π where the angle between the segment and the xy-plane is greater than $\pi/2 - \alpha$. This is a contradiction to the assumption that S is α-safe. □

Lemma 5. *Let $\mathcal{P}$ be an α-safe polyhedron, B an axis-parallel box enclosing $\mathcal{P}$, and let S be an α-safe polyhedral terrain bounded by the silhouette edges of $\mathcal{P}$. Then the mould defined by the parting surfaces $\mathcal{S}_r := \mathcal{S} \cup \mathcal{S}_u$ and $\mathcal{S}_b := \mathcal{S} \cup \mathcal{S}_d$ is α-feasible.*

Proof. Since $\mathcal{P}$ is α-safe, both $\mathcal{S}_u$ and $\mathcal{S}_d$ are α-safe terrains. Since $\mathcal{S}$ is α-safe, both $\mathcal{S}_r$ and $\mathcal{S}_b$ are therefore α-safe. Lemma 4 now implies that the mould is α-feasible. $\qquad\qquad\qquad\qquad\qquad\qquad\qquad\qquad\qquad\qquad\qquad\qquad\qquad$ $\square$

We will now show how to construct a terrain $\mathcal{S}$ as in Lemma 5 by forming the lower envelope of a set of cones. Given a point p on an up-facet of $\mathcal{P}$, the α-*cone* $\mathcal{D}(p)$ of p is the solid vertical upwards oriented cone of angle α with apex p. Formally, if p' is a point vertically above p, then $\mathcal{D}(p) := \{x \mid \angle(xpp') \leq \alpha\}$. Let now $\mathcal{D}_1$ be the union of $\mathcal{D}(p)$ over all points $p \in \mathcal{S}_u$, and let $\mathcal{E}_1$ be the lower envelope of $\mathcal{D}_1$. Clearly, $\mathcal{E}_1$ contains $\mathcal{S}_u$, and so $\mathcal{S} := \mathcal{E}_1 \setminus \mathcal{S}_u$ is bounded by the silhouette edges of $\mathcal{P}$. Since $\mathcal{E}_1$ consists of patches of α-cones, it is clearly α-safe. It follows that $\mathcal{S}$ fulfills the requirements of Lemma 5, except that it is not a polyhedral terrain. We will see below that we can easily "approximate" $\mathcal{S}$ by a polyhedral, α-safe terrain $\mathcal{S}'$ that contains all the linear edges of $\mathcal{S}$ and lies below (or coincides with) $\mathcal{S}$ everywhere.

The construction of $\mathcal{S}$ above appears to require taking the union of an infinite family of cones. We now give an alternative definition of $\mathcal{S}$ as the lower envelope of h constant-complexity objects, where h is the number of silhouette edges of $\mathcal{P}$.

In fact, let pq be a silhouette edge of $\mathcal{P}$. The α-*region* $\mathcal{D}(pq)$ of pq is the convex hull of $\mathcal{D}(p) \cup \mathcal{D}(q)$. The lower envelope of $\mathcal{D}(pq)$ consists of three components: two conic surfaces supported by the α-cones $\mathcal{D}(p)$ and $\mathcal{D}(q)$, and a connecting area consisting of two planar facets.

Let now $\mathcal{D}_2$ be the union of $\mathcal{D}(pq)$, over all silhouette edges pq, and let $\mathcal{E} = \mathcal{E}_2$ be the lower envelope of $\mathcal{D}_2$. It is easy to see that $\mathcal{E}_1$ is in fact the lower envelope of $\mathcal{S}_u$ and $\mathcal{E}_2$, and so $\mathcal{E}_1$ and $\mathcal{E}_2$ coincide "outside" of $\mathcal{P}$. Thus, if we define $\mathcal{S}$ to be the part of $\mathcal{E} = \mathcal{E}_2$ not lying above $\mathcal{S}_u$, we define the same terrain $\mathcal{S}$ as above.

The lower envelope $\mathcal{E}$ consists of *faces*, which are either planar, or supported by a single α-cone $\mathcal{D}(x)$ for a vertex x of $\mathcal{P}$. An edge of $\mathcal{E}$ is either a silhouette edge of $\mathcal{P}$, a straight edge separating a conic patch supported by an α-cone $\mathcal{D}(x)$ from an adjacent planar patch supported by an α-region $\mathcal{D}(xy)$, or is an arc supported by the intersection curve of two α-cones, an α-cone and a plane, or two planes. Such arcs are either straight segments, arcs of parabolas, or arcs of hyperbolas. In all cases, they are contained in a plane.

We can represent $\mathcal{E}$ by its projection on the xy-plane. The projection is in fact a planar subdivision, whose faces are supported by a single plane or α-cone. If we annotate each face with the vertex or silhouette edge of $\mathcal{P}$ whose α-cone or α-region supports it, the resulting map is a complete representation of $\mathcal{E}$.

In general, the lower envelope of m well-behaved, constant-complexity objects can have complexity $\Theta(m^2)$ [11]. We will show in the following that our planar subdivision has in fact *linear* complexity. Roughly speaking, we interpret the planar map as a kind of Voronoi diagram. Our sites are the projections of silhouette edges onto the xy-plane, additively weighted by the "height" of the edge above the xy-plane. (This is, indeed, a strange notion of "weight," as it is not constant for a given site. The concerned reader is asked to wait for the formal definition below.) This diagram does not appear to have been studied before, but it does fit into Klein's framework of *abstract Voronoi diagrams* [7,8], and his results on complexity and computation apply.

Consider a silhouette edge e of $\mathcal{P}$. Let $\bar{e}$ be the projection of e onto the xy-plane. For a point $\bar{p} \in \bar{e}$, let p_z be the z-coordinate of the point $p = (p_x, p_y, p_z) \in e$ whose projection on the xy-plane is $\bar{p}$, and let $w(\bar{p})$ be $p_z \tan \alpha$. We can now define a distance measure in the plane as follows: For $x \in \mathbb{R}^2$ and $\bar{p} \in \bar{e}$, we define

$$d(x, \bar{p}) := |x\bar{p}| + w(\bar{p}) = |x\bar{p}| + p_z \tan \alpha.$$

The distance of a point x to a segment $\bar{e}$ is then

$$d(x, \bar{e}) := \min_{\bar{p} \in \bar{e}} d(x, \bar{p}).$$

Lemma 6. *The vertical projection of the lower envelope $\mathcal{E}$ coincides with the Voronoi diagram of the projected silhouette edges under the distance function defined above.*

Proof. Let x be a point in the plane, and let e be a silhouette edge of $\mathcal{P}$. Let x^* be the point where the vertical line through x intersects the boundary of the α-region $\mathcal{D}(e)$. We observe that $d(x, \bar{e}) = |xx^*| \tan \alpha$. The lemma follows. $\square$

In the following lemma, we show some properties of this Voronoi diagram.

Lemma 7. *Let $\mathcal{P}$ be an α-safe and α-monotone polyhedron. Consider the Voronoi diagram defined by the projections of a subset G' of silhouette edges of $\mathcal{P}$ with the distance function above. It has the following properties:*

1. *A projected silhouette edge $\bar{e}$ lies in its own Voronoi cell.*
2. *Given a point x in the Voronoi cell of $\bar{e}$. Let $y \in \bar{e}$ be the point on $\bar{e}$ minimizing the distance from x. Then the segment xy is contained in the Voronoi cell of $\bar{e}$.*
3. *Each Voronoi cell is simply connected.*
4. *The Voronoi diagram is an abstract Voronoi diagram as defined by by Klein [7].*

Proof. Let G' be a non-empty subset of silhouette edges and vertices, and let $\mathcal{E}'$ be the lower envelope of the α-regions of the silhouette edges in G'.

(1) The claim is identical to stating that the silhouette edge e appears on the lower envelope $\mathcal{E}'$. If it didn't, a point $p \in e$ would have to lie inside the α-region $\mathcal{D}(e')$ of some other silhouette edge e', in contradiction to the assumption that $\mathcal{P}$ is α-monotone.

(2) Assume there is a point $z \in xy$ such that the nearest site point to z is $t \neq y$. Then

$$d(x, t) = |xt| + w(t) \leq |xz| + |zt| + w(t) = |xz| + d(z, t)$$
$$< |xz| + d(z, y) = |xz| + |zy| + w(y) = |xy| + w(y) = d(x, y),$$

in contradiction to the definition of y. So the nearest point on a site is y, for all points on xy, and the segment xy is contained in the Voronoi cell of $\bar{e}$.

(3) Follows from *(1)*, *(2)*, and the fact that the segments considered in *(2)* are all parallel for points x on one side of $\bar{e}$.

(4) The abstract Voronoi diagram framework by Klein [7] assumes a set of (abstract) objects, each pair of which defines a bisector partitioning the plane into two unbounded regions. The system of bisectors has to adhere to a set of four conditions. It is straightforward to verify that the bisectors defined by pairs of silhouette edges do fulfill these conditions, using *(1)*-*(3)* and elementary calculations. □

Figure 2 shows the bisector of two projected silhouette edges $\bar{e}$ and $\overline{e'}$. Note the drop-shaped curves surrounding each edge: these are curves of equal distance from the segment.

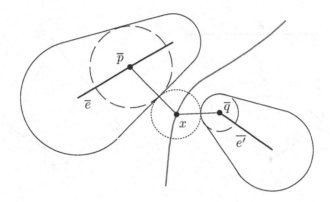

Fig. 2. $d(x, \bar{e}) = d(x, \overline{e'})$

Lemma 8. *Let $\mathcal{P}$ be an α-monotone and α-safe polyhedron with n vertices, and let $\mathcal{E}$ be the lower envelope of the α-regions of its silhouette edges. Then $\mathcal{E}$ has complexity $O(n)$ and can be computed in expected time $O(n \log n)$.*

Proof. From Lemma 6 and Lemma 7 (1)-(3), we can conclude that $\mathcal{E}$ has linear complexity.

We can identify the h silhouette edges of $\mathcal{P}$ in $O(n)$ time by inspecting the normals of all facets. By Lemma 7 (4), the projection of $\mathcal{E}$ onto the xy-plane can be computed in expected time $O(h \log h)$ by the randomized incremental algorithm of Klein et al. [8]. Each face of the Voronoi diagram carries information about the site creating it, and so we can construct the envelope $\mathcal{E}$ in linear time $O(h)$. □

We have now seen how to compute an α-safe terrain $\mathcal{E}$ bounded by the silhouette edges of $\mathcal{P}$ in time $O(n \log n)$. All that remains to be done to fulfill the assumptions of Lemma 5 is to turn $\mathcal{E}$ into a *polyhedral* terrain. We proceed as follows.

The edges of $\mathcal{E}$ consist of a constant number of segments of two types: straight line segments and conic arcs. Let $\delta = v_1 v_2$ be such a conic arc, with endpoints v_1 and v_2. Its projection $\bar{\delta}$ separates two cells of the Voronoi diagram, say of $\bar{e}$ and $\bar{e}'$.

We conceptually add four straight line segments to the graph of the Voronoi diagram by connecting both $\overline{v_1}$ and $\overline{v_2}$ to the nearest point on both $\bar{e}$ and $\bar{e}'$. We do this for all conic arcs of $\mathcal{E}$, adding a linear number of "spokes" to the Voronoi diagram graph. The spokes do not intersect, and so we have increased the complexity of the diagram by a constant factor only. As a result, any conic arc $\bar{\delta}$ is now incident to two constant-complexity faces in the diagram. There are two cases, depicted in Figure 3 (a), depending on whether the spokes meet on one or two sides. Without loss of generality, we can assume that the spokes always meet on $\bar{e}'$.

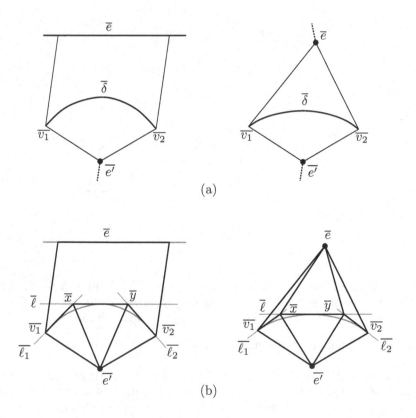

Fig. 3. Approximation of curved surfaces. (a) adding spokes to the diagram: bisector of a vertex and an edge (left) and bisector of two vertices (right). (b) in the new terrain $\mathcal{E}'$, conical facets have been replaced by triangles.

As we have seen before, the conic arc δ is contained in a plane Γ. We now choose a line ℓ in Γ tangent to δ on its convex side, such that its projection $\bar{\ell}$

separates $\bar{\delta}$ from $\bar{e}$. (If Γ is a vertical plane, then $\bar{\delta}$ is a straight segment, and $\bar{\ell}$ contains $\bar{\delta}$.) Let furthermore ℓ_1 and ℓ_2 be the lines in Γ tangent to δ in v_1 and v_2. Let $x := \ell \cap \ell_1$ and $y := \ell \cap \ell_2$.

We now construct a new terrain $\mathcal{E}'$ by replacing the conic arc δ with the polygonal chain $v_1 x y v_2$, and replacing the conic surface patches supported by $\mathcal{D}(e')$ and $\mathcal{D}(e)$ each by three triangles $e'v_1 x$, $e'xy$, $e'yv_2$ (and analogously for e if the spokes meet on both sides). Figure 3 (b) shows the projection of the new terrain $\mathcal{E}'$.

We can perform this operation for all conic arcs of $\mathcal{E}$ simultanously, resulting in a polyhedral terrain $\mathcal{E}'$. Note that the triangles lie on planes that are tangent to α-cones $\mathcal{D}(e)$ or $\mathcal{D}(e')$, and so they are not α-steep. This implies that $\mathcal{E}'$ is α-safe. By Lemma 5, the terrain $\mathcal{E}'$ defines an α-feasible mould, and we have the following result.

Lemma 9. *If a polyhedron $\mathcal{P}$ is α-monotone and α-safe in direction $\vec{d}$, then $\mathcal{P}$ is α-castable in direction $\vec{d}$.*

This concludes the proof of Theorem 2; the theorem follows immediately from Lemmas 3 and 9.

Our proof of Theorem 2 is constructive: Given an α-safe and α-monotone polyhedron, we can compute a feasible cast with uncertainty α in expected time $O(n \log n)$. The construction uses the randomized incremental algorithm by Klein et al. [8], as in Lemma 8. In the full version of this paper we show how to augment Klein's algorithm to correctly recognize if $\mathcal{P}$ is not α-monotone. We have the following theorem.

Theorem 10. *Given a polyhedron $\mathcal{P}$ with n vertices and a direction $\vec{d}$, we can test the α-castability of $\mathcal{P}$ in $\vec{d}$ in expected time $O(n \log n)$. If it is α-castable, then we can construct an α-feasible mould in $O(n \log n)$ expected time. The resulting mould has $O(n)$ vertices.*

4   Computing Feasible Directions

We now describe an algorithm to solve the following problem: Given a polyhedron $\mathcal{P}$ and an angle α, decide whether there is a direction $\vec{d}$ such that $\mathcal{P}$ is α-castable in direction $\vec{d}$. In fact, we will solve the more general problem of finding all directions $\vec{d}$ for which $\mathcal{P}$ is α-castable.

We identify the set of directions with the set of points on the unit sphere $\mathcal{S}^2$ centered at the origin. A point p on $\mathcal{S}^2$ corresponds to the direction $\vec{d}_p$ from the origin o to p. Our goal is to identify the region of $\mathcal{S}^2$ corresponding to directions in which $\mathcal{P}$ is α-castable.

If we imagine the direction $\vec{d}$ changing continuously, there are directions where an up-facet may become a down-facet, or vice versa. The set of these directions forms a collection M of $O(n)$ great circles on $\mathcal{S}^2$. We note that $\mathcal{P}$ is α-safe in a direction $\vec{d}_p$ if and only if p has distance at least α to all great circles in M.

Let C be a cell of the great circle arrangement of M. If $\vec{d}$ varies inside C, the silhouette edges of $\mathcal{P}$ remain the same, but at certain directions the monotonicity of $\mathcal{P}$ changes. In fact, this happens when a line parallel to $\vec{d}$ through a silhouette vertex crosses a silhouette edge. The set of directions for which this occurs forms a collection N of $O(n^2)$ arcs of great circles. We note that $\mathcal{P}$ is α-monotone in direction $\vec{d_p}$ if and only if $\mathcal{P}$ is monotone in direction $\vec{d_p}$ and p has distance at least α to all the arcs in N.

Instead of computing the complete arrangement of $M \cup N$, we can work with a set S of $O(1/\alpha^2)$ sampling points on $\mathcal{S}^2$. The sampling points S are chosen such that any spherical disc of radius α on $\mathcal{S}^2$ contains a point of S.

For each $s \in S$, we first test whether $\mathcal{P}$ is monotone in direction $\vec{d_s}$ in time $O(n \log n)$, using the algorithm by Ahn et al. [1]. If it is, we then construct the cell of the arrangement of M containing s by computing the intersection C_1 of n hemispheres in time $O(n \log n)$. We then compute the $O(n^2)$ arcs of great circles where the monotonicity of $\mathcal{P}$ changes within C_1, and compute the single cell C_2 containing s in their arrangement in time $O(n^2 \log n)$ using the randomized incremental construction algorithm by de Berg et al. [3]. By the observations above, if $p \in C_2$ then $\mathcal{P}$ is α-monotone and α-castable in direction $\vec{d}$ if and only if p has distance at least α to the boundary of C_2. We can compute this set of directions by taking the Minkowski difference [4] of C_2 and a disc of radius α; this is the location of all points p such that the intersection of the disc centered at p and the complement of C_2 is empty.

It remains to argue that all feasible casting directions are found this way. Let $\vec{d_p}$ be a direction in which $\mathcal{P}$ is α-castable. The spherical disc with center p and radius α contains a point $s \in S$, and does not intersect any great circle arc in M or N. This implies that p and s are contained in the same cell of the arrangement of $M \cup N$. Furthermore, $\mathcal{P}$ must be monotone in direction $\vec{d_s}$. It follows that p will be found by our algorithm. We will summarize this result in Theorem 11 below.

Finding the Direction of Maximum Uncertainty. It is desirable that the parting terrain of a cast is as "flat" as possible. So while a relatively small uncertainty α may be given as a minimum requirement for manufacturing, we actually prefer to generate casts with uncertainty as large as possible.

We can easily extend the algorithm described above to solve this problem. Again we are given an angle $\alpha > 0$ and wish to test whether $\mathcal{P}$ is α-castable. If the answer is positive, we now also want to determine the largest $\alpha^* > \alpha$ for which a direction $\vec{d}$ exists such that $\mathcal{P}$ is α^*-castable in direction $\vec{d}$.

We proceed as above: We generate a sampling set S such that any spherical disc of radius α contains a point of S. We then compute, for each $s \in S$, the cell C_2 containing s. The direction of largest uncertainty within C_2 is the center of the maximum inscribed (spherical) disc for C_2, which we compute in $O(n^2 \log n)$ time. The largest inscribed disc, over all cells computed, determines the largest uncertainty for which the object is still castable.

Theorem 11. *Let $\mathcal{P}$ be a polyhedron with n vertices, and $\alpha > 0$. All directions in which $\mathcal{P}$ is castable with uncertainty α can be computed in $O(n^2 \log n/\alpha^2)$*

expected time. If such a direction exists, the largest $\alpha^* > \alpha$ for which $\mathcal{P}$ is castable with uncertainty α^* can be computed within the same time bound.

A Heuristic. If an approximative solution is sufficient, the following heuristic can be applied. It runs in time $O(n \log n)$ for constant α.

Let $\alpha' := (1 - \varepsilon)\alpha$, for some approximation parameter $\varepsilon > 0$. We choose a set S of $O(1)$ sampling directions on $\mathcal{S}^2$, sufficiently dense such that for any spherical disc D of radius α there is a point $s \in S$ such that the disc of radius α' with center s is contained in D.

For each $s \in S$ we test whether $\mathcal{P}$ is α-castable using the algorithm of Section 3. If we are successful, we report $\mathcal{P}$ to be α-castable. If not, we test each direction $s \in S$ again, this time with uncertainty α'. If no feasible casting direction with uncertainty α' is found, we report that $\mathcal{P}$ is not castable with uncertainty α. This is true by the choice of S. If a feasible direction for uncertainty α' is found, we report a "maybe" answer: $\mathcal{P}$ is castable with uncertainty $(1 - \varepsilon)\alpha$, and may or may not be castable with uncertainty α.

The same idea can be used to approximate the largest feasible uncertainty. We can, for instance, set $\alpha' := \alpha/2$, and keep doubling α until $\mathcal{P}$ is no longer α-castable.

References

1. H.-K. Ahn, M. de Berg, P. Bose, S.-W. Cheng, D. Halperin, J. Matoušek, and O. Schwarzkopf. Separating an object from its cast. *Computer-Aided Design*, 34:547–559, 2002.
2. H.-K. Ahn, S.-W. Cheng, and O. Cheong. Casting with skewed ejection direction. In *Proc. 9th Annu. Internat. Sympos. Algorithms Comput.*, volume 1533 of *Lecture Notes Comput. Sci.*, pages 139–148. Springer-Verlag, 1998.
3. M. de Berg, K. Dobrindt, and O. Schwarzkopf. On lazy randomized incremental construction. *Discrete Comput. Geom.*, 14:261–286, 1995.
4. M. de Berg, M. van Kreveld, M. Overmars, and O. Schwarzkopf. *Computational Geometry: Algorithms and Applications*. Springer-Verlag, Berlin, Germany, 2nd edition, 2000.
5. P. Bose, D. Bremner, and M. van Kreveld. Determining the castability of simple polyhedra. *Algorithmica*, 19:84–113, 1997.
6. R. Elliott. *Cast iron technology*. Butterworths, London, UK, 1988.
7. R. Klein. *Concrete and Abstract Voronoi Diagrams*, volume 400 of *Lecture Notes Comput. Sci.* Springer-Verlag, 1989.
8. R. Klein, K. Mehlhorn, and S. Meiser. Randomized incremental construction of abstract Voronoi diagrams. *Comput. Geom. Theory Appl.*, 3:157–184, 1993.
9. J.-C. Latombe. *Robot Motion Planning*. Kluwer Academic Publishers, Boston, 1991.
10. T. Lozano-Pérez, M. T. Mason, and R. Taylor. Automatic synthesis of fine-motion strategies for robots. *Internat. J. Robot. Res.*, 3:3–24, 1984.
11. M. Sharir. Almost tight upper bounds for lower envelopes in higher dimensions. *Discrete Comput. Geom.*, 12:327–345, 1994.
12. C. F. Walton and T. J. Opar, editors. *Iron castings handbook*. Iron Casting Society, Inc., 1981.

Hierarchy of Surface Models and Irreducible Triangulation*

Siu-Wing Cheng[1], Tamal K. Dey[2], and Sheung-Hung Poon[1]

[1] Department of Computer Science, HKUST, Clear Water Bay, Hong Kong.
{scheng,hung}@cs.ust.hk
[2] Department of Computer and Information Sciences, Ohio State University,
Columbus, Ohio, USA.
tamaldey@cis.ohio-state.edu

Abstract. Given a triangulated closed surface, the problem of constructing a hierarchy of surface models of decreasing level of detail has attracted much attention in computer graphics. A hierarchy provides view-dependent refinement and facilitates the computation of parameterization. For a triangulated closed surface of n vertices and genus g, we prove that there is a constant $c > 0$ such that if $n > c \cdot g$, a greedy strategy can identify $\Theta(n)$ topology-preserving edge contractions that do not interfere with each other. Further, each of them affects only a constant number of triangles. Repeatedly identifying and contracting such edges produces a topology-preserving hierarchy of $O(n + g^2)$ size and $O(\log n + g)$ depth. When no contractible edge exists, the triangulation is irreducible. Nakamoto and Ota showed that any irreducible triangulation of an orientable 2-manifold has at most $\max\{342g - 72, 4\}$ vertices. Using our proof techniques we obtain a new bound of $\max\{240g, 4\}$.

1   Introduction

Surface simplification has been a popular research topic in computer graphics [2, 4,8,10,11,15,16]. Most practical surface simplification methods apply to triangulated surface models and are based on local updates including vertex decimation and edge contraction. Vertex decimation removes a vertex together with its incident edges and triangles and then retriangulates the hole left on the surface. Edge contraction collapses an edge to a single vertex (often a new vertex), removing the two incident triangles of the contracted edge and deforming the other triangles touching the contracted edge. If the topology of the surface is not explicitly preserved when applying local updates, the resulting surface might be pinched at a vertex or at an edge. Repeated topology-preserving vertex decimation or edge contraction can produce a hierarchy of models of decreasing level of detail that is useful in many applications. For example, Lee et al. [13] compute a parameterization of the triangulated surface model using such a hierarchy, which can be used for remeshing, texture mapping and morphing. In dynamic

---

* Research of the first and third authors was partially supported by RGC CERG HKUST 6088/99E and HKUST 6190/02E.

virtual environment the hierarchy allows objects to be adaptively refined in a view-dependent manner [4,11,15,16]. Basically, undoing a local update increases the local resolution and redoing a local update reduces the local resolution.

These applications require the local updates to be independent, that is, they do not affect the same triangle. A hierarchy can be conceptually viewed as a directed acyclic graph. The nodes at the topmost level are the triangles in the original surface. When applying a local update, nodes are created for the new triangles and arcs are directed from each old triangle affected to the new triangles created. A new level of detail is obtained by applying a set of independent local updates simultaneously. Each local update should affect a small number of triangles as the time complexity of undoing/redoing the local update is proportional to it [4,16]. Further, the depth of the hierarchy should be small as it bounds the maximum time to obtain a single triangle in the original surface from the model of the lowest level of detail. Given a triangulated surface of n vertices, any hierarchy constructed by repeated applications of independent topology-preserving vertex decimations or edge contractions has depth $\Omega(\log n)$.

For planar subdivisions with straight edges and triangular finite faces, Kirkpatrick [12] and de Berg and Dobrindt [3] showed how to perform independent vertex decimations to construct a hierarchy of $O(\log n)$ depth and $O(n)$ size. Each model in the hierarchy also has straight edges and triangular finite faces. Recently, Duncan et al. [7] showed how to apply planarity-preserving edge contractions to compute a hierarchy of $O(\log n)$ depth for maximal planar graphs. This takes care of triangulated closed surfaces of genus zero as well.

The problem of computing the hierarchy of surface triangulations is related to a mathematical question that has been studied before. An edge is *contractible* if its contraction does not change the surface topology. A triangulation of a 2-manifold is called *irreducible* if no edge is contractible. Is there an upper bound on the number of vertices of an irreducible triangulation in terms of the genus g? Barnette and Edelson [1] first proved that a finite upper bound exists. Later, Nakamoto and Ota [14] proved a bound of $270 - 171\chi$, where χ is the Euler's characteristic. This yields a bound of $342g - 72$ for orientable 2-manifolds.

In this paper, we prove a new upper bound of $240g$ on the number of vertices of an irreducible triangulation. Our proof techniques are different from that of Nakamoto and Ota. By using our techniques and by considering a maximal matching of contractible edges, we prove that for any constant $d > 380$, if $n > \frac{(6008+1310d)g-888-30d}{d-380}$, a greedy strategy can identify at least $\frac{n-1310g+30}{64(d+1)}$ independent topology-preserving edge contractions. Each edge contraction affects at most $d + 2$ triangles. This produces a topology-preserving hierarchy of $O(n+g^2)$ size and $O(\log n+g)$ depth (Theorem 3). These results follow from two topological results about triangulations (Theorem 1 and Theorem 2). Since our topological results are applicable to triangulations with curved edges and curved triangles, we do not assume a piecewise linear embedding of triangulations for our topological results. We may sometime use a piecewise linear embedding as a tool in the proofs and we state this explicitly.

Some proofs and details are omitted due to space limitation. Section 2 provides the basic definitions. Section 3 introduces a family of crossing cycle pairs which is the main tool for obtaining our results. We prove the new upper bound on the number of vertices of an irreducible triangulation in Section 4. Section 5 presents our topological and algorithmic results on constructing a hierarchy.

2   Preliminaries

Triangulated 2-manifolds (without boundaries) are popular representations of object boundaries in solid modeling and computer graphics. The combinatorial structure of a triangulated 2-manifold can be represented using an *abstract simplicial complex* $K = (V, S)$, where V is a set of vertices and S is a set of subsets of V. Each element $\sigma \in S$ has cardinality $k + 1$, $0 \leq k \leq 2$, and σ is called a *k-simplex*. S is required to satisfy the following two conditions. First, for each $v \in V$, $\{v\} \in S$. Second, For each $\sigma \in S$ and $\tau \subset \sigma$, $\tau \in S$. Each proper subset of σ is called a *face* of σ. Two simplices are *incident* if one is a face of the other. For simplicity, we write a 1-simplex $\{u, v\}$ as uv, and a 2-simplex $\{u, v, w\}$ as uvw. We also call the 1-simplices *edges*. The *star* of σ, $\mathrm{St}(\sigma)$, is the collection of simplices $\{\tau : \sigma \subset \tau\}$. If we collect the faces of τ, for all $\tau \in \mathrm{St}(\sigma)$, that are neither σ nor incident to σ, we obtain the *link* of σ denoted as $\mathrm{Lk}(\sigma)$. For each edge uv, its *neighborhood* $\mathrm{N}(uv)$ is $\{\tau \in \mathrm{Lk}(u) \cup \mathrm{Lk}(v) : u \not\subseteq \tau, v \not\subseteq \tau\}$. Figure 1 shows examples of star, link and neighborhood.

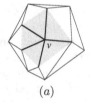 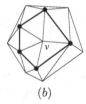 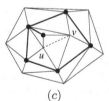

      (a)   (b)   (c)

Fig. 1. In (a), the bold line segments and the shaded triangles are simplices in $\mathrm{St}(v)$. In (b), the black dots and bold line segments are the vertices and edges in $\mathrm{Lk}(v)$. Note that $\mathrm{St}(v) \cap \mathrm{Lk}(v) = \emptyset$. In (c), the dashed line segment is uv and the black dots and bold line segments are vertices and edges in $\mathrm{N}(uv)$. Note that uv is non-contractible.

We use $\mathsf{M_K}$ to denote the underlying space of K, which is a *2-manifold* if the link of each vertex is a simple cycle. The circular ordering of vertices and edges in $\mathrm{Lk}(v)$ of a vertex v induces a circular ordering of edges and 2-simplices in $\mathrm{St}(v)$. A 2-simplex is *oriented* if directions are assigned to its edges so that they form a directed cycle. $\mathsf{M_K}$ is an *orientable 2-manifold* if the 2-simplices of K can be oriented such that each edge is assigned two opposite directions. There are two ways to orient a 2-simplex, so there are two ways to orient K. Orientable 2-manifolds are a popular class of surfaces.

    The *contraction* of an edge uv is a local transformation of K. A new vertex w is introduced to replace uv. $\mathrm{St}(u) \cup \mathrm{St}(v)$ is replaced by a local triangulation:

for each vertex $x \in N(uv)$, we get the edge vx; for every edge $xy \in N(uv)$, we get the 2-simplex vxy. This yields a new abstract simplicial complex. We call a cycle in K *critical* if it consists of three edges and it does not bound a 2-simplex in K. If K is combinatorially equivalent to the boundary of a tetrahedron, no edge can be contracted without changing the topology type of M_K. Otherwise, the contraction of an edge e is topology-preserving if and only if e does not lie on a critical cycle. Dey et al. [5] discussed a more general definition of topology-preserving edge contraction that works for non-manifolds.

3   Family of Cycle Pairs

3.1   Chain, Cycle, and Crossing

We reexamine cycles using concepts from algebraic topology. For $0 \leq k \leq 2$, a k-*chain* is a formal sum of a set of k-simplices with coefficients 0 or 1. The addition is commutative. Terms involving the same k-simplex can be added together by adding their coefficients using modulo 2 arithmetic. The modulo 2 arithmetic implies that a k-simplex appears in the final sum when it appears an odd number of times. The *boundary of a k-simplex* σ is the sum of $(k-1)$-simplices that are faces of σ. The *boundary of a k-chain* is the sum of the boundaries of its k-simplices. We use ∂ to denote the boundary operator.

A 1-chain is a *cycle* if its boundary is empty. The boundary of a 2-chain is always a cycle. The *length* of a cycle is the number of edges in it. We call a cycle *simple* if it is simple in the graph-theoretic sense. Recall that a cycle is *critical* if it consists of three edges and it does not bound a 2-simplex. So a critical cycle is always simple. Two cycles B_1 and B_2 are *homologous* if there exists a 2-chain Σ such that $B_1 = B_2 + \partial \Sigma$.

Let B_1 and B_2 be two simple cycles in K. Suppose that B_1 and B_2 share a vertex v such that the edges of B_1 and B_2 incident to v are distinct. If the edges of B_1 alternate with the edges of B_2 in $St(v)$ (recall that there is a circular ordering of edges and 2-simplices in $St(v)$), we say that B_1 and B_2 *cross at* v. We call v a *crossing of B_1 and B_2*. The above definition of crossing might not be applicable when two cycles share edges. So we will perturb cycle edges in order to proceed further.

Since perturbation is a geometric operation, we need to work with a geometrical realization of K which is a simplicial complex $\widehat{K}$ (embedded without self-intersection in a space of sufficiently high dimension [9]). $\widehat{K}$ is a triangulation of a piecewise linear surface: each edge appears as a line segment and each 2-simplex appears as a triangle. Since K and $\widehat{K}$ have identical combinatorial structure, we do not distinguish corresponding cycles in K and $\widehat{K}$. We would like to emphasize that $\widehat{K}$ is only a tool. Our results are topological and independent of the geometric realization.

Let ξ_1 and ξ_2 be two simple closed curves on the underlying piecewise linear surface of $\widehat{K}$. We say ξ_1 and ξ_2 *cross at a point* p if there is a small region $R(p)$ around p such that $\xi_1 \cap \xi_2 \cap R(p) = \{p\}$ and $\xi_1 \cap R(p)$ contains points on both sides of $\xi_2 \cap R(p)$ locally. We also call p a *crossing of ξ_1 and ξ_2*.

Let B_1 and B_2 be two simple cycles in K. We treat B_1 and B_2 as two simple closed curves on the underlying surface of $\widehat{\mathsf{K}}$. We perturb B_1 to another simple closed curve ξ_1 on $\widehat{\mathsf{K}}$ as follows. Fix the vertices of B_1. For each edge e of B_1, perturb e to a closed curved segment γ such that $\mathrm{int}(\gamma)$ lies in the interior of a triangle of $\widehat{\mathsf{K}}$ incident to e, $\gamma \cap e$ consists of the endpoints of e, and $\mathrm{int}(\gamma)$ does not intersect any curved segment obtained by perturbing other edges of B_1. Consequently, ξ_1 and B_2 intersect only at the vertices of B_1, so the definition of crossings of two simple closed curves is applicable. We use $B_1 \circ B_2$ to denote the *parity of the number of crossings of ξ_1 and B_2*. We can generalize the definition to the case where B_2 is a sum of simple cycles. Let $B_2 = \sum_{j=1}^{q} B_{2j}$, where B_{2j} are simple cycles. Then we define $B_1 \circ B_2 = (\sum_{j=1}^{q} B_1 \circ B_{2j}) \mod 2$.

Lemma 1. *Given a simple cycle B_1 and a sum B_2 of simple cycles in K, $B_1 \circ B_2$ is independent of the sum expression of B_2 and the perturbation of B_1.*

Lemma 2. *Let A, B_1 and B_2 be three simple cycles in K. If B_1 and B_2 are homologous, then $A \circ B_1 = A \circ B_2$.*

Proof. By definition, $B_1 = B_2 + \partial \Sigma$ for some 2-chain Σ. So $A \circ B_1 = (A \circ B_2 + A \circ \partial \Sigma) \mod 2$. Clearly, $A \circ \partial \tau = 0$ for any 2-simplex τ. Thus, $A \circ \partial \Sigma = 0$ which implies that $A \circ B_1 = A \circ B_2$. ▣

3.2   Crossing Cycle Pairs

Let $\ell \geq 3$ be a parameter. Let $\mathcal{F}_\ell$ denote a family of cycle pairs $\{(C_i, D_i) : 1 \leq i \leq |\mathcal{F}_\ell|\}$ that satisfy four conditions: (1) each C_i is a critical cycle, (2) each D_i is a simple cycle of length at most ℓ, (3) for any i, C_i and D_i cross at a vertex called the *anchor* of C_i and C_i does not share any other vertex with D_i, and (4) For $i \neq j$, the anchors of C_i and C_j are different. Note that for $i \neq j$, C_i or D_i may share vertices and edges with C_j and D_j. We will show that $|\mathcal{F}_3|$ is an upper bound on the number of vertices of an irreducible triangulation and we will use $|\mathcal{F}_4|$ to prove our results on constructing a hierarchy. So we want to bound $|\mathcal{F}_\ell|$.

Define a *whisk* to be a collection of mutually homologous C_i's in $\mathcal{F}_\ell$ such that they share a common edge xy and neither x nor y is the anchor of any C_i in the collection. We call xy the *axis* of the whisk. Given a whisk W, we use W^* to denote the set of vertices and edges in W, i.e., the graph formed by the union of the cycles in W. We also call W^* a whisk for convenience.

Lemma 3. *Let W be a whisk. Let xy be the axis of W. Let $\mathcal{Z}$ be a set of whisks such that*

(i)   any two cycles in $W \cup \bigcup_{V \in \mathcal{Z}} V$ are homologous,
(ii) $W \cap V = \emptyset$ for any $V \in \mathcal{Z}$,
(iii) $U^* \cap V^* \subseteq \{x, y\}$ for any two distinct whisks $U, V \in \mathcal{Z}$.

Then $|\mathcal{Z}| \leq \ell - |W|$ and $|W| \leq \ell$.

Proof. Let C_i be a cycle in W. Let D_i be the cycle that pairs up with C_i in $\mathcal{F}_\ell$. By definition, $D_i \circ C_i = 1$. Since C_i and C_j are homologous for any cycle C_j in any whisk in $\mathcal{Z}$, $D_i \circ C_j = D_i \circ C_i = 1$ by Lemma 2. It follows that D_i contains a vertex w of C_j. The vertex w cannot be x or y as D_i does not share an edge with C_i. Since $U^* \cap V^* \subseteq \{x, y\}$ for any two distinct whisks $U, V \in \mathcal{Z}$, each whisk in $\mathcal{Z}$ contributes at least one distinct vertex in D_i. By the same reasoning, D_i must contain the anchors of all cycles in W. Since D_i has length at most ℓ, we conclude that $|\mathcal{Z}| \leq \ell - |W|$. As $|\mathcal{Z}| \geq 0$, rearranging terms yields $|W| \leq \ell$.   ▣

Lemma 4. *There is a subset $\mathcal{F}_\ell' \subseteq \mathcal{F}_\ell$ of cardinality at least $|\mathcal{F}_\ell|/(10\ell^3)$ such that for any two distinct C_i and C_j in $\mathcal{F}_\ell'$, C_i and C_j are non-homologous.*

Proof. We first select a subset $\mathcal{S}_\ell \subseteq \mathcal{F}_\ell$ of cardinality at least $|\mathcal{F}_\ell|/20$ such that for any two distinct C_i and C_j in $\mathcal{S}_\ell$, C_i does not contain the anchor of C_j. Let $\mathcal{H}$ be an equivalence class of mutually homologous C_i's in $\mathcal{S}_\ell$. We first bound $|\mathcal{H}|$. We pick maximal whisks $W_r \subseteq \mathcal{H}$, $1 \leq r \leq m$, in a greedy fashion such that $W_r^* \cap W_s^* = \emptyset$ for $1 \leq r \neq s \leq m$. By Lemma 3 ($W = W_r$ and $\mathcal{Z} = \{W_1, \cdots, W_m\} - \{W_r\}$), $m - 1 \leq \ell - |W_r|$ which implies that $m \leq \ell$ and

$$|W_r| \leq \ell + 1 - m. \tag{1}$$

We partition $\mathcal{H} - \bigcup_{r=1}^{m} W_r$ into a collection $\mathcal{Y}$ of maximal whisks. By the property of $\mathcal{S}_\ell$, no cycle in $\mathcal{S}_\ell$ contains the anchor of another cycle in $\mathcal{S}_\ell$. By greediness, for any $V \in \mathcal{Y}$, $V^* \cap W_r^* \neq \emptyset$ for some $1 \leq r \leq m$. If $V^* \cap W_r^* \neq \emptyset$, the maximality of W_r implies that $V^* \cap W_r^* = \{x\}$ for some endpoint x of the axis of W_r. Take any whisk $V \in \mathcal{Y}$. By Lemma 3 ($W = V$ and $\mathcal{Z} = \{W_1, \cdots, W_m\}$), we have $m \leq \ell - |V|$ which implies that

$$|V| \leq \ell - m, \text{ for any } V \in \mathcal{Y}. \tag{2}$$

Let $x_{r1}x_{r2}$ be the axis of W_r. Let s_{rj}, $1 \leq j \leq 2$, be the number of whisks V in $\mathcal{Y}$ such that $V^* \cap W_r^* = \{x_{rj}\}$. By Lemma 3 ($W = W_r$ and $\mathcal{Z} =$ the set of whisks in $\mathcal{Y}$ that share x_{rj} with W_r^*), we have

$$s_{rj} \leq \ell - |W_r|. \tag{3}$$

Thus, $|\mathcal{H}| \overset{(2)}{\leq} \sum_{r=1}^{m}(|W_r| + (s_{r1} + s_{r2}) \cdot (\ell - m)) \overset{(3)}{\leq} \sum_{r=1}^{m}(|W_r| + 2(\ell - |W_r|)(\ell - m)) = \sum_{r=1}^{m}(2\ell(\ell - m) - (2\ell - 2m - 1)|W_r|)$. If $m = \ell$, then $|\mathcal{H}| \leq \sum_{r=1}^{m}|W_r| \leq \ell$ by (1). If $m < \ell$, then $|\mathcal{H}| < \sum_{r=1}^{m} 2\ell(\ell - m) = 2m\ell(\ell - m)$. This bound is maximized when $m = \ell/2$. So $|\mathcal{H}| < \ell^3/2$.

We pick one C_i from each equivalence class $\mathcal{H}$ of mutually homologous C_i's in $\mathcal{S}_\ell$. Let $\mathcal{F}_\ell' = \{(C_i, D_i) : C_i \text{ picked}\}$. Since $|\mathcal{S}_\ell| \geq |\mathcal{F}_\ell|/20$, $|\mathcal{F}_\ell'| \geq \frac{2}{\ell^3} \cdot \frac{1}{20} \cdot |\mathcal{F}_\ell| = |\mathcal{F}_\ell|/(10\ell^3)$.   ▣

Lemma 5. $|\mathcal{F}_3| \leq 240g$ *and for* $\ell \geq 3$, $|\mathcal{F}_\ell| \leq 20\ell^3 g$.

Proof. If M_K has genus g, K contains at most $2g$ cycles that are mutually non-homologous. Thus, $|\mathcal{F}_\ell'| \leq 2g$. The result then follows from Lemma 4. The proof of the $240g$ bound is omitted.   ▣

4 Irreducible Triangulation

In this section, we prove that any irreducible triangulation of an orientable 2-manifold of positive genus g has at most $240g$ vertices. We need the following lemma about a vertex.

Lemma 6. *Assume that* M_K *has positive genus. Let* A *be a critical cycle passing through vertices* v, x *and* y. *Then one of the following holds.*

(i) There are two contractible edges uv *and* vw *that alternate with* vx *and* vy *in* $\mathrm{St}(v)$.

(ii) A pair of critical cycles cross at v.

Theorem 1. *Any irreducible triangulation of an orientable 2-manifold of genus* g *has at most* $\max\{240g, 4\}$ *vertices.*

Proof. The theorem is clearly true when $g = 0$. Let K be an irreducible triangulation. Assume that $g > 0$. We construct a family $\mathcal{F}_3$ of crossing cycle pairs as follows. Each vertex v in K is incident on a non-contractible edge, so v lies on a critical cycle. Since no edge of K is contractible, Lemma 6(ii) holds and a pair of critical cycles cross at v. We add this cycle pair to $\mathcal{F}_3$. The number of vertices of K is $|\mathcal{F}_3|$ which is at most $240g$ by Lemma 5.   ◻

5 Hierarchy of Surfaces

In this section, we prove that there are linearly many independent topology-preserving edge contractions. Moreover, a simple greedy strategy can be used to find them. Let uv and rs be two edges of K. We say that uv and rs are *independent* if $(\mathrm{St}(u) \cup \mathrm{St}(v)) \cap (\mathrm{St}(r) \cup \mathrm{St}(s)) = \emptyset$. Although $\mathrm{N}(uv)$ and $\mathrm{N}(rs)$ might share vertices and edges, the contractions of uv and rs do not affect the same triangle.

Our proof proceeds in two steps. First, we focus on the contractible edges of K by considering a subgraph G_K that contains *all vertices* of K and the contractible edges of K. (So G_K might be disconnected.) We prove that a maximal matching of G_K has linear size. Second, we prove that any maximal matching of G_K contains an independent subset of edges of linear size. Moreover, they can be found using a greedy strategy.

Lemma 7. *Let* n *be the number of vertices in* K *and let* g *be the genus of* M_K. *Assume that* $g > 0$. *Any maximal matching of* G_K *matches at least* $(n - 1310g + 30)/16$ *vertices.*

Proof. Let $\widehat{K}$ be a geometric realization of K. We use S to denote the underlying surface of $\widehat{K}$, i.e., S is the set of points on $\widehat{K}$ without the triangulation structure. We obtain an embedding of G_K on S by first drawing all vertices and edges of $\widehat{K}$ on S and then erasing all the non-contractible edges. G_K induces a subdivision of S which we denote by $G_K(S)$.

Pick a maximal matching of G_K. Let H_K be the subgraph of G_K (embedded on S) consisting of matched vertices and the edges of G_K between them. So H_K contains all matching edges but H_K may contain some non-matching edges as well. As our argument proceeds, we will create some segments on S, called *purple segments*, that connect matched vertices. The purple segments can be straight or curved. The purple segments will be used later to form a new graph with H_K.

We bound the number of unmatched vertices by charging them to edges in H_K and the purple segments as well as by forming a family $\mathcal{F}_4$ of crossing cycle pairs. We charge for the unmatched vertices one by one in an arbitrary order. Let v be an unmatched vertex. If the degree of v in G_K is at most 1, then Lemma 6(ii) applies and a pair of critical cycles cross at v. We charge for v by adding this cycle pair to $\mathcal{F}_4$.

Suppose that the degree of v in G_K is larger than 1. Since v is unmatched, all neighbors of v in G_K are matched. Let u and w be two consecutive neighbors of v in G_K. Let R be a region in $G_K(S)$ such that uv and vw lie consecutively on the boundary of R. R covers some triangles in $\operatorname{St}(v)$. We pick a subset $T \subseteq \operatorname{St}(v)$ of triangles such that $\bigcup_{t \in T} t \subseteq R$ and uv and vw lie in the boundary of the closure of $\bigcup_{t \in T} t$. Let R_{uvw} denote the closure of $\bigcup_{t \in T} t$. Note that any incident edge of v in $\widehat{K}$ that lies inside R_{uvw} is non-contractible. There are three different ways to charge for v.

Case 1: v is not incident to any edge in $\widehat{K}$ that lies inside R_{uvw}. It follows that $R_{uvw} = uvw$. So uw is an edge in $\widehat{K}$. Since $R_{uvw} \subseteq R$, either uw lies inside R or uw lies on the boundary of R.

Case 1.1: uw is an edge in G_K. It follows that $R = uvw = R_{uvw}$. Since u and w are matched vertices, uw belongs to H_K too. We charge for v by putting a red pebble at uw. Since uw bounds at most two regions in $G_K(S)$, case 1.1 can be applied at most twice to uw producing at most two red pebbles on uw. Thus, each edge of H_K receives at most two red pebbles.

Case 1.2: uw is not an edge in G_K. So uw is non-contractible. If we have created a purple segment γ_{uw} connecting u and w before, we put a green pebble at γ_{uw} to charge for v. Otherwise, we create the straight purple segment $\gamma_{uw} = uw$ and put a green pebble at γ_{uw} to charge for v. It can be shown that the purple segment γ_{uw} receives at most two green pebbles overall.

Case 2: Some edge vx in $\widehat{K}$ lies inside R_{uvw}. Recall that any incident edge of v that lies inside R_{uvw} is non-contractible. So vx lies on a critical cycle A. Let vy and xy be the other two edges of A. If vy lies inside R_{uvw} or on the boundary of R_{uvw}, then Lemma 6(ii) applies, so a pair of critical cycles cross at v. We charge for v by adding this cycle pair to $\mathcal{F}_4$.

Suppose that vy lies outside R_{uvw}. If we have not created a purple segment γ_{uw} connecting u and w before, we create γ_{uw} as follows. If uw is an edge in $\widehat{K}$, we set $\gamma_{uw} = uw$. Otherwise, we draw γ_{uw} as a segment, curved if necessary, inside R_{uvw}. Clearly, γ_{uw} does not cross any edge of H_K. Moreover, by our drawing strategy, γ_{uw} does not cross any other purple segments created before.

After creating γ_{uw} if necessary, we check the number of blue pebbles at γ_{uw}. If γ_{uw} contains less than three blue pebbles, we add a blue pebble to γ_{uw} to charge for v. If γ_{uw} already contains three blue pebbles, these blue pebbles were introduced to charge for three unmatched vertices v_i, $1 \leq i \leq 3$, other than v and each v_i is adjacent to both u and w. We pick v_k such that $v_k \neq x$ and $v_k \neq y$ (recall that x and y are vertices of the critical cycle A passing through v). Let B be the cycle consisting of the edges uv, vw, wv_k, and $v_k u$. Since vx lies inside R_{uvw} and vy lies outside R_{uvw}, A and B cross at v. We add the cycle pair (A, B) to $\mathcal{F}_4$ to charge for v.

By Lemma 5, $|\mathcal{F}_4| \leq 1280g$. It remains to bound the total number of pebbles on the edges of H_{K} and the purple segments. Recall that there is no crossing among the edges of H_{K} and the purple segments. We add the purple segments as edges to H_{K} and we add more edges, if necessary, to obtain a connected graph H^* that is embedded on S without any edge crossing. Let N and E be the number of vertices and edges in H^*. (So N is the number of matched vertices.) By Euler's relation, $E \leq 3N - 6 + 6g$. Since each edge of H_{K} carries at most two red pebbles and each purple segment carries at most two green pebbles and at most three blue pebbles, the total number of pebbles in H^* is at most $5E \leq 15N - 30 + 30g$.

It follows that the number of unmatched vertices is bounded by $1280g + 5E \leq 15N - 30 + 1310g$. Hence, $n \leq N + 15N - 30 + 1310g$ which implies that $N \geq (n - 1310g + 30)/16$. $\square$

Lemma 7 implies the following result.

Theorem 2. *Let n be the number of vertices of K and let g be the genus of $\mathsf{M_K}$. Assume that $g > 0$. For any constant $d > 380$, if $n \geq \frac{(6008 + 1310d)g - 888 - 30d}{d - 380}$, there are at least $\frac{n - 1310g + 30}{64(d+1)}$ independent contractible edges and for each such edge uv, $\mathrm{N}(uv)$ has at most d vertices.*

It is not necessary to compute M first. We initialize an empty output set of edges EDGE_SET. Then we examine the edges of K in an arbitrary order and grow EDGE_SET. For each edge e, we determine whether e is contractible, $\mathrm{N}(e)$ has at most d vertices, and e and the edges in EDGE_SET are independent. If these three conditions are satisfied, we add e to EDGE_SET. In all, we have the following theorem.

Theorem 3. *Given a triangulated closed surface of n vertices and positive genus g, a topology-preserving hierarchy can be constructed by repeated contractions of independent contractible edges. Each edge contraction affects $O(1)$ triangles. The hierarchy has $O(\log n + g)$ depth and $O(n + g^2)$ size.*

The algorithm as described above takes $O(n + g^2)$ time. In practice, the edge contractions should be selected to keep the geometric approximation error small. We suggest using the quadric error proposed by Garland and Heckbert [8] for edge contractions. After sorting the edges, we scan the sorted list using our greedy strategy to select independent contractible edges. Due to sorting, the time complexity of the algorithm increases to $O(n \log n + g^2 \log g)$.

Acknowledgments. The first author thanks Beifang Chen, Herbert Edelsbrunner and Min Yan for helpful discussions.

References

1. D.W. Barnette and A.L. Edelson, All 2-manifolds have finitely many minimal triangulations. *Israel J. Math*, 67 (1989), 123–128.
2. J. Cohen, A. Varshney, D. Manocha, G. Turk, H. Weber, P. Agarwal, F. Brooks, and W. Wright. Simplification Envelopes. *Proceedings of SIGGRAPH 96*, 1996, 119–128.
3. M. de Berg and K.T.G. Dobrindt. On levels of detail in terrains. *Proc. ACM Symposium on Computational Geometry*, 26–27, 1995.
4. L. De Floriani, P. Magillo and E. Puppo. Building and traversing a surface at variable resolution. *Proc. IEEE Visualization '97*, 103–110, 1997.
5. T.K. Dey, H. Edelsbrunner, S. Guha and D.V. Nekhayev, Topology preserving edge contraction, *Publ. Inst. Math. (Beograd) (N.S.)*, 66 (1999), 23–45.
6. A. Dold. *Lectures on Algebraic Topology*, Springer-Verlag, 1972.
7. C.A. Duncan, M.T. Goodrich, and S.G. Kobourov, Planarity-preserving clustering and embedding for large planar graphs, *Proc. Graph Drawing '99*, 186–196.
8. M. Garland and P.S. Heckbert. Surface simplification using quadric error metrics. *Proc. SIGGRAPH '97*, 209–216.
9. P.J. Gilbin. *Graphs, Surfaces and Homology*. Chapman and Hall, 1981.
10. H. Hoppe, Progressive meshes. *Proc. SIGGRAPH '96*, 99–108.
11. H. Hoppe, View-dependent refinement of progressive meshes. *Proc. SIGGRAPH '97*, 189–198.
12. D.G. Kirkpatrick. Optimal search in planar subdivisions. *SIAM Journal on Computing*, 12 (1983), 28–35.
13. A. Lee, W. Sweldens, P. Schröder, L. Cowsar and D. Dobkin, MAPS: multiresolution adaptive parameterization of surfaces. *Proc. SIGGRAPH '98*, 95–104.
14. A. Nakamoto and K. Ota, Note on irreducible triangulations of surfaces. *Journal of Graph Theory*, 10 (1995), 227–233.
15. J.C. Xia, J. El-Sana and A. Varshney. Adaptive real-time level-of-detail-based rendering for polygonal models. *IEEE Transactions on Visualization and Computer Graphics*, 3 (1997), 171–181.
16. J.C. Xia and A. Varshney, Dynamic view-dependent simplification for polygonal models. *Proc. Visualization '96*, 327–334.

Algorithms and Complexity for Tetrahedralization Detections

Boting Yang[1], Cao An Wang[2], and Francis Chin[3]

[1] Department of Computer Science,
University of Regina,
Regina, SK, Canada S4S 0A2
boting@cs.uregina.ca
[2] Department of Computer Science,
Memorial University of Newfoundland,
St.John's, NF, Canada A1B 3X5
wang@cs.mun.ca
[3] Department of Computer Science and Information Systems,
The University of Hong Kong,
Hong Kong
chin@csis.hku.hk

Abstract. Let $\mathcal{L}$ be a set of line segments in three dimensional Euclidean space. In this paper, we prove several characterizations of tetrahedralizations. We present an $O(nm \log n)$ algorithm to determine whether $\mathcal{L}$ is the edge set of a tetrahedralization, where m is the number of segments and n is the number of endpoints in $\mathcal{L}$. We show that it is NP-complete to decide whether $\mathcal{L}$ contains the edge set of a tetrahedralization. We also show that it is NP-complete to decide whether $\mathcal{L}$ is tetrahedralizable.

1 Introduction

Given a set V of points in R^3, a *tetrahedralization* of V is a partition of the convex hull of V into a number of tetrahedra such that (i) each vertex of the tetrahedra belongs to V, (ii) each point of V is the vertex of a tetrahedron, and (iii) the intersection of any two tetrahedra is either empty or a shared face, where a *face* of a tetrahedron is the convex hull of some of its vertices. In particular, a *facet* is the convex hull of three vertices of a tetrahedron. Tetrahedralizations have applications in the finite element method, mesh generation, CAD/CAM, computer graphics, and robotics [4].

Edelsbrunner et al. [8] studied the problem of tetrahedralizing a set of points. They presented several combinatorial results on extremum problems concerning the number of tetrahedra in a tetrahedralization. They also presented an algorithm that tetrahedralizes a set of n points in $O(n \log n)$ time. A similar algorithm was given in [1]. If Steiner points are allowed, then any simple polyhedron of n vertices can be partitioned into $O(n^2)$ tetrahedra. Chazelle [5] showed that this bound is tight in the worst case. Chazelle and Palios [6] described a tetrahedralization algorithm for decomposing a simple polyhedron with n vertices and

P. Bose and P. Morin (Eds.): ISAAC 2002, LNCS 2518, pp. 296–307, 2002.
© Springer-Verlag Berlin Heidelberg 2002

r reflex edges into $O(n + r^2)$ tetrahedra using $O(n + r^2)$ Steiner points. Other algorithms for tetrahedralizing polyhedra with Steiner points are presented in [2,10].

On the issue of NP-completeness, Ruppert and Seidel [15] proved that finding a tetrahedralization of a nonconvex polyhedron without Steiner points is NP-complete; this problem remains NP-complete even for star-shaped polyhedra. It follows that the problem of deciding how many Steiner points are needed to tetrahedralize a polyhedron is also NP-complete. The tetrahedralization with the minimum number of tetrahedra is referred to as a minimal tetrahedralization. Recently, Richter-Gebert proved that finding a minimal tetrahedralization of the boundary of a 4-polytope is NP-complete [14]. Furthermore, Below et al. [3] proved that finding a minimal tetrahedralization of a convex polyhedron is NP-complete.

Given a set of line segments $\mathcal{L}$ in R^3, we investigate the algorithmic complexity of finding tetrahedralizations regarding $\mathcal{L}$. If $\mathcal{L}$ is the edge set of a tetrahedralization, we say that $\mathcal{L}$ *forms* a tetrahedralization. If $\mathcal{L}$ is a superset of the edge set of a tetrahedralization of $V(\mathcal{L})$, we say that $\mathcal{L}$ *contains* a tetrahedralization. If $\mathcal{L}$ is a subset of the edge set of a tetrahedralization of $V(\mathcal{L})$, we say that $\mathcal{L}$ is tetrahedralizable. We present an $O(nm \log n)$ algorithm to determine whether $\mathcal{L}$ forms a tetrahedralization, where m is the number of segments and n is the number of endpoints in $\mathcal{L}$. If $\mathcal{L}$ does not form a tetrahedralization, we prove that it is NP-complete to decide whether $\mathcal{L}$ contains a tetrahedralization of $V(\mathcal{L})$, and we prove that it is also NP-complete to decide whether $\mathcal{L}$ is tetrahedralizable.

This paper is organized as follows. In Section 2, we give some notations and assumptions. In Section 3, we present three characterizations of tetrahedralizations. They describe a tetrahedralization from the facet, vertex, and combinatorial viewpoints, respectively. In Section 4, we describe an $O(nm \log n)$ algorithm to determine whether $\mathcal{L}$ forms a tetrahedralization. In Section 5, we show that the problem of deciding if $\mathcal{L}$ contains a tetrahedralization is NP-complete by a reduction from the two dimensional analog of this problem [12]. In Section 6, we prove that the problem of deciding if $\mathcal{L}$ is tetrahedralizable is NP-complete by a reduction from the Satisfiability problem.

2   Preliminaries

For a set A of geometrical objects (points, edges, facets, or polyhedra), let $V(A)$ be the vertex set, $E(A)$ be the edge set, and $F(A)$ be the facet set. Let also $\text{int}(A)$ and $\text{cl}(A)$ be respectively the interior and closure of A. The convex hull of A, denoted as $CH(A)$, is the smallest convex set containing A. Given arbitrary points a, b, c, d in the Euclidean space, unless otherwise stated, we use ab to denote the line segment (edge) with endpoints a and b, abc to denote the triangle with vertices a, b and c, and $abcd$ to denote the tetrahedron with vertices a, b, c and d.

Definition 1. *For a face or polyhedron P and a line segment set L, if no segment of L intersect the interior of P, then we say that P is L-empty, as shown in Figure 1.*

298   B. Yang, C.A. Wang, and F. Chin

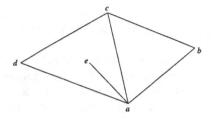

Fig. 1. $L = \{ab, ac, ad, ae, bc, cd\}$; abc is L-empty and acd is not L-empty.

Throughout this paper, let $\mathcal{L}$ be a set of m line segments with n endpoints in R^3; $\Delta_{\mathcal{L}} = \{abc \mid ab, bc, ca \in \mathcal{L}, \operatorname{int}(abc) \cap \mathcal{L} = \emptyset\}$ be the set of $\mathcal{L}$-empty triangles with edges in $\mathcal{L}$; and Δ_{CH} be the set of facets in $CH(\mathcal{L})$. Let $\Delta_I = \Delta_{\mathcal{L}} - \Delta_{CH}$.

For simplicity, we assume throughout this paper that no four vertices of $V(\mathcal{L})$ are coplanar. This assumption implies that each facet of $CH(\mathcal{L})$ is a triangle and no pair of segments in $\mathcal{L}$ intersect.

3   Characterizations of Tetrahedralizations

For each triangle $abc \in \Delta_I$, let abc^+ and abc^- denote its two oriented versions with respect to the opposite normal directions. For the plane H passing through abc, let H^+ and H^- denote its two half spaces corresponding to abc^+ and abc^- respectively. Let $n_t(abc^+)$ (resp. $n_t(abc^-)$) be the number of vertices in $V(\mathcal{L})$ that lie in H^+ (resp. H^-) and form $\mathcal{L}$-empty tetrahedra with abc.

The following theorem gives characterizations of a tetrahedralization from the facet viewpoint.

Theorem 1. (i) *If $\Delta_I = \emptyset$, then $\mathcal{L}$ forms a tetrahedralization if and only if $\mathcal{L}$ forms a tetrahedron.*
(ii) *If $\Delta_I \neq \emptyset$, then $\mathcal{L}$ forms a tetrahedralization if and only if $n_t(abc^+) = 1$ and $n_t(abc^-) = 1$, for each $abc \in \Delta_I$.*

For each vertex $v \in V(\mathcal{L})$, let $\mathcal{L}(v)$ be the edge set induced from vertex v and all its adjacent vertices; let $T(v)$ be a set of tetrahedra in $\mathcal{L}$ with v as a vertex such that each tetrahedron in $T(v)$ is $\mathcal{L}(v)$-empty. For each tetrahedron $t \in T(v)$, the three facets incident to v are called *side facets* of v and the fourth facet is called the *bottom facet* of v. The union of all the bottom facets of v is denoted by $B(v)$. In R^3, the *triangulation surface* is defined as a piecewise linear surface in which each face is a triangle; and the *closed triangulation surface* is defined as the boundary of a simplicial polyhedron.

Given the above, we have the following theorem which is the characterization of a tetrahedralization from the vertex viewpoint.

Theorem 2. *$\mathcal{L}$ forms a tetrahedralization if and only if the following conditions hold:*

1. *for each vertex $v \in V(CH(\mathcal{L}))$, $B(v)$ is a triangulation surface whose boundary edges are contained in $E(CH(\mathcal{L}))$; and*
2. *for each vertex $v \in \mathrm{int}(CH(\mathcal{L}))$, $B(v)$ is a closed triangulation surface with v in its interior.*

Consider the space divided by the empty triangles in $\Delta_{\mathcal{L}}$: $C = R^3 - \{\mathrm{cl}(abc) \mid abc \in \Delta_{\mathcal{L}}\}$. Let n_c be the number of the bounded connected components in C. Each connected component of C is called a *cell*. Each bounded cell is denoted by $C_i(1 \leq i \leq n_c)$, and the unbounded cell is denoted by C_0. So each $\mathrm{cl}(C_i)(0 \leq i \leq n_c)$ can be considered as a polyhedron. Let $|A|$ stand for the cardinality of a set A. Then we have the following theorem which is the characterization of a tetrahedralization from the combinatorial viewpoint.

Theorem 3. *$\mathcal{L}$ forms a tetrahedralization if and only if the following conditions hold:*

1. *$(V(\mathcal{L}), \mathcal{L})$ is a connected graph;*
2. *each $\mathrm{cl}(C_i)(0 \leq i \leq n_c)$ is a simple polyhedron and $\mathrm{cl}(C_0)$ is $\mathcal{L}$-empty; and*
3. *$n_c = |\mathcal{L}| - |V(\mathcal{L})| - |V(CH(\mathcal{L}))| + 3$.*

Similarly, we can prove the two dimensional counterpart of Theorem 3.

Theorem 4. *Let E be a set of line segments in R^2. Then E is a triangulation if and only if E is a connected plane graph and $|\mathsf{E}| = 3|V(\mathsf{E})| - |V(CH(\mathsf{E}))| - 3$ (or $|F(\mathsf{E})| = 2|V(\mathsf{E})| - |V(CH(\mathsf{E}))| - 3$).*

Hopcroft and Tarjan [11] showed that $O(n)$ is sufficient to decide planarity on a conventional random access machine. From Theorem 4 we obtain the following theorem.

Theorem 5. *Given a set of edges E in R^2, whether E is a triangulation can be determined in $O(n)$ time.*

4 Deciding Whether $\mathcal{L}$ Forms a Tetrahedralization

In this section, let us consider a tetrahedralization detection problem — that is, how to design an efficient algorithm to decide if $\mathcal{L}$ forms a tetrahedralization.

For any vertex $v \in V(\mathcal{L})$, let $\mathrm{adj}(v)$ be the set of adjacent vertices of v, and $\mathcal{L}_v = \{ab \mid ab \in \mathcal{L} \text{ and } a, b \in \mathrm{adj}(v)\}$.

Definition 2. *We say that an edge* punctures *a face (facet or polyhedron) if and only if their interiors have a nonempty intersection.*

In summary, the outline of our algorithm can be described as follows:

Algorithm TETRAHEDRALIZATION-DETECTION($\mathcal{L}$) (outline)
Input: A set of line segments $\mathcal{L}$ in R^3.
Output: If $\mathcal{L}$ does not form a tetrahedralization, return (NO). Otherwise, return (YES, T), where T is the set of the tetrahedra in the tetrahedralization $\mathcal{L}$.

Step 1   Compute $CH(\mathcal{L})$. If $E(CH(\mathcal{L})) \nsubseteq \mathcal{L}$, then return (NO).
Step 2   (CHECK-TRIANGULATION) For each vertex v on the boundary of
           $CH(\mathcal{L})$, check if $\mathcal{L}_v$ contains a specified triangulation surface:
Step 2.1   For each segment $pp' \in \mathcal{L}_v$, if there exists a segment $qq' \in \mathcal{L}_v$
                such that qq' punctures face vpp', then delete pp' from $\mathcal{L}_v$.
Step 2.2   If the updated $\mathcal{L}_v$ is the edge set of a triangulation surface
                (denoted as B_v) whose boundary edges are contained in
                $E(CH(\mathcal{L}))$, then let T_v be the set of tetrahedra $vabc$,
                where abc is a bounded face in B_v. Otherwise, return (NO).
Step 3   (CHECK-CLOSED-TRIANGULATION) For each vertex v inside
           $CH(\mathcal{L})$, check if $\mathcal{L}_v$ contains a specified closed triangulation
           surface with v in its interior:
Step 3.1   Same as Step 2.1.
Step 3.2   If the updated $\mathcal{L}_v$ is the edge set of a closed triangulation
                surface (denoted as B_v) with v in its interior, then let T_v
                be the set of tetrahedra $vabc$, where abc is a face in B_v.
                Otherwise, return (NO).
Step 4   Let $T = \cup_{v \in V(\mathcal{L})} T_v$. For each tetrahedron $abcd \in T$, if
           $abcd \in T_a \cap T_b \cap T_c \cap T_d$, then return (YES, T); otherwise
           return (NO).

The following theorem shows the characterization of a tetrahedralization from the algorithmic viewpoint.

Theorem 6. $\mathcal{L}$ *forms a tetrahedralization if and only if* TETRAHEDRALIZATION-DETECTION$(\mathcal{L})$ *(outline) returns (YES, T).*

We have the following main result.

Theorem 7. *If* TETRAHEDRALIZATION-DETECTION *is run on m line segments with n endpoints, then the running time is $O(nm \log n)$ and the running space is $O(n^2)$.*

5   Deciding Whether $\mathcal{L}$ Contains a Tetrahedralization

In this section, we discuss the complexity of deciding if a set of line segments $\mathcal{L}$ in R^3 contains a tetrahedralization (the TETRAD problem). There are two NP-complete problems related to TETRAD. Ruppert and Seidel proved that the problem of deciding if a 3-dimensional nonconvex polyhedron can be tetrahedralized is NP-complete [15] (the NCP problem). Lloyd proved that the problem of deciding if a line segment set E in R^2 contains a triangulation is NP-complete [12] (the TRID problem). Either of the problems can reduce to TETRAD; and the reduction of NCP to TETRAD is easier than that of TRID to TETRAD.

Theorem 8. *The* TETRAD *problem is NP-complete.*

6   Constrained Tetrahedralizations

A *constrained tetrahedralization* for a set $\mathcal{G}$ of vertices and line segments is a tetrahedralization of the vertices and endpoints $V(\mathcal{G})$ such that all the segments $E(\mathcal{G})$ are used as edges in the tetrahedralization. We also say that $\mathcal{G}$ can be tetrahedralized. In this section, we discuss the complexity problem of deciding if a set of vertices and line segments $\mathcal{G}$ in R^3 can be tetrahedralized (briefly, the CT problem).

Ruppert and Seidel proved that it is NP-complete to tetrahedralize a 3-dimensional nonconvex polyhedra [15]. We shall extend their proof to show that the CT problem is also NP-complete. They proved this NP-completeness result by using a transformation from the Satisfiability problem. That is, for any Boolean formula in conjunctive normal form, a nonconvex polyhedron is constructed such that it can be tetrahedralized if and only if the Boolean formula is satisfiable. The main tool in their construction is a gadget, called a *niche*, which is derived from Schönhardt's polyhedron [16]. As illustrated in Figure 2(a), the way to construct Schönhardt's polyhedron is as follows. Start with a triangular prism; fix the bottom face $a_1 b_1 c_1$ so that it cannot move; twist the top face abc so that each rectangular face of the prism folds into two triangles with a reflex edge between them. If the top face abc is removed from the Schönhardt's polyhedron, then the resulting figure that consists of seven triangle faces on six vertices is called a *niche*; and face $a_1 b_1 c_1$ is called the *base* of the niche. In this section, we use S_N to denote a niche, and $\overline{S}_N$ to denote the corresponding Schönhardt's polyhedron.

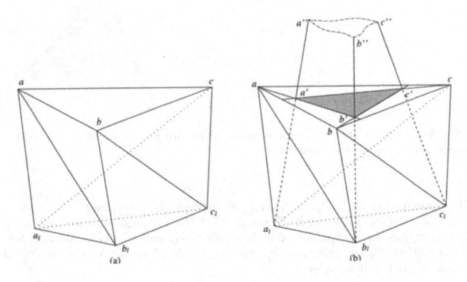

Fig. 2. (a) A Schönhardt's polyhedron that cannot be tetrahedralized. (b) $aa_1 b_1 b'$, $bb_1 c_1 c'$ and $cc_1 a_1 a'$ are plane quadrilaterals. view($a_1 b_1 c_1$) is the truncated cone $a_1 b_1 c_1 a'' b'' c''$; illum($a_1 b_1 c_1$) is the truncated cone $a' b' c' a'' b'' c''$.

Definition 3. *A point p can* **see** *a point q in the base $a_1b_1c_1$ of a niche S_N if the interior of the segment pq intersects the interior of $\overline{S}_N$ but does not intersect any faces of S_N. The set of points that can see the entire base $a_1b_1c_1$ of S_N is called the* **view** *of S_N, denoted as* $\text{view}(a_1b_1c_1)$*. The subset of* $\text{view}(a_1b_1c_1)$ *in the exterior of $\overline{S}_N$ is called the* **illuminant** *of S_N, denoted as* $\text{illum}(a_1b_1c_1)$*.*

Let $S_N = \{a_1b_1c_1, aa_1b_1, bb_1c_1, cc_1a_1, abb_1, bcc_1, caa_1\}$ with base $a_1b_1c_1$. Suppose the three planes passing through the faces aa_1b_1, bb_1c_1, and cc_1a_1, respectively, intersect at point v. They produce eight cones with v as the apex. We are interested in the cone containing base $a_1b_1c_1$. This cone is called the *view cone* of $a_1b_1c_1$. So $\text{view}(a_1b_1c_1)$ and $\text{illum}(a_1b_1c_1)$ can be considered as truncated cones (see Figure 2(b)). Note that each point of $\text{view}(a_1b_1c_1)$ can see the seven faces of S_N from the inside; and if the apex v is below the base $a_1b_1c_1$, then all the points in the symmetric cone can see the seven faces of S_N from the outside as shown in Figure 3(a). If we twist the top face abc of S_N further, $\text{view}(a_1b_1c_1)$ and $\text{illum}(a_1b_1c_1)$ will shrink and $\text{view}(a_1b_1c_1)$ will become a tetrahedron inside $\overline{S}_N$ and $\text{illum}(a_1b_1c_1) = \emptyset$, as shown in Figure 3(b).

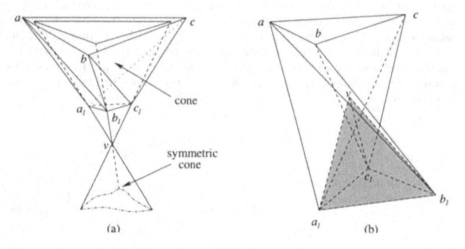

Fig. 3. (a) v is the apex of the double cone. (b) $\text{illum}(a_1b_1c_1) = \emptyset$ and $\text{view}(a_1b_1c_1) = \text{int}(va_1b_1c_1)$ (the shaded tetrahedron).

Lemma 1. *Let $S_N = \{a_1b_1c_1, aa_1b_1, bb_1c_1, cc_1a_1, abb_1, bcc_1, caa_1\}$ be a niche with base $a_1b_1c_1$, as shown in Figure 4, $a_ib_jc_k$ be a triangle such that a_ib_j, a_ic_k puncture face aa_1b_1, b_ja_i, b_jc_k puncture face bb_1c_1, and c_kb_j, c_ka_i puncture face cc_1a_1. Then $S'_N = \{a_ib_jc_k, aa_ib_j, bb_jc_k, cc_ka_i, abb_j, bcc_k, caa_i\}$ is a niche and $\text{illum}(a_ib_jc_k) \subset \text{illum}(a_1b_1c_1)$.*

Theorem 9. *Let $\mathcal{L}$ be a set of segments containing a niche S_N with base $a_1b_1c_1$. If $\text{view}(a_1b_1c_1) \cap V(\mathcal{L}) = \emptyset$, then $\mathcal{L}$ cannot be tetrahedralized.*

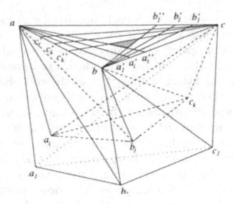

Fig. 4. $\{a_1b_1c_1, aa_1b_1, bb_1c_1, cc_1a_1, abb_1, bcc_1, caa_1\}$ and $\{a_ib_jc_k, aa_ib_j, bb_jc_k, cc_ka_i, abb_j, bcc_k, caa_i\}$ form two niches respectively.

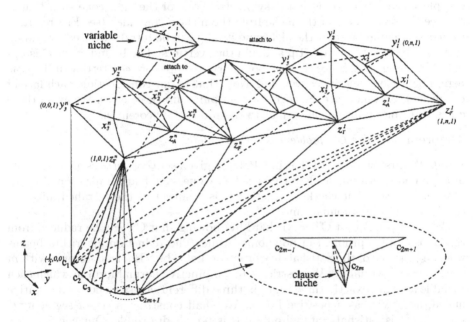

Fig. 5. Polyhedron $\mathcal{P}$ constructed in [15] (not to scale).

Let $\mathcal{P}$ be the nonconvex polyhedron constructed in [15] for showing the NP-completeness of tetrahedralizing nonconvex polyhedra. As illustrated in Figure 5, the starting point for constructing $\mathcal{P}$ is the polyhedron consisting of a row of n squares $y_1^i y_3^i z_T^i z_F^i (1 \le i \le n)$ in plane $z = 1$ and m triangles $c_{2k-1}c_{2k}c_{2k+1}(1 \le k \le m)$ in plane $z = 0$. The coordinates of the squares' vertices are $y_1^i(0, n+1-i, 1)$, $y_3^i(0, n-i, 1)$, $z_T^i(1, n-i, 1)$, and $z_F^i(1, n+1-i, 1)$. Note that $y_3^i = y_1^{i+1}$ and $z_T^i = z_F^{i+1}$. Vertices $c_k(1 \le k \le m)$ lie on a parabola in the xy-plane. The coordinates of c_k are $(\alpha_k, \beta_k, 0)$, where $\alpha_k = 0.5 + 0.005(k-1)/m$ and

$\beta_k = 0.25\alpha_k^2 - 0.25\alpha_k + 0.0625$. To each square $y_1^i y_3^i z_T^i z_F^i$, a roof is attached which contains the variable's three literal vertices x_1^i, x_2^i and x_3^i. As mentioned above, niches are used as gadgets that force any tetrahedralization to have some properties; for example, they can force tetrahedra to appear. For each variable, there is a variable niche attached to $y_1^i y_2^i y_3^i$ of the roof. For each clause, there is a clause niche attached to $c_{2k-1} c_{2k} c_{2k+1}$ on the bottom. Refer to [15] for the detailed placement and coordinates of the roofs and niches.

The ideas of our proof of NP-completeness is as follows. Since the boundary of the constrained tetrahedralization is a convex hull, we intend to construct a convex polyhedron that contains $\mathcal{P}$. If we simply use $CH(\mathcal{P})$ as our convex polyhedron, it is not clear how to tetrahedralize the region between $\mathcal{P}$ and $CH(\mathcal{P})$ because the boundary of $\mathcal{P}$ is too complicated. In order to simplify the boundary of $\mathcal{P}$, we cover each niche by using a *cap* that is defined as a pyramid with base $y_1^i y_2^i y_3^i (1 \le i \le n)$ or $c_{2k-1} c_{2k} c_{2k+1} (1 \le k \le m)$ (see Figure 6 or 7). From the structure of $\mathcal{P}$, we know that the apex of the view cone lies outside $\mathcal{P}$. So we can place the cap's tip inside the symmetric cone of the view cone such that it can see the seven faces of the niche from the niche's "outside" (see Figure 3(a)). Hence, the region between the niche and its cap can be tetrahedralized by adding segments between the cap's tip and all other vertices. The boundary of $\mathcal{P}$ is updated by these caps. Since the new boundary is simple, we can tetrahedralize the region between it and its convex hull. Note that in our construction, each face of the niche is an inner face. So segments can puncture these faces. We avoid these cases by using an argument similar to the proof of Theorem 9.

Theorem 10. *The CT problem is NP-complete.*

Proof. We first show that CT is in NP. For a given set $\mathcal{G}$ of vertices and segments in R^3, a nondeterministic algorithm need only guess a tetrahedralization of $V(\mathcal{G})$. It is easy to see that checking whether $\mathcal{G}$ is a subset of the tetrahedralization can be done in polynomial time.

We next prove that CT is NP-hard by showing that CT can be reduced from the Satisfiability problem in polynomial time. The reduction algorithm begins with an instance of the Satisfiability problem. Let ϕ be a Boolean formula with m clauses over n variables. We restrict ϕ to conjunctive normal form in which each variable appears exactly three times in three different clauses, once as a negative literal, and twice as a positive literal. We shall construct a set of segments $\mathcal{G}$ such that ϕ is satisfiable if and only if $\mathcal{G}$ is tetrahedralizable. Our construction is based on the nonconvex polyhedron $\mathcal{P}$ described above.

For any variable niche with base $q_1^i q_2^i q_3^i (1 \le i \le n)$, let q^i be the apex of the view cone of $q_1^i q_2^i q_3^i$ which lies outside $\mathcal{P}$. Note that in the construction of $\mathcal{P}$ in [15], the variable niche is specified in detail because its base $q_1^i q_2^i q_3^i$ must see the two truth-setting vertices z_T^i and z_F^i, and the three literal vertices x_1^i, x_2^i, x_3^i on the roof of the variable must satisfy the "blocking" condition for the clause niches. From the placement of the variable niche, we know that the x-coordinate of q^i is greater than -1. So we can select a vertex a^i in the plane $x = -1$ such that a^i can see the seven faces of the niche from the outside. Then connect a^i with the six vertices of the niche and the three vertices of face $y_1^i y_2^i y_3^i$ which contains the niche such that the 13 facets of $\mathcal{P}$ bounded by triangle $y_1^i y_2^i y_3^i$

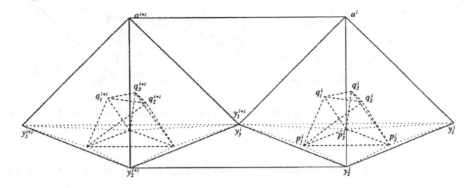

Fig. 6. $a^i y_1^i y_2^i y_3^i$ is the cap of the niche. a^i can see the seven faces of the niche and six faces on face $y_1^i y_2^i y_3^i$.

form 13 tetrahedra with a^i (see Figure 6). So we can update the boundary of $\mathcal{P}$ by using facets $a^i y_1^i y_2^i$, $a^i y_2^i y_3^i$, $a^i y_3^i y_1^i$ $(1 \leq i \leq n)$ to replace the 13 facets in each cap. In particular, from the structure of $\mathcal{P}$, we can select a row of points in the plane $x = -1$ such that segments $a^i a^{i+1}$, $y_2^i y_2^{i+1}$ $(1 \leq i \leq n-1)$ and $a^i y_2^i (1 \leq i \leq n)$ lie on the convex hull of $V(\mathcal{P}) \cup \{a^i \mid 1 \leq i \leq n\}$. Let $\mathcal{P}_1$ be the updated boundary of $\mathcal{P}$.

The construction of the clause niche is much simpler than that of the variable niche. For the kth clause niche, its illuminant contains the segment with end-points $(\alpha_{2k}, 0, 1)$ and $(\alpha_{2k}, n, 1)$, where $\alpha_{2k} = 0.5 + 0.005(2k - 1)/m$. Since this segment intersects $y_1^1 z_F^1$ and $y_3^n z_T^n$, the kth niche must be contained in the region bounded by face $c_{2k-1} c_{2k} c_{2k+1}$ and the planes passing through $z_F^1 c_{2k-1} c_{2k+1}$, $z_T^n c_{2k-1} c_{2k}$ and $z_T^n c_{2k} c_{2k+1}$ respectively. From the illuminant lemma [15] we can construct the niche in this region as small as possible such that $a^i (1 \leq i \leq n)$ cannot see $b_k (1 \leq k \leq m)$ (to be specified later). Similarly to the placement of a^i for the kth clause niche, select a vertex b_k in the plane $z = -0.00001$ such that b_k can see the seven faces of the niche from the outside but b_k cannot see $a^i (1 \leq i \leq n)$. Then connect b_k with the six vertices of the niche and the three vertices of the triangle face to which the niche is attached. Then we update the current boundary of $\mathcal{P}_1$ by using three facets to replace the 13 facets in each cap. In particular, we place $b_k (1 \leq k \leq m)$ in the plane $z = -0.00001$ such that faces $b_k c_{2k-1} c_{2k}$, $b_k c_{2k} c_{2k+1} (1 \leq k \leq m)$ and $b_k b_{k+1} c_{2k+1} (1 \leq k \leq m-1)$ lie on the convex hull of $\mathcal{P} \cup \{a^i \mid 1 \leq i \leq n\} \cup \{b_k \mid 1 \leq k \leq m\}$ (see Figure 7). Let $\mathcal{P}_2$ be the updated boundary of $\mathcal{P}_1$.

It is easy to see that the region between $\mathcal{P}_2$ and $CH(\mathcal{P}_2)$ consists of three parts. The first part is the convex polyhedron with vertex set $\{a^i \mid 1 \leq i \leq n\} \cup \{y_1^i \mid 1 \leq i \leq n\} \cup \{y_3^n, c_1\}$. The second part is the convex polyhedron with vertex set $\{b_k \mid 1 \leq k \leq m\} \cup \{c_{2k+1} \mid 0 \leq k \leq m\} \cup \{y_1^1, z_F^1\}$. So these two parts can be tetrahedralized easily by adding some segments. The third part consists of $n-1$ wedge-shaped polyhedra between the n variable roofs. Since each roof is

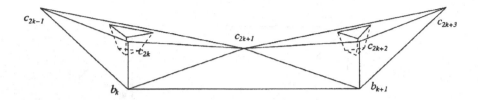

Fig. 7. $b_k c_{2k-1} c_{2k} c_{2k+1}$ is the cap of the niche. b_k can see the seven faces of the niche and six faces inside $c_{2k-1} c_{2k} c_{2k+1}$.

convex, each wedge-shaped polyhedron can be tetrahedralized by adding some segments [9].

Let $\mathcal{G}$ be the segment set consisting of $E(\mathcal{P})$, $E(CH(\mathcal{P}_2))$ and all the segments we have added between $\mathcal{P}$ and $CH(\mathcal{P}_2)$. $\mathcal{G}$ can be considered as the edge set of the tetrahedralization between $\mathcal{P}$ and $CH(\mathcal{P}_2)$. Since $\mathcal{P}$ and $\mathcal{P}_2$ can be constructed in polynomial time, and $n-1$ wedge-shaped polyhedra can be tetrahedralized in polynomial time, $\mathcal{G}$ can be constructed in polynomial time.

We claim that formula ϕ is satisfiable if and only if $\mathcal{G}$ is tetrahedralizable. We first suppose that there exists a constrained tetrahedralization T for $\mathcal{G}$. For any variable niche $p_1 p_2 p_3 q_1 q_2 q_3$ (see Figure 6) in $\mathcal{P}$ with base $q_1 q_2 q_3$ (for simplicity, we drop the i superscripts), it follows from Theorem 1 that there exists a vertex q_4 in the niche side of the plane passing through $q_1 q_2 q_3$ such that $q_1 q_2 q_3 q_4 \in T$. Note that $\text{view}(q_1 q_2 q_3) \cap V(\mathcal{G}) = \{z_T, z_F\}$. If $q_4 \notin \{z_T, z_F\}$, then using an argument similar to the proof of Theorem 9, $q_1 q_4, q_2 q_4$, or $q_3 q_4$ can only puncture the faces $p_1 q_1 q_2, p_2 q_2 q_3$, or $p_3 q_3 q_1$. Without loss of generality, suppose $q_3 q_4$ punctures $p_1 q_1 q_2$. From the structure of $\mathcal{G}$, we know that no other points can form an $\mathcal{L}$-empty tetrahedron with face $q_2 q_3 q_4$. This contradicts Theorem 1. Thus q_4 must lie in the illuminant of the niche. Similarly, for any clause niche in $\mathcal{P}$, the vertex which forms tetrahedron with the base of the niche must lie inside its illuminant. From the proof of Theorem 1 in [15], we can obtain a truth assignment for ϕ. Conversely, suppose ϕ is satisfiable. Using the method of [15], we can add some segments to tetrahedralize $\mathcal{P}$. These adding segments together with $\mathcal{G}$ form a constrained tetrahedralization of $V(\mathcal{G})$. $\qquad\square$

Remark. Note that the NP-completeness of tetrahedralizing nonconvex polyhedra does not depend on coplanarities of faces or other degeneracies. The structure of $\mathcal{P}$ can be modified easily so that $V(\mathcal{P})$ is in a nondegenerate position [15, page 250]. In consequence, we can also modify the structure of $\mathcal{G}$ so that $V(\mathcal{G})$ is in a nondegenerate position. Thus, the NP-completeness is still valid for tetrahedralizing a set of vertices and segments in the general position.

Acknowledgments. We would like to thank Todd Wareham, Donald Craig and Paul Gillard for providing many helpful comments and corrections on the draft of this paper. We would also like to thank the referees who pointed out the relevant work by K. Mehlhorn et al [13] and Devillers et al [7] on checking convex polytopes and planar subdivisions.

This work is partially supported by NSERC grant OPG0041629 and ERA grant in Memorial University of Newfoundland. Part of this work was done when the first author was at Memorial University of Newfoundland.

References

1. D.Avis and H.ElGindy, Triangulating point sets in space, *Discrete and Computational Geometry* , 2(1987)99–111.
2. C.Bajaj and T.K.Dey, Convex decomposition of polyhedra and robustness, *SIAM Journal on Computing* , 21(1992)339–364.
3. A.Below, J.A.De Loera and J.Richter-Gebert, Finding minimal triangulations of convex 3-polytopes is NP-hard, In *Proc. 11th Annual ACM-SIAM Symposium on Discrete Algorithms* , 9–11, 2000.
4. M.Bern and D.Eppstein, Mesh generation and optimal triangulation, In *Computing in Euclidean Geometry* , (D. Du and F. Hwang eds.), World Scientific Publishing Co. 1992.
5. B.Chazelle, Convex partitions of polyhedra: a lower bound and worst-case optimal algorithm, *SIAM Journal on Computing*, 13(1984)488–507.
6. B.Chazelle and L.Palios, Triangulating a nonconvex polytope, *Discrete and computational geometry*, 5(1990)505–526.
7. O.Devillers, G.Liotta, F.P.Preparata and R.Tamassia, Checking the Convexity of Polytopes and the Planarity of Subdivisions, *Computational Geometry: Theory and Applications* , 11(1998)187–208.
8. H.Edelsbrunner, F.Preparata and D.West, Tetrahedralizing point sets in three dimensions, *Journal of Symbolic Computation* , 10(1990)335–347.
9. J.Goodman and J.Pach, Cell Decomposition of Polytopes by Bending, *Israel Journal of Math*, 64(1988)129–138.
10. J.Hershberger and J.Snoeyink, Convex polygons made from few lines and convex decompositions of polyhedra, In *Lecture Notes in Computer Science 621* , 376–387, 1992.
11. J.E.Hopcroft and R.E.Tarjan, Efficient planarity testing, *Journal of ACM* , 21 (1974) 549–568.
12. E.L.Lloyd, On triangulations of a set of points in the plane, In *Proc. 18th FOCS*, 228–240, 1977.
13. K.Mehlhorn, S.Näher, M.Seel, R.Seidel, T.Schilz, S.Schirra and C.Uhrig, Checking Geometric Programs or Verification of Geometric Structures, *Computational Geometry: Theory and Applications* , 12(1999)85–103.
14. J.Richter-Gebert, Finding small triangulations of polytope boundaries is hard, *Discrete and Computational Geometry*, 24(2000)503–517.
15. J.Rupert and R.Seidel, On the difficulty of triangulating three-dimensional nonconvex polyhedra, *Discrete and Computational Geometry*, 7(1992)227–253.
16. E.Schönhardt, Uber die Zerlegung von Dreieckspolyedern in Tetraeder, *Math. Annalen* 98(1928)309–312.

Average-Case Communication-Optimal Parallel Parenthesis Matching

Chun-Hsi Huang[1] and Xin He[2]

[1] Department of Computer Science and Engineering
University of Connecticut, Storrs, CT 06269
huang@cse.uconn.edu
[2] Department of Computer Science and Engineering
State University of New York at Buffalo, Buffalo, NY 14260
xinhe@cse.buffalo.edu

Abstract. We provide the first non-trivial lower bound, $\frac{p-3}{p} \cdot \frac{n}{p}$, where p is the number of the processors and n is the data size, on the average-case communication volume, σ, required to solve the *parenthesis matching* problem and present a parallel algorithm that takes linear (optimal) computation time and optimal expected message volume, $\sigma + p$.

The kernel of the algorithm is to solve the *all nearest smaller values problem*. Provided $\frac{n}{p} = \Omega(p)$, we present an algorithm that achieves optimal sequential computation time and uses only a constant number of communication phases, with the message volume in each phase bounded above by $(\frac{n}{p} + p)$ in the worst case and p in the average case, assuming the input instances are uniformly distributed.

1 Introduction

The *parenthesis matching* problem is defined as follows: A sequence of parentheses is *balanced* if every left (right, respectively) parenthesis has a matching right (left, respectively) parenthesis. Let a balanced sequence of parentheses be represented by $(a_1, a_2, \cdots, a_n)$, where a_k represents the k-th parenthesis. It is required to determine the matching parenthesis of each parenthesis.

The parenthesis matching problem was shown to be reducible to the *all nearest smaller values problem* [3], which is defined as follows: Let $A = (a_1, a_2, \cdots, a_n)$ be an array of elements from a totally ordered domain. For each a_j, $1 \leq j \leq n$, find the nearest element to the left of a_j and the nearest element to the right of a_j that are smaller than a_j.

A typical application of the ANSVP is the merging of two sorted lists [4, 12]. Let $A = (a_1, a_2, \cdots, a_n)$ and $B = (b_1, b_2, \cdots, b_n)$ be two increasing arrays to be merged. The reduction to the ANSVP is done by constructing an array $C = (a_1, a_2, \cdots, a_n, b_n, b_{n-1}, \cdots, b_1)$ and then solving the ANSVP with respect to C. If b_y is the right match of a_x, the location of a_x in the merged list is $x + y$. The locations of b_x's can be found similarly. In addition, the ANSVP is fundamental in that several problems have been shown to be reducible to it [3, 5,10,15].

P. Bose and P. Morin (Eds.): ISAAC 2002, LNCS 2518, pp. 308–319, 2002.
© Springer-Verlag Berlin Heidelberg 2002

1.1   Previous Work and Our Results

A work-optimal CRCW PRAM algorithm for the ANSVP using $O(\log \log n)$ time was proposed in [3]. Katajainen [9] explored the parallel time complexity of the ANSVP on a variation of PRAM, Local Memory PRAM. Kravets and Plaxton [11] presented a parallel algorithm on the hypercube.

These results did not provide communication cost analysis and apply only on the fine-grained machines. Note that communication cost has been shown to be the major bottleneck of the performance of parallel algorithms [7,1]. In this paper, we exploit and extend some observations made in [3] and design a parallel ANSV algorithm that is communication-efficient for any number of processors. We prove that the worst-case message volume is bounded above by $(\frac{n}{p} + p)$ and the average-case message volume is bounded above by p. We then use some novel methods to provide the first non-trivial lower bound, $\frac{p-3}{p} \cdot \frac{n}{p}$, on the expected communication requirement, σ, for the parenthesis matching problem and show that, reducing to the ANSV algorithm in this paper, we have a scalable algorithm that is optimal in both computation cost (linear) and expected communication volume $(\sigma + p)$, provided that $\frac{n}{p} = \Omega(p)$, a condition true for all practical parallel machines.

1.2   Organization of the Paper

In Section 2 we describe the cost model, *Bulk Synchronous Parallel* (BSP), we use for design and analysis. Although a particular parallel model is used to facilitate the description, the algorithms and their analyses in this paper are model independent. A BSP algorithm for the ANSVP and the worst-case and average-case analyses are presented in this section. These analyses establish the foundation of the results in Section 3, which proves the lower bound of the average-case communication volume of the parallel parenthesis matching problem and show the communication optimality when our ANSVP algorithm is used to solve this problem. Section 4 concludes this paper and describes some open problems.

2   The ANSVP Algorithm

2.1   The BSP Model

Interprocessor communication has been shown to become a major bottleneck for parallel algorithm performance. Parallel algorithms should seek to minimize both computation and communication time to be considered practical. For PRAM model, one fundamental assumption is that communication between processors are via common memory, and each memory access consumes one unit computation time. For practical purposes, this assumption does not lead to accurate performance prediction.

Bulk Synchronous Parallel (BSP) model is a model for general-purpose, architecture-independent parallel programming, introduced by Valiant [16] and McColl [13]. The BSP model consists of three components: a set of processors, each with a local memory, a communication network, and a mechanism for globally synchronizing the processors. All existing parallel computers are BSP computers in this sense, but they have very different performance characteristics. These differences are captured by three parameters (in addition to a fourth, the processor speed of the computer), p: the number of processors, g: the ratio of communication throughput to processor throughput, and L: the time required to barrier synchronize all or part of the processors. A BSP program proceeds as a series of *supersteps*. In each superstep, a processor may operate only on values stored in local memory. Values sent through the communication network are not guaranteed to arrive until the end of the current superstep. Parameters p, g, and L can be used to estimate the running time of a BSP program whose communication behavior is known. Similarly, a BSP program that has access to these parameters for the machine it is running on can use them to choose between different algorithms. Our purpose is to design a scalable BSP algorithm, minimizing the number of communication supersteps as well as the local computation time. If the best possible sequential algorithm for a given problem takes $T_s(n)$ time, then ideally we would like to design a BSP algorithm using $O(1)$ communication supersteps, preferably with total message size bounded by $O(\frac{n}{p})$ in each superstep, and $O(\frac{T_s(n)}{p})$ total local computation time.

Throughout the section, we assume $\frac{n}{p} = \Omega(p)$, a condition true for all parallel machines.

2.2   The Algorithm

Given a sequence $A = (a_1, a_2, \cdots, a_n)$ and a p-processor BSP computer with any communication media (shared memory or any interconnection network), each P_i ($1 \leq i \leq p$) stores $(a_{\frac{n}{p}(i-1)+1}, a_{\frac{n}{p}(i-1)+2}, \cdots, a_{\frac{n}{p}i})$. For simplicity, we assume that the elements in the sequence are distinct. We define the nearest smaller value to the right of an element to be its *right match*. The ANSVP can be solved sequentially with linear time using a straightforward stack approach. To find the right matches of the elements, we scan the input, keep the elements for which no right match has been found on a stack, and report the current element as a right match for those elements on the top of the stack that are larger than the current element. The left matches can be found similarly. For brevity and without loss of generality, we will focus on finding the right matches.

Some definitions are given below. Throughout the rest of this paper, we use i ($1 \leq i \leq p$) for processor related indexing and j ($1 \leq j \leq n$) for array element related indexing.

- For any j:
  - $\mathrm{rm}(j)$ ($\mathrm{lm}(j)$, resp.) $\overset{\mathrm{def}}{=}$ the index of the right (left, resp.) match of a_j.
  - $\mathrm{rmp}(j)$ ($\mathrm{lmp}(j)$, resp.) $\overset{\mathrm{def}}{=}$ the index of the processor containing $a_{\mathrm{rm}(j)}$ ($a_{\mathrm{lm}(j)}$, resp.).
- For any i:
  - $\mathrm{min}(i) \overset{\mathrm{def}}{=}$ the index (in A) of the smallest element in P_i.

- rm_min(i) (lm_min(i), resp.) $\overset{\text{def}}{=}$ the index of the right (left, resp.) match of $a_{\min(i)}$ with respect to the array $A_{\min} = (a_{\min(1)}, \cdots, a_{\min(p)})$.
- $\wp_i \overset{\text{def}}{=} \{P_x|\ \text{rm\_min}(x) = i\}$; $\varphi_i \overset{\text{def}}{=} \{P_x|\ \text{lm\_min}(x) = i\}$.

Based on the above definitions, we observe that $P_{\text{rmp}(\min(i))} = P_{\text{rm\_min}(i)}$ ($P_{\text{lmp}(\min(i))} = P_{\text{lm\_min}(i)}$, resp.) Next we prove a lemma used in our BSP algorithm.

Lemma 1. *On a p-processor BSP computer, for any i, if $a_{\text{rm}(\min(i))}$ exists and $\text{rmp}(\min(i)) \neq i + 1$, then there exists a unique processor $P_{k(i)}$, $i < k(i) < \text{rmp}(\min(i))$, such that $\text{lmp}(\min(k(i))) = i$ and $\text{rmp}(\min(k(i))) = \text{rmp}(\min(i))$. (Symmetrically, for any i, if $a_{\text{lm}(\min(i))}$ exists and $\text{lmp}(\min(i)) \neq i - 1$, then there exists a unique processor $P_{k'(i)}$, $\text{lmp}(\min(i)) < k'(i) < i$, such that $\text{lmp}(\min(k'(i))) = \text{lmp}(\min(i))$ and $\text{rmp}(\min(k'(i))) = i$.)*

Proof. We show that P_s, where $a_{\min(s)} = \min\{a_{\min(i+1)},\ a_{\min(i+2)}, \cdots, a_{\min(\text{rmp}(\min(i))-1)}\}$, is the unique processor described in the lemma. For any s' with $i + 1 \leq s' < s$, $\text{rmp}(\min(s'))$ must be $\leq s$. Similarly, for any s' with $s + 1 \leq s' < \text{rmp}(\min(i))$, $\text{lmp}(\min(s'))$ must be $\geq s$. This renders P_s the only candidate processor. We can easily infer that $a_{\min(s)} > a_{\min(i)} > a_{\min(\text{rmp}(\min(i)))}$. Since $a_{\min(s)}$ is the smallest element among those in $P_{i+1}, \cdots, P_{\text{rmp}(\min(i))-1}$, we conclude that P_s is the unique processor $P_{k(i)}$ specified in Lemma 1. (The symmetric part can be proved similarly.)

We next outline our algorithm. To begin with, all processors sequentially find the right matches for their local elements, using the stack approach. Those matched elements require no interprocessor communication. We therefore focus on those elements which are not yet matched. The general idea is to find the right matches for those not-yet-matched elements by reducing the original ANSVP to $2p$ smaller "special" ANSVPs, and solve them in parallel.

Next we compute the right and left matches for all $a_{\min(i)}$'s. To do this, we first solve the ANSVP with respect to the array $A_{\min} = (a_{\min(1)}, \cdots, a_{\min(p)})$. Then, for each processor P_i, we define four sequences, Seq1$_i$, Seq2$_i$, Seq3$_i$ and Seq4$_i$ as follows:

- If $a_{\text{rm}(\min(i))}$ does not exist, then Seq1$_i$ and Seq2$_i$ are undefined.
- If $a_{\text{rm}(\min(i))}$ exists and $\text{rmp}(\min(i)) = i + 1$, then:
  Seq1$_i = (a_{\min(i)}, \cdots, a_{\frac{n}{p}i})$, Seq2$_i = (a_{\frac{n}{p}i+1}, \cdots, a_{\text{rm}(\min(i))})$.
- If $a_{\text{rm}(\min(i))}$ exists and $\text{rmp}(\min(i)) > i+1$, let $P_{k(i)}$ be the unique processor specified in Lemma 1. Then: Seq1$_i = (a_{\min(i)}, \cdots, a_{\text{lm}(\min(k(i)))})$, Seq2$_i = (a_{\text{rm}(\min(k(i)))}, \cdots, a_{\text{rm}(\min(i))})$.
- If $a_{\text{lm}(\min(i))}$ does not exist, then Seq3$_i$ and Seq4$_i$ are undefined.
- If $a_{\text{lm}(\min(i))}$ exists and $\text{lmp}(\min(i)) = i - 1$, then:
  Seq3$_i = (a_{\frac{n}{p}(i-1)+1}, \cdots, a_{\min(i)})$, Seq4$_i = (a_{\text{lm}(\min(i))}, \cdots, a_{\frac{n}{p}(i-1)})$.
- If $a_{\text{lm}(\min(i))}$ exists and $\text{lmp}(\min(i)) < i-1$, let $P_{k'(i)}$ be the unique processor specified in Lemma 1. Then: Seq3$_i = (a_{\text{rm}(\min(k'(i)))}, \cdots, a_{\min(i)})$, Seq4$_i = (a_{\text{lm}(\min(i))}, \cdots, a_{\text{lm}(\min(k'(i)))})$.

Note that Seq1_i and Seq3_i, if they exist, always reside on P_i, Seq2_i, if it exists, always resides on $P_{\text{rmp}(\min(i))}$, and Seq4_i, if it exists, always resides on $P_{\text{lmp}(\min(i))}$. The following observation [3] specifies how to find the right matches for all unmatched elements: The right matches of all not-yet-matched elements in Seq1_i lie in Seq2_i. The right matches of all not-yet-matched elements in Seq4_i, except its first element, lie in Seq3_i.

Each processor P_i therefore is responsible for identifying right matches for not-yet-matched elements in Seq1_i and Seq4_i. Again, we apply the sequential algorithm at each processor P_i with respect to the two concatenated sequences, $\text{Seq1}_i \| \text{Seq2}_i$ and $\text{Seq4}_i \| \text{Seq3}_i$. All elements will be right-matched after the above-mentioned 2p special ANSVPs are solved in parallel [3].

The following technical lemma, Lemma 2, provides the foundation for the communication lower bound in the worst-case analysis.

Lemma 2. *1. Suppose that $\wp_i = \{P_{x_1}, P_{x_2}, \cdots, P_{x_t}\}$ where $x_1 < x_2 < \cdots < x_t$. Then:*

$\text{Seq2}_{x_1} = (a_{\text{rm}(\min(x_2))}, \cdots, a_{\text{rm}(\min(x_1))})$,

$\text{Seq2}_{x_2} = (a_{\text{rm}(\min(x_3))}, \cdots, a_{\text{rm}(\min(x_2))})$,

$\cdots$, $\text{Seq2}_{x_t} = (a_{\frac{n}{p}(i-1)+1}, \cdots, a_{\text{rm}(\min(x_t))})$.

2. Suppose that $\varphi_i = \{P_{y_1}, P_{y_2}, \cdots, P_{y_s}\}$ where $y_1 < y_2 < \cdots < y_s$. Then:

$\text{Seq4}_{y_1} = (a_{\text{lm}(\min(y_1))}, \cdots, a_{\text{lm}(\min(y_2))})$,

$\text{Seq4}_{y_2} = (a_{\text{lm}(\min(y_2))}, \cdots, a_{\text{lm}(\min(y_3))})$,

$\cdots$, $\text{Seq4}_{y_s} = (a_{\text{lm}(\min(y_s))}, \cdots, a_{\frac{n}{p}i})$.

Proof. We only prove Statement 1. The proof of Statement 2 is similar. First observe that, for any $P_x, P_y \in \wp_i$, $x < y$ implies $a_{\min(x)} < a_{\min(y)}$ and $k(x) \leq y$. Based on these observations, we have $k(x_l) = x_{l+1}$ for $1 \leq l < t$ and $x_t = i - 1$. The lemma follows from the definition of Seq2.

The algorithm below finds the right matches and is therefore denoted Algorithm $ANSV_r$.

Algorithm $ANSV_r$:

Input: A partitioned into p subsets of continuous elements. Each processor stores one subset.

Output: The right match of each a_i is computed and stored in the processor containing a_i.

1. Each P_i sequentially solves the $ANSV_r$ problem with respect to its local subset.
2. a) Each P_i computes its local minimum $a_{\min(i)}$.
   b) All $a_{\min(i)}$'s are globally communicated. (Hence each P_i has the array $A_{\min}$.)

3. Each P_i solves the $ANSV_r$ and the $ANSV_l$ problems with respect to $A_{\min}$; and identify the sets $\wp_i$ and φ_i.

4. Each P_i computes $a_{rm(min(x))}$ for every $P_x \in \wp_i$ and $a_{lm(min(y))}$ for every $P_y \in \varphi_i$.
5. Each P_i determines $Seq1_i$, $Seq3_i$ and receives $Seq2_i$, $Seq4_i$ as follows:
   a) Each P_i computes the unique $k(i)$ and $k'(i)$ (as in Lemma 1), if they exist, and determines $Seq1_i$ and $Seq3_i$.
   b) Each P_i determines $Seq2_x$ for every $P_x \in \wp_i$, and $Seq4_y$ for every $P_y \in \varphi_i$ (as in Lemma 2).
   c) Each P_i sends $Seq2_x$, for every $P_x \in \wp_i$, to P_x and $Seq4_y$, for every $P_y \in \varphi_i$, to P_y.

6. a) Each P_i finds the right matches for the unmatched elements in $Seq1_i$ and $Seq4_i$
   b) Each P_i collects the matched $Seq4_y$'s from all P_y's$\in \varphi_i$.

2.3   Cost Analysis

We use the term *h-relation* to denote a routing problem where each processor has at most h words of data to send to other processors and each processor is also due to receive at most h words of data from other processors. In each BSP superstep, if at most w arithmetic operations are performed by each processor and the data communicated forms an *h-relation*, then the cost of this superstep is $w + h * g + L$ (the parameters g and L are as defined in Section 1). The cost of a BSP algorithm using S supersteps is simply the sum of the costs of all S supersteps:

$$BSP\ cost = comp.\ cost + comm.\ cost +\ synch.\ cost = W + H * g + L * S,$$

where H is the sum of the maxima of the *h-relations* in each superstep and W is the sum of the maxima of the local computations in each superstep.

Here we assume the sequential computation time for the $ANSV_r$ problem of input size n is $T_s(n)$, and the sequential time for finding the minimum of n elements is $T_\theta(n)$. Then the BSP cost breakdown of Algorithm $ANSV_r$ can be derived as in Table 1.

Since $\frac{n}{p} = \Omega(p)$ and $T_s(n) = T_\theta(n) = O(n)$, the computation time in each step is obviously linear in the local data size, namely $O(\frac{n}{p})$. Steps 2 (b), 5 (c) and 6 (b) involve communication. Thus the algorithm takes three supersteps. Based on Lemma 2 and the fact that $|\varphi_i| + |\wp_i| \le p$, the communication steps 5 (c) and 6 (b) can each be implemented by an $(\frac{n}{p} + p)$-relation. Therefore we have:

Theorem 1. *The ANSVP with input size n can be solved on a p-processor BSP machine in three supersteps using linear local computation time and at most an $(\frac{n}{p} + p)$-relation in each communication phase, provided $p \le n/p$.*

Due to the page limit, the analysis of expected communication requirement based on uniform distribution of the inputs is omitted. The readers are referred to [17] for the detailed proofs of the following theorem.

Table 1. BSP Cost Breakdown of *Algorithm ANSV$_r$*

Step	Cost		
	comp.	comm.	synch.
1	$T_s(\frac{n}{p})$		
2(a)	$T_\theta(\frac{n}{p})$		
2(b)		pg	L
3	$2T_s(p)$		
4	$\max_i\{T_\theta(\frac{n}{p}+|\varphi_i|)+T_\theta(\frac{n}{p}+|\wp_i|)\}$		
5(a)	$\max_i\{T_\theta(\text{rm\_min}(i)-\text{lm\_min}(i))\}$		
5(b)	$O(\frac{n}{p})$		
5(c)		$\max_i\{(\Sigma_{P_x\in\wp_i}|\text{Seq2}_x| \quad + \Sigma_{P_y\in\varphi_i}|\text{Seq4}_y|)g\}$	L
6(a)	$\max_i\{T_s(|\text{Seq1}_i|+|\text{Seq2}_i|)+T_s(|\text{Seq4}_i|+ |\text{Seq3}_i|)\}$		
6(b)		$\max_i\{\Sigma_{P_x\in\varphi_i}|\text{Seq4}_x|g\}$	L

Theorem 2. *The ANSVP with input size n can be solved on a p-processor BSP machine in 3 supersteps using linear local computation time, with the average-case communication requirement bounded above by a p-relation for each communication phase.*

3   Algorithm *Parenthesis Matching*

3.1   Reduction to ANSVP

The parallel parenthesis matching algorithm will employ a fundamental operation, *prefix sum*, which is defined in terms of a binary, associative operator $\otimes$. The computation takes as input a sequence $< b_1, b_2, \cdots, b_n >$ and produces as output a sequence $< c_1, c_2, \cdots, c_n >$ such that $c_1 = b_1$ and $c_k = c_{k-1} \otimes b_k$ for $k = 2, 3, \cdots, n$.

We will describe the parenthesis matching algorithm first. The details regarding how prefix sum operation can be implemented on BSP will be described next.

Algorithm Parenthesis Matching:

Input: A legal parenthesis list, partitioned into p subsets of continuous elements. Each processor stores one subset.

Output: The mate location of each parenthesis is stored in the processor containing that parenthesis.

1. Each processor assigns 1 for each left parenthesis and -1 for each right parenthesis.
2. Perform BSP prefix sum operation on this array and derive the resulting array A.
3. Perform the *ANSV* algorithm on A.

Note that the BSP prefix sum operation used in step 2 can be implemented in 2 BSP supersteps, provided that $\frac{n}{p} = \Omega(p)$, as below.

Assume that the initial input is $b_1, b_2, \cdots, b_n$ and the processors are labeled $P_1, P_2, \cdots, P_p$. Processor P_i initially contains $b_{\frac{n}{p}(i-1)+1}, b_{\frac{n}{p}(i-1)+2}, \cdots, b_{\frac{n}{p}i}$.

2.1 Each processor computes all prefix sums of $b_{\frac{n}{p}(i-1)+1}, b_{\frac{n}{p}(i-1)+2}, \cdots, b_{\frac{n}{p}i}$ and stores the results in local memory locations $s_{\frac{n}{p}(i-1)+1}, s_{\frac{n}{p}(i-1)+2}, \cdots, s_{\frac{n}{p}i}$.

2.2 Each P_i sends $s_{\frac{n}{p}i}$ to processor P_p.

2.3 Processor P_p computes prefix sums of the p values received in step 2.2 and store these values at $t_1, t_2, \cdots, t_p$.

2.4 Processor P_p then sends t_i to $P_{i+1}, 1 \le i \le p-1$.

2.5 Each processor adds the received value to $s_{\frac{n}{p}(i-1)+1}, s_{\frac{n}{p}(i-1)+2}, \cdots, s_{\frac{n}{p}i}$.

Steps 2.2 and 2.4 are involved with communication and we can easily arrange the prefix sum algorithm in 2 BSP supersteps, each taking a p-relation in the communication phase.

Since step 2 of Algorithm Parenthesis Matching takes only a p-relation. Step 3 therefore becomes dominant in communication requirement. Although the reduction from parenthesis matching to ANSVP is obvious, their average-case communication complexity analyses are very different since these two problems have significantly different problem domains.

In this section we first prove the average-case lower bound of the communication requirement of parallel parenthesis matching and show that our BSP ANSV algorithm actually results in an asymptotically communication-optimal parallel algorithm for this problem.

The Expected Lower Bound. Let S_n be the set of all possible legal parenthesis of length n, where $n = 2m$. We have $|S_n| = C_m = \frac{1}{m+1}\binom{2m}{m}$, the m-th Catalan number [6]. Some definitions in Section 2 are inherited here. In addition, given a sequence $(a_1, a_2, \cdots, a_n) \in S_n$, we define

(1) rm(j) (lm(j), respectively) to be the index of the right (left, respectively) matching parenthesis if a_j is a left (right, respectively) parenthesis;

(2) $M_s(i)$ and $M_r(i)$ to be the total number of right and left parentheses in P_i that are not locally matched, while $(a_1, a_2, \cdots, a_n)$ is evenly partitioned and distributed among the p processors. Note that $M_s(i)$ ($M_r(i)$, respectively) is the size of the message sent (received, respectively) by processor P_i in step 3 of the Algorithm Parenthesis Matching.

Moreover, for a randomly selected sequence from S_n, we denote the event that the j-th element is a right (left, respectively) parenthesis as $a_j = r$ ($a_j = l$, respectively). The following lemma can be derived.

Lemma 3. $Pr[a_j = r] = \left(\sum_{k=0}^{\lfloor \frac{j}{2} \rfloor - 1} C_k C_{m-1-k} \right) / C_m$.

Proof. We count the number of legal sequences where a_j is a right parenthesis. The subsequence between a_j and its left match must be a legal sequence and hence consisting of an even number of parentheses. Thus, the left match of a_j can only occur at locations $j-1, j-3, \cdots$, and $j-(2\lfloor\frac{j}{2}\rfloor-1)$. In addition, the total number of sequences with $a_j = r$ and the left match of a_j being at location $a_{j-(2k-1)}$ is $C_{k-1}C_{m-k}$. (This is because the sequence $a_{j-2k+2}\ldots a_{j-1}$ must be a legal sequence consisting of $(k-1)$ pairs of parentheses and the sequence $a_1 \ldots a_{j-2k}a_{j+1}\ldots a_n$ must be a legal sequence consisting of $(m-k)$ pairs of parentheses). Therefore, the total number of sequences with $a_j = r$ is $\sum_{k=0}^{\lfloor\frac{j}{2}\rfloor-1} C_k C_{m-1-k}$, which completes the proof.

We also observe that:

(1) $P_r[a_j = l] = 1 - P_r[a_j = r]$,
(2) $P_r[a_j = r] = P_r[a_{n-j+1} = l]$, and
(3) $E(M_s(i)) = E(M_r(p-i+1))$.

Observations (2) and (3) can be seen by reversing the sequence and switching the roles of right and left parentheses.

Lemma 4. *Let a_k be a parenthesis in processor P_i. The probability that a_k is a right parenthesis and is not locally matched in P_i is: $P_r[a_k = r] - \sum_{t=0}^{\lfloor\frac{k-\frac{n}{p}(i-1)}{2}\rfloor-1} C_t C_{m-1-t}/C_m$.*

Proof. The probability that a_k is a right parenthesis is $P_r[a_k = r]$. a_k can only be locally matched by a left parenthesis a_{k-2t-1} for $t = 0, 1, \ldots \lfloor\frac{k-\frac{n}{p}(i-1)}{2}\rfloor-1$. By the argument in the proof of Lemma 3, for each fixed t, the probability that a_{k-2t-1} is the matching left parenthesis of a_k is $C_t C_{m-1-t}/C_m$. This proves the lemma.

Lemma 5. *The expected maximum size of messages (sent and received) by any processor is given by $E(M_s(p))$.*

Proof. By Lemma 4, the size of the message sent by processor P_i ($1 \le i \le p$) is: $E(M_s(i)) = \sum_{k=\frac{n}{p}(i-1)+1}^{\frac{n}{p}i} T_k(i)$, where $T_k(i) = P_r[a_k = r] - \sum_{t=0}^{\lfloor\frac{k-\frac{n}{p}(i-1)}{2}\rfloor-1} C_t C_{m-1-t}/C_m$ is the probability that a_k is an unmatched right parenthesis within P_i.

For any fixed $i_1 < i_2$ and k, the second term $(\sum_{t=0}^{\lfloor\frac{k-\frac{n}{p}(i-1)}{2}\rfloor-1} C_t C_{m-1-t}/C_m)$ in $T_k(i_1)$ and $T_k(i_2)$ are identical. The first term $(P_r[a_k = r])$ in $T_k(i_1)$ is less than the first term $(P_r[a_k = r])$ in $T_k(i_2)$. Thus $E(M_s(p)) = max_{1 \le i \le p}\{E(M_s(i))\}$. Thus, the maximum size of the messages sent by any processor is $E(M_s(p))$.

Similarly, the maximum size of the messages received by any processor is $E(M_r(1))$ which, according to Observation (3), equals to $E(M_s(p))$.

Thus, to lower bound the expected maximum message size, we need to lower bound $E(M_s(p))$. In the following lemmas, we will show that, for a random legal sequence, most right parentheses in processor P_p are not locally matched.

Lemma 6. $\dfrac{C_x C_{m-x-1}}{C_m} \geq \dfrac{m^{\frac{3}{2}}}{8\sqrt{\pi}(m-x-1)^{\frac{3}{2}} x^{\frac{3}{2}}}.$

Proof. According to Stirling's approximation $(\sqrt{2\pi n}(\frac{n}{e})^n \leq n! \leq \sqrt{2\pi n}(\frac{n}{e})^{n+\frac{1}{12n}})$, we have

$$
\begin{aligned}
C_x &= \frac{1}{x+1} \frac{(2x)!}{x!x!} \\
&\geq \frac{1}{x+1} \frac{\sqrt{(2\pi)(2x)}(\frac{2x}{e})^{2x}}{[\sqrt{2\pi x}(\frac{x}{e})^{x+\frac{1}{12x}}][\sqrt{2\pi x}(\frac{x}{e})^{x+\frac{1}{12x}}]} \\
&\geq \frac{2^{2x} e^{\frac{1}{6x}}}{(x+1)\sqrt{\pi x}(x^{\frac{1}{6x}})}.
\end{aligned}
$$

Analogously,

$$
C_{m-x-1} \geq \frac{2^{2(m-x-1)} e^{\frac{1}{6(m-x-1)}}}{(m-x)\sqrt{\pi(m-x-1)}(m-x-1)^{\frac{1}{6(m-x-1)}}} \quad \text{and}
$$

$$
C_m \leq \frac{1}{m+1} \frac{1}{\sqrt{\pi m}} \frac{2^{2m+\frac{1}{24m}} m^{\frac{1}{24m}}}{e^{\frac{1}{24m}}}.
$$

Therefore,

$$
\begin{aligned}
\frac{C_x C_{m-x-1}}{C_m} &\geq \frac{\frac{2^{2x} e^{\frac{1}{6x}}}{(x+1)\sqrt{\pi x}(x^{\frac{1}{6x}})} \frac{2^{2(m-x-1)} e^{\frac{1}{6(m-x-1)}}}{(m-x)\sqrt{\pi(m-x-1)}(m-x-1)^{\frac{1}{6(m-x-1)}}}}{\frac{1}{m+1} \frac{1}{\sqrt{\pi m}} \frac{2^{2m+\frac{1}{24m}} m^{\frac{1}{24m}}}{e^{\frac{1}{24m}}}} \\
&\geq \frac{m^{\frac{3}{2}}}{4\sqrt{\pi}x^{\frac{3}{2}}(m-x-1)^{\frac{3}{2}}}(\frac{1}{2}).
\end{aligned}
$$

Lemma 7. $\frac{p-3}{p} \cdot \frac{n}{p} \leq E(M_s(p)) < \frac{n}{p}.$

Proof. By Lemma 3, we have

$$
\begin{aligned}
E(M_s(p)) &= \sum_{k=\frac{n}{p}(p-1)+1}^{n} \left(\Pr[a_k = r] - \sum_{t=0}^{\lfloor\frac{k-\frac{n}{p}(p-1)}{2}\rfloor-1} C_t C_{m-1-t}/C_m \right) \\
&= (C_0 C_{m-1} + C_1 C_{m-2} + \cdots + C_{\lfloor\frac{n}{p}(p-1)+1}{2}\rfloor-1} C_{m-\lfloor\frac{\frac{n}{p}(p-1)+1}{2}\rfloor})/C_m + \\
&\quad (C_1 C_{m-2} + C_2 C_{m-3} + \cdots + C_{\lfloor\frac{\frac{n}{p}(p-1)+2}{2}\rfloor-1} C_{m-\lfloor\frac{\frac{n}{p}(p-1)+2}{2}\rfloor})/C_m + \\
&\quad \cdots + \\
&\quad (C_{\lfloor\frac{n}{p}\rfloor-1} C_{m-\lfloor\frac{n}{p}\rfloor} + C_{\frac{n}{p}} C_{m-\frac{n}{p}-1} + \cdots + C_{m-1} C_0)/C_m \\
&\geq \frac{n}{p} \left[(C_{\lfloor\frac{n}{p}\rfloor-1} C_{m-\lfloor\frac{n}{p}\rfloor} + \cdots + C_{\lfloor\frac{\frac{n}{p}(p-1)+1}{2}\rfloor-1} C_{m-\lfloor\frac{\frac{n}{p}(p-1)+1}{2}\rfloor})/C_m \right].
\end{aligned}
$$

(The last inequality is true because there are n/p lines on the left hand side and each line contains the common expression in the bracket.)

Based on Lemma 6 and the fact that $C_m = \sum\limits_{k=0}^{m-1} C_k C_{m-1-k}$, we conclude the following:

$$E(M_s(p)) \geq \frac{n}{p} \left[(C_{\lfloor \frac{n}{p} \rfloor - 1} C_{m - \lfloor \frac{p}{2} \rfloor} + \cdots + C_{\lfloor \frac{\frac{n}{p}(p-1)+1}{2} \rfloor - 1} C_{m - \lfloor \frac{\frac{n}{p}(p-1)+1}{2} \rfloor}) / C_m \right]$$

$$\geq \frac{n}{p} \left(\frac{1}{m} \left(\lfloor \frac{\frac{n}{p}(p-1)+1}{2} \rfloor - \lfloor \frac{n}{p} \rfloor + 1 \right) \right)$$

$$\geq \frac{n}{p} \left(\frac{p-3}{p} \right).$$

(The second inequality results from the fact that the terms in the right hand side of the first inequality are the larger $(\lfloor \frac{\frac{n}{p}(p-1)+1}{2} \rfloor - \lfloor \frac{n}{p} \rfloor + 1)$ terms of $\sum\limits_{k=0}^{m-1} \frac{C_k C_{m-1-k}}{C_m} = 1$.)

In addition, we can easily infer $E(M_s(p)) < \frac{n}{p}$, since $E(M_s(p)) = \sum\limits_{k=\frac{n}{p}(i-1)+1}^{\frac{n}{p}i} (Pr[a_k = r] - \sum\limits_{t=0}^{\lfloor \frac{k - \frac{n}{p}(i-1)}{2} \rfloor - 1} C_t C_{m-1-t}/C_m) < \frac{n}{p}$. This completes the proof.

We show that $E(M_s(p)) \geq \frac{p-3}{p} \cdot \frac{n}{p}$. Since the BSP algorithm for parenthesis matching employs the prefix sum operation, which introduces an additional p-relation, and the BSP ANSV algorithm, which introduces at most a p-relation (refer to Lemma 2), in addition to the messages required to be sent/received, we conclude that the average-case message volume of our BSP parenthesis matching algorithm is bounded above by $(E(M_s(p)) + \Theta(p))$, which is asymptotically communication-optimal, provided that $\frac{n}{p} = \Omega(p)$.

4   Conclusions and Future Work

We provide the first non-trivial lower bound, $\frac{p-3}{p} \cdot \frac{n}{p}$, where p is the number of the processors and n is the data size, on the average-case communication volume, σ, required to solve the *parenthesis matching* problem and present a parallel algorithm that takes linear (optimal) computation time and optimal expected message volume, $\sigma + p$. The kernel of the algorithm is to solve the *all nearest smaller values problem*. Provided $\frac{n}{p} = \Omega(p)$, we present an algorithm that achieves optimal sequential computation time and uses only a constant number of communication phases, with the message volume in each phase bounded above by $(\frac{n}{p} + p)$ in the worst case and p in the average case.

The parenthesis matching problem is one of the several applications that reduce to the ANSVP. Typical examples are the *binary tree reconstruction* problem [2] and the *monotone polygon triangulation* problem [8,14], which happen to have the same number of problem instances, the n-th Catalan number, as in the parenthesis matching problem. The question regarding whether the ANSV algorithm also yields an expected-case communication-optimal algorithm for them remains open.

References

1. M. Adler and J. W. Byers and R. M. Karp: *Parallel Sorting With Limited Bandwidth*. Proc. ACM Symposium on Parallel Algorithms and Architectures (1995) 129–136
2. Selim G. Akl: *Parallel Computation, Models and Methods*. Prentice Hall, 1997
3. Omer Berkman and Baruch Schieber and Uzi Vishkin: *Optimal Doubly Logarithmic Parallel Algorithms Based on Finding All Nearest Smaller Values*. Journal of Algorithms, Vol 14, 1993, 344–370
4. A. Borodin and J.E. Hopcroft: *Routing, Merging and Sorting on Parallel Models of Computation*. J. Computer and System Science, Vol 30, 1985, 130–145
5. D. Breslauer and Z. Galil: *An Optimal $O(\log\log n)$ Time Parallel String Matching Algorithm*. SIAM J. Computing, Vol 19, 1990, 1050–1058
6. T. H. Cormen and C. E. Leiserson and R. L. Rivest: *Introduction to Algorithms*. McGraw-Hill, 2000
7. F. Dehne and A. Fabri and A. Rau-Chaplin: *Scalable Parallel Geometric Algorithms for Coarse Grained Multicomputers*. Proc. 9th ACM Annual Computational Geometry, 1993, 298–307
8. Joseph JáJá: *An Introduction to Parallel Algorithms*. Addison-Wesley, 1992
9. J. Katajainen: *Finding All Nearest Smaller Values on a Distributed Memory Machine*. Proc. of Conference on Computing: The Australian Theory Symposium, 1996, 100–107
10. Z.M. Kedem and G.M. Landau and K.V. Palem: *Optimal Parallel Prefix-Suffix Matching Algorithm and Applications*. Proc. 1st ACM Symposium on Parallel Algorithms and Architectures, 1989, 388–398
11. D. Kravets and C. G. Plaxton: *All Nearest Smaller Values on the Hypercube*. IEEE Transactions on Parallel and Distributed Systems, Vol 7, No 5, 1996, 456–462
12. C.P. Kruskal: *Searching, Merging and Sorting in Parallel Computation*. IEEE Transactions on Computers, Vol C-32, 1983, 942–946
13. W. F. McColl: *Scalable Computing*. Computer Science Today: Recent Trends and Developments, Edited by J. van Leeuwen, Lecture Notes in Computer Science, Vol 1000, Springer-Verlag, Berlin, 1995, 46–61
14. J.H. Reif: *Synthesis of Parallel Algorithms*. Morgan Kaufmann, 1993
15. B. Schieber and U. Vishkin: *Finding All Nearest Neighbors for Convex Polygons in Parallel: A New Lower Bound Technique and A Matching Algorithm*. Discrete App. Math., Vol 29, 1990, 97–111
16. Leslie G. Valiant: *A Bridging Model for Parallel Computation*. Communications of the ACM, Vol 33, No 8, 1990, 103–111
17. Xin He and Chun-Hsi Huang: *Communication Efficient BSP Algorithm for All Nearest Smaller Values Problem*. Journal of Parallel and Distributed Computing, Vol 61, 2001, 1425–1438

Optimal F-Reliable Protocols for the Do-All Problem on Single-Hop Wireless Networks

(Extended Abstract)

Andrea E. F. Clementi[1], Angelo Monti[2], and Riccardo Silvestri[2]

[1] Dipartimento di Matematica, Università di Roma "Tor Vergata",
via della Ricerca Scientifica, I-00133 Roma, Italy
`clementi@dsi.uniroma1.it`
[2] Dipartimento di Informatica, Università di Roma "La Sapienza",
via Salaria 113, 00198 Roma, Italy
{`monti, silvestri`}`@dsi.uniroma1.it`

Abstract. In the DO-ALL problem, a set of t tasks must be performed by using a synchronous network of p processors. Processors may fail by permanent crashing. We investigate the time and the work complexity of F-*reliable* protocols for the DO-ALL problem on 1-hop wireless networks without collision detection. An F-reliable protocol is a protocol that guarantees the execution of all tasks if at most $F < p$ faults happen during its execution. Previous results for this model are known only for the case $F = p - 1$.
We obtain the following tight bounds.
- The completion time of F-reliable protocols on 1-hop wireless networks without collision detection is

$$\Theta\left(\frac{t}{p-F} + \min\left\{\frac{tF}{p}, F + \sqrt{t}\right\}\right).$$

- The work complexity of F-reliable protocols on 1-hop wireless networks without collision detection is

$$\Theta(t + F \cdot \min\{t, F\}).$$

The two lower bounds hold even when the faults only happen at the very beginning of the protocol execution.

1 Introduction

The problem and the model. Given a synchronous network of p processors (also called nodes or stations), the DO-ALL problem consists in performing t idempotent and independent tasks by using the network processors. Every processor knows the entire list of the tasks to be performed and it is able to execute one task during a time-slot. The DO-ALL problem is trivial when processors do not suffer *faults*. On the other hand, the design of efficient protocols for the DO-ALL problem in presence of unpredictable permanent faults is a fundamental issue in distributed computing. The problem has been introduced by Dwork

P. Bose and P. Morin (Eds.): ISAAC 2002, LNCS 2518, pp. 320–331, 2002.

et al in their seminal paper [4] and it has been the subject of several papers [2,3, 1,5,6]. Recently, the DO-ALL problem has been studied in the model in which processors are stations communicating over a single *multiple-access channel* [2]. The system may be also interpreted as a *single-hop synchronous wireless network* [8] (in the sequel, this model will be called wireless network). Processors fail by *permanent* crashing, i.e., processor restarts are not allowed. A processor which has not failed and has not reached an halt state yet during step s is said to be *operational during step s*. During each step, an operational processor can perform local computations, execute a single task, and it can act as receiver or as transmitter. A message is received by all the processors acting as receivers if and only if it is the only message sent during that step. Furthermore, processors are not able to distinguish between the collision noise and the background noise (so, there is no collision detection).

The fault patterns can be described by means of *adversaries*. An adversary knows the protocol against which it competes. This paper studies protocols which are *reliable* against *F-bounded adversaries*, i.e., adversaries that can decide at most F failures (with $F < p$) and, for each failure, they can decide the processor and the crashing step. We say that a DO-ALL protocol is *F-reliable* if it completes all the tasks against *any* F-bounded adversary.

The performance of the protocols are evaluated with respect to two complexity measures: *completion time* and *work*. The *completion time* of a protocol execution is the number of steps by which all the processors either have halted or have failed. The work measure is the *available processor steps* measure introduced in [10]. The work of a protocol execution is the total number of available-processor steps: each processor contributes a unit for each step when it is *operational*, even when idling.

Previous works. Most of the papers in the literature consider networks in which each node can send a message to any subset of nodes in one step [3,1, 5,6,4]. More recently, Chlebus *et al* [2] studied the work complexity of reliable protocols for the DO-ALL problem on single-hop *wireless networks*.

All the papers cited above consider the model in which protocols must be $(p-1)$-*reliable* (or, simply, reliable): they must complete all the tasks even in the (worst) case in which *only one* processor remains active. This requirement is motivated in situations such as control systems for nuclear reactors or in other applications where high reliability is of vital importance. Such reliable protocols are also useful in local area networks where tasks may be distributed among idle workstations. The scenario in which a set of computations run over idle workstations has been introduced in [11] and extensively studied in the last two decades (see [7] for further references). In this scenario, a fault corresponds to a user reclaiming his machine.

As for wireless networks, Chlebus *et all* [2] provided an $\Omega(t + p\sqrt{t})$ lower bound for the work required by any reliable protocol. Then, they present a work-optimal reliable protocol for the channel with *collision detection* (i.e. stations are able to distinguish between the background noise and the presence of a collision on the channel). As for the model without collision detection, Chlebus *et al* proved the following

Theorem 1 (CKL01).

a) For any reliable DO-ALL *protocol there exists an execution, with at most f faults, that performs work $\Omega(t + p\sqrt{t} + p\min\{f,t\})$.*

b) A reliable DO-ALL *protocol exists such that it performs $O(t + p\sqrt{t} + p\min\{f,t\})$ work, where f is the number of faults really suffered by the network during the execution of the protocol.*

In the sequel, this protocol will be called CKL protocol. The CKL protocol is work optimal.

We emphasize that the proof of the lower bound strongly relies on the fact that protocols must always guarantee the reliability.

Our contribution. We study the complexity of deterministic F-reliable protocols (where F is any positive integer such that $1 \leq F \leq p-1$) for the DO-ALL problem on wireless networks without collision detection. As described before, F-reliability is the natural generalization of reliability. Notice that F is not the number of faults really suffered by the network, rather, it is the maximum number of faults the protocol has to be able to manage. The aim of this paper is to establish lower and upper bounds for the work and the completion time of F-reliable protocols.

We believe that the study of F-reliability does not need further motivations than those mentioned for $(p-1)$-reliability. However, we want to point out that, in several scenarios, the assumption of a bound $F << p-1$ is a more realistic model. Indeed, consider again the case of a distributed control system for the valves of a nuclear reactor: if the system is well-designed and well-implemented, the probability that $p - 1$ processors suffer a fault during the execution of a protocol should be less than the probability that a meteorite hits the reactor. If, on the contrary, this event has a not negligible probability then the same holds for the event "*all* the processors suffer a fault" (provided that $p >> 1$). A further argument supports F-reliability in the idle-workstations application mentioned above. During the set-up phase of the parallel system, it could be convenient to establish a threshold. The success of all the computations (assigned to the idle workstations) is ensured only when the number of available idle workstations is not smaller than the threshold. In this scenario, a weaker adversary becomes relevant: the failures only happen at the very beginning of the execution of the protocol. This corresponds to the situation in which workstation requests from users have real effects only at the end of the protocol execution.

Our contribution can be summarized as follows

Theorem 2 (Completion time).

– *Any F-reliable* DO-ALL *protocol, in the worst case, requires at least*

$$\Omega\left(\frac{t}{p-F} + \min\left\{\frac{tF}{p}, F + \sqrt{t}\right\}\right) \quad \textit{completion time.}$$

It holds even when the faults only happen at the very beginning of the protocol execution.

– *An F-reliable* DO-ALL *protocol exists having* optimal *completion time.*

Theorem 3 (Work complexity).

− *Any F-reliable* DO-ALL *protocol, in the worst case, requires at least*

$$\Omega(t + F \cdot \min\{t, F\}) \ \ work.$$

It holds even when the faults only happen at the very beginning of the protocol execution.
− *An F-reliable* DO-ALL *protocol exists having* optimal *work complexity.*

We remark that the proofs of our lower bounds cannot exploit the techniques of [2]. Indeed, the proof of the lower bound in Theorem 1 strongly relies on the fact that, independently from the number of faults that happen, a $(p-1)$-reliable protocol must be able to assign, at any step, all the remaining tasks to one processor, if this is the only one which is still alive.
On the other hand, our protocols take full advantage of the bound F: in particular, this *a priori* bound allows the use of *redundancy*[1] in a much heavier way than that of the CKL protocol. For some ranges of the parameters, the use of redundancy without any communication yields optimal protocols. As for completion time, this happens whenever the additive term $\frac{tF}{p}$ dominates the other two in our bound. For other ranges, a suitable combination of redundancy and communication is necessary in order to get optimal completion time. A similar remark holds for the term Ft in the work complexity bound.
   A comparison between the bounds in Theorem 1 and our corresponding bounds in Theorems 2-3 is possible by setting $f = F$. As regards the completion time, the F-reliability yields asymptotically faster protocols when F is small and t is not much greater than p (for instance when $F = \mathsf{polylog}(p)$ and $t = O(p)$), or (rather surprisingly) when F is very large, i.e., when $p - F = o(p)$ and $t = O(p^2)$. A more clear situation holds for the work complexity: since our bound in Theorem 3 does not contain p explicitly, then F-reliable protocols are more efficient than $(p-1)$-reliable ones whenever $p > \sqrt{t}$.
   Finally, we emphasize that, thanks to our lower bound in Theorem 3, when $t \geq p^2$ or $t \leq F$, a suitable tuning of the CKL protocol turns out to be a work optimal F-reliable protocol.

Organization of the paper. Section 2 contains the proofs of the two lower bounds. Section 3 provides the optimal protocols and their performance analysis.

2 The Lower Bounds

Both proofs of the two lower bounds for the completion time and the work complexity make use of the following technical lemma.
Let $\mathcal{A}$ be an F-reliable DO-ALL protocol and let $\mathcal{F}$, with $|\mathcal{F}| \leq F$, be a set of processors. We denote by $E_m^{\mathcal{A}}(\mathcal{F})$ the execution of the first m steps of $\mathcal{A}$ when the

---

[1] Redundancy here means making more copies of the same task and distribute them to the processors.

set of processors that fails is $\mathcal{F}$ and all the failures happen at the very beginning of the execution.

A step of a protocol's execution is said to be *silent* if no processor acts as a transmitter in that step.

Lemma 1. *Let $\mathcal{A}$ be a F-reliable DO-ALL protocol and let Q and P be two sets of processors. A set $\overline{P}$ of processors exists such that*

1. *In $E_m^{\mathcal{A}}(Q \cup P \cup \overline{P})$ no transmission succeeds;*
2. *$|\overline{P}| \leq m$.*
3. *If in $E_m^{\mathcal{A}}(Q)$ no transmission succeeds then $\overline{s} \geq s + |\overline{P}|$, where $\overline{s}$ and s are the number of silent steps in $E_m^{\mathcal{A}}(Q \cup P \cup \overline{P})$ and in $E_m^{\mathcal{A}}(Q)$ respectively.*

Proof. The set $\overline{P}$ is obtained by the following procedure.

1. $\overline{P} := \emptyset$;
2. **If** in $E_m^{\mathcal{A}}(Q \cup P \cup \overline{P})$ no communication succeeds **then end**;
3. Let j be the first processor that succeeds to transmit in $E_m^{\mathcal{A}}(Q \cup P \cup \overline{P})$;
4. $\overline{P} := \overline{P} \cup \{j\}$;
5. **go to** 2.

We now show that the procedure terminate in at most m iterations and, consequently, the size of $\overline{P}$ is bounded by m. Let j be any processor in $\overline{P}$ and let $E(j)$ be the execution in the above procedure that has caused the insertion of j in $\overline{P}$. Define $\tau(j)$ as the first step of $E(j)$ in which j succeeds in broadcasting. We claim that for any two distinct processors j and j' in $\overline{P}$ it holds that $\tau(j) \neq \tau(j')$, implying that $|\overline{P}| \leq m$. Indeed, suppose, on the contrary, that $\tau(j) = \tau(j')$. Wlog assume that j has been inserted in $\overline{P}$ before j'. This would imply that during step $\tau(j)$ of $E(j)$, *both* processors j and j' try to transmit and j would not succeed in broadcasting: a contradiction.

If in both $E_m^{\mathcal{A}}(Q)$ and $E_m^{\mathcal{A}}(Q \cup P \cup \overline{P})$ no transmission succeeds then the behavior of the non-faulty processors is the same in both executions. Thus, the silent steps in $E_m^{\mathcal{A}}(Q)$ are also silent in $E_m^{\mathcal{A}}(Q \cup P \cup \overline{P})$.

To prove point 3 it suffices to show that, for every $j \in \overline{P}$, $\tau(j)$ is silent in $E_m^{\mathcal{A}}(Q \cup P \cup \overline{P})$ but it is not silent in $E_m^{\mathcal{A}}(Q)$

In the first $\tau(j)$ steps of $E(j)$ the behavior of non-faulty processors is the same of that of $E_{\tau(j)}^{\mathcal{A}}(Q)$. Thus, at step $\tau(j)$, processor j transmits in both $E(j)$ and $E_m^{\mathcal{A}}(Q)$. Let T be the set of processors that transmit at step $\tau(j)$ in $E_m^{\mathcal{A}}(Q)$. It holds that $j \in T$, $|T| \geq 2$, and $T \subseteq P \cup \overline{P}$. The latter inclusion derives from the facts that j successfully transmits in $E(j)$ and $j \in \overline{P}$. Hence step $\tau(j)$ is not silent in $E_m^{\mathcal{A}}(Q)$ but it is silent in $E_m^{\mathcal{A}}(Q \cup P \cup \overline{P})$ since all the processors in T are faulty. $\square$

2.1 The Completion-Time Complexity

The proof of the lower bound requires two further technical lemmas.

Lemma 2. *For any F-reliable DO-ALL protocol there exists a set of at most F initial faults that forces the completion time m to be such that $m + \frac{mp}{t} > F$.*

Proof. Let $\mathcal{A}$ be a F-reliable DO-ALL protocol. From Lemma 1 with $Q = P = \emptyset$, there exists a set of processors A, with $|A| \leq m$, such that in $E_m^{\mathcal{A}}(A)$ no transmission succeeds. The number of task executions performed in $E_m^{\mathcal{A}}(A)$ is at most mp, therefore there will be a task α executed by at most $\frac{mp}{t}$ processors. Denote the set of processors that execute the task α in $E_m^{\mathcal{A}}(A)$ as $P(\alpha)$.

Let's now apply Lemma 1 with $Q = A$ and $P = P(\alpha)$. Then, there exists a set $\overline{P(\alpha)}$, with $|A \cup \overline{P(\alpha)}| \leq m$, such that in $E_m^{\mathcal{A}}(A \cup P(\alpha) \cup \overline{P(\alpha)})$ no transmission succeeds. There is no broadcast heard in $E_m^{\mathcal{A}}(A \cup P(\alpha) \cup \overline{P(\alpha)})$. Therefore, each operational processor in $E_m^{\mathcal{A}}(A \cup P(\alpha) \cup \overline{P(\alpha)})$ behaves as in $E_m^{\mathcal{A}}(A)$. The task α is not performed in $E_m^{\mathcal{A}}(A \cup P(\alpha) \cup \overline{P(\alpha)})$ since the processors in $P(\alpha)$ have been failed and the remaining ones behave as in $E_m^{\mathcal{A}}(A)$. Hence, it must hold that

$$F \geq |A \cup P(\alpha) \cup \overline{(\alpha)}|$$

Since

$$|A \cup \overline{P(\alpha)} \cup P(\alpha)| \leq |A \cup \overline{P(\alpha)}| + |P(\alpha)| \leq m + \frac{mp}{t}$$

the thesis holds. $\square$

Lemma 3. *For any F-reliable* DO-ALL *protocol there exists a set of at most F initial faults that if $t > 2m^2$ then the completion time m satisfies*

$$\frac{mp}{t - 2m^2} > F.$$

Proof. Let $\mathcal{A}$ be an F-reliable DO-ALL protocol having completion time m so that $\frac{t}{2} > m^2$. Then, define E^0 as the execution of $\mathcal{A}$ when no fault happens. For each step of the execution E^0, if there is only one broadcasting processor, we mark it as *reserved* while if there are two or more broadcasting processors we mark as *reserved* any two of them. Let R be the set of all the *reserved* processors. Since $|R| \leq 2m$, the number of distinct tasks performed by the reserved processors is at most $2m^2$. Hence, there are at least $t - 2m^2 > 0$ distinct tasks that have to be performed by non-reserved processors. So, there will be a task α executed by at most

$$\frac{m(p - |R|)}{t - 2m^2} \leq \frac{mp}{t - 2m^2}$$

non-reserved processors (and, by none of the reserved processors). We define the set $P(\alpha)$ of processors that execute the task α and consider the execution E obtained by failing all the processors in $P(\alpha)$. The task α is not performed in $E_m^{\mathcal{A}}(P(\alpha))$ since each operational processor in E behaves as in E^0. Thus it must be $F < |P(\alpha)|$. It follows that

$$F < |P(\alpha)| \leq \frac{mp}{t - 2m^2}.$$

$\square$

Theorem 4. *Any F-reliable* DO-ALL *protocol in the worst case requires at least*

$$\max\left\{\frac{t}{p-F}, \frac{1}{4}\min\left\{\frac{tF}{p}, F+\sqrt{t}\right\}\right\}$$

completion time. It holds even when the faults only happen at the very beginning of the protocol execution.

Proof. Let $\mathcal{A}$ be an F-reliable DO-ALL protocol. The term $\frac{t}{p-F}$ of the bound follows by considering an execution obtained by failing F processors since the beginning. The remaining $p-F$ processors have to execute t tasks. This requires at list $\frac{t}{p-F}$ steps. Hence we have to show that $\mathcal{A}$ requires, in the worst case, at least

$$\frac{1}{4}\min\left\{\frac{tF}{p}, F+\sqrt{t}\right\}$$

completion time. To this aim, we consider two cases.

- **Case** $\frac{tF}{p} \le F+\sqrt{t}$. For the sake of contradiction, suppose that $m \le \frac{tF}{4p}$ and consider two cases
  - $(F \le \sqrt{t})$. In this case, since $m \le \frac{tF}{4p} \le \frac{F}{4}+\frac{\sqrt{t}}{4} \le \frac{\sqrt{t}}{2}$, we have that $m^2 < \frac{t}{4}$ and

$$\frac{mp}{t-2m^2} \le \frac{mp}{t-\frac{t}{2}} = \frac{2mp}{t} \le \frac{2p}{t}\frac{tF}{4p} = \frac{F}{2} < F.$$

    This contradicts Lemma 3.
  - $(F > \sqrt{t})$. It holds that

$$\frac{mp}{t}+m \le \frac{F}{4}+\frac{tF}{4p} \le \frac{F}{4}+\frac{F}{4}+\frac{\sqrt{t}}{4} \le \frac{3F}{4} < F.$$

    This contradicts Lemma 2.
- **Case** $\frac{tf}{p} > f+\sqrt{t}$. For the sake of contradiction, suppose that $m \le \frac{F+\sqrt{t}}{4}$ and consider two cases
  - $(F \le \sqrt{t})$. Since $m \le \frac{F+\sqrt{t}}{4} \le \frac{\sqrt{t}}{2}$, we have that $m^2 < \frac{t}{4}$ and

$$\frac{mp}{t-2m^2} \le \frac{mp}{t-\frac{t}{2}} = \frac{2mp}{t} < \frac{2p}{t}\frac{tF}{4p} = \frac{F}{2} < F.$$

    This contradicts Lemma 3.
  - $(F > \sqrt{t})$. It holds that

$$\frac{mp}{t}+m < \frac{tF}{4p}\frac{p}{t}+\frac{F}{4}+\frac{\sqrt{t}}{4} \le \frac{3F}{4} < F.$$

    This contradicts Lemma 2.

$\square$

2.2 The Work Complexity

Lemma 1 is used to derive a lower bound on the work required by any F-reliable protocol.

Theorem 5. *Any F-reliable DO-ALL protocol performs in the worst case at least $\max\{t, \frac{F}{6}\min\{t, F\}\}$ work. The lower bound holds even when all the faults happen at the very beginning of the protocol execution.*

Proof. Let $\mathcal{A}$ be a F-reliable DO-ALL protocol. Part t of the bound follows from the fact that each task has to be performed at least once. Now we show that there exists an execution of $\mathcal{A}$ performing at least $F/6\min\{t, F\}$ work. From Lemma 1 with $Q = P = \emptyset$ there exists a set A of at most $F/3$ faults such that in $E_{F/3}^{\mathcal{A}}(A)$ no transmission succeeds. Let w be the work performed in $E_{F/3}^{\mathcal{A}}(A)$. If $w \geq Ft/6$, then the theorem follows. Hence we assume that $w < Ft/6$. In this case, there must be a task α which is executed in $E_{F/3}^{\mathcal{A}}(A)$ by less than $F/6$ processors. Let us denote the set of processors that execute α by $P(\alpha)$. From Lemma 1 with $Q = A$ and $P = P(\alpha)$ there exists a set $\overline{P(\alpha)}$ with $|A \cup \overline{P(\alpha)}| \leq F/3$, such that in $E_{F/3}^{\mathcal{A}}(A \cup P(\alpha) \cup \overline{P(\alpha)})$ no transmission succeeds.
Let P^{op} be the set of processors that are operational in all the steps of

$$E_{F/3}^{\mathcal{A}}(A \cup P(\alpha) \cup \overline{P(\alpha)})$$

From Lemma 1 with $Q = A \cup P(\alpha) \cup \overline{P(\alpha)}$ and $P = P^{op}$ there exists an execution of $\mathcal{A}$ where less than $F/2 + |P^{op}|$ faults occur and the task α is not executed. It turns out that $|P^{op}| > F/2$. Hence in $E_{F/3}^{\mathcal{A}}(A \cup P(\alpha) \cup \overline{P(\alpha)})$ less than $F/2$ faults occur and more than $F/2$ processors are operational for $F/3$ steps. This implies that the work of this execution is at least $F/2 \cdot F/3 = F^2/6$ and the theorem follows.   □

It is interesting to observe that our proof techniques allows to determine also reasonable values for the multiplicative constant hidden by the asymptotical notations.

3 The Upper Bounds

3.1 Completion Time

Theorem 6. *An F-reliable DO-ALL protocol exists that has*

$$O\left(\frac{t}{p-F} + \min\left\{\frac{tF}{p}, F + \sqrt{t}\right\}\right) \quad \text{completion time.}$$

Proof. (sketch) The proof consists in providing two protocols. Protocol A has $O(\frac{t}{p-F} + F + \sqrt{t})$ completion time while Protocol B has $O\left(\frac{tF}{p}\right)$ completion time. The latter is very simple: it just makes $F+1$ copies of each task and, then, it evenly distribute them among the p processors. The distribution is made by avoiding that 2 or more copies of the same task are assigned to the same processor. Every processor executes all the assigned tasks and stops. It is clear that this is a F-reliable DO-ALL protocol having $O(\frac{tF}{p})$ completion time.

Protocol A is executed whenever $F + \sqrt{t} \leq \frac{tF}{p}$, while B is used in the other case. Informally speaking, in Protocol A, a task is considered executed only when a processor p_i has executed it and, afterwards, p_i succeeds in broadcasting. In this case, the task is said *confirmed*. Every processor keeps a list $TASKS$ that contains the tasks which are not yet confirmed. A processor is considered *dead* only when it is scheduled for a broadcast and it fails. Each processor keeps a further list $PROCS$ that contains the indices of the processors which are not dead. The protocol is structured in *phases*. At the beginning of each phase, the tasks in $TASKS$ are evenly distributed among a subset of P processors in $PROCS$. Then, the processors perform the assigned tasks (this is the *task execution* part of the phase). Finally, a *communication part* follows in which the protocol performs a Round Robin of broadcast operations by the P selected processors. This allows the updating of the lists $TASKS$ and $PROCS$ for the next phase.

Protocol A

    Inizialization:
        $TASKS :=$ the list of the t tasks
        $PROCS :=$ the list of the indices of the p processors
        $T := 0;\ P := 0;\ C := 1$

    if $|TASKS| = 0$ **then halt.**
    if $C > T + P$ **then**   { Initialize a new phase }
      if $\sqrt{|TASKS|} \leq |PROCS|$ **then**
        $T := \left\lceil \sqrt{|TASKS|} \right\rceil;\ P := \left\lceil \frac{|TASKS|}{T} \right\rceil$
      else
        $T := \left\lceil \frac{|TASKS|}{|PROCS|} \right\rceil;\ P := |PROCS|$
      end if
      for $j = 1, 2, \ldots, P$ **do**
        Let $ASSIGN_j$ be the list of the tasks in $TASKS$ of positions
        $(j-1) \cdot T + 1, (j-1) \cdot T + 2, \ldots, (j-1) \cdot T + T$
        Let p_j be the index of the processor in $PROCS$ of position j
      end for
      Let k be the position in the list $PROCS$ of the processors
      $C := 1$
    else
      if $C \leq T$ **then**   { Execution part }
        if $k \leq P$ **then execute the** C**-th task in** $ASSIGN_k$
      else
        if $k = C - T$ **then**   { Communication part }
          broadcast 1 bit
          remove from $TASKS$ the tasks in $ASSIGN_k$

```
        else
            if a broadcast is heard then
                remove from TASKS the tasks in ASSIGN_{C-T}
            else
                remove from PROCS the processor p_{C-T}
            end if
        end if
    end if
    C := C + 1
end if.
```

A key issue here is the choice of the number P of selected processors from the list $PROCS$. Indeed, when the number of tasks, that can be assigned to a processor, is not sufficiently large to balance the cost of the Round Robin communication part of the phase, the number P must be smaller than $|PROCS|$. Roughly speaking, this guarantees a good average ratio between the length of a phase and the number of confirmed tasks in that phase.

Claim 1. Protocol A is an F-reliable DO-ALL protocol having $O(\frac{t}{p-F} + F + \sqrt{t})$ completion time.

Proof. (Claim 1). We distinguish two kinds of *phases* of Protocol A. The first kind is that in which $\sqrt{|TASKS|} \leq |PROCS|$. The second kind is when $\sqrt{|TASKS|} > |PROCS|$. In any case, every phase can be split into two components: the task execution component (whose length is T steps) and the communication one (the Round Robin protocol which has length P). We first determine an upper bound on the number of task execution time steps of 1st-kind phases. To this aim, we bound the number of confirmed tasks during each phase. Let f' be the number of failed broadcast in the communication part of the phase. Then, $p - P + f' \leq F$ and, so, $P - f' \geq p - F$. This implies that the number of processors that succeed in broadcasting is at least $p - F$. Since each of such processors confirm T different tasks, at least $T(p - F)$ different tasks will be confirmed. It thus follows that, in every time-step of the task execution part of 1st-kind phases, at least $(p-F)$ different tasks will be confirmed in average. Finally, the overall number of time steps of this part of such phases is $O(\frac{t}{p-F})$.

Notice that, in any phase of the first kind, $T \geq P$. This immediately implies that the overall number of time steps of 1st-kind phases is $O(\frac{t}{p-F})$.

As for phases of second kind, we first evaluate the overall number of communication steps. Failed communication steps can be at most F; so we consider only the "successful" steps. Let t_i be the number of tasks not yet confirmed immediately after the execution of the i-th successful step, starting from the beginning of the execution of Protocol A. Let c_i be the number of tasks which are confirmed during the i-th successful step. By definition, it holds that $t_i \leq t_{i-1} - c_i$ and that $t_1 \leq t$. Observe that $c_i \geq \sqrt{t_{i-1}}$, so we obtain the following inequality

$$t_i \leq t_{i-1} - \sqrt{t_{i-1}}.$$

From the above inequality, it is easy to verify that t_i reaches 0 in $O(\sqrt{t})$ time steps. It thus follows that the overall number of communication steps of 2nd kind phases is $O(F+\sqrt{t})$. Since in phases of the 2nd kind, $P \geq T/2$, the above bound holds for all steps of 2nd-kind phases.

Finally, by combining the above bounds we get the claim.   □

□

3.2   Work Complexity

Theorem 7. *An F-reliable* DO-ALL *protocol exists that has*

$$O\left(t + F \min\{t, F\}\right))\quad work\ complexity.$$

Proof. (sketch) The protocol is a combination of the CKL Protocol having $O(t+p\sqrt{t}+p\min\{F,t\})$ work complexity and the Protocol B in the proof of Theorem 6. The combination depends on the mutual relations among the input parameters. The following scheme provides the correct choice of the protocol for any possible range of the input parameters.

- $(t \geq p^2)$ The CKL protocol performs, in this case, $O(t)$ work.
- $(t < p^2)$
  1. $(\sqrt{t} \geq F)$ it follows that $F \leq \sqrt{t} \leq p$. So $F + \sqrt{t} \leq 2p$. Applying CKL Protocol by using *only* the first $(F + \sqrt{t})/2$ processors, the performed work is $O((F + \sqrt{t})^2) = O(t)$.
  2. $(\sqrt{t} < F)$
     a) $(F < t)$
        i. $(F + \sqrt{t} \leq p)$ Applying CKL Protocol by using *only* the first $(F + \sqrt{t})$ processors, the performed work is $O((F + \sqrt{t})^2) = O(F^2)$.
        ii. $(F + \sqrt{t} > p)$ The CKL protocol has work $O(t + pF) = O(p^2)$. Since $F + \sqrt{t} > p$ and $\sqrt{t} < F$, this means that $F > p/2$. So the work is $O(F^2)$.
     b) $(F \geq t)$ We use our protocol B in the proof of Theorem 6 which performs $O(Ft)$. Notice that in this case, our lower bound is $\Omega(t + Ft) = \Omega(Ft)$.

□

References

1. B.S. Chlebus, D.R. Kowalski (1999). Randomization helps to perform tasks on processors prone to failures. *13th DISC'99*, LNCS, 1693.
2. B.S. Chlebus, D.R. Kowalski, and A. Lingas (2001). The Do-All problem in Broadcast networks, *ACM-PODC'01*.
3. B.S. Chlebus, R. De Prisco, and A.A. Shvartsman (2001). Performing tasks on synchronous restartable message-passing processors. *Distributed Computing*, 14, 49–64.
4. C. Dwork, J. Halpern, and O. Waarts (1998). Performing work efficiently in the presence of faults, *SIAM J. on Computing*, 27, 1457-1491.
5. R. De Prisco, A. Mayer, and M. Yung (1994). Time-optimal message-efficient work performance in the presence of faults, *ACM PODC'94*, 161–172.
6. Z. Galil, A. Mayer, and M. Yung (1995). Resolving message complexity of byzantine agreement and beyond, *IEEE FOCS'95*, 724–733.
7. D. Gelernter and D. Kaminsky (1992). Supercomputing out of recycled garbage: preliminary experience with Piranha. *ACM Int. Conf. on Supercomputing*, 417–427.
8. J.D. Gibson (Ed.) (1996). The Mobile Communications Handbook, CRC Press.
9. P.C. Kanellakis, and A.A. Shavartsman (1992). Efficient Parallel Algorithms Can Be Made Robust, *Distributed Computing*,5, 201-217.
10. P.C. Kanellakis, and A.A. Shavartsman (1997) Fault-Tolerant Parallel Computation, *ISBN 0-7923-9922-6*, Klunder Academic Publishers.
11. J.F. Shoch and J.A. Hupp (1982). The Worm programs – early experience with a distributed computation, *Comm. Assoc. Comput. Mach.*, 25, 95–103.

New Results for Energy-Efficient Broadcasting in Wireless Networks[*]

Ioannis Caragiannis[1], Christos Kaklamanis[1], and Panagiotis Kanellopoulos[1]

Computer Technology Institute and
Dept. of Computer Engineering and Informatics
University of Patras, 26500 Rio, Greece
{caragian, kakl, kanellop}@ceid.upatras.gr

Abstract. Motivated by the problem of supporting energy–efficient broadcasting in ad hoc wireless networks, we study the *Minimum Energy Consumption Broadcast Subgraph* (MECBS) problem. We present the first logarithmic approximation algorithm for the problem which uses an interesting reduction to Node–Weighted Connected Dominating Set. We also show that an important special instance of the problem can be solved in polynomial time, solving an open problem of Clementi et al. [2].

1 Introduction

Wireless networks have received significant attention during the recent years. Especially, *ad hoc wireless networks* emerged due to their potential applications in battlefield, emergency disaster relief, etc. [8]. Unlike traditional wired networks or cellular wireless networks, no wired backbone infrastructure is installed for ad hoc wireless networks.

A node (or station) in these networks is equipped with an omnidirectional antenna which is responsible for sending and receiving signals. Communication in these networks is established by assigning to each station a transmitting power. In the most common power attenuation model [8], the signal power falls as $1/r^\alpha$, where r is the distance from the transmitter and α is a constant which depends on the wireless environment (typical values of α are between 1 and 6). So, a transmitter can send a signal to a receiver if

$$\frac{P_s}{d(s,t)^\alpha} \geq \gamma$$

where P_s is the power of the transmitting signal, $d(s,t)$ is the Euclidean distance between the transmitter and the receiver, and γ is the receiver's power threshold for signal detection which is usually normalized to 1.

So, communication from a node s to another node t may be established either directly if the two nodes are close enough and s uses adequate transmitting

[*] This work was partially supported by the European Union under IST FET Project ALCOM–FT, IST FET Project CRESCCO, and RTN Project ARACNE.

P. Bose and P. Morin (Eds.): ISAAC 2002, LNCS 2518, pp. 332–343, 2002.

power, or by using intermediate nodes. Observe that due to the nonlinear power attenuation, relaying the signal between nodes may result in energy conservation.

A crucial issue in ad hoc wireless networks is to support communication patterns that are typical in traditional networks. These include broadcasting, multicasting, and gossiping (all–to–all communication). The important engineering question to be solved is to guarantee a desired communication pattern minimizing the total energy consumption. In this work, motivated by the problem of supporting energy–efficient broadcasting in ad hoc wireless networks, we study the *Minimum Energy Consumption Broadcast Subgraph* (MECBS) problem.

Consider a complete directed graph $G = (V, E)$ with a symmetric cost function $c : E \to R$ associated with its edges ($c(u, v) = c(v, u)$) and a special node $r \in V$. Given a weight assignment $w : V \to R$ to the nodes of G, the *transmission graph* G_w is the directed graph defined as follows. It has the same set of nodes as G and a directed edge (u, v) belongs to G_w if the weight assigned to node u is at least the cost of the edge (u, v), i.e., $w(u) \geq c(u, v)$. The Minimum Energy Consumption Broadcast Subgraph (MECBS) problem is to assign weights to the nodes of V so that for any node $u \in V - \{r\}$, the transmission graph G_w has a directed path from r to u. The objective is to minimize the sum of weights.

Note that by changing the connectivity requirements for the transmission graph, several interesting combinatorial problems arise. Problems of this kind are examined in [1,4,6]. In these works, the objective is to establish a strongly connected transmission graph (with or without restrictions to its diameter).

MECBS has been proved to be inapproximable within $(1 - \epsilon) \ln n$ unless $\mathcal{NP} \subseteq DTIME\left(n^{O(\log \log n)}\right)$ (where n denotes the number of nodes in the graph) using approximation preserving reductions to SET COVER [2] and CONNECTED DOMINATING SET [7]. However, to our knowledge, no logarithmic approximation algorithms for the problem were known prior to this work.

An important special case of MECBS (which reflects the properties of the engineering problem in practice) is when the nodes of the graph are points in the d–dimensional Euclidean space and the cost function is defined as a α-th power of the distance function. Following the notation of [2], we denote this special case of the problem with MECBS[N_d^α]. Clementi et al. [2] have proved that MECBS[N_d^α] is $\mathcal{NP}$–hard for $d \geq 2$ and $\alpha > 1$. Note that, for $\alpha \leq 1$ and $d \geq 1$, MECBS[N_d^α] has a trivial optimal solution (the transmission graph is a star rooted at r). The complexity of MECBS[N_1^α] for $\alpha > 1$ was left open.

Wieselthier et al. [8] study the behaviour of several greedy algorithms for MECBS[N_2^2]. They present experimental results for three simple algorithms: MST (Minimum Spanning Tree), BIP (Broadcast Incremental Power), and SPT (Shortest Path Tree). By exploring geometric properties of Euclidean spanning trees, Clementi et al. [2] and Wan et al. [7] prove upper bounds on the approximation ratio of algorithm MST on instances of MECBS[N_2^α] for $\alpha \geq 2$. In [2], an upper bound of $\left(2\sqrt{10}\right)^\alpha$ is proved, while an improved bound of 12 is proved in [7]. Wan et al. [7] also prove lower bounds on the performance of algorithms MST, BIP, SPT, and BAIP (Broadcast Average Incremental Power), a greedy algorithm based on the well–known greedy algorithm for SET COVER. They show that even when instances of MECBS[N_2^2] are considered, SPT and BAIP have approximation ratios $\Omega(n)$ and $\Omega(n/\log n)$, respectively. However, it is easy to verify that all the algorithms studied in [2,7,8] have approximation ratios $\Omega(n)$ when applied to general instances of MECBS. Furthermore, none of these algorithms can guarantee optimal solutions to MECBS[N_1^α] for $\alpha > 1$.

In this work, we present (in Section 2) the first logarithmic approximation algorithm for the problem which uses an interesting reduction to the *Node–Weighted Connected Dominating Set* (NWCDS) problem. The Node–Weighted Connected Dominating Set problem is the following: Given a graph $G = (V, E)$ with weights on the nodes, find the smallest weighted subset S of nodes that induce a connected subgraph and each node in $V - S$ is adjacent to at least one node in S. We also show (in Section 3) that MECBS[N_1^α] for $\alpha > 1$ can be solved in polynomial time, solving an open problem of Clementi et al. [2]. Very recently, Andrea Clementi informed us that a similar result was obtained independently in [3].

2   A Logarithmic Approximation Algorithm

In this section, we present a logarithmic approximation algorithm for MECBS. The algorithm uses a new reduction of instances of MECBS to instances of NWCDS.

First, we show how to transform any instance of MECBS to an instance of NWCDS. Let I_{MECBS} be an instance of MECBS which consists of a complete directed graph $G = (V, E)$ with $|V| = n$, a symmetric cost function $c : E \to R$, on the edges of G, and a special node $r \in V$. We will construct an instance I_{NWCDS} of the NWCDS which consists of a node–weighted undirected graph H.

We construct the graph H as follows. For each node v of G, the graph H has a set Z_v of n nodes $Z_{v,1}, Z_{v,2}, ..., Z_{v,n}$ which we call a *supernode*. The supernodes of H corresponding to different nodes of G are disjoint. Let $c_{v,i}$ be the i–th largest cost among the costs of the edges directed out of v in G. The weight associated with the node $Z_{v,i}$, $i = 1, ..., n-1$ of the supernode Z_v is set to $c_{v,i}$. The node $Z_{v,n}$ of the supernode Z_v has infinite weight.

The set of edges in H is defined as follows. Nodes $Z_{v,1}, ..., Z_{v,n-1}$ of the supernode Z_v form the complete graph K_{n-1} while node $Z_{v,n}$ is isolated from the other nodes of its supernode. Node $Z_{v,i}$, for $i = 1, ..., n-1$, is connected to all the nodes of those supernodes Z_u ($v \neq u$) corresponding to nodes u of G which are connected to v by edges of cost no more than the weight of $Z_{v,i}$ (i.e., $c(v,u) \leq c_{v,i}$). The reduction is depicted in Figure 1.

Now, we will show that an approximate solution to NWCDS for instance I_{NWCDS} can be used to obtain an approximate solution to the MECBS for instance I_{MECBS} with similar approximation guarantee. We first show that the cost $OPT(I_{\text{MECBS}})$ of an optimal solution of I_{MECBS} is close to the cost $OPT(I_{\text{NWCDS}})$ of the optimal solution of I_{NWCDS}.

Lemma 1. $OPT(I_{NWCDS}) \leq 2 \cdot OPT(I_{MECBS})$.

Proof. Consider a solution to MECBS for instance I_{MECBS} of minimum cost. Let w be the weight vector corresponding to this solution and G_w the transmission

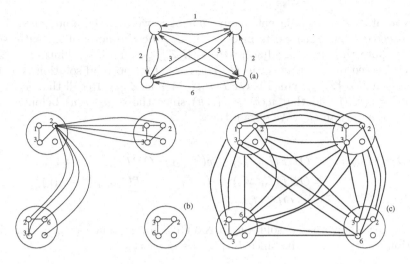

Fig. 1. The reduction to NWCDS. (a) The graph G of an instance of MECBS. (b) The graph H of the corresponding instance of NWCDS. Each large cycle indicates a supernode. Only the edges incident to the node of weight 2 of the upper left supernode are shown. These edges are those which correspond to edges in (a) of cost at most 2, directed out of the left upper node. (c) The graph H of the corresponding instance of NWCDS. Thick edges from a node to a supernode represent edges connecting a node to all the nodes of a supernode.

graph defined by w. Since the solution is optimal, the weight of each node v equals the cost of some edge directed out of v.

We construct a connected dominating set S for instance I_{NWCDS} as follows. For each node v of G_w which has at least one outgoing edge, we add to S the node $Z_{v,i(v)}$ of the supernode Z_v which has weight $w(Z_{v,i(v)}) = w(v) = \max_{(v,u)\in G_w}\{c(v,u)\}$.

Consider the induced subgraph G'_w of G_w which contains the nodes of G_w having at least one outgoing edge in G_w. By construction, each one of the nodes in S belongs to a different supernode of H mapping to a node in G'_w. Also, a directed edge between two nodes u and v in G'_w exists if and only if $w(u) \geq c(u,v)$. This means that the weight of the node $Z_{u,i(u)}$ is at least $w(u)$, and, thus, it is connected with node $Z_{v,i(v)}$ in H. This holds for any directed edge of G'_w and implies that the subgraph of H induced by S is isomorphic to G'_w (if we consider directed edges as undirected ones) and, thus, it is connected.

Also, S dominates all nodes of H except (possibly) for $Z_{r,n}$. The existence of a directed edge (u,v) in G_w implies that the node $Z_{u,i(u)}$ dominates all nodes of Z_v in H. Furthermore, all nodes in Z_r but $Z_{r,n}$ are dominated by $Z_{r,i(r)}$.

If $Z_{r,n}$ is adjacent to a node in S, then S is clearly a connected dominating set in H. The cost of S is

$$\sum_{Z_{v,i(v)}\in S} w(Z_{v,i(v)}) = \sum_{v\in G_w} w(v) = OPT(I_{\text{MECBS}}).$$

Now, if $Z_{r,n}$ is not adjacent to any node of S, we consider a node u such that the directed edge (r, u) exists in G_w. Let $Z_{u,j}$ be the node of Z_u with weight $c(u, r)$. Note that $Z_{u,j}$ is adjacent to both $Z_{u,i(u)}$ and $Z_{r,n}$. Thus, $S \cup \{Z_{u,j}\}$ is a connected dominating set in H. The cost of an optimal solution is at most the cost of $S \cup \{Z_{u,j}\}$ which is $OPT(I_{\text{MECBS}}) + w(Z_{u,j})$. Recall that $w(Z_{u,j}) = c(u, r) = c(r, u)$ and that $w(r) \geq c(r, u)$ since the edge (r, u) belongs to the transmission graph G_w. We obtain that

$$\begin{aligned} OPT(I_{\text{NWCDS}}) &\leq OPT(I_{\text{MECBS}}) + w(Z_{u,j}) = OPT(I_{\text{MECBS}}) + c(u, r) \\ &= OPT(I_{\text{MECBS}}) + c(r, u) \leq OPT(I_{\text{MECBS}}) + w(r) \\ &\leq 2 \cdot OPT(I_{\text{MECBS}}) \end{aligned}$$
□

We now show that a solution to NWCDS for instance I_{NWCDS} can give a solution to MECBS for instance I_{MECBS} of similar cost.

Lemma 2. *Given a solution to NWCDS for instance I_{NWCDS} of cost $COST$ (I_{NWCDS}), we can construct in polynomial time a solution to MECBS for instance I_{MECBS} of cost $COST(I_{\text{MECBS}})$, such that*

$$COST(I_{\text{MECBS}}) \leq 2 \cdot COST(I_{\text{NWCDS}})$$

Proof. Consider a connected dominating set S of H. We may assume that S contains at most one node from each supernode. If this is not the case and there are two nodes $Z_{u,i}$ and $Z_{u,j}$ ($i \neq j$) in S with weights $w(Z_{u,i}) \geq w(Z_{u,j})$, we remove $Z_{u,j}$ from S. By repeatedly executing this procedure, we end up with a connected dominating set having at most one node per supernode. Furthermore, in any solution of finite cost, no node with infinite weight is contained in S. For every supernode Z_u which has a node $Z_{u,i(u)}$ contained in S, we associate a weight $w(Z_{u,i(u)})$ to node u of G. Nodes of G whose corresponding supernodes of H have no node contained in S are assigned zero weight.

Now, consider the transmission graph G_w defined by w. We will first show that the graph G_w is loosely connected, i.e., its undirected counterpart is connected. Consider an edge $(Z_{u,i(u)}, Z_{v,i(v)})$ in H which connects two nodes of S. By the construction of graph H from G and the definition of the transmission graph G_w, we obtain that either (u, v) or (v, u) exists in G_w. This holds for any two nodes of S which are connected with an edge in H and, since the nodes of S are connected, this implies that the subgraph of G_w induced by the nodes corresponding to supernodes of H having a node in S is loosely connected. Also, observe that the nodes of any supernode Z_v containing no node in S are dominated by some node $Z_{u,i(u)}$ in S. By the construction of H from G and the definition of the transmission graph G_w, we obtain that for any node v of G_w with zero weight, there is a directed edge coming from a node with non–zero weight. We conclude that G_w is loosely connected.

We execute a spanning tree algorithm on G_w to compute a loosely connected spanning tree T_w of G_w. Next, we execute the following procedure that transforms T_w to an arborescence T'_w, i.e., a tree T'_w having a directed path from r

to any other node. Starting from r, we compute a Breadth-First-Search (BFS) numbering of the nodes of T_w. We visit the nodes of T_w according to this BFS numbering and, at the step associated with a node u, we transform ingoing edges coming to u from nodes with numbers greater than u's to outgoing ones.

Now, by assigning weights $w'(u) = \max_{(u,v) \in T'_w} c(u,v)$ to the nodes of G, we obtain a transmission graph which contains T'_w as a subgraph and, thus, is a feasible solution to MECBS.

Let l_u be the node such that the edge (u, l_u) belongs to T'_w and has the maximum cost among all edges (u, v) directed out of u in T'_w. The transmission graph G_w contains either (u, l_u) or (l_u, u). If it contains (u, l_u), then $w(u) \geq c(u, l_u)$, otherwise $w(l_u) \geq c(l_u, u) = c(u, l_u)$. In any case,

$$w'(u) = \max_{(u,v) \in T'_w} c(u, l_u) \leq \max\{w(u), w(l_u)\} \leq w(u) + w(l_u).$$

The cost of the solution w' to MECBS is

$$COST(I_{\text{MECBS}}) = \sum_{u \in V} w'(u) \leq \sum_{u \in V} (w(u) + w(l_u)) = \sum_{u \in V} w(u) + \sum_{u \in V} w(l_u)$$

$$\leq 2 \cdot \sum_{u \in V} w(u) = 2 \cdot COST(I_{\text{NWCDS}}) \qquad \square$$

Given an instance I_{MECBS} of MECBS, we first transform it to the corresponding instance I_{NWCDS} as described above. Then, we run an algorithm for the Node–Weighted Connected Dominating Set problem and use the technique described in the proof of Lemma 2 to construct a solution to the original problem. Using Lemmas 1 and 2, we can show that, given a ρ–approximate solution to NWCDS for instance I_{NWCDS} of cost $COST(I_{\text{NWCDS}})$, we can obtain a solution to MECBS for instance I_{MECBS} with cost $COST(I_{\text{MECBS}})$ which is within 4ρ of optimal. Indeed, by Lemma 1, we have $OPT(I_{\text{MECBS}}) \geq OPT(I_{\text{NWCDS}})/2$, while Lemma 2 yields $COST(I_{\text{MECBS}}) \leq 2 \cdot COST(I_{\text{NWCDS}}) \leq 2\rho \cdot OPT(I_{\text{NWCDS}})$. Thus, the approximation ratio we obtain is

$$\frac{COST(I_{\text{MECBS}})}{OPT(I_{\text{MECBS}})} \leq 4\rho.$$

In [5], Guha and Khuller present a $1.35 \ln n$–approximation algorithm for the Node–Weighted Connected Dominating Set problem, where n is the number of nodes in the graph. Given an instance I_{MECBS} of MECBS with n nodes, the corresponding instance I_{NWCDS} has n^2 nodes. Thus, the cost of the solution of I_{NWCDS} is within $4 \cdot 1.35 \ln(n^2) = 10.8 \ln n$ of the optimal solution. The next theorem summarizes the discussion of this section.

Theorem 1. *There exists a $10.8 \ln n$–approximation algorithm for MECBS.*

3   A Polynomial Time Algorithm for MECBS[N_1^α]

Assume that we have n points $x_{u_1}, ..., x_r, ..., x_{u_n}$ located on a line. An instance I of MECBS[N_1^α], for some $\alpha > 1$, consists of a complete directed graph G in which each node u corresponds to a point x_u and the cost of an edge (u, v) in G is defined as

$$c(u, v) = [d(x_u, x_v)]^\alpha,$$

where $d(x_u, x_v)$ is the Euclidean distance of the points x_u and x_v.

We will show that we can compute an optimal solution for the instance I of MECBS[$\mathtt{N}_1^\alpha$] in polynomial time. We start with a few definitions.

We partition the nodes of G into two sets:

- the *left set* $\mathcal{L}$ which contains those nodes of G corresponding to a point at the left of node x_r .
- the *right set* $\mathcal{R}$ which contains those nodes of G corresponding to a point at the right of node x_r.

If one of the sets $\mathcal{L}$ and $\mathcal{R}$ is empty (this means that all the nodes correspond to points located at the right or the left of x_r), then there is a trivial optimal solution. For example, if $r = u_1$ and $\mathcal{R} = \{u_2, ..., u_n\}$, it is easy to verify that $w(u_i) = c(u_i, u_{i+1})$, for $i = 1, ..., n-1$, and $w(u_n) = 0$ is an optimal solution. So, in the following we assume that both $\mathcal{L}$ and $\mathcal{R}$ are non–empty.

Consider an arborescence T of G rooted at r. A node u of T different than r is called *root–crossing* if it belongs to $\mathcal{L}$ and has a child in $\mathcal{R}$ or if it belongs in $\mathcal{R}$ and has a child belonging in $\mathcal{L}$. The node r is called root–crossing if it contains children in both $\mathcal{L}$ and $\mathcal{R}$. Note that, if both $\mathcal{L}$ and $\mathcal{R}$ are non–empty, then any arborescence of G rooted at r contains at least one root–crossing node.

A *chain* from u to v is a directed path from u to v containing only edges between consecutive nodes v' which correspond to point $x_{v'}$ located between the points x_u and x_v on the line. We also say that a node u is a trivial chain from u to u.

An arborescence of G rooted at r is called *single root–crossing* if

- it has exactly one root–crossing node u,
- it contains a chain from r to u, a chain from the node u_L, which is the node among u and its children corresponding to the leftmost point, to u_1, and a chain from the node u_R, which is the node among u and its children corresponding to the rightmost point, to u_n, and
- u is the parent of all nodes corresponding to points between x_{u_L} and x_{u_R} except for nodes in the chain from r to u.

Examples of arborescences are shown in Figure 2.

Let T be an arborescence of G rooted at r. By setting $w(u) = \max_{(u,v)\in T} c(u, v)$, the transmission graph G_w contains T as a subgraph and, thus, it is a feasible solution to the problem with cost $\sum_{u\in V} w(u)$. We will show that given an arborescence T which is not single root–crossing, there exists a feasible solution w' of cost $\sum_{u\in V} w'(u) \leq \sum_{u\in V} w(u)$ such that the transmission graph $G_{w'}$ contains a single root–crossing arborescence T'' as a subgraph. We prove it by constructing from T, a single root–crossing arborescence T'' with

$$\sum_{u\in V} \max_{(u,v)\in T''} c(u, v) \leq \sum_{u\in V} \max_{(u,v)\in T} c(u, v)$$

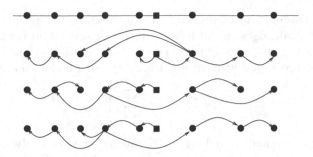

Fig. 2. Points on the line and three arborescences rooted at the node represented by the rectangle. The first arborescence is not single root–crossing because it has two root–crossing nodes. The second and third arborescences have one root–crossing node but, according to the definition, only the third one is single root–crossing.

and setting $w'(u) = \max_{(u,v)\in T''} c(u,v)$. This is proved in Lemmas 3 and 4. In this way, we will obtain that there exists an optimal solution $\hat{w}$ to instance I which defines a transmission graph $G_{\hat{w}}$ containing a single root–crossing arborescence as a subgraph.

Lemma 3. *Let T be an arborescence of G rooted at r containing $k \geq 2$ root–crossing nodes in T. There exists an arborescence T' of G rooted at r with $k-1$ root–crossing nodes such that*

$$\sum_{u\in T'} \max_{(u,v)\in T'} c(u,v) \leq \sum_{u\in T} \max_{(u,v)\in T} c(u,v).$$

Proof. Consider an arborescence T of G rooted at r with $k \geq 2$ root–crossing nodes. Let u_1 and u_2 be two root–crossing nodes such that u_1 is at the same or higher level than u_2 in T. Again, we denote by x_v the point on the line which corresponds to the node v. We call *left* (resp. *right*) children of a node v the children of v in T which belong to set $\mathcal{L}$ (resp. $\mathcal{R}$). Also, we denote by v_L and v_R the node between v and its children which correspond to the leftmost and rightmost point on the line, respectively.

We distinguish between the following four cases

Case 1. If $u_1 = r$, then assume without loss of generality that $u_2 \in \mathcal{R}$. We claim that either all left children of u_2 in T are within distance $\left[\max_{(u_1,v)\in T} c(u_1,v)\right]^{1/\alpha}$ from u_1 or all right children of u_1 in T are within distance $\left[\max_{(u_2,v)\in T} c(u_2,v)\right]^{1/\alpha}$ from u_2. Assume otherwise, i.e., $d(x_{u_1}, x_{u_1^R}) > \left[\max_{(u_2,v)\in T} c(u_2,v)\right]^{1/\alpha}$ and $d(x_{u_2}, x_{u_2^L}) > \left[\max_{(u_1,v)\in T} c(u_1,v)\right]^{1/\alpha}$. Then,

$$d(x_{u_1}, x_{u_1^R}) > \left[\max_{(u_2,v)\in T} c(u_2,v)\right]^{1/\alpha} \geq d(x_{u_2}, x_{u_2^L}) > \left[\max_{(u_1,v)\in T} c(u_1,v)\right]^{1/\alpha},$$

a contradiction since $(u_1, u_1^R) \in T$. So, we can construct an arborescence T' either by removing all edges (u_2, v) from u_2 to its left children (in this way, u_2

340   I. Caragiannis, C. Kaklamanis, and P. Kanellopoulos

is not root–crossing) and adding edges (u_1, v) from u_1 to the left children of u_2 or by removing all edges (u_1, v) from u_1 to its right children (in this way, u_1 is not root–crossing) and adding edges (u_2, v) from u_2 to the right children of u_1. The arborescence T' has at most $k - 1$ root–crossing nodes and, clearly,

$$\sum_{u \in T'} \max_{(u,v) \in T'} c(u,v) \leq \sum_{u \in T} \max_{(u,v) \in T} c(u,v).$$

Case 2. If $u_1 \neq r$ and u_2 belongs to the same set with u_1 (wlog, we assume that $u_1, u_2 \in \mathcal{L}$) then either all right children of u_2 in T are within distance $\left[\max_{(u_1,v) \in T} c(u_1,v)\right]^{1/\alpha}$ from u_1 or all right children of u_1 in T are within distance $\left[\max_{(u_2,v) \in T} c(u_2,v)\right]^{1/\alpha}$ from u_2. Assume otherwise, i.e., $d(x_{u_1}, x_{u_1^R}) > \left[\max_{(u_2,v) \in T} c(u_2,v)\right]^{1/\alpha}$ and $d(x_{u_2}, x_{u_2^R}) > \left[\max_{(u_1,v) \in T} c(u_1,v)\right]^{1/\alpha}$. Then,

$$d(x_{u_1}, x_{u_1^R}) > \left[\max_{(u_2,v) \in T} c(u_2,v)\right]^{1/\alpha} \geq d(x_{u_2}, x_{u_2^R}) > \left[\max_{(u_1,v) \in T} c(u_1,v)\right]^{1/\alpha},$$

a contradiction since $(u_1, u_1^R) \in T$. So, we can construct an arborescence T' by removing all edges (u_1, v) from u_1 to its right children (in this way, u_1 is not root–crossing) and adding edges (u_2, v) from u_2 to the right children of u_1 or by removing all edges (u_2, v) from u_2 to its right children (in this way, u_2 is not root–crossing) and adding edges (u_1, v) from u_1 to the right children of u_2. Again, the arborescence T' has at most $k - 1$ root–crossing nodes and, clearly,

$$\sum_{u \in T'} \max_{(u,v) \in T'} c(u,v) \leq \sum_{u \in T} \max_{(u,v) \in T} c(u,v).$$

Case 3. If $u_1 \neq r$, and u_1, u_2 belong to different sets and u_2 is not a child of u_1, then assume without loss of generality that $u_1 \in \mathcal{L}$ and $u_2 \in \mathcal{R}$. Again, using the same argument with the previous two cases, we can show that either all left children of u_2 in T are within distance $\left[\max_{(u_1,v) \in T} c(u_1,v)\right]^{1/\alpha}$ from u_1 or all right children of u_1 in T are within distance $\left[\max_{(u_2,v) \in T} c(u_2,v)\right]^{1/\alpha}$ from u_2. So, we can construct an arborescence T' either by removing all edges (u_2, v) from u_2 to its left children (in this way, u_2 is not root–crossing) and adding edges (u_1, v) from u_1 to the left children of u_2 or by removing all edges (u_1, v) from u_1 to its right children (in this way, u_1 is not root–crossing) and adding edges (u_2, v) from u_2 to the right children of u_1. Again, we obtain an arborescence T' with at most $k - 1$ root–crossing nodes, such that

$$\sum_{u \in T'} \max_{(u,v) \in T'} c(u,v) \leq \sum_{u \in T} \max_{(u,v) \in T} c(u,v).$$

Case 4. Now, we consider the case where $u_1 \neq r$, and u_1, u_2 belong to different sets and u_2 is a child of u_1 in T. Without loss of generality, we assume that $u_1 \in \mathcal{L}$ and $u_2 \in \mathcal{R}$. If all left children of u_2 are within distance $\left[\max_{(u_1,v) \in T} c(u_1,v)\right]^{1/\alpha}$

from u_1, then we construct T' by removing the edges (u_2, v) from u_2 to every left child v of u_2 and adding edges from u_1 to v. In T', u_2 is not root–crossing and we obtain an arborescence with $k-1$ root–crossing nodes. Again, we have

$$\sum_{u \in T'} \max_{(u,v) \in T'} c(u,v) \le \sum_{u \in T} \max_{(u,v) \in T} c(u,v).$$

Assume now that not all left children of u_2 are within distance at most $\left[\max_{(u_1,v) \in T} c(u_1,v)\right]^{1/\alpha}$ from u_1, i.e., $d(x_{u_1}, x_{u_2^L}) > \left[\max_{(u_1,v) \in T} c(u_1,v)\right]^{1/\alpha}$. Then, u_2^L is at the left of u_1 and

$$\left[\max_{(u_2,v) \in T} c(u_2,v)\right]^{1/\alpha} \ge d(x_{u_1}, x_{u_2^L}) + d(x_{u_1}, x_{u_2}) > \left[\max_{(u_1,v) \in T} c(u_1,v)\right]^{1/\alpha}.$$

Thus, since either u_1^R is at right of u_2 or $u_1^R = u_2$,

$$d(x_{u_2}, x_{u_1^R}) = d(x_{u_1}, x_{u_1^R}) - d(x_{u_1}, x_{u_2})$$
$$\le \max_{(u_1,v) \in T} c(u_1,v)^{1/\alpha} < \max_{(u_2,v) \in T} c(u_2,v)^{1/\alpha}.$$

Now, we construct the arborescence T' as follows. First, we remove from T the edge between the parent of u_1 and u_1 and all edges (u_1, v) from u_1 to its children (in this way, u_1 is not root–crossing). We add to T' an edge from u_2 to u_1 and edges (u_2, v) from u_2 to the children of u_1 in T.

If r has right children in T, then we complete the construction of T' by adding edge (r, u_2). Note that, in this way, we add a right child to r, so r is root–crossing in T' only if it was root–crossing in T. We obtain an arborescence T' with at most $k-1$ root–crossing nodes such that

$$\sum_{u \in T'} \max_{(u,v) \in T'} c(u,v) \le \max_{(r,v) \in T'} c(r,v) + \max_{(u_1,v) \in T'} c(u_1,v) + \max_{(u_2,v) \in T'} c(u_2,v)$$

$$+ \sum_{u \in T' - \{r,u_1,u_2\}} \max_{(u,v) \in T'} c(u,v)$$

$$\le \left(\max_{(r,v) \in T} c(r,v) + d(x_r, x_{u_2})^\alpha\right) + 0 + \max_{(u_2,v) \in T} c(u_2,v)$$

$$+ \sum_{u \in T - \{r,u_1,u_2\}} \max_{(u,v) \in T} c(u,v)$$

$$\le \max_{(r,v) \in T} c(r,v) + d(x_{u_1}, x_{u_2})^\alpha + \max_{(u_2,v) \in T} c(u_2,v)$$

$$+ \sum_{u \in T - \{r,u_1,u_2\}} \max_{(u,v) \in T} c(u,v)$$

$$\le \max_{(r,v) \in T} c(r,v) + \max_{(u_1,v) \in T} c(u_1,v) + \max_{(u_2,v) \in T} c(u_2,v)$$

$$+ \sum_{u \in T - \{r,u_1,u_2\}} \max_{(u,v) \in T} c(u,v)$$

$$= \sum_{u \in T} \max_{(u,v) \in T} c(u,v).$$

If r has only left children in T, then we first remove all edges from r to its children. Then, we add edges (u_2, v) to connect to u_2 all children of r in T that are within distance $\left[\max_{(u_2,v)\in T} c(u_2, v)\right]^{1/\alpha}$ from u_2 while the rest of the children of r in T (if any) are connected to u_2^L. We complete the construction of T' by adding edge (r, u_2) to connect u_2 to r. Note that, now, r has only one (right) child, and, thus, it is not root–crossing. Furthermore, node u_1 has no children in T' and, thus, it is not root–crossing, while node u_2^L is a root–crossing node in T' only if it was a root–crossing node in T. We conclude that T' has at most $k-1$ root–crossing nodes. We denote by r_l the child of r in T which is not within distance $\left[\max_{(u_2,v)\in T} c(u_2, v)\right]^{1/\alpha}$ from u_2 and corresponds to the leftmost point. If no such point exists, then the terms in the following expression refering to r_l can be removed. We have that

$$
\begin{aligned}
\sum_{u\in T'} \max_{(u,v)\in T'} c(u, v) &= \max_{(r,v)\in T'} c(r, v) + \max_{(u_1,v)\in T'} c(u_1, v) + \max_{(u_2,v)\in T'} c(u_2, v) \\
&\quad + \max_{(u_2^L,v)\in T'} c(u_2^L, v) + \sum_{u\in T'-\{r,u_1,u_2,u_2^L\}} \max_{(u,v)\in T'} c(u, v) \\
&\leq d(x_r, x_{u_2})^\alpha + 0 + \max_{(u_2,v)\in T} c(u_2, v) + d(x_{u_2^L}, x_{r_l})^\alpha \\
&\quad + \max_{(u_2^L,v)\in T} c(u_2^L, v) + \sum_{u\in T-\{r,u_1,u_2,u_2^L\}} \max_{(u,v)\in T} c(u, v) \\
&\leq d(x_{u_1}, x_{u_2})^\alpha + \max_{(u_2,v)\in T} c(u_2, v) + d(x_r, x_{r_l})^\alpha \\
&\quad + \max_{(u_2^L,v)\in T} c(u_2^L, v) + \sum_{u\in T-\{r,u_1,u_2,u_2^L\}} \max_{(u,v)\in T} c(u, v) \\
&\leq \max_{(u_1,v)\in T} c(u_1, v) + \max_{(u_2,v)\in T} c(u_2, v) + \max_{(r,v)\in T} c(r, v) \\
&\quad + \max_{(u_2^L,v)\in T} c(u_2^L, v) + \sum_{u\in T-\{r,u_1,u_2,u_2^L\}} \max_{(u,v)\in T} c(u, v) \\
&\leq \sum_{u\in T} \max_{(u,v)\in T} c(u, v)
\end{aligned}
$$

This completes the proof of the lemma.   □

By repeatedly executing the procedure used in the proof of Lemma 3, we obtain an arborescence T' with exactly one root–crossing node. We can also prove the following.

Lemma 4. *Let T' be an arborescence of G rooted at r containing exactly one single root–crossing node. There exists a single root–crossing arborescence T'' such that*

$$
\sum_{u\in T''} \max_{(u,v)\in T''} c(u, v) \leq \sum_{u\in T'} \max_{(u,v)\in T'} c(u, v).
$$

Note that, for any node v, there are $O(n^2)$ single root–crossing arborescences in which v is the root–crossing node. Thus, there are $O(n^3)$ single root–crossing arborescences of G rooted at r. For each possible single root–crossing arborescence T_i, we set $w_i(u) = \max_{(u,v) \in T_i} c(u,v)$. Clearly, w_i is a feasible solution for instance I since the transmission graph G_{w_i} contains the arborescence T_i as a subgraph. The cost of the solution w_i is $\sum_{u \in T_i} w_i(u)$. We select that solution w_i which minimizes $\sum_{u \in T_i} w_i(u)$. By the discussion in this section, this is an optimal solution for I. Since every step in the procedure described above is performed in polynomial time, we have obtained the following theorem.

Theorem 2. *MECBS*$[\mathsf{N}_1^\alpha]$ *can be solved in polynomial time.*

References

1. G. Călinescu, I. Măndoiu, A. Zelikovsky. Symmetric Connectivity with Minimum Power Consumption in Radio Networks. In *Proc. of the 2nd IFIP International Conference on Theoretical Computer Science*, 2002, to appear.
2. A. E. F. Clementi, P. Crescenzi, P. Penna, G. Rossi, P. Vocca. On the Complexity of Computing Minimum Energy Consumption Broadcast Subgraphs. In *Proc. of the 18th Annual Symposium on Theoretical Aspects of Computer Science (STACS '01)*, LNCS 2010, Springer, pp. 121–131, 2001.
3. A. E. F. Clementi, M. Di Ianni, R. Silvestri. The Minimum Broadcast Range Assignment Problem on Linear Multi-Hop Wireless Networks. *Theoretical Computer Science*, to appear.
4. A. E. F. Clementi, P. Penna, R. Silvestri. Hardness Results for the Power Range Assignment Problem in Packet Radio Networks. In *Proc. of Randomization, Approximation, and Combinatorial Optimization (RANDOM/APPROX '99)*, LNCS 1671, Springer, pp. 197–208, 1999.
5. S. Guha, S. Khuller. Improved Methods for Approximating Node Weighted Steiner Trees and Connected Dominating Sets. *Information and Computation*, Vol. 150 (1), pp. 57–74, 1999.
6. L. M. Kirousis, E. Kranakis, D. Krizanc, A. Pelc. Power Consumption in Packet Radio Networks. *Theoretical Computer Science*, Vol. 243(1-2), pp. 289–305, 2000.
7. P.-J. Wan, G. Călinescu, X.-Y. Li, O. Frieder. Minimum-Energy Broadcast Routing in Static Ad Hoc Wireless Networks. In *Proc. of IEEE INFOCOM 2001*, pp. 1162–1171.
8. J. E. Wieselthier, G. D. Nguyen, A. Ephremides. On the Construction of Energy-Efficient Broadcast and Multicast Trees in Wireless Networks. In *Proc. of IEEE INFOCOM 2000*, pp. 585–594.

An Improved Algorithm for the Minimum Manhattan Network Problem

Ryo Kato, Keiko Imai, and Takao Asano

Department of Information and System Engineering
Chuo University, Bunkyo-ku, Tokyo 112-8551, Japan
{asano,imai}@ise.chuo-u.ac.jp

Abstract. For a set S of n points in the plane, a Manhattan network on S is a geometric network $G(S)$ such that, for each pair of points in S, $G(S)$ contains a rectilinear path between them of length equal to their distance in the L_1-metric. The minimum Manhattan network problem is a problem of finding a Manhattan network of minimum length. Gudmundsson, Levcopoulos, and Narasimhan proposed a 4-approximation algorithm and conjectured that there is a 2-approximation algorithm for this problem. In this paper, based on a different approach, we improve their bound and present a 2-approximation algorithm.

1   Introduction

A path connecting two points in the plane is called a *rectilinear path* if it consists of only horizontal and vertical segments. A rectilinear path connecting two points $p = (p.x, p.y), q = (q.x, q.y)$ is called a *Manhattan path* if it is of length equal to their distance in the L_1-metric (i.e., $|p.x - q.x| + |p.y - q.y|$). For a set S of n points in the plane, a *Manhattan network on S* is defined to be a geometric network $G(S)$ such that, for each pair of points in S, there is a Manhattan path between them. A *minimum Manhattan network on S* is a Manhattan network on S of minimum length (Fig.1) and the minimum Manhattan network problem is to find such a network.

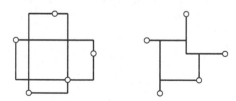

Fig. 1. A Manhattan network (left) and a minimum Manhattan network (right).

Gudmundsson, Levcopoulos, and Narasimhan [5] first introduced this problem in connection with spanners and the Steiner tree problem arising in VLSI

P. Bose and P. Morin (Eds.): ISAAC 2002, LNCS 2518, pp. 344–356, 2002.

applications. The minimum Manhattan network problem can be considered to be the Steiner tree problem in the L_1-metric with an additional constraint that each pair of points in S be connected by a shortest path. For a number $t \geq 1$, a geometric network N is called a t-spanner for S, if for each pair of points p, q in S, there is a path connecting p and q in N of length at most t times the distance between p and q in the L_p-metric. Thus, a Manhattan network $G(S)$ is a 1-spanner for S in the L_1-metric and a minimum Manhattan network is a sparsest 1-spanner. As Gudmundsson, Levcopoulos, and Narasimhan [5] pointed out, spanners have applications in network design, robotics, distributed algorithms and many areas, and have been a subject of considerable research [1,2,3,4,6,9].

The minimum Manhattan network problem, however, is not known whether it is polynomially solvable or NP-hard. By using algorithms for the minimum length partition problem of a rectilinear polygon into rectangles, Gudmundsson, Levcopoulos, and Narasimhan [5] have obtained approximation algorithms for this problem. Actually, they have shown that, if an r-approximation algorithm of $R(n)$ time for the minimum length partition problem is used as a subroutine, a $4r$-approximation algorithm of $O(n \log n + R(n))$ time can be obtained. Since there are an optimal algorithm of $O(n^3)$ time by Lingas et al. [8] and a 2-approximation algorithm of $O(n)$ time by Levcopoulos and Östlin [7] for the minimum length partition problem, their algorithm becomes a 4-approximation algorithm of $O(n^3)$ time and an 8-approximation algorithm of $O(n \log n)$ time. They also conjectured that there is a 2-approximation algorithm for this problem.

In this paper, investigating structural properties of Manhattan networks, we present a 2-approximation algorithm. In Section 2, we present a notion of a sparse generating set Z of pairs of points in S in a sense that $|Z| = O(n)$ and that if we obtain Manhattan paths for the pairs in Z then a Manhattan path for any other pair of points in S is automatically obtained. We also give an algorithm for finding such Z. In Section 3, we present several properties of Z and, in Section 4.1, using these properties, we present an algorithm for constructing a vertical (horizontal) network which uses vertical (horizontal) segments of length at most the length of vertical (horizontal) segments of a minimum Manhattan network on S. The union of these obtained vertical and horizontal networks satisfies nice properties, however, it may contain no Manhattan paths for some pairs of points in S. In Section 4.2, we show that, for such pairs, a set of Manhattan paths of short length can be obtained in polynomial time. Combining the union network with this set of Manhattan paths of short length, we obtain a Manhattan network of length at most twice the length of a minimum Manhattan network on S. Thus, a 2-approximation algorithm is obtained.

2   A Sparse Generating Set

For a pair of points $p = (p.x, p.y), q = (q.x, q.y)$ in S, a (p, q)-Manhattan path is a Manhattan path connecting p and q. A (p, q)-bounding rectangle, denoted by $R(p, q)$, is a rectangle defined by p, q with horizontal and vertical segments, i.e.,

$$R(p, q) = \{(x, y) \mid x_{min} \leq x \leq x_{max}, \ y_{min} \leq y \leq y_{max}\},$$

where $x_{min} = \min\{p.x, q.x\}$, $x_{max} = \max\{p.x, q.x\}$, $y_{min} = \min\{p.y, q.y\}$, $y_{max} = \max\{p.y, q.y\}$. If $p.x = q.x$ ($p.y = q.y$) then a (p, q)-Manhattan path is unique and the vertical (horizontal) segment connecting p and q. Otherwise, there are infinitely many (p, q)-Manhattan paths all of which are entirely contained in (p, q)-bounding rectangle $R(p, q)$.

In a Manhattan network $G(S)$ on S, there is a (p, q)-Manhattan path for each pair of points $p, q \in S$. To construct $G(S)$, however, we need not consider all pairs, since some (p, q)-Manhattan path may be automatically constructed if we construct other Manhattan paths. For example, for three distinct points $p, q, r \in S$, if r is contained in (p, q)-bounding rectangle $R(p, q)$, then a path joining a (p, r)-Manhattan path and an (r, q)-Manhattan path becomes a (p, q)-Manhattan path (Fig.2(left)). In this case we call a pair (p, q) is *decomposable* into pairs (p, r) and (q, r). In general, if a (p, q)-Manhattan path can be constructed by combining (nomempty) parts of *arbitrary* (p_i, q_i)-Manhattan paths $(i = 1, ..., k)$, then we call (p, q) is *decomposable* into $\{(p_i, q_i) | i = 1, ..., k\}$. Fig.2 shows an example of decomposable pairs. For points p, q, s_1, s_2 (t_1, t_2, t_3, t_4) in S, if $(p, s_1), (s_2, q) \in Z$ $((p, t_1), (t_2, q), (t_3, t_4) \in Z)$, then pair (p, q) is decomposable into pairs $(p, s_1), (s_2, q)$ $((p, t_1), (t_2, q), (t_3, t_4))$.

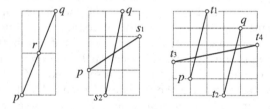

Fig. 2. Example of decomposable pairs (p, q) into pairs shown by solid segments.

Throughout this paper, $\mathcal{P}_2(S)$ denotes the set of unordered pairs of points in S. For a set $Z \subseteq \mathcal{P}_2(S)$, Z is called a *generating set* if, for any pair $(p, q) \in \mathcal{P}_2(S) - Z$, (p, q) is decomposable into pairs in Z. A generating set $Z \subseteq \mathcal{P}_2(S)$ is called *sparse* if $|Z| = O(n)$. A pair $(p, q) \in \mathcal{P}_2(S)$ is called *non-decomposable* if it is not decomposable into pairs in $\mathcal{P}_2(S) - \{(p, q)\}$. A generating set contains all non-decomposable pairs, however, a set of all non-decomposable pairs is not always a generating set. We will give an algorithm for obtaining a sparse generating set below. For simplicity, we assume, throughout this paper, $x_1, x_2, \ldots, x_{k_x}$ $(y_1, y_2, \ldots, y_{k_y})$ are the distinct x-values (y-values) of points in S,

$$x_1 < x_2 < \ldots < x_{k_x}, \quad \text{and} \quad y_1 < y_2 < \ldots < y_{k_y}.$$

We also assume that, for a point p in the plane, $Q_k(p)$ denotes the set of points in $S - \{p\}$ in the k-th quadrant with p being considered to be the origin. Thus, for example, $Q_1(p) = \{q \in S - \{p\} \mid p.x \leq q.x, p.y \leq q.y\}$. Then, based on a simple plane sweep method in computational geometry in x- and y-directions, a sparse generating set $Z = Z_x \cup Z_v \cup Z_y \cup Z_h \cup Z_2$ can be obtained by an algorithm described below. Here, Z_x (Z_y) is the set of adjacent pairs of points in S with same x-values (y-values), Z_v (Z_h, resp.) is the set of adjacent pairs in S with respect to x-values (y-values, resp.) with some slight restriction. Each pair in $Z_x \cup Z_v \cup Z_y \cup Z_h$ is non-decomposable, however, $Z_x \cup Z_v \cup Z_y \cup Z_h$ is not a generating set. Thus we add Z_2 to this (Fig.3).

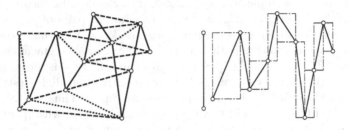

Fig. 3. $Z_x \cup Z_v$ (solid lines), $Z_y \cup Z_h$ (broken lines), Z_2 (dotted lines) in the left figure and the bounding rectangles of pairs in $Z_x \cup Z_v$ in the right figure.

Algorithm A for obtaining a sparse generating set Z

Step 1. Let S_{x_i} be the set of points $p = (p.x, p.y)$ in S with $p.x = x_i$ ($i = 1, 2, ..., k_x$) and let $p_{i1}, p_{i2}, ..., p_{ij_i}$ be the sorted list of points in S_{x_i} in increasing order of their y-values. Then set $Z_x := \cup_{i=1}^{k_x} \{(p_{ij}, p_{ij+1}) | j = 1, ..., j_i - 1\}$.

Step 2. Let S_{y_j} be the set of points $p = (p.x, p.y)$ in S with $p.y = y_j$ ($j = 1, 2, ..., k_y$) and let $p_{1j}, p_{2j}, ..., p_{i_j j}$ be the sorted list of points in S_{y_j} in increasing order of their x-values. Then set $Z_y := \cup_{j=1}^{k_y} \{(p_{ij}, p_{i+1j}) | i = 1, ..., i_j - 1\}$.

Step 3. Set $Z_v := \emptyset$. For each $i = 1, 2, ..., k_x$, set $p_i := p_{i1}$ and $q_i := p_{ij_i}$ (comment: p_i (q_i, resp.) is the point in S_{x_i} with minimum (maximum, resp.) y-value and $p_i.y < q_i.y$ if $|S_{x_i}| > 1$).
   For $i := 1$ to $k_x - 1$ do the following:
      If $[p_i.y, q_i.y] \cap [p_{i+1}.y, q_{i+1}.y] = \emptyset$ then
      if $p_i.y > q_{i+1}.y$ then set $Z_v := Z_v \cup \{(p_i, q_{i+1})\}$
      else (i.e., if $q_i.y < p_{i+1}.y$ then) set $Z_v := Z_v \cup \{(q_i, p_{i+1})\}$.

Step 4. Set $Z_h := \emptyset$. For each $j = 1, 2, ..., k_y$, set $p_j := p_{1j}$ and $q_j := p_{i_j j}$ (comment: p_j (q_j, resp.) is the point in S_{y_j} with minimum (maximum. resp.) x-value and $p_j.x < q_j.x$ if $|S_{y_j}| > 1$).
   For $j := 1$ to $k_y - 1$ do the following:
      If $[p_j.x, q_j.x] \cap [p_{j+1}.x, q_{j+1}.x] = \emptyset$ then
      if $p_j.x > q_{j+1}.x$ then set $Z_h := Z_h \cup \{(p_j, q_{j+1})\}$
      else (i.e., if $q_j.x < p_{j+1}.x$ then) set $Z_h := Z_h \cup \{(q_j, p_{j+1})\}$.

Step 5. Set $Z_2 := \emptyset$ and $Z := Z_x \cup Z_v \cup Z_y \cup Z_h$.
   For $j := 1$ to k_y do the following:
      For $i := 1$ to i_j do the following:

Set $p := p_{ij}$.

For $k := 1$ to 4 do:

if $Q_k(p) \neq \emptyset$ and there is no $q \in Q_k(p)$ with $(p, q) \in Z$ then
let q be the point in $Q_k(p)$ with $|q.x - p.x| = \min\{|q'.x - p.x| \mid q' \in Q_k(p)\}$ (if there are two or more such candidates q'' of q choose q with $|q.y - p.y| = \min\{|q''.y - p.y|\}$ among such q''), and set $Z_2 := Z_2 \cup \{(p, q)\}$ and $Z := Z \cup \{(p, q)\}$.

Fig.3 shows an example of $Z = Z_x \cup Z_v \cup Z_y \cup Z_h \cup Z_2$ obtained by Algorithm A. As described above, Z_x (Z_y) is the set of adjacent pairs of points in S with same x-values (y-values) and each pair (p, q) in Z_x (Z_y) is a non-decomposable pair and a (p, q)-Manhattan path is a vertical (horizontal) segment connecting p and q. Thus, $|Z_x| \leq n-1$ and $|Z_y| \leq n-1$. Similarly, Z_v (Z_h) is the set of adjacent pairs in S with respect to x-values (y-values) with some slight restriction and each pair (p, q) in Z_v (Z_h) is a non-decomposable pair. Thus, $|Z_v| \leq n - 1$ and $|Z_h| \leq n - 1$. If $Q_k(p) \neq \emptyset$ and there is no point q in $Q_k(p)$ with $(p, q) \in Z$, then exactly one point in $Q_k(p)$, say q, is appropriately chosen and a pair (p, q) is added to Z_2 and Z. This (p, q) can be a decomposable pair in general, although it is not decomposable into pairs in $Z - \{(p, q)\}$. Since $|Z_2| \leq 4n$, we have $|Z| = O(n)$. The following lemma establishes that Z is a generating set.

Lemma 1. *Z is a generating set of $\mathcal{P}_2(S)$, that is, for any $(p, q) \in \mathcal{P}_2(S) - Z$, (p, q) is decomposable into some pairs in Z.*

Proof. For $(p, q) \in \mathcal{P}_2(S) - Z$, let $S(p, q)$ be the set of points in $S - \{p, q\}$ contained in (p, q)-bounding rectangle $R(p, q)$. We will show the lemma by induction on $a(p, q) = |S(p, q)|$. By symmetry, we can assume $p.x \leq q.x$ and $p.y \leq q.y$.

Case I. $a(p, q) = 0$: We have $p.x < q.x$ and there is a point r_1 in S such that $p.x \leq r_1.x < q.x$ and $q.y < r_1.y$, or a point r_2 in S such that $p.x < r_2.x \leq q.x$ and $r_2.y < p.y$. Because otherwise we would have $(p, q) \in Z_x \cup Z_v \subseteq Z$ and a contradiction. Similarly, by $(p, q) \notin Z_y \cup Z_h$, we have $p.y < q.y$ and there is a point r'_1 in S such that $q.x < r'_1.x$ and $p.y \leq r'_1.y < q.y$, or a point r'_2 in S such that $r'_2.x < p.x$ and $p.y < r'_2.y \leq q.y$ (Fig.4). If there are several candidates for r_1, let r_1^m (r_1^M) be such a point with minimum (maximum) x-value (if there is just one candidate for r_1 then $r_1^m = r_1^M$). If there are still some candidates for r_1^m (r_1^M), we choose r_1^m (r_1^M) to be the point with minimum y-value (i.e., the point nearest to p) among such candidates. Similarly, if there are some candidates for r_2, let r_2^m (r_2^M) be such a point with minimum (maximum) x-value. If there are still some candidates for r_2^m (r_2^M), we choose r_2^m (r_2^M) to be the point of maximum y-value among such candidates. Thus, we can always assume that r_1^m or r_2^M exists. By exchanging the roles of x- and y-values, we can similarly define $r_1'^m, r_1'^M, r_2'^m, r_2'^M$ and assume that $r_1'^m$ or $r_2'^M$ exists (Fig.4).

We have the following three cases for the existences of r_1^m and r_2^M:

(a) r_1^m exists and no r_2^M exists;   (b) r_2^M exists and no r_1^m exists;

(c) Both r_1^m and r_2^M exist.

Similarly we have three cases for the existences of $r_1'^m$ and $r_2'^M$:

(a') $r_1'^m$ exists and no $r_2'^M$ exists;   (b') $r_2'^M$ exists and no $r_1'^m$ exists;

(c') Both $r_1'^m$ and $r_2'^M$ exist.

We have $(p, r_1^m) \in Z_x \cup Z_v$ in (a), and $(r_2^M, q) \in Z_x \cup Z_v$ in (b). Similarly, we

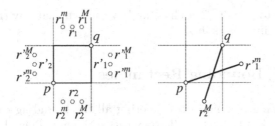

Fig. 4. Examples of r_1, r_2, r_1', r_2' and $(p, r_1'^m), (r_2'^M, q) \in Z$.

have $(p, r_1'^m) \in Z_y \cup Z_h$ in (a') and $(r_2'^M, q) \in Z_y \cup Z_h$ in (b'). Thus, if (b) and (a') occur simultaneously, then we have $(p, r_1'^m), (r_2'^M, q) \in Z_x \cup Z_v \cup Z_y \cup Z_h$ and a (p, q)-Manhattan path can be automatically obtained from $(p, r_1'^m)$- and $(r_2'^M, q)$-Manhattan paths (Fig.4).

By symmetry, if (a) and (b') occur simultaneously, a (p, q)-Manhattan path can be automatically constructed. On the other hand, if (a) and (a') occur simultaneously, then we would have $(p, q) \in Z_2 \subseteq Z$ and a contradiction. Thus, (a) and (a') do not occur simultaneously. For the same reason, (b) and (b') do not occur simultaneously. We consider subcases of (c) and (c') to cope with the remaining cases. (c) can be divided into the following two subcases:

    (c-i) $r_1^m.x < r_2^m.x$;   (c-ii) $r_1^m.x \geq r_2^m.x$.

(c) can also be divided into the following two subcases:

    (c-α) $r_2^M.x > r_1^M.x$;   (c-β) $r_2^M.x \leq r_1^M.x$.

Similarly, (c') can be divided into the following two subcases:

    (c'-i) $r_1'^m.y < r_2'^m.y$;   (c'-ii) $r_1'^m.y \geq r_2'^m.y$.

(c') can also be divided into the following two subcases:

    (c'-α) $r_2'^M.y > r_1'^M.y$;   (c'-β) $r_2'^M.y \leq r_1'^M.y$.

In (c-i), we have $(p, r_1^m) \in Z_x \cup Z_v$, and in (c-$\alpha$), we have $(r_2^M, q) \in Z_x \cup Z_v$. Similarly, $(p, r_1'^m) \in Z_y \cup Z_h$ in (c'-i) and $(r_2'^M, q) \in Z_y \cup Z_h$ in (c'-α). Thus, if (a) and (c'-α) occur simultaneously, then $(p, r_1^m), (r_2'^M, q) \in Z_x \cup Z_v \cup Z_y \cup Z_h$ and a (p, q)-Manhattan path can be obtained automatically. On the other hand, if (a) and (c'-β) occur simultaneously, then we would have $(r_2'^M, q) \notin Z_y \cup Z_h$ and $(p, q) \in Z_2$ and a contradiction. Thus, (a) and (c'-β) cannot occur simultaneously. By symmetry, (b) and (c') can be treated in the same way. (a') and (c) ((b') and (c)) can be treated similarly. The remaining case is the case when (c) and (c') occur simultaneously. By a little complicated case analysis, we can obtain results similar to ones above. Thus, we have shown that $(p, q) \in \mathcal{P}_2(S) - Z$ is decomposable into pairs in Z when $a(p, q) = 0$.

Case II. $a(p, q) = k$ $(k > 0)$: We assume that any $(p', q') \in \mathcal{P}_2(S) - Z$ is decomposable into pairs in Z if $a(p', q') = |S(p', q')| < k$. Let r be any point in $S(p, q)$. Since $a(p, r) + a(r, q) \leq k - 1$, each of (p, r) and (r, q) is in Z or decomposable into pairs in Z. Thus, a (p, q)-Manhattan path can be obtained by combining a (p, r)-Manhattan path and an (r, q)-Manhattan path and (p, q) is decomposable into pairs in Z. (Thus, we have shown that any $(p, q) \in \mathcal{P}_2(S) - Z$ is decomposable into pairs in Z.)

3   Covering Bounding Rectangles

In this section we give a notion of covering all the bounding rectangles of pairs in $Z_x \cup Z_v$ ($Z_y \cup Z_h$) by vertical (horizontal) segments. This plays an important role in our algorithm. Let $\mathcal{R}$ be the set of bounding rectangles of pairs in $Z_x \cup Z_v$, i.e., $\mathcal{R} = \{R(p, q) \mid (p, q) \in Z_x \cup Z_v\}$. Fig.3(right) and Fig.5(left) show $\mathcal{R}$ for $Z_x \cup Z_v$ in Fig.3(left). For $R(p, q) \in \mathcal{R}$ and a horizontal line ℓ, let $R_\ell(p, q)$ be the intersection of ℓ with $R(p, q)$. Similarly, for a set V of vertical segments, let V_ℓ be the intersections of ℓ with V. A set V of vertical segments *covers* $\mathcal{R}$ if, for any horizontal line ℓ, V_ℓ contains a point in each nonempty set $R_\ell(p, q)$ of $\mathcal{R}_\ell \equiv \{R_\ell(p, q) \mid R(p, q) \in \mathcal{R}, R_\ell(p, q) \neq \emptyset\}$. Such a set V is also called a *vertical covering set* for $Z_x \cup Z_v$. A *minimum vertical covering set* is a vertical covering set of minimum length. Fig.5(left) shows an example of a minimum vertical covering set V. In this example, it is easily seen that V covers $\mathcal{R}$. On the other hand, for any v_ℓ in V_ℓ, it can be observed that there is some $R_\ell(p, q) \in \mathcal{R}_\ell$ such that no point in $R_\ell(p, q)$ is contained in $V_\ell - \{v_\ell\}$. Thus, $V_\ell - \{v_\ell\}$ cannot cover $\mathcal{R}_\ell$ and V_ℓ is a minimal set which covers $\mathcal{R}_\ell$. Actually, V_ℓ is a minimum set which covers $\mathcal{R}_\ell$ and thus V is a minimum vertical covering set for $Z_x \cup Z_v$.

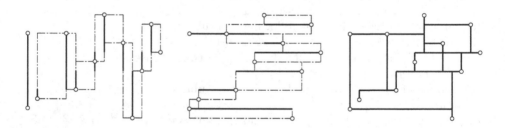

Fig. 5. A minimum vertical covering set (left (thick lines)), a minimum horizontal covering set (center (thick lines)) and the network obtained by combining them (right).

In general, we can assume that a minimum vertical covering set consists of vertical segments on the edges of bounding rectangles in $\mathcal{R}$, since if a vertical covering set contains a segment inside a rectangle $R(p, q) \in \mathcal{R}$, then the segment can be moved in such a way that it is on one of the vertical boundary edges of $R(p, q)$ without losing the vertical covering set property. Let U be the set of vertical edges of rectangles in $\mathcal{R}$, and let U_ℓ be the intersection of ℓ with U. Let $\mathcal{G}_\ell$ be a graph with vertex set U_ℓ and edge set $\mathcal{R}_\ell$ (vertex $u \in U_\ell$ is incident to edge $R_\ell(p, q) \in \mathcal{R}_\ell$ containing u). Then, for any vertical covering set V, V_ℓ is a vertex cover of $\mathcal{G}_\ell$ and if V is a minimum vertical covering set then V_ℓ is also a minimum vertex cover of $\mathcal{G}_\ell$.

We can similarly define a set of horizontal segments covering all the bounding rectangles of pairs $(p, q) \in Z_y \cup Z_h$ and a minimum horizontal covering set (Fig.5(center)). A network in Fig.5(right) is the union of a minimum vertical covering set (Fig.5(left)) and a minimum horizontal covering set (Fig.5(center)).

For any Manhattan network $G(S)$ on S and any pair $(p, q) \in Z_x \cup Z_v$, $G(S)$ contains a (p, q)-Manhattan path which covers $R(p, q)$ vertically. Thus, the set of vertical segments contained in $G(S)$ covers all the rectangles in $\mathcal{R} = \{R(p, q) \mid (p, q) \in Z_x \cup Z_v\}$ and we have the following lemma.

Lemma 2. *Let $G(S)$ be any Manhattan network on S. Then the total length of the vertical (horizontal) segments contained in $G(S)$ is at least the length of a minimum vertical (horizontal) covering set. The union network of a minimum vertical covering set and a minimum horizontal covering set is of length at most the length of a minimum Manhattan network.*

Graph $\mathcal{G}_\ell$ described above with vertex set U_ℓ and edge set $\mathcal{R}_\ell$ consists of paths, since we have the lemma below which is clear from the construction of $Z_x \cup Z_v$. Recall that $Q_k(p)$ is the set of points in $S - \{p\}$ in the k-th quadrant in the plane with p considered as the origin ($k = 1, 2, 3, 4$).

Lemma 3. *For any point $p \in S$, there is at most one point $q \in Q_k(p)$ such that $(p, q) \in Z_x \cup Z_v$ ($k = 1, 2, 3, 4$). Thus, there are at most two points $q \in Q_k(p) \cup Q_{k+1}(p)$ such that $(p, q) \in Z_x \cup Z_v$ (we consider 5=1 for simplicity).*

By Lemma 3, each connected component of $\mathcal{G}_\ell$ is a path and we can find a minimum vertex cover easily. Thus, by a plane sweep method, we can obtain a minimum vertical covering set efficiently. However, a minimum vertical covering set is not sufficient in a sense that we cannot obtain a Manhattan path for some pair in $Z_x \cup Z_v$ even if we use any horizontal segments. We need a minimum vertical covering set which is sufficient for obtaining Manhattan paths for all pairs in $Z_x \cup Z_v$ if we use horizontal segments appropriately. We will call such a set a *sufficient minimum vertical covering set*. Thus, for any pair $(p, q) \in Z_x \cup Z_v$, we can obtain a (p, q)-Manhattan path by adding horizontal segments appropriately to a sufficient minimum vertical covering set for $Z_x \cup Z_v$. Similarly we define a *sufficient minimum horizontal covering set* and, for any pair $(p, q) \in Z_y \cup Z_h$, we can obtain a (p, q)-Manhattan path by adding vertical segments appropriately to a sufficient minimum horizontal covering set for $Z_y \cup Z_h$. Fig.5(left and center) shows such sufficient minimum vertical and horizontal covering sets. An algorithm for constructing such a set will be presented in the next section.

4   2-Approximation Algorithm

Our 2-approximation algorithm consists of two phases. A network is constructed in each phase, and two obtained networks, if combined, lead to a Manhattan network on S of length at most twice the length of a minimum Manhattan network on S. In Phase 1, we will obtain a network (like a network as shown in Fig.5(right)) consisting of a sufficient minimum vertical covering set (Fig.5(left)) and a sufficient minimum horizontal covering set (Fig.5(center)). In Phase 2, we consider pairs in Z for each of which no Manhattan path exists in the network obtained in Phase 1. In this phase, the sufficiency property of a sufficient minimum vertical (horizontal) covering set plays a critical role.

4.1   Phase 1: Constructing Sufficient Minimum Vertical and Horizontal Covering Sets

We will present an algorithm obtaining N_v (Fig.6(left)) from $Z_x \cup Z_v$ which leads to a network with a sufficient minimum vertical covering set (Fig.5(left)). The algorithm is based on the plane sweep method from bottom to top. It will try to join two endpoints of each pair (p, q) in $Z_x \cup Z_v$ using at most two vertical segments and at most one horizontal or oblique segment in $R(p, q)$. However, for some pairs, algorithm may not join their two endpoints. For each pair in Z_x, a vertical segment between its two endpoints is always contained in a minimum vertical covering set and the algorithm joins its two endpoints by a vertical segment. For a pair (p, q) in Z_v, its two endpoints p, q have different x-values and, to join p, q, we have to use a horizontal or oblique segment besides vertical segments. In this case, we can use only a horizontal segment which is assured to be in some minimum horizontal covering set. For the same reason, we can use an oblique segment only if a horizontal segment of the rectangle defined by the oblique segment as a diagonal is assured to be in a minimum horizontal covering set. If there are no such assurances, p and q are joined partly.

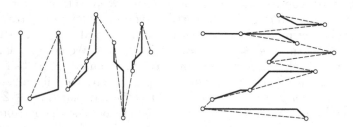

Fig. 6. Example of N_v (left) and N_h (right).

By exchanging the roles of x- and y-coordinates, we can also obtain an algorithm which constructs N_h (Fig.6(right)) from $Z_y \cup Z_h$ leading to a network with a sufficient minimum horizontal covering set (Fig.5(center)). Here we only describe an algorithm for $Z_x \cup Z_v$. Recall that $y_1, y_2, \ldots, y_{k_y}$ are the distinct y-values of points in S and $y_1 < y_2 < \ldots < y_{k_y}$. An intersection of horizontal and vertical segments is called a *Steiner point* if it is not in S.

Algorithm B for constructing N_v
Step 1. For each pair (p, q) in $Z_x \cup Z_v$, join p and q with a straight segment, and construct an edge set E_x from Z_x and an edge set E_v from Z_v.

Step 2. For $\ell := 1$ to $k_y - 1$ do the following Step 3:

Step 3. Let $S_1^\ell = \{p \in S \mid p.y \le y_\ell\}$, $S_2^\ell = \{q \in S \mid y_{\ell+1} \le q.y\}$, and $E^\ell = \{(p,q) \in E_v \mid p \in S_1^\ell, q \in S_2^\ell\}$. Let $G^\ell = (S_1^\ell, S_2^\ell, E^\ell)$ be a bipartite graph with two independent vertex sets S_1^ℓ, S_2^ℓ and edge set E^ℓ. Partition G^ℓ into the connected components $G' = (S_1', S_2', E')$ with $S_1' \subseteq S_1^\ell$, $S_2' \subseteq S_2^\ell$, and $E' \subseteq E^\ell$. For each $G' = (S_1', S_2', E')$ with $E' \ne \emptyset$ do the following Step 4:

Step 4. Let $p_1, p_2, \ldots, p_k$ $(q_1, q_2, \ldots, q_{k'})$ be the points of S_1' (S_2') in increasing order of their x-values. Do one of the following cases (a)–(c):

Case (a) $k < k'$ (Fig.7(a)):

  For $i := 1$ to k do one of the following cases:

  (comment: $q_i.x < p_i.x < q_{i+1}.x$ holds)

  Case (a-i) $q_i.y = q_{i+1}.y = y_{\ell+1}$:

   Let $r_i = (p_i.x, y_{\ell+1})$ be a Steiner point and construct a (p_i, q_i)-Manhattan path and a (p_i, q_{i+1})-Manhattan path as in Fig.7(a-i). Delete edges (p_i, q_i), (p_i, q_{i+1}) from E_v.

  Case (a-ii) either $q_i.y = y_{\ell+1}$ or $q_{i+1}.y = y_{\ell+1}$ (Fig.7(a-ii)):

   If $q_i.y = y_{\ell+1}$ then delete edge (p_i, q_i) from E_v and .
   if $q_{i+1}.y = y_{\ell+1}$ then delete edge (p_i, q_{i+1}) from E_v.

  Case (a-iii) $y_{\ell+1} < q_i.y$ and $y_{\ell+1} < q_{i+1}.y$ (Fig.7(a-iii)): Do nothing.

Case (b) $k' < k$ (Fig.7(b)):

  For $j := 1$ to k' do the following: (comment: $p_j.x < q_j.x < p_{j+1}.x$ holds) Let $r_{j,1} = (p_j.x, y_\ell)$, $r_{j,2} = (q_j.x, y_\ell)$, $r_{j,3} = (p_{j+1}.x, y_\ell)$ be three Steiner points and construct a (p_j, q_j)-Manhattan path with two Steiner points $r_{j,1}$ and $r_{j,2}$ and a (p_{j+1}, q_j)-Manhattan path with two Steiner points $r_{j,2}$ and $r_{j,3}$ as shown in Fig.7(b). Delete edges (p_j, q_j), (p_{j+1}, q_j) from E_v.

Case (c) $k = k'$ (Fig.7(c)):

  By symmetry we assume $p_i.x < q_i.x$ for $i = 1, 2, ..., k$

  (if $p_i.x > q_i.x$ then change the roles of left and right below).

  (comment: $p_i.x < q_i.x < p_{i+1}.x$ holds and it can be shown that there is a point p in S with $p.y = y_\ell$ and $p.x \le p_i.x$)

  Delete edges $\{(p_i, q_{i-1}) \mid 2 \le i \le k\}$ from E_v.

  For $i := 1$ to k do: if there is a point p in S with $p.y = y_\ell$ and $p.x \ge q_i.x$ then do Case (c-i) else do Case (c-ii) or Case (c-iii):

  Case (c-i): Construct a (p_i, q_i)-Manhattan path with two Steiner points $r_{i,1} = (p_i.x, y_\ell)$, $r_{i,2} = (q_i.x, y_\ell)$ as shown in Fig.7(c-i). Delete edge (p_i, q_i) from E_v. For a point $p' \in S$ with $q_i.x < p'.x$ and $y_{\ell+1} < p'.y < q_i.y$, if edge (p', q_i) is in E_v, then delete it.

  Case (c-ii) there is a point q in S with $q.y = y_{\ell+1}$ and $q.x \ge p_i.x$ and no point q' in S with $q'.y = y_{\ell+1}$ and $q'.x \le p_i.x$: Construct a (p_i, q_i)-Manhattan path with two Steiner points $r_{i,1} = (p_i.x, y_\ell)$, $r_{i,2} = (q_i.x, y_{\ell+1})$ as shown in Fig.7(c-ii). Delete edge (p_i, q_i) from E_v. For a point $p' \in S$ with $q_i.x < p'.x$ and $y_{\ell+1} < p'.y < q_i.y$, if edge (p', q_i) is in E_v, then delete it.

  Case (c-iii) otherwise: If $\ell = k_y - 1$ then construct a (p_i, q_i)-Manhattan path with a Steiner point $r_i = (p_i.x, y_{\ell+1})$ as in Fig.7(c-iii), and delete edge (p_i, q_i) from E_v.

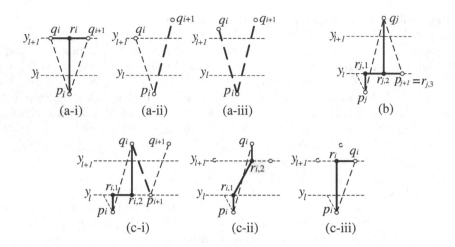

Fig. 7. Step 4 in Algorithm B.

The following lemma holds about the network N_v obtained by the above algorithm B. A similar lemma holds for the network N_h obtained by the algorithm exchanging the roles of x and y.

Lemma 4. *For a horizontal or oblique segment $e = (p, q)$ in N_v (by symmetry, we assume $p.x < q.x$ and $p.y \le q.y$), if $p \notin S$ then there is a point $r_1 \in S$ with $r_1.x < p.x$ and $r_1.y = p.y$, and if $q \notin S$ then there is a point $r_2 \in S$ with $q.x < r_2.x$ and $r_2.y = q.y$.*

Lemma 4 implies that a horizontal segment in N_v is contained in N_h and that an oblique segment $e = (p, q)$ in N_v is consistent with segments in N_h. Similar properties hold for any vertical or oblique segment in N_h. Finally, we consider an oblique segment $e = (p, q)$ in $N_v \cup N_h$ by dividing into three cases: (a) $e \in N_v - N_h$ (Fig.8(a)); (b) $e \in N_h - N_v$ (Fig.8(b)); (c) $e \in N_v \cap N_h$ (Fig.8(c)). We transform each oblique segment $e = (p, q)$ in $N_v \cup N_h$ into solid horizontal or/and vertical segments according to Fig.8 and obtain the union network as shown in Fig.5 (right). The following theorem assures that suffcient minimum vertical and horizonatal covering sets are actually obtained by the above algorithms.

Theorem 1. *The network obtained in Phase 1 consists of a suffcient minimum vertical covering set and a suffcient minimum horizontal covering set and its length is at most the length of a minimum Manhattan network on S.*

4.2   Phase 2: Completing Manhattan Paths

The network obtained in Phase 1 may not be a Manhattan network (Fig.9(left)). This network may have no (p, q)-Manhattan path for some pairs (p, q) in $Z - Z_2$, while it may have a (p', q')-Manhattan path for some pairs (p', q') in Z_2.

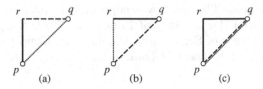

Fig. 8. Transformation of an oblique segment $e = (p, q)$ into (solid) horizontal or/and vertical segments (dotted (broken) lines represent segments in N_v (N_h)).

Now we let Z_3 be the set of pairs in Z for each of which no Manhattan path exists in the network obtained in Phase 1. Let $Z_3' := \emptyset$. For each pair (p, q) in Z_3, we consider its bounding rectangle $R(p, q)$ and we try to modify pair (p, q) to at most two pairs (p, r_p) and (q, r_q) in such a way that a (p, q)-Manhattan path is automatically obtained by combining Manhattan paths for (p, r_p) and (q, r_q) with an (r_p, r_q)-Manhattan path contained in the network obtained in Phase 1. If $R(p, q)$ contains an intersection of vertical and horizontal segments in the network obtained in Phase 1, then let r_p (r_q) be such an intersection in $R(p, q)$ nearest to p (q). If there is no (p, r_p)-Manhattan path ((q, r_q)-Manhattan path) then add (p, r_p) ((q, r_q)) to Z_3'. We repeat this for each pair (p, q) in Z_3 and obtain Z_3' (Fig.9(center)). For these pairs in Z_3', we can obtain all Manhattan paths with short total length efficiently by using a method similar to the $O(n^3)$-time algorithm for the minimum length partition problem [5,8]. Here, we critically use the sufficiency property of a sufficient minimum vertical (horizontal) covering set of the network obtained in Phase 1. Then we can show that the length of newly added segments for the Manhattan paths for Z_3' is at most the length of a minimum Manhattan network on S. The network (Fig.9(right)) obtained by combining the network obtained in Phase 1 with the Manhattan paths for Z_3' is a Manhattan network on S and we have the following theorem.

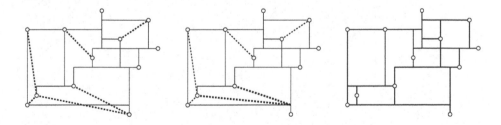

Fig. 9. An example of pairs for which no Manhattan paths exist (left, dotted lines), Z_3' (center, dotted lines) and a Manhattan network (right) obtained by the algorithm.

Theorem 2. *A Manhattan network on S with length at most twice the length of a minimum Manhattan network on S can be obtained in $O(n^3)$ time.*

Acknowledgements. This work was supported in part by Grant in Aid for Scientific Research of the Ministry of Education, Science, Sports and Culture of Japan, The Institute of Science and Engineering, Chuo University, and The Telecommunications Advancement Foundation.

References

1. I. Althöfer, G. Das, D.P. Dobkin, D. Joseph, and J. Soares: On the sparse spanners of weighted graphs, *Discrete Comput. Geom.*, 9 (1993), pp.81–100.
2. S. Arya, G. Das, D.M. Mount, J.S. Salowa, and M. Smid: Euclidean spanners: short, thin, and lanky, *Proc. 27th STOC*, 1995, pp. 489–498.
3. B. Chandra, G. Das, G. Narasimhan, and J. Soares: New sparseness results on graph spanners, *Internat. J. Comput. Geom. Appl.*, 5 (1995), pp.125–144.
4. G. Das and G. Narasimhan: A fast algorithm for constructing sparse Euclidean spanners, *Internat. J. Comput. Geom. Appl.*, 7 (1997), pp.297–315.
5. J. Gudmundsson, C. Levcopoulos, and G. Narasimhan: Approximating minimum Manhattan networks, *Proc. APPROX'99*, 1999, pp.28–37.
6. C. Levcopoulos, G. Narasimhan, and M. Smid: Efficient algorithms for constructing fault-tolerant geometric spanners, *Proc. 30th STOC*, 1998, pp.186–195.
7. C. Levcopoulos and A. Östlin: Linear-time heuristics for minimum weight rectangulation, *Proc. 5th SWAT, LNCS 1097*, 1996, pp.271–283.
8. A. Lingas, R. Pinter, R. Rivest, and A. Shamir: Minimum edge length partitioning of rectilinear polygons, *Proc. 20th Allerton Conf. Commun. Control Comput.*, 1982, pp.53–63.
9. S. B. Rao and W. D. Smith: Improved approximation schemes for geometrical graphs via "spanners" and "banyans", *Proc. 30th STOC*, 1998, pp.540–550.

Approximate Distance Oracles Revisited*

Joachim Gudmundsson[1], Christos Levcopoulos[2], Giri Narasimhan[3], and
Michiel Smid[4]

[1] Department of Computer Science, Utrecht University, P.O. Box 80.089, 3508 TB
Utrecht, The Netherlands. joachim@cs.uu.nl
[2] Department of Computer Science, Lund University, Box 118, 221 00 Lund, Sweden.
christos@cs.lth.se
[3] School of Computer Science, Florida International University, Miami, FL 33199,
USA. giri@fiu.edu
[4] School of Computer Science, Carleton University, 1125 Colonel By Drive, Ottawa,
Ontario, Canada K1S 5B6. michiel@scs.carleton.ca

Abstract. Let G be a geometric t-spanner in $\mathbb{E}^d$ with n vertices and
m edges, where t is a constant. We show that G can be preprocessed
in $O(m \log n)$ time, such that $(1 + \varepsilon)$-approximate shortest-path queries
in G can be answered in $O(1)$ time. The data structure uses $O(n \log n)$
space.

1 Introduction

The *shortest-path* (SP) problem for weighted graphs with n vertices and m edges
is a fundamental problem for which efficient solutions can now be found in any
standard algorithms text. The approximation version of this problem has been
studied extensively, see [1,7,8,9]. In numerous algorithms, query versions frequently appear as subroutines. In such a query, we are given two vertices and
have to compute or approximate the shortest path between them. The latest in
a series of results for undirected weighted graphs is by Thorup and Zwick [14];
their algorithm computes $(2k - 1)$-approximate solutions to the query version of
the SP problem in $O(k)$ time, using a data structure that takes (expected) time
$O(kmn^{1/k})$ to construct and utilises $O(kn^{1+1/k})$ space. It is not an approximation scheme in the true sense because the value k needs to be a positive integer.
Since the query time is essentially bounded by a constant, Thorup and Zwick
refer to their queries as approximate *distance oracles*.

We focus on the geometric version of this problem. A geometric graph has
vertices corresponding to points in $\mathbb{R}^d$ and edge weights from a Euclidean metric. Again, considerable previous work exists on the shortest path and related
problems. A good survey on the topic was written by Mitchell and can be found
in [13], see also [3,4,5,6]. Recently Gudmundsson *et al.* [9] presented the first
data structure that answers approximate shortest-path distance queries in constant time, with $O(n \log n)$ space and preprocessing time for geometric graphs

* J.G. is supported by The Swedish Foundation for International Cooperation in Research and Higher Education and M.S. is supported by NSERC.

P. Bose and P. Morin (Eds.): ISAAC 2002, LNCS 2518, pp. 357–368, 2002.
© Springer-Verlag Berlin Heidelberg 2002

with $O(n)$ edges that are t-spanners for some (possibly large) constant t. This result is *restricted* in the sense that it only supports queries between vertices in G whose Euclidean distance is at least D/n^k, where D is the length of the longest edge in G, and $k > 0$ is an arbitrary integer constant. In this paper we extend the results in [9] so that it holds for *any* two query points, i.e., we remove the restriction on the distance between the query points.

A graph $G = (V, E)$ is said to be a *t-spanner* for V, if for any two points p and q in V, there exists a path in G between p and q of length at most t times the Euclidean distance between p and q. Given a geometric t-spanner $G = (V, E)$ on n vertices and m edges, for some constant t, we show how to build a data structure in time $O(m \log n)$ that answers $(1 + \varepsilon)$-approximate shortest path length queries in G in constant time, for any given real constant $\varepsilon > 0$. The structure uses $O(n \log n)$ space.

We remark that many "naturally occurring" geometric graphs are t-spanners, for some constant $t > 1$, thus justifying the interest in the restricted inputs considered in this paper. In [9] it was shown that an approximate shortest-path distance (ASD) oracle can be applied to a large number of problems, for example, finding a shortest obstacle-avoiding path between two vertices in a planar polygonal domain with obstacles, interesting query versions of *closest pair* problems and, computing the approximate stretch factor of geometric graphs.

In the full paper we also show that the structure presented in this paper easily can be extended such that $(1 + \varepsilon)$-approximate shortest path queries for geometric t-spanners can be answered in time proportional to the number of edges along the path.

2   Preliminaries

Our model of computation is the traditional algebraic computation model with the added power of indirect addressing. We will use the following notation. For points p and q in $\mathbb{R}^d$, $|pq|$ denotes the Euclidean distance between p and q. If G is a geometric graph, then $\delta_G(p, q)$ denotes the Euclidean length of a shortest path in G between p and q. Hence, G is a t-spanner for V if $\delta_G(p, q) \leqslant t \cdot |pq|$ for any two points p and q of V. A geometric graph G is said to be a (t, L)-spanner for V if for every pair of points $p, q \in V$ with $|pq| < L$ it holds that $\delta_G(p, q) \leqslant t \cdot |pq|$. If P is a path in G between p and q having length Δ with $\delta_G(p, q) \leqslant \Delta \leqslant (1+\varepsilon) \cdot \delta_G(p, q)$, then P is a $(1+\varepsilon)$-approximate shortest path for p and q. Finally, a subgraph G' of G is a t'-spanner of G, if $\delta_{G'}(p, q) \leqslant t' \cdot \delta_G(p, q)$ for any two points p and q of V.

The main result of this paper is stated in the following theorem:

Theorem 1. *Let V be a set of n points in $\mathbb{R}^d$, let ε be a positive real constant and let $G = (V, E)$ be a t-spanner for V, for some real constant $t > 1$, having $O(n)$ edges. We can preprocess G in $O(n \log n)$ time using $O(n \log n)$ space, such that for any two points p and q in V, we can compute, in $O(1)$ time, a $(1 + \varepsilon)$-approximation to the shortest-path distance in G between p and q.*

If G has m edges, where m grows superlinearly with n, then we first apply Theorem 3.1 in [9] to obtain a $(1 + \varepsilon)$-spanner of G with $O(n)$ edges. Then we apply Theorem 2 to G'. The preprocessing increases by an additive term of $O(m \log n)$ and the space requirement increases by an additive term of $O(m+n)$, but note that the data structure only uses $O(n \log n)$ space. As described in the theorem we will present a data structure that supports $(1 + \varepsilon)$-ASD queries in a t-spanner graph. As a substructure we will use the data structure in [9] that supports *restricted* $(1+\varepsilon)$-shortest path queries in t-spanner graphs. In the rest of this paper we will refer to the data structure in [9] as the PSP-structure and the construction algorithm of the PSP-structure as the PSP-algorithm.

Fact 2 *(Theorem 4.1 in [9])*
Let V be a set of n points in $\mathbb{R}^d$, and let $G = (V, E)$ be a t-spanner for V, for some real constant $t > 1$, having $O(n)$ edges. Let D be the length of the longest edge in G, let ε be a positive real constant, and let k be a positive real constant. We can preprocess G in $O(n \log n)$ time using $O(n \log n)$ space, such that for any two points p and q in V with $|pq| > D/n^k$, we can compute, in $O(1)$ time, a $(1 + \varepsilon)$-approximation to the shortest-path distance in G between p and q.

We need a slightly modified version of the above fact, stated in the following corollary. The proof is implicit in [9].

Corollary 1. *Let V be a set of n points in $\mathbb{R}^d$, and let $G = (V, E)$ be a (t, L)-spanner for V, having $O(n)$ edges, for some real constant $t > 1$ and some real number L. Let D be the length of the longest edge in G, let ε be a positive real constant, and let k be a positive integer constant. We can preprocess G in $O(n \log n)$ time using $O(n \log n)$ space, such that for any two points p and q in V with $D/n^k < |pq| < L/t$, we can compute, in $O(1)$ time, a $(1+\varepsilon)$-approximation to the shortest-path distance in G between p and q.*

In the rest of this paper, we will apply Corollary 1 with a sufficiently large value of k.

3   A First Result

A straightforward idea is to divide the set E of edges of the t-spanner $G = (V, E)$ into subsets $E_1, \ldots, E_\ell$, and then build a PSP-structure P_i for each subset E_i. In this way, a query can be reduced to the problem of finding an appropriate, to be defined below, PSP-structure. A first step towards our goal is to construct the PSP-structures for each subset of edges and then show that given a query, the correct PSP-structure can be found in constant time.

3.1   Building the PSP-Structures

For each i, $1 \leqslant i \leqslant \ell$, construct the edge set E_i as follows. Initially $E_1' = E$. Let $e_{\min}$ be the edge in E_i' with smallest weight and let $E_i = \{e \in E_i' | wt(e) \leqslant n^k \cdot wt(e_{\min})\}$. Finally, set $E_{i+1}' = E_i' \backslash E_i$. The resulting edge sets are denoted $E_1, \ldots, E_\ell$. We need some straightforward properties of the edge sets that will be needed for the improved algorithm in Section 4.

Observation 3 *Given the subsets of edges $E_1, \ldots, E_\ell$, as described above, the following properties hold.*

1. *If $e_1, e_2 \in E_i$ then $wt(e_1)/wt(e_2) \leqslant n^k$.*
2. *If $e_1 \in E_i$ and $e_2 \in E_{i+1}$ then $wt(e_1) < wt(e_2)$.*
3. *If $e_1 \in E_i$ and $e_2 \in E_{i+2}$ then $wt(e_2) \geqslant n^{2k} \cdot wt(e_1)$.*

When the edge sets are constructed, the ℓ PSP-structures, $P_1, \ldots, P_\ell$, can be computed as follows. For each i, $1 \leqslant i \leqslant \ell$, construct a PSP-structure P_i with the graph $G_i = (\mathcal{V}, \mathcal{E}_i)$ as input, where $\mathcal{V} = V$ and $\mathcal{E}_i = E_1 \cup \ldots \cup E_i$. Since G_i is a (t, L_i)-spanner, where L_i is equal to $1/t$ times the maximum edge weight in $\mathcal{E}_i$, we can apply Corollary 1 to G_i. Hence, the complexity of building the data structure is $\ell \times O(n \log n)$, according to Corollary 2, which can be bounded by $O(n^2 \log n)$. It remains to show that given the ℓ PSP-structures, a $(1 + \varepsilon)$-approximate shortest path query can be answered in constant time, i.e., an appropriate PSP-structure can be found in constant time. Given a query pair (p, q), if i is an integer such that $\min_{e \in E_i} |e| \leqslant t \cdot |pq|$ and $|pq| \leqslant n^{k+1} \min_{e \in E_i} |e|$, then the PSP-structure P_i is said to be *appropriate* for the query pair (p, q). Later in the paper, we will see how this notion can be used to answer ASD-queries in the t-spanner G.

Observation 4 *For every query (p, q) there is an appropriate PSP-structure P_i.*

Proof. Since G is a t-spanner, there is a path in G of length at most $t \cdot |pq|$. Clearly, this path has length at least $|pq|$, and it contains at most n edges. Let e' be the longest edge on this path. Then $|pq|/n \leqslant wt(e') \leqslant t \cdot |pq|$. Let i be the index such that $e' \in E_i$. Then $\min_{e \in E_i} wt(e) \leqslant wt(e') \leqslant t \cdot |pq|$ and $|pq| \leqslant n \cdot wt(e') \leqslant n^{k+1} \min_{e \in E_i} |e|$.

In Section 4, we will show how to improve the time and space bounds to $O(n \log n)$, by using so-called cluster graphs as input to the PSP-algorithm instead of the entire graphs G_i.

3.2   Answering ASD-Queries

Given a query pair (p, q) it suffices to prove, according to Fact 2 and Observation 4, that the appropriate PSP-structure P_i can be found in constant time. For this we build an approximate single-link hierarchical cluster tree, denoted by $\mathcal{T}$. Given a set V of points, a single-link hierarchical cluster tree can be built as follows. Initially, each point is considered to be a cluster. As long as there are two or more clusters, the pair C_1 and C_2 of clusters for which $\min\{|pq| : p \in C_1, q \in C_2\}$ is minimum is joined into one cluster. Constructing the single-linkage hierarchy $\mathcal{T}$ for V just means simulating the greedy algorithm of Kruskal [11] for computing the MST of V. This works fine in the plane but not in higher dimensions. Instead we build an approximate single-link hierarchical cluster tree using an $O(1)$-approximation of the MST of V, that is, a variant of the MST which can be obtained, as in Kruskal's algorithm, by inserting in each step an almost shortest edge which does not induce a cycle with the edges already inserted. Methods for constructing such MST's are given, for example, in [12].

When the approximate single-link hierarchical cluster tree $\mathcal{T}$ is built we pre-process it in $O(n)$ time, such that lowest common ancestors (LCA) queries can be answered in constant time [10].

Observation 5 *Let (u_1, v_1) and (u_2, v_2) be two pairs of vertices that have the same least common ancestor ν. Then $wt(u_1, v_1)/wt(u_2, v_2) = O(n)$.*

Proof. The observation is trivial since an $O(1)$-approximation of an MST is an $O(n)$-spanner of the complete graph.

The final step of the preprocessing is to associate a PSP-structure to each node in $\mathcal{T}$. Let ν be an arbitrary node in $\mathcal{T}$ and let $lc(\nu)$ and $rc(\nu)$ be the left- and right child respectively of ν. Every node ν in $\mathcal{T}$ represents a cluster C. Associated to ν is also a value $d(\nu)$, i.e., the length of the edge connecting C_1 and C_2 where C_1 is the cluster represented by $lc(\nu)$ and C_2 is the cluster represented by $rc(\nu)$. According to Observation 5, this value is an $O(n)$-approximation of the distance between any pair of points (p, q) where $p \in C_1$ and $q \in C_2$. For each node ν in $\mathcal{T}$ we add a pointer to the appropriate PSP-structure. By using binary search on the ℓ PSP-structures one can easily find the appropriate PSP-structure in $O(\log n)$ time, hence in $O(n \log n)$ time in total.

Now we are finally ready to handle a query. Given a query (p, q), compute the LCA ν of p and q in $\mathcal{T}$. The LCA-query will approximate the length of (p, q) within a factor of $O(n)$, according to the above observation, which is sufficient for our needs. Hence, ν will have a pointer to a PSP-structure P_j. One of the PSP-structures P_{j-1} or P_j will be an appropriate PSP-structure, this can be decided in constant time. Finally the query is passed on to the appropriate PSP-structure that answers the query in constant time.

We can now summarize the results presented in this section.

Lemma 1. *Let V be a set of n points in $\mathbb{R}^d$, and let $G = (V, E)$ be a t-spanner for V, for some real constant $t > 1$, having $O(n)$ edges. We can preprocess G in $O(n^2 \log n)$ time and space, such that for any two points p and q in V, we can compute, in $O(1)$ time, a $(1 + \varepsilon)$-approximation to the shortest-path distance in G between p and q, where ε is a positive real constant.*

This is hardly impressive since using Dijkstra's algorithm to compute all-pairs-shortest-paths and then saving the results in a matrix would give a better result. In the next section we will show how to improve both the time and space used to $O(n \log n)$.

4   Improving the Complexity

By considering the above construction, one observes that the main reason why the complexity is so high, is the size of the ℓ graphs constructed as input to the PSP-algorithm. The trivial bound we obtain is that each of the ℓ graphs has size $O(n)$ and $\ell \leqslant n$, thus resulting in an overall $O(n^2 \log n)$ time and space bound. A way to improve the complexity would be to construct "sparser" graphs, for example cluster-graphs. This will improve both the preprocessing time and the space complexity to $O(n \log n)$, as will be shown below. Since each input graph will be a subset of the original input graph, this will complicate the construction and the query structure considerably. This section is organized as follows. First we show how to construct the ℓ cluster graphs, then it is shown that the total space complexity of the input graphs can be bounded by $O(n \log n)$. Finally we consider the problem of answering queries. The correctness of the data structure is proven and finally it is shown how to answer a query in constant time.

4.1   Constructing the Cluster-Graphs

The aim of this section is to produce the ℓ graphs $G_1, \ldots, G_\ell$ that will be the input to the PSP-algorithms. We define the notions of *cluster* and *cluster graph* as follows. Given a graph $G = (V, E)$, a cluster C with cluster center at v is a maximal set of vertices of V such that $\delta_G(v, u) < \infty$ for every vertex u in C. Let $cc(u)$ denote the cluster center of the cluster that u belongs to.

A cluster graph $H_{G,E'}$, of a graph $G = (V, E)$ and a set of edges E', has vertex set $\mathcal{V}$ and edge set $\mathcal{E}$ which are defined as follows. Let $C = \{C_1, \ldots, C_b\}$ be a minimal set of clusters of G. For every edge $(u, v) \in E'$ there is an edge $(cc(u), cc(v)) \in \mathcal{E}$, with weight $|cc(u), cc(v)|$, if and only if u and v are in different clusters. The vertex set $\mathcal{V}$ is the set of cluster centers of C adjacent to at least one edge in $\mathcal{E}$. The cluster graph can be constructed in linear time with respect to the complexity of G and the number of edges in E'. Below we describe, using pseudo code, how this can be done.

ComputeClusterGraph$(G = (V, E), E')$
1.   **for** each edge $(u, v) \in E'$ **do**
2.   add edge $(cc(u), cc(v))$ with weight $|cc(u), cc(v)|$ to $\mathcal{E}$
3.   **while** V not empty **do**
4.   pick an arbitrary vertex $v \in V$
5.   remove every vertex u from V for which $\delta_G(v, u) < \infty$
6.   **if** $\exists u \in V$ such that $(v, u) \in \mathcal{E}$ **then**
7.   add v to $\mathcal{V}$
8.   **output** $H = (\mathcal{V}, \mathcal{E})$

Following the above definition of a cluster graph it is now easy to construct the ℓ cluster graphs $G_1, \ldots, G_\ell$ which will be the input graphs for the construction of the ℓ PSP-structures, $P_1, \ldots, P_\ell$. Let $G_1 = (V, E_1)$ and $G_2 = (V, E_1 \cup E_2)$. For each value of i, $3 \leqslant i \leqslant \ell$, the input graph $G_i = (\mathcal{V}_i, \mathcal{E}_i)$ is the cluster graph of G_{i-2} and the edge set $(E_{i-1} \cup E_i)$.

Lemma 2. *The ℓ PSP-structures uses $O(n \log n)$ space and can be constructed in time $O(n \log n)$.*

Proof. One PSP-structure P_i uses $O((|\mathcal{V}_i| + |\mathcal{E}_i|) \log(|\mathcal{V}_i| + |\mathcal{E}_i|))$ preprocessing time and space, according to Corollary 2. So the total time and space needed is:

$$\sum_{i=1}^{\ell} (|\mathcal{V}_i| + |\mathcal{E}_i|) \log(|\mathcal{V}_i| + |\mathcal{E}_i|). \tag{1}$$

We know from the above definition of $\mathcal{V}_i$ and $\mathcal{E}_i$ that $|\mathcal{V}_i| \leqslant 2|E_i \cup E_{i-1}|$ and that $|\mathcal{E}_i| = |E_i \cup E_{i-1}|$. Hence $|\mathcal{V}_i| + |\mathcal{E}_i| \leqslant 3|E_i \cup E_{i-1}|$. The input graph G has $|E| = O(n)$ edges, which implies that $\sum_{i=1}^{\ell}(|\mathcal{V}_i| + |\mathcal{E}_i|) \leqslant 3\sum_{i=1}^{\ell}|E_i \cup E_{i-1}| = O(n)$ and, hence, (1) will sum up to $O(n \log n)$. The time to compute the ℓ cluster graphs is linear with respect to the total input size, hence a total time of $O(n \log n)$.

PreProcessing$(G = (V, E), \varepsilon)$
1.   construct $E_1, \ldots, E_\ell$
2. $G_1 := (V, E_1)$, $G_2 := (V, E_1 \cup E_2)$
3.   **for** $i := 3$ to ℓ **do**
4. $G_i :=$COMPUTECLUSTERGRAPH$(G, E_{i-1} \cup E_i)$
5. $P_i :=$CONSTRUCTPSP(G_i, ε')
5.   construct an approximate single-linkage hierarchy $\mathcal{T}$ of V
6.   process $\mathcal{T}$ to allow efficient LCA-queries
7.   construct cluster tree L of G
8.   process L to allow efficient level-ancestor queries

Next it will be shown that given a query (p, q) there always exists an appropriate PSP-structure that answers $(1 + \varepsilon)$-ASD queries. Finally, in Section 4.3, we show how this appropriate PSP-structure can be found in constant time.

4.2   There Exists an Appropriate PSP-Structure

Assume that we are given a spanner $G = (V, E)$, a $(1 + \varepsilon)$-ADS query (p, q), and the appropriate PSP-structure P_i for (p, q). As the observant reader already noticed, p and/or q may not be vertices in $\mathcal{V}_i$, which implies that we cannot query P_i with (p, q). Instead it will be shown that querying P_i with $(cc(p), cc(q))$ will help to answer the $(1 + \varepsilon)$-ASD query on G. Let G_i be the cluster graph given as input to the PSP-algorithm that constructed P_i.

Observation 6 *The diameter of a cluster in G_i is at most $|pq|/n^{2k-1}$.*

Proof. According to the requirements the ratio between edge weights in E_i and edge weights in E_{i-2} is at least n^{2k}, and since the longest path in a graph contains at most n edges, we know that the diameter of a cluster in G_i is at most a factor $1/n^{2k-1}$ times longer than the shortest edge in E_i.

The following observation can now be derived.

Observation 7 *Given a $(1 + \varepsilon)$-ASD query (p, q) on a graph G and a cluster graph G_i, where P_i is the appropriate PSP-structure for (p, q) and G constructed with G_i as input, we have*

$$\delta_{G_i}(cc(p), cc(q)) - \frac{2|pq|}{n^{2k-2}} \leqslant \delta_G(p, q) \leqslant \delta_{G_i}(cc(p), cc(q)) + \frac{2|pq|}{n^{2k-2}}.$$

Proof. Let L_1 the path in G between p and q having weight $\delta_G(p,q)$. We shall use the notation $L_1(y,x)$ to denote the vertices of L_1 between vertices y and x, not including y. We construct a cluster path L_2 from p to q in G_i as follows. Let C_0 be the cluster with cluster center $v_0 := cc(p)$. Among all vertices adjacent to v_0 in G_i, let v_1 be the cluster center whose cluster C_1 intersects C_0 in the furthest vertex, say w_1, along $L_1(p,q)$. Add the edge (v_0, v_1) to L_2. Next, among all vertices adjacent to v_1 in G_i, let v_2 be the cluster center whose cluster C_2 intersects C_1 in the furthest vertex, say w_2, along $L_1(w_1, v)$. Add the edge (v_1, v_2) to L_2. This process continues until we reach $cc(q) = v_m$. The two paths are illustrated in Figure 1.

From Observation 6, the difference in length between L_1 and L_2 within one cluster is bounded by $2|pq|/n^{2k-1}$. Since a shortest path can visit at most n clusters and at most n vertices, it follows that the error is bounded by $n \cdot 2|pq|/n^{2k-1} = 2|pq|/n^{2k-2}$, hence the observation follows.

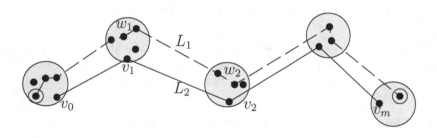

Fig. 1. Illustrating the two paths L_1 and L_2.

Observation 8 *For every graph G and every query (p,q) there is an appropriate PSP-structure P_i.*

Proof. There is a path in G between p and q of length at most $t \cdot |pq|$. This path has length at least $|pq|$, and contains at most n edges. Hence, the longest edge, say e' on this path satisfies $|pq|/n \leqslant wt(e') < t \cdot |pq|$. This means that there is a set E_i such that $\min_{e \in E_i} |e| \leqslant t \cdot |pq|$ and $|pq| \leqslant n^{k+1} \min_{e \in E_i} |e|$.

We summarize this section with the following lemma:

Lemma 3. *A $(1 + \varepsilon)$-ADS query (p,q) on G can be answered by returning the value $\delta + \frac{2|pq|}{n^{2k-2}}$, where δ is the answer obtained by performing a $(1 + \varepsilon')$-ADS query $(cc(p), cc(q))$ on the appropriate PSP-structure for (p,q) on G, where $\varepsilon' \leqslant \varepsilon - \frac{\varepsilon+4}{n^{2k-2}}$.*

Proof. Assume that we are given a $(1 + \varepsilon)$-ASD query (p,q) on G. According to Observation 8 there is an appropriate PSP-structure P_i for (p,q) and G. Now assume that we perform a $(1 + \varepsilon')$-ASD query $(cc(p), cc(q))$ on P_i where we let δ denote the answer of the query. According to Observation 7, we have $\delta_G(p,q) - \frac{2|pq|}{n^{2k-2}} \leqslant \delta_{G_i}(cc(p),cc(q)) \leqslant \delta \leqslant (1+\varepsilon') \cdot \delta_{G_i}(cc(p),cc(q)) \leqslant (1+\varepsilon') \cdot (\delta_G(p,q) - \frac{2|pq|}{n^{2k-2}})$, which gives

$$\delta_G(p,q) \leqslant \delta + \frac{2|pq|}{n^{2k-2}} \leqslant (1+\varepsilon') \cdot \delta_{G_i}(cc(p),cc(q)) + 4(1+\varepsilon')\frac{2|pq|}{n^{2k-2}}.$$

Now choose $\varepsilon' \leqslant \varepsilon - \frac{\varepsilon+4}{n^{2k-2}}$. If a $(1+\varepsilon)$-ASD query (p,q) is performed on G we can answer the query with the value $\delta + \frac{2|pq|}{n^{2k-2}}$ where δ is obtained by performing a $(1+\varepsilon')$-ADS query $(cc(p),cc(q))$ on the appropriate PSP-structure P_i for (p,q).

The result of this section is that if the appropriate PSP-structure for G and (p,q) is P_i then the $(1+\varepsilon)$-ASD query (p,q) on G can be replaced by a $(1+\varepsilon')$-ADS query $(cc(p),cc(q))$ on G_i with $\varepsilon' \leqslant \varepsilon - \frac{\varepsilon+4}{n^{2k-2}}$. By choosing k equal to 2, and assuming that n is sufficiently large, we can take $\varepsilon' = \varepsilon/2$.

4.3   Answering ASD-Queries

Above it was shown that a $(1+\varepsilon)$-ASD query (p,q) can be answered provided that the appropriate PSP-structure and the cluster centers of p and q are found. In this section we will first show how the appropriate PSP-structure P_i can be found in constant time and finally we show how the cluster centers of p and q can be found in constant time.

Build an approximate single-link hierarchical cluster tree, denoted $\mathcal{T}$, as described in Section 3.2. The tree is then preprocessed such that LCA-queries can be answered in constant time [10]. The final step of the preprocessing is to associate a PSP-structure to each node in $\mathcal{T}$, see Section 3.2.

Given a query (p,q) compute the LCA ν of p and q in $\mathcal{T}$. The LCA-query will approximate the length of (p,q) within a factor of $O(n)$, according to Observation 5, which is sufficient for our needs. Hence, ν will have a pointer to a PSP-structure P_i and one of the PSP-structures P_{i-1} or P_i will be the appropriate PSP-structure, which one it is can be decided in constant time.

Finding the cluster centers. Next the query should be passed on to the appropriate PSP-structure. It might be that p and/or q are not points in $\mathcal{V}_i$. Recall that $\mathcal{V}_i$ is the set of input points for the appropriate PSP-structure P_i. Hence, it remains to show that one can find the vertex $cc(p)$ in $\mathcal{V}_i$ that is the cluster center of a query point p in constant time. For this purpose we build the cluster tree L. Let r be the root of L. The parent of a node v in L is denoted $parent(v)$. The depth of v is $d(v) = d(parent(v))+1$, and $d(r) = 0$. The tree has $\ell+1$ levels where the number of nodes at level $1 \leqslant j \leqslant \ell$ is $|\mathcal{V}_{\ell-j+1}|$. At level j in L, every node μ corresponds to a vertex v in $\mathcal{V}_{\ell-j+1}$, hence at leaf level, level ℓ, there are n nodes (the same holds for level $\ell-1$). It holds that $parent(\mu)$ is the node μ' at level $j-1$ that corresponds to the cluster center $cc(v)$ in $G_{\ell-j}$. To obtain a tree we finally add a root at level 0 that connects the nodes at level 1. The cluster tree L can be computed in linear time w.r.t. the number of nodes, and according to the proof of Lemma 2 this is $O(n)$, hence, in total time $O(n\log n)$.

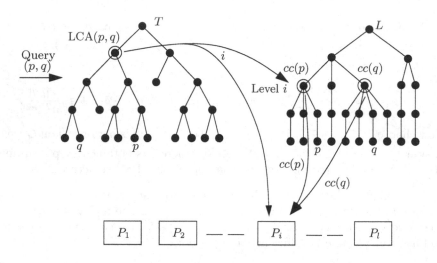

Fig. 2. Processing a query (p, q).

Note that finding the cluster center $cc(p)$ in G_i of a point p in G is equivalent of finding the ancestor of the leaf p at level $j = \ell - i + 1$ in L. We will use a modified variant of the data structure presented by Alstrup and Holm [2]. The structure answers level ancestor queries in constant time. Given a leaf v of a tree T and a level l of T, the level ancestor of v and l is the ancestor of v at level l. Note that if we query the cluster tree $\mathcal{T}$ with a vertex v and a level j the data structure would return the ancestor of v at level j which is the cluster center of v in the cluster graph $G_{\ell-i+1}$.

We briefly describe the modified version of the data structure by Alstrup and Holm [2].

Finding level ancestor. Assume that we are given a tree L with a total of N nodes. A query $LevelAncestor(x, l)$ returns the ancestor y to x at level l in L. The number of nodes in the subtree rooted at v is denoted $s(v)$. The notation $d|a$ means that $a = k \cdot d$ for some positive integer k. Define the *rank* of v, denoted $r(v)$ to be the maximum integer i such that $2^i | d(v)$ and $s(v) \geqslant 2^i$. Note that the rank of the root will be $\lfloor \log_2 n \rfloor$.

Fact 9 *(Observation 1 in [2]) The number of nodes of rank$\geqslant i$ is at most $\lfloor N/2^i \rfloor$.*

The preprocessing algorithm consists of precomputing the depth, size and rank of each node in the tree and then constructing the following three tables:

`logtable`$[x]$: contains the value $\lfloor \log_2 x \rfloor$, for $0 \leqslant x \leqslant N$.
`levelanc`$[v][x]$: contains the x'th ancestor to v, for $0 \leqslant x \leqslant 2^{r(v)}$.
`jump`$[v][i]$: contains the first ancestor to v whose depth is divisible by 2^i, for $0 \leqslant i \leqslant \log_2(d(v) + 1)$.

The idea is then to use the `jump[][]` table a constant number of times in order to reach a node w with a sufficiently large rank to hold the answer to the level ancestor query in its table. The pseudocode looks as follows:

LevelAncestor(v, x)
1. $i := \texttt{logtable}[x+1]$
2. $d := d(v) - x$
3.   **while** $2(d(v) - d) \geqslant 2^i$ **do**
4. $v := \texttt{jump[v][i-1]}$
5.   **return** $\texttt{levelanc[v][d(v)-d]}$

Corollary 2. *(Modified version of Lemma 2 in [2])*
A tree T with N nodes can be preprocessed in $O(N \log N)$ time and space allowing level ancestor queries to be answered in constant time.

Since the size of the L is $O(n)$ we get that L can be constructed in total time $O(n \log n)$ and then processed in time $O(n \log n)$ using $O(n \log n)$ space such that level ancestor-queries can be answered in constant time.

ASD-query(p, q)
1. $\nu := \text{LCA}(\mathcal{T}, p, q)$
2. $P_i := \nu.\text{PSP-structure}$
3. $cc(p) := \text{LEVELANCESTOR}(L, p, i)$
4. $cc(q) := \text{LEVELANCESTOR}(L, q, i)$
5. $\delta := \text{QUERYPSP}(P_i, \varepsilon')$
6.   **return** $(\delta + \frac{2|pq|}{n^{k-2}})$

The following lemma concludes this section.

Lemma 4. *We can preprocess a graph $G = (V, E)$ and a set of PSP-structures of G in time $O(n \log n)$ using $O(n \log n)$ space such that for any two points $p, q \in V$, we can compute in constant time the appropriate PSP-structure P for (p, q) and G, and the cluster centers of p and q in the cluster graph corresponding to P.*

Putting together the results in Lemmas 2, 3 and 4 gives us Theorem 1.

5   Concluding Remarks

We have presented a data structure which supports $(1+\varepsilon)$-approximate shortest distance queries in constant time for geometric t-spanners, hence functions as an approximate distance oracle. This generalises the ASD-oracle for restricted queries presented by the authors in [9] to arbitrary queries. It has been shown that the data structure can be applied to a large number of problems, for example, closest pair queries, shortest paths among obstacles, computing stretch factors of a geometric graph, and so on. We believe that the techniques used are of independent interest.

In the full paper we also show that the structure easily can be extended such that $(1 + \varepsilon)$-approximate shortest path queries for geometric t-spanners can be answered in time proportional to the number of edges along the path.

References

1. D. Aingworth, C. Chekuri, P. Indyk, and R. Motwani. Fast estimation of diameter and shortest paths (without matrix multiplication). *SIAM Journal on Computing*, 28:1167–1181, 1999.
2. S. Alstrup and J. Holm, Improved algorithms for finding level ancestors in dynamic trees. In Proc. *27th International Colloquium on Automata, Languages, and Programming*, 2000.
3. S. Arikati, D. Z. Chen, L. P. Chew, G. Das, M. Smid, and C. D. Zaroliagis. Planar spanners and approximate shortest path queries among obstacles in the plane. In Proc. *4th European Symposium on Algorithms*, LNCS 1136, pp. 514–528, 1996.
4. D. Z. Chen. On the all-pairs Euclidean short path problem. In Proc. *6th ACM-SIAM Symposium on Discrete Algorithms*, pp. 292–301, 1995.
5. D. Z. Chen, K. S. Klenk, and H.-Y. T. Tu. Shortest path queries among weighted obstacles in the rectilinear plane. *SIAM Journal on Computing*, 29:1223–1246, 2000.
6. Y.-J. Chiang and J. S. B. Mitchell. Two-point Euclidean shortest path queries in the plane. In Proc. *10th ACM-SIAM Symposium on Discrete Algorithms*, 1999.
7. E. Cohen. Fast algorithms for constructing t-spanners and paths with stretch t. *SIAM Journal on Computing*, 28:210–236, 1998.
8. D. Dor, S. Halperin, and U. Zwick. All-pairs almost shortest paths. *SIAM Journal on Computing*, 29:1740–1759, 2000.
9. J. Gudmundsson, C. Levcopoulos, G. Narasimhan and M. Smid. Approximate Distance Oracles for Geometric graphs. In Proc. *13th ACM-SIAM Symposium on Discrete Algorithms*, 2002.
10. D. Harel and R. E. Tarjan. Fast algorithms for finding nearest common ancestors. *SIAM Journal on Computing*, 13:338–355, 1984.
11. J. B. Kruskal, Jr. On the shortest spanning subtree of a graph and the traveling salesman problem. Proc. Amer. Math. Soc. 7(1956):48–50, 1956.
12. D. Krznaric, C. Levcopoulos and B. J. Nilsson. Minimum Spanning Trees in d Dimensions. *Nordic Journal of Computing*, 6(4):446–461, 1999.
13. J. S. B. Mitchell. Shortest paths and networks. In *Handbook of Discrete and Computational Geometry*, pp. 445–466. CRC Press LLC, 1997.
14. M. Thorup and U. Zwick. Approximate distance oracles. In Proc. *33rd ACM Symposium on Theory of Computing*, 2001.

Flat-State Connectivity of Linkages under Dihedral Motions[*]

Greg Aloupis[1], Erik D. Demaine[2], Vida Dujmović[1], Jeff Erickson[3],
Stefan Langerman[1], Henk Meijer[4], Joseph O'Rourke[5], Mark Overmars[6],
Michael Soss[7], Ileana Streinu[5], and Godfried T. Toussaint[1]

[1] School of Computer Science, McGill University
{athens, vida, sl, godfried}@uni.cs.mcgill.ca
[2] Laboratory for Computer Science, Massachusetts Institute of Technology
edemaine@mit.edu
[3] Dept. of Computer Science, University of Illinois at Urbana-Champaign
jeffe@cs.uiuc.edu
[4] Dept. of Computing and Information Science, Queen's University
henk@cs.queensu.ca
[5] Dept. of Computer Science, Smith College
{streinu, orourke}@cs.smith.edu
[6] Dept. of Information and Computing Sciences, Utrecht University
markov@cs.uu.nl
[7] Chemical Computing Group Inc.
soss@chemcomp.com

Abstract. We explore which classes of linkages have the property that each pair of their flat states—that is, their embeddings in $\mathbb{R}^2$ without self-intersection—can be connected by a continuous dihedral motion that avoids self-intersection throughout. Dihedral motions preserve all angles between pairs of incident edges, which is most natural for protein models. Our positive results include proofs that open chains with nonacute angles are flat-state connected, as are closed orthogonal unit-length chains. Among our negative results is an example of an orthogonal graph linkage that is flat-state disconnected. Several additional results are obtained for other restricted classes of linkages. Many open problems are posed.

1 Introduction

Motivation: Locked Chains. There has been considerable research on reconfiguration of polygonal chains in 2D and 3D while preserving edge lengths and avoiding self-intersection. Much of this work is on the problem of which classes of chains can *lock* in the sense that they cannot be reconfigured to straight or convex configurations. In 3D, it is known that some chains can lock [4], but the exact class of chains that lock has not been delimited [3]. In 2D, no chains can lock [5,13]. All of these results concern chains with *universal joints*.

---

[*] Research initiated at the 17th Winter Workshop on Computational Geometry, Bellairs Research Institute of McGill University, Feb. 1–8, 2002.

P. Bose and P. Morin (Eds.): ISAAC 2002, LNCS 2518, pp. 369–380, 2002.
© Springer-Verlag Berlin Heidelberg 2002

Motivation: Protein Folding. The backbone of a protein can be modeled as a polygonal chain, but the joints are not universal; rather the bonds between residues form a nearly fixed angle in space. The study of such *fixed-angle* chains was initiated in [11], and this paper can be viewed as a continuation of that study. Although most protein molecules are linear polymers, modeled by open polygonal chains, others are rings (closed polygons) or star and dendritic polymers (trees) [12,6].

The polymer physics community has studied the statistics of "self-avoiding walks" [9,10,15], i.e., non-self-intersecting configurations, often constrained to the integer lattice. To generate these walks, they consider transformations of one configuration to another, such as "pivots" [7] or "wiggling" [8]. Usually these transformations are not considered true molecular movements, often permitting self-intersection during the motion, and perhaps are better viewed as string edits.

In contrast, this paper maintains the geometric integrity of the chain throughout the transformation, to more closely model the protein folding process. We focus primarily on transformations between planar configurations.

Fixed-angle linkages. Before describing our results, we introduce some definitions. A *(general) linkage* is a graph with fixed lengths assigned to each edge. The focus of this paper is *fixed-angle linkages*, which are linkages with, in addition, a fixed angle assigned between each pair of incident edges. We use the term *linkage* to include both general and fixed-angle linkages.

A *configuration* or *realization* of a general linkage is a positioning of the linkage in $\mathbb{R}^3$ (an assignment of a point in $\mathbb{R}^3$ to each vertex) achieving the specified edge lengths. The *configuration space* of a linkage is the set of all its configurations. To match physical reality, of special interest are *non-self-intersecting configurations* or *embeddings* in which no two nonincident edges share a common point. The *free space* of a linkage is the set of all its embeddings, i.e., the subset of configuration space for which the linkage does not "collide" with itself.

A configuration of a fixed-angle linkage must additionally respect the specified angles. The definitions of configuration space, embedding, and free space are the same. A *reconfiguration* or *motion* or *folding* of a linkage is a continuum of configurations. Motions of fixed-angle linkages are distinguished as *dihedral motions*.

Dihedral motions. A dihedral motion can be "factored" into *local dihedral motions* or *edge spins* [11] about individual edges of the linkage. Let $e = (v_1, v_2)$ be an edge for which there is another edge e_i incident to each endpoint v_i. Let Π_i be the plane through e and e_i. A *(local) dihedral motion about e* changes the dihedral angle between the planes Π_1 and Π_2 while preserving the angles between each pair of edges incident to the same endpoint of e. See Fig. 1. The edges incident to a common vertex in a fixed-angle linkage are moved rigidly by a dihedral motion. In particular,

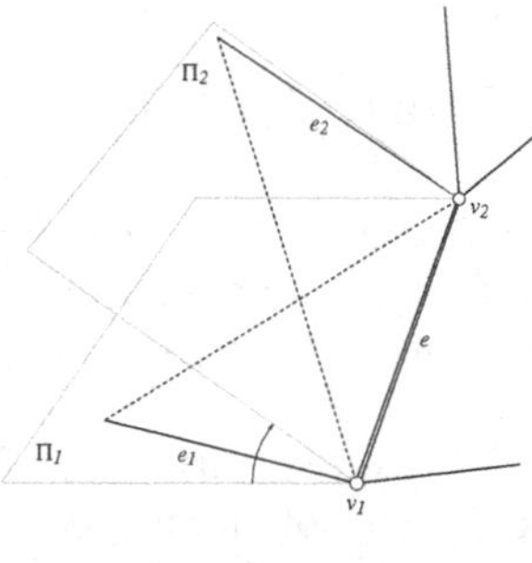

Fig. 1. A local dihedral motion (spin) about edge e.

if the edges are coplanar, they remain coplanar.[1] If we view e and $e_1 \in \Pi_1$ as fixed, then a dihedral motion spins e_2 about e.

Flat-state connectivity. A *flat state* of a linkage is an embedding of the linkage into $\mathbb{R}^2$ without self-intersection. A linkage X is *flat-state connected* if, for each pair of its (distinct, i.e., incongruent) flat states X_1 and X_2, there is a dihedral motion from X_1 to X_2 that stays within the free space throughout. In general this dihedral motion alters the linkage to nonflat embeddings in $\mathbb{R}^3$ intermediate between the two flat states. If a linkage X is not flat-state connected, we say it is *flat-state disconnected.*

Flat-state disconnection could occur for two reasons. It could be that there are two flat states X_1 and X_2 which are in different components of free space but the same component of configuration space. Or it could be that the two flat states are in different components of configuration space. The former reason is the more interesting situation for our investigation; currently we have no nontrivial examples of the latter possibility.

Results. The main goal of this paper is to delimit the class of linkages that are flat-state connected. Our results apply to various restricted classes of linkages, which are specified by a number of constraints, both topological and geometric. The topological classes of linkages we explore include general graphs, trees, chains (paths), both open and closed, and sets of chains. We sometimes restrict all link lengths to be the same, a constraint of interest in the context of protein backbones; we call these *unit-length* linkages. We consider a variety of restrictions on the angles of a fixed-angle linkage, where the angle between two incident links is the smaller of the two angles between them within their common plane. A chain *has a monotone state* if it has a flat state in which it forms a monotone chain in the plane. For sets of chains in a flat state, we *pin* each chain at one of the end links, keeping its position fixed in the plane.

In some cases we restrict the motions of a linkage in one of two ways. First, we may enforce that only certain edges permit local dihedral motion, in which case we call the linkage *partially rigid.* (Such a restriction also constrains the flat states that we can hope to connect, slightly modifying the definition of flat-state connected.) Second, we may restrict the motion to consist of a sequence of 180° edge spins, so that each move returns the linkage to the plane. Most of our examples of flat-state disconnected linkages are either partially rigid or restricted to 180° edge spins.

With the above definitions, we can present our results succinctly in Table 1.

---

[1] Our definition of "dihedral motion" includes rigid motions of the entire linkage, which could be considered unnatural because a rigid motion has no local dihedral motions. However, including rigid motions among dihedral motions does not change our results. For a linkage of a single connected component, we can modulo out rigid motions; and for multiple connected components, we always pin vertices to prevent rigid motions.

Table 1. Summary of results. The '—' means no restriction of the type indicated in the column heading. Entries marked '?' are open problems

Constraints on Fixed-Angle Linkage				Flat-state
Connectivity	Angles	Lengths	Motions	connectivity
Open chain	—	—	—	?
	has a monotone state		—	?
	nonacute	—	—	Connected
	equal acute	—	—	Connected [2]
	each in $(60°, 90°]$	unit	—	Connected [2]
	—	—	180° edge spins	Disconnected
	orthogonal	—	180° edge spins	Connected
Set of chains, each	orthogonal	—	—	Connected
pinned at one end	orthogonal	—	partially rigid	Disconnected
Closed chain	—	—	—	?
	nonacute	—	—	?
	orthogonal	—	—	?
	orthogonal	unit	—	Connected
Tree	—	—	—	?
	orthogonal	—	—	?
	orthogonal	—	partially rigid	Disconnected
Graph	orthogonal	—	—	Disconnected

2   Flat-State Disconnection

It may help to start with negative results, as it is not immediately clear how a linkage could be flat-state disconnected. Several of our examples revolve around the same idea, which can be achieved under several models. We start with partially rigid orthogonal trees, and then modify the example for other classes of linkages.

2.1   Partially Rigid Orthogonal Tree

An *orthogonal tree* is a tree linkage such that every pair of incident links meet at a multiple of 90°. *Partial rigidity* specifies that only certain edges permit dihedral motions. Note that the focus of a dihedral motion is an edge, not the joint vertex.

Fig. 2(a–b) shows two incongruent flat states of the same orthogonal tree; we'll call the flat states $X_{(a)}$ and $X_{(b)}$. All but four edges of the tree are frozen, the four incident to the central degree-4 root vertex x. Call the 4-link branch of the tree containing a the *a-branch*, and similarly for the others. Label the vertices of the a-branch (a, a_1, a_2, a_3), and similarly for the other branches.

We observe three properties of the example. First, as mentioned previously, fixed-angle linkages have the property that all links incident to a particular vertex remain coplanar throughout all dihedral motions. In Fig. 2, this means that $\{x, a, b, c, d\}$ remain coplanar; and we view this as the plane Π of the flat

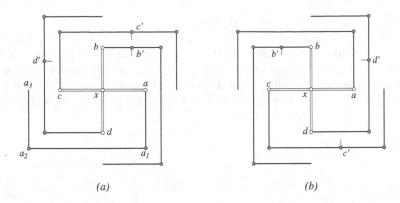

(a)   (b)

Fig. 2. Two flat states of a partially rigid orthogonal tree. The four open edges are the only ones not rigid, permitting dihedral motions.

states under consideration. Note that, for example, a rotation of a about bd would maintain the 90° angles between all edges adjacent consecutively around x, but would alter the 180° angle between xa and xc, and thus is not a fixed-angle motion.

Second, the short links, or "pins," incident to vertices b', c', and d' must remain coplanar with their branch, because they are rigid. For example, the b' pin must remain coplanar with xb, for otherwise the rigid edge bb' would twist.

Third, $X_{(a)}$ and $X_{(b)}$ do indeed represent incongruent flat states of the same linkage. The purpose of the b' pin is to ensure that its relation to (say) the c' pin in the two states is not the same. Without the b' pin, a flat state congruent to $X_{(b)}$ could be obtained by a rigid motion of the entire linkage, flipping it upside-down. It is clear that state $X_{(b)}$ can be obtained from state $X_{(a)}$ by rotating the a-branch 180° about xa, and similarly for the other branches. Thus the two flat states are in the same component of configuration space. We now show that they are in different components of the free space.

Theorem 1. *The two flat states in Fig. 2 of an orthogonal partially rigid fixed-angle tree cannot be reached by dihedral motions that avoid crossing links.*

Proof: Each of the four branches of the tree must be rotated 180° to achieve state $X_{(b)}$. We first argue that two opposite branches cannot rotate to the same side of the Π-plane, either both above or both below. Without loss of generality, assume both the a- and the c-branches rotate above Π. Then, as illustrated in Fig. 3, vertex a_1 must hit a point on the c_1c_2 edge, for the length aa_1 is the same as the distance from a to c_1c_2.

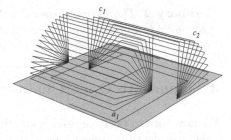

Fig. 3. The a- and c-branches collide when rotated above.

Now we argue that two adjacent branches cannot rotate to the same side of Π. Consider the a- and b-branches, again without loss of generality. As it is more difficult to identify an exact pair of points on the two branches that must collide, we instead employ a topological argument. Connect a shallow rope R from a to a_3 underneath Π, and a rope S from b to b_3 that passes below R. See Fig. 4. In $X_{(a)}$, the two closed loops

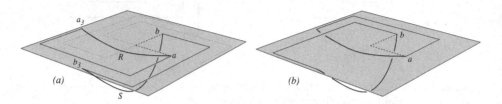

Fig. 4. With the additions of the ropes R and S underneath, the a-chain is not linked with the b-chain in (a), but is linked in (b).

$A = (R, a, a_1, a_2, a_3)$ and $B = (S, b, b_1, b_2, b_3)$ are unlinked. But in $X_{(b)}$, A and B are topologically linked. Therefore, it is not possible for the a- and b-branches to rotate above Π without passing through one another.

By the pigeon-hole principle, at least two branches must rotate though Π. Whether these branches are opposite or adjacent, a collision is forced.   □

2.2   Orthogonal Graphs and Partially Rigid Pinned Chains.

We can convert the partially rigid tree in Fig. 2 to a completely flexible graph by using extra "braces" to effectively force the partial rigidity. We can also convert the tree into four partially rigid chains, each pinned at one endpoint near the central degree-4 vertex. Thus we obtain the following two results:

Corollary 1 *The described orthogonal fixed-angle linkage has two flat states that are not connnected by dihedral motions that avoid crossing links.*

Corollary 2 *The four orthogonal partially rigid fixed-angle pinned chains corresponding to Figure 2 are not connected by dihedral motions that avoid crossing links.*

3   Nonacute Open Chains

We now turn to positive results, starting with the simplest and perhaps most elegant case of a single open chain with nonacute angles. After introducing notation, we consider two algorithms establishing flat-state connectivity, in Sections 3.1 and 3.2.

An abstract polygonal chain C of n links is defined by its fixed sequence of link lengths, $(\ell_1, \ldots, \ell_n)$, and whether it is open or closed. For a fixed-angle chain, the $n-1$ or n angles α_i between adjacent links are also fixed. A realization C of a chain is specified by the position of its $n+1$ vertices: $v_0, v_1, \ldots, v_n$. If the chain is closed, $v_n = v_0$. The links or edges of the chain are $e_i = (v_{i-1}, v_i)$, $i = 1, \ldots, n$, so that the vector along the ith link is $v_i - v_{i-1}$. The plane in which a flat state C is embedded is called Π or the xy-plane.

3.1 Lifting One Link at a Time

The idea behind the first (unrestricted) algorithm is to lift the links of the chain one-by-one into a monotone chain in a vertical plane. Once we reach this *canonical state*, we can reverse and concatenate motions to reach any flat state from any other.

We begin by describing the case of orthogonal chains, as illustrated in Fig. 5, and the algorithm will generalize to arbitrary nonacute chains. The invariant at the beginning of each step i of the algorithm is that we have lifted the chain $e_1, \ldots, e_i$ into a monotone chain in a vertical plane, while the rest of the chain $e_{i+1}, \ldots, e_n$ remains where it began in the xy-plane. Initially, $i = 0$ and the lifted chain contains no links, and we simply lift the first link e_1 to vertical by a 90° edge spin around the second link e_2. For general i, we first spin the lifted chain around its last (vertical) link e_i so that the vertical plane contains the next link to lift, e_{i+1}, and so that the chain $e_1, \ldots, e_{i+1}$ is monotone. Then we pick up e_{i+1} by a 90° edge spin around e_{i+2}. Throughout, the lifted chain remains monotone and contained in the positive-z halfspace, so we avoid self-intersection.

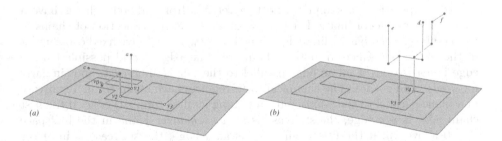

Fig. 5. Picking up a planar orthogonal chain into a monotone canonical state. (a) Lifting edges $e_1 = (v_0, v_1)$ and $e_2 = (v_1, v_2)$: a, b, c. (b) Lifting edges e_3 and e_4: d, e, f.

Nonacute chains behave similarly to orthogonal chains, in particular, the canonical state is monotone, although it may no longer alternate between left and right turns. Now there may be multiple monotone states, and we must choose the state that is monotone in the z dimension. The key property is that, as the chain $e_1, \ldots, e_i$ rotates about e_{i+1}, the chain remains monotone in the z direction, so it does not penetrate the xy-plane.

This algorithm proves the following result:

Theorem 2. *Any nonacute fixed-angle chain is flat-state connected.*

3.2 Lifting Two Links at a Time

The algorithm above makes at most 2 edge spins per link pickup, for a total of $2n$ edge spins to reach the canonical state, or $4n$ edge spins to reach an arbitrary flat state from any other. This bound is tight within an additive constant.

We can reduce the number of edge spins to $1.5n$ to reach the canonical state, or $3n$ to reach an arbitrary flat state, by lifting two edges in each step as follows. As before, in the beginning of each step, we spin the lifted chain $e_1, \ldots, e_i$ about the last link e_i to orient it to be coplanar and monotone with the next link e_{i+1}. Now we spin by $90°$ the lifted chain and the next two links e_{i+1} and e_{i+2} about the following link e_{i+3}, bringing e_{i+1} and e_{i+2} into a vertical plane, and tilting the lifted chain $e_1, \ldots, e_i$ down to a horizontal plane (parallel to the xy-plane) at the top. Then we spin the old chain $e_1, \ldots, e_i$ by $90°$ around e_{i+1}, placing it back into a vertical plane, indeed the same vertical plane containing e_{i+1} and e_{i+2}, so that the new chain $e_1, \ldots, e_{i+2}$ becomes coplanar and monotone. We thus add two links to the lifted chain after at most three motions, proving the $1.5n$ upper bound; this bound is also tight up to an additive constant.

Corollary 3 *Any nonacute fixed-angle chain with n links can be reconfigured between two given flat states in at most $3n$ edge spins.*

4 Multiple Pinned Orthogonal Open Chains

In this section we prove that any collection of open, orthogonal chains, each with one edge pinned to the xy-plane, can be reconfigured to a canonical form, establishing that such chain collections are flat-state connected. We also require a "general position" assumption: no two vertices from different chains have a common x- or y-coordinate. Let C_i, $i = 1, \ldots, k$, be the collection of chains in the xy-plane. Each has its first edge pinned, i.e., v_0 and v_1 have fixed coordinates in the plane; but dihedral motion about this first edge is still possible (so the edge is not frozen). Call an edge parallel to the x-axis an x-*edge*, and similarly for y-*edge* and z-*edge*. The canonical form requires each chain to be a staircase in a plane parallel to the z-axis and containing its first (pinned) edge. If the first chain edge is a y-edge, the staircase is in a yz *quarter plane* in the halfspace $z > 0$ above xy; if the first chain edge is an x-edge, the staircase is in an xz *quarter plane* in the halfspace $z < 0$ below xy.

The algorithm processes independently the chains that are destined above or below the xy-plane, and keeps them on their target sides of the xy-plane, so there is no possibility of interference between the two types of chains. So henceforth we will concentrate on the chains C_i whose first edge is a y-edge, with the goal of lifting each chain C_i into a staircase S_i in a yz quarter plane. At an intermediate state, the staircase S_i is the portion of the lifted chain above the xy-plane, and C_i the portion remaining in the xy-plane. The *pivot edge* of the staircase is its first edge, which is a z-edge. Let $(\ldots, c_i, b_i, a_i)$ be the last three vertices of the chain C_i. Let a_i have coordinates (a_x, a_y); we'll use notation for b_i and c_i. Vertex a_i at the foot of a staircase is its *base vertex* and the last edge of the chain, (b_i, a_i), is the staircase's *base edge*.

After each step of the algorithm, two invariants are reestablished:

1. All staircases for all chains are in (parallel) yz quarter planes;
2. The base edge for every staircase is a y-edge, i.e., is in the plane of the staircase.

We will call these two conditions the *Induction Hypothesis*.

The main idea of the algorithm is to pick up two consecutive edges of one chain, which then ensures that the next edge of that chain is a y-edge. The chain is chosen arbitrarily. To simplify the presentation, we assume without loss of generality that c_i is to the right of b_i. First, the staircases whose pivot's x-coordinates lie in the range $[b_x, c_x]$ are reoriented to avoid crossing above the (b_i, c_i) edge.

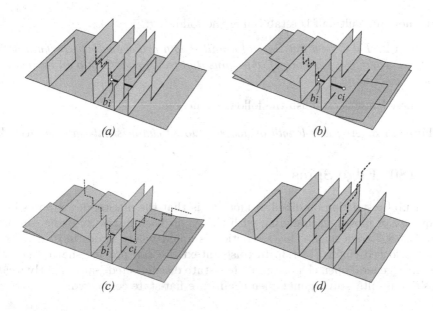

Fig. 6. (a) First, y-edge (a_i, b_i) picked up; (b) Planes parted and flattened in preparation; (c) Two states of staircase shown: Aligned with the second, x-edge (b_i, c_i), and after pickup of that edge; (d) Staircase rotated into vertical plane, and flattened planes made upright.

With S_i aligned with its base y-edge, the (a_i, b_i) edge can be picked up into a vertical plane without collision; see Fig. 6a. We now align S_i with (b_i, c_i), by "parting" the planes at b_x toward the left, laying all planes left of b_x down toward $-x$ (Fig. 6b), and then rotating S_i to be horizontal. Now we pick up (b_i, c_i) into a xz quarter plane, after laying down all planes right of c_x; see Fig. 6(c). Finally, reorient the xz-plane to be vertical and then restore all tilted planes to their yz orientation. We have reestablished the Induction Hypothesis. See Fig. 6(d).

Repeating this process eventually lifts every chain into parallel vertical planes, leaving only the first (pinned) y-edge of each chain in the xy-plane.

5   Unit Orthogonal Closed Chains

Our only algorithm for flat-state connectivity of closed chains is specialized to unit-length orthogonal closed chains. Despite the specialization, it is one of the most complex algorithms, and will only be mentioned in this abstract. We use orthogonally convex polygons as a canonical form, justified by the first lemma:

Lemma 1. *Let C and D be two orthogonally convex embeddings of a unit-length orthogonal closed chain with n vertices. There is a sequence of edge spins that transforms C into D.*

The more difficult half is establishing the following:

Lemma 2. *Let C be a flat state of a unit-length orthogonal closed chain with n vertices. There is a sequence of edge spins that transforms C into an orthogonally convex embedding.*

These lemmas establish the following theorem:

Theorem 3. *Any unit-length orthogonal closed chain is flat-state connected.*

6   180° Edge Spins

A natural restriction on dihedral motions is that the motion decomposes into a sequence of moves, each ending with the chain back in the xy-plane—in other words, 180° edge spins. This restriction is analogous to Erdős flips in the context of locked chains [14,1,3]. In this context, we can provide sharper negative results—general open chains can be flat-state disconnected—and slightly weaker positive results—orthogonal open chains are flat-state connected.

6.1   Restricted Flat-State Disconnection of Open Chains

We begin by illustrating the difficulty in reconfiguring open chains by 180° edge spins; see Fig. 7. Spinning about edge 1 does nothing; spinning about edge 2

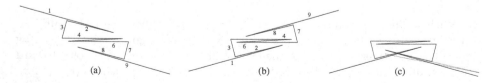

Fig. 7. (a–b) Two flat states of a chain that cannot reach each other via a sequence of 180° edge spins. (c) Attempt at spinning about edge 4.

causes edges 1 and 3 to cross; spinning about edge 3 makes no important change to the flat state; spinning about edge 4 causes edges 2 and 8 to cross as shown in Fig. 7(c); spinning about edge 5 causes edges 4 and 6 to cross (in particular); and the remaining cases are symmetric. This case analysis establishes the following theorem:

Theorem 4. *The two incongruent flat states in Fig. 7(a–b) of a fixed-angle open chain cannot be reached by a sequence of 180° edge spins that avoid crossing links.*

6.2   Restricted Flat-State Connection of Orthogonal Open Chains

The main approach for proving flat-state connectivity of orthogonal chains is outlined in two figures: spin around a convex-hull edge if one exists (Fig. 8), and otherwise decompose the chain into a monotone (staircase) part and an inner part, and spin around a convex-hull edge of the inner part (Fig. 9). Such spins avoid collisions because of the empty infinite strips $R(e_1)$, $R(e_2)$, ... through the edges of the monotone part of the chain. In Fig. 9, the monotone portion of the chain is e_1, e_2, e_3, which terminates with the first edge e_3 that does not have an entire empty strip $R(e_3)$. Each spin of either type makes the chain more monotone in the sense of turning an edge whose endpoints turn in the same direction into an edge whose endpoints turn in opposite directions; hence, the number of spins is at most n. Using a balanced-tree structure to maintain information about recursive subchains, each step can be executed in $O(\log n)$ time, for a total of $O(n \log n)$ time. In addition, we show how the algorithm can be modified to keep the chain in the nonnegative-x halfspace with one vertex pinned against the $x = 0$ plane.

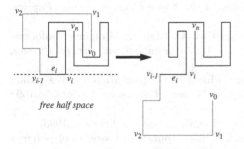

Fig. 8. A dihedral rotation about a convex-hull edge resolves a violation of the canonical form.

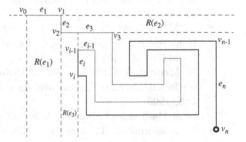

Fig. 9. Determining the chain e_1, e_2, ..., e_{i-1} that can be rotated about e_i.

Theorem 5. *Orthogonal chains are flat-state connected even via restricted sequences of 180° spins that keep the chain in the nonnegative-x halfspace with one vertex pinned at $x = 0$. The sequence of $O(n)$ spins can be computed in $O(n \log n)$ time.*

7 Conclusion and Open Problems

See Table 1 for several open problems. In particular, these three classes of chains seem most interesting, with the first being the main open problem:

1. Open chains (no restrictions).
2. Open chains with a monotone flat state.
3. Orthogonal trees (all joints flexible).

References

1. Oswin Aichholzer, Carmen Cortés, Erik D. Demaine, Vida Dujmović, Jeff Erickson, Henk Meijer, Mark Overmars, Belén Palop, Suneeta Ramaswami, and Godfried T. Toussaint. Flipturning polygons. In *Proc. Japan Conf. Discrete Comput. Geom.*, Lecture Notes in Computer Science, Tokyo, Japan, November 2000. To appear in *Discrete Comput. Geom.*
2. Greg Aloupis, Erik D. Demaine, Henk Meijer, Joseph O'Rourke, Ileana Streinu, and Godfried Toussaint. Flat-state connectedness of fixed-angle chains: Special acute chains. In *Proc. 14th Canad. Conf. Comput. Geom.*, August 2002.
3. T. Biedl, E. Demaine, M. Demaine, S. Lazard, A. Lubiw, J. O'Rourke, M. Overmars, S. Robbins, I. Streinu, G. Toussaint, and S. Whitesides. Locked and unlocked polygonal chains in 3D. *Discrete Comput. Geom.*, 26(3):269–282, 2001.
4. J. Cantarella and H. Johnston. Nontrivial embeddings of polygonal intervals and unknots in 3-space. *J. Knot Theory Ramifications*, 7:1027–1039, 1998.
5. R. Connelly, E. D. Demaine, and G. Rote. Straightening polygonal arcs and convexifying polygonal cycles. In *Proc. 41st Annu. IEEE Sympos. Found. Comput. Sci.*, pages 432–442. IEEE, November 2000.
6. Maxim D. Frank-Kamenetskii. *Unravelling DNA*. Addison-Wesley, 1997.
7. M. Lal. Monte Carlo computer simulations of chain molecules. *Molecular Physics*, 17:57–64, 1969.
8. B. MacDonald, N. Jan, D. L. Hunter, and M. O. Steinitz. Polymer conformations through wiggling. *Journal of Physics A: Mathematical and General Physics*, 18:2627–2631, 1985.
9. N. Madras, A. Orlitsky, and L. A. Shepp. Monte Carlo generation of self-avoiding walks with fixed endpoints and fixed length. *Journal of Statistical Physics*, 58:159–183, 1990.
10. Neal Madras and Gordon Slade. *The Self-Avoiding Walk*. Birkhäuser, 1993.
11. M. Soss and G. T. Toussaint. Geometric and computational aspects of polymer reconfiguration. *J. Math. Chemistry*, 27(4):303–318, 2000.
12. C. E. Soteros and S. G. Whittington. Polygons and stars in a slit geometry. *Journal of Physics A: Mathematical and General Physics*, 21:L857–L861, 1988.
13. I. Streinu. A combinatorial approach to planar non-colliding robot arm motion planning. In *Proc. 41st Annu. IEEE Sympos. Found. Comput. Sci.* IEEE, November 2000. 443–453.
14. G. T. Toussaint. The Erdős-Nagy theorem and its ramifications. In *Proc. 11th Canad. Conf. Comput. Geom.*, pages 9–12, 1999.
15. S. G. Whittington. Self-avoiding walks with geometrical constraints. *Journal of Statistical Physics*, 30(2):449–456, 1983.

Project Scheduling with Irregular Costs: Complexity, Approximability, and Algorithms

Alexander Grigoriev[1] and Gerhard J. Woeginger[2]

[1] University of Maastricht, Faculty of Economics and Business Administration,
Department of Quantitative Economy, P.O.Box 616,
6200 MD, Maastricht, The Netherlands
a.grigoriev@ke.unimaas.nl
http://www.personeel.unimaas.nl/a.grigoriev
[2] University of Twente, Department of Mathematics, P.O.Box 217,
7500 AE Enschede, The Netherlands and
Technische Universität Graz, Institut für Mathematik, Steyrergasse 30,
A-8010 Graz, Austria
g.j.woeginger@math.utwente.nl
http://www.math.utwente.nl/ woeginge/

Abstract. We address a generalization of the classical discrete time-cost tradeoff problem where the costs are irregular and depend on the starting and the completion times of the activities. We present a complete picture of the computational complexity and the approximability of this problem for several natural classes of precedence constraints. We prove that the problem is NP-hard and hard to approximate, even in case the precedence constraints form an interval order. For orders of bounded height, there is a complexity jump: For height one, the problem is polynomially solvable, whereas for height two, it is NP-hard and APX-hard. Finally, the problem is shown to be polynomially solvable for orders of bounded width and for series parallel orders.

1 Introduction

Due to its practical importance, the discrete time-cost tradeoff problem for project networks has been studied in various contexts by many researchers over the last fifty years. For references see survey [1]. In this paper, we look at a generalization of the discrete time-cost tradeoff problem where the costs depend on the exact starting *and* completion times of the activities.

Statement of the Problem. Formally, we consider instances that are called *projects* and that consist of a finite set $\mathcal{A} = \{A_1, \ldots, A_n\}$ of *activities* together with a partial order $\prec$ on $\mathcal{A}$. All activities are available for processing at time zero, and they must be completed before a global project deadline T. Hence, the set of possible starting and completion times of the activities is $\{0, 1, \ldots, T\}$. The set of intervals over $\{0, 1, \ldots, T\}$ (the so-called *realizations* of the activities) is denoted by $\mathcal{R} = \{(x, y) \mid 0 \le x \le y \le T\}$. For every activity A_j, there is a

P. Bose and P. Morin (Eds.): ISAAC 2002, LNCS 2518, pp. 381–390, 2002.

corresponding cost function $c_j : \mathcal{R} \to \mathbb{R}^+ \cup \{\pm\infty\}$ that specifies for every real-ization $(x, y) \in \mathcal{R}$ a non-negative cost $c_j(x, y)$ that is incurred when the activity is started at time x and completed at time y. A *realization* of the project is an assignment of the activities in $\mathcal{A}$ to the intervals in $\mathcal{R}$. A realization is *feasible* if it obeys the precedence constraints: For any A_i and A_j with $A_i \prec A_j$, activity A_j is not started before activity A_i has been completed. The cost of a realization is the sum of the costs of all activities in this realization. The goal is to find a feasible realization of minimum cost. This problem is called *min-cost* project scheduling with irregular costs, or min-cost PSIC for short.

A closely related problem is *max-profit* PSIC. Instead of cost functions c_j for activity A_j, here we have profit functions $p_j : \mathcal{R} \to \mathbb{R}^+ \cup \{\pm\infty\}$ that specify for every realization of A_j the resulting profit. The goal is to find a feasible realization of maximum profit. Clearly, the min-cost and the max-profit version are polynomial time equivalent: The transformations $c_j := \mathrm{const}_1 - p_j$ and $p_j := \mathrm{const}_2 - c_j$ with sufficiently large constants const_1 and const_2 translate one version into the other. However, the two versions seem to behave quite differently with respect to their approximability.

Note that for costs and profits we allow any values from $\mathbb{R}^+ \cup \{\pm\infty\}$. This should be seen as a useful and simple convention: Whenever a cost equals $+\infty$ or a profit equals $-\infty$, then the corresponding realization is forbidden.

Special Cases. Various special cases arise if the cost and profit functions satisfy additional properties. A cost function c is *monotone*, if $[x_1, y_1] \subseteq [x_2, y_2]$ implies $c(x_1, y_1) \geq c(x_2, y_2)$. A profit function p is *monotone*, if $[x_1, y_1] \subseteq [x_2, y_2]$ implies $p(x_1, y_1) \leq p(x_2, y_2)$. The intuition behind these concepts is that short and quick executions should be more expensive than long and slow executions. It is readily seen that the general version of PSIC is equivalent to the monotone version with respect to computational complexity and approximability.

Another interesting special case arises, if $y_1 - x_1 = y_2 - x_2$ implies $c(x_1, y_1) = c(x_2, y_2)$ and $p(x_1, y_1) = p(x_2, y_2)$. In this special case, the cost and the profit of an activity only depend on the length of its realization. This special case actually is equivalent to the DEADLINE problem for the discrete time-cost tradeoff problem: The deadline T is hard, and the goal is to assign lengths to activities such that the overall cost is minimized. Only recently, De, Dunne, Gosh & Wells [2] proved that this problem is NP-hard in the strong sense. Skutella in [3] gives some positive approximability results, and in [4] the authors give some in-approximability results for bicriteria versions. All negative results in this paper are proved for the DEADLINE problem, the weakest variant of PSIC. All positive results in this paper are proved for the most general version of PSIC.

Our Results. We derive several positive and negative statements on the complexity and the approximability of min-cost and max-profit PSIC for several natural classes of precedence constraints. Our results are the following:

(1) Interval orders (Section 2). The min-cost and the max-profit version of the DEADLINE problem (and of their PSIC generalizations) are NP-hard and in-approximable even for interval orders. We establish a close (approximation

preserving) connection of the min-cost DEADLINE problem to minimum vertex cover and of the max-profit DEADLINE problem to maximum independent set. All in-approximability results for these graph problems carry over to the DEADLINE problems. As an immediate consequence, unless P=NP the min-cost DEADLINE problem can not have a polynomial time approximation algorithm with worst case ratio strictly better than 7/6.

(2) Orders of bounded height (Section 3). If the height of the precedence constraints is bounded by 2, then the DEADLINE problems and its PSIC generalizations are NP-hard and in-approximable. However, if the height of the precedence constraints is bounded by 1, then min-cost and max-profit PSIC both can be solved in polynomial time.

(3) Orders of bounded width (Section 4). If the width of the precedence constraints is bounded by some fixed constant d, then min-cost and max-profit PSIC both can be solved in polynomial time $O(n^d T^{2d+1})$.

(4) Series parallel orders (Section 5). For series parallel precedence constraints, min-cost and max-profit PSIC can be solved in polynomial time $O(nT^3)$ by dynamic programming.

2   Interval Orders

In this section we will derive a number of negative results for problem PSIC under interval orders. An *interval order* on a set $\mathcal{A} = \{A_1, \ldots, A_n\}$ is specified by a set of n intervals $I_1, \ldots, I_n$ along the real line. Then $A_i \to A_j$ holds if and only if the interval I_i lies completely to the left of the interval I_j, or if the right endpoint of I_i coincides with the left endpoint of I_j. See e.g. [5].

The central proof in this section will be done by a reduction from the NP-hard INDEPENDENT SET problem in graphs; see [6]: Given a graph $G = (V, E)$ and a bound z, does G contain an independent set (a set that does not induce any edges) of cardinality z? Without loss of generality, we assume that $V = \{1, \ldots, q\}$.

We construct a project with deadline $T = 3q$ for max-profit PSIC. This project contains the activities listed below. For every activity A, we define a so-called *crucial* interval $I(A)$ that will be used to specify the interval order.

- For every vertex $i \in V$, there is a corresponding activity A_i. If A_i is realized by an interval of length zero, then its profit is $-\infty$; for an interval of length 1 or 2 the profit is 0, and for any longer realization the profit is 1. The crucial interval $I(A_i)$ for A_i is $[3i - 3, 3i]$.
- For every edge $\langle i, j \rangle \in E$ with $i < j$, there is a corresponding activity $A_{i,j}$. If $A_{i,j}$ is realized by an interval of length $3j - 3i - 2$ or more then its profit is 0, and for shorter intervals its profit is $-\infty$. The crucial interval $I(A_{i,j})$ is $[3i, 3j - 3]$.
- For $t = 0, \ldots, q$ there are so-called *blocking* activities B_t and C_t. If they are executed for at least $3t$ time units, then they bring profit 0, and for shorter intervals they bring profit $-\infty$. The crucial intervals for them are $I(B_t) = [0, 3t]$ and $I(C_t) = [3q - 3t, 3q]$.

The precedence constraints among these activities are defined as follows: For activities X and Y, $X \prec Y$ holds if and only if the crucial interval $I(X)$ lies completely to the left of the crucial interval $I(Y)$, or if the right endpoint of $I(X)$ coincides with the left endpoint of $I(Y)$. Note that this yields an interval order on the activities. Moreover, for every edge $\langle i, j \rangle \in E$ with $i < j$ this implies $A_i \prec A_{i,j} \prec A_j$.

Lemma 1 *If the graph G has an independent set W, then the constructed project has a feasible realization with profit $|W|$.*

Proof. Let $W \subseteq V$ denote the independent set of cardinality z. If $i \in W$, then process activity A_i with profit 1 during $[3i - 3, 3i]$. If $i \notin W$, then process it with profit 0 during $[3i - 2, 3i - 1]$. All other activities are processed at profit 0: Every blocking activity is processed during its crucial interval. For an edge $\langle i, j \rangle \in E$ with $i < j$ and $i \notin W$, process activity $A_{i,j}$ during $[3i - 1, 3j - 3]$; this puts $A_{i,j}$ after A_i and before A_j exactly as imposed by the precedence constraints. For an edge $\langle i, j \rangle \in E$ with $i < j$ and $i \in W$, process activity $A_{i,j}$ during $[3i, 3j - 2]$. Since $i \in W$, its neighbor j cannot be also in W; hence A_j is processed during $[3j - 2, 3j - 1]$ and after $A_{i,j}$, exactly as imposed by $A_i \prec A_{i,j} \prec A_j$.

Since in this realization activity A_i brings profit 1 if and only if $i \in W$, this realization has profit $|W|$. Moreover it can be verified that all precedence constraints indeed are satisfied. □

Lemma 2 *If the constructed project has a feasible realization with profit $p \geq 1$, then the graph G has an independent set W with $|W| = p$.*

Proof. We first establish three simple claims on such a feasible project realization. The first claim is that (in any feasible realization with positive profit) the processing of every blocking activity must exactly occupy its crucial interval. Indeed, consider the activities B_t and C_{q-t} with their crucial intervals $I(B_t) = [0, 3t]$ and $I(C_t) = [3t, 3q]$. Since the total profit is positive, B_t is processed for at least $3t$ and C_{3q-t} is processed for at least $3q - 3t$ time units. Since B_t is a predecessor of C_{q-t}, they together cover the whole time horizon $[0, 3q]$; this fixes them in their crucial intervals.

The second claim is that every activity A_i is processed somewhere within its crucial time interval $[3i - 3, 3i]$. By our first claim activity B_{i-1} completes at time $3i - 3$ and activity C_{q-i} starts at time $3i$. Since $B_{i-1} \prec A_i \prec C_{q-i}$, activity A_i cannot start before time $3i - 3$ and cannot end after time $3i$.

The third claim is that there exist exactly p activities A_i that exactly occupy their crucial intervals. By construction of the project all the profit results from the activities A_i, and A_i brings positive profit only in case it is executed for at least three time units. By our second claim, A_i cannot be executed for more than three time units. Hence, each activity A_i that brings positive profit occupies its crucial interval $[3i - 3, 3i]$.

Now we are ready to prove the statement in the lemma. Consider the set $W \subseteq V$ that contains vertex i if and only if A_i occupies its crucial interval $[3i - 3, 3i]$. We claim that W is an independent set. Suppose otherwise, and consider $i, j \in W$ with $i < j$ and $\langle i, j \rangle \in E$. Then A_i occupies $[3i - 3, 3i]$, and A_j occupies $[3j - 3, 3j]$, and $A_i \prec A_{i,j} \prec A_j$ holds. Hence, $A_{i,j}$ is processed during the $3j - 3i - 3$ time units between $3i$ and $3j - 3$. But in this case its profit is $-\infty$, and we get the desired contradiction. Hence, W is an independent set, and by our third claim $|W| = p$. □

Theorem 3 *Max-profit project scheduling with irregular costs is NP-hard even for interval order precedence constraints. For any $\varepsilon > 0$, the existence of a polynomial time approximation algorithm for max-profit PSIC for projects with n activities with worst case ratio $O(n^{1/4-\varepsilon})$ implies P=NP (with worst case ratio $O(n^{1/2-\varepsilon})$ implies ZPP=NP).*

Proof. NP-hardness follows from the Lemmas 1 and 2. The constructed reduction preserves objective values. It translates graph instances with independent sets of size z into project instances with realizations of profit z, and thus it is approximation preserving in the strongest possible sense. For a graph with q vertices, the corresponding project consists of $O(q^2)$ activities. Håstad in [7] proved that the clique problem in n-vertex graphs (and hence also the independent set problem in the complement of n-vertex graphs) cannot have a polynomial time approximation algorithm with worst case guarantee $O(n^{1/2-\varepsilon})$ unless P=NP, and it cannot have a polynomial time approximation algorithm with worst case guarantee $O(n^{1-\varepsilon})$ unless ZPP=NP. Since the blow-up in our construction is only quadratic, the theorem follows. □

In the VERTEX COVER problem, the goal is to find a minimum cardinality vertex cover (a subset of the vertices that touches every edge) for a given input graph. Note that vertex covers are the complements of independent sets. We denote by τ_{VC} the approximability threshold for the vertex cover problem, i.e., the infimum of the worst case ratios over all polynomial time approximation algorithms for this problem. Håstad [8] proved that $\tau_{VC} \geq 7/6$ unless P=NP, and it is widely believed that $\tau_{VC} = 2$.

Theorem 4 *Min-cost project scheduling with irregular costs is NP-hard even for interval order precedence constraints. The existence of a polynomial time approximation algorithm for min-cost PSIC with worst case ratio better than τ_{VC} would imply P=NP.*

Proof. By a slight modification of the above construction. For activities $A_{i,j}$ and for blocking activities, we replace low profit $-\infty$ by high cost ∞, and the neutral profit 0 by the neutral cost 0. For activities A_i, we replace low profit $-\infty$ by high cost ∞, profit 0 by cost 1, and profit 1 by cost 0. It can be shown that there exists a realization of cost c for the constructed project, if and only if there exists an independent set of size $q - c$ for the graph, if and only if there exists a vertex cover of size c for the graph. Hence, this reduction preserves objective values. □

Corollary 5 *For the discrete time/cost tradeoff problem, the existence of a polynomial time approximation algorithm with worst case ratio better than τ_{VC} for the DEADLINE problem would imply P=NP.* □

3   Orders of Bounded Height

In this section we will derive a positive result for the project scheduling problem with irregular costs under orders of bounded height. The *height* of an ordered set is the number of elements in the longest chain minus one. Precedence constraints of height 1 are sometimes also called *bipartite* precedence constraints; see [5].

Theorem 6 *Max-profit and min-cost PSIC are NP-hard and APX-hard even when restricted to precedence constraints of height two.*

Proof. Deineko & Woeginger [4] establish APX-hardness for the min-cost DEADLINE problem. Their reduction produces instances of height 2 for min-cost PSIC, and it is straightforward to adapt the construction to max-cost PSIC. □

Let us concentrate on the max-profit PSIC for orders of height 1. Classify the activities into two types: The *A-activities* $A_1, \ldots, A_a$ do not have any predecessors, and the *B-activities* $B_1, \ldots, B_b$ do not have any successors. The only precedence constraints are of the type $A_i \to B_j$. We start with a preprocessing phase that simplifies this instance somewhat.

- If there exists some activity that neither has a predecessor nor a successor, we process this activity at the maximum possible profit, and remove it from the instance. From now on we assume that each activity has at least one predecessor or successor.
- We remove all realizations with profit $-\infty$ from the instance.
- Assume that there is an activity A_i with profit function p_i, and that there are two realizations (x, y) and (u, v) for it with $y \le v$ and $p_i(x, y) \ge p_i(u, v)$. Then the realization (x, y) imposes less restrictions on the successors of A_i and at the same time it comes at a higher profit; so we may disregard this realization (u, v) for A_i. Similarly, we clean up the realizations for B-activities.
- Assume that $A_i \prec B_j$ and that there exists a realization (x, y) of A_i that collides with all surviving realizations of B_j (that is, the endpoint y lies strictly to the right of all possible starting points of B_j). Then we remove realization (x, y) for A_i, since it will always collide with the realization of B_j. Symmetrically, we clean up the realizations of the B-activities.

Lemma 7 *(i) The original instance has a realization with profit p if and only if the preprocessed instance has a realization with profit p.*
(ii) The surviving realizations for A_i can be enumerated as $(x_i^1, y_i^1), \ldots, (x_i^{a(i)}, y_i^{a(i)})$ such that they are ordered by strictly increasing right endpoint and simultaneously by strictly increasing profit for A_i. Similarly, the surviving realizations for B_j can be enumerated as $(u_j^1, v_j^1), \ldots, (u_j^{b(j)}, v_j^{b(j)})$ such that they are ordered by strictly decreasing left endpoint and simultaneously by strictly increasing profit for B_j.
(iii) If the original instance has a realization with non-negative profit, then for every activity A_i (respectively, B_j) there exists a realization in the preprocessed instance that does not collide with any realization of a successor of A_i (respectively, of a predecessor of B_j).

Proof. Statements (i) and (ii) are clear from the preprocessing. To see (iii), consider the realization (x_i^1, y_i^1) that has the smallest left endpoint over all realizations of A_i. Suppose that it collides with some realization (u_j^ℓ, u_j^ℓ) of some successor B_j of A_i. Then this realization of B_j collides with *all* realizations of A_i and would have been removed in the last step of the preprocessing.   □

From now on we assume that the conditions in (iii) in Lemma 7 are satisfied. We translate the preprocessed instance into a bipartite graph with weights on the vertices. The max-profit problem will boil down to finding an independent set of maximum weight in this bipartite graph.

- For every realization (x_i^k, y_i^k) of A_i with profit function p_i, there is a corresponding vertex A_i^k in the bipartite graph. If $k = 1$, then the weight of A_i^k equals $p_i(x_i^1, y_i^1)$. If $k \geq 2$, then the weight of A_i^k equals $p_i(x_i^k, y_i^k) - p_i(x_i^{k-1}, y_i^{k-1})$. Note that all weights are non-negative and that the weight of the first k realizations of A_i equals $p_i(x_i^k, y_i^k)$.
- Symmetrically, the bipartite graph contains for every realization (u_j^ℓ, u_j^ℓ) of activity B_j a corresponding vertex B_j^ℓ. The (non-negative) weights of the vertices B_j^ℓ are defined symmetrically to those of the vertices A_i^k.
- We put an edge between A_i^k and B_j^ℓ if and only if $A_i \prec B_j$ holds and if the interval $[x_i^k, y_i^k]$ does not lie completely to the left of the interval $[u_j^\ell, u_j^\ell]$.

Lemma 8 *The profit of most profitable realization of the preprocessed project equals the weight of the maximum weighted independent set in the bipartite graph.*

Proof. (Only if) Consider the most profitable realization, and consider the following set S of vertices. If activity A_i is realized as (x_i^k, y_i^k), then put the vertices $A_i^1, A_i^2, \ldots, A_i^k$ into S. The weight of these k vertices equals the profit $p_i(x_i^k, y_i^k)$ of realization (x_i^k, y_i^k). If B_j is realized as (u_j^ℓ, u_j^ℓ), then put the vertices $B_j^1, \ldots, B_j^\ell$ into S. The weight of these ℓ vertices equals the profit of the realization of B_j. By construction, the total weight of S equals the total profit p of the considered realization. Moreover, the set S is independent: If in S some A_i^s was adjacent to B_j^t, then $A_i \prec B_j$ and A_i^k and B_j^ℓ would be adjacent. But this would yield a collision in the execution of A_i and B_j, and the realization would be infeasible. (If) Consider an independent set S of maximum weight in the bipartite graph. For an activity A_i, consider the intersection of S with $\{A_i^1, \ldots, A_i^{a(i)}\}$. By Lemma 7.(iii), this intersection is non-empty. Let k denote the largest index such that A_i^k is in S. Since the neighborhood of $A_i^1, \ldots, A_i^{k-1}$ is a subset of the neighborhood of vertex A_i^k, also these $k - 1$ vertices are contained in S. Then we realize activity A_i by (x_i^k, y_i^k); the resulting profit $p_i(x_i^k, y_i^k)$ equals the total weight of the vertices $A_i^1, \ldots, A_i^k$ in S. For activity B_j, we symmetrically compute a realization that is based on the maximum index ℓ for which B_j^ℓ is in S. Since A_i^k and B_j^ℓ are not incident in the bipartite graph, the chosen realizations of A_i and B_j do not collide. Hence, this realization is feasible. By construction, the total profit equals the total weight of S.   □

Theorem 9 *Max-profit and min-cost project scheduling with irregular costs are polynomially solvable when restricted to precedence constraints of height one.*

Proof. By Lemma 8, these problems are polynomial time equivalent to finding a maximum weight independent set in a bipartite graph with non-negative vertex weights. Maximum weight independent set in bipartite graphs is well-known to be polynomially solvable by max-flow min-cut techniques; see [9].   □

4   Orders of Bounded Width

In this section, we will show that if the width of the precedence constraints is bounded by some fixed constant d, then max-profit PSIC is solvable in polynomial time. Without loss of generality, we assume throughout this section that all realizations of length 0 have profit $-\infty$ and hence are forbidden.

In an ordered set, two elements A_i and A_j are called *in-comparable* if neither A_i is a predecessor of A_j nor A_j is a predecessor of A_i. A set of tasks is an *anti-chain*, if its elements are pairwise in-comparable. The *width* of the order is the cardinality of its largest anti-chain. A well-known theorem of Dilworth [10] states that if the width of an ordered set with n elements equals d, then this set can be partitioned into d totally ordered chains $C_1, \ldots, C_d$. Moreover, it is straightforward to compute such a chain partition in $O(n^d)$ time.

For a given instance of max-profit PSIC of width d, we first compute a chain partition $C_1, \ldots, C_d$, and we denote the number of activities in chain C_j by n_j ($j = 1, \ldots, d$). Now let us consider some feasible realization of the project, and let us look at some fixed moment $t + \frac{1}{2}$ in time with $0 \le t \le T$. As the chain C_j is totally ordered, at time $t + \frac{1}{2}$, at most one of its activities is under execution. Chain C_j is called *in-active* at time $t + \frac{1}{2}$ if none of its activities is under execution, and otherwise it is *active* at time $t + \frac{1}{2}$.

For a feasible realization, the *snapshot* S taken at time $t + \frac{1}{2}$ with $0 \le t \le T$ contains the following information:

(S1) For every chain C_j, one bit of information that specifies whether C_j is active or in-active.

(S2) For every in-active chain C_j, a number IN_j with $0 \le \text{IN}_j \le n_j$ that specifies the last activity in C_j that was executed before time $t + \frac{1}{2}$. If no activity has been executed so far, then $\text{IN}_j = 0$.

(S3) For every active chain C_j, a number ACT_j with $1 \le \text{ACT}_j \le n_j$ that specifies the current activity of C_j. Moreover, the starting time x_j of the current activity with $0 \le x_j \le T - 1$.

For the data in (S1) there are at most 2^d possibilities, for all the numbers IN_j and ACT_j in (S2) and (S3) there are at most $O(n^d)$ possibilities, and for all the starting times in (S3) there are at most $O(T^d)$ possibilities. Since d is a fixed constant, this yields that there are at most $O(n^d T^d)$ snapshots at time $t + \frac{1}{2}$.

For any t with $0 \le t \le T$ and for any possible snapshot S, we denote by $F[t; S]$ the maximum possible profit that can be earned on activities completing before time $t + \frac{1}{2}$ in a feasible project realization whose snapshot at time $t + \frac{1}{2}$ equals S. If no such feasible realization exists, then $F[t; S] = -\infty$.

We compute all these values $F[t; S]$ by a dynamic programming that works through them by increasing t. The cases with $t = 0$ are trivial: $F[0; S]$ can only take the values 0 (if there exists a feasible realization with snapshot S at time $\frac{1}{2}$) or $-\infty$ (otherwise). To compute $F[t; S]$ for $t \geq 1$, we check all possibilities for a compatible predecessor snapshot S' at time $t - \frac{1}{2}$ in the following way by considering all the chains separately (the data from snapshots S and S' is represented by un-primed and by primed variables, respectively):

- Chain C_j might be active in S' and in-active in S. Then $\text{IN}_j = \text{ACT}'_j$. The additional profit comes from realizing the ACT'_j-th activity in chain C_j from time x'_j to time t.
- Chain C_j might be in-active in S' and active in S. Then $\text{ACT}_j = \text{IN}'_j + 1$ and $x_j = t$. No additional profit is generated.
- Chain C_j might be in-active in S' and S. Then $\text{IN}_j = \text{IN}'_j$. Since no activity can simultaneously be started and completed at time t, no additional profit is generated.
- Chain C_j might be active in S' and S. There are two cases: If the same activity is executed at time $t - \frac{1}{2}$ and at time $t + \frac{1}{2}$, then $\text{ACT}_j = \text{ACT}'_j$ and $x_j = x'_j$, and no additional profit is generated. And if the executed activities at times $t - \frac{1}{2}$ and $t + \frac{1}{2}$ are distinct, then $\text{ACT}_j = \text{ACT}'_j + 1$ and $x_j = t$ must hold. The additional profit comes from realizing the ACT'_j-th activity in C_j from time x'_j to time t.

If snapshots S and S' are of this form for all d chains, then we say that S' is a *predecessor* of S. Moreover, we denote the total additionally generated profit over all the chains by $\text{profit}(S', S)$. It can be verified that any snapshot S at time $t + \frac{1}{2}$ has at most $O(T^d)$ predecessors at time $t - \frac{1}{2}$. Compute the values

$$F[t; S] := \max \{ F[t - 1; S'] + \text{profit}(S', S) \mid S' \text{ is a predecessor of } S \}. \quad (1)$$

The solution to the instance of max-profit PSIC can be found in $F[T; S^*]$ where S^* is the snapshot at time $T + \frac{1}{2}$ where all chains are in-active and where $\text{IN}_j = n_j$ holds for $j = 1, \ldots, d$. The time complexity of this dynamic programming algorithm is $O(n^d T^{2d+1})$: Since there are $O(n^d T^d)$ snapshots at time $t + \frac{1}{2}$, we altogether compute $O(n^d T^{d+1})$ values $F[t; S]$. Each value can be computed in $O(T^d)$ time by checking all predecessors in (1).

Theorem 10 *Max-profit and min-cost PSIC are solvable in $O(n^d T^{2d+1})$ time when restricted to orders of width bounded by the fixed constant d.* □

5   Series Parallel Orders

Orders are called *series parallel* if (i) they contain a single element, or (ii) they form the series composition of two series parallel order, or (iii) they form the parallel composition of two series parallel orders. Only orders that can be constructed via (i)–(iii) are series parallel. Here the *series composition* of orders $(V_1, \prec_1)$ and $(V_2, \prec_2)$ with $V_1 \cap V_2 = \emptyset$ is the order that results from taking $V_1 \cup V_2$ and making all elements in V_1 predecessors of all elements in V_2, and the *parallel composition* of $(V_1, \prec_1)$ and $(V_2, \prec_2)$ simply is their disjoint union.

It is well known that a series parallel order can be decomposed in polynomial time into its atomic parts according to the series and parallel compositions. Essentially, such a decomposition corresponds to a rooted, ordered, binary tree where all interior vertices correspond to series or parallel compositions and where all leaves correspond to single elements of the order. We associate with every interior vertex v of the decomposition tree the series parallel order $SP(v)$ that is induced by the leaves of the subtree below v. Note that for the root vertex $root$ the corresponding order $SP(root)$ is the whole ordered set.

The usual tool for dealing with series parallel structures is dynamic programming. By a dynamic programming that starts in the leaves of the decomposition tree, and then moves upwards towards the root we can compute all the values $F[v; x, y]$ that are maximum possible profit that can be earned on the activities in $SP(v)$, under condition that all these activities are executed somewhere during the time interval $[x, y]$ and they obey the precedence constraints.

In the end, the solution to the instance of max-profit PSIC can be found in $F[root; 0, T]$. The time complexity of this dynamic programming algorithm is $O(nT^3)$: To compute the values $F[v; x, y]$ for the $O(nT^2)$ leaves, it is sufficient to look once at every possible realization of every activity; this altogether costs $O(nT^2)$ time. And for the inner vertices v, the corresponding $O(nT^2)$ values can be computed in $O(T)$ time per value.

Theorem 11 *Max-profit and min-cost* PSIC *are polynomially solvable in* $O(nT^3)$ *time when restricted to series parallel precedence constraints.*   □

References

1. Brucker, P., Drexl, A., Möhring, R., Neumann, K.,Pesch, E.: Resource-constrained project scheduling: Notation, classification, models, and methods. European Journal of Operational Research. **112** (1999) 3–41.
2. De, P., Dunne, E.J., Gosh, J.B., Wells, C.E.: Complexity of the discrete time-cost tradeoff problem for project networks. Operations Research. **45** (1997) 302–306.
3. Skutella, M.: . Approximation algorithms for the discrete time-cost tradeoff problem. Mathematics of Operations Research. **23** (1998) 909–929.
4. Deineko, V.G., Woeginger, G.J.: Hardness of approximation of the discrete time-cost tradeoff problem. Operations Research Letters. **29** (2001).
5. Möhring, R.H.: Computationally tractable classes of ordered sets. In: Rival, I. (ed.): Algorithms and Order. Kluwer Academic Publishers (1989) 105–193.
6. Garey, M.R., Johnson, D.S.: Computers and Intractability: A Guide to the Theory of NP-Completeness. Freeman, San Francisco (1979).
7. Håstad, J.: Clique is hard to approximate within $n^{1-\epsilon}$. Acta Mathematica. **182** (1999) 105–142.
8. Håstad, J.: Some optimal inapproximability results. In: Proceedings of the 29th ACM Symposium on the Theory of Computing (STOC'1997). (1997) 1–10.
9. Ahuja, R.K., Magnanti, T.L., Orlin, J.B.: Network Flows: Theory, Algorithms, and Applications. Prentice Hall (1993).
10. Dilworth, R.P.: A decomposition theorem for partially ordered sets. Annals of Mathematics. **51** (1950) 161–166.

Scheduling of Independent Dedicated Multiprocessor Tasks*

Evripidis Bampis[1] **, Massimiliano Caramia[2] ***, Jiří Fiala[3†],
Aleksei V. Fishkin[4 ‡], and Antonio Iovanella[5 §]

[1] LaMI, CNRS-UMR 8042, Univerité d'Evry, Boulevard Mitterrand, 91025 Evry
Cedex, France. bampis@lami.univ-evry.fr
[2] Istituto per le Applicazioni del Calcolo "M. Picone" - CNR Viale del Policlinico,
137 - 00161 Roma - Italy. caramia@iac.rm.cnr.it
[3] DIMATIA and ITI, Charles University, Malostranské nám. 2/25, 118 00, Prague,
Czech Republic. fiala@kam.mff.cuni.cz
[4] Institut für Informatik und Praktische Mathematik, Christian-Albrechts-Universität
zu Kiel, Olshausenstrasse 40, 24 098 Kiel, Germany. avf@informatik.uni-kiel.de
[5] Dipartimento di Informatica, Sistemi e Produzione, University of Rome "Tor
Vergata", Via del Politecnico, 1 - 00133 Rome, Italy. iovanella@disp.uniroma2.it

Abstract. We study the off and on-line versions of the well known problem of scheduling a set of n independent multiprocessor tasks with prespecified processor allocations on a set of identical processors in order
to minimize the makespan. Recently, in [12], it has been proven that in
the case when all tasks have unit processing time the problem cannot
be approximated within a factor of $m^{\frac{1}{2}-\varepsilon}$, neither for some $\varepsilon > 0$, unless P= NP; nor for any $\varepsilon > 0$, unless NP=ZPP. For this special case
we give a simple algorithm based on the classical first-fit technique. We
analyze the algorithm for both tasks *arrive over time* and tasks *arrive
over list* on-line scheduling versions, and show that its competitive ratio is bounded by $2\sqrt{m}$ and $2\sqrt{m} + 1$, respectively. Here we also use
some preliminary results on (vertex) coloring of k-tuple graphs. For the
case of arbitrary processing times, we show that any algorithm which
uses the first-fit technique cannot be better than m competitive. Then,
by using our split-round technique, we give a $3\sqrt{m}$-approximation algorithm for the off-line version of the problem. Finally, by using some ideas
from [20], we adapt the algorithm to the on-line case, in the paradigm
of *tasks arriving over time* in which the existence of a task is unknown
until its release date, and show that its competitive ratio is bounded by
$6\sqrt{m}$. Due to the conducted experimental results, we conclude that our
algorithms can perform well in practice.

* Supported by the bilateral French-German project PROCOPE,
** Research supported by EU APPOL II project IST-2001-32007.
*** Research supported by CNR project CNRG007FC1.
† Research supported by EU ARACNE project HPRN-CT-1999-00112, GAČR project
201/99/0242, and by MECR project LN00A056.
‡ Supported by Graduiertenkolleg 357 "Effiziente Algorithmen und Mehrskalenmetho-
den".
§ Research supported by CNR project CNRG007FC1.

P. Bose and P. Morin (Eds.): ISAAC 2002, LNCS 2518, pp. 391–402, 2002.
© Springer-Verlag Berlin Heidelberg 2002

1   Introduction

Optical networks employing wavelength division multiplexing (WDM) are now a viable technology for implementing a next-generation network infrastructure that will support a diverse set of existing, emerging, and future applications [13]. WDM technology initially was deployed in point-to-point links in *wide* area or *metropolitan* area distances [23]. However, the work on WDM *local area* networks[1] has also been currently under way, see e.g. [16,18].

The main future of broadcast WDM local area networks is the so-called *one-to-many transmission* or *multicasting* ability [1]. That is, a transmission by a *node* in such a network on a given channel (wavelength) is received by *all* nodes listening on that channel at that point in time (see [21] for a detailed survey on WDM networks). Since two or more nodes may want to send *data packets* to the same destination node, coordination among nodes that wish to communicate with each other is required. Many *protocols* for coordinating *multicast* data transmissions have been proposed in the literature (see [22] for a survey). The approaches used range from sending a different copy of each data packet to each one of the corresponding destinations (*unicast service*), to transmitting a single copy of the data packet to all the destinations at once (*multicast service with no fanout splitting*).

Due to the relation to the later approach, due to the relation to the later approach, there has been recently a renewed interest to the model of scheduling *dedicated* tasks in the *on-line* context. In this framework, nodes correspond to *processors* and multi-destination data packets to *dedicated tasks*. The goal, minimizing the overall *transmission* time, i.e. the time needed to send all data packets out, corresponds to the schedule *makespan* of dedicated tasks.

In this paper we also address the dedicated variant of the *multiprocessor* tasks scheduling problem (see [8] for a survey on some main results). Formally, we are given a set of n tasks $T = \{1, \ldots, n\}$ and a set of m processors $M = \{1, 2, \ldots, m\}$. Each task $j \in T$ has a processing time p_j, a release date r_j and a prespecified set of processors $\mathrm{fix}_j \subseteq M$. Preemptions are not allowed. Each processor can work on at most one task at a time, each task must be processed simultaneously by all processors of fix_j. The objective is to minimize the *makespan* $C_{\max} = \max_j C_j$, where C_j denotes the completion time of task j.

We call fix_j the *type* and $|\mathrm{fix}_j|$ the *size* of job $j \in T$, and use $\Delta_{\max}$ to denote $\max_j |\mathrm{fix}_j|$ and $r_{\max}$ to denote $\max_j r_j$. To refer to the several variants of the above scheduling problem, we use the standard notation scheme by Graham at al. [15]. For example, $P|\,\mathrm{fix}_j, r_j|C_{\max}$ denotes the above problem itself in the *off-line* case[2]; $P|$ on-line, $\mathrm{fix}_j, p = 1|C_{\max}$ denotes the on-line variant where all

---

[1] These networks are known as *single-hop* WDM networks [19].

[2] In *off-line* scheduling, the scheduler has full information of the problem instance. In contrast, in *on-line* scheduling, information about the problem instance is made available during the course of scheduling.

$p_j = 1$ and the tasks *arrive over list* [3]; and $P \mid$ *on-line*, $fix_j, r_j \mid C_{\max}$ denotes the on-line variant where all p_j are arbitrary and the tasks *arrive over time* [4].

Known results and our contribution. Variants of the multiprocessor tasks scheduling problem have been studied, but the previous research has mainly focused on the off-line case. Hoogeveen et al. [17] showed that already the three-processor problem $P3 \mid fix_j \mid C_{\max}$ is strongly NP-hard, and proved that even if all tasks have unit processing times, there exists no polynomial approximation algorithm for $P \mid fix_j, p_j = 1 \mid C_{\max}$ with performance ratio smaller than $\frac{4}{3}$, unless P=NP. However, in the same work it was shown that if the number of processors m is fixed, i.e. $Pm \mid fix_j, p_j = 1 \mid C_{\max}$, then the problem is solvable in polynomial time. Later, Amoura et al. [2] proposed a polynomial time approximation scheme (PTAS) for problem $Pm \mid fix_j \mid C_{\max}$, and Bampis & Kononov [3] extended this result to a PTAS for $Pm \mid fix_j, r_j \mid C_{\max}$. A $\frac{7}{6}$-approximation algorithm for $P3 \mid fix_j \mid C_{\max}$, which is due to Goemans [14], still remains the best low time complexity approximation algorithm. Recently, Fishkin et al. [12] extended the negative results presented in [17]. It has been proven that $P \mid fix_j, p_j = 1 \mid C_{\max}$ cannot be approximated within a factor of $m^{1/2-\varepsilon}$, neither for some $\varepsilon > 0$, unless P=NP; nor for any $\varepsilon > 0$, unless NP=ZPP.

In the on-line context, there are several results known for the *parallel* variant of the multiprocessor tasks scheduling problem, see e.g. [10,9,5]. Up to our best knowledge, no on-line algorithms with guaranteed competitive ratio are known for the dedicated variant of the multiprocessor model, considered in this paper. Although some algorithms have been recently proposed in the literature, their performances are not evaluated analytically, but by using simulations [7,6].

In this paper we present several results, important from both theoretical and practical point of view. First, we deal with the problems where tasks have unit processing times. Using the so-called *first-fit* technique, we give an on-line algorithm for $P \mid$ *on-line*, $fix_j, p_j = 1 \mid C_{\max}$, which is k-competitive if the maximum task size $\Delta_{\max}$ is bounded by some constant k. It is interesting to point out that our analysis is based on a transformation of our scheduling problem to the classical (vertex) coloring problem of a particular class of graphs, the intersection graphs of k-tuples. We propose simple extensions of this algorithm to a $2\sqrt{m}$-competitive algorithm for $P \mid$ *on-line*, $fix_j, p_j = 1 \mid C_{\max}$ and $(2\sqrt{m} + 1)$-competitive algorithm for $P \mid$ *on-line*, $fix_j, p_j = 1, r_j \mid C_{\max}$. Clearly, the $2\sqrt{m}$-competitive algorithm and the first-fit algorithm are complementary. For instance, if $k = 2$ then the first-fit algorithm is better. From another side, whenever $k = m$ the first algorithm outperforms the second one.

Next, we switch to the general problem with arbitrary processing times. We show that any on-line algorithm which schedules tasks arriving over-list and

---

[3] The tasks of T are ordered in some list (sequence) and presented one by one to the scheduler according to this list. The existence of a job is not known until all its predecessors in the list have already been scheduled.

[4] The tasks j of T arrive at their release dates r_j. The tasks can be started at a time bigger than or equal to their release dates, and at any point of time, the scheduler only has knowledge of the released tasks.

leaves no unnecessary idles cannot be better than m-competitive. In some sense, this leaves no hope for contracting any good on-line algorithm based on the first-fit technique. However, using our *split-round* technique, we obtain an off-line $3\sqrt{m}$-approximation algorithm for $P \mid fix_j \mid C_{\max}$. Then, by using the so-called *active-passive-bins* scheduling technique by Shmoys, Wein & Williamson [20], we give a $6\sqrt{m}$-competitive algorithm for $P \mid on\text{-}line, fix_j, r_j \mid C_{\max}$ in which the existence of a task is unknown until its release date.

Finally, we have conducted some experiments based on random generated instances with $m = 20$ processors; $n = 50$, 100 and 200 tasks; $\Delta_{\max} = 4$, 6, 8, 12 maximum size; $r_{\max} = 6$, 18 and 30 maximum release date. Analyzing the results we observed that our algorithms work much better than it can be expected. For all cases, the output makespan exceeded the lower bound, which is also heuristically obtained, by at most a factor of 3.5, though $\sqrt{20} \approx 4.47$. Furthermore, we have found that all the algorithms have low running time, that is a very important factor for real world applications.

Last notes. We say that an algorithm A is a ρ-*approximation* algorithm for all problem instances it outputs a schedule with the makespan at most $\rho \cdot OPT$, where OPT is the makespan of the optimal schedule. If A is an on-line ρ-approximation algorithm, then we say that it is a ρ-*competitive* algorithm or A is ρ-*competitive*. (We may also that ρ is the approximation and/or competitive ratio of A.) Here, the *competitiveness* of an on-line algorithm is evaluated with respect to an off-line optimal algorithm, and hence it indicates the loss associated with not having complete information of the problem instance. Both approximation algorithms and on-line algorithms are assumed to run in polynomial time.

The paper is organized as follows. In the next section we give some preliminary results. In Section 3.1, we deal with the unit processing times case, and in Section 3.2 with the general case of arbitrary processing times. However, due to space limitation, here we omit the experimental results leaving them to the full version of the paper.

2   Preliminaries

In this section we discuss some general techniques and lemmas that apply throughout our paper. We transform our scheduling problem to the classical (vertex) coloring problem. Then, we use some standard methods from the graph theory to obtain related results. The properties we prove in this section pertain to the case of unit processing times, but the ideas will be used in modified form for arbitrary processing times as well.

2.1   Coloring of the Conflict Graph

Given an undirected graph $G = (V, E)$, a *clique* of G is a set of mutually adjacent vertices. A maximum clique is, naturally, a clique whose number of vertices is at least as large as that for any other clique in the graph. If the vertices have *weights* then a maximum *weighted clique* is a clique with the largest possible sum of vertex weights.

A *(vertex) coloring* of $G = (V, E)$ is an assignment of a *color* to each vertex in V. It is required that the colors on the pair of vertices incident to any edge be different. A minimum coloring of a graph is a coloring that uses as few different colors as possible.

Clique and coloring problems are very closely related. It is straightforward to see that the size of the maximum clique of G is a lower bound on the minimum number of labels needed to color a graph.

Recall our scheduling problem. We are given a task set $T = \{1, \ldots, n\}$, processor set $M = \{1, 2, \ldots, m\}$, and for each task $j \in T$ a processing time p_j, release date r_j, and type $\text{fix}_j \subseteq M$. The goal is to find a schedule that minimizes the makespan $C_{\max} = \max_j C_j$.

Consider the case when all $r_j = 0$ and $p_j = 1$. Then, our scheduling problem can be modeled as clique and coloring problems. It involves forming the so-called *conflict* graph G_T with vertices representing the tasks of T. An edge connects two *incompatible* tasks j and i iff $\text{fix}_j \cup \text{fix}_j \neq \emptyset$. The maximum clique problem is then to find as large a set of pairwise incompatible task as possible. The minimum coloring problem is to assign a color to each task so that every incompatible pair is assigned different colors. Accordingly, having a coloring of the conflict graph G_T one can find a schedule for the tasks of T with the makespan at most the number of colors, and vice versa, having a schedule one can find a coloring with the number of colors at most the schedule makespan. Thus, the maximum clique number of G_T, denoted $\omega(G_T)$, is a lower bound on the schedule makespan of the tasks of T.

Unfortunately, there is no such a simple transformation for the case when all $r_j = 0$, but all p_j are arbitrary. However, we can add a weight p_j to each vertex j of the conflict graph G_T and deal with the maximum weighted clique problem. In this framework, the sum of vertex weights in a maximum weighted clique of G_T, denoted $\omega_p(G_T)$, is a lower bound on the schedule makespan of the tasks of T. (We use this fact only in the last section.)

Consider the case when the maximum task size $\Delta_{\max}$ is bounded by some constant, say k. Then, the conflict graph G_T gets a very specific structure, becoming a *k-tuple graph*. More formally it can be defined as follows. Let X be a finite set. A *k-tuple* of X is a set having k (or less) elements of X. A graph $G = (V, E)$ is a *k-tuple graph* if there exists a set X such that each node of V corresponds to a k-tuple of X and there is an edge in E between any two vertices iff the intersection of the corresponding k-tuples is not empty. (Notice that several vertices can correspond to a single k-tuple.)

Different variants of the graph (vertex) coloring problem have been studied. It is widely known that in the case of general graphs the coloring problem is strongly NP-hard, and there is no ρ-approximation algorithm with $\rho = n^{1/7-\varepsilon}$ unless P=NP, where n is the number of vertices of the graph [4]. However, for restricted graph classes, there can exist either polynomial time exact or approximation algorithms, as well as on-line coloring algorithms (see [11] for a survey on on-line coloring).

For a graph $G = (V, E)$, an on-line coloring algorithm colors the vertices in V one vertex at a time in the externally determined order $v_1 \prec \cdots \prec v_n$, and at the time a color is irrevocably assigned to v_t, the algorithm sees the vertices adjacent to v_t only among $v_1, \ldots, v_{t-1}$. In the following we use a very simple variant of an on-line coloring algorithm:

FIRST FIT COLORING (FFC):
Select vertices v from V in an arbitrary order while coloring with an initial sequence of colors $1, 2, \ldots$. Assign the vertex v the least color that has not already been assigned to any vertex adjacent to v.

Lemma 1. *For a k-tuple graph G, the number of labels used by the algorithm FFC is at most $k \cdot w(G)$, where $w(G)$ is the maximum clique number of G.*

Proof. Let X be a finite set and $G = (V, E)$ be a k-tuple graph with the corresponding k-tuples of X. Let $\{a_1, \ldots, a_k\}$ be the k-tuple of X corresponding to a vertex v. Observe that the neighborhood $N(v) = \{u \in V \mid (u, v) \in E\}$ can be split into at most k sets forming disjoint cliques. Each set of vertices in $N(v)$ whose corresponding k-tuples of X contain a common element, is a clique. Thus, we select elements from $\{a_1, \ldots, a_k\}$ one by one, extracting the corresponding cliques from $N(v)$. These cliques are disjoint and cover $N(v)$. Hence, $|N(v)| \leq k \cdot (\omega(G) - 1)$ and FFC cannot use more than $k \cdot \omega(G)$ colors on G. ∎

3   Scheduling of Dedicated Tasks

In this section we consider the on-line version of the dedicated scheduling problem. In the first part we start from the simplest case of unit processing times. In the last part we consider the general case with arbitrary processing times.

3.1   Unit Processing Times

Consider the case when all $p_j = 1$. Then, we can give the following simple algorithm:

FIRST FIT SCHEDULING (FFS):
Schedule the tasks of T by starting the task j at the earliest interval $[t, t + 1)$ in which the processors of fix_j are not busy.

By using the result of Lemma 1 we have:

Lemma 2. *If the maximum task size $\Delta_{\max}$ is bounded by some constant k, then the algorithm FFS is k-competitive for $P \mid$ on-line, fix_j, $p_j = 1 \mid C_{\max}$.*

Proof. First consider the case when the tasks of T arrive over-list. Let OPT be the optimum makespan for $P \mid \text{fix}_j, p_j = 1 \mid C_{\max}$. Let G_T be the corresponding conflict graph of task set T. One can see that FFC working on G_T cannot outperform FFS working on T, in the worst case analysis. (Each interval $[t, t+1)$ corresponds to color t, the makespan corresponds to the number of colors, and vice versa.) Thus, since $w(G_T)$ is a lower bound on OPT, by Lemma 1 the makespan of the output schedule by FFS is at most $k \cdot OPT$.   □

The above lemma does not guarantee a good performance of FFS on instances with $\Delta_{\max} = m$. However, we can modify the algorithm as follows:

FIRST FIT SCHEDULING+ (FFS+):
Schedule the tasks of T by starting the task j at the earliest interval $[t, t+1)$, $t \geq r_j$ in which the processors of fix_j are not busy and there is no task in $[t, t+1)$

- of size greater than $\sqrt{m}$ if $|\text{fix}_j| \leq \sqrt{m}$, and
- of size at most $\sqrt{m}$ if $|\text{fix}_j| > \sqrt{m}$.

Less formally, the output schedule can be split into two parts — the first part includes the intervals during which the tasks have size at most $\sqrt{m}$, call these intervals "red" colored, and the second part includes the intervals during which the processed tasks have a size greater than $\sqrt{m}$, call these intervals "blue" colored. Accordingly, one can think in terms of assigning the tasks, depending on their sizes, to blue or red intervals.

Lemma 3. *The algorithm* FFS+ *is* $2\sqrt{m}$-*competitive for* $P \mid \text{on-line, fix}_j, p_j = 1 \mid C_{\max}$ *and is* $(2\sqrt{m} + 1)$-*competitive for* $P \mid \text{on-line, fix}_j, p_j = 1, r_j \mid C_{\max}$.

Proof. First consider the case where the tasks of T arrive over-list. Let $T^+ := \{j \in T : |\text{fix}_j| > \sqrt{m}\}$ and $T^- := \{j \in T : |\text{fix}_j| \leq \sqrt{m}\}$. Also, let OPT denote the optimal makespan for $P \mid \text{fix}_j, p_j = 1 \mid C_{\max}$. Then, by Lemma 2 the "red" part of the output schedule corresponding to T^- has a makespan at most $\sqrt{m} OPT$. Regarding the "blue" part of the output schedule corresponding to T^+ observe the following: At least one task appears in each "blue" interval, and since for each $j \in T^+ : |\text{fix}_j| > \sqrt{m}$, one may additionally place at most $(m - \sqrt{m})/\sqrt{m} = \sqrt{m} - 1$ tasks in each "blue" interval. Hence, the blue part has a makespan at most $(\sqrt{m} - 1)OPT + OPT$. Combining the two bounds we get that the output makespan is at most $2\sqrt{m} OPT$.

Now consider the case where the tasks of T arrive over-time. Let OPT' denote the optimal makespan for $P \mid \text{fix}_j, p_j = 1, r_j \mid C_{\max}$. Then, it is clear that the release date $r_j \leq OPT'$ for each task $j \in T$. Furthermore, by the above observation we can estimate the completion time of each task $j \in T$ as follows:

$$C_j \leq r_j + 2\sqrt{m} OPT \leq OPT' + 2\sqrt{m} OPT' \leq (2\sqrt{m} + 1)OPT'.$$

This completes the proof.   □

3.2   Arbitrary Processing Times

Now we consider the general case when all p_j are arbitrary. First we analyze the first-fit technique, and then we go to our split-round technique, presenting the main results of this paper.

First-Fit Technique. Consider m tasks $T_1, T_2 \ldots, T_m$ with processing times $p_k = 1 + \frac{k}{m}\varepsilon$ and types $\text{fix}_k = \{k\}$ $(k = 1, \ldots, m)$, and, in addition, $m-1$ tasks $T_{m+1}, T_{m+2}, \ldots, T_{2m-1}$ with the same type $\text{fix}_k = \{1, \ldots, m\}$ and processing time $p_k = \frac{\varepsilon}{m-1}$ $(k = m+1, \ldots, 2m-1)$.

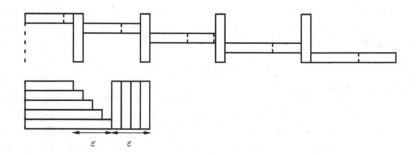

Fig. 1. The output and optimal schedules

   If all these $2m - 1$ tasks arrive in the order $1, (m + 1), 2, (m + 2), \ldots, i,$ $(m+i), \ldots, (m-1), (2m-1), m$, then any deterministic on-line algorithm which leaves no unnecessary idles produces a schedule of makespan greater than $m+2\varepsilon$, while an optimal schedule has a makespan equal to $1 + 2\varepsilon$. (See Figure 1). Thus, m is a lower bound on the competitive ratio of any on-line algorithm which uses the first-fit technique.

   We define $P(T) = \sum_{j \in T} p_j$ to be the total processing time of tasks of T, and $L(T) = \frac{1}{m} \sum_{j \in T} p_j |\text{fix}_j|$ to be the average load of T. Then, it holds that

$$P(T) \le m \cdot L(T) \qquad \text{and} \qquad L(T) \le OPT \le \max_{j \in T} r_j + P(T),$$

where OPT is the minimum makespan for T.

   Consider the following algorithm:

GENERAL FIRST FIT SCHEDULING (GFFS):
Schedule the tasks of T by starting the task j at the earliest time $t \ge r_j$ such that the processors of fix_j are not busy in interval $[t, t + p_j)$.

Sure, the above algorithm uses the first-fit technique and cannot be better than m-competitive. However, it cannot be much worse than that as well.

Lemma 4. *The algorithm* GFFS *is* m*-competitive for* $P \mid$ *on-line,* $\text{fix}_j \mid C_{\max}$ *and* $(m+1)$*-competitive for* $P \mid$ *on-line,* $\text{fix}_j, r_j \mid C_{\max}$.

Proof. First consider the case where the tasks of T arrive over-list. Let OPT be the minimal makespan for $P \mid \text{fix}_j \mid C_{\max}$. Let C denote the makespan of the schedule found by GFFS working on T. Since for each task $j \in T$ the release date $r_j = 0$ and the algorithm GFFS uses the first-fit technique, $P(T)$ is an upper bound on C. Hence, it holds that

$$C \leq P(T) \leq m \cdot L(T) \leq m \cdot OPT,$$

i.e. the output schedule's makespan is at most $m \cdot OPT$.

We complete by using the arguments in the proof of Lemma 3.   □

Split-Round Technique. Let us consider the following off-line algorithm for problem $P \mid \text{fix}_j \mid C_{\max}$.

SPLIT ROUND SCHEDULING (SRS):
Form $T^+ := \{ j \in T : |\text{fix}_j| > \sqrt{m} \}$ and $T^- := \{ j \in T : |\text{fix}_j| \leq \sqrt{m} \}$.

1. Apply GFFS to tasks $j \in T^+$ ordered in an arbitrary way.
2. For each task $j \in T^-$ round p_j to the smallest power of two, say $[p_j]$.[a]
3. Apply GFFS to tasks $j \in T^-$ ordered by non-increasing $[p_j]$'s.

---
[a] Here $[p_j] = 2^a \geq p_j$ and $2^{a-1} < p_j$.

We can state the following result

Theorem 1. *The algorithm* SRS *is a* $3\sqrt{m}$*-approximation algorithm for problem* $P \mid \text{fix}_j \mid C_{\max}$.

Proof. Let OPT be the minimum makespan for $P \mid \text{fix}_j \mid C_{\max}$. Let C^+ be the makespan of the schedule found by GFFS while working on the tasks of T^+ ordered arbitrary (Step 1), and C^- be the makespan of the schedule found by GFFS while working on the rounded tasks $j \in T^-$ ordered by non-increasing $[p_j]$'s (Steps 2 and 3).

Since for each task $j \in T^+$ the size $|\text{fix}_j| > \sqrt{m}$, it holds that

$$L(T^+) = \frac{1}{m} \sum_{j \in T^+} p_j |\text{fix}_j| \geq \frac{\sqrt{m}}{m} \sum_{j \in T^+} p_j = \frac{P(T^+)}{\sqrt{m}}.$$

Thus,

$$C^+ \leq P(T^+) \leq \sqrt{m}\, L(T^+) \leq \sqrt{m}\, OPT.$$

Regarding the tasks of T^- observe the following. First, by applying the rounding procedure we lose at most a factor of 2 in the objective function. Hence, the new minimal makespan, say OPT', is at most $2\,OPT$. Second, the processing time of the tasks of T^- are powers of 2, and GFFS schedules tasks $j \in T^-$ in non-increasing order of $[p_j]$. Hence, a task, say k, is "blocked" by an already scheduled task, say s, if and only if $s : \text{fix}_k \cap \text{fix}_s \neq \emptyset$. (See Figure 2 for an illustration. Notice that if processing times are not powers of 2 we cannot guarantee this property, that is a base element of our analysis.)

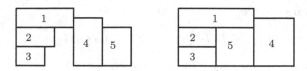

Fig. 2. Schedules for non-rounded and rounded tasks

Consider the schedule found by GFFS while working on the rounded tasks $j \in T^-$ ordered by non-increasing $[p_j]$'s. Let k be a task such that $C^- = C_k$. Accordingly, $S_k = C_k - p_k = C^- - p_k$ is the starting time of task k. Then, by the above observation and by the fact that GFFS uses the first-fit technique, we can find a sequence $s_1, s_2, \ldots, s_q$ of tasks in T^- such that at each moment $t \in [0, S_k)$ for exactly one task $s(t) \in \{s_1, s_2, \ldots, s_q\}$ it holds that $\text{fix}_{s(t)} \cap \text{fix}_k \neq \emptyset$. (See Figure 3). Hence,

$$S_k = \sum_{i=1}^{q} p_{s_i} \quad \text{and} \quad C^- = p_k + \sum_{i=1}^{q} p_{s_i}.$$

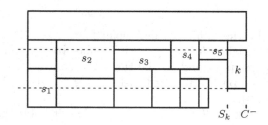

Fig. 3. A task k and a sequence s_1, s_2, s_3, s_4, s_5

Since the size $|\text{fix}_k| \leq \sqrt{m}$ and each task s_i, $i = 1, \ldots, q$ requires at least one of the processors in τ_k, the makespan of any schedule for tasks $s_1, s_2, \ldots, s_q$ and k is at least:

$$L' = \frac{1}{\sqrt{m}} \left(p_k + \sum_{i=1}^{q} p_{s_i} \right) = \frac{C^-}{\sqrt{m}}.$$

Furthermore, since these tasks belong to $T^- \subseteq T$ it also holds that $L' \leq OPT' \leq 2OPT$. Hence, we get

$$C^+ + C^- \leq \sqrt{m}\,OPT + 2\sqrt{m}\,OPT \leq 3\sqrt{m}\,OPT,$$

i.e. the makespan of the output schedule by SRS is at most $3\sqrt{m}\,OPT$.   □

We are ready to present the following main algorithm:

SPLIT-ROUND SCHEDULING+ (SRS+):

1. At each moment we consider two bins, one of which is *active* and the other one is *passive*.
2. At the first release date, collect the tasks into the active bin.
3. Schedule the tasks of the active bin by SRS while collecting the tasks being released into the passive bin.
4. At each moment when SRS has no task to schedule, exchange the activities of bins.

Here we call the following result by Shmoys, Wein & Williamson in [20]: "Let A be a polynomial-time scheduling algorithm that works in an environment in which each job to scheduled is available at time 0 and which always produces a schedule of length at most ρC^*. For the analogous environment in which the existence of a job is unknown until its release date, there exists another polynomial-time algorithm A' that works in this more general setting and produces a schedule of length at most $2\rho C^{*}$". In fact, SRS is similar to the algorithm constructed, and we can give the following result

Theorem 2. SRS+ *is $6\sqrt{m}$-competitive for $P\,|$ on-line, $\mathrm{fix}_j, r_j\,|\, C_{\max}$ in which the existence of a task is unknown until its release date.*

Acknowledgment. We thank to Maxim Sviridenko for giving us useful comments and ideas.

References

1. M. Ammar, G. Polyzos, and S. Tripathi (Eds.). Special issue on network support for multipoint communication. *IEEE Journal Selected Areas in Communications*, 15, 1997.
2. A. K. Amoura, E. Bampis, C. Kenyon, and Y. Manoussakis. Scheduling independent multiprocessor tasks. In *Proceedings 5th European Symposium on Algorithms*, LNCS 1284, pages 1–12. Springer Verlag, 1997.
3. E. Bampis and A. Kononov. On the approximability of scheduling multiprocessor tasks with time dependent processing and processor requirements. In *Proceedings 15th International Parallel and Distributed Processing Symposium*, San Francisco, 2001.
4. M. Bellare, O. Goldreich, and M. Sudan. Free bits, PCPs, and non-approximability — towards tight results. *SIAM Journal on Computing*, 27:804–915, 1998.

5. S. Bischof and E.W. Mayr. On-line scheduling of parallel jobs with runtime restrictions. In *Proceedings 9th Annual International Symposium on Algorithms and Computation*, LNCS 1533, pages 119–129. Springer Verlag, 1998.
6. M. Caramia, P. Dell'Olmo, and A. Iovanella. Lower bound algorithms for multiprocessor task scheduling with ready times, 2001. Personal communications.
7. M. Caramia, P. Dell'Olmo, and A. Iovanella. On-line algorithms for multiprocessor task scheduling with ready times, 2001. Personal communications.
8. M. Drozdowski. Scheduling multiprocessor tasks - an overview. *European Journal on Operations Research*, pages 215–230, 1996.
9. A. Feldmann, M.-Y. Kao, J. Sgall, and S.H. Teng. Optimal online scheduling of parallel tasks with dependencies. In *Proceedings 25th ACM Symposium on the Theory of Computing*, pages 642–651, 1993.
10. A. Feldmann, J. Sgall, and S-H. Teng. Dynamic scheduling on parallel machines. *Theoretical Computer Science*, 130:49–72, 1994.
11. A. Fiat and G. J. Woeginger, editors. *Online algorithms. The state of the art.* LNCS 1442. Springer Verlag, 1998.
12. A. V. Fishkin, K. Jansen, and L. Porkolab. On minimizing average weighted completion time of multiprocessor tasks with release dates. In *Proceedings 28th International Colloquium on Automata, Languages and Programming*, LNCS 2076, pages 875–886, Crete, 2001. Springer Verlag.
13. O. Gerstel, B. Li, A. McGuire, G. N. Rouskas, K. Sivalingam, and Z. Zhang (Eds.). Special issue on protocols and architectures for next generations optical wdm networks. *IEEE Journal Selected Areas in Communications*, 18, October 2000.
14. M. X. Goemans. An approximation algorithm for scheduling on three dedicated processors. *Discrete Applied Mathematics*, 61:49–59, 1995.
15. R. L. Graham, E. L. Lawler, J. K. Lenstra, and A. H. G. Rinnooy Kan. Optimization and approximation in deterministic scheduling: A survey. *Annals of Discrete Mathematics*, pages 287–326, 1979.
16. The NGI Helious project: Regional Testbed Optical Access Network For IP Multicast and Differentiated Services. http://projects.anr.mcnc.org/Helios/, 2000.
17. J. A. Hoogeveen, S. L. Van de Velde, and B. Veltman. Complexity of scheduling multiprocessor tasks with prespecified processor allocations. *Discrete Applied Mathematics*, 55:259–272, 1994.
18. M. Kuznetsov, N. Froberg, S. Henion, H. Rao, J. Korn, K. Rauschenbach, E. Modiano, and V. Chan. A next-generation optical regional access networks. *IEEE Communications Magazine*, 38:66 – 72, January 2000.
19. B. Mukherjee. Wdm-based local lightwave networks Part I: Single-hop systems. *IEEE Network Magazine*, pages 12–27, 1992.
20. D.B. Shmoys, J. Wein, and D.P. Williamson. Scheduling parallel machines on-line. *SIAM Journal on Computing*, 24:1313–1331, 1995.
21. K. M. Sivalingam and S. Subramaniam, editors. *Optical WDM networks: Principles and practice.* Kluwer Academic Publishers, 2000.
22. D. Thaker and G. N. Rouskas. Multi-destination communication in broadcast WDM networks: A survey. Technical Report 2000-08, North Caroline State University, 2000.
23. R. E. Wagner, R. C. Alferness, A. A. M. Saleh, and M. S. Goodman. MONET: Multiwavelength Optical Networking. *Journal of Lightwave Technology*, 14:1349, June 1996.

On the Approximability of Multiprocessor Task Scheduling Problems[*]

Antonio Miranda[1], Luz Torres[1], and Jianer Chen[2]

[1] Department of Computer Science, Bucknell University
Lewisburg, Pennsylvania 17837, USA.
{amiranda,ltorres}@eg.bucknell.edu
[2] Department of Computer Science, Texas A&M University
College Station, TX 77843-3112, USA.
chen@cs.tamu.edu

Abstract. The multiprocessor job scheduling problem has received considerable attention recently. An extensive list of approximation algorithms has been developed and studied for the problem under a variety of constraints. In this paper, we show that from the viewpoint of approximability, the general multiprocessor job scheduling problem has a very rich structures such that by putting simple constraints on the number of processors in the system, we can obtain four versions of the problem, which are NP-hard with a fully polynomial time approximation scheme, strongly NP-hard with a polynomial time approximation scheme, APX-complete (thus with a constant approximation ratio in polynomial time), and with no constant approximation ratio in polynomial time, respectively.

1 Introduction

In recent years we have seen an increasing interest in the approximability of NP-complete problems [18,3]. In particular the multiprocessor scheduling problem has become increasingly studied because processing time is becoming less expensive and systems using several processors are now widely used. Even though the multiprocessor scheduling problem seems to have only to do with the scheduling of tasks on processors, its usefulness goes beyond the boundaries of Computer Science and into several engineering disciplines. For example, in the case of workforce planning, there is a project which has n tasks, each of which must be completed by a team. Several persons can work in parallel, but each person cannot work on more than one task at a time and the total number of persons available is m. We are given the time that each team could spend working on each task. The goal is to find the best way to assign the tasks to the teams in order to minimize the completion time of the entire project.

Over the years several variations of the multiprocessor scheduling problem have been studied. The traditional multiprocessor scheduling problem assumes

---

[*] Supported in part by the National Science Foundation under Grant CCR-0000206.

P. Bose and P. Morin (Eds.): ISAAC 2002, LNCS 2518, pp. 403–415, 2002.

that each task can be processed in a single processor. We will concentrate on the more general version where each job can be processed by several processors working in parallel. This last version has been called multiprocessor job scheduling (or scheduling of multiprocessor tasks) [7]. Two types of multiprocessor job scheduling problems have been extensively studied [12]. The first one is the $P_m|fix|C_{max}$ problem, which consists of a set of tasks, each of which can be processed by a fixed set of processors in a given amount of time. In the second one $P_m|set_j|C_{max}$, each task can be scheduled in different subsets of processors. In both cases, the number of processors is assumed to be fixed. If the number of processors is given as part of the input, these problems are called $P|fix|C_{max}$ or $P|set_j|C_{max}$ respectively. The problem $P_3|fix|C_{max}$ was shown to be NP-hard in the strong sense [10,11] by using a reduction from 3-Partition [21]. $P_3|fix|C_{max}$ subject to the block constraint, where the tasks of the same type requiring two processors must be scheduled consecutively, can be transformed from partition to show that it is NP-hard and can be solved in pseudo-polynomial time [21]. $P_3|fix|C_{max}$ has a polynomial time approximation algorithm with ratio of 7/6 [19]. This result has recently been improved to 9/8 [13]. $P_4|fix|C_{max}$ has a linear-time approximation algorithm of ratio bounded by 1.5 [22]. Comparability graphs [20] were used to show that some instances of $P_m|fix|C_{max}$ can be solved in polynomial time [5,17]. The general problem $P_m|fix|C_{max}$ has a polynomial time approximation scheme [2].

The two processors problem $P_2|set_j|C_{max}$ is also NP-hard [24], can be solved in pseudopolynomial time [7] and has a fully polynomial time approximation scheme [14]. $P_3|set_j|C_{max}$ has an approximation ratio close to $\frac{3}{2}$ [14]. Bianco provides a simple heuristic for $P_m|set_j|C_{max}$ with ratio m [7], which was improved to ratio $m/2$ when the number of processors is fixed [14]. The preemptive case $P_m|set_j, pmtn|C_{max}$ [6] can be solved in polynomial time by using a linear programming approach [7].

Chen and Miranda [15] were able to show that $P_m|set_j|C_{max}$ admits a PTAS. This paper will extend the known results about the general case, showing that $P|set_j|C_{max}$ cannot have a polynomial time algorithm of constant ratio unless $P = NP$, and that there is a problem, obtained as a restriction of $P_m|set_j|C_{max}$, which is MAX-SNP complete. We will use the following definition of PTAS reducibility throughout this paper [16].

Definition 1. *Let A and B be NPO problems A is said to be PTAS-reducible to B, denoted by $A \leq_{PTAS} B$, if there exists a function T, such that*

1. *For any I_A instance of A, $T(I_A)$ is an instance of B, and is computable in time polynomial with respect to $|I_A|$.*
2. *For any instance I_A of A, and for any solution S of $T(I_A)$, a solution $S' = Q(I_A, S)$ to I_A is computable in time polynomial with respect to both $|I_A|$ and $|S|$.*
3. *For any instance I_A of A, and for any solution S of $I_B = T(I_A)$, if $\frac{S}{OPT_B(T(I_A))} \leq \rho$ then $\frac{S'}{OPT_A(I_A)} \leq \rho$, where $\rho \geq 1$. $OPT_A(I_A)$, and $OPT_B(I_B)$ are the optimum solutions to the instances I_A of problem A and I_B of problem B respectively.*

2 Approximability of the General Case with Variable Number of Processors

Formally the problem is stated as follows. We are given m machines $M = \{M_1, M_2, ..., M_m\}$ and a set of jobs $\{J_1, J_2, ..., J_n\}$ that must be processed on those machines. To each job J_i we associate a set of $N(i)$ configurations or processing modes [8] $D_i(k) \subseteq \{M_1, M_2, ..., M_m\}$, $k = 1, ..., N(i)$ where $D_i(k)$ is a machine set on which job J_i requires simultaneous processing for $t_{i,k}$ units of time. In the case of $P_m|fix|C_{max}$, $N(i) = 1$ for all $i = 1, ..., n$, and D_i denotes the processor requirements of task J_i. The main purpose of the problem is to find a schedule for the jobs on the machines where each processor works on at most one job at a time and that minimizes the makespan. This problem is known as $P_m|set_j|C_{max}$ if m is fixed and as $P|set_j|C_{max}$ if m is given as part of the instance [26].

In this section we will show that $P|set_j|C_{max}$ cannot have a polynomial time approximation algorithm with constant ratio unless $P = NP$. To prove this we give a reduction from **minimum graph coloring** which we know cannot be approximable within $|V|^{\frac{1}{7}-\epsilon}$ for any $\epsilon > 0$ [4]. In order to show that Minimum Graph Coloring $\leq_{PTAS} P|set_j|C_{max}$, we take an instance $x = \{G = (V, E)\}$ of **minimum graph coloring** and transform it to $f(x)$, an instance of $P|set_j|C_{max}$ using the algorithm of figure 1 which maps each edge of G to one processor, and each vertex v_i to a task J_i with one processing mode using the processors corresponding to the edges incident on v_i.

Algorithm. C2P

Let s be the number of isolated vertices of G
Let J_i be a job where $i = 1, ..., n$, $n = |V|$, $l = |E| + 1$
Let M_j be a machine where $j = 1, ..., m$ and $m = |E| + s$
For each vertex $v_i \in V$
    if v_i is not an isolated vertex then
        let $e_{i_1}, e_{i_2}, ..., e_{i_k}$ be the k edges incident on v_i
        $D_i(1) = \{M_{i_1}, M_{i_2}, ..., M_{i_k}\}$
    else
        $D_i(1) = \{M_l\}$
        $l = l + 1$
    $N(i) = 1$
    $t_{i,1} = 1$

Fig. 1. Algorithm to create an instance of $P|set_j|C_{max}$ from an undirected graph, instance of graph coloring.

Let y be a solution to the scheduling problem $f(x)$, i.e. y is a schedule where the jobs $J_1, ..., J_n$ have completion times given by $C_1, ..., C_n$ respectively. Let C_{max} be the makespan of the schedule y, and let COLOR be a set of C_{max}

different colors. Since $f(x)$ maps each vertex v_i of G to one task J_i, we can build a coloring of G that assigns to each vertex v_i the color in COLOR corresponding to the completion time of $C_i \leq C_{max}$ of task J_i. Thus we have defined a coloring $g(x,y)$ such that for every vertex $v_i \in V$, $color(v_i) = C_i$. Notice that given two adjacent vertices v_i and v_j, they share one edge. Thus, they require the use of at least one processor in common, which prevents tasks T_i and T_j from having the same completion time, and the same color as stated in the following lemma and corollary.

Lemma 1. *Let y be a schedule for $f(x)$ with completion times $C_i, i = 1, .., n$. If $(v_i, v_j) \in E$, then $color(v_i) \neq color(v_j)$*

Corollary 1. *$g(x,y)$ is a coloring of G.*

Let $m(x, g(x,y))$ be the number of colors required by coloring $g(x,y)$, and let $m(f(x), y)$ be the makespan of schedule y for $f(x)$. Since there is one color for each completion time, the following corollary is true.

Corollary 2. *$m(x, g(x,y)) = m(f(x), y)$*

Now we will show that the optimal solution for x is equal to the optimal solution of $f(x)$. Let y_0 be the schedule that minimizes the makespan for $f(x)$. Let $opt(x)$ be the chromatic number of G and let $opt((f(x))$ be the makespan of y_0, i.e. $opt(f(x)) = m(f(x), y_0)$.

Theorem 1. *$opt(x) = opt(f(x))$*

Proof. We first prove by contradiction that $opt(f(x)) \leq opt(x)$. Suppose that $opt(x) < opt((f(x))$, this implies that $opt(x) < m(f(x), y_0)$. Let $color' : V \to \{1, \ldots, opt(x)\}$ be and optimal coloring function of x. We can use $color'(v)$ to build a schedule y' with smaller makespan than y_0 by making $color'(v_i)$ be the completion time C_i of each task J_i, i.e. for all $v_i \in V$, $C_i = color'(v_i)$. Therefore $opt(f(x)) \leq opt(x)$. Now we show that $opt(f(x)) \geq opt(x)$. Given y_0, the optimal schedule for $f(x)$, from Corollary 1 we can build a coloring $g(x,y)$ for x and from Corollary 2, $m(x, g(x, y_0)) = m(f(x), y_0)$ where $opt(f(x)) = m(f(x), y_0) = m(x, g(x, y_0))$. Therefore, G can be colored with at least $m(x, g(x, y_0))$ colors. Therefore, $opt(f(x)) \geq opt(x)$.

Now we are ready to show that minimum graph coloring is PTAS reducible to $P|set_j|C_{max}$.

Theorem 2. *Minimum Graph Coloring $\leq_{PTAS} P|set_j|C_{max}$.*

Proof. Let x be an instance of minimum graph coloring, and $f(x)$ be the instance of $P|set_j|C_{max}$ obtained by the transformation described in figure 1.

1. Clearly $f(x)$ is computable in polynomial time $O(n + m)$.
2. Given the solution y to $f(x)$, and x, a solution $g(x,y)$ to coloring can be computed in polynomial time.

3. Let $\rho \geq 1$. If $\frac{m(f(x),y)}{opt(f(x))} \leq \rho$, we can use corollary 2 and theorem 1 to conclude that $\frac{m(x,g(x,y))}{opt(f(x))} \leq \rho$.

Therefore **Minimum Graph Coloring** $\leq_{PTAS} P|set_j|C_{max}$

Now we will use theorem 2 to prove the main theorem of this section.

Theorem 3. $P|set_j|C_{max} \notin APX$

Proof. Let GC be the minimum graph coloring problem. From theorem 2 we know that GC $\leq_{PTAS} P|set_j|C_{max}$. However, if $P|set_j|C_{max} \in$ APX then GC $\in$ APX [25]. But we know that GC is not in APX since it cannot be approximated within $|V|^{\frac{1}{7}-\epsilon}$ for any $\epsilon > 0$ unless P=NP [4]. Therefore, $P|set_j|C_{max} \notin$ APX

This shows that if m is given as part of the input, there is no constant ratio approximation algorithm for $P|set_j|C_{max}$.

3   An APX-Complete Version of the Problem

Let $P_{\geq log}|\overline{set_j}|C_{max}$ be the problem $P|Set_j|C_{max}$ subject to the constraints:

1. $m \geq D\lceil \lg n \rceil$ for a fixed constant D.
2. The number of processing modes is bounded by a constant K.
3. Processing modes of different tasks are an equivalence relation with respect to the "can be processed simultaneously" relation (processing modes of the same task are not related).

Notice that the last condition of the definition of $P_{\geq log}|\overline{set_j}|C_{max}$ groups processing modes of different tasks into equivalence classes. This means that all those tasks with a processing mode in a given equivalence class can be processed simultaneously. In the next two sections we will show that $P_{\geq log}|\overline{set_j}|C_{max}$ is APX complete.

3.1 $P_{\geq log}|\overline{set_j}|C_{max}$ Is APX-Hard

In order to show that $P_{\geq log}|\overline{set_j}|C_{max}$ is APX-Complete, we will first show that $P_{\geq log}|\overline{set_j}|C_{max}$ is APX-hard. We will show this result by providing a PTAS reduction from Bounded Degree Vertex Cover (BDVC), which is known to be APX-Complete [1,25].

Let $G = (V, E)$ be a graph with degree bounded by the constant B. We will build an instance of $P_m|Set_j|C_{max}$ with $m = B^2\lceil \lg n \rceil$ from the given graph where each edge will be transformed into a task with two processing modes corresponding to the ends of the edge. Each processing mode will be represented by a binary number with $\lceil \lg n \rceil B^2$ digits, where a 1 in position r means that processor $r+1$ is used to process the corresponding task in that processing mode. The binary numbers used will be selected in such a way that those assigned to one end of those edges incident on the same vertex represent processing modes that can be processed simultaneously while those binary numbers assigned to edges incident on different vertices represent processing modes which cannot be processed in parallel because they have at least one processor in common.

To begin our construction we build B sequences of 2 binary numbers b_j^α with B^2 bits each. We denote by b_{jk}^α the kth bit of the binary number b_j^α , where $1 \le j \le B$, $0 \le k \le B^2 - 1$, and $\alpha \in \{0, 1\}$. b_{jk}^α is defined as follows:

Definition 2.

$$b_{jk}^\alpha = \begin{cases} 1 \; if \; \alpha = 0 \; and \; B(j-1) \le k < Bj \\ 1 \; if \; \alpha = 1 \; and \; (k - j + 1) \; mod \; B = 0 \\ 0 \; otherwise \end{cases}$$

For example, if $B = 3$ we have the following 3 sequences of two binary numbers with 9 bits each.

	$\alpha = 0$	$\alpha = 1$
$j = 1$	$b_1^0 = 000\ 000\ 111$	$b_1^1 = 001\ 001\ 001$
$j = 2$	$b_2^0 = 000\ 111\ 000$	$b_2^1 = 010\ 010\ 010$
$j = 3$	$b_3^0 = 111\ 000\ 000$	$b_3^1 = 100\ 100\ 100$

Now we will use these binary numbers to label each end of the edges of G. We first number the n vertices of G with integers from 0 to $n - 1$ and label each vertex v with the binary representation $l(v)$ corresponding to each integer. Let $l_d(v)$ denote digit d of $l(v)$. The number of binary digits (bits) of each label $l(v)$ will be exactly $\lceil \lg n \rceil$, hence $0 \le d \le \lceil \lg n \rceil - 1$. Next we number the edges incident on vertex v from $j = 1$ to $j = B$ and assign a binary number s_{vj} to each pair of vertex v and edge j as shown in definition 3, where $\tau_1 \tau_2$ denotes concatenation of the binary numbers τ_1 and τ_2, it does not denote product.

Definition 3. $s_{vj} = b_j^{l_{\lceil \lg n \rceil}(v)} \ldots b_j^{l_1(v)} b_j^{l_0(v)}$

Since each edge j is incident on two vertices v and w, each task J_j obtained from each edge j will be given two processing modes s_{vj} and s_{wj}, where a bit equal to 1 in position $0 \le t \le B^2 \lceil \lg n \rceil - 1$ means that processor $t + 1$ is used to process task j in this particular processing mode and a bit equal to 0 in position $0 \le t \le B^2 \lceil \lg n \rceil$ means that processor $t + 1$ is idle while processing task j in this processing mode. The processing time of each processing mode is $t_{ij} = 1$.

Let $T(G)$ be the transformation described above of a given graph G of bounded degree B. The labeling of each edge and vertex requires $O(|E||V|)$ time, and the length of the binary string representing the tasks is $O(\lg |V|)$, therefore, the total time required to compute the transformation is $O(|E||V| \lg |V|)$.

Example 1. In order to illustrate how this transformation works, we now transform the graph of figure 3.1 into the following instance of $P_m|Set_j|C_{max}$ with 4 tasks and 2 processing modes each:

$$T_1 \to \{s_{11}, s_{23}\}$$
$$T_2 \to \{s_{01}, s_{22}\}$$
$$T_3 \to \{s_{21}, s_{32}\}$$
$$T_4 \to \{s_{02}, s_{31}\}$$

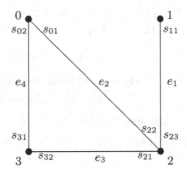

Fig. 2. Graph G to be transformed

where $t_{vj} = 1$ for all v and j, and

$$
\begin{aligned}
s_{11} &= b_1^{l_1(1)} b_1^{l_0(1)} = b_1^0 b_1^1 = 000\ 000\ 111\ 001\ 001\ 001 \\
s_{23} &= b_3^{l_1(2)} b_3^{l_0(2)} = b_3^1 b_3^0 = 100\ 100\ 100\ 111\ 000\ 000 \\
s_{01} &= b_1^{l_1(0)} b_1^{l_0(0)} = b_1^0 b_1^1 = 001\ 001\ 001\ 000\ 000\ 111 \\
s_{22} &= b_2^{l_1(2)} b_2^{l_0(2)} = b_2^1 b_2^0 = 010\ 010\ 010\ 000\ 111\ 000 \\
s_{21} &= b_1^{l_1(2)} b_1^{l_0(2)} = b_1^1 b_1^0 = 001\ 001\ 001\ 000\ 000\ 111 \\
s_{32} &= b_2^{l_1(3)} b_2^{l_0(3)} = b_2^1 b_2^1 = 010\ 010\ 010\ 010\ 010\ 010 \\
s_{02} &= b_2^{l_1(0)} b_2^{l_0(0)} = b_2^0 b_2^0 = 000\ 111\ 000\ 000\ 111\ 000 \\
s_{31} &= b_1^{l_1(3)} b_1^{l_0(3)} = b_1^1 b_1^1 = 000\ 000\ 111\ 000\ 000\ 111
\end{aligned}
$$

Definition 4. *We say that two binary numbers τ_i and τ_j are compatible iff τ_i and τ_j have no 1s in the same position, i.e. $\tau_i \cap \tau_j = 0$ (bitwise operation).*

Definition 5. *A set of binary numbers $T = \{\tau_1, \tau_2, \ldots, \tau_r\}$ is said to be mergeable iff τ_i is compatible with τ_j for every $\tau_i, \tau_j \in T$, $i \neq j$*

Lemma 2. *The sets $B^\alpha = \{b_1^\alpha, b_2^\alpha, \ldots, b_B^\alpha\}$, $\alpha \in \{0, 1\}$ are mergeable*

Proof. (By contradiction) Let $i > j$, and $1 \leq k' \leq B^2$. If $\alpha = 0$ assume that $b_{ik'}^0 = 1$ and $b_{jk'}^0 = 1$. From definition 2, $B(i-1) \leq k' \leq Bi - 1$ and $B(j-1) \leq k' \leq Bj - 1$. Subtracting both inequalities we get $B(i-j) \leq 0 \leq B(i-j)$, which can only be true if $i - j = 0$, but we know that $i \neq j$. Hence, b_i^0 and b_j^0 do not have any 1 in the same position k', which implies that b_i^0 and b_j^0 are compatible. Therefore, B^0 is mergeable. Let $i > j$, and $1 \leq k' \leq B^2$. If $\alpha = 1$ assume that $b_{ik'}^1 = 1$ and $b_{jk'}^1 = 1$. From definition 2, $k' - i + 1 = rB$ and $k' - j + 1 = sB$ for some integers s and r. Since $i > j$ it is clear that $r < s$ and $B \leq B(s - r)$. Subtracting the first two equations, we have $(k' - j + 1) - (k' - i + 1) = sB - rB$, so we have that $i - j = (s - r)B$. On the other hand $1 \leq i \leq B$, and $1 \leq j \leq B$. Subtracting both inequalities we have that $0 < i - j \leq B - 1$, hence

$B \leq B(s-r) = i - j$. Therefore $B \leq i - j$ and $i - j \leq B - 1$, which clearly is a contradiction. This implies that B^1 is mergeable.

Corollary 3. $S_i = \{s_{i1}, s_{i2}, s_{i3}, \ldots, s_{iB}\}$ is mergeable.

Proof. Let $s_{ij} \in S_i$, and $r = \lceil \lg n \rceil - 1$. From definition 3 we have that $s_{ij} = b_j^{l_r(i)} \ldots b_j^{l_1(i)} b_j^{l_0(i)}$ and $s_{ik} = b_k^{l_r(i)} \ldots b_k^{l_1(i)} b_k^{l_0(i)}$. From lemma 2 and definition 3, we know that $b_j^{l_r(i)}$ is compatible with $b_k^{l_r(i)}$, $\ldots$, $b_j^{l_1(i)}$ is compatible with $b_k^{l_1(i)}$, and $b_j^{l_0(i)}$ is compatible with $b_k^{l_0(i)}$. Therefore, s_{ij} is compatible with s_{ik}, hence S_i is mergeable.

Lemma 3. b_i^0 is not compatible with b_j^1 for every $1 \leq i, j \leq B$

Proof. Let $1 \leq i, j \leq B$. b_i^0 has exactly B 1s in bits from $B(i-1)$ to $Bi-1$. From definition 3, b_j^1 has a 1 in position $B(i - 1) + (j - 1)$, and since $1 \leq j \leq B$, we have that $B(i-1) \leq B(i-1) + (j-1) \leq Bi - 1$. Therefore, b_i^0 is not compatible with b_j^1

Corollary 4. s_{vi} is not compatible with s_{uj} for every $u \neq v$ and $1 \leq i, j \leq B$.

Proof. Since the binary representation $l(v)$ of v must differ from the binary representation $l(u)$ of u in at least one digit d, then $b_i^{l_d(v)}$ is not compatible with $b_j^{l_d(u)}$, therefore, s_{vi} is not compatible with s_{uj}.

Notice that corollary 4 guarantees that among all processing modes of a given instance of $P_{\geq log}|\overline{set}_j|C_{max}$, the only compatible ones are those that are derived from the same vertex, i.e. $S_i = \{s_{i1}, s_{i2}, \ldots, s_{iB}\}$ and since S_i is mergeable, then these processing modes form the equivalence classes of the "can be processed simultaneously relation". Therefore, the reduction described above transform an instance of BDVC into an instance of $P_{\geq log}|\overline{set}_j|C_{max}$.

We will now show that if C_{VC}^* is the minimum vertex cover size of $G = (V, E)$ and C_{SCH}^* is the minimum schedule length of $T(G)$ then $C_{VC}^* = C_{SCH}^*$. We first need to show that i) if there is a schedule of $T(G)$ of length C then there is a vertex cover of G of size C and ii) if G has a vertex cover of size C then $T(G)$ has a schedule of length $C' \leq C$. The proof of the following two lemmas is straight forward from lemma 3 and corollary 3.

Lemma 4. If there exists a schedule S of $T(G)$ of length C, then there is a vertex cover A of G of size C.

Lemma 5. If G has a vertex cover of size C then $T(G)$ has a schedule of length $C' \leq C$.

Theorem 4. If C_{VC}^* is the size of a minimum vertex cover of G, and C_{SCH}^* is the minimum makespan of $T(G)$ then $C_{VC}^* = C_{SCH}^*$

Proof. Let C_{VC}^* be the size of a minimum vertex cover of G, and C_{SCH}^* be the minimum makespan of $T(G)$. Using lemma 4, it follows that $C_{VC}^* \leq C_{SCH}^*$, and from lemma 5 it follows that $C_{SCH}^* \leq C_{VC}^*$. Therefore, $C_{VC}^* = C_{SCH}^*$.

Theorem 5. *The transformation $T(G)$described above is a PTAS reduction from bounded degree vertex cover (BDVC) to $P_{\geq log}|\overline{set_j}|C_{max}$.*

Proof. Let G be an instance of BDVC. Let $T(G)$ be the transformed problem, where $T(G)$ is an instance of $P_{\geq log}|\overline{set_j}|C_{max}$.

1. $T(G)$ can be computed in time $O(|E||V|\lg|V|)$, which is polymial with respect to the size of G.
2. Given a schedule of $T(G)$ of length C we can compute, in polynomial time, a vertex cover of G of size C because of lemma 4.
3. Let $C' = Alg(T(G))$ be a solution to $T(G)$. From theorem 4, lemma 4, and lemma 5, it is clear that if for a given $\rho \geq 1$, $\dfrac{C}{OPT_{SCH}(T(G))} \leq \rho$ then $\dfrac{C'}{OPT_{VC}(G)} \leq \rho$.

Therefore $BDVC \leq_{PTAS} P_m|Set_j|C_{max}$.

3.2   Constant Ratio Polynomial-Time Algorithm

Let HVC be the vertex cover problem for the case of hypergraphs, i.e. each edge is allowed to connect several edges and let KHVC be the HVC problem restricted to those hypergraphs where an edge can connect r vertices where $2 \leq r \leq K$ for a fixed K. In this section will provide a PTAS transformation from any instance of $P_{\geq log}|\overline{set_j}|C_{max}$ to an instance of KHVC, and we will show that KHVC has a polynomial time approximation algorithm with ratio K. These two results imply that there is a constant ratio polynomial time approximation algorithm for $P_{\geq log}|\overline{set_j}|C_{max}$, which places $P_{\geq log}|\overline{set_j}|C_{max}$ in APX.

Theorem 6. *KHVC has a polynomial time approximation algorithm A such that given a fixed K and a hypergraph G instance of KHVC, where each edge connects r vertices and $2 \leq r \leq K$, $\dfrac{A(G)}{OPT(G)} \leq K$, where $OPT(G)$ is the size of a minimum hyper vertex cover of G, and $A(G)$ is the size of a hyper vertex cover computed by algorithm A.*

Proof. Let K be a fixed integer greater than or equal to 0, let $G = (V, E)$ be a hypergraph instance of KHVC, where each edge connects r vertices and $2 \leq r \leq K$. We give a polynomial time algorithm where we repeatedly select and remove remove from E all those edges incident on any of the $2 \leq r \leq K$ vertices of e until E is empty. Thus, the size of the vertex cover found by algorithm A is $A(G) \leq K|B|$. Therefore $\frac{A(G)}{OPT(G)} \leq K$.

Let T' be a transformation that takes an instance I of $P_{\geq log}|\overline{set_j}|C_{max}$ and creates a hypergraph $G = T'(I)$. Since the set of processing modes M in I forms an equivalence relation with respect to the "can be processed simultaneously" relation (processing modes of the same task are not related), then the set of all processing modes can be partitioned into equivalence classes. We now build a hypergraph $G = (V, E)$ where there is a vertex $v \in V$ for each one of the equivalence classes of M. The set E is built by adding one hyperedge e_i for each task J_i of I, where e_i links all the vertices that contain processing modes of J_i.

Lemma 6. *If there exists a schedule S of I of length C, then there is a vertex cover A of G of size $C' \leq C$.*

Proof. Let S be a schedule of length C of I. The tasks processed during each time unit $0 \leq t \leq C$ are processed in parallel, hence their corresponding processing modes must belong to the same equivalence class V_t. Let $A = \cup_{t=1}^{C}\{V_t\}$. It is easy to see that A is a vertex cover of G and $C' = |A| \leq C$.

Lemma 7. *If there exists a vertex cover A of size C of $G = T'(I)$ then there exists a schedule S of I with length $C' = C$.*

Proof. Let A be a vertex cover of size C for $G = (V, E)$. Without loss of generality, let $A = \{v_1, v_2, \ldots, v_C\}$ be the first C vertices of V. We build a schedule S where task J_i is processed at time t if the hyperedge i is incident on a vertex $v_t \in A$, where t is the minimum index of those vertices in A. Clearly S is a schedule and since $|A| = C$ then $C' = |S| = C$.

Theorem 7. *If C_{SCH}^* is the minimum makespan of an instance I of $P_{\geq log}|\overline{set_j}|C_{max}$ and C_{VC}^* is the minimum size of a vertex cover of $T'(I)$, then $C_{SCH}^* = C_{VC}^*$*

Proof. Let C_{SCH}^* be the minimum makespan of I. From lemma 7 we can show that $C_{SCH}^* \leq C_{VC}^*$, and from lemma 6 we can show that $C_{VC}^* \leq C_{SCH}^*$. Therefore, $C_{VC}^* = C_{SCH}^*$.

From the above results and using definition 1 the following theorem is proven.

Theorem 8. $P_{\geq log}|\overline{set_j}|C_{max} \leq_{PTAS} KHVC.$

4   Conclusion

The case where the number of processor is given as part of the input ($P_m|set_j|c_{max}$) has been shown to have a PTAS [15]. However, the classification of $P|set_j|C_{max}$ was unknown. In this paper we show that $P|set_j|C_{max}$ is not in APX, which implies that it can have no polynomial time approximation algorithm with fixed ratio unless P=NP. Still with this result it was unknown whether there could be a restricted version of the $P|set_j|C_{max}$ problem which was APX-complete. In this paper we show that if we require the number of processing modes to be bounded by a constant, the number of processors to be in the order of $\lg n$ and the processing modes of all different tasks to form an equivalence relation with respect to the "can be processed simultaneously relation", then this problem is in fact APX-complete.

From the viewpoint of approximability, the general multiprocessor job scheduling problem has a very rich structure. Putting simple constraints on the number of processors in the system we can obtain four versions of the problem which belong to different approximation classes. 1) The problem $P_2|set_j|C_{max}$ is NP-hard but has a fully polynomial time approximation scheme [14]. 2) The $P_m|set_j|C_{max}$ problem was shown to be NP-hard in the strong sense (for any fixed $m \geq 3$) [21]; thus it does not have a fully polynomial time approximation scheme unless P=NP (see also [10,11]), but it admits a polynomial time approximation scheme [15]. 3) The $P_{\geq log}|\overline{set_j}|C_{max}$ problem discussed in this paper is APX-complete, thus with a constant approximation ratio in polynomial time, but cannot have a polynomial time approximation scheme unless P=NP. 4) The general $P|set_j|C_{max}$ discussed in this paper, where the number of processors is given as part of the input is NPO-complete, thus cannot have a polynomial time approximation algorithm with constant ratio unless P=NP.

These results can be summarized in the following table.

Problem	FPTAS	PTAS	APX
$P_2|set_j|C_{max}$	Yes	Yes	Yes
$P_m|set_j|C_{max}$ $m \geq 3$	No	Yes	Yes
$P_{\geq log}|\overline{set_j}|C_{max}$	No	No	Yes
$P|set_j|C_{max}$	No	No	No

References

1. P. ALIMONTI, AND V. KANN, *Hardenss of approximating problems in cubic graphs*, Proc. 34d. Italian Conf. on Algorithms and Complexity, Lecture Notes in Computer Science 1203, Springer-Verlag, pp. 288–298.
2. A. K. AMOURA, E. BAMPIS, C. KENYON, AND Y. MANOUSSAKIS, *Scheduling independent multiprocessor tasks*, Proceedings of the 5th Annual European Symposium, Graz, Austria, 1997, pp. 1–12.
3. S. ARORA, C. LUND, R. MOTWANI, M. SUDAN, AND M. SZEGEDY, *Proof verification and hardness of approximation problems*, Proceedings of the 33rd Annual IEEE Symposium on Foundations of Comput. Sci., IEEE Computer Society, New York, NY, 1992, pp. 14–23.
4. M. BELLARE, O. GOLDREICH, AND M. SUDAN, *Free bits, PCPs and non-approximability-towards tight results*, Proceedings of the 36th Annual IEEE Symposium on Foundations of Comput. Sci., IEEE Computer Society, New York, NY, 1995, pp. 422–431.
5. L. BIANCO, P. DELL'OLMO, AND M. G. SPERANZA, *Nonpreemptive scheduling of independent tasks with prespecified processor allocations*, Naval Res. Logis., 41 (1994), pp. 959–971.
6. L. BIANCO, J. BLAZEWICZ, P. DELL'OLMO, AND M. DROZDOWSKI, *Scheduling preemptive multiprocessor tasks on dedicated processors*, Performance Evaluation, 20 (1994), pp. 361–371.
7. L. BIANCO, J. BLAZEWICZ, P. DELL'OLMO, AND M. DROZDOWSKI, *Scheduling multiprocessor tasks on a dynamic configuration of dedicated processors*, Annals of Operations Research, 58 (1995), pp. 493–517.

8. L. BIANCO, P. DELL'OLMO, AND M.G. SPERANZA, *Scheduling independent tasks with multiple modes*, Discrete Appl. Math., 62 (1995), pp. 35–50.

9. J. BLAZEWICZ, M. DRABOWSKI, AND J. WEGLARZ, *Scheduling multiprocessor tasks to minimize schedule length*, IEEE Trans. Comput., C-35 (1986), pp. 389–393.

10. J. BLAZEWICZ, P. DELL'OLMO, M. DROZDOWSKI, AND M. G. SPERANZA, *Scheduling multiprocessor tasks on three dedicated processors*, Inform. Process. Lett., 41 (1992), pp. 275–280.

11. J. BLAZEWICZ, P. DELL'OLMO, M. DROZDOWSKI, AND M. SPERANZA, *Corrigendum to "Scheduling multiprocessor tasks on three dedicated processors, Inform. Process. Lett., 41 (1992), pp. 275-280"*, Inform. Process. Lett., 49 (1994), pp. 269–270.

12. J. BLAZEWICZ, M. DROZDOWSKI, AND J. WEGLARZ, *Scheduling multiprocessor tasks- a survey*, Microcomputer Applications, 13 (1994), pp. 89–97.

13. J. CHEN, AND J. HUANG, *Semi-normal schedulings: improvement on Goemans' algorithm*, The 9th Annual International Symposium on Algorithms and Computation (ISAAC'01), Lecture Notes in Computer Science 2223, pp. 48–60, 2001.

14. J. CHEN, AND C. -Y. LEE, *General multiprocessor task scheduling*, Naval Res. Logistics 46, pp. 57–74, 1999.

15. J. CHEN, AND A. MIRANDA, *A polynomial time approximation scheme for general multiprocessor job scheduling*, SIAM Journal on Computing, 31 (2001), No. 1, pp. 1–17.

16. P. CRESCENZI, AND L. TREVISAN, *On approximation scheme preserving reducibility and its applications*, Proceedings of the 14th Annual Conf. on Foundations of Software Tech. and Theoret. Comp. Sci.,in Lecture Notes in Comput. Sci. 880, Springer-Verlag, Berlin, Germany, 1994, pp. 330–341.

17. P. DELL'OLMO, AND M. G. SPERANZA, *Graph models for multiprocessor scheduling problems with precedence constraints*, Foundation of Computing and Decisions Sciences, 21 (1996), pp. 17–29.

18. M. GAREY, AND D. JOHNSON, *Computers and Intractability, A Guide to the Theory of NP-Completeness*, W.H. Freeman and Company, San Francisco, CA, 1979.

19. M. X. GOEMANS, *An approximation algorithm for scheduling on three dedicated machines*, Discrete Appl. Math., 61 (1995), pp. 49–59.

20. M. C. GOLUMBIC, *Algorithmic Graph Theory and Perfect Graphs*, Academic Press, New York, NY, 1980.

21. J. A. HOOGEVEEN, S. L. VAN DE VELDE, AND B. VELTMAN, *Complexity of scheduling multiprocessor tasks with prespecified processor allocations*, Discrete Appl. Math., 55 (1994), pp. 259–272.

22. J. HUANG, J. CHEN, AND S. CHEN, *A simple linear time approximation algorithm for multi-processor job scheduling on four processors*, Proceedings of the 8th Annual International Symposium on Algorithms and Computation (ISAAC'00), Lecture Notes in Computer Science 1969, pp. 60–71, 2000.

23. R. M. KARP, *Reducibility among combinatorial problems*, in R. E. MILLER AND J. W. THATCHER, ed., *Complexity of Computer Computations*, Plenum Press, New York, 1972, pp. 85–103.

24. M. KUBALE, *The complexity of scheduling independent two-processor tasks on dedicated processors*, Inform. Process. Lett. (1985), pp. 141–147.

25. C. H. PAPADIMITRIOU, AND M. YANNAKAKIS, *Optimization, approximation, and complexity classes*, Proceedings of the 20th Annual ACM Symposium on the Theory of Computing, ACM, New York, NY 1988, pp. 229–234.

26. B. VELTMAN, B. J. LAGEWEG, AND J. K. LENSTRA, *Multiprocessor scheduling with communication delays*, Parallel Computing, 16 (1990), pp. 173–182.

Bounded-Degree Independent Sets in Planar Graphs

(Extended Abstract)

Therese Biedl* and Dana F. Wilkinson

School of Computer Science, University of Waterloo,
Waterloo, Ontario, Canada, N2L 3G1,
{biedl,d3wilkin}@uwaterloo.ca

Abstract. An independent set in a graph is a set of vertices without edges between them. Every planar graph has an independent set of size at least $\frac{1}{4}n$, and there are planar graphs for which no larger independent sets are possible.

In this paper, similar bounds are provided for the problem of bounded-degree independent set, i.e. an independent set where additionally all vertices have degree less than a pre-specified bound D. Our upper and lower bounds match (up to a small constant) for $D \leq 16$.

1 Introduction

Let $G = (V, E)$ be a graph. An *independent set* of G is a set I of vertices such that no two vertices in I are neighbours. A *D-independent set* is an independent set I_D with the additional constraint that no vertex of I_D has degree D or greater (where the *degree* is the number of incident edges of the vertex in G). A bounded-degree independent set is a D-independent set for some constant D.

A planar graph is a graph that can be drawn without crossings. It is well-known [AH77,RSST97] that every planar graph has a 4-colouring, and in consequence, every planar graph has an independent set of size at least $\frac{n}{4}$.[1] This 4-colouring (and therefore the independent set) can be found in $O(n^2)$ time [RSST97]. No linear-time algorithm to find an independent set of size $\frac{n}{4}$ in a planar graph appears to be known, although there is a linear-time approximation algorithm [Bak94].

The main focus of this paper is on finding bounded-degree independent sets in planar graphs. This problem is motivated by planar point location algorithms (such as those described in [Kir83] or [Iac01]) whose time complexities benefit from the ability to compute a large bounded-degree independent set in linear time. Other applications, such as the reconstruction of Delaunay Triangulations [SvK97], also profit from having a large bounded-degree independent set.

---

* Research partially supported by NSERC.
[1] This fact was conjectured before by Erdös and Vizing (see for example [Alb76]).

P. Bose and P. Morin (Eds.): ISAAC 2002, LNCS 2518, pp. 416–427, 2002.

1.1 Known Results

For a planar graph G, let $\alpha(G)$ be the size of the maximum independent set of G. Let $\alpha(n)$ be the minimum, taken over all planar graphs G with n vertices, of $\alpha(G)$. Thus $\alpha(n)$ defines a guaranteed lower bound on an independent set for any planar graph with n vertices. Similarly define $\alpha_D(n)$ to be the minimum, taken over all planar graphs G with n vertices, of the size of the maximum D-independent set of G.

By the 4-colour theorem, $\alpha(n) \geq \frac{1}{4}n$. On the other hand, by creating a planar graph that consists of many copies of K_4 (the complete graph on four vertices) connected in a planar manner, one can construct a planar graph G for which $\alpha(G) \leq \frac{1}{4}n$, so $\alpha(n) = \frac{1}{4}n$.

This paper contributes similar bounds on $\alpha_D(n)$ for various values of D. The first known result in this area was by Kirkpatrick [Kir83], who showed that there always exists a 13-independent set of size $\frac{1}{24}n$, i.e., $\alpha_{13}(n) \geq \frac{1}{24}n$. In the same paper, he also introduced a general approach for finding a D-independent set: Given a planar graph, delete all vertices of degree D or higher and then find an independent set in the remaining graph. Clearly this yields a D-independent set. It is also easy to see that the maximum D-independent set in the original graph is the same as the maximum independent set in the resulting graph, so in some sense this is the only approach feasible for finding a D-independent set.

Edelsbrunner [Ede87] posed as an exercise to show that $\alpha_D(n) \geq \frac{4(D-6)}{(D+1)^2}n$ for $7 \leq D \leq 13$, and $\alpha_D(n) \geq \frac{1}{7}n$ for $D \geq 13$.[2] Another lower bound on $\alpha_D(n)$ was given by Snoeyink and van Kreveld [SvK97] who showed that every planar graph has a 10-independent set of size $\frac{1}{6}n$, i.e., $\alpha_{10}(n) \geq \frac{1}{6}n$. This bound was later improved by Belleville (private communication) to $\alpha_{10}(n) \geq \frac{4}{21}n + 15$.

No upper bounds on $\alpha_D(n)$ appear to be known. It is obvious that any D-independent set is also an independent set, so a trivial upper bound is $\alpha_D(n) \leq \frac{1}{4}n$.

1.2 Our Results

This paper will provide upper and lower bounds on $\alpha_D(n)$ which match (except for a small constant term) for $7 \leq D \leq 16$. More precisely, it will be shown that:

- $\alpha_D(n) = \theta(1)$ for $D \leq 6$,
- $\alpha_D(n) = \frac{D-6}{4D-18}n + \theta(1)$ for $7 \leq D \leq 16$,
- $\frac{5}{23}n \leq \alpha_D(n) \leq \frac{D-6}{4D-18}n + 2$ for $D = 17, 18$,
- $\frac{5}{23}n \leq \alpha_D(n) \leq \frac{3D-22}{12D-73}n + 2$ for $D = 19$,
- $\frac{D-6}{4D-16}n \leq \alpha_D(n) \leq \frac{3D-22}{12D-73}n + 2$ for $D \geq 20$.

---

[2] Note that these bounds are expressed differently in [Ede87], since our definition of D-independent set *excludes* vertices of degree D in the independent set (as in [Kir83], [Iac01] and [SvK97]), whereas Edelsbrunner's includes them.

Thus, the size of $\alpha_D(n)$ is well-known for $D \leq 16$, whereas a gap remains to be closed for $D \geq 17$.

The paper is structured as follows. In Section 2, an algorithm (which is a modification of the traditional greedy-algorithm) is provided to find an independent set of size at least $\frac{5}{23}n$ in a planar graph in linear time. Using Kirkpatrick's technique, the same algorithm can be used to give a D-independent set of size at least $\min\left(\frac{D-6}{4D-18}n, \frac{5}{23}n\right)$.

This algorithm gives the best-possible bound for D-independent sets for $7 \leq D \leq 16$. To prove this, Section 3 gives graphs for which (up to a small constant) no D-independent set can be larger. Section 4 addresses the case $D \geq 17$, and Section 5 concludes.

Due to space constraints, many details have been left out; see [Bie99] and [Wil02] for more details.

2   A Modified Greedy Algorithm

In this section an algorithm for finding independent sets in planar graphs is described; it is a modification of the well-known greedy algorithm. When finding independent sets, this algorithm guarantees a set of size at least $\frac{5}{23}n$ in linear time.[3] For finding bounded-degree independent sets, combining this algorithm with Kirkpatrick's technique of first deleting vertices of degree D or higher yields a D-independent set of size $\min\left(\frac{D-6}{4D-18}n, \frac{5}{23}n\right)$.

The traditional greedy algorithm for finding an independent set proceeds as follows: repeatedly choose the vertex v of smallest degree, then remove v and all its neighbours from the graph. Next, compute an independent set in the remaining graph recursively, and then add v to it. For simple planar graphs there always exists a vertex of degree 5; hence this greedy algorithm guarantees an independent set of size at least $\frac{1}{6}n$.

This greedy algorithm is modified in two ways: another operation is allowed, and the vertices to be removed are chosen more carefully. These differences are explained in detail below.

2.1   Contracting Vertices

The traditional greedy algorithm uses only one type of operation which consists of deleting a vertex v and all its neighbours. A second operation is now allowed, referred to as *contraction*, in which some of the neighbours of v are retained.

---

[3] It is well-known that such a set can be found in linear time: Baker's approximation algorithm for finding independent sets in planar graphs [Bak94] finds an independent set of size $\frac{k}{k+1}\alpha(G)$ in time $O(k8^k n)$ time. Since every planar graph has $\alpha(G) \geq \frac{1}{4}n$, setting $k = 7$ yields an independent set of size at least $\frac{7}{32}n > \frac{5}{23}n$ in $O(7 \cdot 8^7 n)$ time. However, our algorithm should be substantially faster, since the constant involved is smaller. There is also another approximation algorithm due to Lipton and Tarjan [LT80], but it is even slower than Baker's algorithm.

More precisely, assume that v is a vertex that has two neighbours w_i and w_j that are not adjacent. Then a contraction at v consists of deleting v and all neighbours other than w_i, w_j, then contracting w_i and w_j into one new vertex w^*. Next, recursively find an independent set I in the remaining graph. If w^* is not in the independent set let $I' = I \cup \{v\}$, else let $I' = I - \{w^*\} \cup \{w_i, w_j\}$. One easily verifies that I' is again an independent set in the graph. Note that contracting a vertex of degree d removes only d vertices (as opposed to $d+1$ vertices for a deletion).

In a planar graph, contraction can be applied to any vertex v with $\deg(v) \geq 4$, because not all neighbours of v can be adjacent. Therefore, by repeatedly applying contraction (if possible) and deletion (otherwise) to the vertex of minimum degree in a planar graph, the resulting independent set has size at least $\frac{1}{5}n$.

This technique seems to not have been published independently, but all ideas for it are contained in the paper by Albertson [Alb76].

2.2   Choosing Reduction Vertices

The traditional greedy method always applies the next operation to an arbitrary vertex v of minimum degree. Better bounds can be achieved by choosing vertices more carefully, such that a guaranteed minimum number of edges (relative to the minimum degree of the graph) is deleted.

Assume that graph G has minimum degree d. Then the vertex v and the operation to be applied to it are chosen as follows:

- If $d \leq 2$, then apply deletion at a vertex of degree 2. We call this operation an x_0-*reduction* (x_1-*reduction*, x_2-*reduction*) for $d = 0$ ($d = 1$, $d = 2$).
- If $d = 3$, then apply contraction, if this is possible, at *some* vertex of degree 3. This is referred to as x_3^c-*reduction*. If for all vertices of degree 3 all neighbours are adjacent, then apply a deletion at a vertex of degree 3; this is called an x_3^d-*reduction*.
- If $d = 4$, then let v be a vertex of degree 4. If a deletion at v would delete at least 22 edges, then delete at v (we call this an x_4^d-reduction.) Otherwise, apply contraction at v (this is called an x_4^c-*reduction*.)[4]
- If $d = 5$, then it can be shown that there exists a vertex v of degree 5 such that one of the following three cases is possible:
  - x_5^{19}-reduction: A contraction at v deletes at least 19 edges.
  - x_5^{16}-reduction: A contraction at v deletes at least 16 edges and creates a vertex of degree at most 4. Moreover, the next reduction is not an x_3^d-reduction.
  - x_9-reduction: A contraction at v is followed by an x_3^d-reduction; both reductions together delete at least 25 edges.[5]

The crucial result for this algorithm is the following result.

---

[4] The distinction between these cases is needed only for linear-time complexity, since contraction cannot be done in constant amortized time for arbitrarily high degrees.
[5] The '9' stems from the fact that 9 vertices are deleted during the two reductions.

Lemma 1. *Let G be a planar graph with minimum degree 5. Then it is always possible to do an x_5^{19}-reduction, an x_5^{16}-reduction or an x_9-reduction.*

Proof. (Sketch; see [Bie99] for details) Assume that an arbitrary planar embedding of the planar graph is fixed. Define a *separating triangle* to be a triangle that has vertices both inside and outside it. One can show that there must exist a vertex of degree 5 that is not part of a separating triangle, by taking an "innermost" separating triangle and showing that there is a vertex of degree 5 inside it. Let u be one such vertex; in particular then no two neighbours w_i, w_j of u that are not consecutive in their order around u can be adjacent. This means that a contraction at u removes at least 16 edges. If the neighbours of u have sufficiently high degree, at least 19 edges are, in fact, removed. If this is not the case, then by distinguishing cases by the degrees of the neighbours of u, one can show that either u or one of its neighbours of degree 5 can be chosen as v.

To obtain a D-independent set, apply Kirkpatrick's technique: before running the modified greedy-algorithm, remove all vertices of degree at least D. This operation can be viewed as just another type of reduction, called x_D-*reduction*, applied once for each vertex that had degree D or higher in the original graph.

2.3   Analysis

To analyze the performance of the modified greedy algorithm, we use a linear optimization approach. Each reduction reduces the number of vertices and edges by a certain amount, and increases the size of the independent set by a certain amount. Phrasing this as a constraint system and minimizing the resulting size of the obtained independent set yields a lower bound on this size.

Observe that every reduction is called an x_α^β-reduction for some strings α, β. We will also use x_α^β as a variable to denote the number of x_α^β-reductions; clearly $x_\alpha^\beta \geq 0$. (In fact, x_α^β is an integer, but using this information does not lead to improved bounds.)

The objective function is the size of the resulting independent set; hence it is $\sum i_\alpha^\beta x_\alpha^\beta$, where i_α^β is the number of independent vertices added during an x_α^β-reduction. There are three constraints. Initially there are n vertices, and at the end no vertices are left. So $\sum n_\alpha^\beta x_\alpha^\beta = n$, where n_α^β is the number of vertices deleted during an x_α^β-reduction. Also, initially there are at most $3n - 6$ edges, and each x_α^β-reduction deletes at least a guaranteed number m_α^β of edges. Hence, $\sum m_\alpha^\beta x_\alpha^\beta \leq 3n.^6$ Third, by definition of an x_5^{16}-reduction, the next reduction is at a vertex of degree at most 4, and not an x_3^d-reduction. So we also have $x_5^{16} \leq x_0 + x_1 + x_2 + x_3^c + x_4$.

The values of i_α^β, n_α^β and m_α^β follow from the definition for $\alpha = 5, 9$ and the x_4^d-reduction, and are easily computed for $\alpha \leq 3$ and the x_4^c-reduction. To give just one example: An x_4^c-reduction at v removes 4 vertices and adds one vertex to the independent set so $n_4^c = 4$ and $i_4^c = 1$. Also, since the graph has

---

^{6} We could replace "$3n$" by "$3n - 6$". This would lead to a very small improvement in the constant, but not in the fraction, for the independent set.

minimum degree 4, the two deleted neighbours are incident to at least 8 edges, possibly counting one edge twice. Since contracting v removes at least seven edges incident to the deleted neighbours of v and two more edges during the contraction, $m_4^c = 9$.

To obtain the values for the x_D-reduction, note that the x_D vertices of degree at least D induce a planar graph with at most $3x_D - 6$ edges. Removing these x_D vertices hence removes at least $Dx_D - (3x_D - 6) \geq (D-3)x_D$ edges, since every removed vertex is incident to at least D edges, and at most $3x_D - 6$ edges are counted twice that way. Hence $i_D = 0$, $n_D = 1$ and $m_D = D - 3$ (on average).

Putting it all together, the following linear program is obtained (for the independent set problem, omit all terms containing x_D):

$$
\begin{array}{llllllllllll}
\text{min.} & & x_0 & +x_1 & +x_2 & +x_3^c & +x_3^d & +x_4^c & +x_4^d & +x_5^{16} & +x_5^{19} & +2x_9 \\
\text{s.t.} & x_D & +x_0 & +2x_1 & +3x_2 & +3x_3^c & +4x_3^d & +4x_4^c & +5x_4^d & +5x_5^{16} & +5x_5^{19} & +9x_9 & = & n \\
(D-3)x_D & & & +x_1 & +3x_2 & +5x_3^c & +6x_3^d & +9x_4^c & +22x_4^d & +16x_5^{16} & +19x_5^{19} & +25x_9 & \leq & 3n \\
& & x_0 & +x_1 & +x_2 & +x_3^c & & +x_4^d & +x_4^d & -x_5^{16} & & & \geq & 0 \\
\end{array}
$$

$$x_D, x_0, x_1, x_2, x_3^c, x_3^d, x_4^c, x_4^d, x_5^{16}, x_5^{19}, x_9 \geq 0$$

Solving this linear program to optimality yields the following results:

- For independent set ($x_D = 0$), $x_3^d = \frac{2}{23}n$ and $x_5^{19} = \frac{3}{23}n$ and all other variables 0.
- For a D-independent set and $7 \leq D \leq 16$, $x_3^d = \frac{D-6}{4D-18}n$, $x_D = \frac{6}{4D-18}n$, and all other variables 0.
- For a D-independent set and $D \geq 17$, the results are the same as for general independent set, i.e., $x_3^d = \frac{2}{23}n$ and $x_5^{19} = \frac{3}{23}n$ and all other variables 0.

Thus, the modified greedy-algorithm obtains an independent set of size at least $\frac{5}{23}n$, and a D-independent set of size at least $\min\left(\frac{D-6}{4D-18}n, \frac{5}{23}n\right)$.

2.4   Time Complexity

The modified greedy algorithm can be implemented to run in linear time; we sketch some of the crucial ideas for this below. Finding a vertex v of minimum degree is easily done with a bucket structure. Testing which reduction to do at this vertex can be done in constant time as follows. Note that all that is needed for the analysis is that at least a certain number of edges is deleted. So if v has a neighbour with high degree, then we can always take v. If all neighbours have constant degree, then all tests can be implemented in constant time. The only slightly more complicated problem is how to find the correct vertex for Lemma 1; this can also be done via a bucket structure in amortized constant time.

Reducing the graph at the chosen vertex v takes constant time for a deletion. The time for a contraction is not necessarily constant, but can be made amortized constant time as follows: If the minimum degree is 3 or 4, and if a neighbour of v has degree at least m_4^d, then we do deletion even if contraction is feasible. Thus contraction only occurs if the neighbours have constant degree, and the contraction takes constant time. If the minimum degree is 5, then the neighbours of v to be contracted are chosen in such a way that the neighbour of v with maximum degree gets deleted; the time for contraction is then proportional to the number of deleted edges, and hence is $O(1)$ amortized. To conclude:

Theorem 1. *Any planar graph G has an independent set of size at least $\frac{5}{23}n$, which can be found in $O(n)$ time.*

Theorem 2. *Any planar graph G has a D-independent set of size at least* $\min\left(\frac{D-6}{4D-18}n, \frac{5}{23}n\right)$, *which can be found in $O(n)$ time.*

3   Matching Upper Bounds

By the result of the previous section, every planar graph has a D-independent set of size at least $\min\left(\frac{D-6}{4D-18}n, \frac{5}{23}n\right)$. Note that the minimum is achieved by the first term for $D \leq 16$.

For $D \leq 6$, this bound only states the trivial bound of $\alpha_D(n) \geq 0$. However, this is expected since one can construct arbitrarily large planar graphs with 12 vertices of degree 5 and all other vertices degree 6; hence $\alpha_D(n) = 0$ for $D \leq 5$. For $D = 6$, one can construct planar graphs with 6 vertices of degree 3 (in two triangles) and all other vertices degree 6; hence $\alpha_6(n) \leq 2$ (and, in fact, equality holds for $n \geq 5$.)

The bound of $\frac{D-6}{4D-18}n$ becomes non-trivial for $D \geq 7$. We now show that it is tight for $7 \leq D \leq 16$ (up to a small constant). More precisely, we provide families of graphs where the largest possible D-independent set has size $\frac{D-6}{4D-18}n+2$. These graphs can be built for values of D between 7 and 18.

The solution to the linear program in Section 2.3 suggests what such graphs should look like: they should consist of vertices of degree D and vertices of degree 3 for which all neighbours are adjacent, i.e., K_4's. We now show that such graphs indeed can be constructed.

The construction is given in detail for $D = 8$ and only sketched for other cases. The appropriate graph is built in three steps. The first graph in Figure 1 is a 3×4 grid with diagonals, which has 12 vertices and 12 interior faces. The first step is to insert K_4's into 2 of those faces to get the second graph in Figure 1. As a second step, join an arbitrary number of copies of that second graph together to get the third graph in Figure 1. Notice that in this graph all vertices of the grid have degree 8, unless they are on the outer-face. Finally, take the third graph and identify the vertices on the left with those on the right to form the cylinder-like fourth graph in Figure 1. This graph remains planar and all vertices in the underlying grid with the exception of the top and bottom of the cylinder have degree 8.

Any 8-independent set hence can contain only vertices from the K_4's, as well as one vertex each from the top and the bottom end. To analyze the size of the largest 8-independent set, note that each copy of the second graph adds 14 vertices to the total, but can increase an 8-independent set by at most 2 vertices, one for each K_4. This means that the largest possible 8-independent set is of size $\frac{2}{14}n + 2 = \frac{D-6}{4D-18}n + 2$.

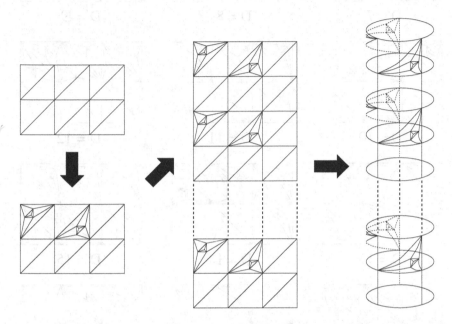

Fig. 1. The lower-bound graph for $D = 8$.

For $D \geq 9$, similarly add $D - 6$ copies of K_4 into the grid in such a manner that after copying the base pattern and rolling it into a cylinder, all vertices of the grid have degree D, except for those at the top end and bottom end of the cylinder. The base patterns are shown in Figure 2.

The case $D = 7$ is special. Ideally, the base pattern would consist of the grid with one K_4 added, but it is not possible to connect it such that all grid-vertices have degree 7. This is resolved by changing the grid pattern as well; see the dashed lines in Figure 2.

In each case, after stacking multiple copies of the pattern and rolling it into a cylinder, the size of the largest D-independent set can be shown to be at most $\frac{D-6}{4D-18}n + 2$.

Theorem 3. *For $7 \leq D \leq 18$, we have $\alpha_D(n) \leq \frac{D-6}{4D-18}n + 2$.*

Combining this with the result of the previous section, therefore $\alpha_D(n) = \frac{D-6}{4D-18}n + \theta(1)$ for $7 \leq D \leq 16$.

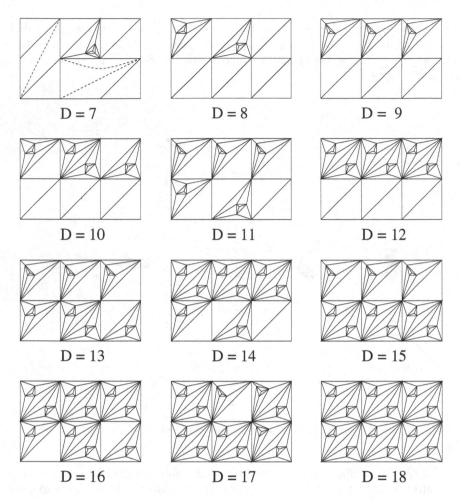

Fig. 2. The base patterns for $8 \leq D \leq 18$.

4   The Case $D \geq 17$

The previous two sections provided upper and lower bounds which together showed that $\alpha_D(n) = \frac{D-6}{4D-18}n$ (plus a small constant) for $7 \leq D \leq 16$. In this section, we now address the case $D \geq 17$.

4.1   Upper Bounds

Recall that the construction of Section 3 works for $7 \leq D \leq 18$, therefore $\alpha_D(n) \leq \frac{D-6}{4D-18}n$ (plus a small constant) even for $D = 17, 18$.

Unfortunately, there seems to be no easy way to extend this bound to higher values of D. Essentially, for each successive value of $7 \leq D \leq 18$ another K_4 was added to the base pattern. For $D > 18$ there are no good places to put additional K_4's, since all 12 interior faces of the grid have been filled in the pattern for $D = 18$ in Figure 2. However, two K_4's can be added to either side of selected edges in the grid and connected such that the degree of each vertex in the grid is increased by five. Figure 3 shows how this can be done to the pattern for $D = 18$. Note that in the resulting graph one face is not a triangle; adding an edge to turn it into two triangles would either add a multiple edge or not contribute to increasing the degrees of vertices in the grid.

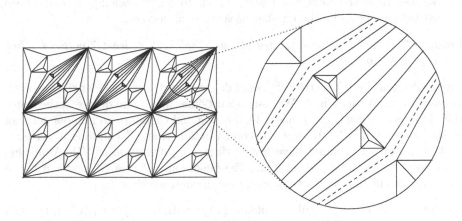

Fig. 3. Extending the pattern for $D = 18$ to a pattern for $D = 23$.

Repeatedly adding two more K_4's to each selected edge yields patterns for $D = 28, 33, 38, \ldots$. Similarly, applying this method to the patterns for $14 \leq D \leq 17$ gives bounds for all values $D \geq 19$. Doing an extended case analysis yields, in the worst case (for $D = 19, 24, 29, \ldots$), an bound on the D-independent set of $\frac{3D-22}{12D-73}n + 2$.

Theorem 4. *For all $D \geq 19$, $\alpha_D(n) \leq \frac{3D-22}{12D-73}n + 2$.*

4.2 Lower Bounds

For $D \geq 17$, the algorithm in Section 2 only guarantees a D-independent set of size $\frac{5}{23}n$, which in particular does not depend on D. Intuitively, one would think that for larger values of D, fewer vertices are removed in the pre-processing step and therefore the lower bound for D-independent set should be closer to $\frac{1}{4}n$. In this section, we show that this is indeed the case, i.e., $\lim_{D \to \infty} \alpha_D(n) = \frac{1}{4}n$.

Note that for studying lower bounds $\alpha_D(n)$ it suffices to study lower bounds on D-independent sets in *triangulated planar graphs* (planar graphs where every face is a triangle.) If G is not triangulated, then one can obtain a triangulated planar graph G' by adding edges to G. Since this can only increase the degrees of vertices, any D-independent set I_D in G' is also a D-independent set in G.

Given such a graph, G, with at least 5 vertices the following is known:

1. G is triconnected, i.e., it cannot be disconnected by removing one vertex (a *cutvertex*) or two vertices (a *cutting pair*).
2. Every vertex in G has degree 3 or larger.
3. The vertices of degree 3 are an independent set. This holds because for any vertex v of degree 3 the neighbours must form a triangle in a triangulated graph. If any neighbour of v had degree 3 as well, then the other two neighbours of v would form a cutting pair, a contradiction.

We also need the following result, which to our knowledge has not been proved before. Let n_i be the number of vertices of degree i.

Lemma 2. *Let G be a triangulated planar graph with at least 7 vertices. Then $\alpha(G) \geq \frac{1}{3}(n_3 + n_4)$.*

Proof. (Sketch) Let V' be the vertices of degree 3 or 4, and consider the graph G' induced by the vertices in V'. We claim that G' has maximum degree 2; the result then follows because then G' has a 3-colouring and therefore an independent set of size at least $|V'|/3$. Assume for contradiction that G' contains a vertex v of degree 3 or 4 (it cannot have higher degree by definition of V'). Analyzing cases considering the degree of v and the neighbours of v in G', one can find a cutvertex or cutting pair of G in each case, a contradiction.

Now let G_D be a triangulated planar graph with $\alpha_D(n) = \alpha_D(G_D)$, i.e., G_D has the smallest possible D-independent set among all graphs with n vertices. Let I_D be a maximum D-independent set of G_D. The above observations yield the following:

- $|I_D| \geq n_3$.
- $3|I_D| \geq n_3 + n_4$.

By definition of n_i, and because every planar triangulated graph G has $3n - 6$ edges, the following also holds:

- $\sum_{i \geq 3} n_i = n$.
- $\sum_{i \geq 3} i \cdot n_i = 6n - 12$.

Denote by n_l the number of vertices with degree less than D. By applying the 4-colouring algorithm to the graph induced by these vertices, one can obtain an independent set of size at least $\frac{1}{4} n_l$.[7] This, together with the definition of n_l, gives two more constraints:

- $n_l = \sum_{i=0}^{D-1} n_i$.
- $|I_D| \geq \frac{1}{4} n_l$.

Computing the minimum possible value of $|I_D|$ under this system of constraints yields $\alpha_D(n) = \alpha(G_D) = |I_D| \geq \frac{D-6}{4D-16} n$.

---

[7] Note that this approach requires the 4-colour theorem. To actually find such a large D-independent set for $D \geq 17$, one must therefore either implement 4-colouring (which may not be practical) or a suitable approximation algorithm for finding independent sets (e.g. [Bak94]).

Theorem 5. *For all $D \geq 7$, $\alpha_D(n) \geq \frac{D-6}{4D-16}n$.*

Note that this bound improves on $\alpha_D(n) \geq \frac{5}{23}n$ for $D \geq 20$. Hence there exists the unusual case that for $D = 16, 17, 18, 19$ the best known lower bound for $\alpha_D(n)$ is the same—$\frac{5}{23}n$. For this reason, it is our conjecture that this bound is not optimal for $D = 17, 18, 19$.

5 Conclusion

In this paper, the problem of bounded-degree independent sets in planar graphs was studied. Both upper and lower bounds were provided for this problem, in terms of the limit, D, on the degree of the vertices in the independent set.

For $7 \leq D \leq 16$, the upper and lower bounds provided in this paper match. For $D \geq 17$, the correct bound remains an open problem. Our conjecture is that our upper bound of $\frac{D-6}{4D-18}n$ for $D = 17, 18$ is tight in these cases as well. Thus one open problem is whether the analysis of the modified greedy algorithm can be tightened to improve the bound for $D = 17, 18$, or whether this algorithm indeed achieves only a set of size $\frac{5}{23}n$ on some graph, and hence a different algorithm would be needed to improve the lower bound.

Another open problem is the case $D \geq 19$. Our suspiscion, in this case, is that both the upper and the lower bound could be improved, and that the true answer lies somewhere in between.

References

[AH77]   K. Appel and W. Haken. Every planar map is four colorable. I. Discharging. *Illinois Journal of Mathematics*, 21(3):429–490, 1977.

[Alb76]   M. Albertson. A lower bound for the independence number of a planar graph. *Journal of Combinatorial Theory (B)*, 20:84–93, 1976.

[Bak94]   B. Baker. Approximation algorithms for NP-complete problems on planar graphs. *Journal of the ACM*, 41(1):153–180, January 1994.

[Bie99]   T. Biedl. Large independent sets in planar graphs in linear time. Technical report, Department of Computer Science, University of Waterloo, 1999.

[Ede87]   H. Edelsbrunner. *Algorithms in combinatorial geometry*. Springer-Verlag, Berlin, 1987.

[Iac01]   J. Iacono. Optimal planar point location. In *Twelfth Annual ACM-SIAM Symposium On Discrete Algorithms*, pages 340–341, New York, 2001. ACM.

[Kir83]   D. Kirkpatrick. Optimal search in planar subdivisions. *SIAM Journal of Computing*, 12(1):28–35, February 1983.

[LT80]   R. Lipton and R. E. Tarjan. Applications of a planar separator theorem. *SIAM Journal of Computing*, 9(3):615–627, August 1980.

[RSST97] N. Robertson, D. Sanders, P. Seymour, and R. Thomas. The four-colour theorem. *JCTB: Journal of Combinatorial Theory, Series B*, 70:2–44, 1997.

[SvK97]   J. Snoeyink and M. van Kreveld. Linear-time reconstruction of Delaunay triangulations with applications. In *Algorithms : 5th Annual European Symposium (ESA '97)*, pages 459–471, Berlin, 1997. Springer-Verlag.

[Wil02]   D. F. Wilkinson. Bounded-degree independent sets. Master's thesis, School of Computer Science, University of Waterloo, May 2002.

Minimum Edge Ranking Spanning Trees of Threshold Graphs

Kazuhisa Makino[1], Yushi Uno[2], and Toshihide Ibaraki[3]

[1] Division of Systems Science, Graduate School of Engineering Science, Osaka University, Toyonaka, Osaka, 560-8531 Japan. makino@sys.es.osaka-u.ac.jp
[2] Department of Mathematics and Information Sciences, College of Integrated Arts and Sciences, Osaka Prefecture University, Sakai, 599-8531 Japan uno@mi.cias.osakafu-u.ac.jp
[3] Department of Applied Mathematics and Physics, Graduate School of Informatics, Kyoto University, Kyoto, 606-8501 Japan. ibaraki@i.kyoto-u.ac.jp

Abstract. Given a graph G, the minimum edge ranking spanning tree problem (MERST) is to find a spanning tree of G whose edge ranking is minimum. However, this problem is known to be NP-hard for general graphs. In this paper, we show that the problem MERST has a polynomial time algorithm for threshold graphs, which have useful applications in practice. The result is also significant in the sense that this is a first non-trivial graph class for which the problem MERST is found to be polynomially solvable.

1 Introduction

Let $G = (V, E)$ be an undirected graph that is simple and connected. An *edge ranking* of a graph G is a labeling $r : E \to \mathbf{Z}_+$, with the property that every path between two edges with the same label i contains an intermediate edge with label $j > i$. Here $\mathbf{Z}_+$ denotes the set of all non-negative integers. An edge ranking is *minimum* if it uses the least number of distinct labels among all edge rankings of G. The *minimum edge ranking spanning tree problem* (MERST) is to find a spanning tree whose edge ranking is minimum. The problem MERST has a number of practical applications in many fields such as parallel processing, integration of distributed data, and so on [10]. It is known [9,10] that the problem MERST is in general NP-hard and linearly solvable if a given graph is a tree. However, the computational complexity of the problem MERST has been still unknown even when a given graph is restricted to be bipartite, chordal, partial k-tree, or threshold, for example.

In this paper, we consider the problem MERST for threshold graphs. This is motivated by efficient parallel joins in a relational database (see Section 2 for the details). Threshold graphs were introduced by Chvátal and Hammer [2], and have been extensively studied (e.g., [4,5]). For example, it is known that threshold graphs are perfect and the recognition can be done in linear time. We show that the problem MERST for threshold graphs can be solved in polynomial time. This is a first result that finds a non-trivial polynomially solvable class for the problem MERST.

P. Bose and P. Morin (Eds.): ISAAC 2002, LNCS 2518, pp. 428–440, 2002.

The rest of the paper is organized as follows. In Section 2, we first give a formal definition of our problem and several notations required, and then overview properties of threshold graphs with emphasis on applications. Section 3 shows that any threshold graph G has a spanning caterpillar whose edge ranking is minimum among all spanning trees of G. We also characterize optimal threshold graphs $G = (V, E)$ with $|V| = 2^k$. Note that such graphs are threshold graphs that have a spanning tree whose minimum edge rank is k, since it is known that any connected graph needs at least $\lceil \log |V| \rceil$ edge ranks. (We assume that the base of the logarithm is 2 throughout the paper.) In Section 4, we reveal several properties of the edge ranking of caterpillars as spanning trees of a threshold graph. In Section 5, we present a linear time algorithm that solves the problem MERST for threshold graphs.

Due to the space limitation, some proofs are omitted.

2   Preliminaries

2.1   Definitions and Notations

Throughout the subsequent sections, we assume that a graph $G = (V, E)$ is simple, undirected and connected. As usual, $V(G)$ and $E(G)$ are used to denote the vertex and edge set of G, respectively. For an edge set $F \subseteq E$, we denote by $V(F)$ the set of all the end vertices of the edges in F. Similarly, for an vertex set $U \subseteq V$, we denote by $E(U)$ the set of all the edges that are incident to the vertices in U. Moreover, for two vertex sets U and W, we use notation $E(U, W) = \{(u, w) \in E \mid u \in U, w \in W\}$. For simplicity, we sometimes write $V(e)$, $E(u)$ and $E(u, W)$ for $V(\{e\})$, $E(\{u\})$ and $E(\{u\}, W)$, for example. Let $G[U]$ (resp., $G[F]$) denote the subgraph induced by a vertex set $U \subseteq V$ (resp., an edge set $F \subseteq E$). If the resulting subgraph induced by U (resp., F) is a tree, we sometimes denote it by $T[U]$ (resp., $T[F]$). For a vertex $u \in V$ and an edge set $F \subseteq E$, the *degree* of u in $G[F]$ is the number of edges in F that are incident to u, and is denoted by $\deg_F(u)$, i.e., $\deg_F(u) = |\{e \mid e = (u, v) \in F\}|$.

A mapping (labeling) $r : E \rightarrow \mathbf{Z}_+$ is called an *edge ranking* if r satisfies that every path between two edges with the same label i contains an intermediate edge with label $j > i$. An integer given to an edge by an edge ranking is called a *rank* of the edge. By definition, every edge ranking r has exactly one edge with the largest rank. An edge ranking by integers $1, 2, \ldots, k$ is called a *k-edge ranking*. A graph G is said to be *k-edge rankable* if it has a k-edge ranking. An edge ranking is *minimum* if the largest rank k in it is the smallest among all edge rankings of G, and such k is called the *minimum edge rank* of G and is denoted by $\mathrm{rank}(G)$. For an edge ranking r of G, we call the multiple set of ranks $\{r(e) \mid e \in E\}$ the *rank set* of r, and denote it by $RS(r)$. Note that $|RS(r)| = |E(G)|$ holds. The *minimum edge ranking problem* (MER) asks to compute a minimum edge ranking r of a given graph G.

As an example, let us consider a graph $G = (V, E)$ in Fig. 1 (a), where $E = \{e_1, e_2, \ldots, e_8\}$. Fig. 1 (b) provides a 4-edge ranking r_1 of G, where $r_1(e_2) = r_1(e_5) = r_1(e_7) = 1$, $r_1(e_1) = r_1(e_6) = 2$, $r_1(e_3) = r_1(e_6) = 3$ and $r_1(e_4) = 4$.

Note that r_1 is in fact a minimum edge ranking of G, i.e., $\mathrm{rank}(G) = 4$. For this r_1, we have $RS(r_1) = \{1, 1, 1, 2, 2, 3, 3, 4\}$.

The problem MER for example has applications in the context of assembling a multi-part product from its components in the smallest number of parallel integration stages, and has been extensively studied (e.g., [1,3,7,8,9,12]). De la Torre, Greenlaw and Schäffer [3] first proposed a polynomial time algorithm for the problem MER if G is a tree, and Lam and Yue [9] finally showed that it can be solved in linear time. On the other hand, Lam and Yue [8] also showed that MER is NP-hard for general graphs.

A spanning tree T of a graph G is called a *minimum edge ranking* spanning tree if it has the minimum $\mathrm{rank}(T)$ among all spanning trees T of G. The minimum edge ranking spanning tree problem (MERST) can be described as follows [10]:

MERST (The minimum edge ranking spanning tree problem)

Input: A graph $G = (V, E)$.
Output: A minimum edge ranking spanning tree T together with a minimum edge ranking of T.

For example, Fig. 1 (c) shows a minimum edge ranking spanning tree $T = (V, E_T)$ of G in Fig. 1 (a), together with a minimum edge ranking r_2 of T.

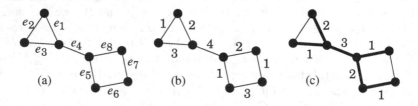

Fig. 1. A graph and edge rankings. (a) a graph $G = (V, E)$, (b) a minimum edge ranking r_1 of G, and (c) a minimum edge ranking spanning tree $T = (V, E_T)$ of G and a minimum edge ranking r_2 of T.

Note that the problem MERST coincides with the problem MER when G is a tree. From a graph theoretical viewpoint, MERST is the problem of minimizing the number of edge contraction steps until G becomes a single vertex (which may have self-loops), under the condition that the edges contracted simultaneously do not share any of their end vertices (i.e., such edges always form a matching) [10], while MER is to minimize the number of edge contraction steps until G becomes a single vertex with no edge, under the same condition but regarding the edge contraction of a self-loop e as the removal of e.

MERST is known to have quite natural practical interpretations, such as join operations in relational databases and scheduling the parallel assembly of a multi-part product [10,11]. In Section 2.2, we briefly discuss one of the applications of MERST for threshold graphs.

Unfortunately, MERST is known to be NP-hard for general graphs. However, it has a simple approximation algorithm with its worst case performance ratio $\min\{(\Delta^*-1)\log n/\Delta^*, \Delta^*-1\}/\log(\Delta^*+1)-1$, where n is the number of vertices in G and Δ^* is the maximum degree of a spanning tree whose maximum degree is minimum [10]. In this paper we present a polynomial time algorithm for MERST assuming that a given graph is threshold.

2.2 Threshold Graphs and Their Applications

A graph $G = (V, E)$ is *threshold* if there exist a weight $a : V \to Z_+$ and a threshold $t \in Z_+$ such that for any subset S of V,

$$S \text{ is stable} \iff \sum_{u \in S} a(u) \leq t$$

Fig. 2 (a) gives a threshold graph with weights to vertices and a threshold $t = 10$.

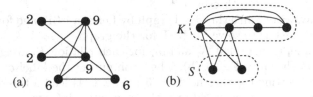

(a)

(b)

Fig. 2. (a) A threshold graph with vertex weights and a threshold $t = 10$, and (b) a vertex partition into a stable set S and a clique K.

It is known [4,5] that threshold graphs can be characterized as follows.

Lemma 1. *For a graph $G = (V, E)$, the following three statements are equivalent.*

(i)   *G is threshold.*
(ii)   *There exists a weight $a : V \to Z_+$ and a threshold t such that for any $u, v \in V$ with $u \neq v$, $(u, v) \in E \iff a(u) + a(v) > t$.*
(iii) *The vertex set V can be partitioned into a clique $K = \{u_1, \ldots, u_{|K|}\}$ and a stable set $S = \{v_1, \ldots, v_{|S|}\}$ such that*

$$(u_i, v_j) \in E \implies (u_{i'}, v_{j'}) \in E \quad (i' \leq i, \ j' \leq j). \tag{1}$$

∎

Fig. 2 (b) shows the vertex partition of the graph G in Fig. 2 (a).

As an example of MERST for threshold graphs, let us consider to compute the join of n relations $R_1, \ldots, R_n$ in a relational database [6,11]. The situation is depicted by a join graph with n vertices corresponding to the n relations, in which each edge represents the join operation of the two relations of its end vertices. To obtain the join of n relations, we select a spanning tree T and perform $n-1$ join operations specified by the edges of T. If these operations are done in parallel, their order can be exactly described by an edge ranking r of T and the number of stages of parallel joins is equal to the rank of r.

In the above process, we may also consider the sizes of intermediate relations formed by join operations. Let relation R_i have N_i tuples, and let m_i be the number of columns used in the join of n relations. Then it can be argued that the number of tuples in the relation resulting from the join of R_i and R_j is approximately proportional to $(N_i/m_i)(N_j/m_j)$. Therefore, it may be reasonable to consider only those edges having small expected numbers of tuples. For this, let $P_i = m_i/N_i$ for all i. Then $(N_i/m_i)(N_j/m_j) < K$ is equivalent to $P_i P_j > 1/K$, i.e., $\log P_i + \log P_j > \log(1/K)$. Therefore, giving weights $a_i = \log P_i$ to vertices R_i, it amounts to consider only those edges satisfying

$$a_i + a_j > t,$$

where $t = \log 1/K$ is the threshold. [1]

In this way, we obtain a threshold graph by Lemma 1 (ii), and for our purpose, the remaining task is to solve MERST for the graph.

For example, assume that N_i and m_i for each relation R_i are given in a pair of (N_i, m_i) as shown in Fig. 3 (b). Then computing the label $a_i = \log P_i + A$ ($A = 12$), and setting the threshold $t = 5$ ($= -7 + A$) lead a threshold graph in Fig. 3 (c), where A is introduced to make a_i positive integers.

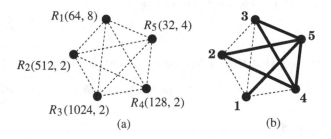

Fig. 3. (a) Five relations with their size and column information, and (b) a graph representing edges that should be processed earlier.

---

[1] If we are concerned with the space to store the relation instead of the number of tuples in it, we can use $P_i = m_i/N_i M_i$, since the space to store the relation resulting from the join of R_i and R_j is approximately proportional to $(N_i M_i/m_i)(N_j M_j/m_j)$. Here M_i denotes the number of columns in R_i.

3   MERST for Threshold Graphs

This section discusses several properties of the edge ranking of spanning tress in threshold graphs. Especially, we show that any threshold graph G has a spanning caterpillar whose edge ranking is minimum among all spanning trees of G. We also characterize threshold graphs $G = (V, E)$ with $|V| = 2^k$ having a spanning tree of minimum edge rank k.

3.1   Threshold Graphs and Their Spanning Caterpillars

A tree $T = (V, E)$ is called a *caterpillar* if V can be partitioned into B and L (i.e., $B \cup L = V$ and $B \cap L = \emptyset$) such that $T[B]$ is a path and L is stable. A vertex in B (resp., L) is called a *body vertex* (resp., a *leg vertex*) of a caterpillar T. By definition, each leg vertex u is adjacent to exactly one body vertex v, where (u, v) is called a *leg* of T. Note that the partition (B, L) of V may not be unique (see Fig 4 (a) and (b)), and hence the names body and/or leg can be used when the partition is specified. We sometimes write a caterpillar $T = (V, E)$ with a vertex partition (B, L) as $T = (B \cup L, E)$.

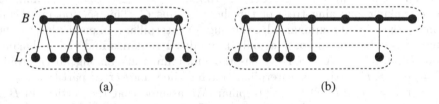

(a)   (b)

Fig. 4. (a) A caterpillar T and a vertex partition (B, L) of V, (b) another partition for the same caterpillar.

We now show two properties of the edge ranking of a spanning tree in a threshold graph.

Lemma 2. *Let $G = (V, E)$ be a threshold graph with a vertex partition (K, S) satisfying Lemma 1 (iii). Let T be a spanning tree of G, and let r be an edge ranking of T. Then G has a spanning tree T^* with an edge ranking r^* such that $T^*[K]$ is connected and $RS(r^*) = RS(r)$ holds.*   ∎

Lemma 3. *Let $G = (V, E)$ be a threshold graph with a vertex partition (K, S) satisfying Lemma 1 (iii). Let T be a spanning tree of G such that $T[K]$ is connected, and let r be an edge ranking of T. Then G has a spanning tree T^* with an edge ranking r^* such that $T^*[K]$ is a path and $RS(r^*) = RS(r)$ holds.*   ∎

Lemmas 2 and 3 immediately imply the following corollary.

Corollary 1. *Let $G = (V, E)$ be a threshold graph with a vertex partition (K, S) satisfying Lemma 1 (iii). Let T be a spanning tree of G, and let r be an edge ranking of T. Then G has a spanning caterpillar $T^* = (B \cup L, E_{T^*})$ with an edge ranking r^* such that $B = K$, $L = S$ and $RS(r^*) = RS(r)$.* ∎

3.2 Optimal Threshold Graphs

It is known that $\mathrm{rank}(G) \geq \lceil \log |V| \rceil$ holds for any connected graph $G = (V, E)$. The following theorem characterizes optimal threshold graphs G satisfying $\mathrm{rank}(G) = \log |V|$ when $|V| = 2^k$.

Theorem 1. *Let $G = (V, E)$ be a threshold graph with $|V| = 2^k$ $(k \geq 0)$ and a vertex partition $(K = \{u_1, \ldots, u_{|K|}\}, S = \{v_1, \ldots, v_{|S|}\})$ satisfying Lemma 1 (iii). Then $\mathrm{rank}(G) = k$ holds if and only if $|K| \geq |S|$ and $(u_{1+j}, v_{|S|-j}) \in E$ for $j = 0, 1, \ldots, |S| - 1$.* ∎

4 Minimum Edge Rankings of Caterpillars

By Theorem 1, we only consider caterpillars $T = (K \cup S, E_T)$ as minimum edge ranking spanning trees of threshold graphs G. This section investigates some properties of minimum edge rankings of caterpillars. In particular, we show that it is sufficient to consider only normalized rankings and that a balanced caterpillar $T^* = (K \cup S, E_{T^*})$ has the minimum rank among all caterpillars $T = (K \cup S, E_T)$ (that is, caterpillars having the same vertex partition).

Let $T = (B \cup L, E)$ be a caterpillar. We assume that the vertices in B are ordered in such a way that $(u_i, u_{i+1}) \in E$ holds for $i = 1, \ldots |B| - 1$. Given a k-edge ranking r of a caterpillar $T = (V, E)$, let $T_p = (V, E_p)$ $(p = 0, 1, \ldots, k)$ denote the spanning subgraph of T consisting of all edges $e \in E$ with $r(e) \leq p$. By definition, T_0 has no edge and $T_k = T$. Furthermore, let $T_{pq} = (V_{pq}, E_{pq})$ $(q = 1, \ldots, c_p)$ be the connected components of T_p, where c_p is the number of connected components. An edge ranking r of a caterpillar $T = (B \cup L, E)$ is *BL-connected* if

$$|V_{pq} \cap B| \geq 2 \implies E(V_{pq} \cap B, L) \subseteq E_{pq}. \tag{2}$$

In other words, if a connected component T_{pq} contains at least two body vertices, then it contains all the legs that are incident to some body vertex in V_{pq}. One can easily see that an edge ranking r is BL-connected if and only if

$$r((u_i, v)) < r((u_{i-1}, u_i)) \quad \text{and} \quad r((u_i, v)) < r((u_i, u_{i+1})) \tag{3}$$

hold for any leg (u_i, v) with $u_i \in B$ and $v \in L$.

Lemma 4. *For any edge ranking r of a caterpillar T, there exists a BL-connected edge ranking r^* of T such that $RS(r^*) = RS(r)$.*

Proof. If r is not BL-connected, then it follows from the discussion above that there exists a leg $(u_i, v) \in E$ such that $u_i \in B$, $v \in L$ and either $r((u_i, v)) > r((u_{i-1}, u_i))$ or $r((u_i, v)) > r((u_i, u_{i+1}))$ holds. We construct a new edge labeling r' from r by exchanging $r((u_i, v))$ with $r((u_{i-1}, u_i))$ if $r((u_i, v)) > r((u_{i-1}, u_i))$; otherwise, with $r((u_i, u_{i+1}))$. Since v is a leaf, it is easy to show that this r' is also an edge ranking. By repeating this argument, we can construct a desired BL-connected edge ranking r^*.   ∎

Note that, for each $u_i \in B$, legs in $E(u_i, L)$ cannot have the same rank. Therefore, any BL-connected edge ranking can be transformed into a "better" BL-connected edge ranking in the sense that the legs in $E(u_i, L)$ $(i = 1, \ldots, |K|)$ receive their ranks from 1 to $|E(u_i, L)|$. We call such a ranking *normalized*. Fig. 5 shows a normalized edge ranking of a caterpillar.

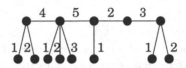

Fig. 5. A normalized (and minimum) edge ranking of a caterpillar originally given in Fig. 4.

In order to compare different edge rankings, we define a *lexicographic order* $\preceq$ on rank sets as follows: for any two rank sets $RS(r)$ and $RS(r')$,

$$RS(r) \preceq RS(r') \iff \left| \{ e \in E \mid r(e) \geq p \} \right| \leq \left| \{ e \in E \mid r(e) \geq p \} \right| \quad \text{for all } p = 1, 2, \ldots.$$

By combining the discussion above with Lemma 4, we have the following lemma.

Lemma 5. *For any edge ranking r of a caterpillar T, there exists a normalized edge ranking r^* of T such that $RS(r^*) \preceq RS(r)$.*   ∎

This lemma shows that, for a caterpillar T, we can always find a minimum edge ranking among normalized ones. Therefore, we can restrict our attention to normalized rankings.

We next consider all the edge rankings of a caterpillar $T = (B \cup L, E)$. The *degree sequence* $\alpha = (\alpha_1, \ldots, \alpha_{|B|})$ of a caterpillar T is the sequence of the numbers of legs incident to $u_i \in B$ (i.e., $\alpha_i = |E(u_i, L)|$) for $i = 1, 2, \ldots, |B|$. Note that the degree sequence α uniquely defines a caterpillar, denoted by T^α. As noticed before, partition (B, L) of V for T is not always unique, and hence a caterpillar can be represented by different degree sequences. For example, for the caterpillar T shown in Fig. 4, there are partitions (B, L) depicted in (a) and (b), which have degree sequences $\alpha = (2, 3, 1, 0, 2)$ and $\beta = (2, 3, 1, 0, 1, 0)$, respectively.

Let α and β be respectively degree sequences of caterpillars T^{α} and T^{β} with the same vertex partition (B, L). Then T^{β} is called the *left-shifted* caterpillar of T^{α} if β is obtained from α by sorting α in the non-increasing order. For example, Fig. 6 shows the left-shifted caterpillar T^{β} of T^{α}.

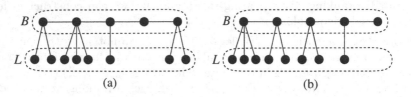

(a)   (b)

Fig. 6. (a) A caterpillar T^{α}, and (b) the left-shifted caterpillar T^{β} of T^{α}. Note that the vertex partition (B, L) is common.

The next lemma shows that the minimum edge rank of the left-shifted caterpillar cannot be worse than that of the original caterpillar.

Lemma 6. *Let r be a normalized edge ranking of a caterpillar T. Then there exists an edge ranking r^* of the left-shifted caterpillar of T such that $RS(r^*) = RS(r)$.*   ∎

We then compare the minimum edge rankings of the left-shifted caterpillars in order to find the best one among those. Similarly to the lexicographic order $\preceq$ on rank sets, we define a lexicographic order $\preceq$ on non-increasing degree sequences $\alpha = (\alpha_1, \ldots, \alpha_{|B|})$ as follows: for $\alpha = (\alpha_1, \ldots, \alpha_{|B|})$ and $\beta = (\beta_1, \ldots, \beta_{|B|})$,

$$\alpha \preceq \beta \iff \sum_{i=1}^{k} \alpha_i \le \sum_{i=1}^{k} \beta_i \quad \text{for all } k = 1, \ldots, |B|.$$

Lemma 7. *Let α and β be two non-increasing degree sequences of caterpillars T^{α} and T^{β} with a common vertex partition (B, L), and let r be a normalized edge ranking of T^{β}. If $\alpha \preceq \beta$, then there exists an edge ranking r' of T^{α} such that $RS(r') \preceq RS(r)$.*   ∎

Let $\alpha^* = (\alpha_1^*, \ldots, \alpha_{|B|}^*)$ be the lexicographically minimum non-increasing degree sequence for a partition (B, L), that is,

$$\alpha_i^* = \begin{cases} \lfloor |L|/|B| \rfloor + 1, \ i = 1, \ldots, p \\ \lfloor |L|/|B| \rfloor, \quad\quad i = p+1, \ldots, |B|, \end{cases}$$

where $p = |L| \bmod |B|$ $(0 \le p < |B|)$.

From Lemmas 5, 6 and 7, we have the following result.

Theorem 2. *Let $\mathcal{T}$ be the set of all caterpillars $T = (V, E)$ with a vertex partition (B, L), and let α^* be defined above. Then for any edge ranking r of T^{α^*}, there exists an edge ranking r^* of T^* such that $RS(r^*) \preceq RS(r)$. In particular, $\mathrm{rank}(T^{\alpha^*})$ is the minimum among all $\mathrm{rank}(T)$, $T \in \mathcal{T}$.*   ∎

5 A Linear Time Algorithm to Solve MERST for Threshold Graphs

This section presents a linear time algorithm that solves MERST for threshold graphs.

5.1 Spanning Caterpillars with the Minimum Degree Sequence

Let $G = (V, E)$ be a threshold graph with a vertex partition (K, S) satisfying Lemma 1 (iii). It follows from Lemmas 1, 5 and 6 that MERST for a threshold graph G can be solved by considering the minimum normalized edge ranking of the spanning left-shifted caterpillar $T = (K \cup S, E_T)$ of G. We note here that a spanning caterpillar $T = (K \cup S, E_T)$ of a threshold graph G can always be transformed into the left-shifted caterpillar, since K is a clique and hence an arbitrary ordering of vertices in K always forms a path. Moreover, if G contains a spanning left-shifted caterpillar $T^\alpha = (K \cup S, E_{T^\alpha})$ with the minimum α^* (with respect to $\preceq$ defined for degree sequences), this T^{α^*} is a minimum edge ranking spanning tree by Lemma 7. We show that this is always the case, and such an α can be computed in linear time.

Note that $K = \{u_1, \ldots, u_{|K|}\}$ and $S = \{v_1, \ldots, v_{|S|}\}$ satisfy condition (1) of Lemma 1, and hence we can restrict spanning caterpillars $T = (K \cup S, E_T)$ to those T that satisfy $E_T(K) = \{(u_i, u_{i+1}) \mid i = 1, \ldots, |K| - 1\}$. In other words, the ordering of the body vertices is the same as in Lemma 1 (iii). This is shown in Fig. 7.

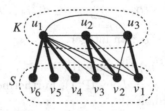

Fig. 7. A threshold graph with a vertex partition ($K = \{u_1, u_2, u_3\}$, $S = \{v_1, \ldots, v_6\}$). Bold edges express the minimum degree sequence $\alpha = (3, 2, 1)$ that can be realized in this graph.

Let $d_i = |E(u_i, S)|$ for $i = 1, \ldots, |K|$. From condition (1), we have $d_1 \geq d_2 \geq \ldots \geq d_{|K|}$. The following procedure computes the minimum degree sequence of (K, S) that can be realized by a spanning caterpillar in G.

Procedure MINIMUM_DS
Input: $d = (d_1, \ldots, d_{|K|})$.
Output: $\alpha = (\alpha_1, \ldots, \alpha_{|K|})$.

begin
    for $i = |K|$ **downto** 1 **do**

$$\alpha_i := \min \left\{ \left\lfloor (d_{i-\ell+1} - \textstyle\sum_{j=i+1}^{|K|} \alpha_j)/\ell \right\rfloor \mid \ell = 1, \ldots, i \right\} \qquad (4)$$

end.

Lemma 8. *Procedure* MINIMUM_DS *correctly computes the minimum degree sequence that can be realized by a spanning caterpillar in a threshold graph* $G = (K \cup S, E)$.

Proof. We separately show the correctness of Procedure MINIMUM_DS as follows:

(i) $\sum_{i=1}^{|K|} \alpha_i = |S|$,
(ii) α is a non-increasing sequence,
(iii) α is realizable by a spanning caterpillar of G.
(iv) α is the minimum degree sequence with respect to the ordering $\preceq$.

Here, note that statement (ii) together with $\alpha_{|K|} \geq 0$ implies that α_i is non-negative for all $i = 1, \ldots, |K|$.

First, we can easily see that α_i computed by Procedure MINIMUM_DS is obtained by solving the following integer programming problem from $i = |K|$ down to 1:

Problem IP(i)

maximize $\quad \alpha_i$

s.t. $\quad \ell\alpha_i + \displaystyle\sum_{j=i+1}^{|K|} \alpha_j \leq d_{i-\ell+1}, \; \ell = 1, \ldots, i$ $\qquad (5)$

α_i: an integer.

Therefore, we show (i) $\sim$ (iv) by using IP(i).

For (i), we have $\sum_{i=1}^{|K|} \alpha_i = d_1$ by IP(1). Furthermore, since G is a connected threshold graph, $d_1 = |S|$ holds.

For (ii), notice that α_{i+1} is the maximum integer that satisfies

$$\ell\alpha_{i+1} + \sum_{j=i+1}^{|K|} \alpha_j \leq d_{i-\ell+1}, \; \ell = 0, 1, \ldots, i.$$

Therefore, α_{i+1} is a feasible solution of the α_i in IP(i). This means that $\alpha_i \geq \alpha_{i+1}$.

For (iii), let us consider the caterpillar $T = (K \cup S, E_T)$ by

$$E_T = \{(u_i, u_{i+1}) \mid i = 1, ..., |K|-1\} \cup \{(u_i, v_{j_i}) \mid i=1, ..., |K|, \; j_i = \sum_{j=i+1}^{|K|} \alpha_j + 1, ..., \sum_{j=i}^{|K|} \alpha_j \}.$$

$$(6)$$

Note that T is a caterpillar defined by α, i.e., $T = T^\alpha$. Moreover, from the constraints (5) for $\ell = 1$, we have $\sum_{j=i}^{|K|} \alpha_j \le d_i$, $i = 1, 2, \ldots, |K|$. This together with Lemma 1 (iii) implies $E_T \subseteq E$, which completes the proof of (iii).

For (iv), suppose that there exists a degree sequence β of (K, S) such that $\beta \not\succeq \alpha$ and β is realizable by a spanning caterpillar of G. Then some k ($2 \le k \le |K| - 1$) satisfies $\sum_{i=1}^{k-1} \alpha_i > \sum_{i=1}^{k-1} \beta_i$, i.e.,

$$\sum_{i=k}^{|K|} \alpha_i < \sum_{i=k}^{|K|} \beta_i. \tag{7}$$

Let k be the largest integer satisfying (7). By definition, we have $\alpha_k < \beta_k$. We can see that β also satisfies (5) (since otherwise non-increasing sequence β is not realizable by a spanning caterpillar of G). In particular, we have $\ell\beta_k + \sum_{j=k+1}^{|K|} \beta_j \le d_{k-\ell+1}$ ($\ell = 1, \ldots, k$). This means

$$\sum_{j=k}^{|K|} \beta_j \le d_{k-\ell+1} - (\ell - 1)\beta_k, \quad \ell = 1, \ldots, k. \tag{8}$$

On the other hand, since α_k is the maximum integer satisfying (5) for $i = k$, some inequality for $\ell = \ell^*$ must be tight for α_k, i.e., $\sum_{j=k}^{|K|} \alpha_j = d_{k-\ell^*+1} - (\ell^* - 1)\alpha_k$. Since $\alpha_k < \beta_k$, we have $\sum_{j=k}^{|K|} \alpha_j \ge \sum_{i=k}^{|K|} \beta_i$, a contradiction to (7)   ∎

For example, by applying this procedure to a threshold graph in Fig. 7, we obtain the minimum degree sequence $\alpha = (3, 2, 1)$, which is also shown in Fig. 7 as bold lines.

5.2   An Algorithm and Its Complexity

Finally we can describe an algorithm that computes a minimum edge ranking spanning tree for a given threshold graph.

Algorithm MERST_THRESHOLD
Input: A threshold graph $G = (K \cup S, E)$.
Output: A minimum edge ranking spanning tree $T = (V, E_T)$ of G with a minimum edge ranking $r : E_T \to Z_+$.
begin

1. Compute the minimum degree sequence α by procedure MINIMUM_DS.
2. Construct a spanning caterpillar $T^\alpha = (K \cup S, E_T)$ by (5.1).
3. Compute a minimum edge ranking r of T^α.

end.

For a threshold graph in Fig. 7, we obtain a minimum edge ranking spanning tree as shown in Fig. 8.

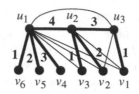

Fig. 8. A threshold graph and a minimum edge ranking spanning tree.

Theorem 3. *Algorithm* MERST_THRESHOLD *correctly computes a minimum edge ranking spanning tree of a threshold graph* G *in* $O(|V| + |E|)$ *time.*

Proof. Since the correctness of the algorithm is clear from the discussion so far, we only examine its time complexity. Note that a vertex partition (K, S) and $d = (d_1, \ldots, d_{|K|})$ can be computed in $O(|V| + |E|)$ time. Procedure MINIMUM_DS computes from (K, S) and d the minimum degree sequence α in $O(|K|^2) = O(|E|)$ time. More careful implementation shows that it can be done in $O(|V|)$ time. Step 2 can be done in $O(|V|)$, and Step 3 is also possible in $O(|V|)$ time [9]. ∎

Note that the proof of Theorem 3 MERST for threshold graphs is solvable in $O(|V|)$ time if $d = (d_1, \ldots, d_{|K|})$ is given in advance.

References

1. H. L. Bodlaender, J. S. Deogun, K. Jansen, T. Kloks, D. Kratsch, H. Müller and Zs. Tuza, Ranking of graphs, *SIAM J. Discrete Mathematics*, **11**, 1 (1998), 168–181.
2. V. Chvátal and P. L. Hammer, Aggregation of inequalities in integer programming, *Annals of Discrete Mathematics*, **1** (1977), 145–162.
3. P. de la Torre, R. Greenlaw and A. A. Schäffer, Optimal edge ranking of trees in polynomial time, *Algorithmica*, **13** (1995), 592–618.
4. M. C. Golumbic, *Algorithmic Graph Theory and Perfect Graphs*, Academic Press, New York, 1980.
5. P. L. Hammer, T. Ibaraki and B. Simeone, Threshold sequences, *SIAM Journal on Algebraic and Discrete Mathematics*, **2**, 39–49, 1981.
6. T. Ibaraki and T. Kameda, On the optimal nesting order for computing N-relational joins, *ACM Transactions on Database Systems*, **9**, 3 (1984), 482–502.
7. A. V. Iyer, H. D. Ratliff and G. Vijayan, On an edge-ranking problem of trees and graphs, *Discrete Applied Mathematics*, **30** (1991), 43–52.
8. T. W. Lam and F. L. Yue, Edge ranking of graphs is hard, *Discrete Applied Mathematics*, **85** (1998), 71–86.
9. T. W. Lam and F. L. Yue, Optimal edge ranking of trees in linear time, *Algorithmica*, **30** (2001), 12–33.
10. K. Makino, Y. Uno and T. Ibaraki, On minimum edge ranking spanning trees, *Journal of Algorithms*, **38** (2001), 411–437.
11. Y. Uno and T. Ibaraki, Complexity of the optimum join order problem in relational databases, *IEICE Transactions*, **E74**, 7 (1991), 2067–2075.
12. X. Zhou, M. A. Kashem and T. Nishizeki, Generalized edge-rankings of trees, *IEICE Transactions*, **E81-A**, 2 (1998), 310–319.

File Transfer Tree Problems

Hiro Ito[1], Hiroshi Nagamochi[2], Yosuke Sugiyama[3], and Masato Fujita[4]

[1] Department of Communications and Computer Engineering, School of Informatics,
Kyoto University, Kyoto 606-8501, Japan, itohiro@kuis.kyoto-u.ac.jp
[2] Department of Information and Computer Sciences, Toyohashi University of
Technology, Aichi 441-8580, Japan, naga@ics.tut.ac.jp
[3] Internet Ware Corporation, sugiyama@intergate.co.jp
[4] Fujitsu Kansai-Chubu Net-Tech Limited, fujita.masato@kcn.fujitsu.com

Abstract. Given an edge-weighted digraph G with a designated vertex
r, and a vertex capacity δ, we consider the problem of finding a shortest
path tree T rooted at r such that for each vertex v the number of children
of v in T does not exceed the capacity $\delta(v)$. The problem has an application in designing a routing for transferring files from the source node
to other nodes in an information network. In this paper, we first present
an efficient algorithm to the problem. We then introduce extensions of
the problem by relaxing the degree constraint or the distance constraint
in various ways and show polynomial algorithms or the computational
hardness of these problems.

1 Introduction

The problem of broadcasting the same contents from a designated node, called
a source, to some other nodes in an information network is an important issue
to find an effective use of the network. We assume that the contents is contained
in a file, and a file can be transmitted through links of the network. A necessary
number of copies of a file can be duplicated at any node. Also assume that a file
containing the same contents is allowed to be transmitted through each link at
most once so that the set of links that transmit files forms a tree rooted at the
source. We call such a routing for transferring files a *file transfer tree* (FTT). As
one of the most important criteria for constructing "effective" FTTs, we employ
the following two measurements:

- Transmission time: the time that a file is transferred from a source to a node
  in an FTT,
- Copy number: the number of copies produced at a node in an FTT.

We assume that each node v has a capacity $\delta(v)$ on the maximum number of
copies of a file that can be produced at v. We formulate the problem of designing
FTTs as a problem of choosing a rooted tree in an edge-weighted digraph G,
where vertices and edges in G represent respectively nodes and links in the
network. Then the above two criteria correspond respectively to the distance
from a root to each vertex in a rooted tree and the number of children at each
vertex in the tree. In this paper, we start with the following problem as the basis
of the study of FTTs with the above criteria.

P. Bose and P. Morin (Eds.): ISAAC 2002, LNCS 2518, pp. 441–452, 2002.

Basic FTT problem:
Input: A digraph G with nonnegative real $\ell(e)$ for each edge e and an integer $\delta(v)$ for each vertex v, and a designated vertex r called a root.
Question: Is there a tree T of G rooted at r that meets the following two constraints?
degree constraint: for each vertex v, the outdegree $d_T^+(v)$ of v in T (i.e., the number of children of v) is at most $\delta(v)$,
distance constraint: for each vertex v, the length of the path in T from r to v is equal to the distance in G from r to v.   □

A tree T is called *degree-bounded* if it satisfies the degree constraint. Note that a tree T that meets the distance constraint is a shortest path tree.

In this paper, we first show that the basic FTT problem is polynomially solvable. Our algorithm checks in $O(m^{1.5})$ time whether or not there is a degree-bounded one among all shortest path trees in a G with m edges and finds one of such trees if any. The key idea of the algorithm is that all shortest path trees are included in the level graph, which is an acyclic digraph, and the problem of constructing a degree-bounded tree on an acyclic digraph can be reduced to the maximum flow problem.

One may need to find a tree that would provide a good FTT even if the answer of the basic FTT problem is "no." For this, we consider designing of algorithms or showing computational hardness of the problem when we relax the degree constraint or the distance constraint. The vertex set and edge set of a graph (or digraph) G are denoted by $V(G)$ and $E(G)$, respectively.

Relaxing the degree constraint. We define the *overflow* of a vertex $v \in V(G)$ in a rooted tree T as

$$\text{overflow}_T(v) := \max\{d_T^+(v) - \delta(v),\ 0\}.$$

A vertex v in a tree T is called *overflowed* if $\text{overflow}_T(v) > 0$, i.e., $d_T^+(v) > \delta(v)$. Instead of meeting the degree constraint rigidly, we introduce the following three types of objective functions to be minimized.

(1) **The total overflow (TO):** $\sum_{v \in V(G)} \text{overflow}_T(v)$,
(2) **The largest overflow (LO):** $\max_{v \in V(G)} \text{overflow}_T(v)$,
(3) **The number of overflowed vertices (NO):** the number of overflowed vertices $v \in V(G)$ (i.e., those v with $\text{overflow}_T(v) > 0$).

We show that the shortest path problem with the objective functions (1) and (2) can be solved in $O(m^{1.5})$ and $O(m^{1.5} \log \Delta)$ time, respectively, where Δ is the maximum degree of G, though the shortest path problem with the objective function (3) is NP-hard even for an acyclic digraph or an undirected graph.

Relaxing the distance constraint. We next relax the distance constraint while meeting the degree constraint. For a graph G and a tree T rooted at r, let $\text{dist}_T(v)$ denote the length of the (unique) path in T from r to a vertex v, and let dist(v) denote the distance (= the length of the shortest path) in G from r to v. The *delay* of v in T is defined as

$$\text{delay}_T(v) := \text{dist}_T(v) - \text{dist}(v).$$

We say that a vertex v is *delayed* in a T if $\text{delay}_T(v) > 0$. We use the following four types of measures as an objective function to be minimized.

(4) **The longest distance (LD):** $\max_{v \in V(G)} \text{dist}_T(v)$,
(5) **The total distance (TD):** $\sum_{v \in V(G)} \text{dist}_T(v)$,
(6) **The largest delay (LY):** $\max_{v \in V(G)} \text{delay}_T(v)$,
(7) **The number of delayed vertices (NY):** the number of delayed vertices $v \in V(G)$ (i.e., those v with $\text{delay}_T(v) > 0$).

(We do not include the one for minimizing "the total delay", which is equivalent to the minimization of the total distance.)

We show that for all objective functions (4)-(7), the problem of finding a degree-bounded spanning tree turns out to be NP-hard even for an acyclic digraph or an undirected graph.

Steiner condition. We further consider a problem in which a subset W of vertices to which a path from a root need to be constructed is prescribed while the degree and distance constraints are being imposed. Thus all vertices in such a subset $W \subseteq V(G)$ should be contained in a tree, but the other vertices in $V(G) - W$ are not necessarily in the tree. Such a condition with a subset W is called *Steiner condition*. From the viewpoint of the application of broadcasting a file, W means the set of nodes that should receive the file and the other nodes should not receive the file (but allowed to transmit the file to other nodes).

We show that the problem with Steiner condition is NP-complete even for an acyclic digraph or an undirected graph.

Related work. The basic FTT problem and its extensions in the above way seem a natural and fundamental problem formulation in the area of network algorithms. Before closing this section, we refer to some problems related our FTT problem. A shortest path tree in a graph with n vertices and m edges can be found in $O(m + n \log n)$ time by the Dijkstra's algorithm with Fibonacci-heap [7]. Finding a spanning tree under the degree constraint is known to be NP-complete [10], even though the minimum cost tree problem is easily solvable [9].

For transferring files in a network, the problem of constructing an FTT, a tree with a similar properties, or a routing more general than trees (such as a file transfer plan) have been studied extensively in many articles [1,2,3,4,8,10,11, 12,13,14,15,17,18,20,22]. Among them, our basic FTT problem has been studied only by Yamagata et al. [22] in which a heuristic algorithm for constructing an FTT has been proposed but any theoretical analysis such as the NP-hardness or polynomial solvability has not been discussed. Kaneko et al. [12,13,14,15] have considered problems finding a minimum cost plan to distribute the required number of copies to each vertex under a restriction on the maximum possible number of copies at each vertex. These problems have some similarity with our basic FTT problem, but allow a choice of non-tree routings differently from our setting. The problem of designing a time-scheduling for transferring files that avoids collisions of file transmitting can be found in [1,2,3,4,20].

This paper organized as follows. Section 2 introduces some definitions. Sections 3 and 4 consider relaxing the degree constraint and the distance constraint, respectively. Section 5 treats the Steiner condition. Finally, section 6 presents concluding remarks.

2   Preliminaries

A graph stands for an undirected graph unless otherwise stated as digraphs. For a graph (or digraph) G, its edges are weighted by nonnegative reals, and the weight of an edge e is denoted by $\ell_G(e)$, where ℓ_G is called a *length function* of G. Also an integer function $\delta : V(G) \to \mathbf{Z}^+$, called a *degree capacity function* is prescribed. We denote $n = |V(G)|$ and $m = |E(G)|$. The *out-degree* of a vertex v in a digraph G is defined by the number of edges outgoing from v and is denoted by $d_G^+(v)$.

A path in a digraph means a directed path. For a vertex $v \in V(G)$ in a digraph G, $E^+(v)$ (resp., $E^-(v)$) denotes the set of edges outgoing from x (resp., coming into x). A spanning subgraph T of a digraph G is called an *outbranching* if T has a root vertex r from which every other vertex in V is reachable along a path of T. A branching T with root r may be simply called a spanning tree of G. Let $\mathcal{ST}(G, r)$ denote the set of all spanning trees of G rooted at r, and $\mathcal{SPT}(G, r)$ denote the set of all shortest path trees of G rooted at a vertex r.

3   Minimizing Overflow in Shortest Path Trees

We have described the basic FTT problem, which asks whether there is a degree-bounded tree among all shortest path trees in $\mathcal{SPT}(G, r)$. To handle the case where there is no such tree T in $\mathcal{SPT}(G, r)$, we define the *overflow* at a vertex v in a tree T by $\max\{d_T^+(v) - \delta(v), 0\}$. In the following sections, we consider the three problems which minimize the total sum of overflows, the maximum of overflows, and the number of vertices with nonzero overflows, respectively.

3.1   Minimizing Total Overflow

In this subsection we consider the problem of finding a shortest path tree that minimizes the total sum of overflows.

Problem 1. **Minimum Total-Overflow Shortest-Path Tree Problem (TO-SPT)**

Input: A digraph G with a root r, an edge length function ℓ and a degree capacity function δ.

Output: A shortest path tree $T \in \mathcal{SPT}(G, r)$ such that $\sum_{v \in V(G)} \max\{d^+{}_T(v) - \delta(v), 0\}$ (i.e., the sum of overflows over all vertices) is minimized.   □

We obtain the following result, where a proof is given after showing some properties on spanning trees in the next subsection.

Theorem 1. *The TO-SPT can be solved in $O(m^{1.5})$ time.*   □

The following corollary is directly obtained from the theorem.

Corollary 1. *The basic FTT problem can be solved in $O(m^{1.5})$ time.*   □

3.2 Algorithm in Acyclic Digraphs

To prove Theorem 1 in the previous subsection, we consider the following problem.

Problem 2. **Minimum Total-Overflow Spanning Tree Problem (TO-ST)**
Input: A digraph $G = (V, E)$ with a root r and a degree capacity function δ.
Output: A spanning tree $T \in \mathcal{ST}(G, r)$ such that $\sum_{v \in V(G)} \max\{d_T^+(v) - \delta(v), 0\}$ (i.e., the sum of total overflows over all vertices) is minimized.   □

Theorem 2. *The TO-ST is NP-hard.*

Proof: The TO-ST such that $\delta(v) = 1$ for all $v \in V(G)$ asks whether G has a Hamilton path starting from r. This implies the NP-hardness of the TO-ST.   □

In what follows, we show that the TO-ST with an acyclic digraph G can be solved in polynomial time. W.l.o.g. assume that G has no edge entering a root r and every vertex can be reachable from r. We convert the TO-ST in an acyclic digraph G into a minimum cost flow problem in an auxiliary graph G', which is defined as follows:

$$V(G') := \{s, t\} \cup \{v_{out}^i \mid v \in V(G), i = 1, 2, \ldots, d_G^+(v)\} \cup \{v_{in} \mid v \in V(G)\},$$
$$E(G') := \{(s, v_{out}^i) \mid v \in V(G), i = 1, 2, \ldots, d_G^+(v)\}$$
$$\cup \{(v_{out}^i, u_{in}) \mid (u, v) \in E(G), i = 1, 2, \ldots, d_G^+(v)\}$$
$$\cup \{(v_{in}, t) \mid v \in V(G) - \{r\}\}.$$

The capacity cap(e) and the cost cost(e) of each edge $e \in E(G')$ in the minimum cost flow problem are given as follows, where we denote $E_s = \{(s, v_{out}^i) \mid v \in V(G), i = 1, 2, \ldots, d_G^+(v)\}$:

$$\text{cap}(e) = 1, \text{cost}(e) = 0 \text{ if } e \in E(G') - E_s, \text{ or } e = (s, v_{out}^i) \in E_s \text{ for some } i \leq \delta(v),$$

$$\text{cap}(e) = 1, \text{cost}(e) = 1 \text{ if } e = (s, v_{out}^i) \in E_s \text{ for some } i > \delta(v).$$

By regarding a path from s to v passing through edge (v_{out}^i, u_{in}) as a choice of edge (v, u) from $E^-(u)$, we observe that a set of $n - 1$ internally disjoint paths from s to t gives a choice of edges from $E(G)$ that forms an outbranching rooted at r and vice versa. Moreover, each of such paths passes through exactly one edge e in E_s incurs unit cost if $e = (s, v_{out}^i)$ for some $i > \delta(v)$ or no cost otherwise. Thus from a minimum cost $n - 1$ disjoint paths (or flow with value $n - 1$), we can obtain an optimal solution to the TO-ST for the original digraph G. Thus we can solve the TO-ST for an acyclic digraph G in polynomial time by an minimum cost flow algorithm.

In the sequel, we present a faster $O(m^{1.5})$ time algorithm for computing the desired minimum cost flow. For a digraph H, we denote by $H + E'$ the digraph obtained from H by adding a set E' of new edges and by $H - E''$ the digraph obtained from H by removing edges in a subset $E'' \subseteq E(H)$. The next result easily follows from a matroidal property on paths [19].

Lemma 1. *Let $G'' = G' - E_s$, and $\mathcal{F} = \{F \subseteq E_s \mid G'' + F$ has $|F|$ internally disjoint paths from s to $t\}$. Then $(E_s, \mathcal{F})$ is a matroid.* □

Theorem 3. *The TO-ST in an acyclic digraph can be solved in $O(m^{1.5})$ time.*

Proof: Since $(E_s, \mathcal{F})$ in Lemma 1 is a matroid, a minimum set of $n - 1$ internally disjoint paths from s to t in G' corresponds to a minimum base B of the matroid with $\text{cost}(e)$, $e \in E_s$. Such $B \subseteq E_s$ can be constructed by the greedy algorithm. Since the cost of any element in E_s is 1 or 0, the greedy algorithm can be executed in the following two phases. Let E_s^0 be the set of edges $e \in E_s$ and $E_s^1 = E_s - E_s^0$. The first phase finds a maximum flow f^0 from s to t in $(G' - E_s) + E_s^0$ (prohibiting use of any edge in E_s^1), and we denote by B^0 the set of edges E_s^0 that are used by the f^0. The second phase computes a maximum flow f^1 from s to t in the digraph $(G' - E_s) + (B^0 \cup E_s^1)$, and we denote by B the set of edges E_s that are used by the f^1. Since $(E_s, \mathcal{F})$ is a matroid, the resulting B is a minimum base of the matroid, implying that $(G' - E_s) + B$ has a set of $n - 1$ internally disjoint paths from s to t. It is not difficult to see that the above algorithm can be executed in $O(m^{1.5})$ time by using by the algorithm due to Even and Tarjan [6].

□

We are ready to prove Theorem 1. Given a digraph G with a root r, we discard all edges that are not on any shortest path from r to some vertex. Then the resulting graph G^* is acyclic, and any outbranching rooted at r in G^* is a shortest path tree of G. Thus, by applying Theorem 3 to the G^*, we can obtain a solution (if any) to the TO-SPT in G in $O(m^{1.5})$ time.

3.3   Minimizing the Largest Overflow

In this subsection, we consider the problem of finding a spanning tree T that minimizes the largest overflow over all vertices.

Problem 3. **Minimum Largest-Overflow Shortest-Path Tree Problem (LO-SPT)**
Input: A digraph G with a root r, an edge length function ℓ and a degree capacity function δ.
Output: A shortest path tree $T \in \mathcal{SPT}(G, r)$ such that $\max_{v \in V(G)} \max\{d_T^+(v) - \delta(v), 0\}$ (i.e., the largest overflow over all vertices) is minimized. □

Given an integer $k > 0$, we can test whether or not there is a shortest path tree $T \in \mathcal{SPT}(G,r)$ such that $d_T^+(v) \leq \delta(v) + k$ for all $v \in V(G)$ by using the algorithm for solving the TO-SPT. Thus, by conducting a binary search, we can find the minimum k for which such a shortest path tree exists.

Theorem 4. *The LO-SPT can be solved in $O(m^{1.5} \log \Delta)$ time, where Δ is the maximum degree of G.* $\square$

We also consider the case where a tree to be chosen is not necessarily a shortest path tree.

Problem 4. **Minimum Largest-Overflow Spanning Tree Problem (LO-ST)**
Input: A graph (or digraph) G with a root r and a degree capacity function δ.
Output: Spanning tree $T \in \mathcal{ST}(G,r)$ such that the maximum overflow over all vertices in T (i.e., $\max_{v \in V(G)} \max\{d_T^+(v) - \delta(v), 0\}$) is minimized. $\square$

Similarly for Theorem 2, we see that the LO-ST is NP-hard. In the case of graphs, we show that the LO-ST can be solved by an approximation algorithm with absolute error 1.

Theorem 5. *For a graph G with a root t and a degree capacity function δ, we can find in $O(n^4 \alpha(n,n) \log n)$ time a spanning tree $T' \in \mathcal{ST}(G,r)$ such that $\max_{v \in V(G)}\{\max\{d_{T'}^+(v) - \delta(v), 0\}\} \leq 1 + \max_{v \in V(G)}\{\max\{d_T^+(v) - \delta(v), 0\}\}$ for all $T \in \mathcal{ST}(G,r)$.* $\square$

To prove the theorem, we review the following problem which has been studied by Fürer and Raghavacari [8]

Problem 5. **Minimum Largest-Degree Spanning Tree Problem (LD-ST)**
Input: A graph G.
Output: A spanning tree T whose maximum degree is minimized over all spanning trees in G. $\square$

For this problem, Fürer and Raghavacari [8] gave an approximation algorithm with absolute error 1. Their algorithm runs in $O(mn\alpha(m,n) \log n)$ time, $\alpha(m,n)$ is a functional inverse of the Ackermann function. We convert a given instance (G,r,δ) of the LO-ST into an instance of the LD-ST so that a solution to the (G,r) can be obtained by applying their algorithm for the LD-ST. We first set a degree constraint that

$$d_T(v) \leq \delta'(v), \ v \in V(G),$$

where $d_T(v)$ denotes the degree of a vertex in a spanning tree T of G and δ' is defined by

$$\delta'(r) := \delta(r), \quad \delta'(v) := \delta(v) + 1 \text{ for } v \in V(G) - \{r\}.$$

Thus, the maximum overflow by a spanning tree T is given by

$$\max_{v \in V(G)} \max\{d_T(v) - \delta'(v),\ 0\}.$$

Let $\delta'_{max} = \max_{v \in V(G)}\{\delta'(v)\}$, and G' be the graph obtained from G by attaching $\delta'_{max} - \delta'(v)$ dummy leaf vertices to each vertex $v \in V(G)$. The G' has $n + \sum_{v \in V(G)}(\delta'_{max} - \delta'(v)) = O(n^2)$ vertices and $m + \sum_{v \in V(G)}(\delta'_{max} - \delta'(v)) = O(n^2)$ edges. Note that any spanning tree T' of G' contains all leaf vertices and satisfies $d_{T'}(v) \geq \delta'_{max} - \delta'(v)$ for all vertices $v \in V(G)$. We then apply the approximation algorithm by Fürer and Raghavacari to the instance G' to obtain a spanning tree T' of G'. We finally construct a spanning tree T of G by discarding dummy edges from T'.

We prove that the resulting tree T is an approximation solution with absolute error 1. Let Δ' be the maximum degree in T'. For each vertex $v \in V(G)$ in G, we have $d_T(v) = d_{T'}(v) - (\delta_{max} - \delta'(v)) \leq \Delta' + \delta'(v) - \delta'_{max}$. Thus, if $\Delta' \leq \delta'_{max}$, then T is a solution to problem 5 by $d_T(v) \leq \delta'(v)$ $(v \in V(G))$. Otherwise, we consider the case of $\Delta' > \delta'_{max}$. For a spanning tree T_0 of G that minimizes the maximum overflow, let T'_0 be the spanning tree of G' obtained by attaching $\delta'_{max} - \delta'(v)$ dummy leaf vertices to each vertex $v \in V(G)$. Since T is a solution with absolute error 1, we have $\Delta' - 1 \leq d_{T'_0}(v) = d_{T_0}(v) + (\delta'_{max} - \delta'(v))$ for all $v \in V(G)$. Hence for each $v \in V(G)$ $\Delta' + \delta'(v) - \delta'_{max} - 1 \leq d_{T_0}(v)$. With this and $d_T(v) \leq \Delta' + \delta'(v) - \delta'_{max}$, we get $d_T(v) \leq d_{T_0}(v) + 1$ $(v \in V(G))$. That is, T is a solution with absolute error 1. ☐

3.4 Minimizing Overflowed Vertices

In this subsection, we consider the following problem.

Problem 6. **Minimum Number Overflowed-Vertices Shortest-Path Tree Problem (NO-SPT)**
Input: A graph (or a digraph) G with a root $r \in V(G)$, a degree capacity function δ, an edge length function ℓ, and a bound $n_0 \geq 0$.
Output: A shortest path tree $T \in \mathcal{SPT}(G, r)$ such that sum of vertices with overflow is at most n_0; Report "no" if such T does not exist. ☐

Theorem 6. *The NO-SPT is NP-complete even if all edge-lengths are equal to 1 and G is a graph or an acyclic digraph.*

We use HITTING SET [10] for proving all NP-completeness in this paper. HITTING SET is defined as follows.

Problem 7. **HITTING SET**
Input: A set S, subsets $S_1, S_2, ..., S_m$ of S and nonnegative integer k.
Output: A subset S' of S which satisfies $|S'| \leq k$, and $S' \cap S_i \neq \phi$ $(i = 1, 2, ..., m)$; Report "no" if such S' does not exist. ☐

Let $(S = \{1, 2, ..., n\},\ S_1, S_2, ..., S_m(S_i \subseteq S),\ k)$ be an instance of the HITTING SET.

Proof of Theorem 6: We construct an instance $(G_0, \delta, \ell, r, n_0)$ of NO-SPT for acyclic digraphs from an instance of the HITTING SET as follows.

$$V(G_0) := \{r\} \cup \{v_i | i = 1, 2, ..., n\} \cup \{s_j | j = 1, 2, ..., m\},$$
$$E(G_0) := \{(r, v_i) | i = 1, 2, ... n\} \cup \{(v_i, s_j) | i \in S_j\},$$
$$\delta(v) := 0 \ (\forall v \in V(G_0)), \quad \ell(e) := 1 \ (\forall e \in E(G_0)), \quad n_0 := k + 1.$$

Observe that G_0 is acyclic.

First, suppose that the instance for the HITTING SET has a feasible solution S', i.e., $|S'| \leq k$ and $S' \cap S_j \neq \phi (j = 1, 2, ..., m)$. We construct a spanning tree T as follows. For $v_i \ (i = 1, 2, ..., n)$, r is the parent. For $s_j \ (j = 1, 2, ..., m)$, select an element $i \in S' \cap S_j$ and let v_i be the parent of S_j. Note that the existence of such i is guaranteed by the assumption that S' is a feasible solution of the instance for the HITTING SET. The T is clearly a shortest path tree. The number of children of $v_i \ (i \notin S')$ and $s_j \ (j \in \{1, 2, ..., m\})$ are zero, and hence they satisfy the degree constraints. Although r and the vertices corresponding to S', i.e., $v_i \ (i \in S')$, may be overflowed, they are at most $|S'| + 1 \leq k + 1 = n_0$. Therefore T is a desired tree.

Next, suppose that NO-SPT has a feasible solution T. We show that $S' = \{i \in S \mid d_T(v_i) > 0\}$ is a feasible solution of the instance of the HITTING SET. Since the root r and all vertices in S' are overflowed, $|S'| + 1 \leq n_0 = k + 1$ and hence $|S'| \leq k$. Every vertex $s_j \ (j = 1, 2, ..., m)$ must have a parent and every vertex $v_i \in S - S'$ can not have a child by the construction of S'. It follows that every $s_j \ (j = 1, 2, ..., m)$ is a child of a $v_i \in S'$. This means that every $S_j \ (j = 1, 2, ..., m)$ has at least one element of S'. Therefore S' is a feasible solution to the instance of the HITTING SET.

The proof for graphs is similar to the preceding one. The instance $G_0' = (V_0, E_0')$ can be obtained from G_0 by replacing all directed edges $(v, w) \in E_0$ by undirected edges (v, w). The remaining part of the proof is almost the same and is omitted from the space limitation.   □

4 Degree-Bounded Trees with Relaxed Distance Constraint

In this section we relax the distance constraint while meeting the degree constraint. Thus we construct only degree-bounded spanning trees (DBSTs for short). We consider the following four problems. From the space limitation, only a constraint characteristic of each problem is shown. Each problem commonly has a graph or a digraph G, a root r, a degree capacity function δ, and an edge length function ℓ for inputs, and requires a degree-bounded spanning tree $T \in \mathcal{ST}(G, r)$ satisfying the characteristic constraint.

Problem 8. **Minimum Longest-Distance DBST Problem (LD-DBST)**
Constraint: $\max_{v \in V(G)} \text{dist}_T(v) \leq \ell_0$. ($\ell_0$ is an input.)   □

Problem 9. **Minimum Total-Distance DBST Problem (TD-DBST)**
Constraint: $\sum_{v \in V(G)} \text{dist}_T(v) \leq \ell_0$. ($\ell_0$ is an input.)   □

Problem 10. **Minimum Largest-Delay DBST Problem (LY-DBST)**
Constraint: $\max_{v \in V(G)} \text{delay}_T(v) \leq \ell_0$. ($\ell_0$ is an input.)   □

Problem 11. **Minimum Number Delayed-Vertices DBST Problem (NY-DBST)**
Constraint: $|\{v \in V | \text{delay}_T(v) > 0\}| \leq n_0$. ($n_0$ is an input.)   □

We show the following theorem.

Theorem 7. *The LD-DBST, the TD-DBST, the LY-DBST, and the NY-DBST are NP-complete even if all edge-lengths are equal to 1 and G is a graph or an acyclic digraph.*

Proof: All proofs are similar to the proof of Theorem 6. Due to the space limitation we show only how to construct their instances.

LD-DBST with acyclic digraphs: We use the graph G_0 constructed in the proof of Theorem 6. The instance $(G_1, \delta, \ell, r, \ell_0)$ is defined as follows. $V(G_1) := V(G_0) \cup \{t\}$, $E(G_1) := E(G_0) \cup \{(s, t)\} \cup \{(t, v_i) | i = 1, 2, ..., n\}$, $\delta(v) := 0$ ($\forall v \in V(G_1)$), $\delta(r) := k + 1$, $\delta(t) := n$, $\delta(v_i) := m$ ($i = 1, 2, ..., n$), $\delta(s_i) := 0$ ($i = 1, 2, ..., m$), $\ell(e) := 1$ ($\forall e \in E(G_1)$), $\ell_0 := 2$.

LD-DBST with graphs: The instance is $(G_1', \delta, \ell, r, \ell_0)$: G_1' is obtained from G_1 by replacing all directed edges $(v, w) \in E(G_1)$ by undirected edges (v, w). The other parameters are the same with the ones for acyclic digraphs.

TD-DBST with acyclic digraphs: The instance is $(G_1, \delta, \ell, r, \ell_0)$: Let G_1, δ, ℓ, r be the same to the instance of the LD-DBST with acyclic digraphs, and let $\ell_0 := 2n + 2m - k + 1$.

TD-DBST with graphs: The instance is $(G_1'', \delta, \ell, r, \ell_0)$: Let G_1' be the instance constructed for the LD-DBST with graphs. G_1'' is obtained from G_1' by adding a vertex t_h' ($h = 1, 2, \ldots, 2n + 2m$) and edges (t, t_h') ($h = 1, 2, \ldots, 2n + 2m$) with $\ell(t, t_h') = 1$. Let $\ell_0 := 6n + 6m - k + 1$.

LY-DBST with acyclic digraphs: The instance $(G_2, \delta, \ell, r, \ell_0)$ is constructed from the instance of the TD-DBST with acyclic digraphs by adding edges (r, s_j), $j = 1, 2, ...m$, of length one and letting $\ell_0 := 1$. The other parameters are the same with the ones in TD-DBST.

LY-DBST with graphs: The instance is $(G_2', \delta, \ell, r, \ell_0)$: G_2' is obtained from G_2 by replacing all directed edges $(v, w) \in E(G_2)$ by undirected edges (v, w), and removing an edge (r, t). Let $\delta(r) := k$, $\delta(t) := n - k - 1$, $l_0 := 2$. Remaining parameters are the same with ones for acyclic digraphs.

NY-DBST with acyclic digraphs: The reduction is the same to one of LD-DBST with acyclic digraphs except for that we set here $n_0 := n - k$.

NY-DBST with graphs: A graph G_1'' of the instance is obtained from the graph G_1 of the instance for acyclic digraphs by replacing all directed edges $(v, w) \in E(G_1)$ by undirected edges (v, w), and removing an edge (r, t). Let $\delta(r) := k$, $\delta(t) := n - k - 1$. Remaining parameters are the same with the ones for acyclic digraphs.   □

5   Steiner Condition

In this section, we consider a problem which has a Steiner condition in addition to the degree constraint and the distance constraint. Thus, a set W of vertices to be included in a tree is prescribed.

Problem 12.   Degree-Bounded Shortest-Path Subtree Problem   (DB-SPST)
Input: A graph (or a digraph) G with a root $r \in V(G)$, a degree capacity function δ, an edge length function ℓ, and a subset $W \subseteq V(G)$.
Output: A degree-bounded tree T of G (not necessarily spanning) such that $W \subseteq V(T)$ and $\mathrm{dist}_T(v) = \mathrm{dist}(v)$ for all $v \in V(T)$. $\square$

This problem is similar to the Steiner tree problem, which requires a minimum weight tree including all vertices of a given subset $W \subseteq V(G)$. The Steiner tree problem is known to be NP-hard [10]. On the other hand, the DB-SPST becomes rather trivial when the degree constraint need not to be met (or $\delta(v) = n - 1$ holds for all $v \in V(G)$); A desired tree can be obtained from a shortest path tree simply by trimming unnecessary edges. Moreover, the DB-SPST in the special case of $W = V(G)$ can be solved in $O(m^{1.5})$ time from Theorem 1, since it is a subproblem of the TO-SPT. However, meeting the degree constraint and the Steiner constraint at the same time makes the problem NP-hard as shown in the following theorem.

Theorem 8. *The DB-SPST is NP-complete even if all edge-lengths are equal to 1 and G is a graph or an acyclic digraph.*

Proof: We construct an instance $(G_0, \delta, \ell, r, W)$ of DB-SPST as follows. Let G_0, ℓ, and r be the same with the instance constructed in the proof of Theorem 6, and let δ and W be $\delta(r) := k$, $\delta(v_i) := m$ $(i = 1, 2, ..., n)$, $\delta(s_i) := 0$ $(i = 1, 2, ..., m)$, $W := \{s_j | j = 1, 2, \ldots, m\}$. The way to show the equivalence of the two instances is almost the same with the proof of Theorem 6. Hence it is omitted.

The proof for graphs is similar to the preceding one. A graph G_0' of the instance can be obtained from G_0 by replacing all directed edges $(v, w) \in E(G_0)$ by undirected edges (v, w). The remaining part is the same, and is omitted. $\square$

6   Concluding Remarks

In this paper, we have presented an $O(m^{1.5})$ time algorithm for solving the basic FTT problem based on a matroidal property of the problem. We have investigated several natural extensions of the basic FTT problem by relaxing the degree constraint or the distance constraint in various ways. As a result, we have seen that in many cases the problem becomes NP-hard when one of the constraint is replaced with an objective function to be minimized except for the minimization of the total overflow or the largest overflow in shortest path trees. An approximation algorithm for minimizing the largest overflow in spanning trees has been obtained. It is the future research to study the approximability of the problems formulated in this paper.

References

1. H. -A. Choi and S. L. Hakimi, Scheduling File Transfers For Trees and Odd Cycles, SIAM J. Comput., Vol. 16, No. 1, pp. 162–168, 1987.
2. H. -A. Choi and S. L. Hakimi, Data Transfers in Networks with Transceivers, Networks, Vol. 17, pp. 393–421, 1987.
3. H. -A. Choi and S. L. Hakimi, Data Transfers in Networks, Algorithmica, Vol. 3, 223–245, 1988.
4. E. G. Coffman. Jr, M. R. Garey, D. S. Johnson and A. S. Lapaugh, Scheduling File Transfers, SIAM J. Comput., Vol. 14, No. 3, pp. 744–780, 1985.
5. E. W. Dijkstra, A Note on Two Problems in Connexion with Graphs, Numerische Mathematik, Vol. 1, pp. 269–271, 1959.
6. S. Even and R. E. Tarjan, Network Flow and Testing Graph Connectivity, SIAM J. Comput., pp. 507–518, 1975.
7. M. L. Fredman and R. E. Tarjan, Fibonacci Heaps and Their Uses in Improved Network Optimization Algorithms, J. ACM, Vol. 34, pp. 596–615, 1987.
8. M. Fürer and B. Raghavachari, Approximating The Minimum Degree Spanning Tree to within One from The Optimal Degree, 3rd ACM-SIAM Symp. on Disc. Algorithms, pp. 317–324, 1992.
9. H. N. Gabow, Z. Galil, T. Spencer, and R. E. Tarjan, Efficient Algorithms for Finding Minimum Spanning Trees in Undirected and Directed Graphs, Combinatorica, Vol. 6, pp. 109–122, 1986.
10. M. Garey and D. Johnson, Computers and Intractability, W. H. Freeman and Company, San Francisco, 1978.
11. D. S. Hochbaum, Approximation Algorithms form NP-Hard Problems, PWS Publishing Company, Boston, 1997.
12. Y. Kaneko, S. Shinoda and K. Horiuchi, A Synthesis of an Optimal File Transfer on a File Transmission Net, IEICE Trans. Fundamentals, Vol. E76-A, No. 3, pp. 377–386, 1993.
13. Y. Kaneko, S. Shinoda and K. Horiuchi, On an Optimal File Transfer on an Arborescence-Net with Constraints on Copying Numbers, IEICE Trans. Fundamentals, Vol. E78-A, No. 4, pp. 517–528, 1995.
14. Y. Kaneko, K. Suzuki, S. Shinoda and K. Horiuchi, A Synthesis of a Forest-Type Optimal File Transfer on a File Transmission Net with Source Vertices, IEICE Trans. Fundamentals, Vol. E78-A, No. 6, pp. 671–679, 1995.
15. Y. Kaneko, R. Tashiro, S. Shinoda and K. Horiuchi, A Liner-Time Algorithm for Designing an Optimal File Transfer Through an Arborescence-Net, IEICE Trans. Fundamentals, Vol. E75-A, No. 7, pp. 901–904, 1992.
16. M. Klein, A Primal Method for Minimal Cost Flows, Management Science, Vol. 14, pp. 205–220, 1967.
17. J. Plesnik, The Complexity of Designing a Network with Minimum Diameter, Networks, Vol. 11, pp. 77–85, 1981.
18. P. I. Rivera-Vega, R. Varadarajan and S. B. Navathe, Scheduling File Transfers in Fully Connected Networks, Networks, Vol. 22, pp. 563–588, 1992.
19. D. J. A. Welsh, Matroid Theory, Academic Press, London, 1976.
20. J. Whitehead, The Complexity Of File Transfer Scheduling with Forwarding, SIAM J. Comput., Vol. 19, No. 2, pp. 222–245, 1990.
21. J. Whitney, On The Abstract Properties of Liner Dependence, American Journal of Mathematics., Vol. 57, pp. 509–533, 1935.
22. T. Yamagata, A. Fujii, and Y. Nemoto, A Multicast Routing Algorithm, IEICE Trans., Vol. J80-D-I, No. 9, pp. 739–744, 1997. (in Japanese)

Approximation Algorithms for Some Parameterized Counting Problems*

V. Arvind and Venkatesh Raman

The Institute of Mathematical Sciences, Chennai 600 113, India
{arvind,vraman}@imsc.ernet.in

Abstract. We give a randomized fixed parameter tractable algorithm to approximately count the number of copies of a k-vertex graph with bounded treewidth in an n vertex graph. As a consequence, we get randomized algorithms with running time $k^{O(k)}n^{O(1)}$, approximation ratio $1/k^{O(k)}$, and error probability $2^{-n^{O(1)}}$ for (a) approximately counting the number of matchings of size k in an n vertex graph and (b) approximately counting the number of paths of length k in an n vertex graph. Our algorithm is based on the Karp-Luby approximate counting technique [8] applied to fixed parameter tractable problems, and the color-coding technique of Alon, Yuster and Zwick [1]. We also show some W-hardness results for parameterized exact counting problems.

1 Introduction

We investigate some counting problems in the paramterized complexity framework of Downey and Fellows [3]: Efficient algorithms are sought for parameterized problems - problems having an input whose size is n, and a fixed parameter k. A parameterized problem is said to be *fixed parameter tractable* (FPT) if there is an algorithm for the problem that takes $O(f(k)n^{O(1)})$ for some function f of k. Examples of fixed parameter tractable problems include VERTEX COVER and UNDIRECTED FEEDBACK VERTEX SET. There is also a hardness theory based on the W-hierarchy [3] and it is known, for example, that the DOMINATING SET and the CLIQUE problems are hard for the parameterized complexity classes $W[2]$ and $W[1]$ respectively. The best known algorithms for these two problems take $n^{O(k)}$ time. In the problems mentioned above, the parameter k is the solution size. (There are problems for which other parameterizations are more natural.)

Recently there has been some attention in studying the parameterized complexity of counting problems [5,6,13]. In particular, Flum and Grohe [6] have shown that (i) exactly counting the number of paths of length k and (ii) exactly counting the number of matchings of size k in an undirected graph, are both hard problems for the class W[1]. Consequently, algorithms with running time $f(k)n^{O(1)}$ for the *exact* counting versions of these problems are unlikely for any computable function $f : \mathcal{N} \to \mathcal{N}$. On the other hand, Alon, Yuster and Zwick

---

* Work supported by a DST-DAAD project (Indo-German Personnel Exchange Programme 2000).

P. Bose and P. Morin (Eds.): ISAAC 2002, LNCS 2518, pp. 453–464, 2002.
© Springer-Verlag Berlin Heidelberg 2002

[1] have shown that the *decision* problem of checking if a given graph has a path of length k is fixed parameter tractable.

In this paper we first consider the following counting problem:

Problem 1.
Input: An n vertex graph G and k-vertex graph H such that H has treewidth b.
Output: Find the number of subgraphs[1] of G isomorphic to H.
Parameter: The parameter for the problem is k, which is the number of vertices in the graph H.
Although we treat b as a *constant*, we will explicitly show the occurrence of b in the running time bound for the main result: the approximation algorithm described in Theorem 2.

This is a generic problem which includes various problems as special cases. E.g. the problems of (i) counting matchings of size k in a graph, (ii) counting paths of length k, and (iii) counting the number of copies of any fixed tree/forest on k nodes in an n-vertex graph.

As our main result, we give a randomized approximate counting algorithm with running time of the form $f(k)n^{b+O(1)}$ for Problem 1 defined above. As a consequence, in one stroke we obtain randomized approximate counting algorithms with similar time bounds for the problems of counting paths of length k, matchings of size k etc.

We now explain the main ingredients of our approximation algorithm. First, we reformulate Karp-Luby's Monte-Carlo sampling method for approximate counting [8] (see also [9]) for parameterized problems. We also suitably adapt the notion of FPRAS (fully polynomial randomized approximation schemes) and define fixed parameter tractable randomized approximation schemes (FPTRAS). We then observe, in Section 2, that under similar conditions as in the case of the approximate counting result of Karp and Luby, FPTRAS exist for parameterized counting problems. Using this and the color coding technique of Alon, Yuster and Zwick [1], we give a randomized approximate counting algorithm taking $O(k^{O(k)}(c^{b^3}k + n^{b+O(1)}2^{b^2/2}))$ time for Problem 1, where the approximation ratio ϵ of the algorithm is $1/k^{O(k)}$ and the error probability δ is $1/2^{n^{O(1)}}$.

It is interesting to note that the factor n^b appears to be a bottleneck in the above-mentioned time bound for the following reason: since a k-clique has treewidth k, it can be easily seen that an $n^{o(b)}f(k)$ algorithm for Problem 1 would imply, in particular, that there is an $f(k)n^{o(k)}$ randomized algorithm for deciding whether a graph on n vertices has a k-clique, which is a long-standing open problem (see e.g. [4] for a detailed discussion on the hardness of finding k-cliques in graphs).

In Section 3, we give some W-hard results on the *exact* counting versions of some fixed parameter tractable problems. Specifically we show that the problem of counting the number of weight k satisfying assignments of a monotone 2-DNF

---
[1] Notice that we are *not* asking for the number of induced subgraphs but for all subgraphs.

formula is $W[1]$-hard. This problem has an FPTRAS algorithm as we show in that Section. We also show that the problems of

- counting the number of cliques *and* independent sets on k vertices in a graph[2], and
- counting the number of satisfying assignments of weight at most k in a bounded CNF formula,

are $W[1]$-hard. The decision version of both these problems are fixed parameter tractable, and the approximate counting versions of these problems are open.

1.1   Definitions and Notation

For a positive integer n, by $[n]$, we mean the set $\{1, 2, \ldots, n\}$. A tree decomposition (see, for example, [11]) of a graph $G = (V, E)$ is a pair $D = (S, T)$ with $S = \{X_i, i \in I\}$ a collection of subsets of vertices of G and $T = (I, F)$ a tree, with one node for each subset in S, such that the following three conditions are satisfied:

1. $\bigcup_{i \in I} X_i = V$,
2. for all edges $(v, w) \in E$, there is a subset $X_i \in S$ such that both v and w are contained in X_i,
3. for each vertex $x \in V$, the set of nodes $\{i | x \in X_i\}$ forms a subtree of T.

The width of a tree decomposition $(\{X_i | i \in I\}, T = (I, F))$ is $\max_{i \in I}(|X_i| - 1)$. The treewidth of a graph G is the minimum width over all tree decompositions of G.

For a subset X of vertices of G, by $G[X]$, we mean the subgraph of G induced by X. I.e., $G[X]$ is the subgraph that contains all the vertices of X and all the edges of G incident on the vertices in X.

For a problem Π, let $\#(I)$ be the number of distinct solutions for an instance I of Π. A fully polynomial randomized approximation scheme (FPRAS) (see, for example [14]) for a counting problem Π is a randomized algorithm A that takes an input instance I with $|I| = n$, and real numbers $\epsilon > 0$ and $0 < \delta < 1$, and in time polynomial in n, $1/\epsilon$, and $\lg 1/\delta$ produces an output $A(I)$ such that

$$\text{Prob}[(1 - \epsilon)\#(I) \le A(I) \le (1 + \epsilon)\#(I)] \ge 1 - \delta.$$

We define an FPTRAS for a parameterized counting problem Π with parameter k as a randomized approximation scheme that takes an input instance I with $|I| = n$, and real numbers $\epsilon > 0$ and $0 < \delta < 1$, and computes an ϵ-approximation to $\#(I)$, with probability at least $1 - \delta$ in time $f(k)g(n, 1/\epsilon, \lg(1/\delta))$ where f is any function of k and g is polynomially bounded in $n, 1/\epsilon$ and $\lg 1/\delta$.

---

[2] Notice that this problem is different from counting just k-cliques *or* just k-independent sets separately. This problem could be potentially easier because the decision version is easier [10]!

2   Randomized Approximate Counting Parameterized Solutions

We first give a parameterized version of the result of Karp-Luby [8] on approximate counting via random sampling.

Theorem 1. *For every positive integer n, and for every integer $0 \leq k \leq n$, let $U_{n,k}$ be a finite universe, whose elements are binary strings of length $n^{O(1)}$. Let $\mathcal{A}_{n,k} = \{A_1, A_2, \ldots A_m\} \subseteq U_{n,k}$ be a collection of m given sets, and let $g : \mathcal{N} \to \mathcal{N}$ be a function and let $d > 0$ be a constant with the following conditions.*

1. *There is an algorithm that computes $|A_i|$ in time $g(k)n^d$, for each i, and every $\mathcal{A}_{n,k}$.*
2. *There is an algorithm that samples uniformly at random from A_i in time $g(k)n^d$, for each i, and every $\mathcal{A}_{n,k}$.*
3. *There is an algorithm that takes $x \in U_{n,k}$ as input and determines whether $x \in A_i$ in time $g(k)n^d$, for each i, and every $\mathcal{A}_{n,k}$.*

Then there is an FPTRAS for estimating the size of $A = A_1 \cup A_2 \cup \ldots A_m$ whenever m is $l(k)n^{O(1)}$ for some function l. In particular, for $\epsilon = 1/g(k)$, and $\delta = 1/2^{n^{O(1)}}$, the running time of the FPTRAS algorithm is $(g(k))^{O(1)}n^{O(1)}$.

Proof. We omit the proof of this theorem as it can be proved exactly on the same lines as original Karp-Luby result [14]. □

Our goal in this section is to use Theorem 1 and design an FPTRAS algorithm for estimating the number of copies of a graph H on k vertices with treewidth bounded by b in a given graph G on n vertices. We first give a high level description of the overall algorithm and explain how we use Theorem 1. Let

$$A = \{K \mid K \text{ is a } k\text{-vertex subgraph of } G \text{ such that } K \text{ is isomorphic to } H\}.$$

Our goal is to estimate $|A|$. We will express A as a union $\bigcup_{i=1}^m A_i$ that fulfills the conditions of Theorem 1 and thereby we will get an FPTRAS algorithm for the problem.

Defining the A_i's

A k-coloring of the graph G is a mapping $f : V(G) \longrightarrow [k]$. Our interest is in k-colorings of G that injectively map k element subsets of $V(G)$. Towards this goal, we consider a family $\mathcal{F}$ of (perfect hash) functions from $[|V(G)|] \longrightarrow [k]$ that satisfies the following crucial property:

$$\forall S \subseteq V(G) \; : \; |S| = k \; \exists f \in \mathcal{F} \; : \; f \text{ is injective on } S.$$

Such a family $\mathcal{F}$ of hash functions with $|\mathcal{F}| = 2^{O(k)} \log^{O(1)} n$, where $n = |V(G)|$ (c.f. [1]).

Suppose G is k-colored by some $f \in \mathcal{F}$. We say that a subset S of vertices of G is *colorful* under f if f restricted to S is injective (i.e. $f(i) \neq f(j)$ for $i \neq j \in S$). A subgraph H of G is *colorful* under f if $V(H)$ is colorful under f. The index set $\mathcal{I}$ defining the collection $\{A_i\}_{i \in \mathcal{I}}$ is the following:

$\mathcal{I} = \{\langle f, \pi \rangle \mid f \in \mathcal{F}$ and π is a k-coloring of H in which each vertex of H is distinctly colored$\}$.

Notice that if we identify $V(H)$ with $[k]$ then π can be seen as a permutation on $[k]$.

We need one more definition: let K_1 and K_2 be two k-colored graphs (by colors from $[k]$). We say that K_1 and K_2 are *color-preserving isomorphic* if there is an isomorphism between them that also preserves the colors. Notice that if both K_1 and K_2 are k-vertex graphs with k-colorings and additionally K_2 is colorful, then there is an efficient algorithm to test if they are color-preserving isomorphic: we simply need to check that K_1 is colorful, and verify that the only color preserving mapping from $V(K_1)$ to $V(K_2)$ is an isomorphism. Now, for each $i = \langle f, \pi \rangle \in \mathcal{I}$ we define the set A_i as follows:

$A_i = \{K \mid K$ is a colorful k-vertex subgraph of G under the coloring f and K is color-preserving isomorphic to H colored by $\pi\}$.

It is easy to verify that $A = \bigcup_{i \in \mathcal{I}} A_i$. Furthermore, $|\mathcal{I}| = 2^{O(k)} \cdot \log^{O(1)} n \cdot k!$. Thus it suffices to verify the three conditions necessary to apply Theorem 1 and we get the required FPTRAS algorithm.

Now, we begin with a lemma to prove the first condition.

Lemma 1. *Let $G = (V, E)$ be a graph on n vertices that is k-colored by some coloring $f : V(G) \longrightarrow [k]$, and let H be a k-vertex graph of treewidth b that is k-colored by some coloring π such that H is colorful. Then there is an algorithm taking time $O(c^{b^3} k + n^{b+2} 2^{b^2/2})$ time to exactly compute the cardinality of the set $\{K \mid K$ is a k-vertex subgraph of G and K is color-preserving isomorphic to $H\}$, where $c > 0$ is some constant.*

Proof. Since H has treewidth b, we can find a rooted tree decomposition $D = (S, T)$ of H, with $S = \{X_i, i \in I\}$ and the rooted tree $T = (I, F)$ satisfying the following properties in time $c^{b^3} k$ (see [2] and [11, Lemma 13.1.3] for details).

1. $|I| \leq 4k$ and every node of T has at most two children.
2. If a node i has two children j and k, then $X_i = X_j = X_k$.
3. If a node i has one child j, then either $X_i = X_j \cup \{x\}$ or $X_j = X_i \cup \{x\}$ for some element $x \in V(H)$.

Furthermore, we know that each $|X_i| \leq b + 1$ for each $i \in I$. Such a tree decomposition is said to be a *nice rooted tree decomposition*, and we will call the set associated with a node i of the tree as a *bag*.

Our goal is to count the number of colorful subgraphs K of G such that K is color-preserving isomorphic to H. We will do this by counting subgraphs of

G that are color-preserving isomorphic to *subgraphs* of H realized by subtrees of the tree decomposition T. As T is a rooted tree, we can do this in a bottom-up fashion. Let the depth of a node in T be the length (the number of edges) in the path from the root to that node (the depth of the root, therefore, is 0). We will process nodes of T inductively in non-increasing order of their depths, i.e. in a bottom up fashion, starting from the leaves of T.

Let y be a node of the tree T and let T_y be a subtree rooted at y. Let H_y be the induced subgraph $H[V_y]$ where $V_y = \bigcup_{x \in T_y} X_x$. Then it is known [11, Lemma 13.1.1] that the nodes of T_y along with their bags form a tree decomposition for H_y. For every node y of T and for each subgraph K of G with $|X_y|$ nodes (which is at most $b + 1$), we define the set

$S(H_y, K) = \{K' \mid K'$ contains K as an induced subgraph, and K' is color-preserving isomorphic to H_y with the nodes of H_y in the bag X_y mapped to the subset $V(K)$ of $V(G)$ by the color-preserving isomorphism$\}$. Let $N(H_y, K)$ denote its cardinality $|S(H_y, K)|$.

At the time of processing a node x of the tree T, we assume inductively that we know $N(H_y, K)$ for every subgraph H_y corresponding to each child y of x in T, and for every subgraph K of G with $|X_y|$ vertices. This assumption is trivially true when processing leaf nodes of T. To prove the inductive step, we make use of the properties of the nice tree decomposition.

At the inductive step, let x be the node of T to be processed and let K be any subgraph of G with $|X_x|$ vertices. Clearly, $N(H_x, K) = 0$ if K is not colorful. If K is colorful then it is easy to check if there is a color-preserving isomorphism from K to $H[X_x]$, the subgraph of H induced by X_x. If there is no color-preserving isomorphism from K to $H[X_x]$ then also $N(H_x, K) = 0$. Thus we need to only consider the case when K is a subgraph of G with $|X_x|$ nodes such that there is a color-preserving isomorphism from K to $H[X_x]$. Then there are two cases depending on whether the node x in T has one or two children.

Case 1: *x has two children y and z.*

In this case, by the property of nice tree decomposition, we have $X_x = X_y = X_z$. By induction hypothesis, we already know $N(H_y, K)$ and $N(H_z, K)$. We claim that $N(H_x, K) = N(H_y, K) \times N(H_z, K)$. To see this it suffices to show that there is a bijective correspondence between $S(H_x, K)$ and the Cartesian product $S(H_y, K) \times S(H_z, K)$. We show this by first injectively mapping $S(H_x, K)$ into $S(H_y, K) \times S(H_z, K)$ and vice-versa.

Let K' be in $S(H_x, K)$, where h is the color-preserving isomorphism from K' to H_x such that h maps X_x to $V(K)$. By the properties of nice tree decompositions we have

$$V(H_x) = V(H_y) \cup V(H_z) \text{ and } X_x = V(H_y) \cap V(H_z)$$

Let K'_1 and K'_2 be the subgraphs of K' induced by $V(H_y)$ and $V(H_z)$ respectively. Then it is clear that h restricted to K'_1 gives a color-preserving isomorphism from K'_1 to H_y, and h restricted to K'_2 gives a color-preserving isomorphism from K'_2 to H_z (the additional properties that X_y is mapped to

$V(K)$ and X_z is mapped to $V(K)$ holds obviously for the restrictions of h as $X_x = X_y = X_z$). The mapping $K \mapsto (K_1', K_2')$ is clearly injective.

Conversely, suppose $(K_1', K_2') \in S(H_y, K) \times S(H_z, K)$. Let h_y be the color-preserving isomorphism from K_1' to H_y, and let h_z be the color-preserving isomorphism from K_2' to H_z such that h_y maps $V(K)$ to X_y and h_z maps $V(K)$ to X_z. Recall that H is colorful under coloring π. Therefore, both h_y and h_z must coincide on $V(K)$. By the tree decomposition property, $V(H_x) = V(H_y) \cup V(H_z)$ and $X_x = V(H_y) \cap V(H_z)$. Thus, the mappings h_y and h_z in fact force $V(K_1') \cap V(K_2') = V(K)$. Now, let K' be the subgraph of G defined as follows: $V(K') = V(K_1') \cup V(K_2')$ and $(x, y) \in E(K')$ if and only if $(x, y) \in E(K_1')$ or $(x, y) \in E(K_2')$. It is easy to see that we can define a color-preserving isomorphism h from $V(K')$ to H_x by letting $h = h_y$ on $V(K_1')$ and $h = h_z$ on $V(K_2')$. Again, the mapping $(K_1', K_2') \mapsto K$ is easily seen to be injective. This completes Case 1.

Case 2: *x has one child y and $X_y = X_x - \{j\}$ for some $j \in V(H)$.*

Firstly, by tree decomposition property, $j \notin V(H_y)$. Thus, $V(H_x) = V(H_y) \cup \{j\}$. Also, tree decomposition guarantees that in H_x, the node j can be adjacent to only nodes in X_x (because the edge (j, k) must lie in some bag, and that bag must be X_x for otherwise j would lie in some other bag forcing it to belong to X_y which is not possible).

Now, let h be the unique color-preserving isomorphism mapping K to the subgraph of H induced by X_x. Let $K_1 = K - \{h^{-1}(j)\}$. Let $K' \in S(H_x, K)$. Clearly, it follows that $K' - \{h^{-1}(j)\}$ is in $S(H_y, K_1)$. Conversely, if $K' \in S(H_y, K_1)$ then there is a color-preserving isomorphism g from K' to H_y that maps $V(K_1)$ to X_y. Since j is adjacent to only nodes in X_x and since only $h^{-1}(j)$ in $V(K)$ can be mapped to $j \in X_x$, it is clear that g can be extended to a color-preserving isomorphism from $K' \cup \{h^{-1}(j)\}$ to H_x by mapping $h^{-1}(j)$ to j.

Thus, $K' \mapsto K' \cup \{h^{-1}(j)\}$ is a bijection from $S(H_y, K_1)$ to $S(H_x, K)$. Therefore, $N(H_x, K)$ is the same as $N(H_y, K - \{h^{-1}(j)\})$ which is already computed by induction hypothesis.

Case 3: *x has one child y and $X_y = X_x \cup \{j\}$ for some $j \in V(H)$.*

By induction hypothesis we already have computed $N(H_y, K')$ for all subgraphs K' with $|X_y|$ vertices of G. Now, let K be a subgraph of G with $|X_x|$ vertices and let h be the color preserving isomorphism from K to the subgraph of H induced by X_x. Let the color of j in H under the coloring π be $c \in [k]$. Notice that we have

$$N(H_x, K) = \sum_{v \in V(G)} N(H_y, K \cup \{v\})$$

This is a direct consequence of $S(H_x, K) = \bigcup_{v \in V(G)} S(H_y, K \cup \{v\})$. In fact, in the above sum, only those $N(H_y, K \cup \{v\})$ are nonzero for which v is colored c in G.

As we have inductively computed $N(X_y, K \cup \{v\})$ for each v, we can add them up to get $N(H_x, K)$.

Finally, notice that we can compute the nice tree decomposition of H in time $c^{b^3}k$ for some constant c (see [2] and [11, Lemma 13.1.3]).

In cases 1 and 2, it takes $O(1)$ time (in the standard RAM model) to compute $N(H_x, K)$ given $N(H_y, K)$ and $N(H_z, K)$ for each subgraph K. So for all subgraphs K on colorful sets of size $|X_x|$, it takes at most $O(n^{b+1}2^{\binom{b+1}{2}})$ time (as the number of such sets can be $O(n^{b+1})$ and the number of subgraphs on a $b+1$ element vertex set is at most $2^{\binom{b+1}{2}}$).

In case 3, it takes $O(n)$ time to compute $N(H_x, K)$ given $N(H_y, K')$. However, in case 3, $|X_x| \le b$ as $|X_y| = |X_x| + 1 \le b + 1$. So to compute $N(H_x, K)$ for all subgraphs K on $|X_x|$ vertices, it would again take $O(n^{b+1}2^{\binom{b}{2}})$ time.

Since T has at most $4k$ nodes, the claimed running time follows. $\square$

Lemma 2. *Let $G = (V, E)$ be a graph on n vertices that is k-colored by some coloring $f : V(G) \longrightarrow [k]$, and let H be a k-vertex graph of treewidth b that is k-colored by some coloring π such that H is colorful. Then there is an algorithm taking time $O(c^{b^3}k + n^{b+O(1)}2^{b^2/2})$ time to sample uniformly at random from the set $\{K \mid K$ is a colorful k-vertex subgraph of G under the coloring f and K is color-preserving isomorphic to H colored by $\pi\}$.*

Proof. Using the algorithm in the proof of Lemma 1, we will first compute $N(H_x, K)$ for each node x of the nice tree decomposition T and for each subgraph K on a colorful subset of $V(G)$ with $|X_x|$ vertices. Let r be the root of T. The uniform random sampler will make use of the rooted tree structure of T (like Lemma 1). For every vertex x of T let k_x denote the number of nodes in H_x. Clearly, $k_r = k$. Now, define S_x as $S_x = \{K \mid K$ is a colorful k_x-vertex subgraph of G under the coloring f and K is color-preserving isomorphic to H_x colored by $\pi\}$. In general, we will explain how to efficiently sample from each S_x (and hence from the desired set S_r). The random sampling proceeds inductively using the tree structure: if x is a leaf node in T, then random sampling from S_x can be easily done by brute force in $n^{b+O(1)}$ time as H_x has at most $b+1$ nodes. In order to sample uniformly from S_x, we pick a subgraph K on a colorful subset of $V(G)$ on $|X_x|$ vertices with probability $\dfrac{N(H_x, K)}{\sum_{K:|V(K)|=|X_x|} N(H_x, K)}$. It is clear that if we can sample uniformly from $S(H_x, K)$ for each K then we have uniform sampling from S_x. We will describe a bottom-up method for sampling uniformly from $S(H_x, K)$ for each K. More specifically, it suffices to show we can efficiently sample uniformly from $S(H_x, K)$, assuming that we can uniformly sample from $S(H_y, K')$ for each child y of x in T and each subgraph K' on $|X_y|$ vertices in G.

We need to consider the three cases for children of x:

If x has two children y and z, then the bijective mapping shown in Lemma 1 for Case 1, between $S(H_x, K)$ and $S(H_y, K) \times S(H_z, K)$, shows immediately that uniformly sampling from $S(H_y, K)$ and $S(H_z, K)$ gives uniform sampling from $S(H_x, K)$.

If x has one child y, with $X_x = X_y \cup \{j\}$ then we argue using Case 2 of Lemma 1. Let h be the unique color-preserving isomorphism mapping K to the subgraph of H induced by X_x. Let $K_1 = K - \{h^{-1}(j)\}$. Let $K' \in S(H_x, K)$. Then, $K' \mapsto K' \cup \{h^{-1}(j)\}$ is a bijection from $S(H_y, K_1)$ to $S(H_x, K)$. Thus, uniform sampling from $S(H_y, K_1)$ yields uniform sampling from $S(H_x, K)$.

Finally, if x has one child y, with $X_y = X_x \cup \{j\}$ then we argue like in Case 3 of Lemma 1. We have that in that case $S(H_x, K) = \bigcup_{v \in V(G)} S(H_y, K \cup \{v\})$, where the union is a *disjoint* union. Thus, if we can uniformly sample from each $S(H_y, K \cup \{v\})$ then we can design a uniform sampling procedure for $S(H_x, K)$ by first randomly picking $v \in V(G)$ with probability $\frac{N(H_y, K \cup \{v\})}{N(H_x, K)}$ and then sampling uniformly from $S(H_y, K \cup \{v\})$.

Thus, putting it together, we have an inductive procedure (following the inductive structure of Lemma 1) that samples uniformly from S_r as desired. It is easy to see that the running time of the algorithm is $O(c^{b^3} k + n^{b+O(1)} 2^{b^2}/2)$. $\square$

Now we are ready to prove our main result.

Theorem 2. *Let G be a graph on n vertices $\{1, 2, \ldots n\}$ and let H be a graph on k vertices $\{1, 2, \ldots k\}$ having tree width b. Then there is an FPTRAS algorithm to approximate the number of copies of H in G. Specifically, the algorithm has running time $k^{O(k)}(c^{b^3} k + n^{b+O(1)} 2^{b^2}/2)$ with $\epsilon = 1/k^{O(k)}$ and $\delta = 1/2^{n^{O(1)}}$.*

Proof. Let $A = \{K \mid K$ is a k-vertex subgraph of G such that K is isomorphic to $H\}$. Our goal is to approximate $|A|$. We have written A as $\bigcup_{i \in \mathcal{I}} A_i$, where $\mathcal{I} = \{\langle f, \pi \rangle \mid f \in \mathcal{F}$ and π is a k-coloring of H in which every vertex of H is distinctly colored$\}$. For each $i = \langle f, \pi \rangle \in \mathcal{I}$:

$A_i = \{K \mid K$ is a colorful k-vertex subgraph of G under the coloring f and K is color-preserving isomorphic to H colored by $\pi\}$.

Notice that by Lemma 1 $|A_i|$ can be computed in time $O(c^{b^3} k + n^{b+2} 2^{b^2}/2)$ for each $i \in \mathcal{I}$, and by Lemma 2 we can uniformly sample from each A_i in time $O(c^{b^3} k + n^{b+O(1)} 2^{b^2}/2)$. Lastly, given any subgraph K it can be checked if K is in A_i in time $O(k^2)$ as we just have to check if K is colorful under coloring f and that K is color-preserving isomorphic to H.

Since $|\mathcal{I}| = 2^{O(k)} \cdot \log^{O(1)} n \cdot k!$ and the three conditions are satisfied by the collection $\{A_i\}_{i \in \mathcal{I}}$, we can apply Theorem 1 to get the desired randomized approximation algorithm for $|A|$. $\square$

The following corollary is an immediate consequence of Theorem 2.

Corollary 1. *Let G be a graph on n vertices and let H be a forest on k vertices. Then there is an FPTRAS that for $\epsilon = 1/k^{O(k)}$ and $\delta = 1/2^{n^{O(1)}}$ has running time $k^{O(k)} n^{O(1)}$ to approximate the number of copies of H in G. As a consequence, for the same ϵ and δ, and with running time $k^{O(k)} n^{O(1)}$, there is an FPTRAS for computing the following:*

1. *The number of matchings of size k in G.*
2. *The number of paths of length k in G.*

We can show the following theorem for directed graphs along the same lines as Theorem 2 (and an immediate corollary).

Theorem 3. *Let G be a directed graph on n vertices $\{1, 2, \ldots n\}$ and let H be a directed graph on k vertices $\{1, 2, \ldots k\}$, such that the underlying undirected graph has treewidth b.[3] Then there is an FPTRAS algorithm for computing the number of copies of H in G.*

Corollary 2. *Given a directed graph G on n vertices, there is an FPTRAS with running time $k^{O(k)} n^{O(1)}$ for $\epsilon = 1/k^{O(k)}$ and δ, inverse exponential in n, for the following problems:*

1. *Given an arborescence H on k vertices, to count the number of copies of H in G.*
2. *The number of directed paths of length k in G.*

Remark on counting vertex covers: Given an undirected graph G, the number of minimal vertex covers of size at most k is at most 2^k. This can be easily seen by a branching algorithm for finding minimal vertex covers (c.f. [3]). Hence we can count the number of vertex covers of size at most k by counting the number of subsets of $V(G)$ of size at most k that contains some minimal vertex cover. This fits into the framework of approximate counting (A_i of Theorem 1 would, for example, be the set of vertex covers containing the i-th minimal vertex cover in some ordering of minimal vertex covers.) and by applying Theorem 1, we can approximately count the number of vertex covers in $2^{O(k)} n^{O(1)}$ time. However, since m, the number of subsets, the size of whose union we are interested in counting, is a function of k alone, we can exactly count the number of vertex covers using inclusion-exclusion in deterministic $2^{O(k^2)}$ time after finding the minimal vertex covers [6]. Rossmanith [16] has recently obtained a $c^k n^{O(1)}$ deterministic algorithm for exactly counting vertex covers of size at most k, for some constant $c < 2$.

3   W-Hard Exact Counting Problems

The exact counting versions of paths of length k or matchings of size k have been proved to be $W[1]$-complete by Flum and Grohe [6]. In this section we give some more examples of problems whose decision versions are FPT, but whose counting versions are hard for some level in the W-hierarchy.

Consider checking for weight k satisfying assignments of a monotone 2-DNF formula. The decision problem is trivial as every 2-DNF has a weight 2 satisfying assignment. Counting the number of weight k satisfying assignments is $W[1]$-hard by the following reduction from the $W[1]$-hard problem [3] which asks whether there is a weight k satisfying assignment for a given *antimonotone* 2-CNF formula (i.e. a 2-CNF formula in which each clause has only negated variables). If F is an antimonotone 2-CNF with m clauses then its complement $\overline{F}$ is a monotone 2-DNF with m terms. The $\binom{n}{k}$ assignments of weight k are partitioned into those that satisfy F and those that satisfy $\overline{F}$. If we can count the number that satisfies $\overline{F}$, then we can clearly decide if there is a weight k assignment that satisfies F. Thus we have the following.

---
[3] Note that this is *not* the notion of directed treewidth [7].

Theorem 4. *Counting the number of weight k satisfying assignments of a monotone 2-DNF formula is $W[1]$-hard.*

On the other hand, randomized approximate counting can be efficiently done by directly applying the Karp-Luby result [14].

Corollary 3. *There is an FPRAS algorithm to approximate the number of weight k-satisfying assignments of DNF formulas.*

Given a c-CNF formula F (i.e., each clause has at most c literals), and integer parameter k, it is FPT to decide if F has a satisfying assignment of weight at most k (see, for example [12]). However, counting the number of satisfying assignments of F with weight at most k is $W[1]$-hard. For, by computing the difference between the number of satisfying assignments of weight at most k and weight at most $k - 1$, we can determine if F has weight k satisfying assignment, which is known to be $W[1]$-hard for antimonotone 2-CNF formulas [3]. Thus we have the following.

Theorem 5. *Counting the number of satisfying assignments of weight at most k in an antimonotone 2-CNF formula is $W[1]$-hard.*

While the CLIQUE and the INDEPENDENT SET problems are $W[1]$-complete, consider the question whether a given undirected graph G has either a clique or an independent set of size k for a given integer parameter k? The decision version of this problem is known to be FPT [10].

We now show that counting the number of solutions – i.e. of k sized cliques and k sized independent sets in a graph is $W[1]$-hard by a reduction from CLIQUE. Let the undirected graph G be an instance of CLIQUE with parameter k. Obtain a graph G' by adding a new vertex and making it adjacent to all vertices of G. Let $CI_k(G)$ be the set of all cliques and independent sets of size k in G. Now G has a clique of size k if and only if $|CI_{(k+1)}(G')| - |CI_{(k+1)}(G)| > 0$. Thus if finding $|CI_k(G)|$ is in FPT then testing whether a given graph has a k-clique is also in FPT. Thus we have the following.

Theorem 6. *It is $W[1]$-hard to find the number of k cliques and k independent sets in an undirected graph G, where k is an integer parameter.*

4 Conclusions and Open Problems

We have obtained a randomized approximate FPT algorithm for counting the number of copies of a bounded treewidth graph in a given graph. Our algorithm is a nice combination of the Karp-Luby approximate counting technique and the color-coding technique of [1]. Two direct applications are FPT approximate counting algorithms for counting paths of length k in a graph and counting matchings of size k. Our results nicely complement the W[1]-hardness of exact counting for the above two problems shown by Flum and Grohe [6]. It would be interesting to find more applications of the combination of Karp-Luby and color coding. Can we get *deterministic* approximate counting for Problem 1 with similar time bounds? The derandomization of the approximate counting algorithm for #-DNF in [8] using the pseudorandom generators [15] is not directly applicable for our algorithm in Theorem 2. In Section 3 we gave examples of FPT problems whose counting versions are W-hard. Can we find randomized approximation algorithms for (a) counting the number of weight at most k satisfying assignments of a 2CNF formula, and (b) counting the number of k cliques and k independent sets in a graph?

Acknowledgment. We are grateful to Johannes Köbler and Rainer Schuler for discussions during the mutual visits supported by a DST-DAAD project.

References

1. N. Alon, R. Yuster and U. Zwick, "Color-Coding", *Journal of the Association for Computing Machinery*, **42**(4) (1995) 844–856.
2. H. Bodlaender, "A Linear Time Algorithm for Finding Tree-Decompositions of Small Treewidth", *SIAM J. Computing* **25** (1996) 1305–1317.
3. R. G. Downey and M. R. Fellows, *Parameterized Complexity*, Springer-Verlag, 1998.
4. U. Feige and J. Kilian, "On Limited versus Polynomial Nondeterminism", *Chicago Journal of Theoretical Computer Science*, March (1997).
5. H. Fernau, "Parameterized Enumeration", to appear *in the Proceedings of CO-COON 2002*.
6. J. Flum and M. Grohe, "The Parameterized Complexity of Counting Problems", To appear in *43rd IEEE Symposium on Foundations of Computer Science* 2002.
7. T. Johnson, N. Robertson, P. D. Seymour, R. Thomas, "Directed Tree-Width", preprint (1998) (available at http://www.math.gatech.edu/#thomas/).
8. R. M. Karp and M. Luby, "Monte-Carlo Algorithms for Enumeration and Reliability Problems", *In Proceedings of the 24th Annual IEEE Symposium on Foundations of Computer Science* (1983) 56-64.
9. R. M. Karp, M. Luby and N. Madras, "Monte-Carlo Approximation Algorithms for Enumeration Problems", *Journal of Algorithms* **10** (1989) 429–448.
10. S. Khot and V. Raman, "Parameterized Complexity of Finding Subgraphs with Hereditary Properties", *Proceedings of the Sixth Annual International Computing and Combinatorics Conference (COCOON*, July 2000, Sydney, Australia, Lecture Notes in Computer Science, Springer Verlag **1858** (2000) 137–147. Full version to appear in *Theoretical Computer Science*.
11. T. Kloks, "Treewidth: Computations and Approximations", Lecture Notes in Computer Science, Springer-Verlag **842** 1994.
12. M. Mahajan and V. Raman, "Parameterizing Above Guaranteed Values: MaxSat and MaxCut", *Journal of Algorithms* **31** (1999) 335–354.
13. C. McCartin, "Parameterized Counting Problems", to appear in *the Proceedings of MFCS 2002 conference*.
14. R. Motwani and P. Raghavan, *Randomized Algorithms*, Cambridge University Press, 1995.
15. N. Nisan, *Using Hard Problems to Create Pseudorandom Generators,* MIT Press (1992).
16. P. Rossmanith, private communication.

Approximating MIN k-SAT[*]

Adi Avidor and Uri Zwick

School of Computer Science, Tel-Aviv University, Tel-Aviv 69978, Israel.
{adi,zwick}@tau.ac.il.

Abstract. We obtain substantially improved approximation algorithms for the
MIN k-SAT problem, for $k = 2, 3$. More specifically, we obtain a 1.1037-
approximation algorithm for the MIN 2-SAT problem, improving a previous 1.5-
approximation algorithm, and a 1.2136-approximation algorithm for the MIN 3-
SAT problem, improving a previous 1.75-approximation algorithm for the problem. These results are obtained by adapting techniques that were previously used
to obtain approximation algorithms for the MAX k-SAT problem. We also obtain
some hardness of approximation results.

1 Introduction

An instance of MIN SAT in the Boolean variables $x_1, \ldots, x_n$ is composed of a collection
of *clauses* $C_1, \ldots, C_m$, and non-negative weights $w_1, \ldots, w_m$ associated with them.
Each clause C_i is of the form $z_1 \vee z_2 \vee \ldots \vee z_{k_i}$ $(k_i \geq 1)$ where each z_j is either
a variable x_l or its negation $\bar{x}_l$. Each such z_j is called a *literal*. The goal is to assign
the variables $x_1, \ldots, x_n$ Boolean values 0 or 1 so that the total weight of the satisfied
clauses is *minimized*. A clause is satisfied if at least one of the literals appearing in it
is assigned the value 1. The more famous MAX SAT problem is defined similarly, with
the goal being to *maximize* the total weight of the satisfied clauses. MIN k-SAT is the
version of MIN SAT in which each clause has *at most* k literals. MIN k-SAT(t) is the
version of MIN k-SAT where each variable appears, negated or unnegated, at most t
times.

MIN SAT was first considered by Kohli *et al.* [KKM94]. They showed that even the
unweighted version of MIN 2-SAT, restricted to instances in which all clauses are Horn
clauses (i.e., containing at most one negated variable), is NP-hard. They also gave a
simple randomized 2-approximation algorithm for the unweighted version of MIN SAT.

Marathe and Ravi [MR96] showed that the MIN SAT problem is in fact equivalent
to the well studied MIN VC, the minimum weight vertex cover problem, in the sense
that a ρ-approximation algorithm for one of them, for any $\rho \geq 1$, gives immediately
a ρ-approximation algorithm for the other. As there are several 2-approximation algorithms for the MIN VC problem, see Bar-Yehuda and Even [BYE81,BYE85], Hochbaum
[Hoc82,Hoc83], Monien and Speckenmeyer [MS85], we get immediately a variety of
2-approximation algorithms for the MIN SAT problem. It is conjectured by many that
there is no better approximation algorithm for the MIN VC problem, and hence also
for the MIN SAT problem, unless P=NP. Dinur and Safra [DS02], improving a result of

---

[*] This research was supported by the ISRAEL SCIENCE FOUNDATION (grant no. 246/01).

Håstad [Hås01], showed that there is no $(10\sqrt{5} - 21 - \varepsilon)$-approximation algorithm for the MIN VC problem, for any $\varepsilon > 0$, unless P=NP. (Note that $10\sqrt{5} - 21 \simeq 1.3606$.)

Furthermore, Marathe and Ravi [MR96] showed that the MIN-VC(Δ) problem, i.e., minimum vertex cover problem for graphs with maximum degree Δ, is equivalent to the MIN Δ-SAT(2) problem, the version of the MIN SAT problem in which each clause is of size at most Δ, and each variable occurs at most twice. Approximation algorithm for the MIN VC(Δ) problem were given by Berman and Fujito [BF99] and by Halperin [Hal00]. Hardness results for this problem were obtained by Berman and Karpinski [BK99].

Marathe and Ravi [MR96] also show that the MIN k-SAT(t) problem is at least as easy as the MIN VC($k(t-1)$) problem. In other words, a ρ-approximation algorithm for MIN VC($k(t-1)$), for any $\rho \geq 1$ can be transformed into a ρ-approximation algorithm for MIN k-SAT(t).

Bertsimas et al. [BTV99] describe a linear programming based approximation algorithm for the MIN k-SAT problem. Their algorithm achieves a performance ratio of $2 - \frac{1}{2^{k-1}}$. For $k = 2$ they thus obtain a ratio of $\frac{3}{2}$, and for $k = 3$ a ratio of $\frac{7}{4}$. Opposed to previous works, where the variables of the linear programming were rounded *independently*, Bertsimias et al. [BTV99] round the variables in a *dependent* manner. Bazgan and Fernandez de la Vega [BdlV99] gave a PTAS for dense instances of MIN 2-SAT. The MIN 2-SAT problem, and other related problems were also considered by Hochbaum [Hoc00] and Hochbaum and Pathria [HP00].

We use semidefinite programming to approximate MIN 2-SAT and MIN 3-SAT. For MIN 2-SAT we improve the approximation ratio from 1.5 to 1.1037. For MIN 3-SAT we improve the approximation ratio from 1.75 to 1.2136. The idea behind our algorithm is as follows. We first assign each Boolean literal x_i a unit vector $v_i \in \mathbb{R}^{n+1}$. In addition, we introduce the vector v_0, that corresponds to the constant false. We then write a semidefinite programming relaxation for the problem. I.e., a minimization program which is linear in the products $v_i \cdot v_j$. We design the semidefinite program in a way that any valid assignment to x_i is a feasible point of the program. This program can then be solved, to any desired precision, in polynomial time. We then have to apply a Rounding Procedure (RP) on the vectors $v_0, \ldots, v_n$, i.e., we need to "round" the vectors to Boolean values. Rounding procedures are usually randomized algorithms.

Geomans and Williamson [GW95] were the first to suggest the semidefinite programming approach for approximating MAX CUT, MAX 2-SAT, and MIN DICUT. They rounded the solution to the semidefinite program by choosing a random hyperplane passing through the origin. The normal of this hyperplane is an $(n + 1)$-dimensional standard normal random variable. Unfortunately, the random hyperplane technique, on its own, cannot be used to obtain approximation algorithms for the MIN k-SAT problem. Feige and Goemans [FG95] extended the random hyperplane technique while studying MAX 2-SAT and MAX DICUT. They *rotated* the vectors towards the vector v_0, and then used random hyperplane rounding. Furthermore, they suggested, but did not explore, another extension to the hyperplane rounding: choose the normal of the hyperplane according to a distribution which is *skewed* towards (or from) the vector v_0. The latter technique was explored for MAX 2-SAT and MAX DICUT first by Matuura and Matsui [MM01a,MM01b], and recently by Lewin et al. [LLZ02].

We show here that these techniques can also be applied to the MIN 2-SAT and MIN 3-SAT problems. Table 1 summarizes the approximability and inapproximability results obtained here, and in previous papers, for the MIN k-SAT problem.

Table 1. Approximability and inapproximability results for MIN SAT. (Results obtained here are marked by [*].)

	Inapproximability ratio	Approximation ratio
MIN 2-SAT	$\frac{15}{14} - \varepsilon$ [*]	1.1037 [*]
MIN 3-SAT	$\frac{7}{6} - \varepsilon$ [*]	1.2136 [*]
MIN k-SAT	$\frac{7}{6} - \varepsilon$ [*]	$2 - \frac{1}{2^{k-1}}$ [BTV99]
MIN SAT	1.3606 [DS02]	2

2 Preliminaries

We let $x_{n+i} = \bar{x}_i$, for $1 \le i \le n$, and we also let $x_0 = 0$. Each clause of a MIN 2-SAT instance is then of the form $x_i \vee x_j$ where $0 \le i, j \le 2n$ (if $i = 0$ the clause is of length 1). An instance of MIN 2-SAT can then be encoded as an array $(w_{ij})_{0 \le i,j \le 2n}$, where w_{ij} is interpreted as the weight of the clause $x_i \vee x_j$. In a similar way, an instance of MIN 3-SAT is encoded as an array $(w_{ijk})_{0 \le i,j,k \le 2n}$, where w_{ijk} is interpreted as the weight of the clause $x_i \vee x_j \vee x_k$, where $0 \le i, j, k \le 2n$. Note that if $i = 0, 0 < j, k \le 2n$, then the clause is of length two, and if $i = j = 0, 0 < k \le 2n$, then the clause is of length one. The goal is then to assign the variables $x_1, \ldots, x_{2n}$ Boolean values, such that $x_{n+i} = \bar{x}_i$ and such that the total weight of the satisfied clauses is minimized.

3 Semidefinite Programming Relaxation for MIN 2-SAT

The semidefinite programming relaxation of the MIN 2-SAT problem is given in Figure 1. In this relaxation, a unit vector $v_i \in \mathbb{R}^{n+1}$ is assigned to each literal x_i, where $0 \le i \le 2n$. As $x_0 = 0$, the vector v_0 corresponds to the constant 0 (false). The vectors

$$\text{Min} \sum_{i,j} w_{ij} \frac{3 - v_0 \cdot v_i - v_0 \cdot v_j - v_i \cdot v_j}{4}$$

$$v_0 \cdot v_i + v_0 \cdot v_j + v_i \cdot v_j \ge -1 \ , \quad 1 \le i, j \le 2n$$
$$v_i \cdot v_{n+i} = -1 \quad , \quad 1 \le i \le n$$
$$v_i \in \mathbb{R}^{n+1} \ , \ v_i \cdot v_i = 1 \quad , \quad 0 \le i \le 2n$$

Fig. 1. A semidefinite programming relaxation of MIN 2-SAT

$v_0, v_1, \ldots, v_{2n}$ are unit vectors on the unit sphere in $\mathbb{R}^{n+1}$. To ensure $x_{n+i} = \bar{x}_i$, we require $v_i \cdot v_{n+i} = -1$, for $1 \leq i \leq n$. This is equivalent to $v_{n+i} = -v_i$, since v_i are unit vectors $(1 \leq i \leq n)$. As the value of a solution of the semidefinite program depends only on the inner products between the vectors, we may assume, with out loss of generality, that $v_0 = (1, 0, \ldots, 0) \in \mathbb{R}^{n+1}$. Note that the semidefinite program in Figure 1 is a relaxation of MIN 2-SAT: Let $x \in \{0, 1\}^{2n+1}$ be a valid assignment (i.e., $x_{n+i} = \bar{x}_i$). Let $v_i = v_0$ if $x_i = 0$, and $v_i = -v_0$ if $x_i = 1$, where $0 \leq i \leq 2n$. Then, it is easy to see that for $0 \leq i, j \leq 2n$ we have

$$\frac{3 - v_0 \cdot v_i - v_0 \cdot v_j - v_i \cdot v_j}{4} = x_i \vee x_j$$

In addition, the constraints $v_0 \cdot v_i + v_0 \cdot v_j + v_i \cdot v_j \geq -1$, called the 'triangle constraints', hold as v_i and v_j are assigned the values $\pm v_0$, for $0 \leq i, j \leq 2n$.

4   Semidefinite Programming Relaxation for MIN 3-SAT

The semidefinite programming relaxation of the MIN 3-SAT problem is given in Figure 2. In a similar way to MIN 2-SAT, a unit vector $v_i \in \mathbb{R}^{n+1}$ is assigned to each literal x_i, where $0 \leq i \leq 2n$. In addition, a scalar z_{ijk} is assigned to each clause, where $0 \leq i, j, k \leq 2n$. Note that the semidefinite program in Figure 2 is a relaxation of MIN 3-SAT: Let $x \in \{0, 1\}^{2n+1}$ be a valid assignment. Let $v_i = v_0$ if $x_i = 0$, and $v_i = -v_0$ if $x_i = 1$, where $0 \leq i \leq 2n$. Then, it is easy to see that for $0 \leq i, j, k \leq 2n$ we have

$$\max \left\{ \begin{array}{c} \frac{3 - v_0 \cdot v_i - v_0 \cdot v_j - v_i \cdot v_j}{4}, \quad \frac{3 - v_0 \cdot v_i - v_0 \cdot v_k - v_i \cdot v_k}{4} \\ \frac{3 - v_0 \cdot v_j - v_0 \cdot v_k - v_j \cdot v_k}{4}, \quad 0 \end{array} \right\} = x_i \vee x_j \vee x_k .$$

Again, the 'triangle constraints' hold.

$$\begin{aligned}
& \text{Min} \sum_{i,j,k} w_{ijk} z_{ijk} \\
z_{ijk} &\geq \frac{3 - v_0 \cdot v_i - v_0 \cdot v_j - v_i \cdot v_j}{4} , \quad 0 \leq i, j, k \leq 2n \\
z_{ijk} &\geq \frac{3 - v_0 \cdot v_i - v_0 \cdot v_k - v_i \cdot v_k}{4} , \quad 0 \leq i, j, k \leq 2n \\
z_{ijk} &\geq \frac{3 - v_0 \cdot v_j - v_0 \cdot v_k - v_j \cdot v_k}{4} , \quad 0 \leq i, j, k \leq 2n \\
z_{ijk} &\geq 0 , \quad 0 \leq i, j, k \leq 2n \\
v_0 \cdot v_i + v_0 \cdot v_j + v_i \cdot v_j &\geq -1 , \quad 1 \leq i, j \leq 2n \\
v_i \cdot v_{n+i} &= -1 , \quad 1 \leq i \leq n \\
v_i \in \mathbb{R}^{n+1}, \; v_i \cdot v_i &= 1 , \quad 0 \leq i \leq 2n
\end{aligned}$$

Fig. 2. A semidefinite programming relaxation of MIN 3-SAT

5   Rounding Procedures

We now describe the rounding procedures we use in this paper, namely $\mathcal{RHP}$, $\mathcal{ROT}_f$ and $\mathcal{THRESH}_g$. For a survey of rounding procedures see Lewin et al. [LLZ02]. The basic rounding technique, introduced by Goemans and Williamson [GW95], is the random hyperplane rounding ($\mathcal{RHP}$) technique. In this rounding procedure, the vectors $v_1, \ldots, v_n$ are rounded using a random hyperplane:

$\mathcal{RHP}$:   Let $r = (r_0, r_1, \ldots, r_n)$ be a standard $(n+1)$-dimensional normal random variable. For $1 \leq i \leq n$, let $x_i \leftarrow 0$ if v_0 and v_i lie on the same side of the hyperplane with the normal vector r, and let $x_i \leftarrow 1$ otherwise.

Since we require $v_{n+i} = -v_i$ for $1 \leq i \leq n$, $\mathcal{RHP}$ preserves consistency, in the sense that x_i and $x_{n+i} = \bar{x}_i$ are always assigned opposite values. As mentioned before, $\mathcal{RHP}$ was generalized by Feige and Goemans [FG95] to the $\mathcal{ROT}_f$ rounding procedure, parameterized by a *rotation* function f, as defined below:

Definition 1. *A continuous function $f : [0, \pi] \rightarrow [0, \pi]$ is called a* rotation *function if $f(\pi - \theta) = \pi - f(\theta)$ for all $\theta \in [0, \pi]$.*

$\mathcal{ROT}_f$:   Let $\theta_i = \arccos(v_0 \cdot v_i)$ be the angle between v_0 and v_i. Rotate v_i into a vector $v_i' \in Span\{v_0, v_i\}$ that forms an angle of $\theta_i' = f(\theta_i)$ with v_0. Round the vectors $v_0, v_1', \ldots, v_n'$ using the $\mathcal{RHP}$ rounding procedure.

Again, since f is a rotation function consistency is preserved. For convenience we let $v_0' = v_0$.

In addition to the rotation technique, Feige and Goemans [FG95] suggested taking the random variable r_0 from an *arbitrary* distribution, rather than the standard normal distribution. In other words, choose the normal of the random hyperplane with a distribution which is *skewed* towards (or away) the vector v_0. The $\mathcal{THRESH}_g$ rounding procedure is a special case of using *rotation* combined with a *skewed* hyperplane distribution. In $\mathcal{THRESH}_g$, the random variable r_0 is assigned a *constant* value. We adopt a different but equivalent definition of $\mathcal{THRESH}_g$ proposed by [LLZ02]. To avoid confusion, we note that $\mathcal{THRESH}_g$ is a member of the family $\mathcal{THRESH}^-$ of [LLZ02]. The $\mathcal{THRESH}_g$ rounding procedure, is defined for an anti-symmetric function $g : [-1, 1] \rightarrow \mathbb{R}$, i.e., a function satisfying $g(-x) = -g(x)$, as follows

$\mathcal{THRESH}_g$:   Let $r = (0, r_1, \ldots, r_n)$, where $r_1, r_2, \ldots, r_n$ are independent standard normal variables. For $1 \leq i \leq n$, let $x_i \leftarrow 1$ if $v_i \cdot r \leq g(v_0 \cdot v_i)$ and $x_i \leftarrow 0$ otherwise.

Again, the anti-symmetry of the function g is used in order to preserve consistency. Lewin et al. [LLZ02] showed that if the rounding procedure is characterized by a rotation function f and the constant random variable $r_0 = a$, then $g(x) = -a \cot f(\arccos x)\sqrt{1 - x^2}$. Though is seems that $\mathcal{THRESH}_g$ is a generalization of $\mathcal{ROT}_f$, the families $\mathcal{ROT}$ and $\mathcal{THRESH}$ are disjoint. This claim can be derived using observations made in [LLZ02].

6 Analysis

In this section we describe the computation of the performance ratio obtained using a given rounding procedure RP. This computation is later used in Section 7 to compute the ratios of our approximation algorithms. In addition, we explain here how to show that a given rounding procedure is an essentially optimal rounding procedures from a given family of rounding procedures.

Let $v_0, v_1, \dots, v_k \in \mathbb{R}^{n+1}$ be vectors corresponding to some clause of MIN k-SAT ($k = 2, 3$). Denote by $prob_{\mathrm{RP}}(v_0, v_1, \dots, v_k)$ the probability that the clause is satisfied when the vectors $v_0, v_1, \dots, v_k$ are rounded using the rounding procedure RP. Denote by $value(v_0, v_1, \dots, v_k)$ the contribution of the clause to the value of the semidefinite program. For MIN 2-SAT we have

$$value(v_0, v_1, v_2) = \frac{1}{4}\left(3 - v_0 \cdot v_i - v_0 \cdot v_j - v_i \cdot v_j\right),$$

and for MIN 3-SAT we have

$$value(v_0, v_1, v_2, v_3) = \max\left\{ \begin{matrix} \frac{3 - v_0 \cdot v_i - v_0 \cdot v_j - v_i \cdot v_j}{4}, & \frac{3 - v_0 \cdot v_i - v_0 \cdot v_k - v_i \cdot v_k}{4} \\ \frac{3 - v_0 \cdot v_j - v_0 \cdot v_k - v_j \cdot v_k}{4}, & 0 \end{matrix} \right\}.$$

Denote

$$ratio_{\mathrm{RP}}(v_0, v_1, \dots, v_k) = \frac{prob_{\mathrm{RP}}(v_0, v_1, \dots, v_k)}{value(v_0, v_1, \dots, v_k)}.$$

We denote by $\alpha_k(RP)$ the performance ratio of the approximation algorithm that uses the rounding procedure RP. We denote by $\Omega_{n+1}^{(k)}$ the vector $(k+1)$-tuples $(v_0, \dots, v_k)$ that satisfy all the constraints of MIN k-SAT semidefinite programming relaxation and have a strictly positive value.

For an instance $(w_{i_1, \dots, i_k})$ of MIN k-SAT, let $v_0, v_1, \dots, v_n$ be an optimal solution to the semidefinite program. As the semidefinite program is a relaxation of MIN k-SAT, the expression $\sum_{i_1, \dots, i_k} w_{i_1, \dots, i_k} \cdot value(v_0, v_{i_1}, \dots, v_{i_k})$ is a lower bound on the value of the optimum solution. We assume that it is bigger than 0. (The case in which it is 0 is easily dealt with.) The expected value of the assignment produced using rounding procedure RP is:

$$\sum_{i_1, \dots, i_k} w_{i_1, \dots, i_k} \, prob_{\mathrm{RP}}(v_0, v_{i_1}, \dots, v_{i_k})$$

Therefore, we get:

$$\begin{aligned}
\alpha_k(RP) &\le \sup_{(w_{i_1, \dots, i_k})} \frac{\sum_{i_1, \dots, i_k} w_{i_1, \dots, i_k} \, prob_{\mathrm{RP}}(v_0, v_{i_1}, \dots, v_{i_k})}{\sum_{i_1, \dots, i_k} w_{i_1, \dots, i_k} \, value(v_0, v_{i_1}, \dots, v_{i_k})} \\
&\le \sup_{(u_0, u_1, \dots, u_k) \in \Omega_{n+1}^{(k)}} \frac{prob_{\mathrm{RP}}(u_0, u_1, \dots, u_k)}{value(u_0, u_1, \dots, u_k)} \\
&= \sup_{(u_0, u_1, \dots, u_k) \in \Omega_{n+1}^{(k)}} ratio_{\mathrm{RP}}(u_0, u_1, \dots, u_k).
\end{aligned}$$

The computation of $prob_{\mathrm{RP}}(v_0, v_1, \dots, v_k)$ for the rounding procedures $\mathcal{ROT}_f$ and $\mathcal{THRESH}_g$ is discussed in details in the full version of the paper. For each of our rounding procedures, the probability $prob_{RP}(v_0, v_1, \ldots, v_k)$ depends only on the inner products $v_i \cdot v_j$, for $0 \leq i, j \leq k$. As we are only interested in the inner products of the vectors $v_0, v_1, \ldots, v_k$, we may assume $v_0, v_1, \ldots, v_k$ lie in $\mathbb{R}^{k+1}$. In other words, we may apply our maximization in $\Omega_{k+1}^{(k)}$, rather than $\Omega_{n+1}^{(k)}$, and get

$$\alpha_k(RP) \leq \beta_k(RP) \stackrel{def}{=} \sup_{(u_0, u_1, \ldots, u_k) \in \Omega_{k+1}^{(k)}} ratio_{RP}(u_0, u_1, \ldots, u_k)$$

For a family $\mathcal{F}$ of rounding procedures (e.g., the family $\bigcup_f \mathcal{ROT}_f$), $\beta_k(RP)$ (where $RP \in \mathcal{F}$) can be bounded from below as follows. Let $C^{(k)} = \{(v_0^{(j)}, v_1^{(j)}, \ldots, v_k^{(j)}) \in \Omega_{k+1}^{(k)} | 1 \leq j \leq J\}$ be a finite set of *configurations*. Then, for all $RP \in \mathcal{F}$

$$\beta_k(RP) \geq \max_{(v_0, \ldots, v_k) \in C^{(k)}} \inf_{R \in \mathcal{F}} ratio_R(v_0, v_1, \ldots, v_k)$$

We use this bounding technique to show that our rounding procedures are essentially the optimal rounding procedures from the $\mathcal{ROT}$ and $\mathcal{THRESH}$ families.

7 Algorithms

In this section we describe our algorithms for MIN 2-SAT and MIN 3-SAT. For both problems we give algorithms, which use a $\mathcal{ROT}_f$ rounding procedure. For MIN 2-SAT we give a better algorithm which uses a $\mathcal{THRESH}_g$ rounding procedure. The performance ratios of our algorithms are calculated by bounding $\beta_k(RP)$ from above. We also obtain lower bounds for $\beta_k(RP)$ among rounding procedures from the $\mathcal{ROT}$ family, and the $\mathcal{THRESH}$ family. These lower bounds show that our choices of rotation functions are almost optimal. The calculation of $\beta_k(RP)$ is a global optimization task. So is the lower bound of $\beta_k(RP)$ for a given family of rounding procedures. These bounds were calculated by using the optimization toolbox of Matlab. We could, in principle, obtain computer-assisted proofs of these bounds, e.g., using $\mathcal{RealSearch}$ [Zwi02], which is a generic software for proving inequalities. However, this would require a huge amount of work.

For both the $\mathcal{ROT}$ and the $\mathcal{THRESH}$ families, we need to conduct a search for a good rotation function. We briefly describe below how this search is conducted. As $\beta_k(RP)$ depends only on the inner product of the vectors, we identify a configuration by the angles between its vectors. In our search, we assume that the rotation function is piecewise linear. We represent our rotation function by a set of p points $\{(\theta_1, f(\theta_1)), \ldots, (\theta_p, f(\theta_p))\}$. It turns out that $\beta_k(RP)$ can be easily bounded from below using a *small* set of configurations, typically three to four configurations. These configurations usually fulfill equality in one (of more) of the triangle inequalities. Therefore, we first bound $\beta_k(RP)$ for all rotation functions by finding these "bad" configurations. In this process we also obtain the optimal rotation values for these angles. These angles, and their rotation values are a good starting point for the search of a good piecewise linear rotation function. At each stage of the search, we obtain a set of "bad" configurations for the current rotation function (typically hundreds of configurations). We then optimize the break points $\{(\theta_1, f(\theta_1)), \ldots, (\theta_p, f(\theta_p))\}$ to perform better on the current "bad" configurations. Next we optimize the "bad" configurations to a new set of configurations that perform worse on the new optimized rotation function. We repeat this process until we find a rotation function which is close enough to our lower bound. The results of our searches are given in Subsections 7.1, 7.2 and 7.3.

7.1   An Approximation Algorithm for MIN 2-SAT Using $\mathcal{ROT}_f$

The algorithm first solves the semidefinite programming of Figure 1. It then rounds the solution using the rounding procedure $\mathcal{ROT}_{f_{2R}}$, which rotates the vectors using the piecewise linear rotation function f_{2R} in Table 2(a). The algorithm achieves a performance ratio of at most 1.115682, and the rotation function f_{2R} is nearly optimal in the $\mathcal{ROT}$ family as

$$1.115674 \; < \; \inf_f \beta_2(\mathcal{ROT}_f) \; < \; \beta_2(\mathcal{ROT}_{f_{2R}}) < 1.115682$$

The lower bound may be obtained in the following way. Let

$$\theta_1 = 1.2897 \;, \; \theta_2 = 1.3688 \;, \; \theta_3 = 1.3889 \;, \; \theta_4 = \pi/2 \;,$$

and denote a configuration Θ_{ij} by $\Theta_{ij} = (\theta_i, \pi - \theta_j, \arccos(1 - \cos(\theta_i) + \cos(\pi - \theta_j)))$. We then optimize the values of any rotation function on the angles $\theta_1, \theta_2, \theta_3, \theta_4$ to minimize the maximum ratio of the four configurations $\{\Theta_{14}, \Theta_{22}, \Theta_{23}, \Theta_{13}\}$. Note that for each configuration Θ_{ij}, one of the triangle constraints is tight.

Table 2. Rotation functions for the MIN 2-SAT approximation algorithms. (a) The piecewise linear rotation function f_{2R} used for the 1.115682-approximation algorithm. (b) The piecewise linear rotation function f_{2T} used for the 1.103681-approximation algorithm.

θ	$f_{2R}(\theta)$
0.00000000000000	0.00000000000000
0.63297009019109	0.00000000000000
1.24730256557798	1.02242252692503
1.33645467789176	1.17771039329272
1.48300834443152	1.42939187408747
1.57079632679490	1.57079632679490

θ	$f_{2T}(\theta)$
0.00000000000000	0.00000000000000
0.27528129161606	0.55427213749512
0.92985503206467	0.97960965696248
1.29587355435084	1.31904539278850
1.44112209825335	1.45517755589338
1.53796647924573	1.54185902561185
1.57079632679490	1.57079632679490

(a)   (b)

7.2   An Approximation Algorithm for MIN 2-SAT Using $\mathcal{THRESH}_g$

The algorithm first solves the semidefinite programming of Figure 1. It then rounds the solution using the rounding procedure $\mathcal{THRESH}_{g_{2T}}$, with $g_{2T}(x) = -a \cot f_{2T}(\arccos x)\sqrt{1 - x^2}$, where $a = 1.5$ and f_{2T} is the piecewise linear function given in Table 2(b). The algorithm achieves a performance ratio of at most 1.103681.

The rounding procedure $\mathcal{THRESH}_{g_{2T}}$ is an essentially optimal rounding procedure from the family $\mathcal{THRESH}$, as

$$1.103639 < \inf_g \beta_2(\mathcal{THRESH}_g) < \beta_2(\mathcal{THRESH}_{g_{2T}}) < 1.103682$$

In our calculation of the lower bound we optimized the values of the function g over the cosines of the four angles

$$\theta_1 = 1.3483 \ , \ \theta_2 = 1.3893 \ , \ \theta_3 = 1.5061 \ , \ \theta_4 = \pi/2 \ .$$

while minimizing the maximum ratio of $\{\Theta_{14}, \Theta_{23}, \Theta_{31}, \Theta_{32}\}$, where the configuration Θ_{ij} is the same as defined in Section 7.1.

7.3   An Approximation Algorithm for MIN 3-SAT Using $\mathcal{ROT}_f$

The algorithm first solves the semidefinite programming of Figure 2. It then rounds the solution using the rounding procedure $\mathcal{ROT}_{f_{3R}}$, which rotates the vectors using the piecewise linear rotation function f_{3R} given in Table 3. The algorithm achieves a performance ratio of at most 1.21358, and the rotation function f_{3R} is nearly optimal in the $\mathcal{ROT}$ family as

$$1.21145 < \inf_f \beta_3(\mathcal{ROT}_f) < \beta_3(\mathcal{ROT}_{f_{3R}}) < 1.21358 \ .$$

We obtain the lower bound by using the three angles

$$\theta_1 = 1.2965 \ , \ \theta_2 = 1.3490 \ , \ \theta_3 = \pi/2 \ ,$$

and the configurations $\{\Theta_{13}, \Theta_{21}, \Theta_{22}\}$. Here each configuration is a 6-tuple $\Theta_{ij} = (\theta_i, \theta_i, \theta''_{ij}, \pi - \theta_j, \theta'_{ij}, \theta'_{ij})$, $\theta'_{ij} = \arccos(1 - \cos(\theta_i) + \cos(\pi - \theta_j))$, and $\theta''_{ij} = \arccos(-1 + 2\min\{|\cos\theta'_{ij}|, |\cos\theta_i|\})$. Note that both angles θ'_{ij}, and θ''_{ij} make at least one of the triangle constraints tight.

Table 3. The piecewise linear rotation function used for the 1.21358 algorithm for MIN 3-SAT.

θ	$f_{3R}(\theta)$
0.00000000000000	0.00000000000000
0.70421578798384	0.00000000000000
1.04856801819789	0.60050742084360
1.24889565726930	0.96416357991621
1.25076553239599	0.98688235088377
1.32051217629965	1.12034955948672
1.38289309655826	1.24098081256639
1.57079632679490	1.57079632679490

8 Hardness Results

Our hardness results are summarized in Theorem 1. We apply a slight modification of the techniques of Trevisan *et al.* [TSSW00]. The results are obtained by using reductions from MAX 3-XOR to MIN k-SAT. The proof of Theorem 1, as well as the proof of optimality of the gadgets used in it, appears in the full version of the paper.

Theorem 1. *For any* $\varepsilon > 0$ *MIN 2-SAT cannot be approximated within a factor of* $15/14 - \varepsilon$, *and MIN 3-SAT cannot be approximated within a factor of* $7/6 - \varepsilon$ *unless* $P = NP$.

9 Concluding Remarks and Open Problems

We obtained substantially improved approximation algorithms for the MIN 2-SAT and the MIN 3-SAT problems. Our algorithms are randomized, but they can be derandomized using the techniques of Mahajan and Ramesh [MR99]. Obtaining further improved approximability and inapproximability results for these problems are challenging open problems.

References

[BdlV99]   C. Bazgan and W. Fernandez de la Vega. A polynomial time approximation scheme for dense min2sat. In *Proc. 12th Int. Symp. on Fundamentals of Computation Theory, Lecture Notes in Comput. Sci. 1684*, pages 91–99. Springer-Verlag, 1999.

[BF99]   P. Berman and T. Fujito. On approximation properties of the independent set problem for low degree graphs. *Theory of Computing Systems*, 32(2):115–132, 1999.

[BK99]   P. Berman and M. Karpinski. On some tighter inapproximability results (extended abstract). In *Proceedings of ICALP'99*, pages 200–209, 1999.

[BTV99]   D. Bertsimas, C. Teo, and R. Vohra. On dependent randomized rounding algorithms. *Oper. Res. Lett.*, 24(3):105–114, 1999.

[BYE81]   R. Bar-Yehuda and S. Even. A linear-time approximation algorithm for the weighted vertex cover problem. *Journal of Algorithms*, 2(2):198–203, 1981.

[BYE85]   R. Bar-Yehuda and S. Even. A local-ratio theorem for approximating the weighted vertex cover problem. *Annals of Discrete Mathematics*, 25:27–45, 1985.

[DS02]   I. Dinur and S. Safra. The importance of being biased. In *Proceedings of STOC'02*, pages 33–42, 2002.

[FG95]   U. Feige and M.X. Goemans. Approximating the value of two prover proof systems, with applications to MAX-2SAT and MAX-DICUT. In *Proceedings of ISTCS'95*, pages 182–189, 1995.

[GW95]   M.X. Goemans and D.P. Williamson. Improved approximation algorithms for maximum cut and satisfiability problems using semidefinite programming. *Journal of the ACM*, 42:1115–1145, 1995.

[Hal00]   E. Halperin. Improved approximation algorithms for the vertex cover problem in graphs and hypergraphs. In *Proceedings of SODA'00*, pages 329–337, 2000.

[Hås01]   J. Håstad. Some optimal inapproximability results. *Journal of the ACM*, 48:798–859, 2001.

[Hoc82]   D.S. Hochbaum. Approximation algorithms for the set covering and vertex cover problems. *SIAM Journal on Computing*, 11(3):555–556, 1982.

[Hoc83]   D.S. Hochbaum. Efficient bounds for the stable set, vertex cover and set packing problems. *Discrete Applied Mathematics*, 6(3):243–254, 1983.

[Hoc00]   D.S. Hochbaum. Instant recognition of polynomial time solvability, half itegrality and 2-approximations. In *Proceedongs of APPROX'00*, pages 2–14, 2000.

[HP00]   D. Hochbaum and A. Pathria. Approximating a generalization of MAX 2SAT and MIN 2SAT. *Discrete Applied Mathematics*, 107(1-3):41–59, 2000.

[KKM94]   R. Kohli, R. Krishnamurti, and P. Mirchandani. The minimum satisfiability problem. *Discrete Mathematics*, 7:275–283, 1994.

[LLZ02]   M. Lewin, D. Livnat, and U. Zwick. Improved rounding techniques for the MAX 2-SAT and MAX DI-CUT problems. In *Proceedings of IPCO'02*, pages 67–82, 2002.

[MM01a]   S. Matuura and T. Matsui. 0.863-approximation algorithm for MAX DICUT. In *Proceedongs of APPROX-RANDOM'01*, pages 138–146, 2001.

[MM01b]   S. Matuura and T. Matsui. 0.935-approximation randomized algorithm for MAX 2SAT and its derandomization. Technical Report METR 2001-03, Department of Mathematical Engineering and Information Physics, the University of Tokyo, Japan, September 2001.

[MR96]   M.V. Marathe and S.S. Ravi. On approximation algorithms for the minimum satisfiability problem. *Information Processing Letters*, 58:23–29, 1996.

[MR99]   S. Mahajan and H. Ramesh. Derandomizing approximation algorithms based on semidefinite programming. *SIAM Journal on Computing*, 28:1641–1663, 1999.

[MS85]   B. Monien and E. Speckenmeyer. Ramsey numbers and an approximation algorithm for the vertex cover problem. *Acta Informatica*, 22(1):115–123, 1985.

[TSSW00]   L. Trevisan, G.B. Sorkin, M. Sudan, and D.P. Williamson. Gadgets, approximation, and linear programming. *SIAM Journal on Computing*, 29:2074–2097, 2000.

[Zwi02]   U. Zwick. Computer assisted proof of optimal approximability results. In *Proceedings of SODA'02*, pages 496–505, 2002.

Average-Case Competitive Analyses for Ski-Rental Problems

Hiroshi Fujiwara and Kazuo Iwama

Graduate School of Informatics, Kyoto University, Kyoto 606-8501, Japan
{fujiwara, iwama}@kuis.kyoto-u.ac.jp

Abstract. Let s be the ratio of the cost for purchasing skis over the cost for renting them. Then the famous result for the ski-rental problem shows that skiers should buy their skis after renting them $(s - 1)$ times, which gives us an optimal competitive ratio of $2 - \frac{1}{s}$. In practice, however, it appears that many skiers buy their skis before this optimal point of time and also many skiers keep renting them forever. In this paper we show that these behaviors of skiers are quite reasonable by using an *average-case competitive ratio*. For an exponential input distribution $f(t) = \lambda e^{-\lambda t}$, optimal strategies are (i) if $\frac{1}{\lambda} \leq s$, then skiers should rent their skis forever and (ii) otherwise should purchase them after renting approximately $s^2 \lambda$ $(< s)$ times. Thus average-case competitive analyses give us the result which differs from the worst-case competitive analysis and also differs from the traditional average cost analysis. Other distributions and related problems are also discussed.

1 Introduction

Suppose that the costs of renting and purchasing skis are \$100 (per ski-tour) and \$1,000, respectively. Then the well-known result for the ski-rental problem says that skiers should rent skis for the first nine $(= (1000 - 100)/100)$ times and should purchase them when going to the tenth. This strategy gives us an optimal competitive ratio $(= 1.9$ for this price setting) [Kar92]. In practice, however, many skiers seem to buy their skis *earlier* than this optimal point of time or to *keep renting* them forever. In this paper, we show that this common behavior of skiers is quite reasonable from a theoretical point of view, using *average-case competitive analyses*. Average-case analyses for the competitive ratio of online problems have rarely appeared in the literature, although they are quite popular in other areas of algorithms. The reason is probably as follows: An online problem can be regarded as a game between an online player and an adversary. The adversary selects the input which attacks weakest points of the online player's strategy. To make this possible, the adversary must have the freedom of selecting inputs without any restriction. Note that average-case analyses assume some input distribution, which is public to the online player. The adversary of course has to follow the distribution and this can be a significant restriction against the adversary's freedom, or even destroy the essence of online games. This seems to be a common perception of most researchers who have had a negative attitude against such a model.

P. Bose and P. Morin (Eds.): ISAAC 2002, LNCS 2518, pp. 476–488, 2002.

1.1   Our Contributions

In this paper, we reconsider this common perception, and claim that average-case competitive analyses occasionally provide us with even more interesting results compared to the conventional worst-case analysis. Our problem here is the ski-rental problem already mentioned. Let X be a random variable for the total number of times the skier goes skiing. We assume an exponential distribution $f(t) = \lambda e^{-\lambda t}$, $\lambda > 0$, for $\Pr(X = t)$. Note that this distribution is equivalent to the following natural model: At each occasion, the skier goes skiing with probability $p = 1 - \lambda$ and quits skiing with probability $1 - p$.

Our results are summarized as follows: Let s be the ratio of the cost for purchasing skis over the cost for renting them.

(A) Both strategies, namely (i) to buy skis after renting them some constant times and (ii) to keep renting them forever, can be optimal depending on the value of λ. Note that case (ii) can never be optimal in the worst-case analysis and case (i) can never be optimal in the average *cost* analysis as described in Section 5.

(B) Case (ii) becomes optimal if $s \geq \frac{1}{\lambda}$ (= the average number of ski tours the skier makes).

(C) Otherwise, i.e. if $s < \frac{1}{\lambda}$, the optimal point of time for buying skis is approximately after renting them $s^2\lambda$ times. (For example, suppose that $s\lambda = \frac{1}{3}$. Then the skier should buy skis after renting them $\frac{1}{3}s$ times.) Thus the optimal point of time for buying skis is shifted to front compared to the worst-case analysis where the skier should buy skis after renting them $s - 1$ times.

Possible merits and demerits of our approach are as follows:

(1) Although it is true that (worst-case) competitive analyses have provided us with a lot of beautiful results which would have been impossible without this new measure, they do have limitations. For example, different algorithms such as LRU, CLOCK, FWF and FIFO for paging have all the same competitive ratio, although people have experienced a lot of difference among their empirical performances [BE98]. Average-case competitive analyses, as well as other attempts including [CN99], might help to overcome these difficulties.

(2) Unfortunately it is usually hard to estimate the input distribution. This is not so serious for the ski-rental problem since its input structure is very simple, but apparently many combinatorial problems have more complicated input structures. Nevertheless, there do exist a good number of interesting problems whose input structure might be a bit more complicated than ski-rental but is still tractable. They include the TCP acknowledgment problem [DGS98], the Bahncard problem [Fle01], and the currency conversion problem [EFKT92]. In this paper, we shall give a brief discussion on how to extend our approach to these problems.

(3) Competitive ratio sometimes seems peculiar. In ski-rental, for example, the always-rent algorithm pays $100 and the immediately-buy algorithm $1,000 if the skier goes skiing only once. If the skier goes 100 times, the former pays $10,000 while the latter $1,000. In both cases the ratio is the same 10 but there is a huge difference in the absolute cost. This might lead us to consider that the absolute cost is a better measure than the competitive ratio and the former

should be used whenever possible. But there is also an opposite viewpoint: Just compare the costs of going to ski tours once and nine times. The always-rent algorithm pays $100 and $900; the latter is nine times as much as the former or the latter has nine times as much weight as the former if we use the absolute cost. However, both are equal to the offline optimal cost and hence to give the same competitive ratio (= 1) seems more reasonable. Thus both competitive ratio and absolute cost have their own characteristics. It should also be noted that interesting phenomena appear when the number of ski tours is relatively small, which might make the first argument going to ski tours 100 times less realistic.

1.2 Related Research

First of all, one should not confuse the average competitive ratio and the *average cost*. (Recall that the latter had been a popular measure before competitive analyses were introduced.) Although details are given later, let us take a look at the following simple example. For the same prices as before ($100 for rental and $1,000 for purchase), consider the following two algorithms: (a) Buy skis at the very beginning, and (b) buy them after renting them six times. Our input distribution is somewhat artificial: People go skiing six times in total with probability 0.5 and ten times with the same probability. Then the average competitive ratio is $(1000/600) \times 0.5 + 1.0 \times 0.5 = 1.33$ for algorithm (a) and $1.0 \times 0.5 + (1600/1000) \times 0.5 = 1.3$ for (b). However, the average cost is $1,000 for (a) and $1,100 for (b). Thus we get completely opposite results depending on which measure is used.

Karlin et al. made a great contribution to online analyses for what they call "the ski-rental-family of problems." [KMMO94] gives a randomized online algorithm for the ski-rental problem whose competitive ratio is optimal $e/(e-1)$. Using a similar analysis, they also show that there exists a deterministic online algorithm such that the ratio of its average-cost over the average offline cost is at most $e/(e-1)$ for any input distribution [FW98,KMMO94]. In their recent paper [KKR01], the randomized, optimal $e/(e-1)$ competitive ratio is extended to other problems of the family such as TCP acknowledgment [DGS98] and Bahncard [Fle01] problems.

Koutsoupias and Papadimitriou [KP94] considered a "partial knowledge" of the input distribution for online players. Raghavan [Rag92] also restricted the power of adversaries, whose input has to satisfy some statistical properties. Also, there are models which allow online players to use several powerful tools or relatively restrict the power of adversaries: Randomization used by online players is a most popular one ([BLS92] and many others). Also included in this category are to allow online players to make "forecasts" [Alb97,IY99] and to allow them to maintain several different solutions and to select (the best) one of them at the end of the game [HIMT00]. The access graph model for paging also restricts the power of adversaries (e.g., [BIRS95]).

2   Average-Case Competitive Analyses

The costs of renting and purchasing skis are denoted by 1 and s, respectively. An online algorithm for this problem is determined completely by deciding how many ($= k$) times the skier should rent skis before buying them, and therefore such an algorithm is denoted by $A(k)$. In this paper, we use a continuous model just because of the ease of calculation. It should be noted that similar results are obtained by using an equivalent discrete model and we often use terminology of the discrete model, such as "in each occasion of going to a ski tour." Let $\text{ALG}(k,t)$ and $\text{OPT}(t)$ denote the cost of the online algorithm and the cost of the optimal offline algorithm, respectively, where t is the total number of times the skier goes skiing. Apparently,

$$\text{ALG}(k,t) = \begin{cases} t & : \ 0 \le t \le k, \\ k+s & : \ k < t, \end{cases} \tag{1}$$

$$\text{OPT}(t) = \min(s,t). \tag{2}$$

The worst-case competitive ratio for the algorithm $A(k)$, denoted by $c(k) = \max_t \frac{\text{ALG}(k,t)}{\text{OPT}(t)}$, becomes minimum when $k = s - 1$, and its value at that point is $2 - \frac{1}{s}$. Now let $f(t)$ be a probability density function for the input distribution. Then we can define the average-case competitive ratio as:

$$\tilde{c}(k) = \boldsymbol{E}\left[\frac{\text{ALG}(k,t)}{\text{OPT}(t)}\right] = \int_0^\infty \frac{\text{ALG}(k,t)}{\text{OPT}(t)} \cdot f(t)dt. \tag{3}$$

As the function $f(t)$, we use a so-called exponential distribution defined by

$$f(t) = \lambda e^{-\lambda t} \quad (\lambda > 0). \tag{4}$$

As mentioned before, this distribution means that in each occasion, the skier continues to go skiing with probability $1 - \lambda$ and quits skiing with λ. Note that its mean value $\int_0^\infty t\lambda e^{-\lambda t}dt$ is equal to $\frac{1}{\lambda}$, which shows how many times in total the skier goes skiing on average.

We calculate (3) for two regions $0 < k \le s$ and $s < k$, for which the value of $\tilde{c}(k)$ is denoted by $\tilde{c}_1(k)$ and $\tilde{c}_2(k)$, respectively. For $0 < k \le s$, one can obtain

$$\tilde{c}_1(k) = 1 - e^{-\lambda k} + (k+s)\int_k^s \frac{1}{t}\lambda e^{-\lambda t}dt + \frac{k+s}{s}e^{-\lambda s}$$

$$= 1 - e^{-\lambda k} + \lambda(k+s)(Ei(-\lambda s) - Ei(-\lambda k)) + \frac{k+s}{s}e^{-\lambda s}, \tag{5}$$

where $Ei(-x) = -\int_x^\infty \frac{e^{-t}}{t}dt$, called the exponential integral, cannot be expressed as an elementary function [MUH56]. On the other hand, for $s < k$, we can obtain

$$\tilde{c}_2(k) = 1 + \frac{1}{\lambda s}e^{-\lambda s} - \left(\frac{1}{\lambda s} - 1\right)e^{-\lambda k}. \tag{6}$$

Differentiating $\tilde{c}_1(k)$, we can get its first derived function:

$$\frac{d\tilde{c}_1(k)}{dk} = -\frac{s}{k}\lambda e^{-\lambda k} + \int_k^s \frac{1}{t}\lambda e^{-\lambda t}dt + \frac{1}{s}e^{-\lambda s}$$

$$= -\frac{s}{k}\lambda e^{-\lambda k} + \lambda(Ei(-\lambda s) - Ei(-\lambda k)) + \frac{1}{s}e^{-\lambda s}. \tag{7}$$

Also, for $s < k$,

$$\frac{d\tilde{c}_2(k)}{dk} = \left(\frac{1}{s} - \lambda\right)e^{-\lambda k}. \tag{8}$$

It is also possible to obtain the second derived functions of $\tilde{c}(k)$, which are the following:

$$\frac{d^2\tilde{c}_1(k)}{dk^2} = \lambda e^{-\lambda k}\frac{1}{k^2}\left(\lambda sk + s - k\right), \tag{9}$$

$$\frac{d^2\tilde{c}_2(k)}{dk^2} = \left(\lambda - \frac{1}{s}\right)\lambda e^{-\lambda k}. \tag{10}$$

3   Optimal Online Strategies

Since we are using the continuous model, $\tilde{c}(k)$ diverges to $+\infty$ as k approaches zero. This is not too important because we are interested in the value of $\tilde{c}(k)$ for $k = $ positive integers. Optimal strategies of the online player differ depending on whether $\frac{1}{\lambda} < s$ or $\frac{1}{\lambda} \geq s$. Note that $\frac{1}{\lambda} < s$ means that the average number of total ski-tours is less than the cost of purchase. *Case 1.* $\frac{1}{\lambda} < s$ (see Table 1). In this case the skier should rent skis forever since $\tilde{c}(k)$ decreases monotonically: One can see that $\frac{d^2\tilde{c}_1(k)}{dk^2}$ is always positive for $0 < k < s$ from (9). In addition,

$$\lim_{k \to s-0} \frac{d\tilde{c}_1(k)}{dk} = \left(s - \frac{1}{\lambda}\right)e^{-\lambda s} < 0, \tag{11}$$

which implies that $\frac{d\tilde{c}_1(k)}{dk}$ is negative and $\tilde{c}_1(k)$ decreases monotonically for $0 < k < s$. It is not hard to see that $\tilde{c}_2(k)$ also decreases monotonically since (8) is negative for $s < k$. Note that

$$\lim_{k \to \infty} \tilde{c}_2(k) = 1 + \frac{1}{\lambda s}e^{-\lambda s}, \tag{12}$$

which is the average-case competitive ratio when the skier does not buy skis. Figure 1 shows the value of $\tilde{c}(k)$ for $s = 10$. The curve for $\lambda = 0.2$ illustrates the current case. As one can see $\tilde{c}(k)$ decreases monotonically and converges to 1.0676676 when $k \to \infty$. Thus the skier following the optimal strategy suffers from only 7% more than the offline cost.

Case 2. $\frac{1}{\lambda} = s$ (see Table 2). In this case the skier should rent skis for the first s times since $\tilde{c}(k)$ decreases monotonically for $0 < k < s$ and is flat for $s < k$ (calculation is just the same as *Case 1*). Note that the skier can buy skis at any

Table 1. *Case 1.* $\frac{1}{\lambda} < s$

k	0	$\cdots$	s	$\cdots$	$+\infty$
$\tilde{c}'(k)$		$-$		$-$	
$\tilde{c}(k)$	$+\infty$	$\searrow$	$1 + e^{-\lambda s}$	$\searrow$	$1 + \frac{1}{\lambda s}e^{-\lambda s}$

Table 2. *Case 2.* $\frac{1}{\lambda} = s$

k	0	$\cdots$	s	$\cdots$	$+\infty$
$\tilde{c}'(k)$		$-$		0	
$\tilde{c}(k)$	$+\infty$	$\searrow$	$1 + e^{-1}$	$\rightarrow$	$1 + e^{-1}$

moment after renting s times and it is also equivalently nice to keep renting skis forever.

Case 3. $\frac{1}{\lambda} > s$ (see Table 3). In this case the skier should buy skis after renting them a certain number of times. As is shown below, $\tilde{c}(k)$ has a minimum peak. Namely, such a unique k, which satisfies $\frac{d\tilde{c}_1(k)}{dk} = 0$, exists between 0 and s. Note that $\tilde{c}_2(k)$ increases monotonically for $s < k$. As mentioned previously, $\frac{d^2\tilde{c}_1(k)}{dk^2}$ is always positive for $0 < k < s$. Then one can see that the minimum peak exists between 0 and s by the following two facts, i.e.,

$$\lim_{k \to s-0} \frac{d\tilde{c}_1(k)}{dk} = \left(s - \frac{1}{\lambda}\right)e^{-\lambda s} > 0, \tag{13}$$

and

$$\lim_{k \to +0} \frac{d\tilde{c}_1(k)}{dk} = -\infty. \tag{14}$$

The reason why (14) holds is the following: By substituting $\frac{1}{t} \le -\frac{1}{sk}(t - s) + \frac{1}{s}$ $(k \le t \le s)$ to the equation of $\frac{d\tilde{c}_1(k)}{dk}$, we can derive

$$\frac{d\tilde{c}_1(k)}{dk} < \frac{1}{\lambda sk}\left\{\lambda s(1 - \lambda s)e^{-\lambda k} + (1 - \lambda k)(e^{-\lambda s} - e^{-\lambda k})\right\}. \tag{15}$$

Setting $k = 0$ in the term surrounded by the brackets in the right-hand side yields

$$e^{-\lambda s} - 1 + \lambda s - (\lambda s)^2. \tag{16}$$

Regarding this equation as a function of λs, it is easy to see that (16) is always finite and negative. Furthermore the right-hand side of (15) is continuous with respect to k and therefore diverges to $-\infty$ as $k \to 0$. Thus, (14) holds (see Table 3). On the other hand, for $s < k$, one can easily see that $\tilde{c}_2(k)$ increases monotonically since (8)> 0. This *Case 3* is illustrated in Fig. 1 as the curves for $\lambda = 0.05$ and $\lambda = 0.03$. One can find a minimum peak for each curve the value of which is much smaller than 1.9 of the worst-case analysis. The value of $\tilde{c}(k)$ for $k = k_0$, namely an optimal competitive ratio, is written as:

Table 3. *Case 3.* $\frac{1}{\lambda} > s$

k	0	$\cdots$	k_0	$\cdots$	s	$\cdots$	$+\infty$
$\tilde{c}'(k)$			$-$		$+$		$+$
$\tilde{c}(k)$	$+\infty$	$\searrow$	$\tilde{c}_1(k_0)$	$\nearrow$	$1 + e^{-\lambda s}$	$\nearrow$	$+\infty$

$$\tilde{c}(k_0) = 1 - \left(1 - \lambda s - \frac{\lambda s^2}{k_0}\right) e^{-\lambda k_0}. \tag{17}$$

Case 4. $\frac{1}{\lambda} \to \infty$. In this case the skier should buy skis at the beginning since $\tilde{c}(k)$ increases monotonically: It turns out that $\tilde{c}(k) \longrightarrow 1 + \frac{k}{s}$ as $\lambda \to 0$ for all range of k. This is obviously true for $0 < k \le s$. For $s < k$, applying the Mean Value Theorem to $\tilde{c}_2(k) = 1 + \frac{1}{\lambda s}\left(e^{-\lambda s} - e^{-\lambda k}\right) + e^{-\lambda k}$, we can conclude that $\tilde{c}_2(k) \longrightarrow 1 + \frac{k}{s}$, namely converging to a linear function of k.

4   Optimal Timing for Purchasing Skis

We consider *Case 3* (i.e. $\frac{1}{\lambda} > s$) in more detail to see the value of k_0 at which the average competitive ratio $\tilde{c}(k)$ becomes minimum. The inequality $k_0 < s$ means that the optimal point of time for purchasing skis is shifted to the front compared to the worst-case. Recall that k_0 is the solution of the equation (7)= 0, which seems to be hard to solve analytically. Let a be defined as $a = s\lambda$ $(0 < a < 1)$ which is the ratio of λ over $\frac{1}{s}$. Figure 2 is the result of numerical calculation using the Newton Method, which suggests the relation between a and k_0. As for upper and lower bounds for k_0, they satisfy the following inequalities (see Fig. 2 where these lower and upper bounds are shown by dotted lines):

$$s\left\{1 - \frac{1}{a} + \sqrt{\left(1 - \frac{1}{a}\right)^2 + 1}\right\} < k_0 < sa. \tag{18}$$

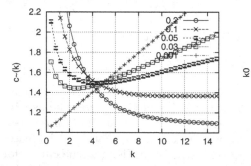
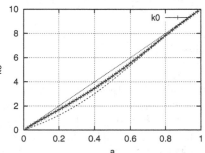

Fig. 1. Average-case CR $\tilde{c}(k)$ ($s = 10$, $\lambda = 0.2, 0.1, 0.05, 0.03, 0.001$)

Fig. 2. Value of k_0 and its upper and lower bounds

We prove the upper bound first. Applying $\frac{1}{t} \geq \frac{1}{s}$ $(s \leq t \leq k)$ to the definite integral term of $\frac{d\tilde{c}_1(k)}{dk}$, we have

$$\frac{d\tilde{c}_1(k)}{dk} > -\frac{s}{k}\lambda e^{-\lambda k} + \int_k^s \frac{1}{s}\lambda e^{-\lambda t}dt + \frac{1}{s}e^{-\lambda s} = \left(\frac{1}{s} - \frac{\lambda s}{k}\right)e^{-\lambda k}. \qquad (19)$$

The last expression becomes zero when $k = s^2\lambda = sa$, which means that $k_0 < sa$ holds for the solution k_0 satisfying $\frac{d\tilde{c}_1(k)}{dk} = 0$, since $\frac{d\tilde{c}_1(k)}{dk}$ increases monotonically as mentioned earlier.

We next prove the lower bound: By using $\frac{1}{t} \leq -\frac{1}{sk}(t-s) + \frac{1}{s}$ $(k \leq t \leq s)$, we have

$$\frac{d\tilde{c}_1(k)}{dk} < \left(\frac{1}{s} - \frac{\lambda s}{k}\right)e^{-\lambda k} + \frac{1}{sk}\int_k^s (s-t)\lambda e^{-\lambda t}dt. \qquad (20)$$

We furthermore use $e^{-\lambda t} < e^{-\lambda k}$ for the definite integral term, which implies

$$\frac{d\tilde{c}_1(k)}{dk} < \frac{\lambda}{2sk}e^{-\lambda k}\left\{k^2 - 2\left(s - \frac{1}{\lambda}\right)k - s^2\right\}. \qquad (21)$$

The last expression yields zero when $k = s - \frac{1}{\lambda} + \sqrt{\left(s - \frac{1}{\lambda}\right)^2 + s^2}$. Thus we can conclude

$$s\left\{1 - \frac{1}{a} + \sqrt{\left(1 - \frac{1}{a}\right)^2 + 1}\right\} < k_0$$ for the same reason as before.

Let us calculate the difference of the upper and lower bounds. Let

$$g(a) = sa - s\left\{1 - \frac{1}{a} + \sqrt{\left(1 - \frac{1}{a}\right)^2 + 1}\right\}. \qquad (22)$$

Then $\frac{dg(a)}{da}$ becomes zero when $a \approx 0.366$, which means the difference becomes maximum there. Since $g(0.366) < 0.099s$, we finally obtain the following theorem.

Theorem 1. *The following strategy provides an optimal average-case competitive ratio for the exponential input distribution $f(t) = \lambda e^{-\lambda t}$: (i) If $\frac{1}{\lambda} \leq s$, then the skier should rent their skis forever. (ii) Otherwise, the skier should purchase their skis after renting k_0 times, where k_0 satisfies $s^2\lambda - \frac{s}{10} < k_0 < s^2\lambda$. Optimal competitive ratios are given by (13) and (18) for cases (i) and (ii), respectively.*

5 Average Cost for Ski-Rental Problem

5-3mm As mentioned in the first section, the average-case competitive ratio is different from the *average cost*. For the same problem and the same input distribution, the average cost $m(k)$ of the algorithm $A(k)$ can be written as:

$$m(k) = \int_0^\infty \mathrm{ALG}(k,t) \cdot f(t)dt = \left(s - \frac{1}{\lambda}\right)e^{-\lambda k} + \frac{1}{\lambda}. \qquad (23)$$

This average cost again shows different properties depending on λ. (i) If $\frac{1}{\lambda} < s$, then $m(k)$ decreases monotonically, (ii) if $\frac{1}{\lambda} = s$, then $m(k)$ is constant, and (iii) if $\frac{1}{\lambda} > s$, then $m(k)$ increases monotonically (see Fig. 3 for $s = 10$ and $\lambda = 0.2, 0.1, 0.05$). Namely, the best strategy of skiers is either to rent their skis forever or to buy them at the very beginning. Thus the strategy of buying skis after a certain times of renting them, which is a heart of competitive analysis and is quite interesting in practice also, can no longer be optimal.

6   Other Input Distributions

Recall that the exponential distribution means the skier continues to go skiing with probability $1 - \lambda$ (quits skiing with probability λ) which is the same at each occasion. A natural modification is to select a distribution such that this probability for continuing skiing decreases or increases as time goes. As an example of the decreasing case, let us consider a *uniform distribution*

$$f_M(t) = \begin{cases} \frac{1}{M} & : \quad 0 \leq t \leq M, \\ 0 & : \quad t > M. \end{cases} \tag{24}$$

Namely the maximum number of ski tours is fixed $(= M)$ and the probability that the skier goes skiing i $(\leq M)$ times in total is the same $(= \frac{1}{M})$ for all $i \leq M$. (As one can see, the probability of continuing skiing at each occasion decreases and becomes zero finally.) Average-case competitive ratio $\tilde{c}_u(k)$ is calculated for $\tilde{c}_{u1}(k)$ $(0 < k \leq s)$ and for $\tilde{c}_{u2}(k)$ $(s < k)$ as before. Although details are omitted, we have

$$\tilde{c}_{u1}(k) = 1 - \frac{s}{M} + \frac{k+s}{M}(\ln s - \ln k) + \frac{k}{s}, \tag{25}$$

$$\tilde{c}_{u2}(k) = 1 + \frac{s}{2M} + \frac{M-s}{Ms}k - \frac{1}{2Ms}k^2. \tag{26}$$

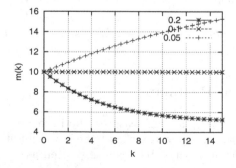

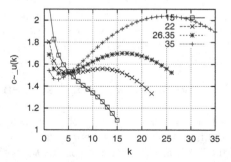

Fig. 3. Average cost $m(k)$

Fig. 4. The average-case CR $\tilde{c}_u(k)$ for the uniform distribution

Figure 4 shows the behavior of $\tilde{c}_u(k)$ for $s = 10$ and $M = 15$, $M = 22$, $M = 26.35$ (in this case, $\tilde{c}_u(k)$ becomes minimum for two different values of k), and $M = 35$. Note that if $M \leq s$ then it is obvious that the skier should not purchase skis. On can see that the general tendency is similar to the case of the exponential distribution: If M is relatively small, then to continue renting skis is better. If M is large, then the skier should purchase skis at some moment, again earlier than the sth ski-tour.

7   Related Problems

7.1   TCP Acknowledgment

Suppose that n packets $P_1, P_2, \cdots, P_n$ arrive at time $a_1, a_2, \cdots, a_n$, respectively, each of which should be acknowledged. However, we do not have to send an acknowledgment for each P_i exactly at time a_i but we can postpone it. If the next packet P_{i+1} arrives while postponing the acknowledgment for P_i, it is enough to send only one acknowledgment packet to acknowledge both P_i and P_{i+1} simultaneously (similarly for three or more packets). As for the cost, we incur (1) a unit cost, called the *acknowledgment cost*, per acknowledging packet and (2) a unit cost, called the *latency cost*, per outstanding packet per unit time. If packet P_i waits 0.7 time units and P_{i+1} 0.4 time units until the acknowledgment packet is sent, for example, then the total cost for these two packets is $0.7 + 0.4 + 1 = 2.1$. (The last 1 is the acknowledgment cost and all the others are the latency cost.) More formally, suppose that k acknowledgments are sent at time $t_1, t_2, \cdots, t_k$. Then the total cost is

$$k + \sum_{1 \leq j \leq k} \text{latency}(j), \tag{27}$$

where

$$\text{latency}(j) = \sum_{i \text{ s.t. } t_{j-1} < a_i \leq t_j} (t_j - a_i). \tag{28}$$

[DGS98] proved that the following natural online algorithm has a competitive ratio of two: The algorithm waits until the latency cost for outstanding packets becomes one, i.e., the same as a unit acknowledgment cost. [KKR01] gave the randomized algorithm which achieves the best possible competitive ratio of $e/(e-1)$. It should be noted that if we know that the number of packets is two and ignore the acknowledgment cost for the second packet (since to acknowledge it immediately is obviously optimal), then the problem becomes exactly the same as the ski-rental problem [DGS98]. Namely, the acknowledgment cost for the first packet corresponds to the ski-purchase cost and its latency cost to the ski-rental cost.

In what follows, we consider the case that the number of packets is three, under the following assumption: (i) The first packet P_1 arrives at time zero. (ii) The second packet P_2 arrives at time t_2 under the distribution that $f(t_2) = \lambda e^{-\lambda t_2}$. (iii) The third packet P_3 arrives at time $t_2 + t_3$ under the same distribution, i.e., $f(t_3) = \lambda e^{-\lambda t_3}$. The final acknowledgment cost for P_3 (and possibly for outstanding P_2 and P_1) is ignored for the same reason as before. (Recall that the two-packet TCP is equivalent to the ski-rental. One can see that the three-packet TCP is equivalent to the *two-person ski-rental* where if the two persons purchase two sets of skis at the same time then they can buy the two sets for the price of one.) Our online algorithm is denoted by $A(k)$, which does not acknowledge until the sum of latency costs for outstanding packets becomes k. (If we set $k = 1$, then, it is the same as [DGS98].) More formally, $A(k)$ operates as follows for our three-packet model: (1) If $t_2 + 2t_3 \leq k$, then $A(k)$ responds for P_1, P_2 and P_3 with a single acknowledgment immediately after P_3 comes and its acknowledgment cost is ignored as mentioned before. (2) If $t_2 + 2t_3 > k$ and $t_2 \leq k$, then $A(k)$ sends two acknowledgments when the latency cost becomes k and when P_3 comes. (3) If $t_2 > k$ and $t_3 \leq k$, then $A(k)$ sends two acknowledgments, one for P_1 and the other for P_2 and P_3. (4) Otherwise (i.e. $t_2 > k$ and $t_3 > k$), $A(k)$ sends three acknowledgments for each packet.

Therefore the cost of $A(k)$, denoted by $\mathrm{ALG}(k, t_2, t_3)$, can be written as follows

$$\mathrm{ALG}(k, t_2, t_3) = \begin{cases} t_2 + 2t_3 & : \ t_2 + 2t_3 \leq k, \\ k+1 & : \ t_2 + 2t_3 > k, \ t_2 \leq k, \\ k+1+t_3 & : \ t_2 > k, \ t_3 \leq k, \\ 2k+2 & : \ t_2 > k, \ t_3 > k. \end{cases} \tag{29}$$

The offline cost $\mathrm{OPT}(t_2, t_3)$ is obtained as

$$\mathrm{OPT}(t_2, t_3) = \min(t_2 + 2t_3, t_2 + 1, t_3 + 1, 2). \tag{30}$$

Now one can see that the average-case competitive ratio $\tilde{c}(k)$ of $A(k)$ can be written as

$$\tilde{c}(k) = \int_0^\infty \int_0^\infty \frac{\mathrm{ALG}(k, t_2, t_3)}{\mathrm{OPT}(t_2, t_3)} f(t_2) f(t_3) dt_2 dt_3. \tag{31}$$

Figure 5 shows the relation between $\tilde{c}(k)$ and k for $\frac{1}{\lambda} = 0.5$, 1 and 5. (Recall that $\frac{1}{\lambda}$ is equal to the average interval of packet arrivals.) If $\frac{1}{\lambda}$ is small, i.e., each packet arrives with short interval, then k should be large, or it is the best to send only one acknowledgment when P_3 comes. If $\frac{1}{\lambda}$ is large, i.e., each packet arrives

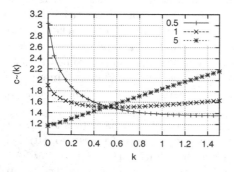

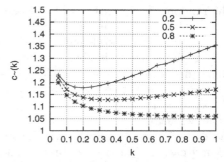

Fig. 5. Average-case CR for TCP acknowledgment

Fig. 6. Average-case CR for Bahncard problem

with long interval, then it becomes better to acknowledge each immediately. In between (e.g., when $\frac{1}{\lambda} = 1.0$ in the figure), there is an optimal value k_0 for k and $\tilde{c}(k_0)$ is much smaller than the worst-case competitive ratio ($= 2.0$).

7.2 Bahncard Problem

If the online player buys a Bahncard with cost C, then subsequently, he/she can buy railway tickets for reduced prices, i.e., β (< 1) times regular prices. The algorithm is determined by the parameter k such that the player buys a Bahncard after he/she has spent a cost of $kC/(1 - \beta)$ for purchasing tickets of regular prices. (Here we assume that Bahncards never expire.) It is known [Fle01] that the optimal algorithm for the worst case is to set $k = 1.0$, which achieves a CR of $2 - \beta$. Average-case analysis uses $f(t) = \lambda e^{-\lambda t}$ for the distribution of the total regular prices of tickets to be purchased. The result is very similar to ski-rental, which is illustrated in Fig. 6. When $\beta = 0.5$, $C = 10$ and $\lambda = 0.01$, the average-case competitive ratio becomes minimum when $k = 0.35$, i.e., Bahncards should be purchased much earlier than the worst-case strategy.

7.3 One-Way Currency Trading

The trader (online player) initially has one dollar and gradually exchanges it to another currency, say yen. In the one-way model, we assume that the exchange rate (= how much yen can be bought for one dollar) is monotonically increasing. However, the rate suddenly drops to the minimum at some unknown point and the game ends at that moment. The trader has to change all the remaining dollar at this minimum rate. The goal is to obtain as much yen as possible when the game ends. A well-known strategy for this game is called *threat-based* [EFKT92], which guarantees the same competitive ratio c, whenever the rate drops. If the distribution $f(t)$ for the time t when the rate drops is known, then we can obviously have more benefit by adopting our trading strategy to the distribution. Details are omitted in this paper, but the improvement of competitive ratio seems moderate compared to the other problems discussed so far. (For example, the ratio 1.92 for threat-based strategy is reduced to 1.86 for the average-case analysis.)

8 Concluding Remarks

Recall that the ski-rental problem has the following structure: (i) Its input is given as a sequence of opportunities for ski tours (as a sequence of 1's formally) whose total number is unknown. (ii) The competitive ratio is relatively independent of the absolute cost (i.e., the CR can be large when the absolute cost is small). If an online problem has such a structure, then the average-case competitive analyses appear to give us some interesting knowledge about optimal strategies. In other words, if (i) is not met, then the analysis becomes hard, and if (ii) is not met, then the result will be similar to the average-cost analysis which is usually easier. The difficulty due to (i) might be bypassed by using, for instance, numerical analyses and/or simulation. Therefore it might be more important, for the future research, to examine online problems from the second point of view, i.e., to study the independency between the competitive ratio and absolute costs.

Acknowledgments. We are grateful to Yasuo Okabe and Gerhard Woeginger for their useful comments.

References

[Alb97] S. al-Binali, "The competitive analysis of risk taking with applications to online trading", *Proc. 38th IEEE FOCS*, pp. 336–344, 1997.
[BE98] A. Borodin and R. El-Yaniv, "Online computation and competitive analysis", Cambridge University Press, 1998.
[BIRS95] A. Borodin, S. Irani, P. Raghavan and B. Schieber, "Competitive paging with locality of reference", *J. Comput. Sys. Sci.*, 50, pp. 244–258, 1995.
[BLS92] A. Borodin, N. Linial and M. Saks, "An optimal online algorithms for metric task systems", *J. of the ACM*, pp. 745–763, 1992.
[CN99] M. Chrobak and J. Noga, "LRU is better than FIFO", *Algorithmica*, Vol.23, No.2, 1999.
[DGS98] D. R. Dooly, S. A. Goldman and S. D. Scott, "TCP Dynamic Acknowledgment Delay: Theory and Practice (Extended Abstract)", *Proc. STOC '98*, pp. 389–398, 1998.
[EFKT92] R. El-Yaniv, A. Fiat, R. M. Karp and G. Turpin, "Competitive analysis of financial games", *Proc. 33rd IEEE FOCS*, pp. 327–333, 1992.
[FW98] A. Fiat and G. J. Woeginger (Eds.), "Online Algorithms", Springer, Chap.16, 1998.
[Fle01] R. Fleischer, "On The Bahncard Problem", *Proc. TCS'01*, pp. 161–174, 2001.
[HIMT00] M. Halldorsson, K. Iwama, S. Miyazaki and S. Taketomi, "Online Independent Sets", *Proc. COCOON'00*, pp. 202–209, 2000.
[IY99] K. Iwama and K. Yonezawa, "Using Generalized Forecasts for Online Currency Conversion", *Proc. COCOON'99*, pp. 409–421, 1999.
[KMMO94] A. R. Karlin, M. S. Manasse, L. McGeogh and S. Owicki, "Competitive Randomize Algorithms for Nonuniform Problems", *Algorithmica*, Vol.11, No.1, January, 1994.
[KKR01] A. R. Karlin, C. Kenyon and D. Randall, "Dynamic TCP Acknowledgement and Other Stories about $e/(e-1)$", *Proc. STOC '01*, pp. 502–509, 2001.
[Kar92] R. Karp, "On-line algorithms versus off-line algorithms: How Much is it Worth to Know the Future?", *Proc. IFIP 12th World Computer Congress*, Vol.1, pp. 416–429, 1992.
[KP94] E. Koutsoupias and C. Papadimitriou, "Beyond competitive analysis," *Proc. 35th IEEE FOCS*, pp. 394–400, 1994.
[MUH56] S. Moriguchi, K. Udagawa and S. Hitotsumatsu, "Mathematics Formulas I", Iwanami Shoten, Publishers, 1956.
[Rag92] P. Raghavan, "A statistical adversary for on-line algorithms," *DIMACS Series in Discrete Mathematics and Theoretical Computer Science*, 7, pp. 79–83, 1992.

On the Clique Problem in Intersection Graphs of Ellipses

Christoph Ambühl and Uli Wagner

Institut für Theoretische Informatik, ETH Zürich
ETH Zentrum, 8092 Zürich, Switzerland
{ambuehl,uli}@inf.ethz.ch

Abstract. Intersection graphs of disks and of line segments, respectively, have
been well studied, because of both, practical applications and theoretically interesting properties of these graphs. Despite partial results, the complexity status of
the CLIQUE problem for these two graph classes is still open.

Here, we consider the CLIQUE problem for intersection graphs of ellipses which
in a sense, interpolate between disc and ellipses, and show that it is $\mathcal{APX}$-hard in
that case. Moreover, this holds even if for all ellipses, the ratio of the larger over
the smaller radius is some prescribed number.

To our knowledge, this is the first hardness result for the CLIQUE problem in
intersection graphs of objects with finite description complexity. We also describe
a simple approximation algorithm for the case of ellipses for which the ratio of
radii is bounded.

1 Introduction

Let $\mathcal{M}$ be a collection of sets. The *intersection graph* of $\mathcal{M}$ is the abstract graph G whose
vertices are the sets in $\mathcal{M}$, and two vertices are connected by an edge if the corresponding
sets intersect; formally,

$$V(G) = \mathcal{M} \text{ and } E(G) = \{\{M, N\} \subseteq \mathcal{M} : M \cap N \neq \emptyset\}.$$

The family $\mathcal{M}$ is called a *representation* of the graph G.

Intersection graphs of various classes of geometric objects have been studied, because of both, practical applications and interesting structural properties of the graphs in
question. Two prominent examples that have received a lot of attention are intersection
graphs of disks (see [15,6]) and of line segments (see [14,11]), respectively.

For instance, intersection graphs of disks, *disk graphs* for short, arise naturally when
studying interference in networks of radio or mobile phone transmitters [1].

Many of these graphs are hard to recognize. For example, recognizing unit disk
graphs is $\mathcal{NP}$-hard [7,10]. Recognizing general disk graphs might be even harder. Only
$\mathcal{PSPACE}$-membership is known [7]. On the other hand, disk contact graphs can be
recognized in linear time, since this class coincides with the class of planar graphs [12].

One reason to study intersection graphs is the hope that they provide classes of graphs
for which optimization problems which are hard for general graphs become tractable. As
an example, CLIQUE is polynomially solvable in unit disk graphs [5]. Since recognition

P. Bose and P. Morin (Eds.): ISAAC 2002, LNCS 2518, pp. 489–500, 2002.

is hard for many of these classes, usually the geometric representation has to be provided in the input.

Even if a problem remains $\mathcal{NP}$-hard in a certain graph class, using its structure might lead to better approximation algorithms or even allow a PTAS, such as for INDEPENDENT SET and VERTEX COVER in the case of disk graphs [9].

In this article, we consider the CLIQUE problem, i.e., the problem of finding a maximal complete subgraph. Its complexity status is unknown for both disk graphs and intersection graphs of line segments.

We do not resolve either of these questions. We consider intersection graphs of ellipses (which contain both of the above classes) and show that the CLIQUE problem is $\mathcal{APX}$-hard in that case. That is, unless $\mathcal{P} = \mathcal{NP}$, there is a constant c such that there is no approximation algorithm with ratio better than c. Hence there is no PTAS. What is more, this is true even if all the ellipses are required to be arbitrarily "round" (or circle-like) or arbitrarily "stretched" (or segment-like). More precisely, given $1 < \rho < \infty$, let ELLIPSE$_\rho$CLIQUE, respectively ELLIPSE$_{\leq \rho}$CLIQUE, be the CLIQUE problem in intersection graphs of ellipses whose ratio of the larger over the smaller radius is exactly ρ, respectively at most ρ.

Theorem 1 *For every $\rho > 1$, the problem* ELLIPSE$_\rho$CLIQUE *is $\mathcal{APX}$-hard.*

This theorem is proved in Section 2 by a reduction from MAX5OCC2SAT, which is the following optimization problem: Given a Boolean formula φ in conjunctive normal form with at most two literals per clause and at most five occurrences of every variable, find an assignment of truth values to the variables that satisfies the maximum number of clauses. MAX5OCC2SATis known to be $\mathcal{APX}$-hard [4].

We would like to stress that the inapproximability ratio in Theorem 1 is independent of the parameter ρ, so it does not matter how close our ellipses are to the "limit cases" $\rho = 1$ (corresponding to circles) or $\rho = \infty$ (corresponding to segments, or to parabolas).

We note that Theorem 1 strengthens a result of Kratochvíl and Kuběna [13], who proved that the CLIQUE problem is $\mathcal{NP}$-complete for intersection graphs of general (compact) convex subsets of the plane. (In fact, they proved a stronger result, namely that every co-planar graph has an (efficiently computable) representation as the intersection graph of some family of convex sets in the plane.) The interesting aspect here is that the proof of Kratochvíl and Kuběna relies in an essential way on the fact that the boundary of convex sets has non-constant description complexity — in technical terms, that convex sets have infinite *VC dimension*. Ellipses, by contrast, have VC dimension 5.

Moreover, if the ratio of radii is bounded, ellipses also have a finite *transversal number*. That is, for every $\rho \geq 1$, there is a number $\tau(\rho) \in \mathbb{N}$ such that, for every family $\mathcal{C}$ of pairwise intersecting ellipses which ratio of radii at most ρ, there is some set S of at most $\tau(\rho)$ points which *pierce* $\mathcal{C}$ in the sense that every $L \in \mathcal{C}$ contains some point $p \in S$. In Section 3, we exploit this to give an approximation algorithm for ELLIPSE$_{\leq \rho}$CLIQUE.

Theorem 2 *For every $1 < \rho < \infty$, the problem* ELLIPSE$_{\leq \rho}$CLIQUE *can be approximated within a factor of* $\min\{9\rho^2, \tau(\rho)/2\}$. *(this also applies when we consider ellipses with their interiors). For* DISKCLIQUE, *the approximation factor can be improved to 2.*

2   Reduction from MAX5OCC2SAT to ELLIPSE$_\rho$CLIQUE

We first recall some facts about ellipses. An ellipse is an affine transformation of the unit circle. That is,

$$E = R \begin{bmatrix} r & 0 \\ 0 & s \end{bmatrix} K + a, \tag{1}$$

where K is the unit circle centered at the origin, R is an orthogonal 2×2 matrix, r, s are positive real numbers, and $a \in \mathbf{R}^2$ (the *center* of E). Then E can also be written as the zero set of a quadratic bivariate polynomial,

$$E = E(A, a) = \{x \in \mathbf{R}^2 : (x - a)^T A(x - a) = 1\}, \tag{2}$$

where "$\cdot^T$" denotes the transpose, and $A = R \begin{bmatrix} 1/r^2 & 0 \\ 0 & 1/s^2 \end{bmatrix} R^T$ is a positive definite symmetric 2×2-matrix (observe that $R^T = R^{-1}$). Thus, A has positive real eigenvalues $\lambda = 1/r^2, \mu = 1/s^2$; in other words, $1/\sqrt{\lambda}$ and $1/\sqrt{\mu}$ are the radii of E.

For computational purposes, we will assume that in instances of ELLIPSE$_\rho$CLIQUE, the ellipses are specified as in (2) with rational coefficients $a \in \mathbf{Q}^2$ and $A \in \mathbf{Q}^{2 \times 2}$.

For the reduction to be polynomial, we also need to ensure that the numbers involved stay polynomial in size. In fact, we will describe a construction involving small algebraic numbers. To complete the reduction, we invoke certain perturbation arguments, which we sketch at the end of this section.

We now start with the description of the reduction. Fix $\rho > 1$ and suppose we are given a formula φ in the variables $x_1, \ldots, x_n$. We begin by introducing ellipses representing the variables and their negations, respectively, in Section 2.1. In Section 2.2, we prove the existence of suitable ellipses which will represent the clauses. In Section 2.3, we combine these building blocks to prove Theorem 1.

2.1   Ellipses Representing the Literals

We introduce ellipses representing the variables and their negations, respectively. We start out with two auxiliary concentric circles C_1 and C_0 of radius r (to be chosen presently) and 1, respectively, with common center c.

Let L be an ellipse with radii $r - 1$ and $\rho(r - 1)$. We place congruent copies $L_1, \ldots, L_{2n}$ of L along the outer circle C_1 such that their centers lie on C_1 and form the vertices of a regular $2n$-gon (with numbering in counterclockwise order), and such that, for each L_i, the main axis corresponding to the radius $r - 1$ is perpendicular to the circle C_1. Thus, each ellipse L_i touches the inner auxiliary circle C_0 in a point p_i.

By choosing r sufficiently large, we may assume that these ellipses pairwise intersect, except for pairs L_i, L_{i+n} of antipodal ellipses, which are disjoint (see Figure 1 for an example with $n = 4$). One can prove that $r = O(n^2)$ is sufficient.

For a literal ξ, let $L(\xi)$ be the ellipse L_i, if ξ is a variable x_i, and L_{i+n} if ξ is a negated variable $\neg x_i$. These ellipses will be called the *literalipses*.

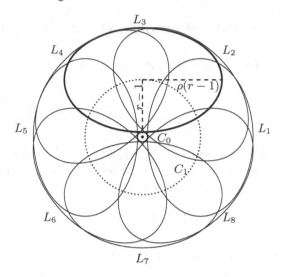

Fig. 1. Rosette of literalipses.

2.2 Ellipses Representing the Clauses

The second building block of our reduction are ellipses which avoid two prescribed literalipses but intersect all others. These are used to represent the clauses of φ, as will be described in Section 2.3.

Lemma 1. *Let $\rho > 1$. For any two literalipses $L(\xi)$ and $L(\omega)$, there is a clause ellipse $E = E(\xi, \omega)$ whose ratio of radii is ρ and which intersects all literal disks except $L(\xi)$ and $L(\omega)$. Moreover, all these clause ellipses intersect one another.*

Note that the lemma also holds if only one literalipse needs to be avoided. The proof of Lemma 1 is based on the upcoming, rather technical, Lemma 2, which is proved in the Appendix. We begin by introducing some notation.

Consider an ellipse L and two points a, b on L. By $(a, b)_L$, we denote the open arc of L that lies to the right of the oriented line $\overrightarrow{ab}$ through a and b.

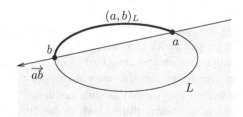

Lemma 2. *Consider a circle K with center c, and four points a, b, a', b' in counter-clockwise order on K such that the arcs $(a, b)_K$ and $(a', b')_K$ are of the same length and disjoint. Let p be the point where the lines ab and $a'b'$ intersect, and let ℓ be the line through p and c (if ab and $a'b'$ are parallel, we take ℓ to be that line through c which is parallel to both of them).*

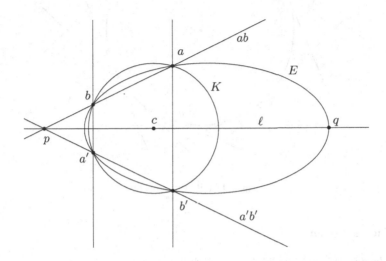

Then, if q is any point on ℓ such that the segment $[p, q]$ intersects K twice, there is a unique ellipse E through the five points a, b, a', b', and q. Moreover, if we move q away from p towards infinity on ℓ, the ratio of radii of E grows monotonically and tends to ∞.

Furthermore, the arcs $(a, b)_E$ and $(a', b')_E$ are completely contained in the interior of K, and the arcs $(b, a')_E$ and $(b', a)_E$ are contained in the intersection of the open half-planes to the left of $\overrightarrow{ab}$ and to the left of $\overrightarrow{a'b'}$. On the other hand, the arcs $(b, a')_K$ and $(b', a)_K$ of K are contained in the interior of E.

Proof (Proof of Lemma 1, using Lemma 2.). Suppose $L(\xi) = L_i$ and $L(\omega) = L_j$. Let $p_{i-1}, p_i, p_{i+1}, p_{j-1}, p_j$, and p_{j+1} be the points at which $L_{i-1}, L_i, L_{i+1}, L_{j-1}, L_j$, and L_{j+1}, respectively, touch the inner circle C_0. Let a_i be the midpoint of the arc $(p_{i-1}, p_i)_{C_0}$ and let b_i be the midpoint of the arc $(p_i, p_{i+1})_{C_0}$. The points a_j and b_j are defined analogously. Finally, let q' be the midpoint of the arc $(b_j, a_i)_{C_0}$. (See Figure 2.) Consider the point $q = q' + t(q' - c)$ for a parameter $t \geq 0$. Let E be the ellipse through a_i, b_i, a_j, b_j, and q whose existence is guaranteed by the preceding lemma. By the containment properties asserted above, E avoids L_i and L_j but intersects all L_l, $l \neq i, j$. Moreover, the ratio of radii of E depends continuously on the parameter t and tends to infinity with t. Thus, since we have ratio 1 for $t = 0$, we can achieve any prescribed ratio. Therefore, E is as advertised.

It is easy to see that all clause ellipses intersect each other since all of them contain point c.

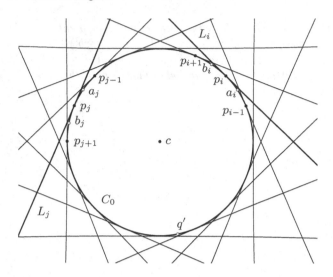

Fig. 2. Detail of the rosette.

2.3 The Reduction

We are now ready to complete the reduction: Fix $\rho > 1$.

Given a Max5Occ2SATformula φ with n variables and m clauses, we construct a collection $\mathcal{L} = \mathcal{L}(\varphi)$ of $14n + 3m$ ellipses, as follows.

1. For each variable x that occurs in φ, we take 7 copies of the ellipse $L(x)$ and 7 copies of the ellipse $L(\neg x)$. These literalipses are arranged into a rosette as described in Section 2.1. We stress that the auxiliary circles C_0 and C_1 are not part of $\mathcal{L}$.
2. For each clause $\kappa = \xi \vee \omega$ of φ, we take the three ellipses $E(\neg\xi, \neg\omega), E(\neg\xi, \omega)$ and $E(\xi, \neg\omega)$ according to Lemma 1. If a clause contains only a single literal ξ, take clause ellipse $E(\neg\xi)$. If there are several clauses $\kappa_1, \ldots, \kappa_l$ that require an ellipse E in this fashion, we take the corresponding number of copies $E_{\kappa_1}, \ldots, E_{\kappa_l}$ of E.

It remains to verify that we have indeed reduced Max5Occ2SAT to the problem Ellipse$_\rho$Clique. This is established by the following:

Lemma 3. *Let φ be an instance of* Max5Occ2SAT *with n variables and m clauses, and let $\mathcal{L}$ be the corresponding* Ellipse$_\rho$Clique *instance just defined. $\mathcal{L}$ contains a clique of size $7n + k$ if and only if there is an assignment of truth values to the variables of φ that satisfies k clauses of φ.*

Proof. We first show how to find a corresponding clique for a given assignment. Fix an assignment $\mathcal{A}$ of truth values. For each literal ξ that is made TRUE by $\mathcal{A}$, take all 7 copies of $L(\xi)$. These form a clique. Moreover, a clause $\xi \vee \omega$ of φ is satisfied by the assignment if and only if one of the following three cases occurs:

1. $\xi = \text{TRUE}$ and $\omega = \text{TRUE}$

2. ξ = TRUE and ω = FALSE
3. ξ = FALSE and ω = TRUE

In the first case, we have already taken 7 copies of $L(\xi)$ and of $L(\omega)$, respectively. Thus, we can enlarge our clique by one element by adding the ellipse $E(\neg\xi, \neg\omega)$ (to be more precise: by adding that copy of it which we have taken into $\mathcal{L}$ on account of the clause $\xi \vee \omega$). We cannot, however, add either of the ellipses $E(\xi, \neg\omega)$ or $E(\neg\xi, \omega)$, which avoid ξ and ω, respectively. The other two cases are treated analogously. Altogether, the clique thus constructed contains $7n$ literal disks (7 for each satisfied literal) and k clause ellipses (one for each satisfied clause).

Conversely, let $\mathcal{C}$ be a clique of size $7n + k$, $k \geq 0$. We may assume that for every variable x, $\mathcal{C}$ contains 7 copies of $L(x)$ or 7 copies $L(\neg x)$. For suppose there is a variable x such that $\mathcal{C}$ does not contain a copy of either $L(x)$ or $L(\neg x)$. Let $\kappa_1^+, \ldots, \kappa_a^+$ and $\kappa_1^-, \ldots, \kappa_b^-$ be the clauses of φ in which x, respectively $\neg x$, occur. We have $a + b \leq 5$. Each κ_i^+ yields two clause ellipses in $\mathcal{L}$ that avoid $L(x)$, and one which avoids $L(\neg x)$. Similarly, each κ_j^- yields two ellipses which avoid $L(x)$, and one which avoids $L(\neg x)$. Therefore, $\mathcal{C}(\subseteq \mathcal{L})$ contains at most $3(a + b) = 15$ clause ellipses which avoid either $L(x)$ or $L(\neg x)$. Thus, for some $\xi \in \{x, \neg x\}$, $L(\xi)$ is avoided by at most $15/2$ ellipses from $\mathcal{C}$, hence in fact by at most 7. But then, if we remove these ellipses from $\mathcal{C}$ and replace them by the 7 copies of $L(\xi)$, we do not decrease $|\mathcal{C}|$.

Therefore, w.l.o.g., $\mathcal{C}$ contains $7n$ literalipses. Then, $\mathcal{C}$ induces a truth value assignment in the obvious fashion: Set variable x to TRUE if $\mathcal{C}$ contains (all 7 copies of) $L(x)$, and to FALSE otherwise.

The remaining k elements of $\mathcal{C}$ are clause ellipses. Consider such an ellipse E. There must be a clause κ that caused $E = E_\kappa(\xi, \omega)$ to be included into the ELLIPSE$_\rho$CLIQUE instance. Call κ the *witness clause* of E (κ could be $\neg\xi \vee \neg\omega$, $\neg\xi \vee \omega$, or $\xi \vee \neg\omega$). Now, E avoids $L(\xi)$ and $L(\omega)$, hence $\mathcal{C}$ must contain all copies of $L(\neg\xi)$ and all copies of $L(\neg\omega)$. Therefore, the assignment induced by $\mathcal{C}$ satisfies the witness clause κ of E. Since this holds for all clause ellipses in $\mathcal{C}$, the assignment satisfies at least k clauses of φ (one for each clause ellipse contained in $\mathcal{C}$).

From the above lemma, it is easy to obtain $\mathcal{APX}$-hardness.

Corollary 1 *Let φ be an instance of* MAX5OCC2SAT *consisting of n variables, m clauses and let $\mathcal{L}$ be the corresponding instance of* ELLIPSE$_\rho$CLIQUE. *Let OPT be the maximum number of satisfied clauses of φ by any assignment of the variables and let OPT' be the size of a maximum clique in $\mathcal{L}$, and let $\epsilon > 0$ and $\gamma > 0$ be constants. Then*

$$OPT \geq (1 - \epsilon)m \Longrightarrow OPT' \geq 7n + (1 - \epsilon)m$$
$$OPT < (1 - \epsilon - \gamma)m \Longrightarrow OPT' < 7n + (1 - \epsilon - \gamma)m.$$

Proof. This follows immediately from Lemma 3. We just have to replace k by $(1 - \epsilon)m$ or $(1 - \epsilon - \gamma)m$ respectively.

In a promise problem of MAX5OCC2SAT, we are promised that either at least $(1-\epsilon)m$ clauses or at most $(1-\epsilon-\gamma)m$ clauses are satisfiable, and we are to find out, which of the two cases holds. This problem is NP-hard for sufficiently small values of $\epsilon > 0$ and $\gamma >$

0 (see [4]). Therefore, Lemma 1 implies that the promise problem for ELLIPSE$_\rho$CLIQUE, where we are promised that the maximum clique is either of size at least $7n + (1 - \epsilon)m$ or at most $7n + (1 - \epsilon - \gamma)m$, is NP-hard as well, for sufficiently small values of $\epsilon > 0$ and $\gamma > 0$. Thus, ELLIPSE$_\rho$CLIQUE cannot be approximated with a ratio of

$$\frac{7n + (1 - \epsilon)m}{7n + (1 - \epsilon - \gamma)m} \geq 1 + \frac{\gamma m}{7n + (1 - \epsilon - \gamma)m}$$

$$\geq 1 + \frac{\frac{n}{2}\gamma}{7n + 5n(1 - \epsilon - \gamma)} = 1 + \frac{\gamma}{14 + 10(1 - \epsilon - \gamma)},$$

where we have used that $m/5 \leq n \leq 2m$. We let $\delta := \frac{\gamma}{14+10(1-\epsilon-\gamma)}$. Since $\delta > 0$, we have shown that ELLIPSE$_\rho$CLIQUE cannot be approximated by any polynomial-time approximation algorithm with an approximation ratio of $1 + \delta$. This proves Theorem 1.

2.4 Perturbations

The reduction produces two kinds of ellipses. The ellipses representing the variables are defined by equation (1), where the entries of a and R are of the form $k \cdot \sin(2\pi/n \cdot i)$ and $k \cdot \cos(2\pi/n \cdot i)$ for $i \in \{0, \ldots n - 1\}$ and $k > 1$ integer. Furthermore, we have $r = \rho s$ with integer s.

The literalipses are defined by five points. Four of them are described by trigonometric expressions similar to the entries of R and one point is of the form $(t, 0)$ where t is the root of a polynomial in ρ, t, and n.

Note that the reduction is stable in the sense that ellipses which intersect do this in such a way that some circle of radius polynomial in $1/n$ fits into the intersection. Conversely, if a pair of ellipses is not to intersect, their distance from each other is of the same form. Note that any dependence on ρ is allowed, since ρ is considered to be a constant.

Thus one can argue that one can approximate all numbers involved by polynomial precision without changing the intersection pattern of the ellipses.

We also note that we actually construct a multiset of ellipses (in other words, the ellipses have nonnegative integer weights). In order to obtain a set of ellipses in which no element occurs more than once, we have to invoke perturbation arguments as above a second time.

3 An Approximation Algorithm for Ellipses of Bounded Ratio

In this section, we consider ellipses with their interiors (the resulting intersection graphs are slightly more general than those of ellipses without interiors). Suppose $\rho \geq 1$, and let ELLIPSE$^\circ_{\leq \rho}$CLIQUE be the CLIQUE problem for intersection graphs of $(\leq \rho)$-ellipses with interiors. We outline an approximation algorithm for this problem, with approximation ratio depending on ρ:

Lemma 4. *Let C be a clique of $(\leq \rho)$-ellipses. Then there is a point p that is contained in at least $|C|/(9\rho^2)$ ellipses from C.*

Proof (Sketch of proof.). This is an adaptation of the proof of Lemma 4.1 of [3]. Let r be the smallest radius of all ellipses in $\mathcal{C}$, and pick $L \in \mathcal{C}$ which has r as its smaller radius and whose larger radius s is minimal among all ellipses having r as smaller radius. We may assume that the center of L is the origin. Furthermore, consider the ellipse $3L$ obtained from L by scaling by a factor of 3. We claim that, for every ellipse $E \in \mathcal{C}$,

$$\text{area}(E \cap 3L) \geq \frac{1}{9\rho^2}\text{area}(3L). \tag{3}$$

It follows that there is some point $p \in \mathbf{R}^2$ which is contained in at least $|\mathcal{C}|/9\rho^2$ ellipses from $\mathcal{C}$. To see why (3) holds, consider an ellipse $E \in \mathcal{C}$ and an arbitrary point p in the intersection of L and the the boundary of E. Let F be the unique ellipse of radii r and r/ρ, respectively, whose shorter axis is parallel to the shorter axis of E and which is tangent to E at p (see Figure 3). Then F is contained in $E \cap 3L$ (proof omitted), which shows (3).

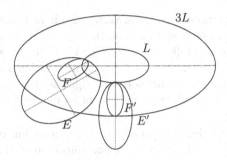

Fig. 3. An area argument for ellipses of bounded ratio.

Having Lemma 4 at our disposal, there is as easy $9\rho^2$-approximation algorithm for ELLIPSE$_{\leq\rho}$CLIQUE:

Algorithm 1. Given ℓ, compute the *arrangement* $\mathcal{A}$ induced by $\mathcal{L}$ and for every cell c, compute the number n_c of ellipses which contain c. (For a family of n ellipses, the arrangement can be computed, for instance, by a randomized incremental algorithm with expected runtime of $O(n \log n + v)$, where $v = O(n^2)$ is the number of vertices of the arrangement, or deterministically with slightly super-quadratic runtime, see [16].) Output the maximum $\max_c n_c$.

Here is an approach for further improvement of the approximation ratio: The proof of Lemma 4 shows that every family $\mathcal{C}$ of pairwise intersecting ellipses has the following property: Every subfamily $\mathcal{L} \subseteq \mathcal{C}$ of cardinality greater than $27\rho^2$, contains 3 distinct ellipses L_1, L_2, L_3 whose intersection $L_1 \cap L_2 \cap L_3$ is non-empty (by (3), there is an ellipse $L \in \mathcal{L}$ such that some point $p \in 3L$ is covered by at least 3 ellipses from $\mathcal{L}$). By the (p, q)-*Theorem* [2], for every ρ, there is some finite number $\tau(\rho)$, called the *transversal number*, such that every clique $\mathcal{C}$ of ($\leq \rho$)-ellipses can be *pierced* by some set of at most $\tau(\rho)$ points (i.e., every $L \in \mathcal{C}$ contains at least one of the points). This suggests the following variant of Algorithm 1.

Algorithm 2. Compute the arrangement induced by $\mathcal{L}$ as above. For every pair $\{c, c'\}$ of cells (there are at most $O(n^4)$), let $\mathcal{L}_{\{c,c'\}}$ be the set of ellipses in $\mathcal{L}$ which contain c, or c', or both. The intersection graph of $\mathcal{L}_{\{c,c'\}}$ is the complement of a bipartite graph on at most n nodes, so we can find a maximum clique in time $O(n^{2.5})$. Output the maximum for all pairs.

The approximation ratio of this algorithm is at least as good as that of the first one, and it is also at most $\tau(\rho)/2$. In general, the bounds for $\tau(\rho)$ implied by the (p, q)-Theorem, are quite large, but in some cases, better bounds are known. For instance, for disks, the transversal number is $\tau(1) = 4$ (see [8]), so we have a 2-approximation in that case (then again, we don't know whether the problem is hard for disks).

4   Discussion

Looking back, the reduction described in Section 2.3 in fact proves $\mathcal{APX}$-hardness for the CLIQUE problem in a more general context.

For $n \in \mathbf{N}$, consider the following graph G_n: $V(G_n)$ contains 7 vertices $v_i^1, \ldots, v_i^7$ for every integer $i \in \{1, \ldots, n\} \cup \{-1, \ldots, -n\}$ and 5 vertices $w_{i,j}^1, \ldots, w_{i,j}^5$ for every pair of such integers. Furthermore, all edges are present in $E(G_n)$ except for those connecting vertices v_i^a and v_{-i}^b, and except for those edges connecting $w_{i,j}^c$ to v_i^a and v_j^b, respectively, $1 \le a, b \le 7$, $1 \le c \le 5$.

By taking suitable induced subgraphs of G_n (depending on the formula φ), our reduction immediately yields the following generalization of Theorem 1.

Theorem 3 *Let $\mathcal{K}$ be a class of sets such that for every n, the graph G_n has a representation (of description size polynomial in n) as a $\mathcal{K}$-intersection graph (i.e., as intersection graph of some subset of $\mathcal{K}$). Then the CLIQUE problem is $\mathcal{APX}$-hard in $\mathcal{K}$-intersection graphs.*

Acknowledgement. We would like to thank Thomas Erlebach for introducing us to the CLIQUE problem in intersection graphs, and for many helpful comments and discussions. We are also indebted to Stephan Eidenbenz for encouraging us to squeeze $\mathcal{APX}$-completeness (instead of mere $\mathcal{NP}$-completeness) from our proof.

References

1. K. I. Aardal, S. P. M. van Hoesel, A. M. C. A. Koster and C. Mannino, and A. Sassano. Models and solution techniques for frequency assignment problems. Technical report, Konrad-Zuse-Zentrum für Informationstechnik Berlin, 2001.
2. N. Alon and D. Kleitman. Piercing convex sets and the Hadwiger-Debrunner (p, q)-problem. *Advances in Mathematics*, 96(1):103–112, 1992.
3. N. Alon, H. Last, R. Pinchasi, and M. Sharir. On the complexity of arrangements of circles in the plane. *Discrete Comput. Geom.*, 26(4):465–492, 2001.

4. S. Arora and C. Lund. Hardness of approximations. In D. S. Hochbaum, editor, *Approximation Algorithms for NP-Hard Problems*. PWS, 1995.
5. C. J. Colbourn B. N. Clark and D. S. Johnson. Unit disk graphs. *Discrete Mathematics*, 86:165–177, 1990.
6. A. Brandstädt, V. B. Le, and J. P. Spinrad. *Graph classes: a survey*. Society for Industrial and Applied Mathematics (SIAM), Philadelphia, PA, 1999.
7. H. Breu and D. G. Kirkpatrick. Unit disk graph recognition is NP-hard. *Computational Geometry*, 9(1-2):3–24, 1998.
8. L. Danzer. Zur Lösung des Gallaischen Problems über Kreisscheiben in der euklidischen Ebene. *Studia Scientiarum Mathematicorum Hungarica*, 21:111–134, 1986.
9. T. Erlebach, K. Jansen, and E. Seidel. Polynomial-time approximation schemes for geometric graphs. In *Proceedings of the 12th ACM-SIAM Symposium on Discrete Algorithms (SODA'01)*, pages 671–679, 2001.
10. P. Hliněný and J. Kratochvíl. Representing graphs by disks and balls (a survey of recognition-complexity results). *Discrete Math.*, 229(1-3):101–124, 2001. Combinatorics, graph theory, algorithms and applications.
11. Petr Hliněný. Contact graphs of line segments are NP-complete. *Discrete Math.*, 235(1-3):95–106, 2001. Combinatorics (Prague, 1998).
12. P. Koebe. Kontaktprobleme der konformen Abbildung. *Berichte über die Verhandlungen der Sächsischen Akademie der Wissenschaften, Math.-Phys. Klasse*, 88:141–164, 1936.
13. J. Kratochvíl and A. Kuběna. On intersection representations of co-planar graphs. *Discrete Math.*, 178(1-3):251–255, 1998.
14. J. Kratochvíl and J. Matoušek. Intersection graphs of segments. *J. Combin. Theory Ser. B*, 62(2):289–315, 1994.
15. T. A. McKee and F. R. McMorris. *Topics in intersection graph theory*. Society for Industrial and Applied Mathematics (SIAM), Philadelphia, PA, 1999.
16. M. Sharir and P.K./ Agarwal. *Davenport-Schinzel Sequences and Their Geometric Applications*. Cambridge University Press, Cambridge, UK, 1995.

Appendix

Proof (Proof of Lemma 2). W.l.o.g., K is the unit circle centered at $c = 0$ ℓ is the x-axis, and, for suitable real parameters r, s, t, $a = (s, \sqrt{1 - s^2})$, $b' = (s, -\sqrt{1 - s^2})$, $b = (r, \sqrt{1 - r^2})$, $a' = (r, -\sqrt{1 - r^2})$, and $q = (t, 0)$.

There is a unique conic E through the five points a, b, a', b', q. Moreover, by the symmetry of these points w.r.t. the x-axis, E is of the form

$$E = \{(x, y) : \lambda(x - m)^2 + \mu y^2 = 1\} \tag{4}$$

for suitable real parameters λ, μ, and m. Here, m is the x-coordinate of the center of E. Moreover, E is an ellipse if λ and μ have the same sign. Then, the radii of E are $1/\sqrt{|\lambda|}$ and $1/\sqrt{|\mu|}$, respectively.

By assumption, $a, b, q \in E$, which yields the three equations

$$\lambda(r - m)^2 + \mu(1 - r^2) = 1, \quad \lambda(s - m)^2 + \mu(1 - s^2) = 1, \quad \lambda(t - m)^2 = 1. \tag{5}$$

Using Maple™, solving these for λ, μ and m yields

$$m = m(r,s,t) = -\frac{1}{2}\frac{-s+st^2-r+rt^2}{rs-st+1-rt} =: -\frac{1}{2}\frac{f(r,s,t)}{g(r,s,t)},$$

$$\lambda = \lambda(r,s,t) = 4\frac{(rs-st+1-rt)^2}{(2trs-st^2+2t-rt^2-s-r)^2} =: 4\frac{g(r,s,t)^2}{h(r,s,t)^2},$$

$$\mu = \mu(r,s,t)$$
$$= 4\frac{r^2t^2-2r^2st+r^2s^2+3st^2r-rt-rt^3+rs-2rs^2t+s^2t^2-st^3-st+t^2}{(2trs-st^2+2t-rt^2-s-r)^2}$$

$$=: 4\frac{i(r,s,t)}{h(r,s,t)^2}.$$

Let us consider the zeros and singularities of these functions: For given r and s,

$$f(r,s,t) = 0 \Leftrightarrow t \in \{-1,+1\},$$
$$g(r,s,t) = 0 \Leftrightarrow t = \frac{rs+1}{r+s},$$
$$h(r,s,t) = 0 \Leftrightarrow t = \frac{1}{2}\frac{2+2rs+2\sqrt{(1-r^2)(1-s^2)}}{r+s} \quad \text{(double zero)},$$
$$i(r,s,t) = 0 \Leftrightarrow t \in \{r,s,\frac{rs+1}{r+s}\}.$$

By symmetry, we may assume that the point p lies to the left of K, i.e. that $r \leq -|s|$. Fix such r and s. Straightforward calculations show that $t = s$ is largest among the roots of $g(r,s,t)$, $h(r,s,t)$, and $i(r,s,t)$. Therefore, on the interval $s < t < \infty$, $\lambda(r,s,t)$ and $\mu(r,s,t)$ are continuous functions of t that do not change signs. Since $\mu(r,s,1) = \lambda(r,s,1) = 1$, we see that $\lambda(r,s,t), \mu(r,s,t) > 0$ for $t \in (s,\infty)$, hence E is an ellipse in that range of t. Note also that the ratio $\mu(r,s,t)/\lambda(r,s,t) \to \to \infty$ as $t \to \infty$, and that it grows monotonically for

$$t > \max\{(rs+1)/(r+s), (2+2rs+2\sqrt{(1-r^2)(1-s^2)})/(2(r+s))\},$$

in particular for $t \geq 1$.

For the claimed containment properties, it suffices to observe that K and E have no points of intersection except a,b,a',b', and that there are no points of intersection of E and ab except a and b, and analogously for $a'b'$.

A Geometric Approach to Boolean Matrix Multiplication[*]

Andrzej Lingas

Department of Computer Science, Lund University,
Box 118, S-22100 Lund, Sweden,
Andrzej.Lingas@cs.lth.se

Abstract. For a Boolean matrix D, let r_D be the minimum number of rectangles sufficient to cover exactly the rectilinear region formed by the 1-entries in D. Next, let m_D be the minimum of the number of 0-entries and the number of 1-entries in D.

Suppose that the rectilinear regions formed by the 1-entries in two $n \times n$ Boolean matrices A and B totally with q edges are given. We show that in time $\tilde{O}(q + \min\{r_A r_B, n(n + r_A), n(n + r_B)\})^1$ one can construct a data structure which for any entry of the Boolean product of A and B reports whether or not it is equal to 1, and if so, reports also the so called witness of the entry, in time $O(\log q)$.

As a corollary, we infer that if the matrices A and B are given as input, their product and the witnesses of the product can be computed in time $\tilde{O}(n(n + \min\{r_A, r_B\}))$. This implies in particular that the product of A and B and its witnesses can be computed in time $\tilde{O}(n(n + \min\{m_A, m_B\}))$.

In contrast to the known sub-cubic algorithms for Boolean matrix multiplication based on arithmetic $0 - 1$-matrix multiplication, our algorithms do not involve large hidden constants in their running time and are easy to implement.

1 Introduction

In 1969, Strassen published the first substantially sub-cubic (i.e., of order approximately $n^{2.81}$) algorithm for arithmetic matrix multiplication [1,19]. At present, the best asymptotic upper bound on the number of arithmetic operations necessary to multiply two $n \times n$ matrices is $O(n^{2.376})$ due to Coppersmith and Winograd [9]. As Boolean matrix multiplication is trivially reducible to arithmetic $0 - 1$-matrix multiplication [1], the same asymptotic upper bound holds in the Boolean case. Unfortunately, the aforementioned substantially sub-cubic algorithms for arithmetic matrix multiplication rely on algebraic approaches which involve large hidden constants in the running time and are not easy to implement. The fastest known combinatorial algorithm for Boolean matrix multiplication, based on a refinement of the idea of the "Four Russian's" algorithm [3], is due to Bash, Khanna and Motwani [4]. It runs in time $O(n^3/\log^2 n)$. On the other hand, it has been known for long time that Boolean matrix multiplication requires $\Omega(n^3)$ operations in the model of monotone circuits [14,16].

---

[*] Research supported in part by TFR grant 221-99-344.

[1] $\tilde{O}(f(n,q))$ stands for $O(f(n,q) \cdot \log^c(n+q))$, for any constant $c > 0$

In [17], Schnorr and Subramanian have shown that the Boolean product of two $n \times n$ Boolean matrices, drawn uniformly at random, can be determined by a simple combinatorial algorithm with high probability in time $\tilde{O}(n^2)$. Their algorithm is based on the following simple observation (cf. A-columns-B-rows approach in [17]).

Let A, B be the input $n \times n$ Boolean matrices. For any index k, $1 \leq k \leq n$, let $A1(k)$ be the set of the positions (given as row numbers) of 1's in the k-th column in A, and similarly let $B1(k)$ be the set of the positions (given as column numbers) of 1's in the k-th row in B. Then, we can set to 1 each entry $C_{i,j}$ of the product C of A and B for which $i \in A1(k)$ and $j \in B1(k)$ since $A_{i,k} = 1$ and $B_{k,j} = 1$.

By this observation, the product C can be computed in time $O(n^2 + \sum_{k=1}^{n} |A1(k)||B1(k)|)$. This upper time is quite useful, especially in sparse cases. For instance, it yields rather immediately an $O(n(n+s))$-time bound on Boolean matrix multiplication in case one of the matrices has only s 1-entries [7].

Consequently, Schnorr and Subramanian raised the question of whether or not there exists a substantially sub-cubic combinatorial algorithm for Boolean matrix multiplication[2].

In [7], another evidence that such a combinatorial algorithm might exist has been given. The combinatorial algorithm for Boolean matrix product presented in [7] is substantially sub-cubic in case the rows of the first $n \times n$ matrix or the columns of the second one are highly clustered, i.e., their minimum spanning tree in the Hamming metric has low cost. More exactly, the algorithm runs in time $\tilde{O}(n(n+c))$ where c is the minimum of the costs of the minimum spanning trees for the rows and the columns, respectively, in the Hamming metric.

Unfortunately, the problem of whether or not there exists a substantially sub-cubic combinatorial algorithm for Boolean matrix multiplication seems to be very hard. During the last two decades no essential progress as for upper time bounds in terms of n could be reported. For this reason and because of the practical and theoretical importance of Boolean matrix multiplication, it seems of interest to investigate special cases of structured and random matrices and derive partial results, even if they are not too complicated (e.g., [7,19]). (For comparison, the vast literature on sorting includes several papers on sorting presorted files.) It might happen that a combination of such partial results could eventually lead to a substantially sub-cubic combinatorial algorithm for Boolean matrix multiplication.

In this paper, we follow the aforementioned suggestion by considering the complexity of Boolean $n \times n$ matrix multiplication in a more geometric setting. Namely, we study this problem in terms of three parameters: the size q of the rectilinear regions formed by the 1-entries in the input matrices, the minimum number of rectangles necessary to cover these regions and n. We design algorithms which initially find such rectangular coverings within $O(\log q)$ of the minimum by using the fast sweep-line method of Franzblau [11] and then utilize the coverings to compute the Boolean product substantially faster in case at least one of the input matrices admits a substantially sub-quadratic covering of its 1's. Interestingly, our algorithms rely on efficient standard geometric data structures for orthogonal object intersection [10,13]. By combining them, we obtain the following main result, where for a Boolean matrix D, r_D stands for the minimum number of

---

[2] The question has been also recently addressed as an open challenging problem by Zwick in his invited talk on graph distance problems [20].

rectangles sufficient to cover exactly the rectilinear region formed by the 1-entries in D^3.

Suppose that the rectilinear regions formed by the 1-entries in two $n \times n$ Boolean matrices A and B totally with q edges are given. In time $\tilde{O}(q+\min\{r_A r_B, n(n+r_A), n(n+r_B)\})$, one can construct a data structure which for any entry of the Boolean product of A and B reports whether or not it is equal to 1 in time $O(\log q)$.

In case the input matrices are given in a standard way (i.e., the size of the input is $\Omega(n^2)$), we obtain the following result as a corollary.

The product of two $n \times n$ Boolean matrices A and B can be computed in time $\tilde{O}(n(n + \min\{r_A, r_B\}))$.[4]

It is not difficult to observe that whenever the rectilinear region formed by the 1-entries in a Boolean matrix D has q edges then $r_D \leq q$. It follows immediately that if in at least one of the matrices A and B, the 1-region has a substantially sub-quadratic number of edges then the product of A and B can be computed in substantially sub-cubic time.

For a Boolean matrix D, let m_D stand for the minimum of the number of 0-entries and the number of 1-entries in D. We observe that $r_D = O(m_D)$ which yields a useful application of our results to sparse and/or dense instances. Simply, we can substitute m_A and m_B for r_A and r_B in the statements of our main results. This substitution yields in particular the $\tilde{O}(n(n + \min\{m_A, m_B\}))$ time bound on computing the product of the $n \times n$ matrices A and B. This is useful in case at least one of the input matrices is sparse or conversely dense in terms of 1's. The sparse variant of this upper bound is known, e.g., by the aforementioned column-row approach while the dense variant could be also deduced from the upper bound given in [7] but only in a randomized setting. Summarizing, taking into account the aspect of determinism, our upper bound in terms of m_A and m_B seems new.

If an entry $C_{i,j}$ of the Boolean product C of two Boolean matrices A and B is equal to 1 then any index k such that $A_{i,k}$ and $B_{k,j}$ are equal to 1 is a *witness* to this. More recently, Alon and Naor [2] and Galil and Margalit [12] (see also [18]) have shown that the witnesses for the Boolean matrix product of two $n \times n$ Boolean matrices (i.e., for all its nonzero entries) can be computed in time $\tilde{O}(n^{2.376})$ by repeatedly applying the aforementioned algorithm of Coppersmith and Winograd for arithmetic matrix multiplication [9]. The geometric data structures and algorithms for Boolean matrix multiplication presented in this paper yield the witnesses directly without any extra asymptotic time-cost.

Our paper is structured as follows. The next section presents facts on two basic geometric data structures for intervals and rectangles used by our algorithms. The main third section shows our geometric algorithms and data structures for Boolean matrix multiplication and their analysis leading to our upper time bounds on Boolean matrix multiplication in terms of minimum rectangular coverings of the 1-regions in the input matrices. The fourth section contains rather immediate corollaries from our main upper time bound, in particular the upper time bound devoted to the sparse and/or dense cases.

---

[3] Note that the minimum number of rectangles into which the region can be partitioned might be substantially larger than r_D, in extreme cases, almost quadratic in r_D.

[4] A substantially weaker bound $O(n^2 \min\{r_A, r_B\})$ easily follows from the observation that the product of A and B can be computed in time $O(n^2 \min\{rank(A), rank(B)\})$ [6].

In the fifth section, we compare the size of minimum rectangular covering of 1's in a Boolean square matrix with the weight of minimum spanning tree for the rows of the matrix in the context of our method for Boolean matrix multiplication and that from [7]. We exhibit an infinite family of matrices for which the size of minimum covering is substantially smaller than the weight of the aforementioned tree.

2   Geometric Data Structures

Our geometric algorithms for Boolean matrix multiplication utilize efficient data structures for interval and rectangle intersection.

An interval tree is a leaf-oriented binary search tree that supports intersection queries for a set S of closed intervals on the real line in logarithmic time.

Fact 1 [13]. *Suppose that the left endpoints of the intervals in S belong to a subset U of R of size N and $|S| = n$. An interval tree T of depth $O(\log N)$ for S can be constructed in time $O(N + n \log Nn)$ using space $O(n + N)$. The insertion or deletion of an interval with left endpoint in U into T takes time $O(\log n + \log N)$. The intersection query is supported by T in time $O(\log N + r)$, where r is the number of the reported intervals.*

We shall also use the standard data structure for reporting the intersection of orthogonal objects from [10], restricted to 2-dimensional space, rectangles and point (i.e., degenerate rectangle) queries respectively.

Fact 2 [10]. *Let S be an arbitrary collection of n rectangles in 2-dimensional space. In time $O(n \log^2 n)$, one can construct a static data structure such that for an arbitrary query point x, all the t rectangles in S that include x can be reported in time $O(\log n + t)$.*

3   The Algorithms

Throughout the paper, A and B denote two $n \times n$ Boolean matrices, and C stands for their Boolean product. Additionally, we need to introduce or recall and/or formalize the following concepts and notation.

1. For an Boolean matrix D, and $1 \le i_1 \le i_2 \le n, 1 \le k_1 \le k_2 \le n, D(i_1, i_2, k_1, k_2)$ is the rectangular sub-matrix of D composed of all the entries $D_{i,k}$ where $i \in [i_1, i_2]$ and $k \in [k_1, k_2]$. We shall identify it with the rectangle with the corners (i_1, k_1), (i_1, k_2), (i_2, k_1), (i_2, k_2) on the $n \times n$ integer grid.
2. For an 1-entry $C_{i,j}$ of the Boolean matrix product C of A and B, an index $k \in \{1, 2, ..., n\}$ such that $A_{i,k} = 1$ and $B_{k,j} = 1$ is called a *witness*. An $n \times n$ matrix W such that $W_{i,j}$ is a witness of $C_{i,j}$ is called a *witness matrix* for the product C of A and B.
3. For a Boolean matrix D, r_D is the minimum number of its rectangular sub-matrices whose entries are completely filled with 1's such that each entry $D_{i,j} = 1$ is included into at least one of the sub-matrices.

The idea of our first geometric algorithm for Boolean matrix product relies on the two following observations.

A Geometric Approach to Boolean Matrix Multiplication

Lemma 1. *If each entry in $A(i_1, i_2, k_1, k_2)$ and $B(k'_1, k'_2, j_1, j_2)$ is set to 1 and $[k_1, k_2] \cap [k'_1, k'_2] \neq \emptyset$ then for any $i \in [i_1, ..., i_2], j \in [j_1, ..., j_2], C_{i,j} = 1$ holds and any element in $[k_1, k_2] \cap [k'_1, k'_2]$ is a witness of $C_{i,j}$.*

Proof. For any $k \in [k_1, k_2] \cap [k'_1, k'_2], i \in [i_1, ..., i_2], j \in [j_1, ..., j_2]$, we have $A_{i,k} = 1$ and $B_{k,j} = 1$.

Lemma 2. *Let C_A, C_B, be some sets of rectangular sub-matrices of A and B respectively such that for each 1-entry in A or B there is a rectangular sub-matrix in C_A or C_B respectively that includes it. If $C_{i,j} = 1$ and k is a witness of $C_{i,j}$ then there is $A(i_1, i_2, k_1, k_2) \in C_A$ and $B(k'_1, k'_2, j_1, j_2) \in C_B$ such that $i \in [i_1, i_2], j \in [j_1, j_2]$ and $k \in [k_1, k_2] \cap [k'_1, k'_2]$.*

Proof. Since $A_{i,k} = 1$, there is a rectangular sub-matrix $A(i_1, i_2, k_1, k_2)$ in C_A that includes it. Symmetrically, there is a rectangular sub-matrix $B(k'_1, k'_2, j_1, j_2) \in C_B$ including $B_{k,j}$ since $B_{k,j} = 1$. Now, it is easy to see that such two sub-matrices satisfy the lemma.

Algorithm 1
Input: the rectilinear regions formed by the 1-entries in two square Boolean matrices A and B of total size q.
Output: a data structure T (for reporting point inclusion in a set of rectangles) which will be proved to satisfy Theorem 7.

1. Find a covering C_A of the rectilinear polygonal region formed by the 1-entries in A with rectangles (i.e., rectangular sub-matrices) which is at most $O(\log q)$ times as large as minimum rectangular covering;
2. Find a covering C_B of the rectilinear polygonal region formed by the 1-entries in B with rectangles which is at most $O(\log q)$ times as large as minimum rectangular covering;
3. Initialize an interval tree S on the k-coordinates of the rectangles in C_A and C_B. For each rectangle $A(i_1, i_2, k_1, k_2)$ insert $[k_1, k_2]$, with a pointer to $A(i_1, i_2, k_1, k_2)$, into S.
4. For each rectangle $B(k_1, k_2, j_1, j_2)$ in C_B report all intervals $[k'_1, k''_2]$ in S that intersect $[k_1, k_2]$. For each such an interval $[k'_1, k''_2]$, with pointer to $A(i_1, i_2, k'_1, k''_2)$, insert the rectangle with the corners $(i_1, j_1), (i_1, j_2), (i_2, j_1), (i_2, j_2)$, and a pointer to an element k in $[k_1, k_2] \cap [k'_1, k''_2]$, into a two-dimensional data structure T for reporting point inclusion in rectangles.

Lemma 3. *Algorithm 1 runs in time $\tilde{O}(q + r_A r_B)$*

Proof. To implement steps 1 or 2, we use the sweep-line approximation algorithm for covering a rectilinear polygon with holes with a minimum number of axis-parallel rectangles due to Franzblau [11]. The algorithm runs in time time $O(q \log q)$ where q is the size of the input rectilinear figure and produces a rectangular covering within $O(\log q)$ of the minimum. Note here that whenever the rectilinear region formed by the 1-entries in a Boolean matrix D has at most q edges then $r_D \leq q$. Simply, produce a rectangular partition of the region by drawing a horizontal line-segment from each reflex vertex of the region to the opposite boundary. Hence, we have $r_A \leq q$ and $r_B \leq q$. By Fact 1, step 3 can be implemented in time $O(q \log q + r_A \log q \log q)$. The queries of S in step 4 take time $O(r_B \log q \log q + r_A \log q r_B \log q)$ by Fact 1. The construction of the data structure T in step 4 takes time $O(r_A \log q r_B \log q \log^2 q)$ by Fact 2. Thus, the total time taken by step 4 is $O(r_A r_B \log^4 q)$.

Lemma 4. *Suppose that the rectilinear regions formed by the 1-entries in two square Boolean matrices A and B with totally q edges are given. In time $\tilde{O}(q + r_A r_B)$, one can construct a data structure which for any entry of the Boolean product of A and B reports whether or not it is equal to 1, and if so, reports also a witness of the entry, in time $O(\log q)$.*

Proof. Run Algorithm 1 to construct the data structure T. To decide whether or not the entry $C_{i,j}$ of the product of A and B is set to 1, test the point (i,j) for containment in the rectangles in T. It follows immediately from Lemmata 1, 2 that if (i,j) is contained in at at least one rectangle then $C_{i,j}$ is equal to 1 and the witness associated with such a rectangle is also a witness of $C_{i,j}$.

Our second algorithm computes directly the Boolean matrix product of the input matrices.

Algorithm 2
Input: $n \times n$ Boolean matrices A and B.
Output: the Boolean product C of A and B and the witness matrix W for the product.

1. Find a covering C_A of the rectilinear polygonal region formed by the 1-entries in A with rectangles which is at most $O(\log n)$ times as large as minimum rectangular covering;
2. Initialize an interval tree S on the point set $\{1, ..., n\}$. For each rectangle $A(i_1, i_2, k_1, k_2)$ insert $[k_1, k_2]$, with a pointer to $A(i_1, i_2, k_1, k_2)$, into S.
3. For $j = 1, ..., n$, set S' to S and for $k = 1, ..., n$, whenever $B_{k,j} = 1$, report all intervals $[k', k'']$ in S' that contain k. For each such an interval $[k', k'']$ with a pointer to $A(i_1, i_2, k', k'')$ insert $[i_1, i_2]$, with a pointer to the witness k, into the interval tree S_j on the point set $\{1, ..., n\}$ and remove $[k', k'']$ from S'.
4. For $j = 1, ..., n$, compute the union U_j of the intervals in S_j assigning to each element in S_i a witness of an interval in S_j containing it.
5. For $j = 1, ..., n$, for each i in U_j, set $C_{i,j}$ to 1 and $W_{i,j}$ to the witness assigned to i.

Lemma 5. *Algorithm 2 runs in time $\tilde{O}(n(n + r_A))$.*

Proof. Steps 1, 2 take time $\tilde{O}(n^2 + r_A n)$ by the analysis of steps 1,3 of Algorithm 1 with $q = n^2$. In step 3, for each j, each rectangle $A(i_1, i_2, k', k'')$ is considered only once. Therefore, the total time taken by this step is $O(n^2 \log n + n r_A \log n)$. Step 4 can be implemented for $j = 1, ..., n$, by querying the interval tree S_j with $[1],...,[i],...,[n]$. Whenever an intersection is reported, we add i with the witness pointer of the intersected interval to the union U_j. Totally, it takes time $O(n^2 \log n)$ by Fact 1. Finally, step 5 takes time $O(n^2)$.

The correctness of Algorithm 2 is also implied by Lemmata 1, 2 (if one sees each 1-entry in B as covered by a unit rectangle) combined with the following observation: For a given j and $A(i_1, i_2, k_1, k_2)$ it is sufficient to find the smallest k such that $B_{k,j} = 1$ and $k \in [k_1, k_2]$ in order to know that $C_{i,j} = 1$ for $i \in [i_1, i_2]$. Hence, we obtain the following lemma by Lemma 5.

Lemma 6. *The Boolean product of two $n \times n$ Boolean matrices A and B as well as the witnesses of the product can be computed in time $\tilde{O}(n(n + r_A))$.*

Now, we can simply combine these two algorithms under the assumption that the rectilinear regions formed by the 1-entries in the matrices A and B with totally q edges are given. First, we run steps 1, 2 of Algorithm 1 in order to estimate the approximate values of $r_A r_B$, $n(n + r_A)$ and $n(n + r_B)$. Next, depending on which of them is approximately the smallest one, we either continue Algorithm 1, or run Algorithm 2, or run Algorithm 2 for the transposed matrices B^t and A^t (clearly, $r_{B^t} = r_B$). In this way, we obtain our main result.

Theorem 7. *Suppose that the rectilinear regions formed by the 1-entries in two $n \times n$ Boolean matrices A and B with totally q edges are given. In time $\tilde{O}(q + \min\{r_A r_B, n(n + r_A), n(n + r_B)\})$, one can construct a data structure which for any entry of the Boolean product of A and B reports whether or not it is equal to 1, and if so, reports a witness of the entry, in time $O(\log q)$.*

If one cannot assume any succinct representation of the 1-regions in the input matrices and the matrices are given in a standard way, the $q = n^2$ term jointly with $n(n + r_A)$ and $n(n + r_B)$ will unfortunately shade off $r_A r_B$. Therefore, in this standard input case, we need use just Algorithm 2 for A and B or B^t and A^t. This yields our second main result (which in fact is the special case of Theorem 7 for $q = n^2$).

Theorem 8. *The Boolean product of two $n \times n$ Boolean matrices A and B as well as the witnesses of the product can be computed in time $\tilde{O}(n(n + \min\{r_A, r_B\}))$.*

4   Corollaries

We have observed in the proof of Lemma 3 that whenever the rectilinear region formed by the 1-entries in a Boolean matrix D has q edges then $r_D \leq q$. Hence, we obtain immediately the following corollary from Theorem 8.

Theorem 9. *Let A and B be two $n \times n$ Boolean matrices. If the rectilinear region formed by the 1-entries in one of the matrices has at most q edges then the product of A and B and its witnesses can be computed in time $\tilde{O}(n(n + q))$.*

To state another interesting corollary from Theorem 8, we need to recall that for a Boolean matrix D, m_D stands for the minimum of the number of 0-entries and the number of 1-entries in D.

Lemma 10. *For any Boolean matrix D, the equality $r_D = O(m_D)$ holds.*

Proof. Let d_0, d_1 be the number of 0-entries and the number of 1-entries in D, respectively. The 1-entries can be trivially covered by d_1 unit rectangles. On the other hand, the rectilinear region formed by them has at most $4d_0$ reflex corners. Draw a horizontal line segment from each reflex corner of the rectilinear region within the region up to the region boundary. A partition into at most $4d_0 + 1$ rectangles of the region follows. Thus, the minimum number of rectangles sufficient to cover the rectilinear region formed by 1-entries in D is $O(m_D)$.

By Lemma 10, we can substitute m_A and m_B for r_A and r_B respectively in Theorems 7, 8. In this way, we obtain the following counterpart of the aforementioned theorems.

Theorem 11. *Let A and B be two $n \times n$ Boolean matrices, and let C be their product.*

1. *If the rectilinear regions formed by the 1-entries in A and B with totally q edges are given then in time $\tilde{O}(q + \min\{m_A m_B, n(n + m_A), n(n + m_B)\})$, one can construct a data structure which for any entry of C reports whether or not it is equal to 1, and if so, reports a witness of the entry, in time $O(\log q)$.*
2. *The product C and its witnesses can be computed in time $\tilde{O}(n(n+\min\{m_A, m_B\}))$.*

Our second corollary from Theorem 8 demonstrates that if for a square Boolean matrix D the value of r_D is very low then a moderate Boolean power of D still can be computed in substantially sub-cubic time.

The following rather immediate consequence of Lemmata 1, 2 reveals the idea of the second corollary.

Lemma 12. *For any two square Boolean matrices A and B, the inequality $r_{A \times B} \leq r_A r_B$ holds.*

Now, we are ready to state our second corollary.

Theorem 13. *For $i \leq 1$, the 2^i-th power of an $n \times n$ Boolean matrix D can be computed in time $\tilde{O}(n(n + r_D^{2^{i-1}}))$.*

Proof. We may assume $i = O(\log n)$ w.l.o.g. Assume inductively that the 2^{i-1}-th power of D can be computed in time $\tilde{O}(n(n + r_D^{2^{i-2}}))$ and $r_{D^{2^{i-2}}} \leq r_D^{2^{i-2}}$ holds. Then, the theorem immediately follows from Theorem 8 by Lemma 12.

5   Rectangular Covering versus Hamming MST

The main drawback of the geometrical method of Boolean matrix multiplication presented in this paper is that it is not invariant to the permutations of rows in A and columns in B respectively. For instance, if the first half of the rows in A have 1's at the even positions and the second half of them have 1's at the odd positions then the 1-entries can be covered with n rectangles trivially. On the other hand, for the matrix A' obtained by a random permutation of the rows in A, the expected value of r_A would be $\Omega(n^2)$.

The method of Boolean matrix multiplication based on the Hamming minimum spanning tree (MST) for the rows of A or the columns of B presented in [7] is invariant to such permutations. On the other hand, it is difficult to compute such a MST in substantially sub-cubic time deterministically. One is forced to apply a Monte Carlo method to achieve a substantial speed up [7]. Furthermore, there exists matrices A_i for which r_{A_i} is substantially smaller than the weight of the Hamming MST for the rows of A_i. For such matrices A_i, our "geometric" method of Boolean matrix multiplication is correspondingly substantially faster than the Monte Carlo method from [7].

Theorem 14. *For any positive constant $\epsilon < 1$, there is an infinite sequence of Boolean $O(n_i) \times O(n_i)$ matrices A_i such that $r_{A_i} = O(n_i^{2-2\epsilon})$ and the weight of the minimum spanning tree for the rows of A_i in the Hamming metric is $\Omega(n_i^{2-\epsilon})$.*

Proof. By [8], there exists an infinite sequence of the so called Hadamard matrices H_i of size $\lceil n_i^{1-\epsilon} \rceil \times \lceil n_i^{1-\epsilon} \rceil$ where the Hamming distance between each pair of rows in H_i is $\Omega(n_i^{1-\epsilon})$. Let A_i be the $O(n_i) \times O(n_i)$ matrix resulting from substituting the $\lceil n_i^\epsilon \rceil \times \lceil n_i^\epsilon \rceil$ Boolean matrix completely filled with 1's for each 1 and the $\lceil n^\epsilon \rceil \times \lceil n^\epsilon \rceil$ Boolean matrix completely filled with 0's for each 0 in H_i. Clearly, we have $r_{A_i} = O(n^{2-2\epsilon})$. On the other hand, any MST for the rows of A_i has to have the weight $\Omega(n_i^{1-\epsilon} \times n_i^{1-\epsilon} \times n_i^\epsilon)$.

6 Final Remarks

The question of whether or not there exists a substantially sub-cubic deterministic algorithm for Boolean matrix product is a challenging open problem. Although the upper time bounds presented in this paper are still cubic in the worst case, in the various applications of Boolean matrix multiplication the situation where at least one of the input matrices is structured is not unusual. If the 1's in such a matrix can be separated from the 0's by a substantially sub-quadratic number of axis-parallel edges then our relatively simple geometric algorithms yield the product in the correspondingly sub-cubic time by Theorem 9.

Acknowledgments. The author is grateful to Anders Björklund for inspiration and to Christos Levcopoulos for valuable comments.

References

1. A.V. Aho, J.E. Hopcroft and J.D. Ullman. The Design and Analysis of Computer Algorithms (Addison-Wesley, Reading, Massachusetts, 1974).
2. N. Alon and M. Naor. Derandomization, Witnesses for Boolean Matrix Multiplication and Construction of Perfect hash functions. Algorithmica 16, pp. 434–449, 1996.
3. V.L. Arlazarov, E.A. Dinic, M.A. Konrod and L.A. Faradzev. On economical construction of the transitive closure of a directed graph. Doklady Acad. Nauk SSSR, 194, pp. 487–488, 1970 (in Russian).
4. J. Basch, S. Khanna and R. Motwani. On Diameter Verification and Boolean Matrix Multiplication. Technical Report, Standford University CS department, 1995.
5. P. Berman and B. Dasgupta. Approximating Rectilinear Polygon Cover Problems. Algorithmica 17(4), 1997, pp. 331–356.
6. A. Björklund. Fast Multiplication of Low Rank Matrices. Unpublished manuscript, Lund University, 2002.

7. A. Björklund and A. Lingas. Fast Boolean matrix multiplication for highly clustered data. Proc. 7th International Workshop on Algorithms and Data Structures (WADS 2001), Lecture Notes in Computer Science, Springer Verlag.

8. P.J. Cameron. Combinatorics. Cambridge University Press 1994.

9. D. Coppersmith and S. Winograd. Matrix Multiplication via Arithmetic Progressions. J. of Symbolic Computation 9 (1990), pp. 251–280.

10. H. Edelsbrunner and H. Maurer. On the intersection of orthogonal objects. IPL vol. 13 (1981), pp. 177–181.

11. D.S. Franzblau. Performance Guarantees on a Sweep Line Heuristic for Covering Rectilinear Polygons with Rectangles. SIAM J. Discrete Math., Vol. 2, 3, 1989, pp. 307–321.

12. Z. Galil and O. Margalit. Witnesses for Boolean Matrix Multiplication and Shortest Paths. Journal of Complexity, pp. 417–426, 1993.

13. K. Mehlhorn. *Data Structures and Algorithms 3: Multi-dimensional Searching and Computational Geometry*. EATCS Monographs on Theoretical Computer Science, Springer Verlag, Berlin, 1984.

14. K. Mehlhorn and Z. Galil. Monotone switching circuits and Boolean matrix product. Computing 16(1-2), pp. 99–111, 1976.

15. M.H. Overmars. Efficient Data Structures for Range Searching on a Grid. J. Algorithms, 9(2), pages 254–275, 1988.

16. M.S. Paterson. Complexity of monotone networks for Boolean matrix product. Theoretical Computer Science 1(1), pp. 13–20, 1975.

17. C.P. Schnorr and C.R. Subramanian. Almost Optimal (on the average) Combinatorial Algorithms for Boolean Matrix Product Witnesses, Computing the Diameter. Randomization and Approximation Techniques in Computer Science. Second International Workshop, RANDOM'98, Lecture Notes in Computer Science 1518, pp. 218–231.

18. R. Seidel. On the all-pairs-shortest-path problem in unweighted undirected graphs. Journal of Computer and System Sciences, 51, pp. 400–403, 1995.

19. V. Strassen. Gaussian elimination is not optimal. Numerische Mathematik 13, pp. 354–356, 1969.

20. U. Zwick. Exact and Approximate Distances in Graphs - A Survey. Proc. 9th Annual European Symposium on Algorithms (ESA 2001), Lecture Notes in Computer Science, Springer Verlag.

The Min-Max Voronoi Diagram of Polygons and Applications in VLSI Manufacturing

Evanthia Papadopoulou[1] and D.T. Lee[2]

[1] IBM TJ Watson Research Center, Yorktown Heights, NY 10598, USA
evanthia@watson.ibm.com
[2] Institute of Information Science, Academia Sinica, Nankang, Taipei, Taiwan
dtlee@iis.sinica.edu.tw

Abstract. We study the *min-max Voronoi diagram* of a set S of polygonal objects, a generalization of Voronoi diagrams based on the *maximum* distance between a point and a polygon. We show that the min-max Voronoi diagram is equivalent to the Voronoi diagram under the Hausdorff distance function. We investigate the combinatorial properties of this diagram and give improved combinatorial bounds and algorithms. As a byproduct we introduce the *min-max hull* which relates to the min-max Voronoi diagram in the way a convex hull relates to the ordinary Voronoi diagram.

1 Introduction

Given a set S of polygonal objects in the plane their *min-max* Voronoi diagram is a subdivision of the plane into regions such that the Voronoi region of a polygon $P \in S$ is the locus of points t whose maximum distance from (any point in) P is less than the maximum distance from any other object in S. The min-max Voronoi region of P is subdivided into finer regions by the farthest point Voronoi diagram of the vertex set of P. This structure generalizes both the ordinary Voronoi diagram of points and the farthest-point Voronoi diagram. The ordinary Voronoi diagram is derived if shapes degenerate to points and the farthest-point Voronoi diagram appears in the case where the set of points is the set of vertices of a single polygon ($|S| = 1$). The *min-max* Voronoi diagram can be defined equivalently on a collection S of sets of points instead of polygonal objects. It is equivalent to the Voronoi diagram of sets of points or polygonal objects under the Hausdorff metric (see Section 2).

The min-max Voronoi diagram problem was formulated in [11] where the *critical area* computation problem for *via-blocks* in VLSI designs was addressed via the L_∞ min-max Voronoi diagram of disjoint rectilinear shapes. This diagram had also been considered in [4] where it was termed the *Voronoi diagram of point clusters*, and in [1] where it was termed the *closest covered set diagram* for the case of disjoint convex shapes and arbitrary convex distance functions. In [4] general combinatorial bounds regarding this diagram were derived by means of envelopes in three dimensions. It was shown that the size of this diagram is $O(n^2 \alpha(n))$ in general and linear in the case of clusters of points with disjoint

P. Bose and P. Morin (Eds.): ISAAC 2002, LNCS 2518, pp. 511–522, 2002.
© Springer-Verlag Berlin Heidelberg 2002

convex hulls. The latter was also shown in [1] for disjoint convex shapes and arbitrary convex distances. Using a divide and conquer algorithm for computing envelopes in three dimensions, [4] concluded that the cluster Voronoi diagram can be constructed in $O(n^2\alpha(n))$ time. In [1] the problem for disjoint convex sets was reduced to abstract Voronoi diagrams and the randomized incremental construction of [6] was proposed for its computation. This approach results in an $O(kn \log n)$ -time algorithm, where k is the time to construct the bisector of two convex polygons.

In this paper we provide tighter combinatorial bounds and algorithms that improve in an output sensitive fashion the time complexity bounds given in [4] and [1]. Specifically we show that the size of the diagram is $O(m)$ where m is the size of the *intersection graph* of S which reflects the number of relevant intersections among the shapes of S in addition to the number of convex hull edges of each shape in S (see Def. 7). In the worst case m is $O(n^2)$. Furthermore we show that in case of *non-crossing* polygons (not necessarily disjoint or convex, see Def. 5) the min-max Voronoi regions are connected and the size of the min-max Voronoi diagram is linear in the number of vertices on the convex hulls of the polygons in S. Thus, the connectivity and linearity of the diagram is maintained for a more general class of polygons than the ones shown in [4,1]. We present a divide and conquer algorithm of time complexity $O((n + M + N + K) \log m)$ where n is the total number of points in S, M is the total number of vertices of *crossing* pairs of shapes (see Def. 5), K is the total number of vertices of *interacting*[1] pairs of shapes and $N = O(M \log m)$ is the sum for all dividing lines of the size of pairs of crossing shapes that intersect a dividing line in the divide and conquer scheme. As a byproduct we introduce the *min-max hull* of a set of shapes which relates to the min-max Voronoi diagram in the way a convex hull relates to the ordinary Voronoi diagram (see Def. 9). The special properties of the min-max Voronoi diagram are heavily exploited to maintain efficiency as the *merge curve* of the standard divide-and-conquer technique contains cycles. The algorithm simplifies to $O((n+N+K) \log n)$ for *non-crossing* polygons which is the case of interest to our VLSI manufacturing application. In our application, VLSI contact shapes tend to be small in size and well spaced in which case N, K are both negligible compared to n. A plane sweep approach of time complexity $O((n + K') \log n)$ for a non-crossing S is presented in a companion paper [10].

Our motivation for studying the min-max Voronoi diagram comes from an application in VLSI *yield* prediction, in particular the estimation of *critical area*, a measure reflecting the sensitivity of a VLSI design to spot defects during manufacturing [7,8,9,11,13]. In [12,11] the critical area computation problem for *shorts*, *opens*, and *via-blocks* was reduced to variations of L_∞ Voronoi diagrams of segments. In particular, the L_∞ min-max Voronoi diagram of disjoint rectilinear shapes was introduced as a solution to the critical area computation problem for via-blocks [11]. The L_∞ metric corresponds to a square defect model. The square defect model has been criticized [8] for the case of via-blocks and thus the Euclidean version of the problem needs to be investigated. The construction of

---

[1] A shape $Q \in S$ is *interacting* with $P \in S$ if Q is entirely enclosed in the minimum enclosing circle of P (Def. 11).

the Euclidean min-max Voronoi diagram remains realistic as robustness issues are similar to the construction of Voronoi diagrams of points and not to those of segments.

2 Preliminaries

Let $d(p, q)$ denote the ordinary distance between two points in the plane p and q. The ordinary bisector between p and q, denoted as $b(p, q)$, is the locus of points equidistant from p and q. The bisector $b(p, q)$ partitions the plane into two *half-planes* $H(p, q)$ and $H(q, p)$, where $H(p, q)$ denotes the half plane associated with p. The *farthest distance* of a point p from a shape Q is $d_f(p, Q) = \max\{d(p, q), \forall q \in Q\}$. It is well known that $d_f(p, Q) = d(p, q)$ for some vertex q on the convex hull of Q. The convex hull of Q is denoted as $CH(Q)$. It is also well known that $d_f(p, Q)$ can be determined by the *farthest point* Voronoi diagram of the vertex set of Q, denoted as *f-Vor(Q)*. Let *freg(q)* denote the farthest Voronoi region of $q \in CH(Q)$, where q is on $CH(Q)$. The farthest distance between two polygons P and Q is $d_f(P, Q) = \max\{d(p, q), \forall(p, q), p \in P, q \in Q\}$. Clearly $d_f(P, Q) = d(p, q)$ where p and q are vertices of $CH(P)$ and $CH(Q)$ respectively. For two arbitrary sets of points A, B $d_f(A, B)$ is defined equivalently.

Definition 1. *The farthest bisector, denoted $b_f(P, Q)$, between P and Q is the locus of points equidistant from P and Q according to $d_f(P, Q)$, i.e., $b_f(P, Q) = \{y \mid d_f(y, P) = d_f(y, Q)\}$. Any point r such that $d_f(r, P) < d_f(r, Q)$ is said to be closer to P than to Q. The locus of points closer to P than to Q is denoted by $H(P, Q)$.*

$H(P, Q)$ need not be connected in general. We now show that $b_f(P, Q)$ is equivalent to the bisector of P, Q under the Hausdorff metric. The (directed) Hausdorff distance from P to Q is $d_h(P, Q) = \max_{p \in P} \min_{q \in Q} d(p, q)$. The (undirected) Hausdorff distance between P and Q is $D_h(P, Q) = \max\{d_h(P, Q), d_h(Q, P)\}$. The Hausdorff bisector between P and Q is $b_h(P, Q) = \{y \mid D_h(y, P) = D_h(y, Q)\}$. But for any point y, $d_h(y, P) = d(y, P)$, where $d(y, P) = \min_{p \in P}\{d(y, p)\}$, and $d_h(P, y) = d_f(y, P)$. But $d_f(y, P) \geq d(y, P)$. We thus conclude that $b_f(P, Q)$ and $b_h(P, Q)$ are equivalent.

Definition 2. *The* min-max *Voronoi diagram of S, denoted as mM-Vor(S), is a subdivision of the plane into regions such that the min-max Voronoi region of a polygon P, denoted as mM-reg(P), is the locus of points closer to P according to d_f, than to any other shape in S i.e., $mM\text{-}reg(P) = \{y \mid d_f(y, P) \leq d_f(y, Q), \forall Q \in S, Q \neq P\}$. The min-max Voronoi region of P is subdivided into finer regions by the farthest point Voronoi diagram of the vertex set of P. That is, $mM\text{-}reg(p) = mM\text{-}reg(P) \cap freg(p)$ i.e., $mM\text{-}reg(p) = \{y \mid d(y, p) = d_f(y, P) \leq d_f(y, Q), \forall Q \in S, Q \neq P\}$. The Voronoi edges on the boundary of mM-reg(P) are called inter-bisectors. The bisectors in the interior of mM-reg(P) are called intra-bisectors. Since $b_f(P, Q)$ is equivalent to $b_h(P, Q)$ the min-max Voronoi diagram is also called the Hausdorff Voronoi diagram.*

Figure 1 illustrates the min-max Voronoi diagram of $S = \{P_1, P_2, P_3\}$. The shaded regions depict $mM\text{-}reg(P_1)$ and $mM\text{-}reg(P_3)$; the unshaded portion corresponds to $mM\text{-}reg(P_2)$. Inter-bisectors are shown in solid lines and intra-bisectors are depicted in dashed lines. Figure 2 illustrates the min-max Voronoi diagram of two intersecting segments P, Q; $mM\text{-}reg(P)$ is illustrated shaded. Both an intra- and an inter-bisector correspond to an ordinary bisector $b(p, q)$ between two points p, q. For an inter-bisector, p and q belong to different shapes, $H(p, q)$ consists of all points closer to p than q, and p is located in $H(p, q)$. For an intra-bisector, p, q are points of the same shape, $H(p, q)$ consists of points farther from p than q and p is not located in $H(p, q)$. Similarly we distinguish between three types of vertices: *inter-vertices* where at least three inter-bisectors meet, *intra-vertices* where at least three intra-bisectors meet, and *mixed-vertices* where one intra-bisector and two inter-bisectors meet. The term *intra-bisector* is used to denote any bisector in $f\text{-}Vor(P)$, the farthest point Voronoi diagram of the vertex set of P.

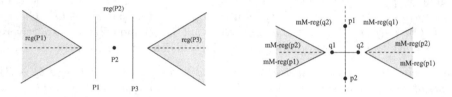

Fig. 1. The min-max Voronoi diagram of $S = \{P_1, P_2, P_3\}$.

Fig. 2. The min-max Voronoi diagram of two intersecting segments.

Definition 3. *The circle K_y, centered at an intra-bisector point y of radius $d_f(y, P)$ is called a P-circle. A P-circle is empty if it contains no shape other than P in its interior.*

Definition 4. *A supporting line of a convex polygon P is a straight line l passing through a vertex v of P such that the interior of P lies entirely on one side of l. Vertex v is called a supporting vertex. l and v are called left (resp. right) supporting) if P lies to the right (resp. left) of l. The portion of the common supporting line between the supporting vertices of two convex polygons such that both polygons lie on the same side of l is called a supporting segment.*

Definition 5. *Two polygons are called non-crossing if their convex hulls admit at most two supporting segments. Otherwise they are called crossing.*

An example of non-crossing and crossing polygons can be seen in Figure 3. In Figures 3(a) and 3(b) all polygons are non-crossing; in Figure 3(c) polygons C and D are crossing. In our application, VLSI via-shapes are disjoint and thus, their convex hulls must be non-crossing.

Definition 6. *A chord $\overline{p_i p_j} \in P$ is a diagonal or convex hull edge of P that induces an intra-bisector of $f\text{-}Vor(P)$. Two chords $\overline{p_i p_j} \in P$ and $\overline{q_i q_j} \in Q$, are* *called* non-crossing *if they do not intersect or if they intersect but at least one of their endpoints lies strictly in the interior of $CH(P \cup Q)$. Otherwise they are called* crossing. *Shape Q is called* crossing *with chord $\overline{p_i p_j} \in P$ if there is a chord $\overline{q_i q_j} \in Q$ that is crossing with $\overline{p_i p_j}$. Otherwise it is called* non-crossing.

Definition 7. *The* intersection graph *of S, $G(S)$, has a vertex for every chord of a shape $P \in S$ and an edge for any two crossing chords $\overline{p_i p_j} \in P, \overline{q_i q_j} \in Q$. The size of the intersection graph is denoted by m. The total number of vertices of all crossing pairs of shapes is denoted by M.*

Definition 8. *A* chain *is a simple polygonal line. A chain C is* monotone *with respect to a line l if every line orthogonal to l intersects C in at most one point.*

3   Structure and Properties

In this section we list the structural properties of $mM\text{-}Vor(S)$. Unless we explicitly mention otherwise these properties have not been identified in [4,1].

Property 1. The farthest bisector $b_f(P,Q)$ is a subgraph of $f\text{-}Vor(P \cup Q)$ consisting of edge disjoint monotone chains. If a chain has just one edge, this is a straight line; otherwise its two extreme edges are semi-infinite rays.

Corollary 1. *Any semi-infinite ray of $b_f(P,Q)$ corresponds to the perpendicular bisector induced by a distinct supporting segment between $CH(P)$ and $CH(Q)$. The number of chains constituting $b_f(P,Q)$ is derived by the number of supporting segments between $CH(P)$ and $CH(Q)$.*

Property 2. For any point $x \in mM\text{-}reg(p)$ the segment $\overline{px} \cap freg(p)$ lies entirely in $mM\text{-}reg(p)$. $mM\text{-}reg(p)$ is said to be essentially star-shaped.

Property 3. The boundary of any connected component of $mM\text{-}reg(p)$, $p \in P$, $p \neq P$, consists of a sequence of outward convex chains, each one corresponding to an inter-bisector $b_f(p, Q_i)$, $Q_i \in S, Q_i \neq P$, and an inward convex chain corresponding to the intra-bisector $b_f(p, P)$. (Convexity is characterized as seen from the interior of $mM\text{-}reg(p)$).

Let's now define the *min-max hull*, or *mM-hull* for short, of S, denoted as $mMH(S)$. The mM-hull is related to the $mM\text{-}Vor(S)$ as the ordinary convex hull is related to the ordinary Voronoi diagram. An example is depicted in Figure 3.

Definition 9. *A shape $P \in S$, in particular a vertex $p \in P$, is said to be on the mM-hull of S, if and only if p admits a supporting line ℓ such that $CH(P)$ lies totally on one side of ℓ and none of the rest of shapes in S lie totally on the same side, except possibly from a shape having its boundary common to ℓ and its interior lying on the same side of ℓ as $CH(P)$. A segment $\overline{pq}$ joining two mM-hull vertices $p \in P, q \in Q$ such that $\overline{pq}$ is a supporting segment of P, Q is called an mM-hull* supporting segment. *An mm-hull edge is either a convex hull edge joining mM-hull vertices of one shape or an mM-hull supporting segment joining mM-hull vertices of different shapes.*

Fig. 3. The min-max hull.   **Fig. 4.** $Q_1 \in \mathcal{K}_y^r$ and $Q_2 \in \mathcal{K}_y^f$.

By Definition 9, a supporting segment $\overline{pq}$, $p \in P$ and $q \in Q$, is an edge of $mMH(S)$ if and only if $CH(P)$ and $CH(Q)$ lie totally on one side of the line $\ell_{\overline{pq}}$ passing through $\overline{pq}$ and no other shape in S lies totally on the same side of $\ell_{\overline{pq}}$ (except possibly from a shape having its boundary common to $\ell_{\overline{pq}}$). Furthermore, a convex hull edge $\overline{pr}$, $p, r \in P$, is an mM-hull edge if and only if $CH(P)$ is the only shape in S lying entirely on one side of the underlying line $\ell_{\overline{pr}}$. Thus, the boundary of the mM-hull consists of a sequence of supporting segments, interleaved with a subsequence of convex chains. The order of traversal, say clockwise, of the mM-hull edges satisfies the property that each edge defines a supporting line of a shape P on the mM-hull: a right supporting line for mM-hull supporting segments and a left supporting line for convex hull chains. Figure 3 illustrates the mM-hulls of (a) disjoint, (b) non-crossing, and (c) crossing shapes respectively. Supporting segments are illustrated in arrows according to a clockwise traversal.

Property 4. Region mM-$reg(p)$ is unbounded if and only if vertex p lies on the mM-hull of S.

Corollary 2. *All unbounded bisectors of mM-Vor(S) (both inter- and intra-bisectors) are cyclically ordered in the same way as the edges are ordered on the boundary of $mMH(S)$.*

Let $T(P)$ denote the tree of intra-bisectors of P i.e., the tree induced by f-$Vor(P)$. $T(P)$ is assumed to be rooted at the point of minimum weight i.e., the center of the minimum enclosing circle of P. Let $\overline{y_j y_k} \in T(P)$ be the intra-bisector segment of chord $\overline{p_i p_j}$, where y_j is the parent of y_k, and let y_i be a point on $\overline{y_j y_k}$. Point y_i partitions $T(P)$ in at least two parts. Let $T(y_i)$ denote the part containing the descendents of y_i in the rooted $T(P)$ i.e., the subtree of $T(P)$ rooted at y_i that contains segment $\overline{y_i y_k}$, and let $T_c(y_i)$ denote the complement of $T(y_i)$. $T(y_i)$ is referred to as the *subtree of $T(P)$ rooted at y_i*. Let $\mathcal{K}_{y_i}$ be the P-circle centered at y_i. Chord $\overline{p_i p_j}$ partitions $\mathcal{K}_{y_i}$ in two parts $\mathcal{K}_{y_i}^f$ and $\mathcal{K}_{y_i}^r$, where $\mathcal{K}_{y_i}^r$ is the part enclosing the portion of $CH(P)$ inducing $T(y_i)$ and $\mathcal{K}_{y_i}^f$ is the part enclosing the portion of $CH(P)$ inducing $T_c(y_i)$. Figure 4 depicts $\mathcal{K}_{y_i}^f$ shaded.

Definition 10. *A shape $Q \in S$ is called* limiting *with respect to chord $\overline{p_i p_j} \in P$, if Q is enclosed within a P-circle $\mathcal{K}$ passing through p_i, p_j, and Q is non-crossing with $\overline{p_i p_j}$. (Note that Q may be limiting but still be crossing with P). Q is called* forward limiting *if $Q \in \mathcal{K}_{y_i}^f \cup CH(P)$ or* rear limiting *if $Q \in \mathcal{K}_{y_i}^r \cup CH(P)$.*

Note that any shape enclosed in a P-circle must be forward limiting, rear limiting or crossing with P. In Figure 4 shapes Q_1 and Q_2 are rear and forward limiting respectively with respect to y and chord $\overline{p_1 p_2}$.

Lemma 1. *Let $Q \in S$ be a limiting shape with respect to chord $\overline{p_i p_j} \in P$ and let $y \in T(P)$ be the center of the P-circle enclosing Q. If Q is forward (resp. rear) limiting then the whole $T(y)$ (resp. $T_c(y)$) is closer to Q than to P.*

Proof. Sketch. It is not hard to see that $\mathcal{K}_y^f \cup CH(P) \subset \mathcal{K}_{y_k}$ for any $y_k \in T(y)$, and $\mathcal{K}_y^r \cup CH(P) \subset \mathcal{K}_{y_j}$ for any $y_j \in T_c(y)$.

The following property gives a sufficient condition for a shape to have an empty Voronoi region and it is directly derived from Lemma 1. In the case of non-crossing shapes the condition is also necessary. For the subset of disjoint convex shapes property 3 had also been identified in [1].

Property 5. *$mM\text{-}Vor(P) = \emptyset$ if there exist two limiting shapes Q, R, enclosed in the same P-circle $\mathcal{K}_y$, $y \in b(p_i, p_j)$, $\overline{p_i p_j} \in P$, such that $Q \in \mathcal{K}_y^f \cup CH(P)$ and $R \in \mathcal{K}_y^r \cup CH(P)$. In case of a non-crossing S the condition is also necessary (except from the trivial case where $CH(P)$ entirely contains another shape).*

Theorem 1. *The min-max Voronoi diagram of an arbitrary set of polygons S has size $O(m)$, where m is the size of the intersection graph of S. In case of a non-crossing S, there is at most one connected Voronoi region for each $P \in S$ and $mM\text{-}Vor(S)$ has size $O(n)$, where n is the total number of vertices in S.*

Proof. Sketch. It is enough to show that the number of mixed Voronoi vertices is $O(m)$. For any P at most two mixed Voronoi vertices can be induced by limiting shapes. Any other mixed Voronoi vertex on $b(p_i, p_j)$, $p_i, p_j \in P$ must be induced by some $Q \in S$ that is crossing $\overline{p_i p_j}$. But for $\overline{p_i p_j}$ at most two mixed Voronoi vertices can be induced by Q. Thus, the bound is derived.

For a non-crossing S any shape enclosed in a P-circle must be limiting. By property 3 the boundary of any connected component of $mM\text{-}reg(p)$ for any $p \in P$ must contain a single intra-bisector chain of $T(P)$. Thus, by lemma 1 and the non-crossing property, $T(P) \cap mM\text{-}reg(P)$ must consist of a single connected component (if not empty). Hence, $mM\text{-}reg(P)$ must be connected.

In the case of disjoint convex shapes the connectivity and linearity of $mM\text{-}Vor(S)$ had also been established in [4] and [1]. Our proof however is much simpler and extends the linearity property to the more general class of non-crossing polygons.

4   A Divide and Conquer Algorithm

Let S_l and S_r be the sets of shapes in S to the left and to the right respectively of a vertical dividing line $\mathcal{L}$, where a shape P is said to be to the *left* (resp. *right*) of $\mathcal{L}$ if the leftmost x-coordinate of P is to the left (resp. right) of $\mathcal{L}$. Assume that $mM\text{-}Vor(S_l)$ and $mM\text{-}Vor(S_r)$ have been computed. We shall compute $mM\text{-}Vor(S)$

by merging $mM\text{-}Vor(S_l)$ and $mM\text{-}Vor(S_r)$. Let $\sigma(S_l, S_r)$ denote the *merge curve* between $mM\text{-}Vor(S_l)$ and $mM\text{-}Vor(S_r)$ i.e., the collection of bisectors $b(p_l, p_r)$ in $mM\text{-}Vor(S)$ such that $p_l \in S_l$ and $p_r \in S_r$. Let $S_\mathcal{L}$ consist of the shapes in S_l that intersect the dividing line and the shapes in S_r that are crossing with shapes in S_l. In case of non-crossing shapes $S_\mathcal{L} \subseteq S_l$.

Lemma 2. *The merge curve $\sigma(S_l, S_r)$ is a collection of edge disjoint unbounded chains and cycles. The cycles can only enclose regions of shapes in $S_\mathcal{L}$ that is, regions of shapes in S_l that intersect the dividing line and regions of shapes in S_r that are crossing with shapes in S_l (if any).*

Figure 5 shows $mM\text{-}Vor(S)$, $S = \{P_1, P_2, Q_1, Q_2\}$ with $mM\text{-}reg(P_2)$ depicted shaded. Dividing S as $S_l = \{P_1, P_2\}$ and $S_r = \{Q_1, Q_2\}$ shows that $\sigma(S_l, S_r)$ can have cycles enclosing regions of shapes in S_l. In Figure 6 the shaded regions depict $mM\text{-}reg(Q)$. Dividing S as $S_l = \{P_1, P_2\}$ and $S_r = \{Q\}$ shows that, in case of crossing shapes, a cycle in $\sigma(S_l, S_r)$ can enclose regions of shapes in S_r.

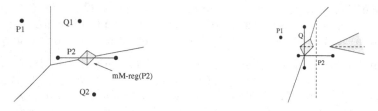

Fig. 5. $mM\text{-}Vor(S)$, $S = \{P_1, P_2, Q_1, Q_2\}$.   **Fig. 6.** $mM\text{-}Vor(S)$, $S = \{P_1, P_2, Q\}$.

To trace the merge curve $\sigma(S_l, S_r)$ we need to identify a starting point on every component. Then each component can be traced as in the ordinary Voronoi diagram case [14]. By Corollary 2 the unbounded portions of $\sigma(S_l, S_r)$ are induced by the mM-hull supporting segments between vertices of S_l and S_r. We first concentrate on identifying those mM-hull supporting segments. We then show how to identify a point on every cycle of $\sigma(S_l, S_r)$.

4.1 Merging Two mM-Hulls

The mM-hull of S is represented by an ordered (e.g. clockwise) list of its vertices. An alternative representation can be obtained by the *Gaussian Map* [3] of S, for short *GMap(S)*, onto the unit circle. In the Gaussian Map every mM-hull edge e_i is mapped to a point $\nu(e_i)$ on the circumference of a unit circle $\mathcal{K}_o$ as obtained by the outward pointing unit normal. That is, point $\nu(e_i)$ represents a normal vector pointing in the half-plane bordered by the supporting line ℓ_{e_i} away from the mM-hull i.e., towards the interior of the shape(s) where the endpoints of e_i belong. We will use the same notation to denote both the vector and the corresponding point in the GMap. Figure 7 illustrates two mM-hulls and their respective Gaussian maps. In the Gaussian maps the supporting segments of the mM-hull are represented by longer arrows and are marked by the names of the two shapes they support.

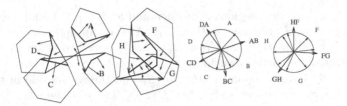

Fig. 7. The Gaussian Map of $mM\text{-}hull(\{A, B, C, D\})$ and $mM\text{-}hull(\{F, H, G\})$.

Lemma 3. *The normal vectors in GMap(S) appear in the same cyclic order (e.g. clockwise) as the respective cyclic traversal of the boundary of mM-hull(S).*

By Lemma 3 merging two mM-hulls corresponds to merging their Gaussian Maps. An edge e of $mMH(S_l)$ or $mMH(S_r)$ is called *valid* if e remains on the $mMH(S)$, otherwise e is called *invalid*. To merge $GMap(S_l)$ and $GMap(S_r)$ we merge their valid portions in cyclic order, and add vectors of the new supporting segments between $mMH(S_l)$ or $mMH(S_r)$. We will only provide the self-explanatory Figure 8 which illustrates the merging process of the two mM-hulls appearing in Figure 7. In particular, Figure 8(a) illustrates $mMH(S), S = S_l \cup S_r, S_l = \{A, B, C, D\}, S_r = \{G, F, H\}$, Figure 8(b) illustrates $GMap(S)$, and Figures 8(c) and 8(d) illustrate $GMap(S_l)$ and $GMap(S_r)$ respectively. In Figures 8(c) and 8(d) the invalid vectors are shown crossed out. The new vectors corresponding to the new supporting segments between $mMH(S_l)$ and $nMH(S_r)$ are indicated by thicker arrows.

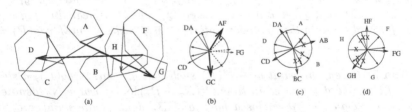

Fig. 8. Merging $mMH(S_l)$ and $mMH(S_r)$.

Theorem 2. *Merging $mMH(S_l)$ and $mMH(S_r)$ into $mMH(S)$ can be computed in linear time.*

4.2 Tracing the Cycles of $\sigma(S_l, S_r)$

Our goal is to identify a starting point on every cycle of $\sigma(S_l, S_r)$. Let $S_P = S_l$ and $S_Q = S_r$ (resp. $S_P = S_r, S_Q = S_l$) for any $P \in S_l \cap S_{\mathcal{L}}$ (resp. $P \in S_r \cap S_{\mathcal{L}}$). Let $mM\text{-}reg'(p), p \in P$, denote the region of p in $mM\text{-}Vor(S_P)$ or $mM\text{-}Vor(S_Q)$ and $mM\text{-}reg(p)$ denotes the region of p in $mM\text{-}Vor(S)$. Consider a connected component σ of $\sigma(S_l, S_r)$. It partitions the plane into two parts such that one side of $\sigma(S_l, S_r)$ borders regions of S_l, denoted as $H(S_l, S_r)$, and the other side borders regions of S_r, denoted as $H(S_r, S_l)$. A region of mM-$Vor(S_l)$ (resp. mM-$Vor(S_r)$) that falls in $H(S_r, S_l)$ (resp. $H(S_l, S_r)$) is said to be on the *wrong* side of σ and it may be partially enclosed by a cycle of $\sigma(S_l, S_r)$.

Let $P \in S_{\mathcal{L}}$ have a region in mM-$Vor(S_P)$ on the wrong side of the components of $\sigma(S_l, S_r)$ computed so far. Let $T'(P) \subseteq T(P)$ be the intra-bisector tree derived by one connected component of mM-$reg'(P)$ in $H(S_Q, S_P)$. We assume that $T'(P)$ is rooted at the point of minimum weight. Note that in the general crossing case there may be more than one connected component of mM-$reg'(P)$ but each one is treated independently.

Consider an intra-bisector segment $\overline{y_i y_j} \in T'(P)$ where y_i is an ancestor of y_j in $T'(P)$ (i.e., $d_f(y_i, P) < d_f(y_j, P)$). Let p_i and p_j be the vertices inducing $\overline{y_i y_j}$ (i.e.,$\overline{y_i y_j} \in b(p_i, p_j)$). Let $q_i \in Q_i$ and $q_j \in Q_j$ be the owners of y_i and y_j respectively in mM-$Vor(S_Q)$ (i.e., $y_i \in mM$-$reg'(q_i)$ and $y_j \in mM$-$reg'(q_j)$). Q_i and Q_j may or may not be the same shape. Comparing $d(y_k, p_k)$ and $d(y_k, q_k)$ for $k = i, j$ we can easily determine whether y_k is closer to P or Q_k (i.e., whether $y_k \in mM$-$reg(P)$ or $y_k \in mM$-$reg(q_k)$ in mM-$Vor(S)$). The following lemma can be derived from lemma 1 and property 3.

Lemma 4. *For an intra-bisector segment $\overline{y_i y_j} \in T'(P)$, and a merge curve cycle τ enclosing a portion of $\overline{y_i y_j}$ (if any) we have the following:*

1. *If $y_i \in mM$-$reg(P)$ and $y_j \in mM$-$reg(P)$, then τ exists but if S is non-crossing then τ cannot intersect $\overline{y_i y_j}$.*
2. *If exactly one of y_i, y_j belongs in mM-$reg(P)$ then τ must intersect $\overline{y_i y_j}$.*
3. *If $y_i \in mM$-$reg(Q_i)$ and $y_j \in mM$-$reg(Q_j)$ and if $Q_i = Q_j$, or if Q_i is forward limiting or if Q_j is rear limiting, then τ cannot intersect $\overline{y_i y_j}$.*
4. *If $y_i \in mM$-$reg(Q_i)$ and $y_j \in mM$-$reg(Q_j)$ but Q_i is rear limiting and Q_j is forward limiting then either τ intersects $\overline{y_i y_j}$ twice or mM-$reg(P) = \emptyset$.*

Lemma 5. *Given any point $y_i \in T'(P)$ such that $y_i \in mM$-$reg(P)$, a starting point of τ can be determined in time $O((N_\tau + B_\tau)\log n)$ where N_τ denotes the number of regions in S_P enclosed by τ and B_τ denotes the number of intra-bisectors of shapes in S_Q that get eliminated by τ.*

Proof. Sketch. Consider segment $\overline{q_i y_i}$ ($y_i \in mM$-$reg'(q_i)$). Let x be the intersection point of $\overline{q_i y_i}$ with the chain B of intra-bisectors bounding $reg'(q_i)$. If x is closer to q_i than the owner of x in mM-$Vor(S_P)$ then τ must intersect segment $\overline{y_i x}$. Determine the intersection point by walking on segment $\overline{y_i x}$, starting at y_i, and considering intersections with the regions of mM-$Vor(S_P)$ (if any). Otherwise, if x is closer to its owner in mM-$Vor(S_P)$, walk along chain B until either the whole chain is determined to be closer to S_P or a point equidistant from q_i i.e., a point on τ, is found. In the former case this connected component of mM-$reg(q_i)$ must be empty; update y_i to the intersection point, if any, of $\overline{y_i y_j}$ and mM-$reg(q_i)$ and repeat the process. The method maintains time complexity.

Lemma 6. *For any segment $\overline{y_i y_j} \in T'(P)$ induced by chord $\overline{p_i p_j}$ such that $y_i \in mM\text{-}reg(Q_i)$, $y_j \in mM\text{-}reg(Q_j)$ and Q_i is rear limiting or crossing with $\overline{p_i p_j}$ and Q_j is forward limiting or crossing with $\overline{p_i p_j}$, the first intersection x of τ with $\overline{y_i y_j}$ (if any) can be determined in time $O(M_\tau \log n)$ where M_τ denotes the total size of rear limiting or crossing shapes shapes enclosed in the P-circles centered along $\overline{y_i x}$.*

Proof. Sketch. We traverse segment $\overline{y_i y_j}$, starting at y_i, considering the intersection points of $\overline{y_i y_j}$ with $mM\text{-}Vor(S_Q)$ until either a point closer to P is encountered, or y_j is reached, or a point t on some inter-bisector $b(q_k, q_l)$ ($q_k \in Q_k, q_l \in Q_l$) of $mM\text{-}Vor(S_Q)$ is reached such that Q_l is forward limiting. In the first case we can easily determine the starting point of a cycle τ. In the second and third case we conclude that τ does not intersect $\overline{y_i y_j}$. For non-crossing shapes the latter observation implies $reg(P) = \emptyset$.

A starting point of τ in the non-crossing case can be found as follows:
Non-crossing case: Identify $S_{\mathcal{L}} \subset S_l$. For every $P \in S_{\mathcal{L}}$ such that $mM\text{-}reg'(P)$ is on the *wrong* side of the unbounded portion of $\sigma(S_l, S_r)$ identify $T'(P)$. Locate the root of $T'(P)$ in $mM\text{-}Vor(S_r)$. If the root remains in $mM\text{-}reg(P)$ identify a starting point of τ as explained in lemma 5. Otherwise visit the vertices of $T'(P)$ in order of increasing weight until either Case 2 or Case 4 of Lemma 4 is encountered. In case 2 follow lemma 5 and in case 4 follow lemma 6.

We now generalize to arbitrarily crossing shapes. Let $T(S_l)$ (resp. $T(S_r)$) denote the collection of all intra-bisector trees induced by the regions of shapes in $S_{\mathcal{L}} \cap S_l$ (resp. $S_{\mathcal{L}} \cap S_r$) that fall on the *wrong side* of the merge curves computed so far.
General case: Locate the vertices of $T(S_l)$ and $T(S_r)$ into $mM\text{-}Vor(S_r)$ and $mM\text{-}Vor(S_l)$ respectively. Process first $T(S_l)$ until empty and then repeat for $T(S_r)$. For each tree $T'(P)$ in $T(S_l)$ or $T(S_r)$ do:

1. For any vertex $y_i \in T(P)$ such that $y_i \in mM\text{-}reg(Q_i)$ and Q_i is a forward (resp. rear) limiting shape, eliminate $T(y_i)$ (resp. $T_c(y_i)$).
2. For any adjacent vertices $y_i, y_j \in T(P)$ such that $y_i \in mM\text{-}reg(Q_i)$ and $y_j \in mM\text{-}reg(Q_j)$ but $Q_i = Q_j$ eliminate segment $\overline{y_i y_j} \in T(P)$.
3. For any vertex $y_i \in T(P)$ such that $y_i \in mM\text{-}reg(P)$, determine a starting point of the cycle enclosing y_i following lemma 5.
4. For any adjacent vertices $y_i, y_j \in T(P)$ such that $y_i \in mM\text{-}reg(Q_i)$ and $y_j \in mM\text{-}reg(Q_j)$ determine the first intersection of $\overline{y_i y_j}$ and a merge cycle (if any) as explained in lemma 6. If a forward (resp. rear) limiting shape is identified during the traversal eliminate $T(y_i)$ (resp. $T_c(y_i)$). Eliminate any portion of $\overline{y_i y_j}$ encountered during the traversal.
5. For any starting point identified in Steps 3 and 4 trace the merge cycle.
6. Update $T(S_P)$ and $T(S_Q)$.
7. Repeat for every tree in $T(S_l)$ until $T(S_l) = \emptyset$. Repeat for every tree in $T(S_r)$. Repeat until both $T(S_l)$ and $T(S_r)$ are empty. (During the processing of $T(S_r)$, $T(S_l)$ may become non-empty and vice versa).

By lemma 1 all rear limiting shapes with respect to a chord of P are enclosed within the minimum enclosing circle of P.

Definition 11. *A shape $Q \in S$ is said to be* interacting *with $P \in S$ if Q is entirely enclosed in the minimum enclosing circle of P i.e., the minimum radius P-circle. Pair (P, Q) is called an* interacting pair. *The total number of vertices of interacting pairs of shapes is denoted by K.*

Theorem 3. *The min-max Voronoi diagram mM-$Vor(S)$ can be computed by divide and conquer in time $O((n + M + N + K) \log m)$ where n is the number of input points, M is the number of vertices of crossing pairs of shapes, m is the size of the intersection graph $G(S)$, K is the number of vertices of interacting pairs of shapes, and $N = \Sigma_{\mathcal{L}} |S_{\mathcal{L}}|$ for all dividing lines $\mathcal{L}$ (N is $O(M \log m)$). For a non-crossing S, the algorithm simplifies to $O((n + N + K) \log n)$.*

References

1. M. Abellanas, G. Hernandez, R. Klein,V. Neumann-Lara, and J. Urrutia, *Discrete Computat. Geometry* 17, 1997, 307–318.
2. F. Aurenhammer, "Voronoi diagrams: A survey of a fundamental geometric data structure," *ACM Comput. Survey*, 23, 345–405, 1991.
3. L.L. Chen, S.Y. Chou, and T.C. Woo, 1993, "Optimal parting directions for mold and die design," *Computer-Aided Design*, Vol. 26, No. 12, pp.762–768.
4. H. Edelsbrunner, L.J. Guibas, and M. Sharir, "The upper envelope of piecewise linear functions: algorithms and applications", *Discrete Computat. Geometry* 4, 1989, 311–336.
5. R. Klein, "Concrete and Abstract Voronoi Diagrams", vol. 400, *Lecture Notes in Computer Science*, Springer-Verlag, 1989.
6. R. Klein,, K. Melhorn, S. Meiser, "Randomized Incremental Construction of Abstract Voronoi diagrams", *Computational geometry: Theory and Aplications* 3,1993, 157–184
7. W. Maly, "Computer Aided Design for VLSI Circuit Manufacturability," *Proc. IEEE*, vol.78, no.2, 356–392, Feb. 90.
8. C.H. Ouyang, W.A. Pleskacz, W. Maly, "Extraction of Critical Areas for Opens in Large VLSI Circuits", *IEEE Trans. on Computer-Aided Design*, vol. 18, no 2, 151-162, February 1999.
9. B. R. Mandava, "Critical Area for Yield Models", IBM Technical Report TR22.2436, East Fishkill, NY, 12 Jan 1982.
10. E. Papadopoulou, "Plane sweep construction for the min-Max (Hausdorff) Voronoi diagram", Manuscript in preparation.
11. E. Papadopoulou, "Critical Area Computation for Missing Material Defects in VLSI Circuits", *IEEE Transactions on Computer-Aided Design*, vol. 20, no.5, May 2001, 583–597.
12. E. Papadopoulou and D.T. Lee, "Critical Area Computation via Voronoi Diagrams", *IEEE Trans. on Computer-Aided Design*, vol. 18, no.4, April 1999,463–474.
13. E. Papadopoulou and D.T. Lee, "The L_∞ Voronoi Diagram of Segments and VLSI Applications", *International Journal of Computational Geometry and Applications*, Vol. 11, No. 5, 2001, 503—528.
14. Preparata, F. P. and M. I. Shamos, *Computational Geometry: an Introduction*, Springer-Verlag, New York, NY 1985.

Improved Distance Oracles for Avoiding Link-Failure[*]

Rezaul Alam Chowdhury and Vijaya Ramachandran

Department of Computer Sciences
The University of Texas at Austin
Austin, Texas 78712, USA
{shaikat, vlr}@cs.utexas.edu

Abstract. We consider the problem of preprocessing an edge-weighted directed graph to answer queries that ask for the shortest path from any given vertex to another avoiding a failed link. We present two algorithms that improve on earlier results for this problem. Our first algorithm, which is a modification of an earlier method, improves the query time to a constant while maintaining the earlier bounds for preprocessing time and space. Our second result is a new algorithm whose preprocessing time is considerably faster than earlier results and whose query time and space are worse by no more than a logarithmic factor.

1 Introduction

Given an edge-weighted directed graph $G = (V, E, w)$, where w is a weight function on E, the *distance sensitivity problem* asks for the construction of a data structure called the *distance sensitivity oracle* that supports any sequence of the following two queries:

- distance(x, y, u, v): return the shortest distance from vertex x to vertex y in G avoiding the edge (u, v).
- path(x, y, u, v): return the shortest path from vertex x to vertex y in G avoiding the edge (u, v).

This problem as formulated above was first addressed by Demetrescu and Thorup in [DT02] for directed graphs with nonnegative real valued edge weights. In an earlier paper, King and Sagert [KS99] addressed a variant of this problem related to reachability in directed acyclic graphs.

As in [DT02], in this paper we concentrate on answering distance queries in digraphs with nonnegative real valued edge weights under the failure of a single link. The goal is to preprocess the graph in order to answer distance queries quickly when a link failure is detected. It is assumed that the time gap between two successive link failures is long enough to permit us to compute a new data

[*] This work was supported in part by Texas Advanced Research Program Grant 003658-0029-1999 and NSF Grant CCR-9988160. Chowdhury was also supported by an MCD Graduate Fellowship.

structure in the background that will assist us in answering distance queries when the next link failure is detected.

In this paper we present two new algorithms for the distance sensitivity problem. An overview of our results is given in Section 4. In Section 2 we describe the notation we use, and we summarize earlier results in Section 3. In Sections 5 and 6 we present details of our two new methods.

2   Notations

Most of the notation we use is from [DT02], which we include below for convenience. By n and m we denote respectively the number of vertices and the number of edges in G, i.e., $n = |V|$ and $m = |E|$. By $T(x)$ we denote the shortest-path tree rooted at vertex x of G and by $\pi_{x,y}$ we denote the unique x to y path in $T(x)$. As in [DT02] we assume that all shortest paths in G are unique, since if not, we can make them unique by adding small random fractions to the edge weights. By $h_{x,y}$ we denote the number of edges in $\pi_{x,y}$. The weight of any edge (x,y) will be represented by $w_{x,y}$ and the weighted shortest distance from x to y will be represented by $d_{x,y}$. We denote by $\hat{G}$ the graph obtained by reversing the orientation of the edges in G and $\hat{\pi}$ and $\hat{T}$ respectively represent π and T related to $\hat{G}$. Since we assume that all shortest paths are unique, $\pi_{x,y}$ and $\hat{\pi}_{y,x}$ will have the same set of edges and for any vertex z on the shortest path from x to y, $\pi_{x,z}, \pi_{z,y} \subseteq \pi_{x,y}$.

Let u, v be vertices on a shortest path from x to y, with u being closer to x than v. By $\pi_{x,y}^{u,v}$ we denote a shortest path from vertex x to vertex y in G that avoids the sub-path from u to v (this is a slight generalization of the notation used in [DT02]). Let a be a vertex on $\pi_{x,u}$ and b a vertex on $\pi_{v,y}$ such that $\pi_{x,y}^{u,v} = \pi_{x,a} \cdot \pi_{a,b}^{u,v} \cdot \pi_{b,y}$ and $\pi_{a,b} \cap \pi_{a,b}^{u,v} = \emptyset$, where $\cdot$ denotes the path concatenation operator. So $\pi_{a,b}^{u,v}$ is the *best detour* one can follow in order to go from x to y while avoiding the subpath from u to v, and $\pi_{x,a}$ and $\pi_{b,y}$ represent the longest common portions of $\pi_{x,y}$ and $\pi_{x,y}^{u,v}$ (see Fig. 1). Now consider vertices l_1, l_2, r_1, r_2 that lie on $\pi_{x,y}$ at increasing distances from x. Let d be an upper bound on the length of $\pi_{x,y}^{l_2,r_1}$. The value d is said to *cover* $[l_1, l_2] \times [r_1, r_2]$ if $d \le$ shortest distance from x to y avoiding the interval $[l_2, r_1]$ when using a detour with a in π_{l_1,l_2} and b in π_{r_1,r_2}.

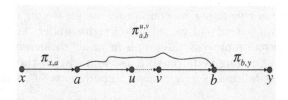

Fig. 1. Optimal detour [DT02].

Let P be a set of shortest paths in G. The paths in P are *independent* in G w.r.t. x if for any $\pi_1, \pi_2 \in P$ no edge of π_1 is a descendent of π_2 in $T(x)$ and vice versa. In our algorithms we will make use of two procedures *exclude* and *exclude-d* from [DT02]. If P is a set of shortest paths in G, both *exclude*(G, x, P) and *exclude-d*(G, x, P) compute, for each path $\pi \in P$ the shortest distance from vertex x to all other vertices in G avoiding the edges on π. The procedure *exclude*(G, x, P) runs in $O(|P|(m+n \log n))$ worst-case time using a fast Dijkstra computation [FT87] for each $\pi \in P$. On the other hand, if the shortest paths in P are independent, *exclude-d* can perform the same computation in $O(m + n \log n)$ worst-case time.

3   Existing Algorithms

A straight-forward approach to solving the distance sensitivity problem is to use a recent dynamic all pairs shortest paths (APSP) algorithm [DI01,K99,KT01] and delete a *specific* failed link. The time required is quite high ($\tilde{O}(n^{2.5})$ [1] amortized) even for unweighted graphs though after that the queries can be answered in $O(1)$ time assuming failure of that specific link. In contrast, the distance sensitivity problem asks for a preprocessing of the graph so that queries can be answered quickly under failure of *any* one link.

Another straightforward approach is to construct a table of size $O(n^3)$ that will store for each vertex pair (x, y) the shortest distance from x to y avoiding each of the $O(n)$ edges on the shortest x to y path in the original graph. The total time required to construct the table is $\tilde{O}(mn^2)$ [T01] and the query time is $O(1)$. The problem with this approach is the excessive space requirement (and its relatively large preprocessing time).

The first two non-trivial algorithms for solving the distance sensitivity problem were presented by Demetrescu and Thorup [DT02]. For convenience we name these two algorithms as DT-1 and DT-2, and summarize their complexities in Table 1. In DT-1 each x to y shortest path is divided into $O(\log n)$ segments and the shortest distance from x to y avoiding each of these $O(\log n)$ segments is stored in a table. The procedure *exclude* is used to compute these distances. The query algorithm progressively decreases an upper bound on the shortest distance from x to y avoiding a failed link (u, v) and returns the correct answer in $O(\log n)$ iterations. In DT-2, each shortest-path tree $T(x)$ is divided into $O(\sqrt{n})$ bands of independent paths. For each of the bands in each $T(x)$ the procedure *exclude-d* is used to compute the shortest distances from x to all y's avoiding the paths in that band. After this preprocessing, queries can be answered in constant time.

4   Our Results

In this paper, we present two algorithms: CR-1 and CR-2. The algorithm CR-1 is a simple but useful modification of DT-1 which brings down the query

---

[1] $\tilde{O}(f(n))$ is used to denote $O(f(n)\text{polylog}(n))$.

Table 1. Distance Sensitivity Oracles in [DT02]

Algorithm	Preprocessing Time	Extra Space	Query Time
DT-1	$O(mn^2 \log n + n^3 \log^2 n)$	$O(n^2 \log n)$	$O(\log n)$
DT-2	$O(mn^{1.5} \log n + n^{2.5} \log^2 n)$	$O(n^{2.5})$	$O(1)$

Table 2. Our Contribution

Algorithm	Preprocessing Time	Extra Space	Query Time
CR-1	$O(mn^2 \log n + n^3 \log^2 n)$	$O(n^2 \log n)$	$O(1)$
CR-2	$O(mn \log^2 n + n^2 \log^3 n)$	$O(n^2 \log^2 n)$	$O(\log n)$

time to constant while leaving the other costs unchanged. The second algorithm (CR-2) is based on a new approach for finding bands of independent paths that reduces the preprocessing time significantly without increasing the other costs by more than a logarithmic factor. More precisely, when compared with DT-1, CR-2 reduces the preprocessing time by a factor of $\frac{n}{\log n}$ at the cost of increasing the space requirement by a factor of $\log n$. The query time is unchanged. When compared with DT-2, CR-2 reduces the preprocessing time by a factor of $\frac{\sqrt{n}}{\log n}$ at the cost of increasing the query time by a factor of $\log n$. In this case the space requirement is also reduced significantly — by a factor of $\frac{\sqrt{n}}{\log^2 n}$.

5 The Algorithm CR-1

CR-1 is a simple modification of DT-1 which improves the query time to $O(1)$ while keeping the preprocessing time and the space requirement unchanged ($O(mn^2 \log n + n^3 \log^2 n)$ and $O(n^2 \log n)$ respectively). The only difference in preprocessing between CR-1 and DT-1 is the use of two additional matrices, sl and sr, in CR-1. The query algorithm in CR-1 uses these two matrices to reduce the query time to constant (from $\Theta(\log n)$ in DT-1). The matrix sl is similar to the matrix dl in DT-1: while $dl(x, y, i)$ stores the shortest distance from vertex x to y in G without the edges of the sub-path of $\pi_{x,y}$ between level $\lfloor 2^{i-1} \rfloor$ and level 2^i in $T(x)$, $sl(x, y, i)$ stores the x to y shortest distance avoiding only the first edge (i.e., edge nearest to x) on that sub-path. A similar relation holds between matrices dr and sr. Combined with the data stored in dl and dr, the sl and sr matrices allow us to answer the query $distance(x, y, u, v)$ in constant time.

Data Structure

We maintain each $d_{x,y}$ and $h_{x,y}$ using $O(n^2)$ space. We use $O(n^2 \log n)$ space in order to maintain six matrices dl, dr, sl, sr, vl and vr of size $n \times n \times \log n$ each defined for any x, y and for any i, $0 \leq i < \log n$, as follows:

- $dl[x, y, i]$ = shortest distance from vertex x to vertex y in G without the edges on the sub-path of $\pi_{x,y}$ between level $\lfloor 2^{i-1} \rfloor$ and level 2^i in $T(x)$;

- $dr[x, y, i]$ = shortest distance from vertex y to vertex x in $\hat{G}$ without the edges on the sub-path of $\hat{\pi}_{y,x}$ between level $\lfloor 2^{i-1} \rfloor$ and level 2^i in $\hat{T}(y)$;
- $sl[x, y, i]$ = shortest distance from vertex x to vertex y in G without the edge of $\pi_{x,y}$ between level $\lfloor 2^{i-1} \rfloor$ and level $\lfloor 2^{i-1} \rfloor + 1$ in $T(x)$;
- $sr[x, y, i]$ = shortest distance from vertex y to vertex x in $\hat{G}$ without the edge of $\hat{\pi}_{y,x}$ between level $\lfloor 2^{i-1} \rfloor$ and level $\lfloor 2^{i-1} \rfloor + 1$ in $\hat{T}(y)$;
- $vl[x, y, i]$ = vertex of $\pi_{x,y}$ at level $\lfloor 2^{i-1} \rfloor$ in $T(x)$;
- $vr[x, y, i]$ = vertex of $\hat{\pi}_{y,x}$ at level $\lfloor 2^{i-1} \rfloor$ in $\hat{T}(y)$;

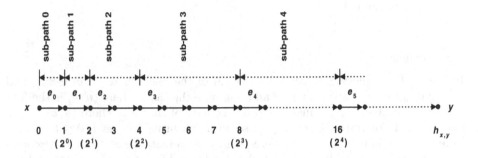

Fig. 2. Edge e_i is the edge on $\pi_{x,y}$ that is excluded in calculating the distance $sl(x, y, i)$. Sub-path i is the sub-path excluded in calculating $dl(x, y, i)$.

Preprocessing

As in [DT02], we initialize the distances $d_{x,y}$ and $h_{x,y}$ and the matrices vl and vr from the shortest-path trees of G, and the matrices dl and dr by calling the procedure *exclude*.

Let $B_{T(x)}(j, k)$ denote the band of paths in $T(x)$ that connect vertices at level j with vertices at level $k > j$ in the tree assuming that x is at level 0 in $T(x)$. For each x and for each i, $0 \le i < \log n$, we compute $sl[x, y, i]$ by calling procedure $exclude\text{-}d(G, x, B_{T(x)}(\lfloor 2^{i-1} \rfloor, \lfloor 2^{i-1} \rfloor + 1))$. We compute $sr[x, y, i]$ by calling procedure $exclude\text{-}d(\hat{G}, y, B_{\hat{T}(y)}(\lfloor 2^{i-1} \rfloor, \lfloor 2^{i-1} \rfloor + 1))$, for each y and for each i, $0 \le i < \log n$.

Query

The query algorithm is as follows:

```
function distance(x, y, u, v): ℜ
1.      if d_{x,u} + w_{u,v} + d_{v,y} > d_{x,y} then return d_{x,y} fi
2.      if x = u then return sl[x, y, 0] fi
```

```
3.      if y = v then return sr[x, y, 0] fi
4.      l := ⌈log₂ hₓ,ᵤ⌉
5.      if l = log₂ hₓ,ᵤ then return sl[x, y, l + 1] fi
6.      r := ⌈log₂ hᵥ,ᵧ⌉
7.      if r = log₂ hᵥ,ᵧ then return sr[x, y, r + 1] fi
8.      p := vr[x, u, l],  q := vl[v, y, r]
9.      d := min(dₓ,ₚ + sl[p, y, l], sr[x, q, r] + dᵩ,ᵧ)
10.     if hₓ,ᵤ ≤ hᵥ,ᵧ then
11.         d := min(d, dl[x, y, l])
12.     else
13.         d := min(d, dr[x, y, r])
14.     fi
15.     return d
```

Correctness

In line 1 of the query algorithm, we get rid of the case where $(u, v) \notin \pi_{x,y}$ and return $d_{x,y}$ as the answer. Lines 2 and 3 handle the cases where (u, v) is the first or the last edge on $\pi_{x,y}$. Lines 4 and 5 take care of the case where (u, v) is at a distance 2^l from vertex x on $\pi_{x,y}$ for some nonnegative integer l, $0 \leq l < \log n$. Lines 6 and 7 handle the case where (u, v) is at a distance 2^r from vertex y on $\hat{\pi}_{y,x}$ for some nonnegative integer r, $0 \leq l < \log n$. Lines 8 to 15 take care of the remaining cases.

Lines 2 to 7 handle the following four trivial cases: (1) $h_{x,u} = 0$, (2) $h_{v,y} = 0$, (3) $h_{x,u} = 2^l$ for some nonnegative integer l, $0 \leq l < \log n$, and (4) $h_{v,y} = 2^r$ for some nonnegative integer r, $0 \leq r < \log n$. So in order to prove the correctness of *distance* we need only to prove the correctness of the code segment of lines 8 to 14 that handles the nontrivial case when none of the above four conditions hold.

Let $d_1 = d_{x,p} + sl[p, y, l]$. The distance d_1 covers $[p, u] \times [v, y]$ and the distance $d_2 = sr[x, q, r] + d_{q,y}$ covers $[x, u] \times [v, q]$. Now let us consider the case when $h_{x,u} \leq h_{v,y}$. In this case $0 \neq h_{x,p} < h_{p,u} = 2^{l-1} \leq h_{v,q}$. $dl[x, y, l]$ is the distance from x to y avoiding a sub-path of length 2^{l-1} at a distance of 2^{l-1} from x on $\pi_{x,y}$. Let the endpoints of that sub-path be p' and q', and $h_{x,p'} < h_{x,q'}$. So $[p', q']$ contains (u, v) and $[p, q]$ contains $[p', q']$. Hence, $dl[x, y, l]$ covers $[x, p] \times [q, y]$. But $[p, u] \times [v, y] \cup [x, u] \times [v, q] \cup [x, p] \times [q, y] = [x, u] \times [v, y]$. So d covers $[x, u] \times [v, y]$. Similar argument holds for the case when $h_{x,u} > h_{v,y}$. Hence we conclude

Claim 1. *The query function distance(x, y, u, v) for algorithm CR-1 correctly computes the shortest distance from vertex x to vertex y in G avoiding the edge (u, v).*

Preprocessing, Query, and Space Bounds

Claim 2. *In algorithm CR-1, the preprocessing requires $O(mn^2 \log n + n^3 \log^2 n)$ worst-case time, the data structure requires $O(n^2 \log n)$ space, and any distance query can be answered in $O(1)$ worst-case time.*

Proof. The entries of the matrices dl and dr are computed as in [DT02] in $O(mn^2 \log n + n^3 \log^2 n)$ worst-case time. For each x and for each i, $0 \leq i < \log n$, we can compute $sl[x, y, i]$ by calling procedure *exclude-d*$(G, x, B_{T(x)}(\lfloor 2^{i-1} \rfloor, \lfloor 2^{i-1} \rfloor + 1))$ since the single-edge paths in $B_{T(x)}(\lfloor 2^{i-1} \rfloor, \lfloor 2^{i-1} \rfloor + 1)$ are trivially independent. Since *exclude-d*$(G, x, B_{T(x)}(\lfloor 2^{i-1} \rfloor, \lfloor 2^{i-1} \rfloor + 1))$ runs in $O(m + n \log n)$, the total time required to compute the matrix sl is $O(mn \log n + n^2 \log^2 n)$. Similarly the matrix sr can be computed in $O(mn \log n + n^2 \log^2 n)$. Hence the preprocessing time is dominated by the time to compute the dl and dr matrices and requires $O(mn^2 \log n + n^3 \log^2 n)$ worst-case time.

The query algorithm does not contain any loops and it does not make any function call either. Therefore it runs in $O(1)$ worst-case time.

The d and h matrices have size $n \times n$, and each of the matrices dl, dr, sl, sr, vl and vr has size $n \times n \times \log n$. Therefore the total space requirement is $O(n^2 \log n)$.   □

6   The Algorithm CR-2

This is a new algorithm with preprocessing time $O(mn \log^2 n + n^2 \log^3 n)$, query time $O(\log n)$ and space $O(n^2 \log^2 n)$. In this algorithm we divide each shortest-path tree $T(x)$ into $O(\log n)$ bands of independent paths. In order to achieve a logarithmic query time we further subdivide each path in a band into $O(\log n)$ segments, thus producing a total of $O(\log^2 n)$ sub-bands (or *segments*) of independent paths in each $T(x)$. For each sub-band of independent paths we use *exclude-d* to compute the required distances to be stored in our data structure. Compared to DT-1, DT-2 and CR-1, this strategy significantly reduces the preprocessing time without increasing the query time or space requirement by more than a factor of $\log n$. (Note that DT-2 divides each $T(x)$ into $O(\sqrt{n})$ bands of independent paths.)

Data Structure

In the following, each of the matrices d, h, bcl and bcr has size $O(n^2)$, and each of the matrices vl, vr, dl and dr has size $O(n^2 \log^2 n)$.

- $bcl[x, y] =$ index i such that the ith segment of $\pi_{x,w}$ contains y where w is a descendent of y in $T(x)$. Defined only for the internal nodes of $T(x)$;
- $bcr[x, y] =$ index i such that the ith segment of $\hat{\pi}_{y,w}$ contains x where w is a descendent of x in $\hat{T}(y)$. Defined only for the internal nodes of $\hat{T}(y)$;
- $vl[x, y, i] =$ first vertex (i.e., vertex nearest from x) of $\pi_{x,y}$ contained in the ith segment of $\pi_{x,y}$ in $T(x)$;
- $vr[x, y, i] =$ first vertex (i.e., vertex nearest from y) of $\hat{\pi}_{y,x}$ contained in the ith segment of $\hat{\pi}_{y,x}$ in $\hat{T}(y)$;
- $dl[x, y, i] =$ shortest distance from vertex x to vertex y in G without the edges on the ith segment of $\pi_{x,y}$ in $T(x)$;
- $dr[x, y, i] =$ shortest distance from vertex y to vertex x in $\hat{G}$ without the edges on the ith segment of $\hat{\pi}_{y,x}$ in $\hat{T}(y)$;

Preprocessing

Distances $d_{x,y}$ and $h_{x,y}$ and matrices vl and vr are easily initialized from shortest-path trees of G. However, before describing how the other matrices are initialized we will first describe how the shortest paths are segmented.

A *chain* in a directed graph G is a path $\langle v_1, v_2, \ldots, v_k \rangle$ such that each v_i has exactly one incoming edge and one outgoing edge in G. A *maximal chain* is one that cannot be extended. As defined in [R97] a *leaf chain* $\langle v_1, v_2, \ldots, v_{l-1}, v_l \rangle$ in a rooted tree T consists of a maximal chain $\langle v_1, v_2, \ldots, v_{l-1} \rangle$, together with a vertex v_l, which is a leaf and the unique child of v_{l-1} in T. For convenience we modify this definition slightly by augmenting the leaf chain $\langle v_1, v_2, \ldots, v_l \rangle$ to $\langle v_0, v_1, \ldots, v_l \rangle$ where v_0 is the predecessor of v_1 in T and call this $\langle v_0, v_1, \ldots, v_l \rangle$ a leaf chain instead of $\langle v_1, v_2, \ldots, v_l \rangle$.

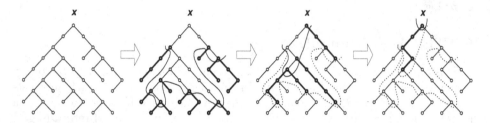

Fig. 3. Repeated applications of *Shrink* create bands of independent paths in $T(x)$. In the figure the dark paths below each new solid curve represent a collection of independent paths.

Claim 3. *The set of all leaf chains of a shortest-path tree $T(x)$ form a band of independent paths with respect to x.*

Proof. By definition, no two leaf chains share an edge. It also follows from the definition that no edge in one leaf chain is a descendent of another edge in another leaf chain. Therefore the set of all leaf chains of $T(x)$ form a band of independent paths with respect to x. □

Let us consider a tree operation *Shrink* [R87] that removes all leaf chains. The following Lemma (with a slightly different wording) has been proved in [R87] (see also [NNS89]).

Lemma 1. *Any n node tree can be transformed into a single vertex with $O(\log n)$ applications of the Shrink operation.*

This leads to the following corollary:

Corollary 1. *Each shortest-path tree $T(x)$ of G can be partitioned into $O(\log n)$ bands of independent paths with respect to x.*

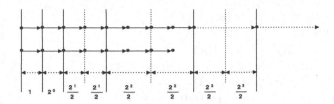

Fig. 4. Segmentation of leaf chains and creation of sub-bands in a band.

Now consider any leaf chain $\langle v_0, v_1, \ldots, v_l \rangle$ in a band. We divide the chain into $O(\log l)$ segments where the ith $(i \geq 0)$ segment is the sub-chain $\langle v_{s(i)}, v_{s(i)+1}, \ldots, v_{\min(e(i),l)} \rangle$ and this segment exists iff $s(i) < l$. Here $s(i)$ and $e(i)$ are defined as follows:

$$\langle s(i), e(i) \rangle = \begin{cases} \langle i, i+1 \rangle, & if\ i \leq 1 \\ \langle 2^{\frac{i}{2}}, 3 \times 2^{\frac{i}{2}-1} \rangle, & if\ i > 1\ and\ even \\ \langle 3 \times 2^{\frac{i-3}{2}}, 2^{\frac{i+1}{2}} \rangle, & if\ i > 1\ and\ odd \end{cases}$$

Since we know that splitting any band of independent paths yields again bands of independent paths, we can split each of the $O(\log n)$ bands created using the *Shrink* operation, into $O(\log n)$ sub-bands of independent paths where the ith $(i \geq 0)$ sub-band includes the ith segment (if it exists) of each of the sub-paths in that band (see Fig. 4).

Each shortest-path tree $T(x)$ has $O(\log^2 n)$ sub-bands of independent paths and for convenience we will number them from top to bottom with consecutive integers starting from 0. By $\beta_{T(x),i}$ we denote the set of paths in the ith sub-band of $T(x)$. In order to compute $dl[x,y,i]$ for each vertex x and for each of the $O(\log^2 n)$ values of i, we call $exclude\text{-}d(G, x, \beta_{T(x),i})$ once. Similarly, for each vertex y and for each of the $O(\log^2 n)$ values of i, we compute $dr[x,y,i]$ by calling $exclude\text{-}d(\hat{G}, y, \beta_{\hat{T}(y),i})$ once.

It is also clear that any shortest path $\pi_{x,y}$ in $T(x)$ can have $O(\log^2 n)$ segments. For convenience we number these segments from x to y with consecutive integers starting from 0.

If a vertex u on $\pi_{x,y}$ lies on the boundary of two consecutive segments in $T(x)$ we assume that the lower numbered segment contains u. Any edge (u,v) on $\pi_{x,y}$ is assumed to be contained in the same segment that contains vertex u.

Once the paths are segmented, it is straightforward to compute the entries in the matrices bcl, bcr, vl and vr.

Query

The query algorithm is as follows:

```
function distance(x, y, u, v): ℜ
1.    if d_{x,u} + w_{u,v} + d_{v,y} > d_{x,y} then return d_{x,y} fi
2.    i := bcl[x,u],  d := dl[x,y,i]
```

```
3.      l := vl[x, y, i],  r := vl[x, y, i + 1]
4.      while h_{l,r} > 1 do
5.          if h_{l,v} ≤ ⌈h_{l,r}/2⌉ then
6.              i := bcr[v, r],  d := min(d, dr[x, r, i] + d_{r,y})
7.              t := vr[x, r, i + 1],  r := vr[x, r, i]
8.              if h_{x,l} > h_{x,t} then l := t fi
9.          else
10.             i := bcl[l, u],  d := min(d, d_{x,l} + dl[l, y, i])
11.             t := vl[l, y, i + 1],  l := vl[l, y, i]
12.             if h_{r,y} > h_{t,y} then r := t fi
13.         fi
14.     end while
15.     return d
```

Correctness

In line 1, we get rid of the case when $(u, v) \notin \pi_{x,y}$ and return $d_{x,y}$ as the answer. If $h_{x,l} \leq 1$ before entering the *while* loop, d trivially covers $[x, u] \times [v, y]$ and the answer is returned without executing the loop. However, if $h_{x,l} > 1$, we have $h_{l,r} \leq \frac{h_{x,l}}{2} < \frac{h_{x,y}}{2}$. $[l, r]$ contains (u, v) and d covers $[x, l] \times [r, y]$.

Now, consider what happens in an iteration of the *while* loop. Let us examine the case when $h_{l,v} \leq \lceil \frac{h_{l,r}}{2} \rceil$. Let $p = vr[x, r, i + 1]$, $q = vr[x, r, i]$ and l', r' respectively be the new values of l, r in the next iteration.

After the assignment in line 9, d covers $[x, p] \times [q, r] \cup [x, l] \times [r, y]$.

If $h_{x,p} < h_{x,l}$ then $l' = p$ and $r' = q$. So, $h_{l',r'} = h_{p,q} \leq \frac{h_{q,r}}{2} < \frac{h_{l,r}}{2} < \frac{3}{4}h_{l,r}$

If $h_{x,p} \geq h_{x,l}$ then $l' = l$ and $r' = q$. Let m be a vertex on $\pi_{l,r}$ such that $h_{l,m} = \lceil \frac{h_{l,r}}{2} \rceil$.

Now, if $h_{l,q} \leq h_{l,m}$, then $h_{l',r'} = h_{l,q} \leq h_{l,m} \leq \lceil \frac{h_{l,r}}{2} \rceil < \frac{3}{4}h_{l,r}$.

Otherwise,

$$
\begin{aligned}
h_{l',r'} &= h_{l,q} \\
&= h_{l,p} + h_{p,q} \\
&\leq \frac{h_{l,r}}{2} + h_{p,q} && [h_{l,p} \leq h_{l,u} < h_{u,r} \Rightarrow h_{l,p} \leq \frac{h_{l,r}}{2}] \\
&\leq \frac{h_{l,r}}{2} + \frac{h_{q,r}}{2} \\
&< \frac{h_{l,r}}{2} + \frac{h_{l,r}}{4} && [h_{l,q} > h_{l,m} \Rightarrow h_{q,r} < h_{m,r} \Rightarrow h_{q,r} < \frac{h_{l,r}}{2}] \\
&= \frac{3}{4}h_{l,r}
\end{aligned}
$$

So, when $h_{l,v} \leq \lceil \frac{h_{l,r}}{2} \rceil$, we always have $h_{l',r'} < \frac{3}{4}h_{l,r}$. Moreover in each of the above cases $[x, l'] \times [r', y] \subseteq [x, p] \times [q, r] \cup [x, l] \times [r, y]$, i.e., d covers $[x, l'] \times [r', y]$. Using similar arguments we can prove the same results for $h_{l,v} > \lceil \frac{h_{l,r}}{2} \rceil$, too.

Note that $[l, r]$ always contains (u, v) and $h_{l,r}$ is reduced by a factor of at least $\frac{4}{3}$ in each iteration of the *while* loop until it becomes equal to 1. So, at the end of the algorithm $[l, r] = [u, v]$. Therefore, d covers $[x, u] \times [v, y]$ eventually. Hence we obtain

Claim 4. *In algorithm CR-2, the query function distance(x, y, u, v) correctly computes the shortest distance from vertex x to vertex y in G avoiding the edge (u, v).*

Preprocessing, Query, and Space Bounds

Claim 5. *In algorithm CR-2, preprocessing requires $O(mn \log^2 n + n^2 \log^3 n)$ worst-case time, any distance query can be answered in $O(\log n)$ worst-case time, and the data structure requires $O(n^2 \log^2 n)$ space.*

Proof. In order to initialize the matrices dl and dr the function *exclude-d* is called a total of $O(n \log^2 n)$ times and *exclude-d* runs in $O(m + n \log n)$. It is trivial to see that the cost of computing dl and dr dominates the preprocessing cost. Therefore, preprocessing requires $O(mn \log^2 n + n^2 \log^3 n)$ worst-case time.

The $O(\log n)$ worst-case query time follows from the fact that $h_{l,r} < n$ initially, and is reduced by a factor of at least $\frac{4}{3}$ in each iteration of the while loop in the query algorithm.

Each of the matrices d, h, bcl and bcr has size $O(n^2)$, and each of the matrices vl, vr, dl and dr has size $O(n^2 \log^2 n)$. Therefore, the total space requirement is $O(n^2 \log^2 n)$. □

7   Further Extensions and Concluding Remarks

We have presented two improved distance sensitivity oracles for directed graphs with real edge weights, that answer queries asking for the shortest distance from any given vertex to another one avoiding a failed link. The query algorithms can be easily modified to handle the following situations without increasing the query time:

Arbitrary change in a link weight. If the weight of a link (u, v) changes from $w_{u,v}$ to $w'_{u,v}$ any distance query asking for the shortest distance from any vertex x to another vertex y in the changed graph can be answered as $\min(\text{distance}(x, y, u, v), d_{x,u} + w'_{u,v} + d_{v,y})$.

At most one link fails and a constant number of links are restored. If a link (u, v) fails and another link (u_1, v_1) gets restored then the shortest distance from a vertex x to another vertex y can be answered as $\min(\text{distance}(x, y, u, v), \text{distance}(x, u_1, u, v) + w'_{u_1,v_1} + \text{distance}(v_1, y, u, v))$ where w'_{u_1,v_1} is the weight of the restored link (u_1, v_1). If no more than a constant number of links are restored we will need to consider only a constant number of permutations of the restored links along the shortest x to y path in the changed graph. So the query time increases by no more than a constant factor.

There are a number of avenues for further research in this area. Designing efficient oracles that are able to answer distance queries when several links fail at the same time is one direction for further research. Designing more efficient pre-processing algorithms, for instance by using incremental calculations, is another avenue.

It may be possible to obtain more efficient oracles for special types of graphs such as undirected graphs or unweighted graphs. For instance, for undirected graphs the computation may be simplified by avoiding the use of the graph $\hat{G}$. Further, for sparse undirected graphs, the preprocessing time can be improved by using the undirected shortest path algorithm in [PR02] if the ratio r of the maximum to minimum edge weight is not too large (i.e., $r \in 2^{n^{o(1)}}$). By using the algorithm in [PR02] in place of Dijkstra's algorithm the preprocessing time is improved to $O(mn^2\alpha(m,n)\log n + n^3 \log n \log \log r)$ for CR-1 and $O(mn\alpha(m,n)\log^2 n + n^2 \log^2 n \log \log r)$ for CR-2.

References

[DI01]   C. Demetrescu, G. F. Italiano. Fully dynamic all pairs shortest paths with real edge weights. In *Proc. of the 42nd IEEE Annual Symposium on Foundations of Computer Science (FOCS'01), Las Vegas, Nevada*, pp. 838–843, 2001.

[DT02]   C. Demetrescu, M. Thorup. Oracles for distances avoiding a link-failure. In *Proc. of the 13th IEEE Annual ACM-SIAM Symposium on Discrete Algorithms (SODA'02), San Fransisco, California*, pp. 838–843, 2002.

[FT87]   M. L. Fredman, R. E. Tarjan. Fibonacci heaps and their uses in improved network optimization algorithms. *Journal of the ACM*, 34:596–615, 1987.

[KS99]   V. King, G. Sagert. A fully dynamic algorithm for maintaining the transitive closure. In *Proc. of the 31st ACM Symposium on Theory of Computing (STOC'99)*, pp. 492–498, 1999.

[KT01]   V. King, M. Thorup. A space saving trick for directed fully dynamic transitive closure and shortest path algorithms. In *Proc. of the 7th Annual International computing and Combinatorics Conference (COCOON), LNCS 2108*, pp. 268–277, 2001.

[K99]   V. King. Fully dynamic algorithms for maintaining all-pairs shortest paths and transitive closure in digraphs. In *Proc. of the 40th IEEE Annual Symposium on Foundations of Computer Science (FOCS'99)*, pp. 81–99, 1999.

[NNS89]   J. Naor, M. Naor, A. A. Schaffer, "Fast parallel algorithms for chordal graphs," *SIAM J. Computing*, Vol. 18, pp. 327–349, 1989.

[PR02]   S. Pettie, V. Ramachandran. Computing shortest paths with comparisons and additions (extended abstract). In *Proc. of the 13th Annual ACM-SIAM Symposium on Discrete Algorithms (SODA)*, pp. 267–276, 2002.

[R87]   V. Ramachandran. Parallel algorithms for reducible flow graphs. *Journal of Algorithms*, 23:1–31, 1997. (Preliminary version in *Princeton Workshop on Algorithm, Architecture, and and Technology*, S.K. Tewksbury, B.W. Dickinson, S.C. Schwartz, ed., 1987, pp. 117–138. Plenum Press.)

[T01]   M. Thorup. Fortifying OSPF/IS-IS against link-failure, 2001. Manuscript.

Probabilistic Algorithms for the Wakeup Problem in Single-Hop Radio Networks*

Tomasz Jurdziński[1,2] and Grzegorz Stachowiak[1]

[1] Institute of Computer Science, Wrocław University
[2] Institute of Computer Science, Technical University of Chemnitz

Abstract. We consider the problem of waking up n processors of a completely broadcast system. We analyze this problem in both globally and locally synchronous models, with or without n being known to processors and with or without labeling of processors. The main question we answer is: how fast we can wake all the processors up with probability $1 - \varepsilon$ in each of these eight models. In [10] a logarithmic waking algorithm for the strongest set of assumptions is described, while for weaker models only linear and quadratic algorithms were obtained. We prove that in the weakest model (local synchronization, no knowledge of n or labeling) the best waking time is $O(n/\log n)$. We also show logarithmic or poly-logarithmic waking algorithms for all stronger models, which in some cases gives an exponential improvement over previous results.

1 Introduction

We concentrate on the effects of synchronization level in broadcast systems such as the Ethernet or radio networks (RN). The system is assumed to be synchronous, i.e. processors send messages in rounds. It is assumed that the processors succeed in hearing a message in a given round if and only if exactly one processor sends a message in that round (the situations in which more than one processor or none of the processors send a message are indistinguishable). Hence the communication model is equivalent to the radio model [1,2,3,6,14,16], in a complete graph. This case in which every processor is directly accessible by the other ones is called a *single-hop* radio network. The model considered here is called single-hop radio networks without collision detection (no-CD RN), as opposite to the model in which processors are able to distinguish between the situations in which more than one processor or none of the processors sends a message (CD RN). Because of diversity of technical capabilities of existing technologies, both models (noCD RN and CD RN) are investigated in the literature.

A central problem in such systems is to establish the pattern of access to the shared communication media that allows messages to go through with a small delay, i.e., avoid or efficiently resolve message collisions. Another important problem considered is to design algorithms working without complete information about the network (for example topology, size) and/or an information which makes possible to distinguish its elements (e.g. different labels) [6,16,9,2,5]. This direction is motivated by the mobility of radio networks, changes of the topology and the set of participants.

---

* The first author was supported by DFG grant GO 493/1-1, and the second by KBN grant 8T11C 04419.

P. Bose and P. Morin (Eds.): ISAAC 2002, LNCS 2518, pp. 535–549, 2002.

Problem statement. In this paper, we consider the fundamental problem of waking up all of n processors (or „radio stations") in a completely broadcast system or single-hop radio network [10] (see [10] for motivations). Some processors wake up spontaneously, in different rounds, while others have to be woken up. Only awake processors can send messages. All sleeping processors wake up non-spontaneously upon hearing a message. This happens in the first round when exactly one processor sends a message. Time complexity of the execution of the algorithm is the number of rounds executed. We consider here the worst case time complexity of the wakeup problem, measured by the number of rounds elapsing from the time the first processor wakes up (spontaneously) to the time all processors are woken up (i.e. up to the first round in which exactly one processor sends a message). The notion „worst case complexity" means here that an adversary controls which processors wake up spontaneously and when; it also knows the algorithm used by processors, but does not know in advance the values of random choices made by processors.

Following [10], we consider two modes of synchronization. In *globally synchronous* systems, all processors have access to a global clock showing the current round number. In *locally synchronous* systems, all clocks tick at the same rate, one tick per round. However, no global counter is available, the local clock of each processor starts counting rounds when processor wakes up. In both models, we distinguish the following assumptions concerning the knowledge of processors (see e.g. [16,10,18] for motivations):

- the size of the system, n, is known or unknown to processors,
- processors are labeled by different numbers from the set $\{1, \ldots, n\}$ or they are unlabeled
- the acceptable probability of error for randomized algorithms, ϵ, is given explicitly as a constant parameter, or only as a function of n.

In [5], authors pointed out that algorithms for RN should be as simple as possible (because „processors" are very often small hand-held devices with limited resources). Therefore they defined a notion of *uniform* algorithms in the model of radio networks. A randomized algorithm for a model with a global clock is called *uniform* if in every round all awake processors send a message with the same probability (independently of the history of communication). Our goal is also to construct efficient uniform algorithms for the wakeup problem.

2   Previous and Related Work

The wakeup problem has been studied in other contexts of distributed computations [7,8, 15]. Collision detection and resolution, and access management algorithms for the model considered here were studied mainly assuming a probability distribution on the arrival rate of messages at the different processors, cf. [12,11]. Unlike in these papers we consider worst-case analysis of the noCD RN model, introduced in [10]. It was proved ([10]) that deterministic algorithms for the wake-up problem require time $\Omega(n)$. For globally synchronized model, the upper bound matching the lower bound was obtained. For local synchronization, it was shown that there exists a deterministic algorithm achieving time $O(n \log^2 n)$. A constructive version of this result (with slightly worse time performance) was presented by Indyk [13]. Results (from [10]) concerning probabilistic algorithms are summarized and compared with our work in the next section.

Recently many other problems in *multi-hop* and *single-hop* radio networks (as broadcasting, gossiping, initialization i.e. labeling, leader election) were considered [6,5,2,1, 16] (one can find more pointers to previous work on the topic in [4]).

3   Our Results

In this paper, we concentrate on probabilistic algorithms. Previous authors consider the model with an allowed error probability ϵ given as a constant parameter. They obtained an algorithm waking up all processors in time $O(\log n \log(1/\epsilon))$ with probability $1 - \epsilon$ in this model, but only in case of global clock, and n known for all processors. When global clock or the knowledge of n is missing, their algorithms require at least linear time. We present algorithms waking up all processors in time $O(\log n \log(1/\epsilon))$ with probability $1 - \epsilon$ even for weakest assumptions considered in [10] (i.e. with local synchronization, unknown n, but labeled processors). These algorithms show that also in this case randomization gives, in comparison with determinism, almost exponential improvement in time complexity. Moreover, in the model with global synchronization, unknown n and no processor labeling we obtain a polylogarithmic algorithm. This result establishes an almost exponential gap between global and local synchronization, because a lower bound $\Omega(n/\log n)$ holds in the model with local synchronization, without labels, and with unknown n. Further, the lower $\Omega(n/\log n)$ compared with our upper bounds show that in locally synchronous environment the knowledge of the size of the network (or at least its approximation given by labels) is crucial for the complexity of the problem. In tables presented above we make more detailed comparison with previous results. It is assumed here that ϵ is given explicitly as a parameter.

Previous results			
n known	Labels	Global clock	Local clock
Yes	Yes	$O(\log n \log(1/\epsilon))$	$O(n \log(1/\epsilon))$
Yes	No	$O(\log n \log(1/\epsilon))$	$O(n \log(1/\epsilon))$
No	Yes	–	$O(n^2 \log(1/\epsilon))$
No	No	not considered	$O(n^2 \log(1/\epsilon))$

Our results			
n known	Labels	Global clock	Local clock
Yes	Yes	$O(\log n \log(1/\epsilon))$ [uniform]	$O(\log n \log(1/\epsilon))$
Yes	No	$O(\log n \log(1/\epsilon))$ [uniform]	$O(\log n \log(1/\epsilon))$
No	Yes	$O(\log n \log(1/\epsilon))$	$O(\log n \log(1/\epsilon))$
No	No	$O(\log^2 n(\log \log n)^3 \log(1/\epsilon(n)))$ [uniform]	$O\left(\frac{n \log(1/\epsilon)}{\log n}\right)^\star$

$\star$ – matches the lower bound

Algorithm's name	Minimal requirements				Time
	n known	Global	Labels	ϵ explicit	
Repeated Prob. Decrease	Yes	Yes	No	No	$O(\log n \log(1/\epsilon))$
Probability Increase	Yes	No	No	No	$O(\log n \log(1/\epsilon))$
Increase from Square	No	No	Yes	Yes	$O(\log n \log(1/\epsilon))$
Use Factorial Representation	No	Yes	No	No	$O(f(n, \epsilon))$
Decrease From Half	No	No	No	Yes	$O(n \log(1/\epsilon)/\log n)$
Decrease Slowly	No	No	No	No	$O(n \log(1/\epsilon))$

where $f(n, \epsilon) = \log^2 n (\log \log n)^3 \log(1/\epsilon)$

We show also a lower bound $\Omega\left(\frac{\log n \log(1/\epsilon)}{\log \log n + \log \log(1/\epsilon)}\right)$ for uniform algorithms working with known n and global clock. This bound almost matches our upper bound for these assumptions: $O(\log n \log(1/\epsilon))$.

Further, we consider the case that ϵ is not given as a constant, but as a function of n only. Note that this assumption makes difference only when n is unknown for processors. For the global clock, mentioned earlier polylogarithmic algorithm for unlabeled networks can work with unknown ϵ. When the global clock and the value of n are not available, we can wake up all processors with probability $1 - \epsilon$ in time $O(n \log(1/\epsilon))$, even without labels (but without labels the function ϵ should be polynomially decreasing).

Our upper bounds are obtained by a sequence of algorithms, presented in direction of decreasing amount of knowledge and synchronization needed. The above table (see page 3) summarizes time bounds and minimal requirements of these algorithms.

Aside from the fact that our results establish exponential gaps between global and local synchronization, known and unknown (approximation of) n as well as determinism and randomization, they may be helpful in construction of efficient probabilistic algorithms for more general problems on multi-hop networks without global synchronization. This is because of large time efficiency of presented solutions for the wake-up problem, which maybe used as a tool in more general settings. From a technical point of view, maybe the most interesting result was obtained in Section 8 where some interesting properties of a so called *factorial representation* of natural numbers were showed and used for the construction of a very simple but efficient algorithm. The additional advantage of this algorithm is that it has also small expected time (opposite to some other algorithms presented here).

In the next section we present some basic mathematical facts and a very useful Sums Lemma. Sections 5–9 are devoted to our randomized algorithms and appropriate lower bounds. We present them in direction of decreasing amount of resources needed. In some cases, our algorithms need an explicit knowledge of allowed error probability, ϵ. Then, we present also other solutions that work for ϵ given as a function of n only.

4   Basic Lemmata

We say that a round of computation is *successful* if and only if exactly one processor is broadcasting in this round. Let *success probability* in a round be a probability that the round is successful. The *broadcast probability* of a processor i in its round r is the probability that i broadcasts a message in the round r. The *broadcast sum* of a round is a sum of broadcast probabilities (in this round) of all processors that are awake in that round.

Let us recall some basic mathematical facts:

Lemma 4.1.   *1.* $(1-p)^{1/p} \geq 1/4$ *for every* $0 < p \leq 1/2$.
2. $(1-p)^{1/p} \leq 1/e$ *for every* $0 < p \leq 1$.
3. $\log n! = \Theta(n \log n)$.
4. $\sum_{i=1}^{\infty} 1/i^2 = \pi^2/6$.
5. $\frac{\log n}{2} \leq \sum_{i=2}^{n} \frac{1}{i} \leq \log n$ *and* $\frac{p \log n}{4} \leq p \sum_{i=1}^{n} \frac{1}{2p+i} \leq p \log(2n)$ *for every* $n > 16$
that is a natural power of two and $1 \leq p < n^{1/4}$.

Lemma 4.2 (Sums Lemma). *Assume that there are k awake processors in a round, with broadcast probabilities $p_1, \ldots, p_k$ such that $p_i \leq 1/2$ for $i = 1, \ldots, k$. Then, the probability that the round is successful is not smaller than* $\left(\sum_{i=1}^{k} p_i\right) \left(\frac{1}{4}\right)^{\sum_{i=1}^{k} p_i}$.

Proof. The probability that exactly one processor is broadcasting is $\sum_{i=1}^{k} (p_i \prod_{j=1; j \neq i}^{k} (1 - p_j)) \geq \sum_{i=1}^{k} (p_i \prod_{j=1}^{k} (1 - p_j)) = (\sum_{i=1}^{k} p_i)(\prod_{j=1}^{k} (1 - p_j))$. Observe that, by Fact 4.1, $(1 - p_j) = ((1 - p_j)^{1/p_j})^{p_j} \geq (1/4)^{p_j}$. Thus, $\prod_{j=1}^{k} (1 - p_j) = \prod_{j=1}^{k} ((1 - p_j)^{1/p_j})^{p_j} \geq \prod_{j=1}^{k} (1/4)^{p_j} = (1/4)^{\sum_{j=1}^{k} p_j}$. So, the success probability is not smaller than $(\sum_{i=1}^{k} p_i) \cdot (1/4)^{\sum_{j=1}^{k} p_j}$.   □

Corollary 4.3. *Assume that there are k awake processors in a round, with broadcast probabilities $p_1, \ldots, p_k$ such that $p_i \leq 1/2$ for $i = 1, \ldots, k$.*

1. If $1/2 \leq \sum_{i=1}^{k} p_i \leq y$ then the success probability is not smaller than $\frac{1}{2} \left(\frac{1}{4}\right)^y \geq \left(\frac{1}{16}\right)^y$.
2. If $1/2 \leq \sum_{i=1}^{k} p_i \leq 2$ then the success probability is not smaller than $1/32$.
3. If $y \leq \sum_{i=1}^{k} p_i \leq 1/2$ then the success probability is not smaller than $y/2$.

5   The Globally Synchronous Model with Known n

In this section we consider the globally synchronous model where the number of processors, n, is known to all of them. Recall that in the case when n is known, there is no difference whether ϵ is given explicitly or as a function. Gąsieniec et al. [10] presented algorithm Repeated-Decay (based on the algorithm Decay from [2]) for this model. This algorithm achieves time complexity $O(\log n \log(1/\epsilon))$ with probability $1 - \epsilon$. We present a modification of this algorithm that is *uniform* and achieves the same time complexity.

Algorithm Repeated Probability Decrease (RPD)
Let $l = 2\lceil \log n \rceil$. Each processor which wakes up spontaneously broadcasts a wakeup message in the round s (for every s) with probability $2^{-1-(s \bmod l)}$.

Theorem 5.1. *The algorithm Repeated Probability Decrease succeeds in waking up the system in time $O(\log n \log(1/\epsilon))$ with probability $1 - \epsilon$.*

Proof. Consider rounds $s, s + 1, \ldots, s + l - 1$ for any s such that $s \bmod l = 0$ and at least one processor is awake in the round s. So, at least one processor is awake in the round s and at most n are awake in the round $s + l - 1$. It means that there exists $1 \leq i \leq l$ such that the number of awake processors in round $s + i + 1$ is greater or equal to 2^{i-1} and smaller or equal to 2^i. For any woken up processor the probability of broadcasting in round $s + i + 1$ is equal to $1/2^i$, so the broadcast sum is not smaller than $2^{i-1} \cdot \frac{1}{2^i} = \frac{1}{2}$ and not bigger than $2^i \frac{1}{2^i} = 1$. By Corollary 4.3, the success probability in this round is not smaller than $1/32$. Let a *phase* be a part of computation which consists of rounds $s, s + 1, \ldots, s + l - 1$ for any s such that $(s \bmod l = 0)$ and at least one processor is awake in the round s. Thus, the probability of success in one phase is at least $1/32$. Hence, the probability that none of the first $\log(1/\epsilon(n))/\log(32/31)$ phases succeeds is at most $(31/32)^{\log(1/\epsilon)/\log(32/31)} = \epsilon$. $\qquad\square$

Kushilevitz and Mansour showed ([14], Lemma 3) that if l labeled processors are woken up spontaneously at the same round (for $l \in \{2^0, 2^1, \ldots 2^{\lceil \log n \rceil}\}$) then the expected number of rounds until successful round is $\Omega(\log n)$ (the expectation is taken over l and probabilistic choices), even in case of nonuniform algorithm, with known n and global clock. Thus, our algorithms that work with probability $1 - \epsilon$ in time $O(\log n \log(1/\epsilon))$ are optimal for constant ϵ and their time bound differ only by a factor $\log(1/\epsilon)$ from this lower bound in general case.

In the *uniform* case, the above algorithm almost matches the following lower bound that for $1/\epsilon$ growing with n is tighter than the bound $\Omega(\log n)$ from [14].

Theorem 5.2. *Any uniform probabilistic algorithm (for globally synchronous model and known system size) requires $\Omega\left(\frac{\log n \log(1/\epsilon)}{\log\log n + \log\log(1/\epsilon)}\right)$ rounds in order to succeed in waking up the system with probability at least $1 - \epsilon$.*

Proof. Assume that $\epsilon = \epsilon(n) < 1/2 - c$ for every large enough n and some constant c (otherwise, we can apply Lemma 3 from [14]). We analyze only scenarios in which $m \geq 10$ processors are woken up spontaneously in the first step and no processor is woken up spontaneously later (all considerations below concern this scenario). Because of uniformity, in every round of the computation the broadcast probabilities of all awake processors are equal. Let m be the number of woken-up processors and p be the broadcast probability in a round. Then, the success probability in that round is $mp(1 - p)^{m-1} \leq 3/4$, what we show by considering two cases:

Case 1: $p \leq 1/2$.
Then $mp(1 - p)^{m-1} \leq (1/(1 - p))mp(1/e)^{mp} \leq 2 \cdot mp(1/e)^{mp}$. Observe that $0 < mp \leq m/2$. For a function $f(x) = x(1/e)^x$ we have $f'(x) = (1/e)^x(1 - x)$. So, the only local extremum of $f(x)$ is obtained for $x = 1$. Thus, by simple calculations one can see that the maximal value of the success probability is smaller than $2 \cdot 1 \cdot (1/e)^1 < 3/4$.

Case 2: $p > 1/2$.
Then $mp(1 - p)^{m-1} \leq mp(1/2)^{m-1} \leq m(1/2)^{m-1} < 1/2$ for $m > 10$.
We say that a given round is *lost* if the success probability in that

round is smaller than $1/(\log n \log(1/\epsilon))$. Observe that the probability of unsuccessful work in $(1/2)\log n \log(1/\epsilon)$ lost rounds is not smaller than $\left(1 - \frac{1}{\log n \log(1/\epsilon)}\right)^{(1/2)\log n \log(1/\epsilon)} \geq \left(\frac{1}{4}\right)^{1/2} = \frac{1}{2}$. Thus, in order to wake up all processors with probability $1 - \epsilon$, the probability of unsuccessful work in rounds that are not lost should be at most 2ϵ. Recall, the success probability is in every round not bigger than $3/4$. So, in order to get the probability of unsuccessful work smaller than 2ϵ, we need at least $\log_{3/4}(2\epsilon) = \Theta(\log(1/\epsilon))$ rounds that are not lost.

Let $x = \lceil 2(\log \log n + \log \log(1/\epsilon)) \rceil$, $m_i = \lfloor x/2 \rfloor + (i-1)x$ for $i = 1, 2, \ldots, \lfloor \log n/x \rfloor - 1$. As shown in following claim, there is no round r in the uniform computation such that r is not the lost round when 2^{m_i} processors are woken-up and r is not the lost round when 2^{m_j} processors are awake for $i \neq j$.

Claim. Let p be a broadcast probability in a round, let m be a number of awake processors. If $|\log m - \log(1/p)| \geq x$ and $m \geq \max\{10, \lceil \log \log n + \log \log(1/\epsilon) \rceil\}$ then the round is lost.

Proof: Let p_s be a success probability in a round considered. First, assume that $p \geq 1/2$. Then $p_s = mp(1-p)^{m-1} \leq m(1/2)^{m-1} \leq (1/2)^{m/2} \leq 1/(\log n \log(1/\epsilon))$ for every $m \geq \max\{10, \lceil \log \log n + \log \log(1/\epsilon) \rceil\}$.

Now, consider the case $p \leq 1/2$. Then, $p_s = mp((1-p)^{1/p})^{mp}/(1-p) \leq 2mp(1/e)^{mp}$. Note that $mp = 2^{\log m + \log p} = 2^{\log m - \log(1/p)}$. If $\log m - \log(1/p) > x$ then $mp > 2^x = (\log n \log(1/\epsilon))^2$, so $p_s \leq 2mp \cdot (1/e)^{mp} \leq 2 \cdot (1/e)^{mp/2} \leq 1/(\log n \log(1/\epsilon))$. If $\log(1/p) - \log m > x$ then $mp < 2^{-x} \leq \frac{1}{(\log n \log(1/\epsilon))^2}$. Thus, $p_s \leq 2mp(1/e)^{mp} \leq 2mp \leq 1/(\log n \log(1/\epsilon))$. $\square$ *Claim*

Thus, for every $i = 1, 2, \ldots, \lfloor \log n/x \rfloor - 1$ we need $\Theta(\log(1/\epsilon))$ rounds that are not lost. On the other hand, each round is not lost for not more than one value of i. So, time needed in order to obtain the probability of success $1 - \epsilon$ is $\Omega(\log n \log(1/\epsilon)/x) = \Omega\left(\frac{\log n \log(1/\epsilon)}{\log \log n + \log \log(1/\epsilon)}\right)$. $\square$

6 Only the Number of Processors Is Known

In this section we consider the case when the number of processors, n, is known but processors are unlabeled and no global clock is available (recall that there is no difference between ϵ given as a constant parameter and as a known function of n in this case). As shown in following algorithm, one can obtain efficient solutions in these settings, despite the lack of the global synchronization.

Algorithm Probability Increase
Let $k = \lceil \log(1/\epsilon(n))/\log(32/31) \rceil$, $l = \lceil \log n \rceil + 1$. After waking up spontaneously, every processor works in l phases numbered in decreasing order $l, l-1, \ldots, 1$. Each phase lasts k rounds. In each of these rounds the processor broadcasts a wakeup message with probability $1/2^j$, where j is the number of the phase.

Theorem 6.1. *Algorithm Probability Increase succeeds in waking up the system in time* $O(\log n \log(1/\epsilon))$ *with probability at least* $1 - \epsilon$.

Due to limited space, we omit the proof of the above theorem (it is similar to the more general proof of Theorem 7.1).

7 Only Labels Are Available

In this section we consider the model with labeled processors, without a global clock, where the number of processors, n, is unknown to them. We present an efficient algorithm that requires a direct access to a parameter ϵ. The main observation that enables to achieve efficient solution in this model is that one can use (in appropriate way) labels of processors as „local approximations" of the size of the network.

Algorithm Increase From Square (IFS)
Let $k = \lceil \log(1/\epsilon)/\log(32/31) \rceil$, let $y = \pi^2/6$. Upon waking up spontaneously, the processor i performs the following:

1. $p \leftarrow \frac{1}{2^{\lceil \log(2y(i+1)^2) \rceil}}$
2. if $p \leq 1/2$ then in k consecutive rounds the processor works as follows: randomly set a bit b with probabilities $\mathbf{P}[b = 1] = p$, $\mathbf{P}[b = 0] = 1 - p$; if $b = 1$, broadcast a wakeup message.
3. $p \leftarrow 2p$
4. Goto 2.

Theorem 7.1. *The algorithm Increase From Square succeeds in waking up all processors in time $O(\log n \log(1/\epsilon))$ with probability at least $1 - \epsilon$.*

Proof. Consider the first round in which at least one processor is awake. The broadcast sum is not bigger than $\sum_{i=1}^{\infty} 1/(2^{\lceil \log(2yi^2) \rceil}) \leq \sum_{i=1}^{\infty} 1/(2yi^2) = 1/(2y) \sum_{i=1}^{\infty} 1/i^2 = 1/2$ in this round. Starting from this round, the broadcast sum is growing. Let r be the last round in which this sum is smaller than $1/2$. Let $s \leq 1/2$ be the value of this sum in the round r. Observe that the sum of probabilities of all processors that were woken up in the round r is not greater than $2s \leq 1$ and not smaller than $1/2$ in every round in the sequence $r + 1, r + 2, \ldots, r + k$. Moreover, the sum of broadcast probabilities of all processors woken up spontaneously in rounds $r + 1, r + 2, \ldots, r + k$ is not greater than $\sum_{i=1}^{\infty} 1/(2^{\lceil \log(2yi^2) \rceil}) \leq 1/2$ in all these rounds. So, the broadcast sum is in rounds $r + 1, r + 2, \ldots, r + k$ not smaller than $1/2$ and not bigger than $3/2$. By Corollary 4.3, the probability of success in each of these rounds is at least $1/32$. So, the probability that every round in this sequence was unsuccessful is at most $(1/32)^k \leq \epsilon$.

Finally, observe that the largest possible distance between the first round in which at least one processor is woken up (spontaneously) and the last round in which the broadcast sum is smaller than $1/2$, is $O(k \log(2^{\lceil \log(2y(n+1)^2) \rceil})) = O(\log n \log(1/\epsilon))$. $\qquad \square$

8 Only Global Clock Is Available

Now, we consider the model with global clock, unlabeled processors, where the number of processors, n, is unknown to them. We present a fast algorithm for this scenario that does not require an explicit information about ϵ. First, we introduce a very useful notion of a factorial representation of natural numbers.

Definition 8.1 (Factorial representation). *Let* $t > 0$ *be a natural number. The factorial representation of* t *is equal to the sequence* $t_1, t_2, \ldots$ *such that* $t = \sum_{i=1}^{\infty} t_i \cdot i!$ *and* $0 \le t_i \le i$ *for every* $i \in \mathbb{N}$ *(and only finite number of* t_i*'s are not equal to 0).*

Lemma 8.2. *For every natural number* t *there exists exactly one factorial representation of* t.

Definition 8.3. *Let* t *be a natural number,* $\{t_i\}_{i \in \mathbb{N}}$ *be a factorial representation of* t. *Let* $j(t) = \min\{i \mid t_i = 0\}$, *and:*

- $y(t) = 1$ *for* $t \in \{0, 1\}$,
- $y(t) = y((j(t) - 1)! - 1) + 1 + \sum_{l=2}^{j(t)-1} (t_l - 1)(l - 1)!$ *for* $t > 1$.

Lemma 8.4. *The functions* y *and* j *satisfy following conditions:*

1. $y((j(t) - 1)! - 1) \le y(t) \le y(j(t)! - 1)$ *for every natural number* $t > 1$,
2. $y(k! - 1) = 1! + 2! + \ldots + (k - 1)!$ *for every* $k > 1$.

Proof. 1. Let $\{t_i\}_{i \in \mathbb{N}}$ be a factorial represention of t. The inequality $y((j(t)-1)!-1) \le y(t)$ follows directly from the definition. So, we concentrate on the second inequality. Let $j(t) = k$. Observe that the factorial representation of $k! - 1$ is equal to $1, 2, 3, \ldots, (k-1), 0, 0, 0 \ldots$ Thus, $j(t) = j(k!-1) = k$ and further $y(k!-1) = y((k-1)!-1)+1+ \sum_{l=2}^{k-1}(l-1)(l-1)! \ge y((k-1)!-1)+1+\sum_{l=2}^{k-1}(t_l - 1)(l-1)! = y(t)$, because $t_l \le l$ for every l.

2. We prove this property by induction. For the base step, observe that $y(2!-1) = 1 = 1!$ Now, assume that the property is true for $k-1$. Recall that $j(k!-1) = k$ and the factorial representation of $k! - 1$ is equal to $1, 2, 3, \ldots, (k-1), 0, 0, 0, \ldots$ Then, $y(k! - 1) = y((k-1)!-1)+1+\sum_{l=2}^{k-1}(l-1)(l-1)!$ Thus, using inductive hypothesis, $y(k!-1) = (1!+2!+\ldots+(k-2)!)+1+\sum_{l=2}^{k-1}(l-1)(l-1)! = 1!+2!+\ldots+(k-2)!+(k-1)!$ $\square$

We propose very simple but efficient algorithm based on properties of factorial representations:

Algorithm Use Factorial Representations (UFR)
Each woken processor performs the following in round t (for every t): Randomly set a bit b with probabilities $\mathbf{P}[b = 1] = 1/2^{y(t)}$, $\mathbf{P}[b = 0] = 1 - 1/2^{y(t)}$. If $b = 1$, broadcast a wakeup message.

Theorem 8.5. *Algorithm UFR is uniform and succeeds in waking up a globally synchronous system with probability* $1 - \epsilon$ *in time* $O(\log^2 n (\log \log n)^3 \log(1/\epsilon))$, *even if the number* n *and a parameter* ϵ *are unknown and processors are unlabeled.*

Proof. First, we show some useful facts that are crucial for the analysis of our algorithm:

Claim 1. For any $y \in \mathbb{N}$, there exists a natural number t such that $y(t) = y$; moreover, the smallest t such that $y(t) = y$ is not greater than $c_1 y \log y$ for a constant c_1.

Proof: Let y be a natural number, $(y_i)_{i \in \mathbb{N}}$ be a sequence denoting the factorial representation of $y - 1$, k be the largest index such that $y_k \neq 0$. First, we show that there exists a *modified factorial* representation $(y'_1, \ldots, y'_{k'})$ of $y - 1$ such that $y - 1 = \sum_{l=1}^{k'} y'_l l!$, where $0 < y'_i \leq i + 1$ for each $i \in \{1, \ldots, k'\}$, and $k' \leq k$. If $y_i > 0$ for each $i = 1, \ldots, k$ then the sequence $y_1, \ldots, y_k$ satisfies conditions of „modified" representation. Otherwise, we can transform the sequence $y_1, \ldots, y_k$ in order to obtain appropriate modified factorial representation. To this aim we can use the following algorithm that given a factorial representation $y_1, \ldots y_k$ (where $y_i = 0$ for $i > k$), generates a modified factorial representation $y'_1, \ldots, y'_{k'}$.

```
y₀ ← 1
sub ← false;
For l = 1, 2, ..., k do
    if (yₗ > 0) and (not sub) then y'ₗ ← yₗ
    else if yₗ = 0 then begin
            sub ← true;
            if yₗ₋₁ > 0 then y'ₗ ← l + 1 else y'ₗ ← l
    end
    else begin{i.e.when yₗ > 0 and sub =true}
            if (yₗ > 1) or (l = k) then
            (y'ₗ ← yₗ - 1; sub ← false)
            else (y'ₗ ← l + 1; yₗ ← 0; sub ← true)
    end
```

In other words, every subsequence of zeroes, say $y_a, \ldots, y_b$ (such that $y_{a-1} \neq 0$ and $y_{b+1} \neq 0$) is replaced into a sequence $y'_a, \ldots, y'_b$ such that $y'_a = a + 1$ and $y'_c = c$ for $c = a + 1, \ldots, b$. This is done by „borrowing" one occurrence of $(b + 1)!$ from y_{b+1}, so then $y'_{b+1} = y_{b+1} - 1$ (and possibly consecutive „borrowing" is needed if $y_{b+1} = 1$ and $b + 1 < k$).

Using Lemma 8.4, we show that $y = y(t)$ for $t = 1 + \sum_{l=1}^{k'} y'_l (l + 1)!$. The factorial representation of t is equal to $1, y'_1, y'_2, \ldots, y'_{k'}, 0, 0, \ldots$ Thus, $j(t) = k' + 2$ and $y(t) = y((k' + 1)! - 1) + 1 + \sum_{l=1}^{k'} (y'_l - 1)l! = (\sum_{l=1}^{k'} l!) + 1 + (\sum_{l=1}^{k'} (y'_l - 1)(l)!) = (\sum_{l=1}^{k'} y'_l l!) + 1 = (y - 1) + 1 = y$. For the second part of the claim, observe that $t = 1 + \sum_{l=1}^{k'} y'_l (l + 1)! \leq 1 + (k' + 1) \sum_{l=1}^{k'} y'_l l! \leq 1 + (k' + 1)(y - 1) \leq (k' + 1)y$, and $y \geq (k')!$. Thus, $k' \leq d \log y$ (for some constant d, see Fact 4.1.3), so $t \leq (k' + 1)y \leq 2dy \log y$.   □ *Claim 1*

Claim 2. For each y, the difference $t_2 - t_1$ between every two consecutive numbers $t_1 < t_2$ such that $y(t_1) = y(t_2) = y$ is not bigger than $c_2 y \log^3 y$ (for a constant c_2).

Proof: Let y be a natural number, let t be any number that satisfies $y(t) = y$, let $j = j(t)$. First, observe that $y(t) = y(t + (j + 1)!)$. Indeed, first j numbers in the factorial representation of $t' = t + (j + 1)!$ are equal to the first j numbers in factorial representation of t. So, $j(t') = j(t) = j$ and $y(t') = y(t)$. Now, we must only show

that $(j + 1)! = O(y \log^3 y)$. By Lemma 8.4, $y((j - 1)! - 1) \leq y \leq y(j! - 1)$. Thus (also by Lemma 8.4), $1! + 2! + \ldots + (j - 2)! \leq y \leq 1! + 2! + \ldots + (j - 1)!$ It means that $y \geq (j - 2)!$ and $\log y \geq e(j - 2) \log(j - 2)$ for some constant e. In consequence, $j + 1 \leq f \log y$ (for a constant f). Finally, $(j + 1)! \leq (j - 2)!(j + 1)^3 \leq f^3 y \log^3 y$.
□ *Claim 2*

Now, we apply these properties to the analysis of our algorithm. Observe that in every round all awake processors have the same broadcast probability. If this probability is $1/2^y$ and the number of awake processors is in the range $\langle 2^{y-1}, 2^{y+1} \rangle$ then the broadcast sum is not smaller than $2^{y-1} \cdot \frac{1}{2^y} = \frac{1}{2}$ and not bigger than $2^{y+1} \cdot \frac{1}{2^y} = 2$. Thus, the probability that the round is successful is not smaller than $1/32$, by Corollary 4.3. After $\Theta(\log(1/\epsilon(n))$ such rounds we obtain appropriate probability of success. Let $y = \lceil \log n \rceil$. For any $m = 1, 2, \ldots, \lceil \log n \rceil$ and for every $y \log^3 y = O(\log n (\log \log n)^3)$ consecutive rounds there is a round in which each (awake) processor has broadcast probability $1/2^m$ (by Claims 1 and 2). Let us split the computation into blocks of length $y \log^3 y$ (starting from the first round in which at least one processor is woken up). In each block, at least one of the following two conditions is satisfied:

(1) There is at least one round t in the block such that $y(t) = x$ and the number of woken up processors in t is in the range $\langle 2^{x-1}, 2^{x+1} \rangle$.
(2) The number of awake processors at the end of the block is at least twice larger than the number of awake processors at the beginning of the block.

There are at most $\lceil \log n \rceil$ blocks of type (2). Moreover, the probability that none of b blocks of type (1) is successful is at most $(1 - 1/32)^b$. So, we get required probability of success after $\log n + \log_{32/31}(1/\epsilon(n)) = O(\log n + \log(1/\epsilon(n))$ blocks. Time consumed by these blocks is $O(\log n(\log \log n)^3(\log n + \log(1/\epsilon(n)))) = O(\log^2 n(\log \log n)^3 \log(1/\epsilon(n)))$.   □

9   The Weakest Model

In this section we consider the weakest model, without global clock, with unlabeled processors, where the number of processors, n, is unknown to them. First, we show an almost linear lower bound which contrasts to polylogarithmic algorithms obtained for stronger models.

Theorem 9.1. *If a global clock is not available, the size n of the system is not known to processors and processors do not have labels then every probabilistic algorithm needs at least $\Omega(n/\log n)$ rounds in order to wake up the system of n processors with probability $1 - \epsilon$.*

Proof. In this model, the algorithms of all processors are identical and may be described by a (infinite) sequence $p_1, p_2, p_3, \ldots$ such that p_i is the broadcast probability in the ith round after spontaneous wake-up of the processor. If $p_i = 0$ for every $i \in \mathbb{N}$ then the algorithm is incorrect (an adversary may never wake-up all processors). Let $j = \min\{i \mid p_i > 0\}$, let $p_j = 1/p$ (n is unknown, thus we can assume that p is constant with respect to n). For a natural number n, let $r_n = \lfloor n/(\lceil 4p \log n \rceil) \rfloor$. We take any (large enough) n that satisfies inequalities $(1/n)^4 \leq 1/(2n)$, $8 \log n/n^2 \leq$

$1/(2n)$ and $r_n \leq n\log(1/\epsilon)/2$. Let us call a round in which first processor is woken up spontaneously as the round 1. Assume that the adversary wakes up spontaneously $x = \lceil 4p\log n\rceil$ processors in each round in the sequence $1, 2, \ldots, \lfloor n/x\rfloor = \Theta(n/\log n)$. The probability of success in rounds $1, \ldots, j-1$ is equal to zero. Let U_k be a set of processors woken up in the round k. We show that for any $k \in \{j, j+1, \ldots, \lfloor n/x\rfloor\}$, the probability that at least two processors from U_{k-j+1} send broadcast message in the round k is not smaller than $1-1/n$. First, observe that the probability that none of the processors from U_{k-j+1} broadcasts is $(1-1/p)^x \leq (1-1/p)^{4p\log n} \leq (1/2)^{4\log n} \leq (1/n)^4 \leq 1/(2n)$. Moreover, the probability that exactly one element of U_{k-j+1} broadcasts is $\frac{x}{p}(1-1/p)^{x-1} \leq \frac{8p\log n}{p}(1-1/p)^{2p\log n} \leq (8\log n)/n^2 \leq 1/(2n)$. Concluding, the probability that more than one processor from U_{k-j+1} is broadcasting in the round k is at least $1 - 1/(2n) - 1/(2n) = 1 - 1/n$.

Thus, the probability of success in each round from the sequence $j, j+1, \ldots, \lfloor n/x\rfloor$ is smaller than $1/n$. The probability of unsuccessful work in rounds $1, \ldots, \lfloor n/x\rfloor$ is not smaller than $(1-1/n)^{r_n} > (1-1/n)^{n\log(1/\epsilon)/2} \geq (1/4)^{log(1/\epsilon)/2} = \epsilon$.   □

Now, we present an algorithm working in the model with known and constant ϵ that matches above lower bound.

Algorithm Decrease From Half

Let $k = \lceil \log(1/\epsilon)/\log(32/31)\rceil$. Upon waking up spontaneously, each processor performs the following (until all processors are woken up):

1. $p \leftarrow 1/2$
2. In k consecutive rounds: randomly set a bit b with probabilities $\mathbf{P}[b = 1] = p$, $\mathbf{P}[b = 0] = 1 - p$; if $b = 1$, broadcast a wakeup message.
3. $p \leftarrow p/2$
4. Goto 2.

Theorem 9.2. *The algorithm Decrease From Half succeeds in waking up all processors in time* $O\left(\frac{n\log(1/\epsilon)}{\log n}\right)$ *with probability* $1 - \epsilon$.

The proof of this theorem is omitted, it is similar to the proof of Theorem 9.3.

Finally, we present an algorithm for the weaker scenario, when ϵ is given as a function of n, not as a constant parameter. This algorithm works for any $\epsilon(n)$ that polynomially goes to zero as n is growing.

Algorithm Decrease Slowly

Assume that $\epsilon(n) \geq 1/n^r$ for a constant r. Let $q = 16r\ln 2$. Any processor, after waking up spontaneously performs the following:

1. $i \leftarrow 0$
2. $p \leftarrow q \cdot \frac{1}{2q+i}$
3. In one round: randomly set a bit b with probabilities $\mathbf{P}[b = 1] = p$, $\mathbf{P}[b = 0] = 1-p$; if $b = 1$, broadcast a wakeup message.
4. $i \leftarrow i + 1$
5. Goto 2.

Theorem 9.3. *Algorithm Decrease Slowly wakes up a system of n processors in time* $O(n\log(1/\epsilon(n)))$ *with probability at least* $1 - \epsilon(n)$.

Proof. Assume that n is a natural power of two. Let us analyze the work of the algorithm during the first $2m$ rounds (starting from the first round in which at least one processor is woken up), for $16qn \leq m < n^2/2$. Note that the sum of broadcast probabilities of one processor in all these $2m$ rounds is not bigger than $q\sum_{i=0}^{n^2} 1/(2q + i) \leq q\log(2n^2) \leq q(2\log n + 1) \leq 4q\log n$. Thus the sum of broadcast probabilities of all processors in first $2m$ rounds of the algorithm is not bigger than $4qn\log n$. This implies that in at least m of the first $2m$ rounds, the broadcast sum is not bigger than $\frac{4qn\log n}{m} \leq \frac{\log n}{4}$ (indeed, otherwise the sum over all rounds would be bigger than $4qn\log n$). Let rounds with broadcast sum smaller than $\log n/4$ be called *light*. Let us split *light* rounds into two categories:

Category 1: The rounds with broadcast sum bigger than $1/2$. By Corollary 4.3.1, the success probability in each such round is bigger than $(1/16)^{(\log n)/4} = 1/n$. If the number of rounds of category 1 is bigger than $n\ln(1/\epsilon(n))$ then the probability of unsuccessful work is smaller than $(1 - 1/n)^{n\ln(1/\epsilon(n))} \leq (1/e)^{\ln(1/\epsilon(n))} = \epsilon(n)$.

Category 2: The *light* rounds with broadcast sums not bigger than $1/2$. First note that:

Claim 1

(a) Assume that the broadcast sum in the round j is not smaller than $q/(2q + i)$. Then the broadcast sum in the round $j + 1$ is not smaller than $q/(2q + i + 1)$.

(b) The broadcast sum of the first light round of Category 2 is not smaller than $q/(2q+1)$.

Proof: (a) Let U_j be the set of processors that are awake in the round j, let s_j be the broadcast sum in the round j. Then $s_j = q\sum_{l\in U_j} 1/(2q + i_l)$, where i_l are natural numbers. So, $s_{j+1} \geq q\sum_{l\in U_j} 1/(2q+i_l+1)$. If there exists $k \in U_j$ such that $i_k \leq i$ then $s_{j+1} \geq q/(2q+i_k+1) \geq q/(2q+i+1)$. Otherwise, $s_{j+1} \geq q\sum_{l\in U_j} \frac{1}{2q+i_l} \cdot \frac{2q+i_l}{2q+i_l+1} \geq q\sum_{l\in U_j} \frac{1}{2q+i_l} \cdot \frac{2q+i}{2q+i+1} = \frac{2q+i}{2q+i+1}s_j \geq \frac{2q+i}{2q+i+i} \cdot \frac{q}{2q+i} = \frac{q}{(2q+i+1)}.$

(b) Note that the broadcast sum of the first round during the work of the algorithm is not smaller than $q \cdot \frac{1}{2q}$. The rest follows from (a).   □ *Claim 1*

One can show by induction that the broadcast sum of the ith *light* round of Category 2 is not smaller than $q/(2q + i)$. By above claim, it is true for $i = 0$. For the inductive step, let us only comment the case when the ith and the $(i+1)$st *light* round of Category 2 are not consecutive in the execution of the algorithm. Note that then the broadcast sums of the rounds between them are bigger than the broadcast sum in the round i. So, our statement is satisfied by the application of Claim 1.

Thus by Corollary 4.3.3 the broadcast probability in the ith *light* round of Category 2 is not smaller than $\frac{1}{2} \cdot \frac{q}{2q+i}$. So the probability of unsuccessful work in first x *light* rounds of Category 2 is not bigger than $\prod_{i=1}^{x} \left(1 - \frac{1}{2} \cdot \frac{q}{2q+i}\right) \leq \left(\frac{1}{e}\right)^{\frac{q}{2}\sum_{i=1}^{x}\frac{1}{2q+i}} \leq \left(\frac{1}{e}\right)^{\frac{1}{2} \cdot \frac{q\log x}{4}} = x^{-2r}$. In order to bound this probability by $\epsilon(n)$ we need $x = (\epsilon(n))^{-1/(2r)} = \sqrt{n}$ *light* rounds of Category 2.

Finally, if $m \geq 2 \cdot \max(n\ln(1/\epsilon(n)) + \sqrt{n}, 16qn)$ then we get required success probability.   □

10 Conclusions

We presented efficient probabilistic algorithms waking up all processors in polylogarithmic time with probability $1 - \epsilon$ if global clock or knowledge of n or labeling is accessible. These algorithms substantially improve previous results, and show that efficient solutions are possible even without the knowledge of the basic parameters of the network. Further, we showed an almost linear lower bound for the weakest model and we presented an algorithm that matches this bound.

Some interesting problems remain open. First, there is a gap of size $\log(1/\epsilon)$ between lower bounds and our algorithms for the „standard" model with labels and ϵ given as a parameter. Second, is it possible to construct polylogarithmic (or even sublinear) randomized algorithm for the model with local clocks, n unknown, known labels and ϵ given as a function? The best solution known for us is a protocol in which each processor broadcasts in each step with probability equal to $1/l$, where l is the label of the processor. However, this protocol requires $\Theta(n \log(1/\epsilon))$ steps for waking up all processors with probability $1 - \varepsilon(n)$.

References

1. N. Alon, A. Bar-Noy, N. Linial, D. Peleg, *A Lower Bound for Radio Broadcast*, Journal of Computer Systems Sciences 43, (1991), 290–298.
2. R. Bar-Yehuda, O. Goldreich, A. Itai, *On the Time-Complexity of Broadcast in Multi-hop Radio Networks: An Exponential Gap Between Determinism and Randomization*. JCSS 45(1), (1992), 104-126.
3. I. Chlamtac, S. Kutten, *On Broadcasting in Radio Networks – Problem Analysis and Protocol Design*, IEEE Trans. on Communications 33, 1985.
4. B. S. Chlebus, *Randomized communication in radio networks*, a chapter in „Handbook on Randomized Computing" P. M. Pardalos, S. Rajasekaran, J. H. Reif, J. D. P. Rolim, (Eds.), Kluwer Academic Publishers, to appear.
5. B. Chlebus, L. Gąsieniec, A. Lingas, A.T. Pagourtzis, *Oblivious gossiping in ad-hoc radio networks*, in Proc., DIALM 2001, 44–51.
6. M. Chrobak, L. Gąsieniec, W. Rytter, *Fast Broadcasting and Gossiping in Radio Networks*, in Proc. FOCS, 2000, 575-581.
7. S. Even, S. Rajsbaum, *Unison, canon, and sluggish clocks in networks controlled by a synchronizer*, Mathematical Systems Theory 28, (1995), 421–435.
8. M.J. Fischer, S. Moran, S. Rudich, G. Taubenfeld, *The Wakeup Problem*, SIAM Journal of Computing 25, (1996), 1332–1357.
9. P. Fraigniaud, A. Pelc, D. Peleg, S. Perennes, *Assigning labels in unknown anonymous networks*, Distributed Computing 14(3), 163-183 (2001).
10. L. Gąsieniec, A. Pelc, D. Peleg, *The wakeup problem in synchronous broadcast systems*, PODC 2000, 113–121.
11. J. Goodman, A. G. Greenberg, N. Madras, P. March, *On the stability of Ethernet*, 17th ACM-STOC, 1985, 379–387.
12. J. Hastad, T. Leighton, B. Rogoff, *Analysis of Backoff Protocols for Multiple Access Channels*, 19th ACM-STOC, 1987, 241–253.
13. P. Indyk, *Explicit constructions of selectors and related combinatorial structures, with applications*, Proc. SODA, 2002.

14. E. Kushilevitz, Y. Mansour, *An $\Omega(Dlog(N/D))$ Lower Bound for Broadcast in Radio Networks*, SIAM J. Comput. 27(3), (1998), 702–712.
15. J. Mazoyer, *On optimal solutions to the firing squad synchronization problem*, Theoretical Computer Science 168, (1996), 367–404.
16. K. Nakano, S. Olariu, *Energy-Efficient Initialization Protocols for Single-Hop Radio Networks with no Collision Detection*, IEEE Trans. on Parallel and Distributed Systems, Vol. 11, No. 8, pp. 851–863, Aug. 2000.
17. D. Peleg, *Deterministic radio broadcast with no topological knowledge*, a manuscript, 2000.
18. D. E. Willard, *Log-logarithmic selection resolution protocols in multiple access channel*, SIAM Journal on Computing 15 (1986), 468–477.

A Simple, Memory-Efficient Bounded Concurrent Timestamping Algorithm

Vivek Shikaripura and Ajay D. Kshemkalyani

Computer Science Department, Univ. of Illinois at Chicago,
Chicago, IL 60607, USA

Abstract. Several constructions have been proposed for implementing a Bounded Concurrent Timestamp System (BCTS). Some constructions are based on a recursively defined *Precedence Graph*. Such constructions have been viewed as hard to understand and to prove correct. Other constructions that are based on the *Traceable Use* abstraction first proposed by Dwork and Waarts have been regarded as simple and have therefore been preferred. The Dwork-Waarts (DW) algorithm, however, is not space-efficient. Haldar and Vitanyi (HV) gave a more space-efficient construction based on Traceable Use, starting with only safe and regular registers as building blocks. In this paper, we present a new algorithm by making simple modifications to DW. Our algorithm is simple and is more memory-efficient than the DW and HV algorithms.

1 Introduction

We consider the problem of constructing a Bounded Concurrent Timestamp System (BCTS), which can be described briefly as follows. *There is a set of n asynchronous processors which can perform two types of operations, **Label()** and **Scan()**. The **Label()** operation corresponds to some event and generates a new timestamp (label, hereafter) for the processor performing it. The **Scan()** operation enables a processor to obtain a list of the current labels and to determine the order among them, consistent with the real-time order in which the labels were selected. The **Label()**/ **Scan()** operations of a processor may be interleaved or concurrent with a **Label()**/ **Scan()** operation of another processor. The problem is to construct such a system by using shared variables as building blocks but with the restriction that the size and number of all labels/shared variables/registers so used be bounded. The system must also be wait-free [5].*

Bounded Concurrent Timestamping is a well-studied problem. The reader is referred to [2,5] for an introduction. Bounded Concurrent Timestamping abstracts a large class of problems in concurrency control, notably Lamport's first-come first-served mutual exclusion [11], randomized consensus [1], MRMW atomic register construction [16] and fifo-l-exclusion [6].

There are many BCTS algorithms [3,4,5,8,10]. The basic technique used is to first construct an Unbounded Concurrent Timestamp System (UBCTS) in which processors assume monotonically increasing label values (a potentially unbounded number of them) and then convert the UBCTS into a Bounded

P. Bose and P. Morin (Eds.): ISAAC 2002, LNCS 2518, pp. 550–562, 2002.

Concurrent Timestamp System (BCTS) by employing a mechanism to recycle the labels. So far, two recycle mechanisms (and their variants) have been used.

1. *Precedence Graph*: In this mechanism, processors choose labels from the domain of a recursively defined, nested digraph called precedence graph. The domain consists of the set of labels (node names) in the precedence graph. To execute a label operation, a processor first collects all the node labels which have been chosen by other processors and then chooses its new label by selecting a free node such that it is the lowest node "dominating" the collected nodes. A scanner determines the order among the labels by following the edges of the nodes corresponding to the labels. Owing to some special properties of the graph, successive labeling operations are able to continuously choose new nodes without ever running out of nodes. This mechanism, first used by Israeli and Li [9] for sequential timestamp systems, was later extended and used in [4,3,7,10] for concurrent timestamp systems.

2. *Traceable Use*: In the precedence graph, there is a global pool of values from which processors choose their labels. In the Traceable Use mechanism, each processor maintains a separate, local pool of private values. To execute the labeling operation, a processor first collects the private value of every processor, chooses a new, unused private value for itself from its local pool and finally obtains a vector clock from these private values. This vector clock is the new label of the processor. So as to allow other processors to know which one of their private values has been collected, the labeler records in shared memory, every private value that it collects. This enables a processor to trace out which of its private values are not in use at any time by checking shared memory. With this arrangement, a labeler can safely recycle private values. A scanner determines the order among the labels by comparing the vector clocks. This mechanism, known as Traceable Use, is the basis of the algorithms in [5,8].

While the precedence graph mechanism has allowed construction of efficient algorithms, it has been regarded as hard to understand [8,7] and having high conceptual complexity [4,12,14]. The Traceable Use mechanism has been regarded as "simple", but it has not enabled space-efficient constructions so far, since long registers and large amount of shared memory space are required in its implementation. Despite this drawback, Traceable Use mechanism has enjoyed popularity since it renders simple, elegant constructions that are easy to understand and prove correct. We are thus motivated to seek improvements in algorithms [5,8], while retaining the simplicity with which they model BCTS. The presented work is such an attempt.

Several criteria have been used to compare BCTS algorithms.

- *Label/Register size*: The maximum size in bits of a register owned by any processor.
- *Label time complexity*: The maximum number of shared variable accesses in a single execution of the **Label**() operation.
- *Scan time complexity*: The maximum number of shared variable accesses in a single execution of the **Scan**() operation.

– *Memory overhead*: Every processor accesses two types of memory – shared memory and local memory. Shared memory includes all shared variables. Local memory includes local variables and private pools of values used by a processor, which are kept in local space.

Tables 1 and 2 compare the performance of the BCTS algorithms.

Table 1. Comparison of BCTS algorithms based on the precedence graph [8].

Algorithm	Register size (bits)	Label time complexity	Scan time complexity	Total shared space (bits)
Dolev-Shavit [3]	$O(n)$	$O(n)$	$O(n^2 \log n)$	$O(n^3)$
Gawlick et. al. [7]	$O(n^2)$	$O(n \log n)$	$O(n \log n)$	$O(n^4)$
Israeli-Pinhasov [10]	$O(n^2)$	$O(n)$	$O(n)$	$O(n^4)$
Time-lapse [4]	$O(n)$	$O(n)$	$O(n)$	$O(n^3)$

Table 2. Comparison of BCTS algorithms based on the Traceable Use mechanism.

Algorithm	Register size (bits)	Label time complexity	Scan time complexity	Total shared space (bits)	Local space (bits)
Dwork-Waarts [5]	$O(n \log n)$	$O(n)$	$O(n)$	$O(n^5 \log n)$	$O(n^2 \log n)$
Haldar-Vitanyi [8]	$O(n \log n)$	$O(n)$	$O(n)$	$O(n^3 \log n)$	$O(n^2 \log n)$
This paper	$O(n \log n)$	$O(n)$	$O(n)$	$O(n^3 \log n)$	$O(n \log n)$

The DW algorithm [5] has three main drawbacks: (i) each processor uses $O(n^2)$ SWMR registers, so that the total overhead of shared memory is $O(n^5 \log n)$ SWSR safe bits, (ii) an extra, expensive garbage collection mechanism is required for tracking down private values not in use, and (iii) long registers of size $O(n \log n)$ bits are used in the construction. Haldar and Vitanyi (HV) [8] improved the first two drawbacks by giving an independent construction using only regular and safe registers as building blocks.

In the present work, we improve the algorithm in [5] by proposing a new, modified design for the **Label()** and **Scan()** operations. Unlike [5,8], while our **Label()** operation uses the RWRWR handshake technique of [5,13] to collect private values, our **Scan()** operation collects processor labels by capturing a snapshot of the labels as they existed at a single point in time [4,7]. Every processor in our construction uses two pools of integers, a private value pool and a tag pool, each of size $O(n)$ integers. The size of the private pool of values is $3n - 1$ integers in our case, which is better than the $2n^2$ in [8] and the $22n^2$ in [5]. Our algorithm is inspired by a detailed study of existing BCTS algorithms; we integrate many of the ideas embedded in the algorithms of [4,5,7] into our construction. As can be seen from the tables, although the Time-lapse algorithm

[4] based on the precedence graph is more efficient, it does not enjoy the simplicity of approach of our construction based on the Traceable Use abstraction.

2   The Dwork-Waarts Construction

We first review the structure of shared memory in the DW algorithm. Figure 1 (top) sketches the shared variables associated with any processor b. X_b, known as the principal shared variable, is the label (timestamp) of processor b. It is a vector of length n whose components are the private values collected from every processor. The $A1-X_{b*}$, $A2-X_{b*}$, $A3-X_{b*}$, and $B-X_{b*}$ registers are auxiliary shared variables needed for the construction. The p_{b*}, q_{b*} and tog_b registers implement the RWRWR handshake mechanism of [5,13] which can detect concurrent writes while the processor b is reading. The Ord_{b*} registers maintain the ordering information among the pool of private values of b. As the shared variables X_b, tog_b and handshake bits p_{b*} are fields of one SWMR atomic register r_b, all of them can be read/written in one atomic operation. Similarly, $A1-X_{b*}$, $A2-X_{b*}$, $A3-X_{b*}$ and $B-X_{b*}$ can be read/written atomically from r_{b*}.

The central idea of the DW algorithm is as follows: Each processor b maintains an integer pool of private values $- v_{b(1)}, \ldots \ldots \ldots, v_{b(11n^2)}$ from which it draws values for its label. The relative order between $v_{b(i)}$, $v_{b(j)}$ (i.e., which of the two was taken up before the other), for all i, j, is stored in Order registers (Ord_{b*} in Figure 1) by b. Whenever b executes a labeling operation, it assumes a new label, a vector clock consisting of n components – the current private value collected from each processor plus a new private value v drawn from its own local pool (assuming it is not empty). The value v so drawn is now written in the Order registers Ord_{b*} as the largest of b's values. The key to the algorithm is to ensure that when b takes up v, v is indeed "new" in the sense that it does not currently appear and will not appear later in the label of any processor in the system. To see why this is critical, consider the following scenario. A scanner c collects labels from all processors but goes to sleep temporarily before determining their relative ordering. Meanwhile an updater b wishing to assume a new label, cannot take up any of its private values v_k which c may have obtained from b (or indirectly from another processor d). If b, before c woke up, were indeed to take up v_k and update register Ord_{b*} with v_k as the largest value, any of c's subsequent computations making reference to newly updated Ord_{b*} would be erroneous. Hence, b would have to make sure that it does not take up a value which is already in use. The problem of determining which values are in use in the system at any time is not trivial. The DW algorithm solves this problem by requiring that every private value read in an operation, either **Label()** or **Scan()**, be recorded in shared memory somewhere, hence, *all private values consumed in read operations are recorded in shared variables*. A processor can trace out which of its values are in use at any time, by scanning all the shared variables (this is known as garbage collection). Only values not found in shared memory (and hence not in use) are valid new values. It turns out that in order

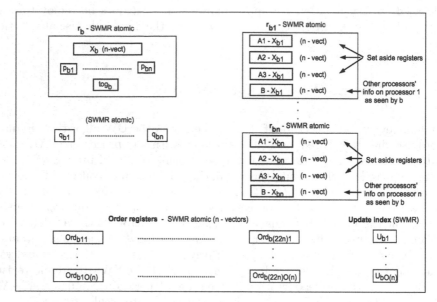

Shared variables and registers at Processor b in the DW algorithm [1]

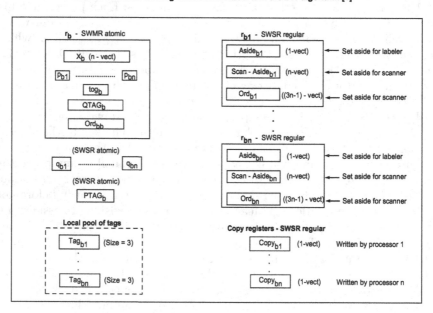

Shared variables and registers at Processor b in proposed algorithm

Fig. 1. Data structures for the DW algorithm [5] and the proposed algorithm.

Label() procedure, code for processor b

1. For all c, $val[]_c = $ **traceable-collect**$(X_c[c])$
   // collect private values from all other processors except b in $val[]_c$
2. Choose $v_b = $ smallest element in b's pool of available private values in Ord_{bb}
3. $new-l = (val_1, val_2, \ldots, v_b, \ldots, val_n)$
4. **traceable-write**$(X_b, new-l)$ // write $new-l$ into X_b

Scan() procedure, code for processor b

1. The set of labels is $S = $ **traceable-collect**(X)
2. To decide the order between two labels lab_i and lab_j:
   Read the Order register Ord_{kb} where k is the most significant index in
   which lab_i differs from lab_j.
   Do the same for all labels

traceable-collect(X) of RWRWR, code for processor b

Perform handshake

1. For each c, Collect p_{cb}
2. For each c, Write $q_{bc} = p_{cb}$

Remainder of the first Read-Write-Read

3. For each c, Collect $X_c[c]$, tog_c
4. For each c, Write $B-X_{bc} = X_1[c], X_2[c], \ldots, X_n[c]$
5. For each c, Collect $X_c[c]$, p_{cb}, tog_c
6. For each c, if $q_{bc} = p_{cb}$ and tog_c is unchanged since line 3
   then $Y_c = X_c$
   else $Y_c = null$

Remainder of the second Read-Write-Read

7. For each c for which $Y_c = null$
   Write $B-X_{bc} = X_1[c], X_2[c], \ldots, X_n[c]$
8. For each c, Collect $X_c[c]$, p_{cb}, tog_c
9. For each c for which $Y_c = null$
   if $q_{bc} = p_{cb}$ and tog_c is unchanged since line 5
   then $Y_c = X_c$

10. For each c such that $Y_c = null$
    Read $Y_c = A1-X_{cb}$
11. Return $(Y_1, Y_2, \ldots, Y_n)$

Fig. 2. **Scan()**, **Label()** and **traceable-collect** procedures, DW algorithm [5].

to implement this mechanism, known as Traceable Use, every processor must maintain a separate, private pool of size $22n^2$ integers.

Though simple, the DW algorithm has the following drawbacks identified in [8]. (1) If b takes up a new private value every time it executes a **Label()**, its local pool will soon become empty unless it stops once in a while to perform garbage collection, allowing it to return those private values which are not in use, back into its private pool. To avoid this situation, garbage collection is performed as an ongoing, background operation, a constant number of steps during every **Label()** operation on one half of the pool, while exhausting the other half. That is, garbage collection which is of $O(n^2)$ complexity, is amortized over **Label()**

operation, which is of $O(n)$ complexity. This implies that **Label()** is not truly
linear. (2) To avoid having to use a long register of size $O(n^2 \log n)$ bits for
maintaining the ordering information among its $22n^2$ integers, b uses $22n$ SWMR
Order registers $Ord_{b(1)}$ to $Ord_{b(22n)}$, each having n values. But distributing the
ordering information over $22n$ separate SWMR Order registers does not allow
b to update its Order registers atomically, which can be a problem if there is
a concurrent read of Order registers by a scanner c. Hence, b maintains $O(n)$
sets of the Order registers and update indices $U_{b1} \ldots \ldots U_{bO(n)}$ to maintain the
status (i.e., whether being read currently) of the corresponding Order register.
Such an arrangement requires every processor b to use $O(n^2)$ SWMR registers.
Implementing this system using fundamental level SWSR safe bits is not space-
efficient. An analysis in [8] reveals that since it takes $O(n^2 \log n)$ SWSR safe bits
to construct a single SWMR atomic register of size $O(n \log n)$, each processor
b would need $O(n^4 \log n)$ SWSR safe bits and hence the whole system would
need $O(n^5 \log n)$ SWSR safe bits. (3) Each processor needs to maintain a pool
of values of size $22n^2$ integers in local memory, which means that there is an
overhead of $O(n^2 \log n)$ bits of local memory on every processor.

3   A Memory-Efficient Algorithm

3.1   Intuition and Basic Idea

As discussed in the previous section, a processor b can reuse a private value
v_k only if it determines that v_k is not in use by any processor in the system.
The Traceable Use Abstraction used in DW enables b to track down any of its
private values v_k which are in use by another processor d, even if v_k propagated
through many levels of indirection via other processors to d. Hence DW allows
arbitrary levels of indirection of propagation for v_k and if v_k is detected to be
in use, it will not be recycled. But as noted in [8], the propagation of private
values is restricted to only one level of indirection in BCTS and not to arbitrary
levels. Hence, the complete power of Traceable Use is redundant for BCTS. For
example, in the scenario described in Section 2, for b to recycle v_k, it is enough
to check for the following – v_k does not appear currently in any existing label, v_k
will not appear in a label that some processor might write in the future, v_k does
not appear in a label that is returned by any concurrent or future scan of b's
label. It is not necessary for b to keep track of more than one level of indirection
of propagation of its private values.

We now make an observation. In Figure 2, both the **Label()** and **Scan()**
procedures collect labels from other processors. While **Scan()** collects whole
labels, **Label()** collects only the private value of the processors. However, both
operations employ a common **traceable-collect** routine for collecting values.
Intuitively, from the point of view of a processor b, while it is necessary for b
to keep track of its private values that other labelers consumed from it (which
would go towards making up new labels), it is not necessary for b to keep track
of private values that scanners consumed from it (as long as the scanners are
able to determine, in some other way, the same order imposed on the labels by

Label() procedure, code for processor b

1. For each $c \neq b$, $val_c = \textbf{traceable-collect}(X_c)$ // collect private values
2. Read $localOrder = Ord_{bb}$.   // copy b's order into local variable $localOrder$

$$y[] = Scan-Aside_{b*}[b] \bigcup Aside_{b*} \bigcup Copy_{b*} \bigcup X_b[b]$$

//garbage collect private values

Choose v_b = smallest element in $localOrder$ not in $y[]$

Reorder elements in $localOrder$ with v_b as the largest element

3. $new-l = (val_1, val_2, \ldots, v_b, \ldots, val_n)$ // this is new label of b
4. $\textbf{traceable-write}(X_b, new-l)$ // write new label into X_b

Fig. 3. Label() procedure.

traceable-collect(X) of RWRWR, code for processor b // similar to that of [5]

Perform handshake

1. For each c, Collect p_{cb}
2. For each c, Write $q_{bc} = p_{cb}$

Remainder of the first Read-Write-Read

3. For each c, Collect $X_c[c]$, tog_c
4. For each c

   Write $Copy_{cb} = X_c[c]$
5. For each c, Collect $X_c[c]$, p_{cb}, tog_c
6. For each c, if $q_{bc} = p_{cb}$ and tog_c is unchanged since line 3

   then $Y_c = X_c[c]$

   else $Y_c = null$

Remainder of the second Read-Write-Read

7. For each c for which $Y_c = null$

   Write $Copy_{cb} = X_c[c]$
8. For each c, Collect $X_c[c]$, p_{cb}, tog_c
9. For each c for which $Y_c = null$

   if $q_{bc} = p_{cb}$ and tog_c is unchanged since line 5

   then $Y_c = X_c[c]$

10. For each c such that $Y_c = null$

   Read $Y_c = Aside_{cb}$
11. Return $(Y_1, Y_2, \ldots, Y_n)$

Fig. 4. traceable-collect RWRWR procedure.

the labelers). This is because *in the **Label()** operation, the collected values go back into the system as the new label of a processor but in **Scan()**, the collected labels do not go back.* While the **Label()** operation introduces a new label into the system and hence changes the system, the **Scan()** operation only makes an inference about the order of labeling events and hence does not change the system. Therefore, the approach taken in [5,8], of using the same **traceable-collect** routine for tracking down private values in both **Label()** and **Scan()** operations, does not seem natural.

We propose a new, modified design. In our algorithm, a processor does not keep track of labels that scanners consumed from it. Rather, the scanner notes

traceable-write(X_b, new-X) of RWRWR, code for processor b

1. For all c, Read q_{cb}
    // read each processor's b-th q to check for overlapping Label operation
2. For all $c \neq b$, Read $qtag_b[c] = PTAG_c[b]$
    // read each processor's b-th component of $PTAG$ to check for overlapping Scan
3. For all c, if $p_{bc} = q_{cb}$ // if labeling operation of processor c overlaps, do as follows
    Write $Aside_{bc} = X_b[b]$ // set aside private value for overlapping labeler c
4. For all $c \neq b$, if $qtag_b[c] \neq QTAG_b[c]$
    // if Scan() operation of processor c overlaps, do as follows
    Write $Scan-Aside_{bc} = X_b$ //set aside label for overlapping scanner c
    Write $Ord_{bc} = Ord_{bb}$ // set aside order for overlapping scanner c
5. Atomically write to r_b: // update all fields of r_b atomically
    X_b = new-X
    $tog_b = \mathbf{not}(tog_b)$
    for all c, $p_{bc} = \mathbf{not}(q_{cb})$
    $Ord_{bb} = localOrder$
    $QTAG_b = qtag_b$

Fig. 5. traceable-write procedure.

Scan() procedure, code for processor b // steps 1-3 collect Time-Lapse snapshot [4]
1. **Produce-tag()**
2. For all c, atomically Read from r_c: // copy X_c, $QTAG_c[b]$, Ord_{cc} into local var
    $value_b[c] = X_c$
    $qtag_b[c] = QTAG_c[b]$
    $order_b[c] = Ord_{cc}$
3. For all c
    If $qtag_b[c] = ptag_b[c]$
    // if some processors c have updated X_c, read values set aside during traceable-write
        Read $value_b[c] = Scan-Aside_{cb}$
        Read $order_b[c] = Ord_{cb}$
4. The labels and respective orders are $value_b$, $order_b$.   // determine real-time order
    To decide the order between the labels $value_b[i][1\ldots\ldots n]$ and $value_b[j][1\ldots\ldots n]$:
    For $i = 1$ to n
        For $j = 1$ to n
            Let k be the most significant index in which
            $value_b[i][1\ldots k\ldots n]$ differs from $value_b[j][1\ldots k\ldots n]$
            Determine the relative order using $order_b[k]$

Fig. 6. Scan() procedure.

down from each processor, at the same time as it collects the labels, the ordering information necessary for it to determine the order. Each processor uses two pools – a private value pool and a tag pool. Private values are used by labelers for making up labels. Tags are used by scanners to collect a snapshot of labels in the system. As in DW, **Label()** invokes **traceable-collect**, but in **Scan()**, a different approach is used. The scanner collects two things from each processor – its label and the ordering information of its private value pool. To get a

Produce-tag() procedure, code for processor b // produces new tag for every proc
1. Read $ptag_b = PTAG_b$
2. For all $c \neq b$, $x[c]$ = **garbage collect**($PTAG_b[c]$, $QTAG_c[b]$) //garbage collect tags
3. For all $c \neq b$, choose from Tag_{bc}, the local pool of tags for c, a tag $ptag_b[c] \notin x[c]$
4. Write $PTAG_b = ptag_b$ // this is new set of tags produced for other processor

Fig. 7. Produce-tag() procedure.

consistent view of the system, the scanner captures a snapshot, i.e., it collects labels and orders as they existed at a single point in time during the interval of the **Scan**()'s execution. The ordering information collected in the snapshot is used to determine the relative order among the labels. Consider the scenario described in Section 2. In the DW algorithm, all the labels collected by c during **Scan**() – totally $O(n^2)$ private values, $O(n)$ values for each of the n processors – are recorded in shared variables. Labeler b cannot recycle any of these private values since c may refer to those values again when it wakes up. In fact, b can reuse these values only after c performs a subsequent **Scan**(), writing new values in shared memory, in place of the earlier values. Hence, every processor needs to have a minimum pool size of $O(n^2)$ private values for the $O(n)$ scanners. In our algorithm, the scanner collects the labels and corresponding ordering information as they existed at a single point in time, without recording the private values in shared memory. Only the labeler, when it updates, sets aside values in shared memory, for those processors from which it noticed a concurrent **Scan**(). Hence, every processor needs to maintain a pool of only $O(n)$ private values. The scanner can determine the order among the labels by referring to the ordering information it has already noted down from each processor. Conceptually, the algorithm invokes **traceable-collect** from **Label**() and the Time-Lapse snapshot [4] from **Scan**(). This approach is comparable to [7,4] where atomic snapshot and Time-Lapse snapshot respectively are invoked from **Scan**(). While the precedence graph forms the backbone recycling mechanism in [7,4], our algorithm uses the Traceable Use mechanism for recycling private values at the back-end and is hence simpler.

Figure 1 illustrates the main differences between our construction and the Dwork-Waarts construction [5] at the implementation level. X_b is the label of processor b. The boolean p_{b*}, q_{b*} and tog_b registers implement the RWRWR handshake mechanism to coordinate between the **traceable-collect** of one labeler and the **traceable-write** of another labeler. $Aside_{bi}$ is written by b in **traceable-write** if b notices a concurrent **traceable-collect** operation by processor i. The boolean $PTAG_b$ and $QTAG_b$ registers enable b to detect scans from other processors, concurrent with its Label's **traceable-write**. $Scan-Aside_{bj}$ and Ord_{bj} are written by b in **traceable-write** if b notices a concurrent scan from processor j. The $Copy_{bi}$ registers, written by the corresponding processor i, enable i to make it known to b which of b's private values it read from label X_b in **traceable-collect**. Ord_{bb}, which is a field of r_b, is updated atomically with X_b by b. The local pool of tags used by b (tags Tag_{b1}, ..., Tag_{bn}) is a collection of $(n-1)$ pools, one pool of three tags for every processor. Since a scanner b needs to produce a new tag for every processor i, it needs to choose a tag other than the ones which i and b are currently holding and therefore three tags suffice for every processor. Hence the size of the pool of tags is bounded by $3n - 3$. From Figure 1, it should be clear that private values of processor b that may be in use at any time are recorded in registers $Scan-Aside_{b*}[b]$, $Aside_{b*}$, $Copy_{b*}$ and $X_b[b]$. Hence the private pool size is bounded by $3(n-1) + 1 + 1 = 3n - 1$. It is also clear from the figure that since there is only one long SWMR atomic register of length $O(n \log n)$ bits and the remaining long registers r_{b*} are all SWSR regular, the total shared space associated with b is $O(n^2 \log n)$ SWSR safe bits. Hence the total shared space of the system is bound by $O(n^3 \log n)$ SWSR safe bits. The size of the two pools being $O(n)$ integers, the local memory is bounded by $O(n \log n)$ bits.

3.2   Algorithm Description

The pseudo-code of the proposed algorithm is given in Figure 3 to Figure 7. Local variables such as $localorder$, $y[]$, v_b, $new-l$, $value_b$, $qtag_b$, $ptag_b$, $order_b[]$, Y_c, etc. used in the routines are assumed to be persistent.

In **Label()**, line 1 invokes the **traceable-collect** routine and collects private values of other processors. Line 2 corresponds to the **garbage collect** routine of [5]. In our algorithm, garbage collection is embedded in **Label()**. Private values of b which could be in use in the system, are collected in $y[]$ by scanning the $Scan-Aside$, $Aside$, $Copy$ & X_b registers. After choosing a new private value $v_b \in y[]$ from $localOrder$, b writes the new order in $localOrder$. The new label is written in X_b during **traceable-write** in line 4.

The **Scan()** operation collects a snapshot of $label_c$ and $order_c$ from all c, as they existed at a single point in time. In line 1, b invokes **Produce-tag()** to choose a new tag for each processor from its local pool of tags. Tags are used by a scanner i to determine if a processor d updated itself after i began its **Scan()**. If so, i discards the label and order collected from d in favor of the $Scan-Aside$ & Ord_{db} values, which d set aside for it. Line 2 does an atomic collect of the label and order from every processor. Updates that may have occurred during the interval of **Scan()** by some processors are determined by checking $qtag$ in line 3; for each such updater, the values set aside for it by the updater are taken. The total order on the collected labels consistent with the real-time order of their corresponding labeling operations is determined in line 4.

The **traceable-collect** routine invoked by **Label()** is identical to the one in [5]. Over lines (1,3,5), the **traceable-write** sets aside in $Aside$ its own component of label if **traceable-collect** of another labeler is detected to overlap using the RWRWR mechanism. Over lines (2,4,5), the **traceable-write** sets aside the label with the associated order in $Scan-Aside$ and Ord registers if **Scan()** of a scanner is detected to overlap using the $PTAG$ and $QTAG$ registers. The **Produce-tag()** routine used by **Scan()** produces a new set of tags for every other processor. Line 2 performs garbage collection to determine a new, unused tag that the scanner can produce again. The pools of tags used in our construction correspond to the pools of colors used in [4].

4   Remarks

We presented a simple, linear-time algorithm for constructing a Bounded Concurrent Timestamp System (BCTS) by giving new, modified designs for the **Label**() and **Scan**() operations. The time complexity of our algorithm matches the best know BCTS algorithms [4,5,8] while the total shared memory required in terms of the fundamental building blocks (SRSW safe registers) is less than that of [5] by two orders of magnitude. No expensive garbage collection is necessary. The pool of private values used by each processor is smaller than [5,8] so that the local memory overhead on each processor is correspondingly smaller. A full and formal version of the paper, including the proof of correctness of the construction, is given in [15]. The main drawback of our construction, and also of [5,8], is that we use long registers of size $O(n \log n)$ bits.

References

1. K. Abrahamson, On achieving consensus using a shared memory, *Proc. 7th ACM Symposium on Principles of Distributed Computing*, 291–302, 1988.
2. H. Attiya and J. Welch, Distributed computing: Fundamentals, simulations and advanced topics, *McGraw-Hill Publishing Company, London, UK*, 1998.
3. D. Dolev and N. Shavit, Bounded concurrent time-stamping, *SIAM Journal of Computing*, 26(2): 418–455, 1997.
4. C. Dwork, M. Herlihy, S. Plotkin and O. Waarts, Time-lapse snapshots, *SIAM Journal of Computing*, 28(5): 1848–1874, 1999.
5. C. Dwork and O. Waarts, Simple and efficient bounded concurrent timestamping and the traceable use abstraction, *Journal of the ACM*, Vol. 46, pp. 633–666, 1999.
6. M. Fischer, N. Lynch, J. Burns and A. Borodin, Distributed Fifo allocation of identical resources using small shared space, *ACM Transactions on Programming Language Systems*, 11(1): 90–114, 1989.
7. R. Gawlick, N. Lynch, N. Shavit, Concurrent timestamping made simple, *Proc. Israeli Symp. on Computing and Systems*, 171–183, LNCS 601, Springer, 1992.
8. S. Haldar and P. Vitanyi, Bounded concurrent timestamp systems using vector clocks, *Journal of the ACM*, Vol. 49, pp. 101–126, 2002.
9. A. Israeli and M. Li, Bounded timestamps, *Proc. 28th IEEE Symposium on Foundations of Computer Science*, pp. 371–382, 1987.
10. A. Israeli and M. Pinhasov, A concurrent timestamp scheme which is linear in time and space, *Proc. Workshop on Distributed Algorithms*, 95–109, LNCS 647, Springer-Verlag, 1992.
11. L. Lamport, A new solution to Dijkstra's concurrent programming problem, *Communications of the ACM*, 17, 1974.
12. M. Li, J. Tromp and P. Vitanyi, How to share concurrent wait-free variables, *Journal of the ACM*, 43(4): 723–746, 1996.
13. G. Peterson, Concurrent reading while writing, *ACM Transactions on Programming Language Systems*, 5(1): 46–55, 1983.
14. T. Petrov, A. Pogosyants, S. Garland, V. Luchangco and N. Lynch, Computer-assisted verification of an algorithm for concurrent timestamps, *Formal Description Techniques and Protocol Specification, Testing and Verification, FORTE/PSTV'96*, IFIP Procs., pp. 29–44, 1996.

15. V. Shikaripura, A. Kshemkalyani A simple memory-efficient bounded concurrent timestamping algorithm, Technical Report UIC-CS-02-04, June 2002.
16. P. Vitanyi and B. Awerbuch, Shared register access by asynchronous hardware, *Proc. 27th IEEE Symp. on Foundations of Computer Science*, pp. 233–243, 1986.

Crossing Minimization for Symmetries[*]

Christoph Buchheim[1] and Seok-Hee Hong[2]

[1] Institut für Informatik, Universität zu Köln, Germany
buchheim@informatik.uni-koeln.de
[2] School of Information Technologies, University of Sydney, Australia
shhong@it.usyd.edu.au

Abstract. We consider the problem of drawing a graph with a given symmetry such that the number of edge crossings is minimal. We show that this problem is NP-hard, even if the order of orbits around the rotation center or along the reflection axis is fixed. Nevertheless, there is a linear time algorithm to test planarity and to construct a planar embedding if possible. Finally, we devise an $O(m \log m)$ algorithm for computing a crossing minimal drawing if inter-orbit edges may not cross orbits, showing in particular that intra-orbit edges do not contribute to the NP-hardness of the crossing minimization problem for symmetries. From this result, we can derive an $O(m \log m)$ crossing minimization algorithm for symmetries with an orbit graph that is a path.

1 Introduction

In Automatic Graph Drawing, the display of symmetries is one of the most desirable aims [16]. Each symmetry displayed in the drawing of a graph reduces the complexity of the drawing for the human viewer; see Fig. 1. Furthermore, symmetric drawings are regarded as aesthetically pleasing in general. Finally, the existence of a symmetry can point to an important structural property of the data being displayed.

The usual approach for drawing graphs symmetrically consists of two steps: first, try to find an abstract symmetry, i.e., an automorphism of the graph that can be represented by a symmetric drawing in two or three dimensions; second, compute such a drawing, while trying to keep an eye on other criteria characterizing a nice and readable graph layout. The symmetry detection problem to be solved in the first step is NP-hard in general [14]. Nevertheless, exact algorithms for general graphs have been devised in [4] and [1]. In planar graphs, planar symmetries can be detected in linear time [9,10,11,12].

In this paper, we focus on the second step and on two-dimensional drawings. More specifically, we consider the problem of minimizing the number of edge crossings over all two-dimensional drawings displaying a given symmetry π. Obviously, the aims of displaying π and minimizing the number of edge crossings

[*] The second author is supported by a grant from the Australian Research Council. This paper was partially written when the first author was visiting the University of Sydney.

P. Bose and P. Morin (Eds.): ISAAC 2002, LNCS 2518, pp. 563–574, 2002.
© Springer-Verlag Berlin Heidelberg 2002

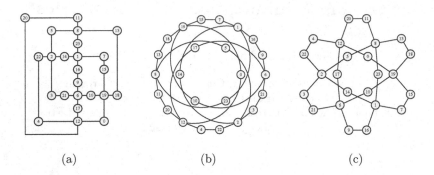

(a)   (b)   (c)

Fig. 1. Three drawings of the same abstract graph. The drawing in (a) is planar but does not display any symmetry. In (b) and (c), the same symmetries are displayed; the number of edge crossings in (b) is 18, while the number of edges crossings in (c) is 6, which is minimal for the given symmetries

interfere with each other; see Fig. 1 again. We always assume that the symmetry π has to be displayed in any case. Finally observe that we do not consider the case of two symmetries that can be displayed simultaneously.

As far as we know, the problem of crossing minimization for symmetries has not been examined before, so we first introduce some new notation in Sect. 2. In Sect. 3, we show that crossing minimization for symmetries is NP-hard. Next, we show that planarity can be tested in linear time; see Sect. 4. In Sect. 5, we investigate the impact of intra-orbit edges on complexity. In Sect. 6, we give some ideas of how the results presented in this paper can be used in order to develop heuristic crossing minimization algorithms for symmetries. Sect. 7 concludes.

2   Preliminaries

In the following, a *graph* is always a simple graph with n nodes and m edges. For runtime evaluations, we assume $m \geq n$. A *drawing* of a graph is a representation of its nodes by different points of the plane and of its edges by arbitrary curves in the plane that connect the corresponding points but do not traverse any other node point. An *edge crossing* in a drawing of a graph is a crossing of two curves.

A *symmetry* or *geometric automorphism* of a graph G is a permutation of the nodes of G that is induced by a symmetric drawing of G, i.e., by a drawing of G that is fixed by some non-trivial isometry of the plane. By [6], there are two different types of symmetries: a *reflection* or *axial symmetry* is a symmetry induced by a reflection of the plane at an axis, the *reflection axis*; a *rotation* is a symmetry induced by a rotation of the plane around a center point, the *rotation center*. The *order* of a symmetry π is the smallest positive integer k such that π^k equals the identity. In particular, all reflections have order one or two. For a node v, the *orbit* of v under π consists of the nodes $\pi^k(v)$ for all integers k; the orbit of an edge is defined analogously. An edge is called *intra-orbit edge* if the nodes connected by this edge belong to the same orbit, otherwise it is called *inter-orbit edge*. A *diagonal* of a rotation π of even order k is an edge $(v, \pi^{k/2}(v))$. For technical reasons, we consider diagonals as inter-orbit edges in this paper. A node v of G is a *fixed node* of π if $\pi(v) = v$.

The *orbit graph* G/π of π is the graph resulting from G when identifying each node $v \in V$ with its image $\pi(v)$ and deleting multiple edges and loops afterwards. Hence every node of G/π represents a π-orbit of G and every edge of G/π corresponds to a set of orbits of inter-orbit edges of G; see Fig. 2.

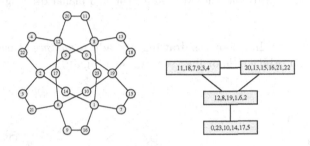

Fig. 2. A rotation of order six and its orbit graph

A *drawing* of a symmetry π of G is a drawing of G such that π is induced by an isometry of the plane fixing this drawing. In any drawing of a rotation π, every curve representing a diagonal has to cross the rotation center. By adding a fixed node to π, if none exists, and by splitting up every diagonal at the fixed node, we may assume throughout this paper that the considered rotations never have diagonals. For every drawing of a rotation of order k, there is an integer $e \in \{1, \ldots, k\}$ with $\gcd(e, k) = 1$ such that rotating the drawing by $360/k$ degrees in clockwise order around the rotation center maps each node v to $\pi^e(v)$. We call e the *exponent* of the drawing; see Fig. 3. The exponent of a reflectional drawing is always one. Different exponents correspond to what is called different *representations* in [1].

In a drawing of a rotation, all nodes of a common orbit have the same distance to the rotation center; the corresponding circle is called the *orbit circle*. For a drawing of a reflection, the nodes of a common orbit lie on the same line orthogonal to the reflection axis; this line is called the *orbit line*.

A graph is *planar* if it has a *plane drawing*, i.e., a two-dimensional drawing without edge crossings. A symmetry π of a graph G is *planar* if there is a drawing of π that is planar as a drawing of G. We call a drawing of a symmetry *loopless* if all orbit circles or lines are distinct and if no edge crosses more orbit circles or lines than necessary. More precisely, in a loopless drawing of a rotation, every curve connecting two orbits may only cross the circles of the orbits in-between, and each only once; see Fig. 4. Analogously, in a loopless drawing of a reflection, every curve connecting two orbits may only cross the lines of orbits in between,

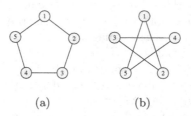

(a)   (b)

Fig. 3. Two drawings displaying the same symmetry $\pi = (12345)$. The exponent of the planar drawing (a) is one; the exponent of the non-planar drawing (b) is three

and each only once. In a loopless drawing, we add a *dummy* at each crossing of an edge with an orbit circle or line; see Fig. 4(a) again.

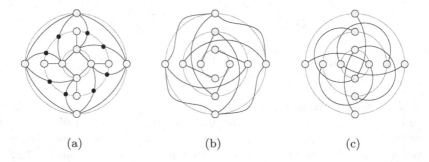

(a)   (b)   (c)

Fig. 4. The drawing (a) is loopless, the dummies are represented by filled circles. In both drawings (b) and (c), all edges are loops

In this paper, we will mainly consider the following problems:

Problem (SCM): Given a symmetry π, compute a drawing of π with a minimal number of edge crossings.

Problem (SCM+): Given a symmetry π, compute a loopless drawing of π with a minimal number of edge crossings.

Problem (SCM+1): Given a symmetry π and a fixed order of its orbits, compute a loopless drawing of π respecting this order such that the number of edge crossings is minimal.

Problem (SCM+2): Given a symmetry π, a fixed order of its orbits, and a fixed order of nodes and dummies on each orbit circle or line, compute a loopless drawing of π respecting these orders such that the number of edge crossings is minimal.

3   Crossing Minimization for Symmetries Is NP-Hard

In the following, we prove that the problems (SCM), (SCM+), and (SCM+1) defined in the last section are NP-hard. In fact, the proofs show that the crossing minimization problem for symmetries can be considered as a generalization of the crossing minimization problem for graphs.

Lemma 1. *The problem (SCM) can be reduced to (SCM+) in $O(m^2)$ time, also if both problems are restricted to reflections or rotations of fixed order.*

Proof. Let π be a symmetry of G. Add m new nodes on each edge of G and let π' be the resulting symmetry. We claim that for every drawing of π there is a loopless drawing of π' with the same number of edge crossings; for a proof, see the extended version of this paper [3]. Hence, a crossing minimal loopless drawing of π' gives rise to a crossing minimal general drawing of π.   □

Theorem 1. *The problems (SCM) and (SCM+) are NP-hard, even if restricted to reflections or rotations of fixed order.*

Proof. We can easily reduce the NP-hard problem of crossing minimization for graphs [8] to the crossing minimization problem for reflections or rotations of fixed order k. For this, let G be any graph. Construct a new graph G' as the disjoint sum of k copies of G. Define π by mapping the copies cyclically to each other. Obviously, drawing π with a minimal number of edge crossings is equivalent to drawing G' without intersections between the copies of G and drawing the copies of G with a minimal number of edge crossings. Hence an optimal drawing of π according to (SCM) gives rise to a drawing of G with a minimal number of edge crossings. Finally, the NP-hardness of (SCM+) follows from Lemma 1.   □

Theorem 2. *The problem (SCM+1) is NP-hard, even if restricted to reflections or rotations of fixed order.*

Proof. We reduce the NP-hard fixed linear crossing minimization problem [15] to (SCM+1). Consider an instance of this problem, given by an ordered sequence of nodes $(v_1, \ldots, v_n)$ and a set of edges E. Define an instance for (SCM+1) as follows: start with $G = (\{v_1, \ldots, v_n\}, E)$ and add nodes $v_{i,j}$ for $i = 1, \ldots, n-1$ and $j = 1, \ldots, m^2$. Define a node order by $v_i < v_{i,j} < v_{i+1}$ and $v_{i,j} < v_{i,j+1}$. Add new edges $(v_i, v_{i,j})$ and $(v_{i,j}, v_{i+1})$ for each $i = 1, \ldots, n-1$ and $j = 1, \ldots, m^2$. Finally, define G' and π as in the proof of Theorem 1.

Now consider an optimal drawing of π according to (SCM+1) with the given order of orbits. We claim that the center of the rotation lies outside of each component of G' then. Indeed, the new edges would induce at least m^2 edge crossings otherwise, but this is not possible since a drawing with less than m^2 edge crossings obviously exists. By the same reason, no edge crosses the new edges of its own component. In summary, the solution of (SCM+1) gives rise to a solution of the fixed linear crossing minimization problem.   □

We do not know whether problem (SCM+2) is NP-hard in general. If no dummies exist, (SCM+2) can be solved in $O(m \log m)$ time by Theorem 4 below.

4   Testing Planarity of Symmetries in Linear Time

We now consider a more promising candidate:

Problem (SPL): Given a symmetry π, decide whether there is a drawing of π without edge crossings. If there is such a drawing, compute one.

Theorem 3. *The problem (SPL) can be solved in $O(n)$ time.*

Proof. See [3] again. The outline of this algorithm is similar to the one presented in [9,10,11,12]. However, it cannot be derived directly from the results given there. First, the problem can be reduced to biconnected planar graphs. Using SPQR-trees [5], it can be further reduced to the triconnected case. In this case, the result follows from Lemma 3.1 and Theorem 4.1 in [12].   □

In [3], we also show that every planar symmetry admits a planar loopless drawing as well as a planar straight-line drawing. Hence, planarity of symmetries does not depend on whether we require straight lines or not. This is not true for the crossing number of a symmetry, which follows from the corresponding result for graphs [2] but is obvious even for symmetries with a single orbit; see Fig. 5.

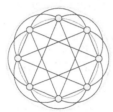 

Fig. 5. Intra-orbit edges may be drawn inside or outside their orbit circle. In general, requiring all edges to be drawn on the same side, for example in straight-line drawings, increases the number of necessary edge crossings

5   Crossing Minimization and Intra-Orbit Edges

In Sect. 3, we showed that the symmetric crossing minimization problems (SCM), (SCM+), and (SCM+1) are NP-hard. In the reductions of the proofs, we did not make use of intra-orbit edges at all. These problems thus remain NP-hard even for symmetries containing only inter-orbit edges. In this section, we show that the corresponding statement for intra-orbit edges is not true. More generally, we show that (SCM+2) can be solved in $O(m \log m)$ time if no dummies exist. Since no information about intra-orbit edges is given in (SCM+2), we derive that intra-orbit edges do not make crossing minimization significantly harder.

Lemma 2. *Let k be a positive integer and $(a_1, \ldots, a_r)$ a sequence of integers. Let $I = \{1, \ldots, r\}$. Then an integer sequence $(c_1, \ldots, c_r)$ minimizing the sum*

$$\sum_{\substack{i,j \in I \\ i<j}} |a_j + c_j k - a_i - c_i k|$$

as well as the minimal objective value can be computed in $O(r \log r)$ time.

Proof. See [3] again. Showing that $a_i + c_i k - a_j - c_j k < k$ for all $i, j \in I$ in any optimal solution, we can restrict our attention to r remaining candidates that can be evaluated and compared in a total runtime of $O(r \log r)$. $\qquad \square$

Lemma 3. *Let I be a set of r non-negative integers and $t_1, t_2 \geq 0$. Then a partition $I = I_1 \oplus I_2$ minimizing*

$$\sum_{\substack{i,j \in I_1 \\ i<j}} i + \sum_{\substack{i,j \in I_2 \\ i<j}} i + t_1 \sum_{i \in I_1} i + t_2 \sum_{i \in I_2} i$$

as well as the minimal objective value can be computed in $O(r \log r)$ time.

Proof. Assuming $t_1 \leq t_2$, we claim that an optimal partition is found by traversing the set I in descending order, adding the first $\lfloor t_2 - t_1 \rfloor$ elements to I_1 and the following elements to I_1 and I_2 in turn. For a proof, see [3]. $\qquad \square$

Theorem 4. *The problem (SCM+2) can be solved in $O(m \log m)$ time for symmetries without dummies. The corresponding number of edge crossings can be computed in $O(m/k \cdot \log m)$ time, where k is the order of the symmetry.*

Proof. First note that this statement is trivial for reflections. Hence for the rest of this proof we consider a rotation π of order k. Consider the plane polar coordinate function φ mapping a point (α, d) in the plane to $(d \sin \alpha, d \cos \alpha)$. For simplicity, we will not construct the required drawing D directly but a drawing D' such that $\varphi(D') = D$; see Fig. 6.

First observe that the placement of nodes is essentially determined by the data given in (SCM+2): the α-th node on the d-th orbit is placed to the point $(2\pi\alpha/k, d)$, where we count orbits from inside to outside, the number of the first non-trivial orbit being one and the number of an eventual fixed orbit being zero.

For drawing edges, we first consider the inter-orbit edges independently. If an edge e connects two nodes placed to (α_1, d_1) and (α_2, d_2), the optimal way to draw e is by a straight line. However, e does not have to connect (α_1, d_1) to (α_2, d_2) but may connect it to $(\alpha_2 + 2\pi c, d_2)$ for any integer c. This corresponds to the possibility of wrapping e around the inner orbit circle. The resulting number of edge crossings is influenced by the choice of c in general; see Fig. 7. Below we explain how to choose the wrapping coefficients c optimally.

After drawing the inter-orbit edges, we consider intra-orbit edges. For an intra-orbit edge e connecting (α_1, d_1) and (α_2, d_2), the optimal coefficient c is

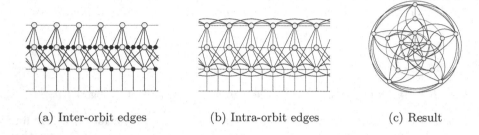

(a) Inter-orbit edges   (b) Intra-orbit edges   (c) Result

Fig. 6. Constructing an optimal drawing

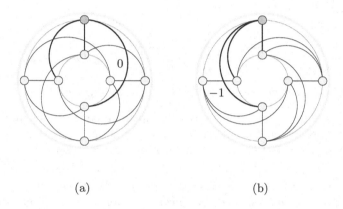

(a)   (b)

Fig. 7. Different wrapping coefficients c induce different numbers of edge crossings. The number of crossings in (a) is 12, the number of crossings in (b) is 4

obviously the one minimizing $|\alpha_2 - \alpha_1 + 2\pi c|$. But now we have to decide for each orbit of intra-orbit edges whether its edges are drawn inside or outside the orbit circle; see Fig. 5. The algorithm to decide this is explained below. For a given partition into inside and outside edges, we proceed as follows: the edge e is drawn as an arc. The angle of the arc when leaving a node is the same for all arcs; it is equal to the minimum angle of a line representing an inter-orbit edge incident to the same node, but small enough to ensure that the distance of the longest arc from the line of the orbit is at most $1/3$. If e is chosen to be an inside edge, the corresponding arc is drawn below the orbit line, otherwise above.

In summary, for fixed wrapping coefficients and a fixed partition into inside and outside edges, all crossings involving intra-orbit edges are unavoidable by the data given in (SCM+2). In particular, the optimal partition into inside and outside edges depends only on the given orders but not on the drawing of the inter-orbit edges computed before. Since the number of crossings between inter-orbit edges in the drawing created by this algorithm is minimal, this justifies our strategy to consider inter-orbit edges and intra-orbit edges separately. So to prove the first statement of Theorem 4, it remains to compute optimal wrapping coefficients and an optimal partition of intra-orbit edges into inside and outside edges in $O(m \log m)$ time.

For the first part, we may assume that π has exactly two orbits, and that both orbits have k nodes, since if the inner orbit contains only one fixed node, the problem of determining optimal wrapping coefficients is trivial. First we explain how to count the number of inter-orbit edge crossings depending on the choice of wrapping coefficients. Let v be any node on the outer orbit and number the nodes on the inner orbit $0, \ldots, k-1$ according to the order given in (SCM+2), starting at any node. Let $e_1, \ldots, e_r$ be the inter-orbit edges incident to v, sorted by ascending x-coordinate of the corresponding node on the inner orbit. Let $I = \{1, \ldots, r\}$. Note that each orbit of inter-orbit edges has exactly one representing edge in $\{e_1, \ldots, e_r\}$; hence $r \leq m/k$. For $i \in I$, let a_i be the number of the node adjacent to e_i on the inner orbit, and let c_i be the unique integer such that the x-coordinate of a_i minus the x-coordinate of the node with number 0 is contained in $[2\pi c_i, 2\pi(c_i + 1))$. Then for $i < j$ the number of edge crossings between the orbits of e_i and e_j is

$$k(|a_j + c_j k - a_i - c_i k| - 1) .$$

Since inter-orbit edges of the same orbit do not cross each other, the total number of crossings between inter-orbit edges is

$$k \sum_{\substack{i,j \in I \\ i<j}} |a_j + c_j k - a_i - c_i k| - k \cdot \#\{i, j \in I \mid i < j\} .$$

By Lemma 2, the optimal coefficients c_i can be computed in $O(m/k \cdot \log m)$ time.

Next we determine the optimal partition of intra-orbit edges into inside and outside edges. For this, consider a fixed orbit p. Assume that the nodes on p are numbered $0, 1, \ldots, k-1$ according to the order given in (SCM+2). Depending on a given partition, we can compute the number of edge crossings involving intra-orbit edges of orbit p as follows: consider the set

$$I = \{i \in \{1, 2, \ldots, \lfloor k/2 \rfloor\} \mid (0, i) \in E\} .$$

Let $I = I_1 \oplus I_2$ be the partition of I representing the chosen partition of edges into inside and outside edges; i.e., let the set $\{(0, i) \mid i \in I_1\}$ contain exactly one representing edge for each orbit of inside edges. Observe that the total sum of elements of I for all orbits is at most m/k, since all edge orbits contain exactly k edges (recall that we excluded diagonals). Furthermore, the number of crossings between inside edges is given by the sum of the numbers of crossings between the orbits of $(0, i)$ and $(0, j)$ for $i, j \in I_1$ and $i \leq j$ (the case $i = j$ corresponds to crossings between edges of the same inside edge orbit). In an optimal drawing as described above, this crossing number for i and j is $2k(i - 1)$. So the total number of crossings between inside edges is

$$\sum_{\substack{i,j \in I_1 \\ i \leq j}} 2k(i-1) \, .$$

The situation for outside edges is symmetric. Additionally, we have

$$\sum_{i \in I_1} kt_1(i-1) + \sum_{i \in I_2} kt_2(i-1)$$

crossings between an intra-orbit edge of p and inter-orbit edges incident to p, where t_1 is the number of edges connecting node 0 with an orbit inside of p and t_2 is the number of edges connecting node 0 with an orbit outside of p. So the total number of crossings involving at least one intra-orbit edge is

$$2k \left(\sum_{\substack{i,j \in I_1 \\ i<j}} (i-1) + \sum_{\substack{i,j \in I_2 \\ i<j}} (i-1) + \frac{t_1}{2} \sum_{i \in I_1} (i-1) + \frac{t_2}{2} \sum_{i \in I_2} (i-1) \right) + 2k \sum_{i \in I} (i-1)$$

and can be minimized by Lemma 3 after decreasing each element of I by one. The total runtime for all orbits is $O(m/k \cdot \log m)$.

It remains to explain how to compute the total number of edge crossings in $O(m/k \cdot \log m)$ time. This number is given as the sum of the two formulas derived above for the number of crossings between inter-orbit edges and the number of crossings involving intra-orbit edges. Hence it can be computed in $O(m/k \cdot \log m)$ time by Lemma 2 and 3. □

Corollary 1. *The problem (SCM+) can be solved in $O(m \log m)$ time if the orbit graph of π is a path.*

Proof. Obviously, the optimal order of orbits is one of the two orders given by the path; for rotations with a fixed node, this node has to be the innermost orbit. Hence it suffices to show that (SCM+1) can be solved in $O(m \log m)$ time if all inter-orbit edges connect neighboring orbits. Let k be the order of the symmetry to be drawn. Observe that there are at most k possible orders of nodes and dummies on the orbits, since we have no dummies and the order of nodes on the orbits is determined by the exponent of the drawing. By Theorem 4, we can compute the minimal number of edge crossings for each possible exponent in $O(m/k \cdot \log m)$, since after fixing the exponent we have an instance of (SCM+2) without dummies. Hence we can determine the best exponent in $O(m \log m)$ time. Finally, we can compute an optimal drawing for this exponent in $O(m \log m)$ time by Theorem 4 again. □

Corollary 2. *The problem (SCM) can be solved in $O(m \log m)$ time for symmetries without inter-orbit edges.*

Observe that Corollary 1 and Lemma 1 imply that the problem (SCM) can be solved in $O(m^2 \log m)$ time if the orbit graph of π is a path.

6 Heuristic Crossing Minimization for Symmetries

From the results given in the last section, we can derive some ideas for a first approach to heuristic crossing minimization for symmetries. Corollary 1 suggests the following algorithmic framework:

1. Test planarity of π by Theorem 3. If π is planar, draw it without crossings by Theorem 3 and stop.
2. Compute the orbit graph G/π.
3. Compute a layering of G/π with a small number of dummy nodes, assigning only one node to each layer.
4. For every edge in G/π connecting non-neighboring layers, delete the corresponding edge orbits in G. Let π' be the resulting symmetry.
5. The orbit graph of π' is a path; draw it optimally using Corollary 1.
6. Reinsert the deleted edge orbits one after another, each one optimally.

In step 3, we have an optimal linear arrangement problem, since the number of dummies equals the total length of all edges minus m. This problem is NP-hard [7], but many heuristic and exact algorithms have been devised; see [13] for an overview. In step 6, we can add a single edge optimally in linear time by searching for a shortest path in the dual graph. After that, we can insert all other edges of the same orbit symmetrically.

This algorithmic framework is similar to the one used for the planarization method for graphs. In both methods, edges of the graph are removed until the resulting graph or symmetry is tractable. After solving the reduced instance, the deleted edges are reinserted. However, note that we do not have to delete edges until the symmetry is planar but only until Corollary 1 is applicable. Finally note that the number of edges in the orbit graph G/π is at most m/k, where k is the order of π. Loosely speaking, the more symmetry we have, the smaller G/π is, and the fewer edges we have to delete in step 4.

7 Conclusion

In this paper, we started the discussion of crossing minimization for graphs with a given symmetry. This is a generalization of the usual crossing minimization problem for graphs and hence a hard problem both theoretically and practically. The presented results allow to restrict the attention to symmetries without intra-orbit edges, since these do not increase the complexity significantly. In the case of a single orbit we have a crossing minimization problem that is interesting on its own and that can be solved in $O(m \log m)$ time.

This is a first approach to symmetric crossing minimization, so there is much left for future research; heuristic crossing minimization algorithms, for example in the framework of Sect. 6, have to be developed and evaluated. Furthermore, all results should be generalized to the case where not only one single symmetry can be considered but also two symmetries that can be displayed simultaneously (the dihedral case). Finally, it is open how much exactly has to be fixed for the drawing of a symmetry to make the crossing minimization problem polynomial time solvable.

References

1. D. Abelson, S. Hong, and D. Taylor. A group-theoretic method for drawing graphs symmetrically. Technical Report IT-IVG-2002-01, University of Sydney, 2002.
2. D. Bienstock and N. Dean. Bounds for rectilinear crossing numbers. *Journal of Graph Theory*, 17:333–348, 1993.
3. C. Buchheim and S. Hong. Crossing minimization for symmetries. Technical Report zaik2002-440, ZAIK, Universität zu Köln, 2002.
4. C. Buchheim and M. Jünger. Detecting symmetries by branch & cut. In P. Mutzel, M. Jünger, and S. Leipert, editors, *Graph Drawing 2001*, volume 2265 of *Lecture Notes in Computer Science*, pages 178–188. Springer-Verlag, 2001.
5. G. Di Battista and R. Tamassia. On-line maintenance of triconnected components with SPQR-trees. *Algorithmica*, 15:302–318, 1996.
6. P. Eades and X. Lin. Spring algorithms and symmetry. *Theoretical Computer Science*, 240(2):379–405, 2000.
7. M. Garey and D. Johnson. *Computers and Intractability – A Guide to the Theory of NP-Completeness*. W. H. Freeman, 1979.
8. M. Garey and D. Johnson. Crossing number is NP-complete. *SIAM Journal on Algebraic and Discrete Methods*, 4:312–316, 1983.
9. S. Hong and P. Eades. Drawing planar graphs symmetrically II: Biconnected graphs. Technical Report CS-IVG-2001-01, University of Sydney, 2001.
10. S. Hong and P. Eades. Drawing planar graphs symmetrically III: Oneconnected graphs. Technical Report CS-IVG-2001-02, University of Sydney, 2001.
11. S. Hong and P. Eades. Drawing planar graphs symmetrically IV: Disconnected graphs. Technical Report CS-IVG-2001-03, University of Sydney, 2001.
12. S. Hong, B. McKay, and P. Eades. Symmetric drawings of triconnected planar graphs. In *SODA 2002*, pages 356–365, 2002.
13. S. Horton. *The optimal linear arrangement problem: Algorithms and approximation*. PhD thesis, Georgia Institute of Technology, 1997.
14. J. Manning. *Geometric Symmetry in Graphs*. PhD thesis, Purdue University, 1990.
15. S. Masuda, K. Nakajima, T. Kashiwabara, and T. Fujisawa. Crossing minimization in linear embeddings of graphs. *IEEE Transactions on Computers*, 39(1):124–127, 1990.
16. H. Purchase. Which aesthetic has the greatest effect on human understanding? In Giuseppe Di Battista, editor, *Graph Drawing '97*, volume 1353 of *Lecture Notes in Computer Science*, pages 248–261. Springer-Verlag, 1997.

Simultaneous Embedding of a Planar Graph and Its Dual on the Grid*

Cesim Erten and Stephen G. Kobourov

Department of Computer Science University of Arizona
{cesim,kobourov}@cs.arizona.edu

Abstract. Traditional representations of graphs and their duals suggest the requirement that the dual vertices should be placed inside their corresponding primal faces, and the edges of the dual graph should cross only their corresponding primal edges. We consider the problem of simultaneously embedding a planar graph and its dual on a small integer grid such that the edges are drawn as straight-line segments and the only crossings are between primal-dual pairs of edges. We provide an $O(n)$ time algorithm that simultaneously embeds a 3-connected planar graph and its dual on a $(2n - 2) \times (2n - 2)$ integer grid, where n is the total number of vertices in the graph and its dual.

Keywords. Graph drawing, planar embedding, simultaneous embedding, convex planar drawing.

1 Introduction

In this paper we address the problem of simultaneously drawing a planar graph and its dual on a small integer grid. The *planar dual* of an embedded planar graph G is the graph G' formed by placing a vertex inside each face of G, and connecting those vertices of G' whose corresponding faces in G share an edge. Each vertex in G' has a corresponding primal face and each edge in G' has a corresponding primal edge in the original graph G. The traditional manual representations of a graph and its dual, suggest two natural requirements. One requirement is that we place a dual vertex inside its corresponding primal face and the other is that we draw a dual edge so that it only crosses its corresponding primal edge. We provide a linear-time algorithm that simultaneously draws a planar graph and its dual using straight-line segments on the integer grid while satisfying these two requirements.

1.1 Related Work

Straight-line embedding a planar graph G on the grid, i.e., mapping the vertices of G on a small integer grid such that each edge can be drawn as a straight-line segment and that no crossings between edges are created, is a well-studied graph

---

* A full version of this extended abstract is at www.cs.arizona.edu/~cesim/dual.ps.

P. Bose and P. Morin (Eds.): ISAAC 2002, LNCS 2518, pp. 575–587, 2002.

drawing problem. The first solution to this problem was given by de Fraysseix, Pach and Pollack [6] who provide an algorithm that embeds a planar graph on n vertices on the $(2n - 4) \times (n - 2)$ integer grid. Later, Schnyder [13] present another method that requires grid size $(n - 2) \times (n - 2)$. Also, several restrictions of this problem have been considered. Harel and Sardas [7] provide an algorithm to embed a biconnected graph on the $(2n - 4) \times (n - 2)$ grid without triangulating the graph initially. The algorithm of Chrobak and Kant [5] embeds a 3-connected planar graph on a $(n - 2) \times (n - 2)$ grid so that each face is convex. Miura, Nakano, and Nishizeki [11] further restrict the graphs under consideration to 4-connected planar graphs with at least 4 vertices on the outer face and present an algorithm for straight-line embedding of such graphs on a $(\lceil n/2 \rceil - 1) \times (\lfloor n/2 \rfloor)$ grid.

In a paper dating back to 1963, Tutte [14] shows that there exists a simultaneous straight-line representation of any planar graph and its dual in which the only intersections are between corresponding primal-dual edges. However, a disadvantage of this representation is that the area required by the algorithm can be exponential in the number of vertices of the graph.

Brightwell and Scheinerman [2] show that every 3-connected planar graph G can be represented as a collection of circles, one circle representing each vertex and each face, so that, for each edge of G, the four circles representing the two endpoints and the two neighboring faces meet at a point. Moreover, the vertex-circles cross the face-circles at right angles. This result implies that one can represent a 3-connected planar graph and its dual simultaneously in the plane with straight-line edges so that the primal edges cross the dual edges at right angles (provided that the vertex corresponding to the unbounded face is located at infinity). Mohar [12] extends the results of [2] by presenting an approximation algorithm that given a 3-connected planar graph $G = (V, E)$ and a rational number $\epsilon > 0$ finds an ϵ-approximation for the radii and the coordinates of the centers for the primal-dual circle representation for G and its dual. Mohar's algorithm runs in time polynomial in $|E(G)|$ and $\log(1/\epsilon)$ and the angles of the primal-dual edge crossings are arbitrarily close to $\pi/2$.

Bern and Gilbert [1] address a variation of the simultaneous planar-dual embedding problem: finding suitable locations for dual vertices, given a straight-line planar embedding of a planar graph, so that the edges of the dual graph are also straight-line segments and cross only their corresponding primal edges. They present a linear time algorithm for the problem in the case of convex 4-sided faces and show that the problem is NP-hard for the case of convex 5-sided faces.

1.2   Our Results

The simultaneous embedding in [2] guarantees right angles for the primal-dual edge crossings where the unbounded face needs to be handled in a special way by creating a vertex at infinity. Even without considering the unbounded face, the methods in [2] and [12] do not provide bounds on the area required for the simultaneous embedding and they are less practical than our approach.

In this paper we present an algorithm for embedding a given planar graph G and its dual simultaneously so that following conditions are met:

- The primal graph is drawn with straight-line segments without crossings.
- The dual graph is drawn with straight-line segments without crossings.
- Each dual vertex lies inside its primal face.
- A pair of edges cross if and only if the edges are a primal-dual pair.
- Both the primal and the dual vertices are on the $(2n - 2) \times (2n - 2)$ grid, where n is the number of vertices in the primal and dual graphs.
- The running time of the algorithm is $O(n)$.

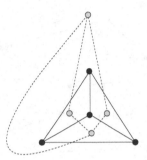

Fig. 1. A drawing of K_4 and its dual. If the vertex corresponding to the outer face is drawn explicitly, then one of its adjacent edges must have a bend.

Similar to most primal-dual representation methods, the unbounded (outer) face must be treated differently. If the vertex corresponding to the unbounded face is not explicitly drawn in the plane, then all of the conditions above are met. However, if it is drawn explicitly, then one of the dual edges emanating from it cannot be a straight-line segment; see Fig. 1. In our grid-embedding algorithm, we provide an option for not drawing the vertex representing the outer face explicitly, or if it is drawn, then one of the edges emanating from it has one bend (that also is on the grid). Note, that if the embedding is done on the surface of a sphere, the edges emanating from this vertex are arcs of great circles and the unbounded face does not require special treatment.

In section 2 we describe the algorithm in detail and in section 3, we briefly discuss the implementation and present several drawings of primal-dual graphs produced by our algorithm.

2 Algorithm for Embedding a Graph and Its Dual

Let G_1 be a 3-connected planar graph. We construct a new graph G_2 that combines information about both the planar graph G_1 and its dual. For this construction we make some changes in G_1. We introduce a new vertex v_i' corresponding to a face $\mathcal{F}_i'$ of G_1, for all $1 \leq i \leq f$, where f is the number of faces of G_1. We connect each newly added vertex v_i' to each vertex v_j of $\mathcal{F}_i'$ with a single new edge and delete all the edges that originally belonged to G_1. Fig. 2 shows a sample construction. We call the resulting planar graph G_2 *fully-quadrilateralated (FQ)*, i.e., every face of G_2 is a quadrilateral. Since the original graph G_1 is 3-connected, the resulting graph G_2 is also 3-connected (proven formally in [14]).

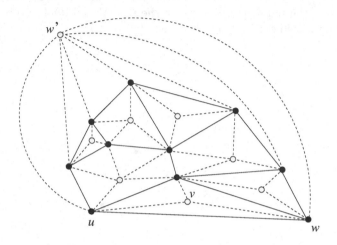

Fig. 2. Creating graph G_2: the original 3-connected graph G_1 is drawn with solid lines and filled-in circles; we insert the dual vertices (drawn as empty circles) and add the edges connecting primal and dual vertices (drawn as dashed lines). To obtain G_2 we remove the original edges of G_1 (drawn with solid lines).

Observation: If we can embed the graph G_2 on the grid so that each inner face of G_2 is strictly convex and the outer face of G_2 lies on a strictly concave quadrilateral, then we can embed the initial graph G_1 and its dual so that we meet all the problem requirements with the only exception that one edge of the primal graph G_1(or its dual) is drawn with one bend.

The requirement that the edges of the dual graph be straight and cross only their corresponding primal edges is guaranteed by the strict convexity of the quadrilateral faces. Let the outer face of the graph G_2 be (u, v, w, w'), where u, w are primal vertices and v, w' are dual vertices, as shown in Fig. 2. The exception arises from the fact that we need to draw (u, w) and (v, w'), while both of these edges can not lie inside the quadrilateral (u, v, w, w'). In order to get around this problem we embed the quadrilateral (u, v, w, w') so that it is strictly concave. This way only one bend for one of the edges (u, w) or (v, w') will be sufficient. As a result all the edges in the primal and the dual graph are straight-line edges, except for one edge. In fact, it is easy to choose the exact edge we need (either from the primal or from the dual).

Hence, the original problem can be transformed into a problem of straight-line embedding an FQ-3-connected planar graph G on the grid so that each internal face of G is strictly convex and the outer face of G lies on a strictly concave quadrilateral. Note that this problem can be solved by the algorithm of Chrobak *et al.* [4]. However, the area guaranteed by their algorithm is $O(n^3) \times O(n^3)$, whereas our algorithm guarantees a drawing on the $(2n-2) \times (2n-2)$ grid, which is stated in the main theorem in this paper:

Theorem 1. *Given a 3-connected planar graph G_1, we can embed G_1 and its dual on a $(2n-2) \times (2n-2)$ grid, where n is the number of vertices in G_1 and its dual, so that each dual vertex lies inside its primal face, each dual edge crosses only its primal edge and every edge in the overall embedding is a straight-line segment except for one edge which has a bend placed on the grid. Furthermore, the running time of the algorithm is $O(n)$.*

2.1 Overview of the Algorithm

Given a 3-connected graph G_1, we summarize our algorithm to simultaneously embed G_1 and its dual as follows:

- Find a topological embedding of G_1 using [8].
- Apply the construction described above to find G_2.
- Let $G = G_2$, where G is an FQ-3-connected planar graph.
- Find a suitable canonical labeling of the vertices of G.
- Place the vertices of G on the grid one at a time using this ordering.
- Remove all the edges of G and draw the edges of G_1 and its dual.

Note that our method works only for 3-connected graphs. A commonly used technique for drawing a general planar graph is to embed the graph after fully triangulating it by adding some extra edges and then to remove the extra edges from the final embedding. Using the same idea, we could first fully triangulate any given planar graph. Then after embedding the resulting 3-connected planar graph and its dual, we could remove the extra edges that were inserted initially. However, the problem with this approach is that after removing the extra edges there could be faces with multiple dual vertices inside. Thus the issue of choosing a suitable location for the duals of such faces remains unresolved. In fact, depending on the drawing of that face, it could as well be the case that no suitable location for the dual exists [1]. In the rest of the paper we consider only 3-connected graphs.

2.2 The Canonical Labeling

We present the canonical labeling for the type of graphs under consideration. It is a simple restriction of the canonical labeling of [9], which in turn is based on the ordering defined in [6].

Let G be an FQ-3-connected planar graph with n vertices. Let (u, v, w, w') be the outer face of G s.t. u, w are primal vertices and v, w' are dual vertices. Then there exists a mapping δ from the vertices of G onto v_i, $1 \leq i \leq m$ such that δ maps u and v to v_1, w' to v_m and satisfies the following invariants for every $3 \leq k \leq m$:

1. The subgraph $G_{k-1} \subseteq G$, induced by the vertices labeled v_i, $1 \leq i \leq k-1$ is biconnected and the boundary of its exterior face is a cycle C_{k-1} containing the edge (u, v).
2. Either one vertex or two vertices can be labeled v_k.
   a) Let z_0 be the only vertex labeled v_k. Then z_0 belongs to the exterior face of G_{k-1}, has at least two neighbors in G_{k-1} and at least one neighbor in $G - G_k$.
   b) Let z_0, z_1 be the two vertices labeled v_k, where (z_0, z_1) is an edge in G. Then z_0, z_1 belong to the outer face of G_{k-1}, each has exactly one neighbor in G_{k-1} and at least one neighbor in $G - G_k$.

Since G is FQ, all the faces created by adding v_k, $3 \leq k \leq m$, have to be quadrilaterals; see Fig. 3.

Note that assigning the mappings onto v_1 and v_m as above provides us the embedding where all the edges of both the primal and the dual graph are straight except for one primal edge, (u, w), which has a bend. Alternatively assigning v and w to map onto v_1, and u to map onto v_m would choose a dual edge, (v, w'), to have a bend.

Lemma 1. *Every FQ-3-connected planar graph has a canonical labeling as defined above.*

Kant [9] provides a linear-time algorithm to find a canonical labeling of a general 3-connected planar graph. It is easy to see that the canonical labeling definition of [9] when applied to FQ-3-connected planar graphs, gives us the labeling defined above.

2.3   The Placement of the Vertices

The main idea behind most of the straight-line grid embedding algorithms is to come up with a suitable ordering of the vertices and then place the vertices one at a time using the given order, while making sure that the newly placed vertex (or vertices) is (are) visible to all the neighbors. In order to realize this last goal, at each step, a set of vertices are shifted to the right without affecting the planarity of the drawing so far. Our placement algorithm is similar to the algorithm of Chrobak and Kant [5], with some changes in the invariants that we maintain to guarantee the visibility together with strict convexity of the faces.

Let the canonical labeling, δ, that maps the vertices of G onto $v_1, v_2, ...v_m$ be defined as in the previous section. Let $\mathcal{U}(g_i)$ denote the vertices *under* g_i. $\mathcal{U}(g_i)$ should be shifted to the right whenever the vertex g_i is shifted to the right. $\mathcal{U}(g_i)$ is initialized to $\{g_i\}$ for every vertex g_i of G. Let $\delta(g_i) = v_{i'}$ and $\delta(g_j) = v_{j'}$. Then we define $Low(g_i, g_j) = i$ if $i' < j'$, $Low(g_i, g_j) = j$ if $j' < i'$. If $i' = j'$ then let $Low(g_i, g_j)$ be the one that is placed to the left. Let $x(g_i)$, $y(g_i)$ respectively denote the x and y coordinates of the vertex g_i.

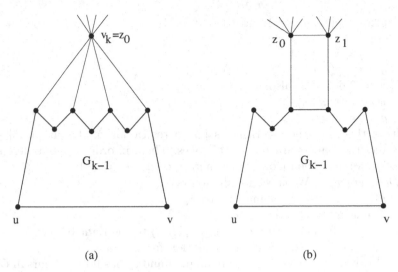

Fig. 3. (a) One vertex, z_0, is labeled v_k; (b) Two vertices, z_0 and z_1, are labeled v_k.

● *Embed the First Quadrilateral Face:*   We begin by placing the vertices mapped onto v_1 and v_2. The ones that are mapped onto v_1 are u and v. We place u at $(0,0)$ and v at $(3,0)$. Note that two vertices should be mapped to v_2. We place the vertex that is mapped to v_2 and that has an edge with u at $(1,1)$ and the other at $(2,1)$.

Then, for every k, $3 \le k \le m$, we do the following:

● *Update $\mathcal{U}(g_i)$:*   Let $C_{k-1} = (u = c_1, c_2, ..., c_r = v)$. Let $c_p, c_q \in C_{k-1}$, respectively be the first and the last neighbor of the vertex(vertices) mapped to v_k. If only one vertex, z_0, is mapped to v_k, we update $\mathcal{U}(c_p), \mathcal{U}(c_q)$ and $\mathcal{U}(z_0)$ as follows:

$$Low(c_p, c_{p+1}) = p + 1 \Longrightarrow \mathcal{U}(c_p) = \mathcal{U}(c_p) \cup \mathcal{U}(c_{p+1})$$

$$Low(c_{q-2}, c_{q-1}) = q - 2 \Longrightarrow \mathcal{U}(c_q) = \mathcal{U}(c_q) \cup \mathcal{U}(c_{q-1})$$

$$\mathcal{U}(z_0) = \mathcal{U}(z_0) \cup \bigcup_{i=Low(c_p, c_{p+1})+1}^{Low(c_{q-2}, c_{q-1})} \mathcal{U}(c_i)$$

We do not change $\mathcal{U}(g_i)$ if two vertices, z_0 and z_1, are mapped to v_k.

● *Shift to the right:*   We then perform the necessary shifting. We shift each vertex $g_i \in \bigcup_{i=q}^r \mathcal{U}(c_i)$ to the right by one if only one vertex is mapped to v_k, by two otherwise.

● *Locate the New Vertices:*   Finally we locate the vertex(vertices) mapped to v_k on the grid. Let $|v_k|$ denote the number of vertices mapped to v_k. Then we have:

If c_p has no neighbors in $G - G_k$
$$x(z_0) = x(c_p)$$

$$y(z_0) = y(c_q) + x(c_q) - x(c_p) - |v_k| + 1$$
otherwise
$$\cdot \quad x(z_0) = x(c_p) + 1$$
$$y(z_0) = y(c_q) + x(c_q) - x(c_p) - |v_k|$$
If $|v_k| = 2$ define z_1 also:
$$x(z_1) = x(z_0) + 1$$
$$y(z_1) = y(z_0)$$

Up to this step, the algorithm is just a restriction of the one in [5] and it guarantees the convex drawing of the faces. Then, in order to guarantee strict-convexity, we note the following degenerate cases; see Fig. 4:

• *Degeneracies:* We check for the following:

If only one vertex, z_0, is mapped to v_k

(d_1) If $x(z_0) = x(c_{p+1}) = x(c_{p+2})$
Shift each vertex $g_i \in \bigcup_{i=p+1}^{r} \mathcal{U}(c_i)$ to the right by one.
Perform the location calculation for z_0 again.

(d_2) If $k < m$ and z_0, c_q, c_{q+1} are aligned and c_q has no neighbors in $G - G_k$
Shift each vertex $g_i \in \bigcup_{i=q+1}^{r} \mathcal{U}(c_i)$ to the right by one.

If two vertices, z_0 and z_1 are mapped to v_k

(d_3) If $y(z_0) = y(z_1) = y(c_p)$
Shift each vertex $g_i \in \bigcup_{i=q}^{r} \mathcal{U}(c_i)$ to the right by one.
Perform the location calculation for z_0 and z_1 again.

(d_4) If $k < m$ and z_1, c_q, c_{q+1} are aligned and c_q has no neighbors in $G - G_k$
Shift each vertex $g_i \in \bigcup_{i=q+1}^{r} \mathcal{U}(c_i)$ to the right by one.

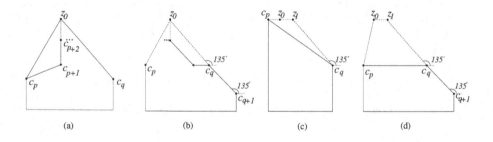

(a) $\qquad\qquad$ (b) $\qquad\qquad$ (c) $\qquad\qquad$ (d)

Fig. 4. Four possible degenerate cases: type d_2, type d_2, type d_3, type d_4.

2.4   Proof of Correctness

Lemma 2. *Let $C_k = (u = c_1, c_2, ..., c_r = v)$ be the exterior face of G_k after the kth placement step. Let $\alpha(c_j, c_{j+1})$ denote the angle of the vector $\mathbf{c_j c_{j+1}}$, for $1 \leq j \leq r - 1$. The following holds for $2 \leq k \leq m - 1$:*

1. $\alpha(c_j, c_{j+1})$ *lies in* $[-45°, \arctan -1/2] \cup \{0\} \cup [45°, 90°]$. *It can not lie in* $(-45°, \arctan -1/2]$ *if* c_j *has a neighbor in* $G - G_k$.
2. *If* $c_j \in C_k, c_j \notin \{c_1, c_r\}$ *s.t.* c_j *does not have a neighbor in* $G - G_k$, *then:*
   a) *If* $Low(c_{j-1}, c_j) = j - 1$ *then* $\alpha(c_j, c_{j+1}) = 90°$; *else* $\alpha(c_{j-1}, c_j) = -45°$.
   b) *If* $\alpha(c_j, c_{j+1}) = 90°$ *then* $\alpha(c_{j-1}, c_j) \neq 90°$.
   c) *If* $\alpha(c_j, c_{j+1}) = -45°$ *then* $\alpha(c_{j-1}, c_j) \neq -45°$.

Proof Sketch: Due to space limitations the proof of this lemma is left out of this extended abstract.[1]   □

Preserving Planarity. Let only one vertex, z_0, be mapped to v_k. If (z_0, c_j) is an edge in G_k for some $c_j \in C_{k-1}$, then the placement algorithm and the previous lemma guarantees that $-90 < \alpha(z_0, c_j) < -45$, for $j \neq p, j \neq q$. Then no crossing is created between a new edge (z_0, c_j) and the edges of C_{k-1}. Because such a crossing would imply that there exists $j' < j$ s.t. $c_{j'} \in C_k$ and $\alpha(c_{j'}, c_j) < -45$. But this is impossible by the first part of the above lemma. The same idea applies to the case where $|v_k| = 2$. Then the following corollary holds:

Corollary 1. *Insertion of the vertex(vertices) mapped to* v_k, *at the* k^{th} *placement step, where* $2 \leq k \leq m$ *preserves planarity.*

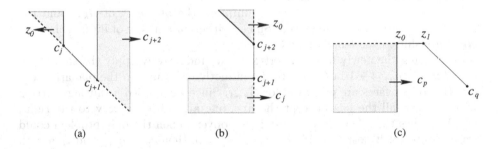

(a)   (b)   (c)

Fig. 5. The vertices pointed to by the arrows must lie in the indicated area. The dashed lines indicate open boundaries that are not included in the area.

Strictly Convex Faces. Let $|v_k| = 1$ and z_0 be the vertex mapped to v_k. Let $\mathcal{F}_j = (c_j, c_{j+1}, c_{j+2}, z_0)$ be a quadrilateral face created after the insertion of z_0. If $Low(c_j, c_{j+1}) = j + 1$, then by the previous lemma $\alpha(c_j, c_{j+1}) = -45°$. Fig. 5(a) shows the area where z_0 and c_{j+2} must lie. If $Low(c_j, c_{j+1}) = j$, then $\alpha(c_{j+1}, c_{j+2}) = 90°$. Fig. 5(b) shows the area where z_0 and c_{j+2} must lie in this case. Both cases imply that $\mathcal{F}_j = (c_j, c_{j+1}, c_{j+2}, z_0)$ is strictly convex.

If $|v_k| = 2$ and z_0, z_1 are mapped to v_k, the placement algorithm requires that c_p must lie in the area shown in Fig. 5(c), which implies that the newly created face is strictly convex. The following corollary holds:

---

[1] The proof of this lemma can be found in the full version of the paper, available at www.cs.arizona.edu/~cesim/dual.ps.

Corollary 2. *The newly created faces after the insertion of the vertex(vertices) mapped to v_k, at the kth placement step, where $2 \le k \le m$, are strictly convex.*

Shifting Preserves Planarity and Strictly Convex Faces. The above discussion shows that after the insertion of the vertex(vertices) at the kth placement step, no new edge crossings are created and all the newly added faces are strictly convex. In order to complete the proof of correctness we only need to prove that the same holds for shifting also:

Lemma 3. *Let $C_k = (u = c_1, c_2, ..., c_r = v)$ be the exterior face of G_k after the kth placement step, where $2 \le k < m$. For any given j, where $1 \le j \le r$, shifting the vertices in $\bigcup_{i=j}^{r} \mathcal{U}(c_i)$, to the right by s units preserves the planarity and the strictly convex faces of G_k.*

Proof Sketch: The claim holds trivially for $k = 2$. Assume it holds for $k' = k-1$, where $2 \le k' < m - 1$. We assume $|v_k| = 1$. The case where $|v_k| = 2$ is similar. Let z_0 be the vertex mapped to v_k and $c_p, c_q \in C_{k-1}$, respectively be the first and the last neighbor of z_0 in G_{k-1}.

If $j \le p$ then by the inductive assumption the planarity of G_{k-1} and the strictly convex faces of G_{k-1} are preserved. The faces introduced by z_0 shifts rigidly to the right, which, by the previous corollaries, implies that G_k is planar and all its faces are strictly convex.

If $j > q$, then by the inductive assumption the planarity of G_{k-1} and the strictly convex faces are preserved. Since neither z_0 nor any of its neighbors in G_{k-1} are shifted the lemma follows.

If shifting the newly inserted vertex z_0, we inductively apply the shifting to $j' = Low(c_p, c_{p+1}) + 1$ in G_{k-1}. By the inductive assumption the planarity and strictly convex faces are preserved for G_{k-1}. Since we applied a shifting starting with j' then, all the faces except the first one are shifted rigidly to the right, which implies that those faces are strictly convex. Then the only problem could arise with the leftmost face. If $Low(c_p, c_{p+1}) = p$, then c_{p+1}, c_{p+2} and z_0 are all shifted to the right by the same amount. Since initially the face $(c_p, c_{p+1}, c_{p+2}, z_0)$ was strictly convex, it continues to be so after shifting those three vertices also. In the case where $Low(c_p, c_{p+1}) = p + 1$, the only shifted vertices are z_0 and c_{p+2}. Again shifting those two vertices does not violate the convexity of the face.

If $j = q$, the situation is very similar to the previous case, except now the only deformed face is the rightmost face, instead of the leftmost one. The same idea applies to this case also, i.e., given that initially the face is strictly convex, it remains so after shifting.   □

2.5   Grid Size

Lemma 4. *The algorithm requires a grid of size at most $(2n - 4) \times (2n - 4)$.*

Proof Sketch: If no degeneracies are created then the exact grid size required is $(n - 1) \times (n - 1)$. We show that each degenerate case can be associated with a newly added quadrilateral face of G.

Degenerate case of type d_1 is associated with the face $(c_p, c_{p+1}, c_{p+2}, z_0)$. Degenerate case of type d_2 at some step k of the algorithm, is associated with a face (z_0, c_q, c_{q+1}, g_i), where g_i is a vertex that will be added at some step $k' > k$ of the algorithm. We know that such a face exists, since $k < m$, c_q has no neighbors in $G - G_k$ and each face under consideration is a quadrilateral. Similar argument holds for degenerate case of type d_4. Finally degenerate case of type d_3 is associated with the face (c_p, c_q, z_1, z_0). Fig. 4 shows all four types of degeneracies that can occur. Note that each quadrilateral face is associated with at most one degeneracy.

Since an FQ graph G with n vertices has $n - 3$ inside faces, the placement algorithm requires grid size of at most $(2n - 4) \times (2n - 4)$. □

Final Shifting. Let (u, v, w, w') be the outer face of G. The placement algorithm and Lemma-2 imply that the outer face is the isosceles right triangle $\triangle uvw'$ and that w lies on the line segment (v, w'). One final right shift is needed to guarantee that the outer face (u, v, w, w') lies on a strictly concave quadrilateral. For this we just shift v to the right by one. As a result we can draw the edge (v, w') as a straight-line segment. In order to draw the edge (u, w), we place a bend point at $(x(w') - 1, y(w') + 2)$, where $x(w')$ and $y(w')$, respectively denote the x and y coordinates of the vertex w'. We connect the bend point with u and w. Then the total area required is $(2n - 2) \times (2n - 2)$ and Theorem-1 follows.

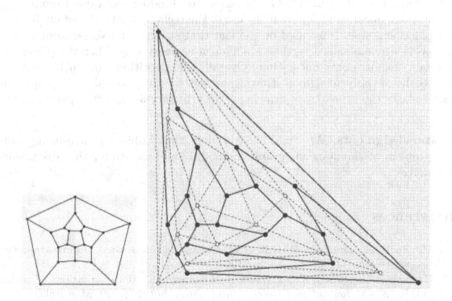

Fig. 6. Dodecahedral graph and its dual representation. The filled-in vertices and solid edges represent the primal graph; the empty circles and dashed edges represent the dual.

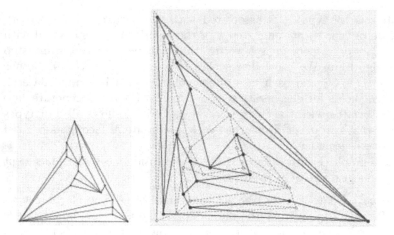

Fig. 7. A random 3-connected planar graph with 16 vertices and its dual representation.

3   Implementation

We have implemented our algorithm[2] to visualize 3-connected planar graphs and their duals using the LEDA/AGD libraries [10]. Finding a suitable canonical labeling takes linear time [9]. We make use of the technique introduced by [3] to do the placement step. It is based on the fact that storing relative x-coordinates of the previously embedded vertices is sufficient at every step. Then the placement step also requires only linear time. Overall, the algorithm runs in linear time. Fig. 6 shows the primal/dual drawing we get for the dodecahedral graph and Fig. 7 shows the primal/dual drawing of an arbitrary 3-connected planar graph.

Acknowledgments. We would like to thank Anna Lubiw for introducing us to the problem of simultaneous graph embedding and for stimulating discussions about it.

References

1. Marshall Bern and John R. Gilbert. Drawing the planar dual. *Information Processing Letters*, 43(1):7–13, August 1992.
2. Graham R. Brightwell and Edward R. Scheinerman. Representations of planar graphs. *SIAM Journal on Discrete Mathematics*, 6(2):214–229, May 1993.
3. M. Chrobak and T. Payne. A linear-time algorithm for drawing planar graphs. *Inform. Process. Lett.*, 54:241–246, 1995.
4. Marek Chrobak, Michael T. Goodrich, and Roberto Tamassia. Convex drawings of graphs in two and three dimensions. In *Proc. 12th Annu. ACM Sympos. Comput. Geom.*, pages 319–328, 1996.

---

[2] The C++ code is publicly available at www.cs.arizona.edu/~cesim/dual.tar.gz.

5. Marek Chrobak and Goos Kant. Convex grid drawings of 3-connected planar graphs. *International Journal of Computational Geometry and Applications*, 7(3):211–223, 1997.
6. H. de Fraysseix, J. Pach, and R. Pollack. How to draw a planar graph on a grid. *Combinatorica*, 10(1):41–51, 1990.
7. David Harel and Meir Sardas. An algorithm for straight-line drawing of planar graphs. *Algorithmica*, 20(2):119–135, 1998.
8. J. Hopcroft and R. E. Tarjan. Efficient planarity testing. *Journal of the ACM*, 21(4):549–568, 1974.
9. G. Kant. Drawing planar graphs using the canonical ordering. *Algorithmica*, 16:4–32, 1996. (special issue on Graph Drawing, edited by G. Di Battista and R. Tamassia).
10. K. Mehlhorn and St. Näher. *The LEDA Platform of Combinatorial and Geometric Computing*. Cambridge University Press, 1999.
11. Kazuyuki Miura, Shin-Ichi Nakano, and Takao Nishizeki. Grid drawings of 4-connected plane graphs. *Discrete and Computational Geometry*, 26(1):73–87, 2001.
12. Bojan Mohar. Circle packing of maps in polynomial time. *European Journal of Combinatorics*, 18:785–805, 1997.
13. Walter Schnyder. Embedding planar graphs on the grid. In *Proceedings of the 1st ACM-SIAM Symposium on Discrete Algorithms (SODA)*, pages 138–148, 1990.
14. William T. Tutte. How to draw a graph. *Proceedings London Mathematical Society*, 13(52):743–768, 1963.

Meaningful Information*

(Extended Abstract)

Paul Vitányi[1]**

Centrum voor Wiskunde en Informatica (CWI),
Kruislaan 413, 1098 SJ Amsterdam, The Netherlands.
Paul.Vitanyi@cwi.nl, http://www.cwi.nl/~paulv

Abstract. The information in an individual finite object (like a binary string) is commonly measured by its Kolmogorov complexity. One can divide that information into two parts: the information accounting for the useful regularity present in the object and the information accounting for the remaining accidental information. There can be several ways (model classes) in which the regularity is expressed. Kolmogorov has proposed the model class of finite sets, generalized later to computable probability mass functions. The resulting theory, known as Algorithmic Statistics, analyzes the algorithmic sufficient statistic when the statistic is restricted to the given model class. However, the most general way to proceed is perhaps to express the useful information as a recursive function. The resulting measure has been called the "sophistication" of the object. We develop the theory of recursive functions statistic, the maximum and minimum value, the existence of absolutely nonstochastic objects (that have maximal sophistication—all the information in them is meaningful and there is no residual randomness), determine its relation with the more restricted model classes of finite sets, and computable probability distributions, in particular with respect to the algorithmic (Kolmogorov) minimal sufficient statistic, the relation to the halting problem and further algorithmic properties.

1 Introduction

The information contained by an individual finite object (like a finite binary string) is objectively measured by its Kolmogorov complexity—the length of the shortest binary program that computes the object. Such a shortest program contains no redundancy: every bit is information; but is it meaningful information? If we flip a fair coin to obtain a finite binary string, then with overwhelming probability that string constitutes its own shortest program. However, also with overwhelming probability all the bits in the string are meaningless information,

---

* First electronic version published November, 2001, on the LANL archives http://xxx.lanl.gov/abs/cs.CC/0111053 .
** Partially supported by the EU fifth framework project QAIP, IST–1999–11234, the NoE QUIPROCONE IST–1999–29064, the ESF QiT Programmme, and the EU Fourth Framework BRA NeuroCOLT II Working Group EP 27150. Also affiliated with the University of Amsterdam.

P. Bose and P. Morin (Eds.): ISAAC 2002, LNCS 2518, pp. 588–599, 2002.

random noise. On the other hand, let an object x be a sequence of observations of heavenly bodies. Then x can be described by the binary string pd, where p is the description of the laws of gravity, and d the observational parameter setting: we can divide the information in x into meaningful information p and accidental information d. The main task for statistical inference and learning theory is to distil the meaningful information present in the data. The question arises whether it is possible to separate meaningful information from accidental information, and if so, how.

We summarize a discussion in [6]: In statistical theory, every function of the data is called a "statistic" of the data. The central notion in probabilistic statistics is that of a "sufficient" statistic, introduced by the father of statistics R.A. Fisher [4]: "The statistic chosen should summarise the whole of the relevant information supplied by the sample. This may be called the Criterion of Sufficiency ... In the case of the normal curve of distribution it is evident that the second moment is a sufficient statistic for estimating the standard deviation." This approach deals with averages, and one cannot improve on sufficiency in the probabilistic setting. For traditional problems, dealing with frequencies over small sample spaces, this approach is appropriate. But for current novel applications, average relations are often irrelevant, since the part of the support of the probability density function that will ever be observed has about zero measure. This is the case in, for example, complex video and sound analysis. There arises the problem that for individual cases the selection performance may be bad although the performance is good on average. There is also the problem of what probability means, whether it is subjective, objective, or exists at all.

A.N. Kolmogorov in 1974 proposed to found statistical theory on finite combinatorial principles independent of probabilistic assumptions, as the relation between the individual data and its explanation (model). In contrast to the situation in probabilistic statistical theory, the algorithmic relation of (minimal) sufficiency is an absolute relation between the individual model and the individual data sample. To simplify matters, and because all discrete data can be binary coded, we consider only data samples that are finite binary strings. The basic idea is to study separation of meaningful versus accidental information in an initially limited setting where this information be represented by a finite set of which the object (the data sample) is a typical member. Especially interesting is the case where the two-part description of the finite set, together with the index of the object in that set, is as concise as the shortest one-part description. The finite set statistic models the regularity present in the object (since it is a typical element of the set). This approach has been generalized to computable probability mass functions. The combined theory has been developed in detail in [6] and called "Algorithmic Statistics."

Here we study the most general form of algorithmic statistic: recursive function models. In this setting the issue of meaningful information versus accidental information is put in its starkest form; and in fact, has been around for a long time in various imprecise forms unconnected with the sufficient statistic approach: The issue has sparked the imagination and entered scientific popularization in [8] as "effective complexity" (here "effective" is apparently used in the sense of "producing an effect" rather than "constructive" as is customary in the theory of computation). It is time that it receives formal treatment. Formally, we study the minimal length of a total recursive function that leads to an optimal length two-part code of the object being described. This minimal length has been called the "sophistication" of the object in [12,13] in a different, but related, setting of compression and prediction properties of infinite sequences. That treatment is technically sufficiently vague so as to have no issue for the present work. We develop the notion based on prefix Turing machines, rather than on a variety of monotonic Turing machines as in the cited papers. Below we describe related work in detail and summarize our results. Subsequently, we formulate our problem in the formal setting of computable two-part codes.

1.1   Related Work

At a Tallinn conference in 1973, A.N. Kolmogorov formulated the approach to an individual data to model relation, based on a two-part code separating the *structure* of a string from meaningless *random* features, rigorously in terms of Kolmogorov complexity (attribution by [15,2]). Cover [2,3] interpreted this approach as a (sufficient) statistic. The "statistic" of the data is expressed as a finite set of which the data is a "typical" member. Following Shen [15] (see also [19, 16,18]), this can be generalized to computable probability mass functions for which the data is "typical." Related aspects of "randomness deficiency" were formulated in [10,11] and studied in [15,19]. Despite its evident epistemological prominence in the theory of hypothesis selection and prediction, only selected aspects of algorithmic statistic were studied initially, [15,2,19,16,20], while [6] can be considered as a comprehensive investigation. Here we extend the latter work to the most general form of algorithmic statistic in terms of total recursive functions. This idea was pioneered by [12,13] who, unaware of a statistic connection, coined the cute word "sophistication." The algorithmic (minimal) sufficient statistic was related to an applied form in [18,7]: the well-known "minimum description length" principle [1] in statistics and inductive reasoning. In an important paper [17] (following the present paper) we vindicate, for the first time, the rightness of the original "structure function", proposed by Kolmogorov in 1974. This resulted in the first rigorous proof that—in complexity-restricted model classes—ML, MDL, and related methods in model selection, *always* give a best possible model, and not only with high probability. Following Kolmogorov we analyzed a canonical setting where the models are finite sets. As Kolmogorov himself pointed out, this is no real restriction: the finite sets model class is equivalent, up to a logarithmic additive term, to the model class of probability density functions, as studied in [15,6]. The analysis is valid, up to logarithmic additive terms, also for the model class of total recursive functions, as studied here.

1.2   This Work

It will be helpful for the reader to be familiar with initial parts of [6]. That being said, we define the notion of sophistication (minimal sufficient statistic in the total recursive function model class). We distinguish two types, "sophistication" and "typical sophistication," based on description of the function concerned. The notions are demonstrated to be meaningful (existence and nontriviality). We then establish lower and upper bounds on the sophistication; in particular we show that there are objects for which all description types of sophistication are maximal. In fact, these are objects in which all information is meaningful and there is (almost) no accidental information. That is, the simplest explanation of such an object is the object itself. In the simpler setting of finite set statistic the analogous objects were called "absolutely non-stochastic" or "negatively random" by Kolmogorov. Such objects can only be a random outcome of a "complex" random process, and Kolmogorov questioned whether such objects can occur in nature. Notice that these are not the "random" strings of maximal Kolmogorov complexity; those are very unsophisticated (with sophistication about 0). Such objects are outcomes of "simple" random processes and were therefore called "positively" random by Kolmogorov. We subsequently establish the equivalence between sophistication and the algorithmic minimal sufficient statistics of the finite set class and the probability mass function class. Finally, we investigate the algorithmic properties of sophistication: nonrecursiveness, upper semicomputability, and intercomputability relations of Kolmogorov complexity, sophistication, halting sequence.

Structure Function: "Sophistication" is the algorithmic version of "minimal sufficient statistic" for data x in the model class of total recursive functions. However, the full stochastic properties of the data can only be understood by considering the Kolmogorov structure function $\lambda_x(\alpha)$ (mentioned earlier) that gives the length of the shortest two-part code of x as a function of the maximal complexity α of the total function supplying the model part of the code. This function has value about $l(x)$ initially, is nonincreasing, and drops to the line $K(x)$ at complexity $\alpha_0 = \mathrm{soph}(x)$, after which it remains constant, $\lambda_x(\alpha) = K(x)$ for $\alpha \geq \alpha_0$, everything up to a logarithmic additive term. A comprehensive analysis has been given in [17] for the model class of finite sets containing x, but it is shown that all results extend to the model class of computable probability distributions and the model class of total recursive functions, up to an additive logarithmic term.

2   Background on Two-Part Codes

For definitions, notation, and an introduction to Kolmogorov complexity, see [14]. It is a deep and useful fact that the shortest effective description of an object x can be expressed in terms of a *two-part code*: the first part describing an appropriate Turing machine and the second part describing the program that interpreted by the Turing machine reconstructs x. The essence of the the theory is the Invariance Theorem, that can be informally stated as follows: For a fixed reference universal prefix Turing machine U the length of the shortest program to compute x is $\min\{l(p) : U(p) = x\}$. Let $T_1, T_2, \ldots$ be the standard enumeration of Turing machines, U be a standard Universal Turing machine satisfying $U(\langle i, p \rangle) = T_i(p)$ for all indices i and programs p, where $\langle \cdot, \cdot \rangle$ is a standard pairing function. Notice that we can define $U(\langle \epsilon, p \rangle) = U(0p) = U(p)$. From the definitions it then follows that

$$K(x) = \min\{l(T) + l(p) : T(p) = x\} \pm 1,$$

where $l(T)$ is the length of a self-delimiting encoding for a prefix Turing machine T (rather, with $T = T_i$, of the index i identifying T). This provides an alternative definition of prefix Kolmogorov complexity (similarly, for conditional Kolmogorov complexity). For any string x let x^* be the shortest program from which the universal machine U computes x (if there is more than one of them then x^* is the first one in standard enumeration). The above expression for Kolmogorov complexity can be rewritten as

$$K(x) \stackrel{+}{=} \min\{K(T) + K(x \mid T^*) : T \in \{T_0, T_1, \dots\}\}, \qquad (1)$$

that emphasizes the two-part code nature of Kolmogorov complexity: using the regular aspects of x to maximally compress. In the example

$$x = 101010101010101010101010$$

we can encode x by a small Turing machine which computes x from the program "13." Intuitively, the Turing machine part of the code squeezes out the *regularities* in x. What is left are irregularities, or *random aspects*, of x relative to that Turing machine. The minimal-length two-part code squeezes out regularity only insofar as the reduction in the length of the description of random aspects is greater than the increase in the regularity description.

This interpretation of $K(x)$ as the shortest length of a two-part code for x, one part describing a Turing machine, or *model*, for the *regular* aspects of x and the second part describing the *irregular* aspects of x in the form of a program to be interpreted by T, has profound applications. We can interpret this as that the regular, or "valuable," information in x is constituted by the bits in the "model" while the random or "useless" information of x constitutes the remainder.

The "right model" is a Turing machine T among the ones that reach the minimum description length in (1). This T embodies the amount of useful information contained in x. The main remaining question is which such T to select among the ones that satisfy the requirement. Following Occam's Razor we opt here for the shortest one—a formal justification for this choice is given in [18]. The problem is how to separate a shortest program x^* for x into parts $x^* = pq$ such that p represents an appropriate T.

3   Recursive Statistic

Instead of finite set statistic, or computable probability density functions $P : \{0,1\}^* \to [0,1]$ statistic, as in [6], we consider the most general form of algorithmic statistic: recursive functions. For convenience we consider data in the form of finite binary strings.

Notation 1 From now on, we will denote by $\overset{+}{<}$ an inequality to within an additive constant, and by $\overset{+}{=}$ the situation when both $\overset{+}{<}$ and $\overset{+}{>}$ hold. We will also use $\overset{*}{<}$ to denote an inequality to within an multiplicative constant factor, and $\overset{*}{=}$ to denote the situation when both $\overset{*}{<}$ and $\overset{*}{>}$ hold.

Definition 1. *A statistic for data x is a total recursive function p such that there exists an input d and $p(d) = x$ and $K(x \mid p^*) \overset{+}{=} l(d)$. Note that, by definition, $K(x \mid p^*) \overset{+}{<} K(d \mid p^*) \overset{+}{<} l(d)$, so it is the other inequality, $K(x \mid p^*) \overset{+}{>} l(d)$, that needs to be satisfied. Formally, we call x typical for p and, conversely, p typical for x, if*

$$K(x \mid p^*) \overset{+}{=} l(d), \quad p(d) = x. \tag{2}$$

That is, there a program p^* (shortest self-delimiting program to compute p) and input d satisfying the displayed equation. We can use a simple program q of a constant number of bits, to parse the following self-delimiting constituent programs p^* and d, retrieve the objects p and d, and then compute $U(p, d)$. That is, $U(qp^*d) = U(p, d)$.

3.1 Sufficient Statistic

The most important concept in this paper is the sufficient statistic. For an extensive discussion see [6].

Definition 2. *The (prefix-) complexity $K(p)$ of a recursive partial function p is defined by*

$$K(p) = \min_i \{K(i) : \text{Turing machine } T_i \text{ computes } p\}.$$

A recursive function p is a sufficient statistic *for data x if*

$$K(x) \overset{+}{=} K(p) + l(d), \quad p(d) = x.$$

3.2 Description Mode of Statistic

The complexities involved depend on what we know about the length of the argument d in the computation $p(d) = x$ in the case x is typical for p. (Related distinctions with respect to description format of finite set statistics or computable probability mass function statistics appear in [6].) The description format of a total recursive function is:

- *typical* if p is a total recursive function such that we can compute the length of an argument resulting in x satisfying (2)—x is typical— from p^*, in the sense that $K(l(d) \mid p^*) \overset{+}{=} 0$.
- *Unrestricted* if it may lead to arguments, resulting in a typical value, of unknown length.

Example 1. (i) Let $p(d) = x$, with $x \in \{0,1\}^n$ and d is the position of x in the lexicographical order. To ensure totality of p let $p(d) = 0$ otherwise. Then, $K(p) \overset{+}{=} K(n)$ and $K(d \mid p^*) \overset{+}{=} K(x \mid p^*)$ for all $x \in \{0,1\}^n$ with $p(d) = x$. Moreover, there are n-bit x's, namely those of maximal complexity $\overset{+}{=} n + K(n)$, such that $K(x) \overset{+}{=} K(p) + l(d)$. That is, $K(l(d) \mid p^*) \overset{+}{=} 0$ since $K(d \mid p^*) \overset{+}{=} l(d) \overset{+}{=} n$. This means that we can compute $l(d)$ exactly from p. Hence p is typical.

(ii) A function like $p(d) = d$ is total, but not typical.

As apparent from the example, the explicit descriptions of finite sets and of computable probability mass functions in [6] are special cases of typical descriptions of total functions.

3.3   Minimal Sufficient Statistic

Let $T_1, T_2, \ldots$ be the standard enumeration of prefix Turing machines. The *reference* prefix Turing machine U is chosen such that $U(i, p) = T_i(p)$ for all i and p. Looking at it slightly more from a programming point of view, we can define a pair (p, d) is a *description* of a finite string x if $U(p, d)$ prints x and p is a *total* self-delimiting program with respect to x.

Definition 3. *A minimal sufficient statistic for data x is a sufficient statistic for x with minimal prefix complexity. Its length is known as the* sophistication *of x, and is defined by* $\mathrm{soph}(x) = \min\{l(p) : (p, d)$ *is a description of x*\}.

For these notions to be nontrivial, it should be impossible to shift, for every x with description (p, d), information from p to d and write (q, pd), where q is a $O(1)$-bit program that considers the first self-delimiting prefix of the data as a program to be simulated. The key to this definition should be to force a separation of a program embodying nontrivial regularity from its data embodying accidental information. It turns out that the key to nontrivial separation is the requirement that the program witnessing the sophistication be *total*.

Lemma 2. *Assume for the moment we allow all partial recursive programs as statistic. Then, the sophistication of all data x is $\overset{+}{=} 0$.*

Proof. Let $T_1, T_2, \ldots$ be the standard enumeration of (prefix) Turing machines. Let the index of U (the reference prefix Turing machine) in the standard enumeration $T_1, T_2, \ldots$ of prefix Turing machines be u. Suppose that (p, d) is a description of x. Then, (u, pd) is also a description of x since $U(u, pd) = U(p, d)$.

This shows that unrestricted partial recursive statistics are uninteresting. Apart from triviality, a class of statistics can also possibly be vacuous by having the length of the minimal sufficient statistic exceed $K(x)$. Our first task is to determine whether the definition is non-vacuous. We will distinguish sophistication in different description modes:

Notation 3 *The sophistication of x is denoted by "$\mathrm{soph}(x)$" with respect to unrestricted total recursive function statistic, "$\mathrm{sophtyp}(x)$" with respect to typical total recursive function statistic.*

Lemma 4 (Existence). *For every finite binary string* x*, the sophistication satisfies* $\mathrm{soph}(x), \mathrm{sophtyp}(x) \stackrel{+}{<} K(x)$.

Proof. By definition of the prefix complexity there is a program x^* of length $l(x^*) = K(x)$ such that $U(x^*, \epsilon) = x$. This program x^* can be partial. But we can define another program $x_s^* = sx^*$ where s is a program of a constant number of bits that tells the following program to ignore its actual input and compute as if its input were ϵ. Clearly, x_s^* is total and is a typical sufficient statistic of the total recursive function type, that is, $\mathrm{soph}(x), \mathrm{sophtyp}(x) \le l(x_s^*) \stackrel{+}{<} l(x^*) = K(x)$.

The previous lemma gives an upper bound on the sophistication. This still leaves the possibility that the sophistication is always $\stackrel{+}{=} 0$, for example in the most liberal case of unrestricted totality. But this turns out to be impossible.

Theorem 1. *For every* x*, with* p *a minimal sufficient statistic for* x*, with witness* $p(d) = x$*, we have*

(i) $I(x : p) \stackrel{+}{=} l(p^*) = \mathrm{sophtyp}(x) \stackrel{+}{>} K(K(x))$ *and* $l(d) \stackrel{+}{<} K(x) - K(K(x))$.
(ii)

$$\liminf_{l(x) \to \infty} \mathrm{soph}(x) \stackrel{+}{=} 0. \tag{3}$$

(iii) For every n *there exists an* x *of length* n *such that* $\mathrm{sophtyp}(x) \stackrel{+}{>} n$.

For every n *there exists an* x *of length* n *such that* $\mathrm{soph}(x) \stackrel{+}{>} n - \log n - 2 \log \log n$.

Proof. (i) If p is a sufficient statistic for x, with $p(d) = x$, then

$$k = K(x) \stackrel{+}{=} K(p) + K(d \mid p^*) \stackrel{+}{=} K(p) + l(d). \tag{4}$$

By symmetry of information, a deep result of [5],

$$K(x, y) \stackrel{+}{=} K(x) + K(y \mid x^*) \stackrel{+}{=} K(y) + K(x \mid y^*). \tag{5}$$

The *mutual information* of x about y is $I(x : y) = K(y) - K(y \mid x^*)$. By rewriting according to symmetry of information we see that $I(x : y) \stackrel{+}{=} I(y : x)$. Now, $K(x) + K(p \mid x^*) \stackrel{+}{=} K(p) + K(x \mid p^*)$. Since p is a sufficient statistic we see that $K(x) \stackrel{+}{=}$ the left-hand side of the last equation, therefore

$$K(p \mid x^*) \stackrel{+}{=} 0. \tag{6}$$

Then, $I(x : p) \stackrel{+}{=} K(p) - K(p \mid x^*) \stackrel{+}{=} K(p) \stackrel{+}{=} l(p^*)$. From p^* in *typical* description format we can retrieve a value $\stackrel{+}{=} l(d)$ and also $K(p) \stackrel{+}{=} l(p^*)$. Therefore, we can retrieve $K(x) \stackrel{+}{=} K(p) + l(d)$ from p^*. That shows that $K(K(x)) \stackrel{+}{<} K(p)$. This proves the first statement. The second statement follows by (4) since $p(d) = x$ and p is a sufficient statistic using d.

(ii) An example of very unsophisticated strings are the individually random strings with high complexity: x of length $l(x) = n$ with complexity $K(x) \stackrel{+}{=} n + K(n)$. Then, the identity program i with $i(d) = d$ for all d is total, has complexity $K(i) \stackrel{+}{=} 0$, and satisfies $K(x) \stackrel{+}{=} K(i) + l(x^*)$. Hence, i witnesses that $\mathrm{soph}(x) \stackrel{+}{=} 0$. This shows (3).

(iii) By reduction to the similar results for finite sets in [6]. Omitted in this version.

The useful (6) states that for every x there are only finitely many sufficient statistics and there is a constant length program that generates all of them from x^*. In fact, there is a slightly stronger statement from which this follows:

Lemma 5. *There is a universal constant c, such that for every x, the number of p^*d such that $p(d) = x$ and $K(p) + l(d) \stackrel{+}{=} K(x)$, and a fortiori with $l(p^*) \geq \mathrm{soph}(x)$ or $l(p^*) \geq \mathrm{sophtyp}(x)$, is bounded above by c.*

Proof. Since $U(p, d) = x$ and $K(p) + l(d) \stackrel{+}{=} K(x)$, the combination p^*d (with self-delimiting p^*) is a shortest prefix program for x. From [14], Exercise 3.3.7 item (b) on p. 205, it follows that the number of shortest prefix programs is upper bounded by a universal constant.

4   Algorithmic Properties

We investigate the recursion properties of the sophistication function. An important and deep result is the Kolmogorov complexity quantification of the uncomputability of the Kolmogorov complexity, due to [5], was stated in (7). For every length n there is an x of length n such that:

$$\log n - \log\log n \stackrel{+}{<} K(K(x) \mid x) \stackrel{+}{<} \log n. \tag{7}$$

Note that the right-hand side holds for every x by the simple argument that $K(x) \leq n + 2\log n$ and hence $K(K(x)) \stackrel{+}{<} \log n$. But there are x's such that the length of the shortest program to compute $K(x)$ almost reaches this upper bound, even if the full information about x is provided. In fact, (7) quantifies the uncomputability of $K(x)$ (the bare uncomputability can be established in a much simpler fashion). It is natural to suppose that the sophistication function is not recursive either. The lemma's below suggest that the complexity function is more uncomputable than the sophistication.

Theorem 2. *The functions $\mathrm{soph}, \mathrm{sophtyp}$ are not recursive.*

Proof. Given n, let x_0 be the least x such that $\mathrm{soph}(x), \mathrm{sophtyp}(x) > n - 2\log n$. By Theorem 1 we know that $\mathrm{soph}(x), \mathrm{sophtyp}(x) \to \infty$ for $x \to \infty$, hence x_0 exists. Assume by way of contradiction that either sophistication function is computable. Then, we can find x_0, given n, by simply computing the successive values of the function. But then $K(x_0) \stackrel{+}{<} K(n)$, while by Lemma 4 $K(x_0) \stackrel{+}{>} \mathrm{soph}(x_0), \mathrm{sophtyp}(x_0)$ and by assumption $\mathrm{soph}(x_0), \mathrm{sophtyp}(x_0) > n - 2\log n$, which is impossible.

The *halting sequence* $\chi = \chi_1\chi_2\ldots$ is the infinite binary characteristic sequence of the halting problem, defined by $\chi_i = 1$ if the reference universal prefix Turing machine halts on the ith input: $U(i) < \infty$, and 0 otherwise. It turns out that the sophistication can be computed from the Kolmogorov complexity, which in turn can be computed from the halting sequence. It is not known, and seems doubtful, whether the Kolmogorov complexity or the halting sequence can be computed from the sophistication. However, both of the former are computable from the object itself plus the witness program of which the sophistication is the length.

Lemma 6. *Let p witness the (typical) sophistication of x and hence $p = p^*$. Then,*

(i) *In the typical case, sophtyp$(x) = l(p)$, we can compute $K(x)$ from p, which implies that $K(K(x) \mid p) \stackrel{+}{=} 0$.*

(ii) *In the unrestricted total case, soph$(x) = l(p)$, we can compute $K(x)$ from p and x, which implies that $K(K(x) \mid p, x) \stackrel{+}{=} 0$.*

Proof. (i) follows directly from the definitions of typical description, and the assumption that p is a sufficient statistic of x.

(ii) We are given p^* such that $p(d) = x$, $K(p) + l(d) \stackrel{+}{=} K(x)$, and p is total. Run $p(e)$ on all strings e in lexicographical length-increasing order. Since p is total we will find a shortest string e_0 such that $p(e_0) = x$. Set $d = e_0$.

Theorem 3. *Given an oracle that on query x answers p, witnessing the (typical) sophistication of x by $l(p) = (\text{sophtyp}(x))\ \text{soph}(x)$. Then, we can compute the halting sequence χ.*

Proof. By Lemma 6 we can compute the function K given the oracle in the statement of the theorem. In [14], Exercise 2.2.7 on p. 175, it is shown that if we can solve the halting problem for plain Turing machines, then we can compute the (plain) Kolmogorov complexity, and *vice versa*. The same holds for the halting problem for prefix Turing machines and the prefix Turing complexity. This proves the theorem.

Lemma 7. *There is a constant c, such that for every x there is a program (possibly depending on x) of at most c bits that computes soph(x), sophtyp(x) and the witness programs p, p', respectively, from $x, K(x)$. That is, $K(p, p' \mid x, K(x)) \stackrel{+}{=} 0$. With some abuse of notation we can express this as $K(\text{soph}, \text{sophtyp} \mid K) \stackrel{+}{=} 0$.*

Proof. By definition of sufficient statistic p, we have $p(d) = x$ and $K(p) + K(d \mid p^*) \stackrel{+}{=} K(x)$. By (6) the number of sufficient statistics for x is bounded by an independent constant, and we can generate all of them from x by a $\stackrel{+}{=} 0$ length program (possibly depending on x). Then, we can simply determine the least length of a sufficient statistic, say p_0. Then, $l(p_0) = \text{soph}(x)$ or $l(p_0) = \text{sophtyp}(x)$, depending on the description mode considered.

There is a subtlety here: Lemma 7 is nonuniform. While for every x we only require a fixed number of bits to compute the sophistication from $x, K(x)$, the result is nonuniform in the sense that these bits may depend on x. But how do we know if we have the right bits? We can simply try all programs of length up to the upper bound, but then we don't know if they halt or if they halt they halt with the correct (total) program. The question arising is if there is a single program that computes the sopistication and its witness program for all x. In [17] this much more difficult question is answered in a strong negative sense: there is no algorithm that for every x, given $x, K(x)$, approximates the sophistication of x to within precision $l(x)/(10 \log l(x))$.

Theorem 4. *For every x of length n and p the program that witnesses the sophistication of x we have $K(p \mid x) \overset{+}{<} \log n$. For every length n there are strings x of length n such that $K(p \mid x) \overset{+}{>} \log n - \log \log n$.*

Proof. Omitted.

Definition 4. *A function f from the rational numbers to the real numbers is* upper semicomputable *if there is a recursive function $H(x,t)$ such that $H(x,t+1) \leq H(x,t)$ and $\lim_{t \to \infty} H(x,t) = f(x)$. Here we interpret the total recursive function $H(\langle x,t \rangle) = \langle p,q \rangle$ as a function from pairs of natural numbers to the rationals: $H(x,t) = p/q$. If f is upper semicomputable, then $-f$ is* lower semicomputable. *If f is both upper-a and lower semicomputable, then it is* computable.

When everything is over the natural numbers then computability turns into recursivity. Since $K(\cdot)$ is upper semicomputable, [14], and from $K(\cdot)$ we can compute $\mathrm{soph}(x), \mathrm{sophtyp}(x)$, we have the following:

Lemma 8. *The functions $\mathrm{soph}(x), \mathrm{sophtyp}(x)$ are upper semicomputable.*

Proof. Omitted.

Theorem 5. *Given the halting sequence χ we can compute $\mathrm{soph}(x), \mathrm{sophtyp}(x)$ from x. With some abuse of notation this implies $K(\mathrm{soph}, \mathrm{sophtyp} \mid \chi) \overset{+}{=} 0$.*

Proof. From χ we can compute the prefix Kolmogorov complexity function K, and from K we can compute $\mathrm{soph}, \mathrm{sophtyp}$ by Lemma 7.

Acknowledgment. The author thanks Luis Antunes and Lance Fortnow for their comments and for stimulating this investigation.

References

1. A.R. Barron, J. Rissanen, and B. Yu, The minimum description length principle in coding and modeling, *IEEE Trans. Inform. Theory*, IT-44:6(1998), 2743–2760.
2. T.M. Cover, Kolmogorov complexity, data compression, and inference, pp. 23–33 in: *The Impact of Processing Techniques on Communications*, J.K. Skwirzynski, Ed., Martinus Nijhoff Publishers, 1985.
3. T.M. Cover and J.A. Thomas, *Elements of Information Theory*, Wiley, New York, 1991.
4. R. A. Fisher, On the mathematical foundations of theoretical statistics, *Philosophical Transactions of the Royal Society of London, Ser. A*, 222(1922), 309–368.
5. P. Gács, On the symmetry of algorithmic information, *Soviet Math. Dokl.*, 15 (1974) 1477–1480. Correction: ibid., 15 (1974) 1480.
6. P. Gács, J. Tromp, and P. Vitányi, Algorithmic statistics, *IEEE Trans. Inform. Theory*, 47:6(2001), 2443–2463.
7. Q. Gao, M. Li and P.M.B. Vitányi, Applying MDL to learn best model granularity, *Artificial Intelligence*, 121(2000), 1–29.
8. M. Gell-Mann, *The Quark and the Jaguar*, W. H. Freeman and Company, New York, 1994.
9. A.N. Kolmogorov, Three approaches to the quantitative definition of information, *Problems Inform. Transmission* 1:1 (1965) 1–7.
10. A.N. Kolmogorov, On logical foundations of probability theory, Pp. 1–5 in: *Probability Theory and Mathematical Statistics*, Lect. Notes Math., Vol. 1021, K. Itô and Yu.V. Prokhorov, Eds., Springer-Verlag, Heidelberg, 1983.
11. A.N. Kolmogorov and V.A. Uspensky, Algorithms and Randomness, *SIAM Theory Probab. Appl.*, 32:3(1988), 389–412.
12. M. Koppel, Complexity, depth, and sophistication, *Complex Systems*, 1(1987), 1087–1091
13. M. Koppel, Structure, *The Universal Turing Machine: A Half-Century Survey*, R. Herken (Ed.), Oxford Univ. Press, 1988, pp. 435–452.
14. M. Li and P. Vitanyi, *An Introduction to Kolmogorov Complexity and Its Applications*, Springer-Verlag, New York, 1997 (2nd Edition).
15. A.Kh. Shen, The concept of (α, β)-stochasticity in the Kolmogorov sense, and its properties, *Soviet Math. Dokl.*, 28:1(1983), 295–299.
16. A.Kh. Shen, Discussion on Kolmogorov complexity and statistical analysis, *The Computer Journal*, 42:4(1999), 340–342.
17. N.K. Vereshchagin and P.M.B. Vitányi, Kolmogorov's structure functions and an application to the foundations of model selection, *Proc. 47th IEEE Symp. Found. Comput. Sci.*, 2002. Full version: http://xxx.lanl.gov/abs/cs.CC/0204037
18. P.M.B. Vitányi and M. Li, Minimum Description Length Induction, Bayesianism, and Kolmogorov Complexity, *IEEE Trans. Inform. Theory*, IT-46:2(2000), 446–464.
19. V.V. V'yugin, On the defect of randomness of a finite object with respect to measures with given complexity bounds, *SIAM Theory Probab. Appl.*, 32:3(1987), 508–512.
20. V.V. V'yugin, Algorithmic complexity and stochastic properties of finite binary sequences, *The Computer Journal*, 42:4(1999), 294–317.

Optimal Clearing of Supply/Demand Curves[*]

Tuomas Sandholm[1] and Subhash Suri[2]

[1] Computer Science, Carnegie Mellon University, Pittsburgh, PA 15213
sandholm@cs.cmu.edu
[2] Computer Science, University of California, Santa Barbara, CA 93106
suri@cs.ucsb.edu

Abstract. Markets are important coordination mechanisms for multiagent systems, and market clearing has become a key application area of
algorithms. We study optimal clearing in the ubiquitous setting where
there are multiple indistinguishable units for sale. The sellers and buyers
express their bids via supply and demand curves. Discriminatory pricing
leads to greater profit for the party who runs the market than non-
discriminatory pricing. We show that this comes at the cost of computation complexity. For piecewise linear curves we present a fast polynomial-
time algorithm for nondiscriminatory clearing, and show that discriminatory clearing is $\mathcal{NP}$-complete (even in a very special case). We then show
that in the more restricted setting of linear curves, even discriminatory
markets can be cleared fast in polynomial time. Our derivations also un-
cover the elegant fact that to obtain the optimal discriminatory solution,
each buyer's (seller's) price is incremented (decremented) *equally* from
that agent's price in the quantity-unconstrained solution.

1 Introduction

Commerce is moving online to an increasing extent, and there has been a significant shift to dynamic pricing via auctions (one seller, multiple buyers), reverse
auctions (one buyer, multiple sellers), and exchanges (multiple buyers, multiple sellers). These market types have also become key coordination methods in
multiagent systems. These trends have led to an increasing need for fast market
clearing algorithms. Recent electronic commerce server prototypes such as *eMediator* [10] and *AuctionBot* [14] also have demonstrated a wide variety of new
market designs, leading to the need for new clearing algorithms.

There has been a recent surge of interest (e.g., [11,2,4]) in clearing combinatorial auctions where bids can be submitted on bundles of distinguishable items,
potentially multiple units of each [10,6,13]. There has also been recent work on
clearing combinatorial reverse auctions [10,13] and combinatorial exchanges [10,
13]. The clearing problem in a combinatorial market is $\mathcal{NP}$-complete [9], in-
approximable [11], and in certain variants even finding a feasible solution is
$\mathcal{NP}$-complete [13]. On the other hand, markets where there is only one unit of
one item for sale are trivial to clear.

---

[*] This work was funded by, and conducted at, CombineNet, Inc., 311 S. Craig St.,
Pittsburgh, PA 15213.

P. Bose and P. Morin (Eds.): ISAAC 2002, LNCS 2518, pp. 600–611, 2002.

In this paper we study a setting which is in between. We study the ubiquitous market setting where there are multiple *indistinguishable* units of an item for sale. This setting is common in markets for stocks, bonds, electricity, bandwidth, oil, pork bellies, memory chips, CPU time, etc. We study the problem where the bids are known up front. This is the case in many business-to-business markets. The algorithms can also be used in markets where bids arrive over time (and the market is cleared periodically, for example, every 5 minutes, or after some number of bids have arrived since the last clearing) and in multi-stage markets (where a tentative clearing is carried out after each round of bidding).

The naive approach to bidding in a multi-unit market would require the bidders to express their offers as a list of points, for example ($2 for 1 unit) XOR ($5 for 2 units) XOR ($6 for 3 units), etc. The mapping from quantities to prices can be represented more compactly by allowing each bidder to express his offer as a price-quantity curve (supply curve for a seller, demand curve for a buyer). Such curves are natural ways of expressing preferences, are ubiquitous in economics [8], and are becoming common in electronic commerce as well [10, 5,7].

In classic economic theory of supply and demand curves (called partial equilibrium theory [8]), the market is cleared as follows. The supply curves of the sellers and the demand curves of the buyers are separately aggregated. The market is cleared at a per-unit price for which supply equals demand (there may be multiple solutions). This way of clearing the market maximizes social welfare.

However, it turns out that the *auctioneer* (that is, the party who runs the market—who is neither a buyer nor a seller) will achieve greater (or equal) profit from the same supply/demand curves by reducing the number of units traded, and charging one per-unit price to the buyers while paying a lower per-unit price to the sellers.[1] We call such pricing *non-discriminatory* because each buyer pays the same amount per unit, and each seller gets paid the same amount per unit. The auctioneer's profit can be further improved by moving to *discriminatory* pricing where each seller and each buyer can be cleared at a different per-unit price.

Interestingly, the pricing scheme and the shape of the supply/demand curves significantly impact the computational complexity of clearing the market. We show that markets with piecewise linear curves are clearable in polynomial time under non-discriminatory pricing, but are $\mathcal{NP}$-complete to clear under discriminatory pricing. With linear curves, even discriminatory markets can be cleared in polynomial time.[2]

---

[1] The profit could be allocated entirely to the party who runs the market, or it could be divided among the market participants. How it is divided can affect the bidders' incentives for revealing their preferences, but we do not address incentives in this paper.

[2] These issues have been settled for auctions (and part of reverse auctions) [12]. We settle the general case: multiple buyers and sellers.

2   The Market Model

The market has n sellers and m buyers. Without loss of generality, we assume that no agent is both a buyer and a seller (if an agent is both, we treat him as two separate agents). Each seller expresses his willingness to sell via a *supply curve* $s : R^+ \to R^+$ from non-negative *unit* prices to non-negative supply quantity. Thus, if the unit price is p, the seller is willing to supply $s(p)$ units of the good.[3] Similarly, each buyer submits a *demand curve* $d : R^+ \to R^+$. We assume that supply/demand curves are *piecewise linear*. Such curves can approximate any curve arbitrarily closely. We make the natural assumption that *supply curves* are *upward sloping* and *demand curves* are *downward sloping*. This is economically reasonable in that higher prices increase supply and decrease demand. We do not assume that the curves are continuous—they could have discrete "jumps".

In this paper, we study clearing where the objective is to maximize the auctioneer's profit. We study two pricing schemes: *non-discriminatory* and *discriminatory*.

2.1   Non-discriminatory Pricing

In a non-discriminatory market, there are two clearing prices, one shared by the sellers and one shared by the buyers. Specifically, suppose the market clears the sellers at the unit price p^*_{ask}, and the buyers at the unit price p^*_{bid}. These prices uniquely determine the quantity supplied by each seller, and the quantity bought by each buyer, using their supply/demand curves. In particular, suppose s_i is the supply curve of seller i, and d_j is the demand curve of buyer j. Then, for the solution to be feasible, supply must equal demand: $\sum_{i=1}^{n} s_i(p^*_{ask}) = \sum_{j=1}^{m} d_j(p^*_{bid})$. Subject to this, the goal is to maximize the auctioneer's profit: $p^*_{bid} \sum_{j=1}^{m} d_j(p^*_{bid}) - p^*_{ask} \sum_{i=1}^{n} s_i(p^*_{ask})$. The computational problem is to determine the clearing prices p^*_{ask} and p^*_{bid}.

2.2   Discriminatory Pricing

In a discriminatory market, the market can clear each seller and each buyer at a distinct unit price. Specifically, suppose seller i is cleared at unit price p^*_i, and buyer j is cleared at unit price p^*_j. Then, the feasibility condition of supply meeting demand is $\sum_{i=1}^{n} s_i(p^*_i) = \sum_{j=1}^{m} d_j(p^*_j)$, and the auctioneer's profit to be maximized is $\sum_{j=1}^{m} p^*_j d_j(p^*_j) - \sum_{i=1}^{n} p^*_i s_i(p^*_i)$. The computational problem is to determine the clearing price for each seller and buyer. The profit generated under discriminatory pricing is greater (or equal) than that under non-discriminatory pricing.[4]

---

[3] The model also covers the possibility that the seller (buyer) is willing to accept a higher (lower) price for the same quantity. However, given our objective, in any optimal solution, each party is cleared exactly on his supply/demand curve.

[4] Non-discriminatory markets offer fairness. Discriminatory markets offer a weak form of *ex ante* fairness: they are anonymous in the sense that had two players swapped their bids, their allocations would also have been swapped.

3 Clearing Preliminaries

We begin with the simple case of a *one seller, one buyer* market, where the seller has an upward sloping linear supply curve $q = a_s p - b_s$, and the buyer has a downward sloping linear demand curve $q = -a_d p + b_d$, where $a_s, a_d > 0$, and $b_s, b_d \geq 0$. The following elementary lemmata will be useful throughout the paper. (Observe that in a 1-seller, 1-buyer market, non-discriminatory and discriminatory pricing are identical.) Figure 1 illustrates the clearing in this setting. Due to lack of space, we omit proofs of most of the lemmas and theorems, but they can be found in the full version of the paper available on the web at www.cs.ucsb.edu/~suri/pubs.html.

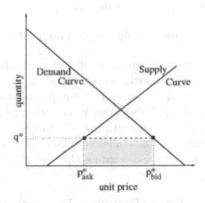

Fig. 1. 1-seller, 1-buyer market with linear supply/demand curves. The profit equals the area of the shaded rectangle. The clearing occurs at quantity q^* which is half the height of the triangle formed by the supply and demand lines.

Lemma 1 (1-Seller, 1-Buyer Unconstrained Trading). *Consider a market of one seller, with upward sloping supply $q = a_s p - b_s$, and one buyer, with downward sloping demand $q = -a_d p + b_d$, where $a_s, a_d > 0$, and $b_s, b_d \geq 0$. Then, the profit-maximizing trade occurs at quantity*

$$q^* = \frac{1}{2}\left(\frac{a_s b_d - a_d b_s}{a_s + a_d}\right)$$

The clearing prices for the seller and the buyer are

$$p^*_{ask} = \frac{1}{2}\left(\frac{b_s}{a_s} + \frac{b_s + b_d}{a_s + a_d}\right), \qquad p^*_{bid} = \frac{1}{2}\left(\frac{b_d}{a_d} + \frac{b_s + b_d}{a_s + a_d}\right)$$

Proof. The proof uses elementary calculus, and is omitted due to lack of space.

In general, when participants put price or quantity constraints on their curves, the trade will not occur at the "unconstrained" optimal quantity q^* determined by Lemma 1. The following lemma determines the effect of moving the traded quantity away from q^*. We consider the same one-seller, one-buyer market as above, and compute the profit achieved when trade occurs at some quantity $q^* + \varepsilon$, where ε can be positive or negative. (We assume $|\varepsilon| \leq q^*$, which ensures that the trade is feasible and produces non-negative profit.)

Lemma 2. *In the 1-seller, 1-buyer market, if the trade occurs at quantity $q^* + \varepsilon$, then the profit is*

$$q^*(p_{bid}^* - p_{ask}^*) - \varepsilon^2 \left(\frac{1}{a_s} + \frac{1}{a_d} \right).$$

That is, the profit shrinks (quadratically) with $|\varepsilon|$.

Proof. The proof uses simple calculus, and is omitted due to lack of space.

Our next lemma states a corollary of Lemma 2 for the case where the buyer and seller curves are quantity-constrained. Suppose the buyer's curve is the downward-sloping linear function $q = -a_d p + b_d$, but restricts quantity to the range $[q_d', q_d'']$, and the seller's curve is the upward-sloping linear function $q = -a_d p + b_d$, but restricts the quantity to the range $[q_s', q_s'']$. What trade maximizes the profit? The only feasible trades that can occur are those in the quantity range that is common to both. So, assume that $[q', q'']$ is the intersection of the intervals $[q_d', q_d'']$ and $[q_s', q_s'']$.

Lemma 3 (1-Seller, 1-Buyer Bounded Trading). *Consider the 1-seller, 1-buyer market, with linear supply/demand curves, where the buyer and the seller can only trade in the quantity range $[q', q'']$. The profit-maximizing trade occurs either at q^* (if $q^* \in [q', q'']$), or at that endpoint of the range $[q', q'']$ which is closer to q^*.*

Proof. If the unconstrained trade quantity q^* is in the feasible range, profit is maximized at q^*. Otherwise, Lemma 2 shows that the profit shrinks quadratically with the deviation ε from q^*. Thus, the optimal trade occurs at the feasible point closest to q^*, which is an endpoint of the range $[q', q'']$.

4   Non-discriminatory Markets

In this section, we show how to clear a market with multiple buyers and sellers, each with a *piecewise linear* demand or supply curve, under non-discriminatory pricing. Our approach is to *aggregate* the demand and the supply curves separately, and then reduce the problem to the 1-seller, 1-buyer case. The single seller is the aggregate of all sellers, and the single buyer is the aggregate of all buyers. The key idea here is that since *all buyers are cleared at the same price* p_{bid}^*, we can infer quantities sold to each individual buyer by evaluating their curves at p_{bid}^*. The same holds for the sellers.

Let us consider the aggregation of buyer curves; seller curves are handled in the same way. Consider a set of piecewise linear demand curves $d_1, d_2, \ldots, d_n$. Their *aggregate curve* is a piecewise linear function $D : R^+ \to R^+$ such that $D(p)$ is the total demand at unit price p. That is, $D(p) = d_1(p) + d_2(p) + \ldots + d_n(p)$, where $d_i(p)$ is the demand by curve i at unit price p. The aggregation of *linear* functions leads to a linear function. Thus, if a price interval $[p_1, p_2]$ does not contain the *breakpoints* of any of the demand curves, then the aggregate curve in the interval $[p_1, p_2]$ has the form $q = (\sum_i a_i)p + \sum_i b_i$, where a_i and b_i are the coefficients of the component linear curves.

The breakpoints of D are the union of the breakpoints of the component curves—the aggregate demand curve changes only when one of the component curves changes. Thus, given a set of n piecewise linear curves each of which has at most k pieces, their aggregate curve D has at most nk breakpoints. Given n piecewise linear curves, their aggregate is easily computed in $O(nk \log(nk))$ time, by a sweep-line algorithm [1].

Despite the fact that the input demand curves are downward sloping, the aggregate demand curve need not be downward sloping. (This can happen because the demand curves may have disconnected pieces.) Similarly, the aggregate supply curve need not be upward sloping. This can lead to the problem of having multiple prices for a given quantity. We rectify this by computing the *rightmost* envelope of the aggregate demand curve, and *leftmost* envelope of the aggregate supply curve. In other words, for each quantity q, we just keep the point with the maximum demand price, and the minimum supply price. This does not affect the solution space since in a profit-maximizing market, all clearings occur at these *envelope* prices.

Our non-discriminatory clearing algorithm can now be described as follows (see Figure 2.):

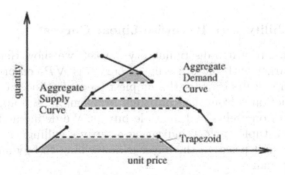

Fig. 2. Non-discriminatory market. The figure shows decomposition of the feasible region into four trapezoids. Each trapezoid corresponds to a 1-seller, 1-buyer market. The aggregate demand curve is intentionally drawn to be discontinuous and not downward sloping (although each piece is downward sloping).

Algorithm ND-Market

1. Compute the piecewise linear aggregate demand curve D, and the aggregate supply curve S.
2. Let the *feasible space* denote the set of points (p, q) for which $D(p)$ and $S(p)$ both exist and $S(p) \le D(p)$; that is, there is both aggregate demand and aggregate supply for q, and the aggregate demand price is no smaller than the aggregate supply price.
3. Decompose the feasible region into trapezoids, by "drawing" horizontal lines through each breakpoint of D or S.
4. The market clearing problem for each trapezoid corresponds to the 1-seller, 1-buyer bounded trade (cf. Lemma 2), so it can be solved in $O(1)$ time. Since there are $O(K)$ trapezoids, the time complexity of this step is $O(K)$.
5. The maximum-profit solution over all trapezoids is the optimal solution. Once the clearing prices, p^*_{bid} and p^*_{ask}, are determined, we can evaluate each seller curve at p^*_{ask} and each buyer's curve at p^*_{bid} to determine the quantity sold by each seller, and bought by each buyer.

We summarize this result in the following theorem.

Theorem 1. *Consider a 1-item, multi-unit market with multiple sellers and buyers, where each seller (buyer) has an upward (downward) sloping piecewise linear curve. Then, a profit-maximizing clearing using non-discriminatory pricing can be determined in $O(K \log K)$, where K is the total number of pieces in all of the piecewise linear curves.*

5 Discriminatory Markets

We now study the complexity of discriminatory markets.

5.1 Intractability with Piecewise Linear Curves

In sharp contrast to a non-discriminatory market, we show that clearing a discriminatory market with piecewise linear curves is $\mathcal{NP}$-complete. In fact, this complexity jump occurs even for the simplest piecewise linear curves: *step functions*. The reduction is from the **knapsack** problem [3], and applies even to the restricted case of one seller and multiple buyers. We define a *step function* demand curve as a tuple (p_j, q_j), indicating a buyer's willingness to buy q_j units at or below the unit price p_j; the buyer is not willing to buy any units at price strictly greater than p_j.

Theorem 2. *Consider a discriminatory market where each participant submits a step function supply or demand curve. Determining a profit-maximizing clearing of the market is $\mathcal{NP}$-complete. This holds even if one side of the market has only one participant—who submits a constant curve.*

Proof. Reduction from 0-1 Knapsack problem. Due to limited space, we omit the details.

5.2   Fast Algorithms for Linear Curves

In this section we show that in the more restricted setting where supply and demand curves are **linear**, even discriminatory markets can be cleared in polynomial time. As we show, the discriminatory market clearing problem is a convex quadratic program with linear constraints, which could be solved in polynomial time using general techniques. However, we present fast ($O(N \log N)$) and simple specialized algorithms. In addition, our algorithms lend insight into the structure of the problem, such as closed-form expressions for prices. We begin with the simpler problems: *reverse auctions* where one buyer is matched with multiple sellers, and *auctions* where one seller is matched with multiple buyers.

Reverse Auction: One Buyer, Multiple Sellers. Say n sellers bid in a reverse auction to sell multiple indistinguishable units of an item. There is one buyer in the market. He wants to acquire at least Q units, at minimum total cost. Each seller has an upward sloping supply curve: $q = a_i p - b_i$, where $a_i > 0$ and $b_i \geq 0$. The clearing problem is

$$\min \ \sum_{i=1}^{n} p_i q_i \quad \text{s.t.} \quad q_i = a_i p_i - b_i \ \text{and} \ \sum_{i=1}^{n} q_i \geq Q$$

Eliminating p_i's from the objective yields the quadratic function $\sum q_i \left(\frac{q_i + b_i}{a_i} \right)$, leaving $\sum q_i \geq Q$ as the only constraint. Since the seller curves are upward sloping, it is clear that the quantity constraint is tight on optimality. Thus, we can use the method of Lagrangian multipliers to optimize

$$\min \ \left(\frac{q_i^2}{a_i} + \frac{b_i q_i}{a_i} \right) + \lambda \left(Q - \sum_{i=1}^{n} q_i \right).$$

Setting each partial derivative with respect to q_i to zero gives $2q_i = a_i \lambda - b_i$. Since $\sum_{i=1}^{n} q_i = Q$, we get

$$\lambda = \frac{2Q + \sum b_i}{\sum a_i} \tag{1}$$

Substituting the value of λ into q_i yields the clearing quantities and prices:

$$q_i = -\frac{b_i}{2} + \frac{a_i}{2} \left(\frac{2Q + \sum b_i}{\sum a_i} \right), \quad p_i = \frac{b_i}{2a_i} + \frac{1}{2} \left(\frac{2Q + \sum b_i}{\sum a_i} \right) \tag{2}$$

There is only one difficulty in this solution: some of the quantities q_i may be negative (because we ignored the non-negativity constraint from the mathematical program). In particular, if the quantity $a_i \lambda$ is smaller than b_i, then $q_i < 0$. Such a solution could easily arise even in the case of two sellers, where it might be advantageous to *buy some extra units from one, and sell to the other the excess over Q*. For a simple example, consider two sellers, with curves $q = 100p - 100$ and $q = p - 100$, and suppose $Q = 50$. In this case, the Lagrangian method gives $q_1 = 98.5$ and $q_2 = -48.5$, which is clearly infeasible. We show below how to control the Lagrangian solution to keep it feasible. Basically, we show that sellers whose clearing quantity in Eq. (2) is negative must sell zero quantity in any optimal solution.

Algorithm D-ReverseAuction

1. Index the sellers by increasing value of their smallest feasible price, namely, b_i/a_i. Let S_i denote the set of sellers $\{1, 2, \ldots, i\}$.
2. For $i = 1, 2, \ldots, n$ do
   a) Compute clearing prices and quantities for the set S_i given by Eq. (2).
   b) If :any $q_j < 0$, terminate, and output the clearing computed for set S_{i-1}.
   c) If the *maximum* clearing price among all sellers is less than b_{i+1}/a_{i+1}, or if $i = n$, terminate and output the solution.

Of course, if none of the clearings involve negative quantities, then the Lagrangian method ensures the correctness of the solution. If all sellers in S_i clear for less than b_{i+1}/a_{i+1}, which is the minimum feasible price for $i+1$, then we can obviously terminate. What remains to be shown is that as soon as some quantity becomes negative, say, for the set S_j, we can disregard the sellers j through n.

Lemma 4. *Suppose in the algorithm* **D-ReverseAuction**, *the first occurrence of a negative clearing quantity is for* S_j, *then all the sellers* $j, j+1, \ldots, n$ *sell zero quantity in an optimal solution of the one-sided 1-buyer, multi-seller market.*

Proof. Due to limited space, we omit the proof.

Finally, algorithm D-ReverseAuction can be implemented to run in $O(n \log n)$ time, for n sellers, as follows. The key is to maintain the Lagrangian multiplier λ_i at each iteration of the algorithm, which takes $O(1)$ time to update. Since the sellers are sorted in increasing order of b_i/a_i, the highest clearing price belongs to the most recently added seller i. Similarly, since we only need to check if i's quantity is negative, we only need to compute q_i in round i of the algorithm, which takes $O(1)$ time. Thus, the run time of the algorithm is dominated by the initial sorting. Put together, we have:

Theorem 3. *Consider a reverse auction with n sellers, where each seller has an upward sloping linear supply curve. We can determine the minimum-cost discriminatory-price clearing for buying Q units in total time $O(n \log n)$.*

Auction: One Seller, Multiple Buyers. A similar solution holds when the market has one seller, with Q units to sell, and many buyers.[5] The goal is to maximize the seller's revenue. Let the (downward sloping) demand curve of the jth buyer be $q = -a_j p + b_j$, for $j = 1, 2, \ldots, n$, where $a_j > 0, b_j \geq 0$. The *unconstrained* solution for the market is to sell exactly $\frac{1}{2} b_j$ units to buyer j. However, if the total number of units available is insufficient, that is, $Q <$

---

[5] This case was recently solved [12]. We summarize some of the key results here because we will use them as components for deriving the algorithm for clearing discriminatory exchanges.

$\frac{1}{2}\sum_j b_j$, then we solve the problem using the Lagrangian multiplier method, and obtain the following clearing quantities and prices:

$$q_i = \frac{b_i}{2} - \frac{a_i}{2}\left(\frac{\sum b_i - 2Q}{\sum a_i}\right), \quad p_i = \frac{b_i}{2a_i} + \frac{1}{2}\left(\frac{\sum b_i - 2Q}{\sum a_i}\right) \tag{3}$$

In other words, if the unconstrained solution is quantity-infeasible, we increase the price for each buyer **by the same amount** until the demand reduces to Q. In increasing the price, if some buyer's curve reaches a point of infeasibility (its demand goes to zero), then we remove that buyer from the set, and recalculate the Lagrangian multiplier. To summarize:

Theorem 4 ([12]). *Consider an auction with n buyers each with a downward sloping linear demand curve. We can determine the revenue-maximizing discriminatory-price clearing for selling (at most) Q units in time $O(n \log n)$.*

Exchange: Multiple Sellers, Multiple Buyers. We now describe how to clear discriminatory-price exchanges using auctions and reverse auctions as building blocks. We plot the "aggregate quantity vs. aggregate revenue" curve for both the sellers and the buyers. On the demand side, let $D(q)$ denote the maximum revenue achieved by selling exactly q units to the buyers. On the supply side, let $S(p)$ denote the minimum cost of procuring q units from the sellers. Assume that both the sellers and buyers are sorted in increasing order of $\frac{b_i}{a_i}$ (these are the roots of the supply and demand curves, that is, the *maximum* feasible prices for buyers, and the *minimum* feasible prices for sellers).

For the D curve, the starting point is the quantity $q_{\max} = \sum_{j=1}^{m} b_j/2$, which is the optimal quantity sold when quantity is *unconstrained*, as was discussed in Section 5.2. The corresponding revenue is $D(q_{\max}) = \sum \frac{b_j^2}{4a_j}$. To determine the aggregate revenue as we decrease the aggregate demand, we use the clearing expressions in Eq. (3). So, the clearing price of each buyer is uniformly increased by $\frac{1}{2}\lambda_0$, where $\lambda_0 = \frac{1}{2}\left(\frac{\sum_{j=1}^{m} b_j - 2q}{\sum_{j=1}^{m} a_j}\right)$ is a function of the quantity q being sold. Consequently, the revenue $D(q)$ is a *quadratic* function of q.

The parameter λ changes when the first buyer's price reaches the upper bound (his quantity goes to zero); the first buyer is the one with the smallest b_j/a_j term. We then update λ to the new value $\lambda_1 = \frac{1}{2}\left(\frac{\sum_{j=2}^{m} b_j - 2q}{\sum_{j=2}^{m} a_j}\right)$, and recompute the quadratic revenue function, and so on. Thus, $D(q)$ consists of m quadratic pieces, starts at quantity $q_{\max}$ and ends at the origin. See Figure 3.

Similarly, we determine the aggregate supply curve $S(q)$, which starts at the origin and ends at $q_{\max}$. We determine the maximum quantity that can be purchased from the first seller subject to the price being at most the root of the second seller's supply curve which is b_2/a_2 (below this root only the first seller's curve is active). This quantity is determined using Equations (2), and the revenue is obtained by multiplying the quantity by the price. The first piece of $S(q)$ is a quadratic curve that starts from the origin and ends at this point.

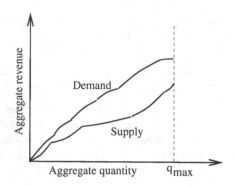

Fig. 3. Discriminatory exchange clearing.

The curve is obtained from Equations (2). The second piece of $S(q)$ extends from the end point of the first to the quantity that can be purchased using the first *two* sellers at maximum price b_3/a_3 (where the 3rd seller's curve enters), and so on. We stop when the quantity purchased reaches q_{max}.

Once we have $D(q)$ and $S(q)$, the former with m pieces and the latter with n pieces, the maximum profit is the maximum *vertical* distance between them, which can be computed easily in $O(n+m)$ time. Computing the curves $D(q)$ and $S(q)$ takes $O(m \log m)$ and $O(n \log n)$ time, respectively. The following theorem summarizes our result.

Theorem 5. *In a multi-buyer, multi-seller discriminatory-price exchange with linear supply/demand curves, a profit-maximizing clearing can be determined in $O(N \log N)$ time, where N is the number of participants.*

6   Conclusions and Future Research

We studied profit-maximizing clearing of markets in the ubiquitous setting where there are multiple indistinguishable units for sale. The sellers and buyers express their bids via supply and demand curves. We focused on the natural and classical economic setting where each seller's supply increases and each buyer's demand decreases as the price increases. Discriminatory pricing leads to greater (or equal) profit for the auctioneer than non-discriminatory pricing. However, we showed that this comes at the cost of computational complexity.

Future research includes studying other restrictions besides linearity that may allow polynomial-time clearing of discriminatory markets, such as continuity assumptions and conditions on second (or higher) derivatives of the supply/demand curves. Also, the question of polynomial-time clearability under other types of supply/demand curves besides piecewise linear ones remains open for both discriminatory and non-discriminatory markets. Finally, we plan to design incremental clearing algorithms for these settings.

References

1. M. de Berg, M. van Kreveld, M. Overmars and O. Schwarzkopf. *Computational Geometry: Algorithms and Applications.* Springer, 2000.
2. Y. Fujishima, K. Leyton-Brown, and Y. Shoham. Taming the computational complexity of combinatorial auctions: Optimal and approximate approaches. In *Proceedings of the Sixteenth International Joint Conference on Artificial Intelligence (IJCAI)*, pages 548–553, Stockholm, Sweden, August 1999.
3. M. R Garey and D. S Johnson. *Computers and Intractability.* W. H. Freeman and Company, 1979.
4. H. Hoos and C. Boutilier. Solving combinatorial auctions using stochastic local search. In *Proceedings of the National Conference on Artificial Intelligence (AAAI)*, pages 22–29, Austin, TX, August 2000.
5. R. Lavi and N. Nisan. Competitive analysis of incentive compatible on-line auctions. In *Proceedings of the ACM Conference on Electronic Commerce (ACM-EC)*, pages 233–241, Minneapolis, MN, 2000.
6. K. Leyton-Brown, M. Tennenholtz, and Y. Shoham. An algorithm for multi-unit combinatorial auctions. In *Proceedings of the National Conference on Artificial Intelligence (AAAI)*, Austin, TX, August 2000.
7. W. A Lupien and J. T Rickard. Crossing network utilizing optimal mutual satisfaction density profile. US Patent 5,689,652, granted Nov. 18 to Optimark Technologies, 1997.
8. A. Mas-Colell, M. Whinston, and J. R. Green. *Microeconomic Theory.* Oxford University Press, 1995.
9. M. H Rothkopf, A. Pekeč, and R. M Harstad. Computationally manageable combinatorial auctions. *Management Science*, 44(8):1131–1147, 1998.
10. T. Sandholm. eMediator: A next generation electronic commerce server. In *Proceedings of the Fourth International Conference on Autonomous Agents (AGENTS)*, pages 73–96, Barcelona, Spain, June 2000.
11. T. Sandholm. Algorithm for optimal winner determination in combinatorial auctions. *Artificial Intelligence*, 135:1–54, January 2002.
12. T. Sandholm and S. Suri. Market clearability. In *Proceedings of the Seventeenth International Joint Conference on Artificial Intelligence (IJCAI)*, pages 1145–1151, Seattle, WA, 2001.
13. T. Sandholm, S. Suri, A. Gilpin, and D. Levine. Winner determination in combinatorial auction generalizations. In *International Conference on Autonomous Agents and Multi-Agent Systems (AAMAS)*, Bologna, Italy, July 2002.
14. P. R Wurman, M. P Wellman, and W. E Walsh. The Michigan Internet AuctionBot: A configurable auction server for human and software agents. In *Proceedings of the Second International Conference on Autonomous Agents (AGENTS)*, pages 301–308, Minneapolis/St. Paul, MN, May 1998.

Partitioning Trees of Supply and Demand

Takehiro Ito, Xiao Zhou, and Takao Nishizeki

Graduate School of Information Sciences, Tohoku University
Aoba-yama 05, Sendai, 980-8579, Japan.
take@nishizeki.ecei.tohoku.ac.jp {zhou,nishi}@ecei.tohoku.ac.jp

Abstract. Assume that a tree T has a number n_s of "supply vertices" and all the other vertices are "demand vertices." Each supply vertex is assigned a positive number called a supply, while each demand vertex is assigned a positive number called a demand. One wish to partition T into exactly n_s subtrees by deleting edges from T so that each subtree contains exactly one supply vertex whose supply is no less than the sum of demands of all demand vertices in the subtree. The "partition problem" is a decision problem to ask whether T has such a partition. The "maximum partition problem" is an optimization version of the partition problem. In this paper, we give three algorithms for the problems. First is a linear-time algorithm for the partition problem. Second is a pseudo-polynomial-time algorithm for the maximum partition problem. Third is a fully polynomial-time approximation scheme (FPTAS) for the maximum partition problem.

Keywords: algorithm, approximation, demand, maximum partition problem, partition problem, FPTAS, supply, tree

1 Introduction

Let $G = (V, E)$ be a graph with vertex set V and edge set E. Set V is partitioned into two sets V_d and V_s. Let $|V| = n$, $|V_d| = n_d$ and $|V_s| = n_s$, then $n = n_d + n_s$. Each vertex $v \in V_d$ is called a *demand vertex* and is assigned a positive real number $d(v)$, called a *demand of v*, while each vertex $u \in V_s$ is called a *supply vertex* and is assigned a positive real number $s(u)$, called a *supply of u*. We wish to partition G into exactly n_s connected components by deleting edges from G so that each component C contains exactly one supply vertex whose supply is no less than the sum of demands of all demand vertices in C. For example, the graph in Fig. 1 (a) has such a partition. In Fig. 1 (a) the deleted edges are drawn by thick dotted lines. Not every graph has such a partition. The *partition problem* is a decision problem to ask whether G has such a partition. If G has no such partition, then we wish to partition G into connected components so that each component C either has no supply vertex or has exactly one supply vertex whose supply is no less than the sum of demands of all demand vertices in C. The *maximum partition problem* is an optimization problem to find one of these partitions that maximize the "fulfillment," that is, the sum of demands of the demand vertices in all components with supply vertices. Clearly, the maximum partition problem is a generalization of the partition problem. Figure 1 (b)

P. Bose and P. Morin (Eds.): ISAAC 2002, LNCS 2518, pp. 612–623, 2002.

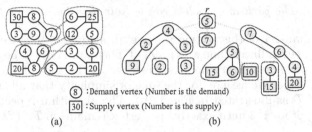

8 : Demand vertex (Number is the demand)

30 : Supply vertex (Number is the supply)

(a)   (b)

Fig. 1. (a) Partition of a graph, and (b) partition of a tree with maximum fulfillment.

illustrates a solution of the maximum partition problem for a tree G. The partition problem and the maximum partition problem have some applications in the power supply problem for power delivery networks and VLSI circuits [2,6,7,11]. Let G be the graph of a power delivery network. Each supply vertex v represents a "feeder," which can supply at most an amount $s(v)$ of electrical power. Each demand vertex v represents a "load," which requires an amount $d(v)$ of electrical power supplied from exactly one of the feeders through a network. Each edge of G represents a cable segment, which can be "turned off" by a switch. Then the partition problem for G represents a simplified version of the "power delivery problem," while the maximum partition problem represents the "power supply switching problem" to minimize the sum of all loads that cannot be supplied powers in a network "reconfigured" by turning off some cable segments. Several types of graph partition problems related to ours have been studied in [3,4,8].

The partition problem is NP-complete even for complete bipartite graphs, because the "0/1 multiple knapsack problem" [9] can be easily reduced to the partition problem for a complete bipartite graph in polynomial time. On the other hand, the maximum partition problem is NP-hard even for a particular tree, i.e., a "star" with exactly one supply vertex, because the "maximum subset sum problem" [5] can be easily reduced in linear time to the maximum partition problem for such a tree. Thus it is very unlikely that these problems can be solved in polynomial time in general. However, it has been desired to obtain efficient algorithms for a restricted class of graphs, say trees, which often appear in the power delivery problem. Several works have been done for related problems on trees [1,3,10].

In this paper we give three algorithms for trees. First is a linear-time algorithm for the partition problem. Second is a pseudo-polynomial-time algorithm to solve the maximum partition problem in time $O(F^2 n)$ if all demands and supplies are positive integers and $F = \min\{\sum_{v \in V_d} d(v), \sum_{v \in V_s} s(v)\}$. Thus the algorithm takes polynomial time if F is bounded by a polynomial in n. Third is a fully polynomial-time approximation scheme (FPTAS) for the maximum partition problem.

2   Linear Algorithm for the Partition Problem

The maximum partition problem is NP-hard even for a star with exactly one supply vertex. However, in this section, we have the following theorem on the partition problem.

Theorem 1. *The partition problem can be solved in linear time for trees.*

In the remainder of this section we give an algorithm to solve the partition problem for trees in linear time as a proof of Theorem 1. Let T be a given "free" tree. We choose an arbitrary demand vertex r as the root of T, and regard T as a "rooted" tree. We assume for the sake of simplicity that all supply vertices are leaves of T as illustrated in Fig. 1 (b). The algorithm repeatedly executes the following Steps 1–3 until exactly one vertex remains in T. **(Step 1)** Choose

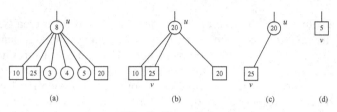

Fig. 2. Steps 1–3.

any internal vertex u of T whose all children are leaves. (See Fig. 2 (a).)
(Step 2) Let D be the set of all children of u that are demand vertices. Let $d(u) := d(u) + \sum_{w \in D} d(w)$, delete all vertices in D from T, and let T be the resulting tree, in which all children of u are supply vertices. (See Figs. 2 (a) and (b). All demand vertices in D and u should be supplied from the same supply vertex.)
(Step 3) If u has a child, then execute the following operations. Find a child v of u whose supply is maximum among all children of u, delete all the other children of u from T, and let T be the resulting tree. (See Figs. 2 (b) and (c). One may assume that u should be supplied from v if u were supplied from a child of u.) If $d(u) \le s(v)$, then delete u from T, join v with the parent of u, let $s(v) := s(v) - d(u)$, and let T be the resulting tree in which v is a leaf. (See Figs. 2 (c) and (d). One may assume that u is supplied from v.) If $d(u) > s(v)$, then delete the supply vertex v from T and let T be the resulting tree in which v is a leaf. (The supply vertex v cannot supply power to any demand vertex.)

If the vertex in T is a supply vertex when the algorithm terminates, then there is a desired partition. Otherwise, there is no desired partition.

One can easily observe that the algorithm correctly solves the partition problem in linear time, and that the algorithm can be easily modified so that it actually finds a partition of a tree if there is.

3   Algorithm for the Maximum Partition Problem

The maximum partition problem is NP-hard even for a star with exactly one supply vertex. However, in this section, we give a pseudo-polynomial-time algorithm to solve the maximum partition problem for trees. The main result of this section is the following theorem.

Theorem 2. *The maximum partition problem can be solved for a tree $T = (V, E)$ in time $O(F^2 n)$ if the demands and supplies are integers and $F = \min\{\sum_{v \in V_d} d(v), \sum_{v \in V_s} s(v)\}$.*

In the remainder of this section we give an algorithm to solve the maximum partition problem in time $O(F^2 n)$ as a proof of Theorem 2. The algorithm takes polynomial time if F is bounded by a polynomial in n.

In this section and the succeeding section, a *partition* P of a graph G is to partition G into n_s or more connected components by deleting edges from G so that

(a) each component contains at most one supply vertex; and
(b) if a component C contains a supply vertex, then the supply is no less than the sum of demands of all demand vertices in C.

The *fulfillment $f(P)$ of a partition* P is the sum of demands of the demand vertices in all components with supply vertices. Thus $f(P)$ corresponds to the maximum sum of all loads that are supplied electrical power from feeders through a network reconfigured by cutting off some edges. The *maximum partition problem* is to find a partition of G with the maximum fulfillment. The *maximum fulfillment $f(G)$ of a graph* G is the maximum fulfillment $f(P)$ among all partitions P of G. Clearly $f(G) \leq F$.

Assume for the sake of simplicity that T is a rooted tree with a demand vertex r as the root, and all supply vertices are leaves of T. Let m_d be the maximum demand, and let m_s be the maximum supply, that is, $m_d = \max\{d(v)|v \in V_d\}$ and $m_s = \max\{s(u)|u \in V_s\}$.

For a vertex v of T, we denote by T_v the maximum subtree of T rooted at v. For a non-negative real number z, we denote by R_z the set of all real numbers $x \leq z$, and denote by R_z^+ the set of all real numbers x, $0 \leq x \leq z$. Our idea is to introduce functions $g_{\text{out}}(T_v, R_F; x)$, $g_{\text{in}}(T_v, R_F; x)$ and $g_0(T_v, R_F; x)$, $x \in R_F$, for each vertex v, and to compute them from leaves to the root of T by dynamic programming.

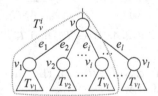

Fig. 3. Subtree T_v^i of T_v.

Let v be a vertex of T, let v_1, v_2, $\cdots$, v_l be the children of v, and let e_i, $1 \leq i \leq l$, be the edge joining v and v_i. T_{v_i}, $1 \leq i \leq l$, is the maximum subtree of T rooted at v_i. We denote by T_v^i the subtree of T which consists of the vertex v, the edges e_1, e_2, $\cdots$, e_i and the subtrees T_{v_1}, T_{v_2}, $\cdots$, T_{v_i}. (See Fig. 3.) Clearly $T_v = T_v^l$. We denote by T_v^0 the subtree of a single vertex v.

Let P be a partition of a rooted subtree T_v of T, and let $C(P)$ be the set of all vertices in the connected component containing the root v of T_v in P. We now define j-out, j-in and isolated partitions as follows.

(a) A partition P of T_v is called a *j-out partition* for $j \in R_F^+$ if $C(P)$ contains a supply vertex w and $s(w) \geq j + \sum_{u \in C(P) - \{w\}} d(u)$. A j-out partition of T_v corresponds to a partition of T in which v is supplied from a supply vertex in T_v. If a virtual demand vertex v_d with $d(v_d) = j$ is joined to the root v of T_v, then v_d and all demand vertices in $C(P)$ can be supplied power from w in the resulting tree T_v^+. (See Fig. 4 (a).) A j-out partition P of T_v induces a partition P^+ of T_v^+ such that $f(P^+) = f(P) + j$.

(b) A partition P of T_v is called a *j-in partition* for $j \in R_F^+$ if $C(P)$ contains no supply vertex and $\sum_{u \in C(P)} d(u) \leq j$. Thus, if P is a j-in partition, then v is a demand vertex and $j \geq d(v) > 0$. A j-in partition of T_v corresponds to a partition of T in which v is supplied from a supply vertex outside T_v. All vertices in $C(P)$ are not supplied power in a j-in partition of T_v, but if a virtual supply vertex v_s with $s(v_s) = j$ is joined to the root v of T_v then all (demand) vertices in $C(P)$ can be supplied from v_s in the resulting tree T_v^*. (See Fig. 4 (b).) For a j-in partition P, let $f^*(P) = f(P) + \sum_{u \in C(P)} d(u)$. Clearly, a j-in partition P of T_v induces a partition P^* of T_v^* such that $f(P^*) = f^*(P)$.

(c) A partition P of T_v is called an *isolated partition* if $C(P)$ consists of a single demand vertex v. An isolated partition of T_v corresponds to a partition P' of T in which v is not supplied from any supply vertex in T. Such a partition P' of T does not always induce an isolated partition P of T_v, but T has a partition P'' such that $f(P'') = f(P')$ and P'' induces an isolated partition P of T_v.

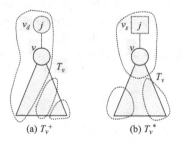

(a) T_v^+ (b) T_v^*

Fig. 4. Trees T_v^+ and T_v^*.

We define the "margin" $g_{\mathrm{out}}(T_v, R_F; x)$, the "deficiency" $g_{\mathrm{in}}(T_v, R_F; x)$, and $g_0(T_v, R_F; x)$, $x \in R_F$, as follows:

$$g_{\mathrm{out}}(T_v, R_F; x) = \max \ \{j \in R_F^+ \mid T_v \text{ has a } j\text{-out partition } P$$
$$\text{such that } f(P) \geq x \ \}; \quad (1)$$

$$g_{\mathrm{in}}(T_v, R_F; x) = \min \ \{j \in R_F^+ \mid T_v \text{ has a } j\text{-in partition } P$$
$$\text{such that } f^*(P) \geq x\}; \quad (2)$$

and

$$g_0(T_v, R_F; x) = \begin{cases} 0 & \text{if } T_v \text{ has an isolated partition } P \text{ such that } f(P) \geq x; \\ +\infty & \text{otherwise.} \end{cases}$$
(3)

If T_v has no j-out partition P with $f(P) \geq x$ for any $j \in R_F^+$, then let $g_{\text{out}}(T_v, R_F; x) = -\infty$. If T_v has no j-in partition P with $f^*(P) \geq x$ for any $j \in R_F^+$, then let $g_{\text{in}}(T_v, R_F; x) = +\infty$. Clearly, for any negative real number $x < 0$, $g_{\text{out}}(T_v, R_F; x) = g_{\text{out}}(T_v, R_F; 0)$, $g_{\text{in}}(T_v, R_F; x) = g_{\text{in}}(T_v, R_F; 0)$, and $g_0(T_v, R_F; x) = g_0(T_v, R_F; 0)$.

Let $f_{\text{out}}(T_v, R_F)$ be the maximum fulfillment $f(P)$ taken over all j-out partitions P of T_v. Thus

$$f_{\text{out}}(T_v, R_F) = \max\{x \in R_F \mid g_{\text{out}}(T_v, R_F; x) \neq -\infty\},$$
(4)

and if there does not exist any $x \in R_F$ such that $g_{\text{out}}(T_v, R_F; x) \neq -\infty$, that is, T_v has no j-out partition P with $f(P) \geq x$ for any $j \in R_F^+$, then let $f_{\text{out}}(T_v, R_F) = -\infty$. On the other hand, let $f_0(T_v, R_F)$ be the maximum fulfillment $f(P)$ taken over all isolated partitions P of T_v. Thus

$$f_0(T_v, R_F) = \max\{x \in R_F \mid g_0(T_v, R_F; x) = 0\},$$
(5)

and if there does not exist any $x \in R_F$ such that $g_0(T_v, R_F; x) = 0$, that is, v is a leaf and is a supply vertex, then let $f_0(T_v, R_F) = -\infty$. Note that

$$f(T_v) = \max\{f_{\text{out}}(T_v, R_F), f_0(T_v, R_F)\},$$
(6)

and that $f_0(T_v, R_F) = f(T_v - \{v\}) = \sum_{i=1}^{l} f(T_{v_i})$ if v is an inner vertex.

Our algorithm computes $g_{\text{out}}(T_v, R_F; x)$, $g_{\text{in}}(T_v, R_F; x)$ and $g_0(T_v, R_F; x)$ for each vertex v of T from leaves to the root r of T by means of dynamic programming. Since $T = T_r$ for the root r of T, one can compute the fulfillment $f(T)$ of T from $g_{\text{out}}(T, R_F; x)$ and $g_0(T, R_F; x)$ by Eqs. (4)–(6).

We first compute $g_{\text{out}}(T_v^0, R_F; x)$, $g_{\text{in}}(T_v^0, R_F; x)$ and $g_0(T_v^0, R_F; x)$ for each vertex v of T as follows. Since T_v^0 consists of a single vertex v, $T_v = T_v^0$ if v is a leaf. If v is a demand vertex, then

$$g_{\text{out}}(T_v^0, R_F; x) = -\infty$$
(7)

for any $x \in R_F$, and

$$g_{\text{in}}(T_v^0, R_F; x) = \begin{cases} d(v) & \text{if } x \leq d(v); \\ +\infty & \text{if } d(v) < x \leq F, \end{cases}$$
(8)

and

$$g_0(T_v^0, R_F; x) = \begin{cases} 0 & \text{if } x \leq 0; \\ +\infty & \text{if } 0 < x \leq F. \end{cases}$$
(9)

If v is a supply vertex, then

$$g_{\text{out}}(T_v^0, R_F; x) = \begin{cases} s(v) & \text{if } x \leq 0; \\ -\infty & \text{if } 0 < x \leq F, \end{cases}$$
(10)

and

$$g_{\text{in}}(T_v^0, R_F; x) = +\infty \tag{11}$$

and

$$g_0(T_v^0, R_F; x) = +\infty \tag{12}$$

for any $x \in R_F$.

We next compute $g_{\text{out}}(T_v^i, R_F; x)$, $g_{\text{in}}(T_v^i, R_F; x)$ and $g_0(T_v^i, R_F; x)$, $1 \le i \le l$, for each internal vertex v of T from the counterparts for T_v^{i-1} and T_{v_i}, where l is the number of the children of v. Note that $T_v = T_v^l$.

We first explain how to compute $g_{\text{out}}(T_v^i, R_F; x)$. There are the following four possibilities (a)–(d), and we define $g_{\text{out}}^a(T_v^i, R_F; x, y)$, $g_{\text{out}}^b(T_v^i, R_F; x, y)$, $g_{\text{out}}^c(T_v^i, R_F; x, y)$ and $g_{\text{out}}^d(T_v^i, R_F; x, y)$ for (a), (b), (c) and (d), respectively. (In Fig. 5 an arrow represents the direction of power supply, a dotted line represents a deleted edge, and a shaded circle represents a demand vertex supplied power.)

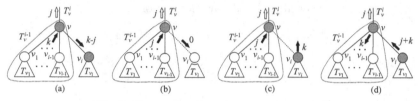

Fig. 5. Power flow in a j-out partition of T_v^i.

(a) A j-out partition P of T_v^i with $f(P) \ge x$ can be obtained by merging a k-out partition P_1 of T_v^{i-1} and a $(k - j)$-in partition P_2 of T_{v_i} such that $f(P_1) \ge y$ and $f^*(P_2) \ge x - y$ for some $k \ge j$ and $y \in R_x$. (See Fig. 5 (a).) Then let

$$g_{\text{out}}^a(T_v^i, R_F; x, y) = g_{\text{out}}(T_v^{i-1}, R_F; y) - g_{\text{in}}(T_{v_i}, R_F; x - y). \tag{13}$$

(b) A j-out partition P of T_v^i with $f(P) \ge x$ can be obtained by merging a j-out partition P_1 of T_v^{i-1} and an isolated partition P_2 of T_{v_i} such that $f(P_1) \ge y$ and $f(P_2) \ge x - y$ for some $y \in R_x$. (See Fig. 5 (b).) Then let

$$g_{\text{out}}^b(T_v^i, R_F; x, y) = g_{\text{out}}(T_v^{i-1}, R_F; y) - g_0(T_{v_i}, R_F; x - y). \tag{14}$$

(c) A j-out partition P of T_v^i with $f(P) \ge x$ can be obtained by merging a j-out partition P_1 of T_v^{i-1} and a k-out partition P_2 of T_{v_i} such that $f(P_1) \ge y$ and $f(P_2) \ge x - y$ for some $k \in R_{m_s}^+$ and $y \in R_x$. (See Fig. 5 (c).) Then let

$$g_{\text{out}}^c(T_v^i, R_F; x, y) = \begin{cases} g_{\text{out}}(T_v^{i-1}, R_F; y) & \text{if } g_{\text{out}}(T_{v_i}, R_F; x - y) \ge 0; \\ -\infty & \text{if } g_{\text{out}}(T_{v_i}, R_F; x - y) = -\infty. \end{cases} \tag{15}$$

(d) A j-out partition P of T_v^i with $f(P) \ge x$ can be obtained by merging a k-in partition P_1 of T_v^{i-1} and a $(j + k)$-out partition P_2 of T_{v_i} such that $f^*(P_1) \ge y$ and $f(P_2) \ge x - y$ for some $k \in R_{m_s}^+$ and $y \in R_x$. (See Fig. 5 (d).) Then let

$$g_{\text{out}}^d(T_v^i, R_F; x, y) = g_{\text{out}}(T_{v_i}, R_F; x - y) - g_{\text{in}}(T_v^{i-1}, R_F; y). \tag{16}$$

From g_{out}^a, g_{out}^b, g_{out}^c and g_{out}^d above one can compute $g_{\text{out}}(T_v^i, R_F; x)$ as follows:

$$g_{\text{out}}(T_v^i, R_F; x) = \max \ \{g_{\text{out}}^a(T_v^i, R_F; x, y), \ g_{\text{out}}^b(T_v^i, R_F; x, y),$$
$$g_{\text{out}}^c(T_v^i, R_F; x, y), \ g_{\text{out}}^d(T_v^i, R_F; x, y) \mid y \in R_x\}. \ (17)$$

We next explain how to compute $g_{\text{in}}(T_v^i, R_F; x)$. There are the following three possibilities (a), (b) and (c), and we define $g_{\text{in}}^a(T_v^i, R_F; x, y)$, $g_{\text{in}}^b(T_v^i, R_F; x, y)$ and $g_{\text{in}}^c(T_v^i, R_F; x, y)$ for (a), (b) and (c), respectively.

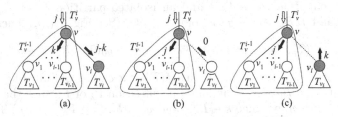

Fig. 6. Power flow in a j-in partition of T_v^i.

(a) A j-in partition P of T_v^i with $f^*(P) \geq x$ can be obtained by merging a k-in partition P_1 of T_v^{i-1} and a $(j-k)$-in partition P_2 of T_{v_i} such that $f^*(P_1) \geq y$ and $f^*(P_2) \geq x - y$ for some $k \in R_j^+$ and $y \in R_x$. (See Fig. 6 (a).) Then let

$$g_{\text{in}}^a(T_v^i, R_F; x, y) = g_{\text{in}}(T_v^{i-1}, R_F; y) + g_{\text{in}}(T_{v_i}, R_F; x - y). \qquad (18)$$

(b) A j-in partition P of T_v^i with $f^*(P) \geq x$ can be obtained by merging a j-in partition P_1 of T_v^{i-1} and an isolated partition P_2 of T_{v_i} such that $f^*(P_1) \geq y$ and $f(P_2) \geq x - y$ for some $y \in R_x$. (See Fig. 6 (b).) Then let

$$g_{\text{in}}^b(T_v^i, R_F; x, y) = g_{\text{in}}(T_v^{i-1}, R_F; y) + g_0(T_{v_i}, R_F; x - y). \qquad (19)$$

(c) A j-in partition P of T_v^i with $f^*(P) \geq x$ can be obtained by merging a j-in partition P_1 of T_v^{i-1} and a k-out partition P_2 of T_{v_i} such that $f^*(P_1) \geq y$ and $f(P_2) \geq x - y$ for some $k \in R_{m_s}^+$ and $y \in R_x$. (See Fig. 6 (c).) Then let

$$g_{\text{in}}^c(T_v^i, R_F; x, y) = \begin{cases} g_{\text{in}}(T_v^{i-1}, R_F; y) & \text{if } g_{\text{out}}(T_{v_i}, R_F; x - y) \geq 0; \\ +\infty & \text{if } g_{\text{out}}(T_{v_i}, R_F; x - y) = -\infty. \end{cases} \qquad (20)$$

From g_{in}^a, g_{in}^b and g_{in}^c above one can compute $g_{\text{in}}(T_v^i, R_F; x)$ as follows:

$$g_{\text{in}}(T_v^i, R_F; x) = \min \ \{g_{\text{in}}^a(T_v^i, R_F; x, y), \ g_{\text{in}}^b(T_v^i, R_F; x, y),$$
$$g_{\text{in}}^c(T_v^i, R_F; x, y) \mid y \in R_x\}. \qquad (21)$$

We finally explain how to compute $g_0(T_v^i, R_F; x)$. There are the following two possibilities (a) and (b), and we define $g_0^a(T_v^i, R_F; x, y)$ and $g_0^b(T_v^i, R_F; x, y)$ for (a) and (b), respectively.

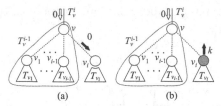

Fig. 7. Power flow in an isolated partition of T_v^i.

(a) An isolated partition P of T_v^i with $f(P) \geq x$ can be obtained by merging an isolated partition P_1 of T_v^{i-1} and an isolated partition P_2 of T_{v_i} such that $f(P_1) \geq y$ and $f(P_2) \geq x - y$ for some $y \in R_x$. (See Fig. 7 (a).) Then let

$$g_0^a(T_v^i, R_F; x, y) = g_0(T_v^{i-1}, R_F; y) + g_0(T_{v_i}, R_F; x - y). \qquad (22)$$

(b) An isolated partition P of T_v^i with $f(P) \geq x$ can be obtained by merging an isolated partition P_1 of T_v^{i-1} and a k-out partition P_2 of T_{v_i} such that $f(P_1) \geq y$ and $f(P_2) \geq x - y$ for some $k \in R_{m_s}^+$ and $y \in R_x$. (See Fig. 7 (b).) Then let

$$g_0^b(T_v^i, R_F; x, y) = \begin{cases} g_0(T_v^{i-1}, R_F; y) & \text{if } g_{\text{out}}(T_{v_i}, R_F; x - y) \geq 0; \\ +\infty & \text{if } g_{\text{out}}(T_{v_i}, R_F; x - y) = -\infty. \end{cases} \qquad (23)$$

From g_0^a and g_0^b above one can compute $g_0(T_v^i, R_F; x)$ as follows:

$$g_0(T_v^i, R_F; x) = \min\{g_0^a(T_v^i, R_F; x, y), \; g_0^b(T_v^i, R_F; x, y) \mid y \in R_x\}. \qquad (24)$$

Suppose now that the supplies and demands of all vertices in T are integers. Then $f(P)$ and $f^*(P)$ are integer for any partition of T_v. For $z \in R_F$ we denote by I_z^+ the set of all integers m, $0 \leq m \leq z$. Define $g_{\text{out}}(T_v, I_F^+; x)$, $g_{\text{in}}(T_v, I_F^+; x)$ and $g_0(T_v, I_F^+; x)$ for integers $x \in I_F^+$ similarly as those for R_F in Eqs. (1)–(3), then they are integers. Furthermore $g_{\text{out}}(T_v, R_F; x) = g_{\text{out}}(T_v, I_F^+; \lceil x \rceil)$, $g_{\text{in}}(T_v, R_F; x) = g_{\text{in}}(T_v, I_F^+; \lceil x \rceil)$ and $g_0(T_v, R_F; x) = g_0(T_v, I_F; \lceil x \rceil)$ for any real number $x \in R_F$. Define integral values $f_{\text{out}}(T_v, I_F^+)$ and $f_0(T_v, I_F^+)$ similarly as those for R_F in Eqs. (4) and (5). Then $f_{\text{out}}(T_v, I_F^+) = f_{\text{out}}(T_v, R_F)$ and $f_0(T_v, I_F^+) = f_0(T_v, R_F)$, and hence

$$f(T_v) = \max\{f_{\text{out}}(T_v, I_F^+), f_0(T_v, I_F^+)\}. \qquad (25)$$

Therefore, it suffices to compute values $g_{\text{out}}(T_v, I_F^+; x)$, $g_{\text{in}}(T_v, I_F^+; x)$ and $g_0(T_v, I_F^+; x)$ only for integers $x \in I_F^+$.

One can compute $g_{\text{out}}(T_v^0, I_F^+; x)$, $g_{\text{in}}(T_v^0, I_F^+; x)$ and $g_0(T_v^0, I_F^+; x)$ for a vertex v of T and all integers $x \in I_F^+$ in time $O(F)$ by the counterparts of Eqs. (7)–(12). One can recursively compute $g_{\text{out}}(T_v^i, I_F^+; x)$, $g_{\text{in}}(T_v^i, I_F^+; x)$ and $g_0(T_v^i, I_F^+; x)$ of an internal vertex v of T in time $O(F^2)$ by the counterparts of Eqs. (13)–(24), and from them for $i = l$ one can compute $f_{\text{out}}(T_v, I_F^+)$, $f_0(T_v, I_F^+)$ and $f(T_v)$ by Eq. (25) and the counterparts of Eqs. (4) and (5) in time $O(F)$. Since $f(T) = f(T_r)$ for the root r of T, the number of vertices in T is n and the number of edges is $n-1$, one can compute $f(T)$ in time $O(F^2 n)$. This completes a proof of Theorem 2.

4 Approximation Scheme for the Maximum Partition Problem

The main result of this section is the following theorem.

Theorem 3. *There is a fully polynomial-time approximation scheme for the maximum partition problem on trees.*

In the remainder of this section, we give an algorithm to find a partition P of a tree T with $f(P) \geq (1 - \varepsilon)f(T)$ in time $O\left(n^5/\varepsilon^2\right)$ for any ε, $0 < \varepsilon < 1$, as a proof of Theorem 3. Assume in this section that the supplies and demands of all vertices in a tree T are positive real numbers, and assume for the sake of simplicity that all supply vertices are leaves of T. The algorithm is similar to the algorithm in Section 3. For a positive real number t, we define a set R_F^t of real numbers as $R_F^t = \{\cdots, -2t, -t, 0, t, 2t, \cdots, \lfloor F/t \rfloor\, t\}$, and define a set R_F^{t+} of non-negative real numbers as $R_F^{t+} = \{0, t, 2t, \cdots, \lfloor F/t \rfloor\, t\}$. In this section we
(i) recursively define and compute the "digitized" functions $\bar{g}_{\mathrm{out}}(T_v, R_F^t; x)$, $\bar{g}_{\mathrm{in}}(T_v^i, R_F^t; x)$ and $\bar{g}_0(T_v^i, R_F^t; x)$, $x \in R_F^t$, by the counterparts of Eqs. (7)–(24);
(ii) define and compute values $\bar{f}_{\mathrm{out}}(T_v, R_F^t)$ and $\bar{f}_0(T_v, R_F^t)$ by the counterparts of Eqs. (4) and (5);
(iii) define and compute

$$\bar{f}(T_v) = \max\{\bar{f}_{\mathrm{out}}(T_v, R_F^t), \bar{f}_0(T_v, R_F^t)\}; \text{and} \qquad (26)$$

(iv) define and compute $\bar{f}(T) = \bar{f}(T_r)$ where r is the root of T.
One can prove that T has a partition P such that $f(P) \geq \bar{f}(T)$. Therefore $\bar{f}(T)$ is an approximate solution of the maximum partition problem for T.

The "digitized" functions $\bar{g}_{\mathrm{out}}$, $\bar{g}_{\mathrm{in}}$ and $\bar{g}_0$ approximate the original functions g_{out}, g_{in} and g_0 as follows. Note that $\bar{g}_{\mathrm{out}}(T_v, R_F^t; x) = \bar{g}_{\mathrm{out}}(T_v, R_F^t; 0)$, $\bar{g}_{\mathrm{in}}(T_v, R_F^t; x) = \bar{g}_{\mathrm{in}}(T_v, R_F^t; 0)$ and $\bar{g}_0(T_v, R_F^t; x) = \bar{g}_0(T_v, R_F^t; 0)$ for any negative number $x \in R_F^t$.

Lemma 1. *Let $s(T_v^i)$ be the size of T_v^i, that is, the number of vertices and edges in T_v^i. Then the following (a), (b) and (c) hold:*

(a) (i) *$\bar{g}_{\mathrm{out}}(T_v^i, R_F^t; x) \leq g_{\mathrm{out}}(T_v^i, R_F; x)$ for any $x \in R_F^t$;*
   (ii) *$\bar{g}_{\mathrm{out}}(T_v^i, R_F^t; x)$ is non-increasing; and*
   (iii) *for any $x \in R_F$, there is an integer α such that $0 \leq \alpha \leq s(T_v^i) - 1$ and $\bar{g}_{\mathrm{out}}(T_v^i, R_F^t; \lfloor x/t \rfloor\, t - \alpha t) \geq g_{\mathrm{out}}(T_v^i, R_F; x)$,*
(b) (i) *$\bar{g}_{\mathrm{in}}(T_v^i, R_F^t; x) \geq g_{\mathrm{in}}(T_v^i, R_F; x)$ for any $x \in R_F^t$;*
   (ii) *$\bar{g}_{\mathrm{in}}(T_v^i, R_F^t; x)$ is non-decreasing; and*
   (iii) *for any $x \in R_F$, there is an integer β such that $0 \leq \beta \leq s(T_v^i)$ and $\bar{g}_{\mathrm{in}}(T_v^i, R_F^t; \lceil x/t \rceil\, t - \beta t) \leq g_{\mathrm{in}}(T_v^i, R_F; x)$,*
and
(c) (i) *$\bar{g}_0(T_v^i, R_F^t; x) \geq g_0(T_v^i, R_F; x)$ for any $x \in R_F^t$;*
   (ii) *$\bar{g}_0(T_v^i, R_F^t; x)$ is non-decreasing; and*
   (iii) *for any $x \in R_F$, there is an integer γ such that $0 \leq \gamma \leq s(T_v^i)$ and $\bar{g}_0(T_v^i, R_F^t; \lceil x/t \rceil\, t - \gamma t) \leq g_0(T_v^i, R_F; x)$,*

Proof. Omitted in this extended abstract. $\qquad \square$

By Lemma 1 (a), there is an integer α such that $0 \leq \alpha \leq s(T_v^i) - 1$ and $\bar{g}_{\text{out}}(T_v^i, R_F^t; \lfloor x/t \rfloor t - \alpha t) \geq g_{\text{out}}(T_v^i, R_F; x)$. Clearly $\lfloor x/t \rfloor t - \alpha t + s(T_v^i)t \geq x$. Therefore by Eq. (4) and its counterpart we have

$$\bar{f}_{\text{out}}(T_v^i, R_F^t) + s(T_v^i)t \geq f_{\text{out}}(T_v^i, R_F). \tag{27}$$

Similarly, by Lemma 1 (c), there is an integer γ such that $0 \leq \gamma \leq s(T_v^i)$ and $\bar{g}_0(T_v^i, R_F^t; \lceil x/t \rceil t - \gamma t) \leq g_0(T_v^i, R_F; x)$. Clearly $\lceil x/t \rceil t - \gamma t + s(T_v^i)t \geq x$. Therefore by Eq. (5) and its counterpart we have

$$\bar{f}_0(T_v^i, R_F^t) + s(T_v^i)t \geq f_0(T_v^i, R_F). \tag{28}$$

Thus by Eqs. (6) and (26)–(28) we have $\bar{f}(T_v^i) + s(T_v^i)t \geq f(T_v^i)$. Since $T = T_r$ for the root r and $s(T) < 2n$, we have

$$\bar{f}(T) + 2nt \geq f(T). \tag{29}$$

One may assume that T has a path Q going from the demand vertex v with the maximum demand $d(v) = m_d$ to some supply vertex u such that Q passes through only demand vertices except u and $s(u)$ is no less than the sum of demands on Q, because otherwise u cannot be supplied from any supply vertex and hence it suffices to solve the problem for a forest obtained from T by deleting v. Thus one may assume that

$$f(T) \geq m_d. \tag{30}$$

Let
$$t = \frac{\varepsilon m_d}{2n}. \tag{31}$$

Then by Eqs. (29)–(31) we have

$$f(T) \leq \bar{f}(T) + 2n\frac{\varepsilon m_d}{2n} \leq \bar{f}(T) + \varepsilon f(T),$$

and hence $(1 - \varepsilon)f(T) \leq \bar{f}(T)$.

One can observe that the algorithm takes time $O\left(\left|R_F^{t+}\right|^2 n\right) = O\left(n^5/\varepsilon^2\right)$, because $|R_F^{t+}| = \lfloor F/t \rfloor + 1$, $F \leq nm_d$ and hence by Eq. (31) $F/t \leq 2n^2/\varepsilon$. This completes a proof of Theorem 3.

5   Conclusions

In this paper we obtained three algorithms for trees. First is a linear-time algorithm for the partition problem. Second is a pseudo-polynomial-time algorithm to solve the maximum partition problem in time $O(F^2n)$. The algorithm takes polynomial time if F is bounded by a polynomial in n. Third is a fully polynomial-time approximation scheme (FPTAS) for the maximum partition problem. It is easy to modify all algorithms so that they actually find a partition of a tree.

One can solve the maximum partition problem of a given tree in time $O(m_s^2 n)$ by using the "inverse" functions of g_{out} and g_{in} if the demands and supplies of all vertices in a tree T are positive integers. One can similarly obtain FPTAS for the "minimum partition problem" to find a partition of a tree that minimizes the sum of demands in all components having no supply vertices. We finally remark that both the partition problem and the maximum partition problem can be solved for series-parallel graphs similarly as for trees.

Acknowledgments. We thank M. Kabakura, T. Tanaka and H. Saito for fruitful discussions with them.

References

1. I. Bárány, J. Edmonds and L. A. Wolsey. Packing and covering a tree by subtrees. *Combinatorica*, Vol. 6, No. 3, pp. 221–233, 1986.
2. N. G. Boulaxis and M. P. Papadopoulos. Optimal feeder routing in distribution system planning using dynamic programming technique and GIS facilities. *IEEE Trans. on Power Delivery*, Vol. 17, No. 1, pp. 242–247, 2002.
3. M. E. Dyer and A. M. Frieze. On the complexity of partitioning graphs into connected subgraphs. *Discrete Applied Math.*, Vol. 10, pp. 139–153, 1985.
4. E. Györi. On division of connected subgraphs. in *Proc. 5th Hungarian Combinatorial Coll.*, pp. 485–494, 1978.
5. M. R. Garey and D. S. Johnson. Computers and Intractability: A Guide to the Theory of NP-Completeness. *Freeman, San Francisco, CA*, 1979.
6. T. Lengauer. Combinatorial Algorithms for Integrated Circuit Layout. *John Wiley and Sons*, Chichester, 1990.
7. A. B. Morton and I. M. Y. Mareels. An efficient brute-force solution to the network reconfiguration problem. *IEEE Trans. on Power Delivery*, Vol. 15, No. 3, pp. 996–1000, 2000.
8. S. Nakano, M. S. Rahman and T. Nishizeki. A linear-time algorithm for four-partitioning four-connected planar graphs. *Inf. Proc. Let.*, Vol. 62, pp. 315–322, 1997.
9. D. Pisinger. An exact algorithm for large multiple knapsack problems. *European Journal of Operation Research*, Vol. 114, pp. 528–541, 1999.
10. Y. Perl and S. R. Schach. Max-min tree partitioning. *J. Asso. Comput. Mach.*, Vol. 28, No. 1, pp. 5–15, 1981.
11. J-H. Teng and C-N. Lu. Feeder-switch relocation for customer interruption cost minimization. *IEEE Trans. on Power Delivery*, Vol. 17, No. 1, pp. 254–259, 2002.

Maximizing a Voronoi Region: The Convex Case

Frank Dehne[1], Rolf Klein[2], and Raimund Seidel[3]

[1] Carleton University, Ottawa. frank@dehne.net
[2] Institut für Informatik I, Universität Bonn. rolf.klein@uni-bonn.de
[3] Universität des Saarlandes, Saarbrücken. rseidel@stone.cs.uni-sb.de

Abstract. Given a set S of s points in the plane, where do we place a new point, p, in order to maximize the area of its region in the Voronoi diagram of S and p? We study the case where the Voronoi neighbors of p are in convex position, and prove that there is at most one local maximum.

Keywords: Computational geometry, locational planning, optimization, Voronoi diagram.

1 Introduction

Suppose that we want to place a new supermarket where it wins over as many customers as possible from the competitors that already exist.

Let us assume that customers are equally distributed and that each customer shops at the market closest to her residence. Our task then amounts to finding a location, p, for the new market amidst the locations p_i of the existing markets, such that the Voronoi region of p, that is, the set of all points in the plane that are closer to p than to any p_i, has a maximum area.

Not much seems to be known about this problem. The area of Voronoi regions has been addressed in the context of games, where players can in turn move their existing sites, or insert new sites, such as to end up with a large total area of their Voronoi regions; see the Hotelling game described in Okabe et al. [6], or recent work by Cheong et al. [3] and Ahn et al. [1]. But none of these papers gives an explicit method for maximizing the region of a new site.

In this paper we take the first, nontrivial step towards a solution of the area maximization problem. Let the Voronoi region of the new point, p, in the Voronoi diagram $V(S \cup \{p\})$ consist of parts of the former regions of certain sites $p_1, \ldots, p_n$ in $V(S)$; these sites form the set N of Voronoi neighbors of p in $V(S \cup \{p\})$. In general, this set N spans a polygon that is star-shaped as seen from p.[1] We show that if the set N is in *convex* position then there can be at most one local maximum for the Voronoi area of p, in the interior of the locus of all positions that have N as their neighbor set. The proof is based on a delicate analysis of certain rational functions; it will be given in Section 3.

---

[1] A set P is called *star-shaped* as seen from one of its points, p, if any line segment connecting p to a point in P is fully contained in P.

P. Bose and P. Morin (Eds.): ISAAC 2002, LNCS 2518, pp. 624–634, 2002.

In Section 4 we describe an overall algorithm for determining the location of p that attains a maximum Voronoi area. Finally, we discuss some directions for future work in Section 5. Section 2 contains some preliminaries, among them a tractable formula for the area of a Voronoi region.

For general properties of Voronoi diagrams see the monograph by Okabe et al. [6] or the surveys by Fortune [4] and Aurenhammer and Klein [2].

2   Preliminaries

First, we restate some basic definitions and facts. Let S be a set of s point sites in the plane that are in general position, that is, no four of them are co-circular, no three of them co-linear. By $V(S)$ we denote the Voronoi diagram of the set S. It consists of Voronoi regions $VR(q, S)$, one to each point q of S, containing all points in the plane that are closer to q than to any other site in S. The planar dual of $V(S)$ is the Delaunay triangulation, $DT(S)$, of S. It consists of all triangles with vertices in S whose circum- (or: Delaunay) circle does not contain a site of S in its interior. Both, $V(S)$ and $DT(S)$, are of complexity $O(s)$ and can be constructed in optimal time $O(s \log s)$.

If we add a new point site, p, to S, it will be connected to a site $q \in S$ by an edge of $DT(S \cup p)$ if, and only if, there exists a Delaunay triangle with vertex q in $DT(S)$ whose circumcircle contains p. The set N of such Voronoi or Delaunay neighbors q of p forms a polygon, $P(N)$, that is star-shaped as seen from p. The locus of all placements of p that have N as their neighbor set is denoted by C_N. Its shape will be discussed in Section 4.

In this section we derive some useful formulae for the area of the Voronoi region of a new site p with neighbor set N, assuming that $P(N)$ is convex. It is based on computing the *signed areas* of certain triangles. Let (v_0, v_1, v_2) be the vertices of some triangle, D, where $v_i = (a_i, b_i)$ in Cartesian coordinates. Then,

$$\text{SignedArea}(D) := \frac{1}{2} \sum_{i=0}^{2} (a_i b_{i+1} - a_{i+1} b_i)$$

gives the positive area of D if (v_0, v_1, v_2) appear in counterclockwise order on the boundary of D; otherwise, we obtain the negative value. Here, indices are counted mod 3.

Now let p_i, p_{i+1} be two consecutive vertices on the boundary of $P(N)$, in counterclockwise order. Unless p is co-linear with p_i and p_{i+1}, these three point sites define a Voronoi vertex v_i that may or may not be contained in $P(N)$; see Figure 1.

Let D_i denote the triangle (p_i, v_i, p_{i+1}); its *signed area* is positive if and only if these vertices appear on D_i in counterclockwise order, that is, if and only if v_i lies outside the convex polygon $P(N)$.

Lemma 1. *With the notations from above we have the following identity.*

$$Area(VR(p, S \cup \{p\})) = \frac{1}{2}\left((Area(P(N)) + \sum_{i=1}^{n} SignedArea(D_i)\right)$$

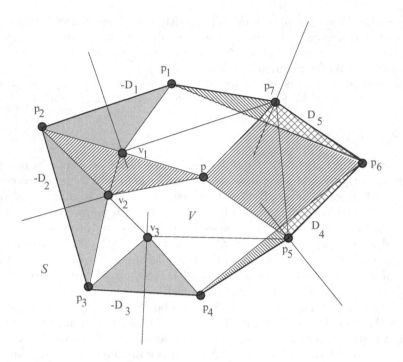

Fig. 1. Decomposing the area of the Voronoi region of site p.

Proof. The area of $\mathrm{VR}(p, S \cup \{p\})$ equals the sum of the areas of the triangles (v_{i+1}, p, v_i); each of them is the reflected image of the triangle (v_i, p_{i+1}, v_{i+1}). The union of all these triangles equals $P(N)$ minus those triangles D_j that are contained in $P(N)$, plus those D_i not contained in $P(N)$; see Figure 1.

Lemma 1 reduces the problem of maximizing the area of the Voronoi region of p to maximing the sum of the signed areas of the triangles D_i, assuming N is fixed. Thus, two vertices of D_i are the given points p_i, p_{i+1}, while only the third, v_i, can move, and its movement is constrained to the bisector of p_i, p_{i+1}, depending on the placement of p.

Next, we express the signed area of D_i as a function of p. To this end, let $p_i = (s_i, t_i)$, and let $m_i = (\frac{s_i + s_{i+1}}{2}, \frac{t_i + t_{i+1}}{2})$ be the midpoint of $p_i p_{i+1}$. We put $b_i = |p_i m_i|$ and $l_i = |p m_i|$. Finally, let α_i be the angle at p in the triangle $F_i = (p_i, p_{i+1}, p)$; see Figure 2 for an illustration.

Lemma 2. *Let $p = (X, Y)$ be the new point site. Then the following identities hold.*

$$SignedArea(D_i) = b_i^2 \frac{l_i^2 - b_i^2}{2\,SignedArea(F_i)} \qquad (1)$$

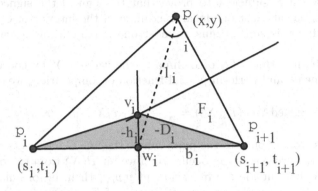

Fig. 2. Computing the signed area of the triangle D_i. In this case, the sign is negative.

$$= b_i^2 \frac{(X - \frac{s_i + s_{i+1}}{2})^2 + (Y - \frac{t_i + t_{i+1}}{2})^2 - b_i^2}{X(t_i - t_{i+1}) + Y(s_{i+1} - s_i) + s_i t_{i+1} - s_{i+1} t_i} \qquad (2)$$

Proof. Let h_i denote the signed height of the triangle $D_i = (p_i, v_i, p_{i+1})$, so that $\mathrm{SignedArea}(D_i) = b_i h_i$ holds. The Voronoi vertex v_i can be expressed as a vector sum by

$$\mathbf{v_i} = \mathbf{m_i} - h_i \mathbf{e_i},$$

where $\mathbf{e_i} = \frac{1}{2b_i}(t_i - t_{i+1}, s_{i+1} - s_i)$ denotes the unit vector along the bisector of p_i, p_{i+1}. On the other hand, $p = (X, Y)$ lies on a circle of radius $\sqrt{h_i^2 + b_i^2}$ centered at v_i. Plugging the cartesian coordinates of v_i into the equation of this circle, and solving for h_i, leads to formula (2), since the coefficient of h_i reduces to zero. The numerators and denominators in formulae (1) and (2) are identical. We observe that the denominator, that is, the sign of the area of F_i, is positive as long as p stays inside the polygon $P(N)$. It becomes 0 when p hits the line through p_i and p_{i+1}. The numerator of formula (2) is the equation of the circumcircle of the line segment $p_i p_{i+1}$. Thus, if $p \in \{p_i, p_{i+1}\}$ holds then the denominator's zero cancels out, and the area of D_i is zero because the Voronoi vertex v_i equals m_i.

3   Uniqueness of the Local Maximum

In this section we assume that N, the set of Voronoi neighbors of the new site, p, consists of n points in convex position. Now we state our main result.

Theorem 1. *For a convex set, N, of n points, there is at most one interior position in $P(N)$, intersected with the locus C_N of all locations with neighbor set N, where the area of the Voronoi region of p has a local maximum.*

Proof. By Lemma 1 it is sufficient to prove that the sum of the signed areas of the triangles D_i has at most one local maximum in the interior of C_N. It is enough to show that this sum attains at most one maximum along each line through $P(N)$.

If we substitute, in formula (2) of Lemma 2, the variable Y by coordinates $eX + f$ of some line G, and perform partial fraction decomposition, we obtain

$$-\mathrm{SignedArea}(D_i(X)) = \frac{A_i}{X - a_i} + c_i X + d_i.$$

The pole at $X = a_i$ corresponds to the point where the line G intersects the line G_i through p_i, p_{i+1}. Three cases can occur when we hit $P(N)$ from the outside. If the point $G \cap G_i$ lies outside the line segment $p_i p_{i+1}$ then, in formula (1) of Lemma 2, we have $l_i > b_i$, while the sign of the area of F_i changes from $-$ to $+$. Consequently, the sign of $-D_i(X)$ changes from $-$ to $+$. But if G intersects the interior of $p_i p_{i+1}$ then $l_i < b_i$, so that $-D_i(X)$ changes from $+$ to $-$. Finally, if G happens to run through one of p_i, p_{i+1} then there is no pole at a_i, i. e., $A_i = 0$ holds, as we noted at the end of the proof of Lemma 2.

Let us assume that line G equals the X-axis, and let $a_1 \leq a_2 \leq \ldots \leq a_m \leq l < r \leq b_1 \leq \ldots \leq b_k$ denote the n poles that correspond to its intersections with the lines G_i. By the convexity of $P(N)$, the two intersections of the X-axis with the boundary of $P(N)$ must be consecutive in this sequence; they are denoted by l and r.

Figure 3 shows the behavior of

$$f(X) := -\sum_{i=1}^{n} \mathrm{SignedArea}(D_i) =$$

$$= \sum_{i=1}^{m} \frac{A_i}{X - a_i} - \frac{L}{X - l} + \frac{R}{X - r} - \sum_{i=1}^{k} \frac{B_i}{X - b_i} + cX + d$$

as a function of X. By the above discussion, we have $A_i, L, R, B_i \geq 0$.

We want to prove that $f(X)$ has at most one local minimum in the interval (l, r). Since f comes from, and returns to, $-\infty$ at l resp. r it is sufficient to show that its second derivative

$$2f''(X) = \sum_{i-=1}^{m} \frac{A_i}{(X - a_i)^3} - \frac{L}{(X - l)^3} + \frac{R}{(X - r)^3} - \sum_{i=1}^{k} \frac{B_i}{(X - b_i)^3}$$

has at most two zeroes in (l, r). We split function $2f''$ into two constituent parts,

$$g(X) := \sum_{i=1}^{m} \frac{A_i}{(X - a_i)^3} - \frac{L}{(X - l)^3} \quad \text{and}$$

$$h(X) := \sum_{i=1}^{k} \frac{B_i}{(X - b_i)^3} - \frac{R}{(X - r)^3},$$

such that $2f'' = g - h$ holds, and discuss g and h independently.

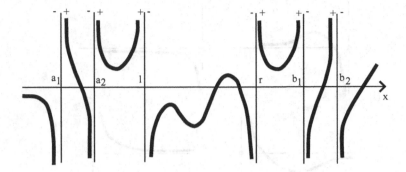

Fig. 3. Discussing the number of minima of $f(X)$ between l and r.

Lemma 3. *Each of the functions g and g'' has at most one zero in (l, ∞), and each of h, h'' has at most one zero in $(-\infty, r)$.*

Proof. Let $x_1 \neq x_0 \in (l, \infty)$ be such that x_0 is a zero of g. Then,

$$0 = g(x_0) = \sum_{i=1}^{m} \frac{A_i}{(x_0 - a_i)^3} - \frac{L}{(x_0 - l)^3} \tag{3}$$

$$= \sum_{i=1}^{m} \frac{A_i}{(x_0 - a_i)^3} \frac{(x_0 - l)^3}{(x_1 - l)^3} - \frac{L}{(x_1 - l)^3} \tag{4}$$

$$= \sum_{i=1}^{m} \frac{A_i}{(x_1 - a_i)^3} \left(\frac{(x_1 - a_i)^3}{(x_0 - a_i)^3} \frac{(x_0 - l)^3}{(x_1 - l)^3} \right) - \frac{L}{(x_1 - l)^3} \tag{5}$$

$$< \sum_{i=1}^{m} \frac{A_i}{(x_1 - a_i)^3} - \frac{L}{(x_1 - l)^3} = g(x_1), \text{ if } x_1 > x_0 \tag{6}$$

$$> g(x_1), \text{ if } x_1 < x_0; \tag{7}$$

observe that formula (4) follows from (3) by multiplying both sides by $\frac{(x_0 - l)^3}{(x_1 - l)^3}$. The alternatives (6) or (7) follow from (5) because $a_i < l < x_0, x_1$ implies that

$$\frac{(x_1 - a_i)^3}{(x_0 - a_i)^3} \frac{(x_0 - l)^3}{(x_1 - l)^3}$$

is of value < 1 if $x_1 > x_0$ holds, and of value > 1, otherwise. Consequently, g has at most one zero in (l, ∞). The other claims are proven analogously.

As a consequence of Lemma 3, the function g has at most one zero and at most one turning point to the right of l. Since g has a negative pole at l and tends to 0 for large values of X, its graph has one of the two possible shapes shown in Figure 4 (i). The possible shapes of the graph of h are shown in (ii).

Our next lemma implies that $2f'' = g - h$ has at most two zeroes in the interval (l, r).

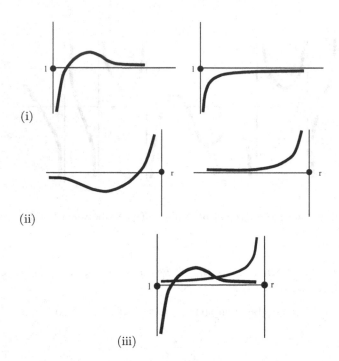

Fig. 4. (i) The possible shapes of the function $g(x)$. (ii) Possible shapes of $h(x)$. (iii) $g(x) = h(x)$ holds for at most two points between l and r.

Lemma 4. *The graphs of the functions g and h have at most two points of intersection over (l, r).*

Proof. If neither g nor h have a zero in (l, r) their graphs do not intersect; see Figure 4. Suppose that h has a zero in (l, r), and assume that p_1 and p_2 are the leftmost points of intersection of the two graphs to the right of l.

We argue that p_2 must be situated to the right of the minimum, m, of h. Indeed, m lies below the X-axis, where g is increasing, and h is decreasing to the left of m, so that only p_1 could lie to the left of m. If p_2 lies to the left of the maximum, M, of function g, or if g does not have a maximum, then the two graphs are separated by their tangents at p_2. If p_2 lies to the right of M then, to the right of p_2, function g is decreasing while function h is increasing. In either case, no third point of intersection can exist.

Now we have shown that the function f takes on at most one local minimum for all points p on L inside the polygon $P(N)$. In the interior of $C_N \subset P(N)$, we have, by Lemma 1,

$$\text{Area}(\text{VR}(p, S \cup \{p\})) = \frac{1}{2}(\text{Area}(P(N)) - f(X)).$$

This completes the proof of Theorem 1.

To give an example, let us assume that n points are evenly placed on the boundary of the unit circle. For $n \leq 4$ there is no local maximum of the Voronoi area. In fact, there is a unique local minimum at the center for $n = 3$; for $n = 4$, the cross formed by the four point sites consists of minimal positions. But for $n \geq 5$ we have a unique local maximum at the center of the circle.

4   Computing the Maximum

In this section we describe a general algorithm for computing, for a new site p, a location of maximum Voronoi area amidst s existing sites. As p is moved over the plane, three events may happen. First, the set N of p's Voronoi neighbors can change.

Let C_N denote a maximal connected region in the plane such that all placements of p inside C_N have N as their set of Voronoi neighbors; we call such a set C_N a *neighborship cell* of N with respect to S. The nature of these cells is quite simple; the proof of the following lemma follows from standard facts on the Delaunay triangulation. Observe that for two neighboring sites, q and r, on the convex hull of S we define, as their Delaunay triangle and circumcircle, the halfplane defined by the line through q, r that does not contain a site of S.

Lemma 5. *Let S be a set of s point sites in the plane.*

1. *The neighborship cells with respect to S are the cells of the arrangement of the Delaunay circles of S. Each cell C has, as its neighbor set, all sites that span a Delaunay circle containing C. The total complexity of all neighborship cells is in $O(s^2)$.*
2. *Let $N \subset S$ be such that $P(N)$ is star-shaped. Then the neighborship cells C_N can be obtained as the intersection of the circumcircles of those Delaunay triangles that are contained in, and share an edge with the boundary of, $P(N)$, minus the union of all Delaunay circles passing through points of $S \setminus N$.*

Lemma 5 is illustrated by Figure 5. Figure 6 shows an example where many neighborship cells are associated with a set, N, of sites.

The arrangement of $O(s)$ many circles can be constructed in time $O(s\lambda_4(s))^2$ by a deterministic algorithm, or in expected time $O(s \log s + k)$, where k denotes the complexity of the arrangement; see Sharir and Agarwal [8].

Another event happens when p hits the boundary of the convex hull of the site set S. At this point, the region of p becomes unbounded. To exclude this phenomenon[3] we assume that a certain feasability domain, F, is given, that consists of neighborship cells contained in the interior of the convex hull of S, and that the placement of p is restricted to F.

Finally, the position of the new site, p, could coincide with one of the existing sites, $p_i \in S$. At these points the area function fails to be continuous; in fact, the

[2] As usual, $\lambda_t(s)$ denotes the maximum length of a Davenport-Schinzel sequence of order t over s characters.

[3] Far out of town there are no customers to win.

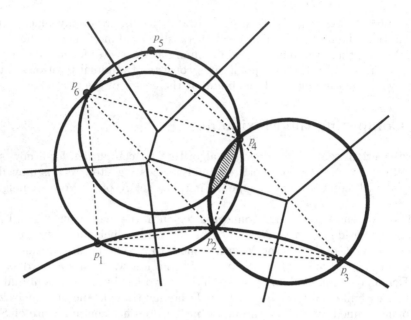

Fig. 5. The shaded area is the neighborship cell of $p_1, \ldots, p_6$. Only the circumcircles of those Delaunay triangles contribute to its boundary that are contained in the star-shaped polygon and share one of its edges.

former region of p_i is split among p and p_i by a bisector through $p = p_i$ whose slope is perpendicular to the direction in which p has approached p_i. But apart from these points, the area function is smooth, as was shown independently by Okabe and Aoyagi [5] and by Piper [7] who generalized work by Sibson [9].

In order to find the optimum placement of p within the whole feasibility domain F, we inspect each cell C of F in turn, and compute the optimal placement of p within the closure of C. Within the interior of C we apply some Newton-based approximation algorithm, which is possible thanks to the smoothness of the area function. If the neighbor set N is convex, we even know that there is at most one local maximum, by Theorem 1, so that following the gradient leads straight to the maximum (or to the boundary of C). Next, we have to check for maxima the boundary of C, which consists of circular arcs, by Lemma 5. This includes checking all placements of p on top of some site p_i; for each of them it takes time proportional to its Delaunay degree to find the optimum slope of the bisector. The solution to our problem is then the maximum of these $O(s^2)$ many cell maxima, together with the corresponding placement of p.

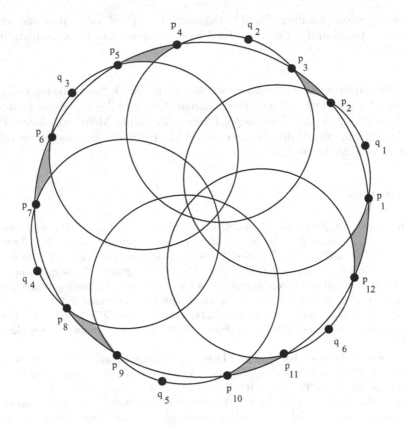

Fig. 6. Each of the shaded cells has $p_1, \ldots, p_{12}$ as neighbor set.

5 Conclusions

In this paper we have shown that the Voronoi area of a new site has at most one local maximum in the interior of each neighborship cell, if the Voronoi neighbors are in convex position. This result gives rise to many further questions.

The obvious open problem is if the maximum is still unique if the neighbors are in star-shaped position. The main difference to the convex case is the following. The line G, along which the new site p was supposed to move in the proof of Theorem 1, can now intersect edge extensions of the neighbor polygon $P(N)$ *inside* $P(N)$, too. Consequently, the functions g and h in the proof of Lemma 3 become more complicated. We expect that considerably more (mathematical) effort will be necessary in order to settle this problem.

Other questions concern the customer model. One could specify bounded populated areas, together with population densities, instead of the uniform distribution, with or without defining a feasibility domain F. Also, it would be interesting to study metrics different from the Euclidean, that are frequently used in location planning. From a theoretical point of view, it would also be interesting to minimize the area of a Voronoi region, and to investigate higher dimensions.

Acknowledgement. The authors would like to thank the following colleagues for fruitful discussions: Franz Aurenhammer, Otfried Cheong, Horst Hamacher, Christian Icking, Elmar Langetepe, Lihong Ma, Kurt Mehlhorn, Belen Palop, Emo Welzl, and Jörg Wills. Also, we would like to thank the anonymous referees for their valuable comments.

References

1. Hee-Kap Ahn, Siu-Wing Cheng, Otfried Cheong, Mordecai Golin, and René van Oostrum. Competitive facility location along a highway. *Proc. 7th Annu. Int. Conf. (COCOON 2001)*, Lecture Notes Comput. Sci.(2108):237–246, 2001.
2. Franz Aurenhammer and Rolf Klein. Voronoi diagrams. In Jörg-Rüdiger Sack and Jorge Urrutia, editors, *Handbook of Computational Geometry*, pages 201–290. Elsevier Science Publishers B.V. North-Holland, Amsterdam, 2000.
3. Otfried Cheong, Sariel Har-Peled, Nathan Linial, and Jiří Matoušek. The one-round Voronoi game. *Proc. 18th Annu. ACM Symp. on Computational Geometry*, 2002.
4. S. Fortune. Voronoi diagrams and Delaunay triangulations. In Jacob E. Goodman and Joseph O'Rourke, editors, *Handbook of Discrete and Computational Geometry*, chapter 20, pages 377–388. CRC Press LLC, Boca Raton, FL, 1997.
5. Atsuyuki Okabe and M. Aoyagi. Existence of equilibrium configurations of competitive firms on an infinite two-dimensional space. *J. of Urban Economics*, 29:349–370, 1991.
6. Atsuyuki Okabe, Barry Boots, Kokichi Sugihara, and Sung Nok Chiu. *Spatial Tessellations: Concepts and Applications of Voronoi Diagrams*. John Wiley & Sons, Chichester, UK, 2000.
7. B. Piper. Properties of local coordinates based on dirichlet tessellations. *Computing Suppl.*, 8:227–239, 1993.
8. Micha Sharir and P. K. Agarwal. *Davenport-Schinzel Sequences and Their Geometric Applications*. Cambridge University Press, New York, 1995.
9. R. Sibson. *A Brief Description of the Natural Neighbor Interpolant*. In: D.V. Barnett (ed.) Interpolation Multiariate Data. Wiley, 1981.

Random Tries

Luc Devroye

School of Computer Science
McGill University
3480 University Street, Suite 318
Montréal, Québec
Canada, H3A 2A7
luc@cs.mcgill.ca

Abstract. We look at the properties of random tries, random Patricia tries, and random level compacted tries and show by means of Talagrand-style concentration inequalities that most parameters, such as the height and the profile of these tries, are stable (i.e., close to their expected value) in a universal manner.

P. Bose and P. Morin (Eds.): ISAAC 2002, LNCS 2518, p. 635, 2002.
© Springer-Verlag Berlin Heidelberg 2002

Expected Acceptance Counts for Finite Automata with Almost Uniform Input

Nic holas Pippenger

Department of Computer Science, University of British Columbia, 2366 Main Mall,
V ancouver, BC V6T 1Z4, Canada

Abstract. If a sequence of independent unbiased random bits is fed into
a finite automaton, it is straightforward to calculate the expected number of acceptances among the first n prefixes of the sequence. This paper
deals with the situation in which the random bits are neither independent
nor unbiased, but are nearly so. We show that, under suitable assumptions concerning the automaton, if the the difference between the entrop y
of the first n bits and n conv erges to a constant exponentially fast, then
the change in the expected number of acceptances also converges to a
constan texponentially fast. We illustrate this result with a variety of
examples in which numbers follo wing the reciprocal distribution, whid
governs the significands of floating-point n umbers, are recoded in the
execution of various multiplication algorithms.

1 Introduction

Consider a finite automaton M with binary input alphabet $\{0,1\}$. It is a straightforward exercise to calculate the number P_n of input strings of length n accepted
b y M. The number of directed paths of length n betw een the initial state and
accepting states is given by a linear recurrence with constant coefficients, and
thus its generating function

$$\Phi(x) = \sum_{n \geq 1} P_n x^n$$

is a rational function (see Chomsky and Miller [C]).

If the input to M is a sequence of n independent unbiased random bits, then
all 2^n input strings are equally likely, the probability of acceptance is $p_n = P_n/2^n$,
and the generating function

$$\varphi(x) = \sum_{n \geq 1} p_n x^n$$

for these probabilities is $\varphi(x) = \Phi(x/2)$.

If M is strongly connected (that is, if each state can be taken to every other
state b y a suitable input string), then with random input it will spend a positie

---

[1] This research was supported by an NSERC Research Grant and a Canada Research
Chair.

P. Bose and P. Morin (Eds.): ISAAC 2002, LNCS 2518, pp. 636–646, 2002.

fraction of its time in accepting states, and the series $\sum_{n\geq 1} p_n$ will divgerge. This implies that the rational function $\varphi(x)$ has a pole at $x = 1$. The automaton that accepts all strings has $1/(1 - x)$ as its probability generating function, so we have $\varphi(x) \leq 1/(1-x)$. This implies that the pole of $\varphi(x)$ at $x = 1$ is simple. If $\rho = \lim_{x\to 1}(1 - x)\,\varphi(x)$ is the residue of $\varphi(x)$ at the pole at $x = 1$, then the expected number

$$e_n = \sum_{1\leq k\leq n} p_k$$

of accepted prefixes of the first n bits satisfies

$$e_n \sim \rho n$$

as $n \to \infty$. If in addition M is aperiodic (that is, if there is an integer k such that each state can be taken to every other state by a suitable input string of length k), then we have the stronger conclusion that $p_n \to \rho$ as $n \to \infty$.

If, for example, M accepts all strings ending with 1, then $\varphi(x) = x/2(1-x)$, $p_n = 1/2$ for all n, and $e_n = n/2$. If, however, M accepts all strings of even length, then $\varphi(x) = x^2/(1-x^2)$, p_n alternates between 0 and 1, and $e_n = \lfloor n/2\rfloor$. Thus p_n does not tend to a limit, but $e_n \sim n/2$.

Our goal in this paper is to study what happens when the distribution of the random input to M is not uniform, but is in some sense close to uniform. Our motivating example is the case in which the input bits $X_1, X_2, \ldots$ are the successive bits in the binary expansion of a real number $X = \sum_{n\geq 1} X_n 2^{-n}$ that is distibuted on the interval $[1/2, 1)$ according to the reciprocal distribution (also kno wn as the logarithmic distribution):

$$\Pr[x \leq X \leq y] = \frac{1}{\log 2}\int_x^y \frac{dX}{X} = \log_2\left(\frac{y}{x}\right).$$

Hamming [H] has argued that the reciprocal distribution is the appropriate one to use for the significand (also known as the mantissa) of a random floating-point number. In particular, Hamming sho ws among other things that the product of a large number of independent indentically distributed numbers (with mild assumptions concerning their distribution) has a significand with the reciprocal distribution. (This phenomenon is related to the logarithmic distribution of leading digits observed by Newcomb [N].)

Consider first feeding the bits of a reciprocally distributed significand into a finite automaton M_1 that accepts all strings ending with 1, so that e_n is simply the expected number of 1s among the first n bits of the significand. A simple calculation shows that

$$p_n = \frac{1}{\log 2}\sum_{2^{n-1}<k<2^n}\int_{k/2^n}^{(k+1)/2^n}\frac{dX}{X}$$

$$= \log_2\left(\frac{2^{n-1}+2}{2^{n-1}+1}\cdots\frac{2^n}{2^n-1}\right),$$

so that

$$e_n = \log_2 \left(\frac{2}{1} \cdots \frac{2^n}{2^n - 1} \right)$$

$$= \frac{1}{2} \log_2 \left(\frac{2}{1} \cdots \frac{2^n}{2^n - 1} \right)^2$$

$$= \frac{n}{2} + \frac{1}{2} \log_2 \left(\frac{2}{1} \cdot \frac{2}{3} \cdots \frac{2^n - 2}{2^n - 1} \cdot \frac{2^n}{2^n - 1} \right).$$

Using Wallis's [W1] formula

$$\frac{\pi}{2} = \frac{2}{1} \cdot \frac{2}{3} \cdots \frac{2^n - 2}{2^n - 1} \cdot \frac{2^n}{2^n - 1} \cdots ,$$

together with the estimate

$$\frac{k}{k-1} \cdot \frac{k}{k+1} = 1 + O\left(\frac{1}{k^2} \right),$$

which yields

$$\frac{k}{k-1} \cdot \frac{k}{k+1} \cdot \frac{k+2}{k+1} \cdot \frac{k+2}{k+3} \cdots = 1 + O\left(\frac{1}{k} \right),$$

we obtain

$$e_n = \frac{n}{2} + \log_2 \left(\frac{\pi}{2} \right) + O\left(\frac{1}{2^n} \right).$$

This result differs from the result $n/2$ for uniform input by a constant, $\frac{1}{2} \log_2 \frac{\pi}{2}$, and an exponentially small error term. (Note that the first bit of a significand is always 1, which raises the expected number of acceptances by $1/2$. Each remaining bit, how ever, is more likely to be 0 than 1, so the limit of the number of extra acceptances, $\frac{1}{2} \log_2 \frac{\pi}{2} = 0.3257 \ldots$, is less than $1/2$.)

Consider next feeding the bits of a reciprocally distributed significand into a finite automaton M_2 that accepts all strings that either (1) consist of the single bit 1, or (2) contain two or more bits, the last two of which are diferent. A simple calculation shows that $p_1 = \log_2 \frac{2}{1}$, $p_2 = \log_2 \frac{3}{2}$ and, for $n \geq 3$,

$$p_n = \log_2 \left(\frac{2^{n-1} + 3}{2^{n-1} + 1} \cdots \frac{2^n - 1}{2^n - 3} \right),$$

so that

$$e_n = \log_2 \left(\frac{3}{1} \cdot \frac{7}{5} \cdots \frac{2^n - 5}{2^n - 7} \cdot \frac{2^n - 1}{2^n - 3} \right)$$

$$= \frac{1}{2} \log_2 \left(\frac{3}{1} \cdot \frac{7}{5} \cdots \frac{2^n - 5}{2^n - 7} \cdot \frac{2^n - 1}{2^n - 3} \right)^2$$

$$= \frac{n}{2} + \frac{1}{2} \log_2 \left(\frac{3}{1} \cdot \frac{3}{5} \cdots \frac{2^n - 1}{2^n - 3} \cdot \frac{2^n - 1}{2^n + 1} \right) + O\left(\frac{1}{2^n} \right) \qquad (1.1)$$

where we have used $\frac{1}{2} \log_2 (2^n + 1) = \frac{n}{2} + O(1/2^n)$. We shall use the formula

$$\prod_{k \geq 1} \frac{(k + \alpha_1) \cdots (k + \alpha_t)}{(k + \beta_1) \cdots (k + \beta_t)} = \frac{\Gamma(1 + \beta_1) \cdots \Gamma(1 + \beta_t)}{\Gamma(1 + \alpha_1) \cdots \Gamma(1 + \alpha_t)}, \tag{1.2}$$

where $\Gamma(\cdots)$ denotes Euler's Gamma-function (see Whittaker and Watson [W2], Section $12 \cdot 13$). (Note that Wallis's formula is the special case of (1.2) in which $t = 2$, $\alpha_1 = \alpha_2 = 0$, $\beta_1 = -1/2$ and $\beta_2 = 1/2$, since

$$\prod_{k \geq 1} \frac{2k}{2k - 1} \cdot \frac{2k}{2k + 1} = \prod_{k \geq 1} \frac{k}{k - \frac{1}{2}} \cdot \frac{k}{k + \frac{1}{2}},$$

and $\Gamma(1) = 1$, $\Gamma(1/2) = \pi^{1/2}$ and $\Gamma(3/2) = \pi^{1/2}/2$.) Applying (1.2) to (1.1), we have

$$
\begin{aligned}
e_n &= \frac{n}{2} + \frac{1}{2} \log_2 \left(\frac{1 - \frac{1}{4}}{1 - \frac{3}{4}} \cdot \frac{1 - \frac{1}{4}}{1 + \frac{1}{4}} \cdots \frac{(2^{n-2} - \frac{1}{4})}{(2^{n-2} - \frac{3}{4})} \cdot \frac{(2^{n-2} - \frac{1}{4})}{(2^{n-2} + \frac{1}{4})} \right) + O\left(\frac{1}{2^n} \right) \\
&= \frac{n}{2} + \frac{1}{2} \log_2 \left(\frac{\Gamma(\frac{1}{4}) \Gamma(\frac{5}{4})}{\Gamma(\frac{3}{4})^2} \right) + O\left(\frac{1}{2^n} \right) \\
&= \frac{n}{2} + \frac{1}{2} \log_2 \left(\frac{\Gamma(\frac{1}{4})^4}{8\pi^2} \right) + O\left(\frac{1}{2^n} \right),
\end{aligned}
$$

where we have used $\Gamma(5/4) = \Gamma(1/4)/4$ and $\Gamma(3/4) = 2^{1/2}\pi/\Gamma(1/4)$ (see Whittaker and Watson [W2], Section $12 \cdot 14$). This result again differs from the result $n/2$ for uniform input by a constant, in this case $\frac{1}{2} \log_2 \left(\Gamma(\frac{1}{4})^4/8\pi^2 \right)$, and an exponentially small error term. (Note that the first bit of a significand is always 1, which raises the expected number of acceptances by $1/2$. The second bit is more likely to be 0 than 1, and thus more likely to be different from than the same as the first bit, which further raises the expected number of acceptances by $\log_2(3/2) - (1/2) = 0.08496\ldots$. Each remaining bit is also more likely to be 0 than 1, and thus more likely to be the same as than different from the previous bit, but by rapidly diminishing amounts, so the limit of the number of extra acceptances, $\frac{1}{2} \log_2 \left(\Gamma(\frac{1}{4})^4/8\pi^2 \right) = 0.5649\ldots$, is greater than $1/2$.)

The acceptances of the automaton M_1 correspond to the 1s in its input, and reflect the additions performed by the standard shift-and-add algorithm for multiplication. The acceptances of M_2 reflect the additions and subtractions performed by a multiplication algorithm that recodes the multiplier in a manner suggested by Booth [B1]. (This recoding differs from what has become known as "Booth recoding", which we shall mention later.) This recoding replaces a substring of the form $0^k 1^l$ in the input by the string $0^{k-1} 1 0^{l-1} \bar{1}$, where $\bar{1}$ calls for a subtraction rather than an addition. When applying this algorithm to a finite input string, it is necessary to append 0s to the beginning and end of the string, and this adds a further $1/2$ in the uniform case (or $1/2 + O(1/2^n)$ in the reciprocal case) to the expected number of additions and subtractions. Thus this recoding actually increases the number of operations required for a random multiplier. Its advantage, however, lies in eliminating almost all of the operations corresponding to leading 1s when the multiplier is a small negative number represented in 2s-complement form.

At this point we could give many other examples of feeding sequences of bits with various distributions into various finite automata. For most of the examples that arise in analysis of arithmetic algorithms with natural input distributions, the results are similar to those presented above: the expected number of acceptances for the given input distribution differs from that for the uniform input distribution by a constant and an exponentially small error term. Our goal in the next section will be to determine reasonably general conditions under which this type of result holds.

2   Main Theorem

Our main result will relate the asymptotic behaviour of the acceptance counts to that of the entropy of the input sequence. Following Shannon [S], we define the *entropy* of a random variable Ξ to be

$$H(\Xi) = - \sum_{\xi} \Pr[\Xi = \xi] \log_2 \Pr[\Xi = \xi].$$

Let $\Xi_n = X_1 \cdots X_n$ denote the first n bits of the input sequence $X_1, X_2, \ldots$. Define

$$h_n = H(\Xi_n)$$

to be the entropy of these n bits.

Theorem 2.1 *Suppose the entropy of $X_1, X_2, \ldots$ satisfies*

$$h_n = n - A + O(a^n)$$

for some constant $A \geq 0$ and some $a < 1$. If the input $X_1, X_2, \ldots$ is fed into a strongly-connected and aperiodic finite automaton, then

$$p_n = \rho + O(b^n)$$

for some constant $b < 1$, and thus

$$e_n = \rho n + B + O(b^n)$$

for some constant B.

Proof. Let $\Xi_n = YZ$, where $|Y| = m = \lfloor n/2 \rfloor$ and $|Z| = l = \lceil n/2 \rceil$. A simple calculation shows that

$$H(\Xi_n) = H(Y) + H(Z \mid Y),$$

where the *conditional entropy* of Z with respect to Y is given by

$$H(Z \mid Y) = - \sum_{\eta} \Pr[Y = \eta] \sum_{\zeta} \Pr[Z = \zeta \mid Y = \eta] \log_2 \Pr[Z = \zeta \mid Y = \eta].$$

Thus

$$H(Z \mid Y) = h_n - h_m$$
$$= l + O(a^{n/2}),$$

by the hypothesis of the theorem. We can write this as

$$\sum_{\eta} \Pr[Y = \eta](l - H(Z \mid Y = \eta)) = O(a^{n/2}), \tag{2.1}$$

where the *entropy* of Z conditioned on the event $Y = \eta$ is given by

$$H(Z \mid Y = \eta) = -\sum_{\zeta} \Pr[Z = \zeta \mid Y = \eta] \log_2 \Pr[Z = \zeta \mid Y = \eta].$$

Since Z takes on at most 2^l different values, $H(Z \mid Y = \eta) \leq l$. Thus the quantity in parentheses in (2.1) is non-negative, which implies that there exists a set V such that

$$\Pr[Y \notin V] = O(a^{n/4}) \tag{2.2}$$

and, for every $\eta \in V$, we have

$$H(Z \mid Y = \eta) = l + O(a^{n/4}).$$

We shall now show that this implies that there exists a set W_η such that

$$\Pr[Z \notin W_\eta \mid Y = \eta] = O(a^{n/12}), \tag{2.3}$$

$$\frac{|\{0,1\}^l \setminus W_\eta|}{2^l} = O(a^{n/12}), \tag{2.4}$$

and, for every $\zeta \in W_\eta$, we have

$$\Pr[Z = \zeta \mid Y = \eta] = \frac{1}{2^l}(1 + O(a^{n/12})). \tag{2.5}$$

This will follow immediately from the following lemma, whose proof is straightforward.

Lemma 2.2 *Let U be a random variable taking values in $\{0,1\}^l$. Suppose that there is a set $T \subseteq \{0,1\}^l$ such that either*

$$\Pr[U \in T] \geq \varepsilon$$

or

$$\frac{|T|}{2^l} \geq \varepsilon,$$

and, for every $v \in T$, we have either

$$\Pr[U = v] \geq \frac{1 + \varepsilon}{2^l}$$

or

$$\Pr[U = v] \leq \frac{1 - \varepsilon}{2^l}.$$

Then

$$H(U) \leq l - \Omega(\varepsilon^3).$$

Let the automaton be $M = (Q, \Sigma, \iota, \delta, F)$, where Q is the set of states, $\Sigma = \{0, 1\}$ is the input alphabet, $\iota \in Q$ is the initial state, $\delta : Q \times \Sigma \to Q$ is the transition function, and $F \subseteq Q$ is the set of final (or accepting) states. We shall extend δ to a function $\delta : Q \times \Sigma^* \to Q$ in the usual way.

We have

$$p_n = \sum_{\beta \in F} \sum_{\substack{|\xi| = n \\ \delta(\iota, \xi) = \beta}} \Pr[\Xi_n = \xi].$$

From this we obtain

$$p_n = \sum_{\substack{\alpha \in Q \\ |\eta| = m \\ \delta(\iota, \eta) = \alpha}} \Pr[Y = \eta] \sum_{\substack{\beta \in F \\ |\zeta| = l \\ \delta(\alpha, \zeta) = \beta}} \Pr[Z = \zeta \mid Y = \eta].$$

Taking account of (2.2), (2.3) and (2.5), we have

$$p_n = \sum_{\substack{\alpha \in Q \\ \eta \in V \\ \delta(\iota, \eta) = \alpha}} \Pr[Y = \eta] \sum_{\substack{\beta \in F \\ \zeta \in W_\eta \\ \delta(\alpha, \zeta) = \beta}} \frac{1}{2^l} + O(a^{n/12}).$$

Taking account of (2.4), we have

$$p_n = \sum_{\substack{\alpha \in Q \\ \eta \in V \\ \delta(\iota, \eta) = \alpha}} \Pr[Y = \eta] \sum_{\substack{\beta \in F \\ |\zeta| = l \\ \delta(\alpha, \zeta) = \beta}} \frac{1}{2^l} + O(a^{n/12}).$$

The innermost sum is the probability $q(l, \alpha, \beta)$ that a uniformly distributed sequence of l bits takes M from state α to state β. From the hypotheses of the theorem, a standard coupling argument shows that this probability is almost independent of α: there exists a constant $c < 1$ such that for all l, α and β,

$$|q(l, \alpha, \beta) - q(l, \iota, \beta)| = O(c^l).$$

This implies

$$p_n = \sum_{\substack{\alpha \in Q \\ \eta \in V \\ \delta(\iota, \eta) = \alpha}} \Pr[Y = \eta] \sum_{\beta \in F} q(l, \iota, \beta) + O(c^{n/2}) + O(a^{n/12}).$$

The innermost sum is now the probability that M accepts a uniformly distributed sequence of l bits. The hypotheses of the theorem again show that, with an appropriate choice of $c < 1$,

$$\sum_{\beta \in F} q(l, \iota, \beta) = \rho + O(c^{n/2}).$$

Thus we obtain

$$p_n = \rho \sum_{\alpha \in Q} \sum_{\substack{\eta \in V \\ \delta(\iota, \eta) = \alpha}} \Pr[Y = \eta] + O(c^{n/2}) + O(a^{n/12}).$$

Again taking account of (2.2), we have

$$p_n = \rho \sum_{\alpha \in Q} \sum_{\substack{|\eta| = m \\ \delta(\iota, \eta) = \alpha}} \Pr[Y = \eta] + O(c^{n/2}) + O(a^{n/12}).$$

The double sum is 1, since every sequence η must take e the initial state to some state α. This completes the proof of the theorem.   □

Let us now consider what probability distributions meet the condition of Theorem 2.1.

Theorem 2.3 *Suppose that $X_1, X_2, \ldots$ are the successive bit in the binary expansion of a real number $X = \sum_{n \geq 1} X_n 2^{-n}$ that is distributed on the interval $[0, 1)$ according to the density function f, which satisfies the followoing conditions: there exist $j \geq 0$ and breakpoints $0 = a_0 < a_1 < \cdots < a_j < a_{j+1} = 1$ such that, for each $0 \leq i \leq j$, f satisfies the Lipschitz condition $|f(x) - f(y)| = O(|x - y|)$ for $a_i < x < y < a_{i+1}$. Then*

$$h_n = n - \nu + O\left(\frac{n}{2^n}\right),$$

where

$$\nu = \int_0^1 f(X) \log_2 f(X) \, dX.$$

Pr of. We shall deal with the case with $j = 0$ breakpoints; the general case follo ws merely by elaborating the notation. We have

$$h_n = - \sum_{0 \leq k < 2^n} \int_{k/2^n}^{(k+1)/2^n} f(X) \, dX \log_2 \int_{k/2^n}^{(k+1)/2^n} f(X) \, dX. \qquad (2.6)$$

F rom the Lipschitz condition we have

$$\int_{k/2^n}^{(k+1)/2^n} f(X) \, dX = \frac{1}{2^n} f\left(\frac{k}{2^n}\right) \left(1 + O\left(\frac{1}{2^n}\right)\right).$$

Substituting this into (2.6), and using the fact that the Lipschitz condition ensures that f, and therefore also

$$g(X) = f(X) \log_2 f(X)$$

is bounded, yields

$$h_n = n - \frac{1}{2^n} \sum_{0 \le k < 2^n} f\left(\frac{k}{2^n}\right) \log_2 f\left(\frac{k}{2^n}\right) + O\left(\frac{n}{2^n}\right).$$

Estimating the sum by an integral, using the fact that the Lipschitz condition ensures that f, and therefore also g, has bounded total variation, we obtain

$$h_n = n - \int_0^1 f(X) \log_2 f(X) \, dX + O\left(\frac{n}{2^n}\right).$$

This completes the proof of the theorem.   □

As an example, consider the reciprocal distribution, which corresponds to

$$f(x) = \begin{cases} 0, & \text{for } 0 \le x < 1/2; \\ \dfrac{1}{x \log 2}, & \text{for } 1/2 \le x < 1. \end{cases}$$

For this distribution we obtain $\nu = \frac{1}{2} + \log_2 \log 2e = 1.0287\ldots$ bits. Since the first bit is always 1, it alone accounts for 1 bit of entropy loss; and this first bit can be omitted from the representation of the significand as a "hidden" bit. The remaining $n - 1$ bits thus have just $-\frac{1}{2} + \log_2 \log 2e = 0.0287\ldots$ bits less entropy than $n - 1$ uniformly distributed bits.

3   Conclusion

We have seen in Theorem 2.1 that if the input to a strongly connected and aperiodic finite automaton has entropy that differs from that of a unifomly distibuted input by a constant andan exponen tially small error term, then the expected acceptance count also differs from that for a uniformly distributed input by a constant and an exponentially small error term. Our motivating example has been the reciprocal distribution, which governs the significands of floating-point numbers. As Theorem 2.3 shows, how ever, many other natually arising distibutions satisfy this condition. We may take, for further examples, the stationary distributions governing the partial remainders when a dividend (with mild assumptions concerning its distribution) is divided by a positive integer using the original S-R-T division algorithm (see F reiman [F1]). These distributions are piecewise constant, with breakpoints at dyadic rational n umbers; as a result, they have, for all sufficiently large n,

$$h_n = n - \nu,$$

with no error term!

Theorem 2.1 has a corollary that co vers the case in which the automaton is strongly connected but periodic. In this case we can find a positive integer d such that the automaton with input alphabet $\{0, 1\}^d$ that accepts d input bits and produces d output bits at a time is both strongly connected and aperiodic.

Analysis similar to that in the proof of Theorem 2.1 then applies to the original automaton for each equivalence class of n modulo d. For each suc h class $0 \leq c \leq d - 1$, there will be an acceptance rate ρ_c, and the ov erall accceptance rate ρ will be the average of these:

$$\rho = \frac{\rho_0 + \cdots \rho_{d-1}}{d}.$$

The conclusion is then that

$$p_n = \rho_c + O(b^n)$$

as $n \to \infty$ through integers congruent to c modulo d, and that

$$e_n = \rho n + B + O(b^n)$$

as $n \to \infty$ in any fashion. Examples of strongly connected but periodic automata are those whose acceptances correspond to the additions and subtractions per-formed by multiplication algorithms that recode pairs or triplets of bits at each step (see MacSorley [M]). (These algorithms have become know as "Booth re-coding" algorithms; they were first published by MacSorley, though Booth [B2] attributes the triplet version to K. D. Tocher!)

Finally, w eshould mention that though Theorem 2.1 deals with finite au-tomata processing in successive bits of a real number from left to right (that is, from most significant to least significant), it is also possible to apply it to multiplication algorithms that recode the multiplier from right to left. Examples of these are certain algorithms that recode variable-length substrings of 0s and 1s (see MacSorley [M]). The analysis using Theorem 2.1 is made possible by con-sidering the recoding to be done left-to-right using the "on-the-fly" conv ersion tec hnique described by Frougny [F2]. The sequential machine that performs this conv ersion is not a finite automaton, but the number of additions and subtrac-tions called for by the recoding can be determined by the acceptances of a finite automaton that processes the input bits from left to right.

References

[B1]   Booth, A. D.: A signed binary multiplication technique. Quart. J. Mech. Appl. Math., **4** (1951) 236–240

[B2]   Booth, A. D.: Review of "A proof of the modified Booth's algorithm for multi-plication" by Louis P. Rubinfeld. Math. Rev., **53** #4610

[C]   Chomsky, N., Miller, G. A.: Finite-state languages. Inform. and Control, **1** (1958) 91–112

[F1]   F reiman, C. V.: Statistical analysis of certain binary division algorithms. Proc. IRE, **49** (1961) 91–103

[F2]   F rougn yₚ.: On-the-fly algorithms and sequential machines. IEEE Trans. on Computers, **49** (2000) 859–863

[H]   Hamming, R. W.: On the distribution of numbers. Bell System Tech. J., **49** (1970) 1609–1625

[M]   MacSorley, O. L.: High-speed arithmetic in binary computers. Proc. IRE, **49** (1961) 67–91

[N]   Newcomb, S.: Note on the frequency of use of the different digits in natural numbers,. Amer. J. Math., **4** (1881) 39–40

[S]   Shannon, C. E.: A mathematical theory of communication. Bell System Tech. J., **27** (1948) 379–423, 623–655

[T]   Tocher, K. D.: T echniques ofm ultiplicationand division for automatic binary computers. Quart. J. Mech. Appl. Math., **11** (1958) 364–384

[W1] Wallis, J.: Arithmetica infinitorum. Oxford, 1656

[W2] Whittaker, E. T., Watson, G. N.: A course of modern analysis. Cambridge, 1963

Monotone Drawings of Planar Graphs*

János Pach and Géza Tóth

[1] City College, CUNY and Courant Institute of Mathematical Sciences,
New York University, New York, NY 10012, USA
pach@cims.nyu.edu
[2] Rényi Institute of the Hungarian Academy of Sciences,
H-1364 Budapest, P.O.B.127, Hungary
geza@renyi.hu

Abstract. Let G be a graph drawn in the plane so that its edges are represented by x-monotone curves, any pair of which cross an even number of times. We show that G can be redrawn in such a way that the x-coordinates of the vertices remain unchanged and the edges become non-crossing straight-line segments.

1 Introduction

A *drawing* $\mathcal{D}(G)$ of a graph G is a representation of the vertices and the edges of G by points and by possibly crossing simple Jordan arcs between them, resp. When it does not lead to confusion, we make no notational or terminological distinction between the vertices (resp. edges) of the underlying abstract graph and the points (resp. arcs) representing them. Throughout this paper, we assume that in a drawing

1. no edge passes through any vertex other than its endpoints;
2. no three edges cross at the same point;
3. if two edges of a drawing share an interior point p, then they properly cross at p, i.e., one arc passes from one side of the other arc to the other side.

A drawing is called *x-monotone* if every vertical line intersects every edge in at most one point. For simplicity, we assume that no two vertices in an x-monotone drawing have the same x-coordinate. We call a drawing *even* if any two edges not incident to the same vertex cross an even number of times.

Hanani (Chojnacki) [Ch34] (see also [T70]) proved the remarkable theorem that if a graph G permits an even drawing, then it is *planar*, i.e., it can be redrawn without any crossing. On the other hand, by Fáry's theorem [F48], [W36], every planar graph has a straight-line drawing. We can combine these two facts by saying that every even drawing can be *"stretched"*

---

* Work on this paper by János Pach has been supported by NSF grant CCR-00-98246, by PSC-CUNY Research Award 63382-0032. Work by Géza Tóth has been supprted by Hungarian Science Foundation grant OTKA T-038397.

The aim of this note is to show that if we restrict our attention to *x-monotone* drawings, then every even drawing can be stretched without changing the *x*-coordinates of the vertices.

Consider an *x*-monotone drawing $\mathcal{D}(G)$ of a graph G. If the vertical ray starting at $v \in V(G)$ and pointing upward (resp. downward) crosses an edge $e \in E(G)$, then v is said to be *below* (resp. *above*) e. Two drawings of the same graph are called *equivalent*, if the above-below relationships between the vertices and the edges coincide.

In the next two sections we establish the following two results.

Theorem 1. *For any x-monotone even drawing of a connected graph, there is an equivalent x-monotone drawing in which no two edges cross each other and the x-coordinates of the corresponding vertices are the same.*

Theorem 2. *For any non-crossing x-monotone drawing of a graph G, there is an equivalent non-crossing straight-line drawing, in which the x-coordinates of the corresponding vertices are the same.*

2   Proof of Theorem 1

We follow the approach of Cairns and Nikolayevsky [CN00]. Consider an *x*-monotone drawing $\mathcal{D}$ of a graph on the *xy*-plane, in which any two edges cross an even number of times. Let u and v denote the leftmost and rightmost vertex, respectively. We can assume without loss of generality that $u = (-1, 0)$ and $v = (1, 0)$. Introduce two additional vertices, $w = (0, 1)$ and $z = (0, -1)$, each connected to u and v by arcs of length $\pi/2$ along the unit circle C centered at the origin, and suppose that every other edge of the drawing lies in the interior of C. Denote by G the underlying abstract graph, including the new vertices w and z.

For each crossing point p, attach a *handle* (or bridge) to the plane in a very small neighborhood $N(p)$ of p. Assume that (1) these neighborhoods are pairwise disjoint, (2) $N(p)$ is disjoint from every other edge that does not pass through p, and that (3) every vertical line intersects every handle only at most once. For every p, take the portion belonging to $N(p)$ of one of the edges that participate in the crossing at p, and lift it to the handle without changing the *x*- and *y*-coordinates of its points. The resulting drawing $\mathcal{D}_0$ is a crossing-free embedding of G on a surface S_0 of possibly higher genus.

Let S_1 be a very small closed neighborhood of the drawing $\mathcal{D}_0$ on the surface S_0. Note that S_1 is a compact, connected surface, whose boundary consists of a finite number of closed curves. Attaching a disc to each of these closed curves, we obtain a surface S_2 with no boundary. According to Cairns and Nikolayevsky [CN00], S_2 must be a 2-dimensional *sphere*. To verify this claim, consider two closed curves, α_2 and β_2, on S_2. They can be deformed into closed walks, α_1 and β_1, respectively, along the edges of $\mathcal{D}_0$. The projection of these two walks into the (x, y)-plane are closed curves, α and β, in $\mathcal{D}$, which must cross an even number of times. Every crossing between α and β occurs either at a vertex of $\mathcal{D}$

or between tw o of its edges. By the assumptions, any two edges in $\mathcal{D}$ cross an even number of times. (The same assertion is trivially true in $\mathcal{D}_0 \subset S_2$, because there no two edges cross.) Using the fact that in $\mathcal{D}_0 \subset S_2$ the cyclic order of the edges incident to a vertex is the same as the cyclic order of the corresponding edges in $\mathcal{D}$, we can conclude that α_1 and β_1 cross an even n umber of times, and the same is true for α_2 and β_2. Thus, S_2 is a surface with no boundary, in which any tw o closed curves cross an even number of times. This implies that S_2 is a sphere. Consequently, $\mathcal{D}_0$, a crossing-free drawing of G on S_2, corresponds to a plane drawing.

Next, we argue that $\mathcal{D}_0$ can also be regarded as an x-*monotone* plane drawing of G, in which the x-coordinates of the vertices are the same as the x-coordinates of the corresponding vertices in $\mathcal{D}$.

For any point q (either in the plane or in 3-space), let $x(q)$ denote the x-coordinate of q. As before, every boundary curve of S_1 corresponds to a cycle of G. Since in the original drawing the cycle $vwuz$ encloses all other edges and vertices of G, one of the boundary curves of S_1, say γ, corresponds to the cycle $vwuz$. Consider another boundary curve, $\kappa \neq \gamma$, which corresponds to a cycle $v_1 v_2 \ldots v_i$ in G. Let D_κ be a closed disc in the plane bounded by an equivalent non-crossing x-monotone drawing of the cycle $v_1 v_2 \ldots v_i$. Glue the boundary of D_κ to κ so that a boundary point b of D_κ will be glued to a point $k \in \kappa$ if and only if b and k correspond to the same point of the cycle $v_1 v_2 \ldots v_i$ in $\mathcal{D}$. Consequently, $x(b) = x(k)$. Repeat the same procedure for each $\kappa \neq \gamma$.

Finally, w e obtain a new surface S con taining $\mathcal{D}_0$, which is topologically isomorphic to the unit disk bounded by C, and a natural extension of the x-coordinate function from S_1 to S, whic h is a con tin uous real function with no local minimum or maximum. Therefore, $\mathcal{D}_0$ can be regarded as a crossing-free x-monotone drawing of G, equivalent to $\mathcal{D}$. This completes the proof of Theorem 1.

Remark. Theorem 1 cannot be extended to disconnected graphs. To see this, consider a pair of edges, e_1 and e_2, intersecting twice, and place a vertex below e_1 and above e_2, and another one above e_1 and below e_2. Clearly, there exists no equivalen t crossing-free x-monotone drawing. On the other hand, if we drop the condition that the new drawing must be equivalen t to the original one, then the connected components can be treated separately and their drawings can be shifted in the vertical direction so as to avoid any crossing between them.

3   Proof of Theorem 2

Let $\mathcal{D} = \mathcal{D}(G)$ be a non-crossing x-monotone drawing of a graph G. First, we show that it is sufficient to prove Theorem for *triangulated* graphs. Deleting all vertices (points) and edges (arcs) of $\mathcal{D}$ from the plane, the plane falls into connected components, called *faces*. The x-coordinate of any vertex v will be denoted by $x(v)$.

Lemma 3.1. *By the addition of further edges, if necessary, every non-crossing x-monotone drawing $\mathcal{D}$ can b e extendd to an x-monotone triangulation.*

Proof. Consider a face F, and assume that it has more than 3 vertices. It is sufficient to show that one can always add an x-monotone edge between two non-adjacent vertices of F, which does not cross any previously drawn edges.

For the sake of simplicity, we outline the argument only for the case when F is a bounded face; the proof in the other case is very similar.

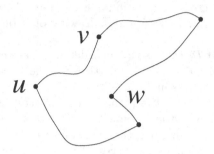

Fig. 1. The vertex w is extreme, u and v are not.

A vertex w of F is called *extreme* if it is not the left endpoint of any edge or not the right endpoint of any edge in $\mathcal{D}$, and a small neighborhood of w on the vertical line through w belongs to F. In particular, if the boundary of F is not connected, the leftmost (and the rightmost) vertex of each component of the boundary other than the exterior component, is extreme. See Fig. 1.

Suppose first that F has an extreme vertex w. We may assume, by symmetry, that w is not the right endpoint of any edge in $\mathcal{D}$. Starting at w, draw a horizontal ray in the direction of the negative x-axis. Let p be the first intersection point of this ray with the boundary of F. If p is a vertex, then the segment wp can be added to $\mathcal{D}$. Otherwise, one can add an x-monotone edge joining w to the left endpoint of the edge that p belongs to.

Suppose next that none of the vertices of F are extreme. In this case, the boundary of F is connected and any two vertices of F can be joined by an x-monotone curve inside F. However, an edge can be added to $\mathcal{D}$ only if the corresponding two vertices do not induce an edge in the exterior of F. Clearly, letting v_1, v_2, v_3, and v_4 denote four consecutive vertices of F, at least one of the pairs (v_1, v_3) and (v_2, v_4) has this property. $\square$

Now we turn to the proof of Theorem 2. The proof is by induction on the number of vertices. If G has at most 4 vertices, the assertion is trivial. Suppose that G has $n > 4$ vertices and that we have already established the theorem for graphs having fewer than n vertices. By Lemma 3.1, we can assume without loss of generality that the original x-monotone drawing $\mathcal{D}$ of G is triangulated.

CASE 1. There is a triangle $T = v_1 v_2 v_3$ in $\mathcal{D}$, which is not a face.

Then there is at least one vertex of $\mathcal{D}$ in the interior and at least one vertex in the exterior of T. Consequently, the drawings $\mathcal{D}_{\text{in}}$ and $\mathcal{D}_{\text{out}}$ defined as the part of $\mathcal{D}$ induced by v_1, v_2, v_3, and all vertices *inside* T and *outside* T, resp., have fewer than n vertices. By the induction hypothesis, there exist straight-line drawings $\mathcal{D}'_{in}$ and $\mathcal{D}'_{out}$, equivalent to $\mathcal{D}_{in}$ and $\mathcal{D}_{out}$, resp., in which all vertices have the same x-coordinates as in the original drawing. Notice that there is an affine transformation A of the plane, of the form

$$A(x,y) = (x, ax + by + c),$$

which takes the triangle induced by v_1, v_2, v_3 in $\mathcal{D}_{in}$ into the triangle induced by v_1, v_2, v_3 in $\mathcal{D}_{out}$. Since the image of a drawing under any affine transformation is equivalen t to the original drawing, we conclude that $A\left(\mathcal{D}'_{in}\right) \cup \mathcal{D}'_{out}$ meets the requirements.

In the sequel, we can assume that $\mathcal{D}$ has no triangle that is not a face. Fix a vertex v of $\mathcal{D}$ with minimum degree. Since every triangulation on $n > 4$ vertices has $3n - 6$ edges, the degree of v is $3, 4$, or 5. If the degree of v is 3, the neighbors of v induce a triangle in $\mathcal{D}$, which is not a face, contradicting our assumption.

There are tw o more cases to consider.

CASE 2. The degree of v is 4.

Let v_1, v_2, v_3, v_4 denote the neighbors of v, in clockwise order. There are three substantially different subcases, up to symmetry. See Fig. 2.

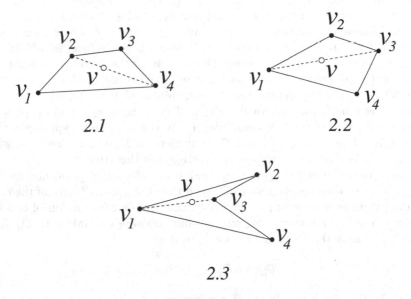

Fig. 2. CASE 2.

SUBCASE 2.1: $x(v_1) < x(v_2) < x(v_3) < x(v_4)$

Clearly, at least one of the inequalities $x(v) > x(v_2)$ and $x(v) < x(v_3)$ is true. Suppose without loss of generality that $x(v) < x(v_3)$. If v_1 and v_3 were connected by an edge, then vv_1v_3 would be a triangle with v_2 and v_4 in its interior and in its exterior, resp., contradicting our assumption. Remove v from $\mathcal{D}$, and add an x-monotone edge between v_1 and v_3, running in the interior of the face that contains v. Applying the induction hypothesis to the resulting drawing, we obtain that it can be redrawn by straight-line edges, keeping the x-coordinates fixed. Subdivide the segment v_1v_3 by its (uniquely determined) point whose x-coordinate is $x(v)$. In this drawing, v can also be connected by straight-line segments to v_2 and to v_4. Thus, we obtain an equivalent drawing which meets the requirements.

SUBCASE 2.2: $x(v_1) < x(v_2) < x(v_3) > x(v_4) > x(v_1)$
SUBCASE 2.3: $x(v_1) < x(v_2) > x(v_3) < x(v_4) > x(v_1)$

In these two subcases, the above argument can be repeated *verbatim*. In Subcase 2.3, to see that $x(v_1) < x(v) < x(v_3)$, we have to use the fact that in $\mathcal{D}$ both vv_2 and vv_4 are represented by x-monotone curves.

CASE 3. The degree of v is 5.

Let v_1, v_2, v_3, v_4, v_5 be the neighbors of v, in clockwise order. There are four substantially different cases, up to symmetry. See Fig. 3.

SUBCASE 3.1: $x(v_1) < x(v_2) < x(v_3) < x(v_4) < x(v_5)$
SUBCASE 3.2: $x(v_1) < x(v_2) < x(v_3) < x(v_4) > x(v_5) > x(v_1)$
SUBCASE 3.3: $x(v_1) < x(v_2) < x(v_3) > x(v_4) < x(v_5) > x(v_1)$
SUBCASE 3.4: $x(v_1) < x(v_2) > x(v_3) > x(v_4) < x(v_5) > x(v_1)$

In all of the above subcases, we can assume, by symmetry or by x-monotonicity, that $x(v) < x(v_4)$. Since $\mathcal{D}$ has no triangle which is not a face, we obtain that v_1v_3, v_1v_4, and v_2v_4 cannot be edges. Delete from $\mathcal{D}$ the vertex v together with the five edges incident to v, and let $\mathcal{D}_0$ denote the resulting drawing. Furthermore, let $\mathcal{D}_1$ (and $\mathcal{D}_2$) denote the drawing obtained from $\mathcal{D}_0$ by adding two non-crossing x-monotone diagonals, v_1v_3 and v_1v_4 (resp. v_2v_4 and v_1v_4), which run in the interior of the face containing v. By the induction hypothesis, there exist straight-line drawings $\mathcal{D}_1'$ and $\mathcal{D}_2'$ equivalent to $\mathcal{D}_1$ and $\mathcal{D}_2$, resp., in which the x-coordinates of the corresponding vertices are the same.

Apart from the edges v_1v_3, v_1v_4, and v_2v_4, $\mathcal{D}_1'$ and $\mathcal{D}_2'$ are non-crossing straight-line drawings equivalent to $\mathcal{D}_0$ such that the x-coordinates of the corresponding vertices are the same. Obviously, the convex combination of two such drawings is another non-crossing straight-line drawing equivalent to $\mathcal{D}_0$. More precisely, for any $0 \le \alpha \le 1$, let $\mathcal{D}_\alpha'$ be defined as

$$\mathcal{D}_\alpha' = \alpha \mathcal{D}_1' + (1-\alpha)\mathcal{D}_2'.$$

That is, in $\mathcal{D}_\alpha'$, the x-coordinate of any vertex $u \in V(G) - v$ is equal to $x(u)$, and its y-coordinate is the combination of the corresponding y-coordinates in $\mathcal{D}_1'$ and $\mathcal{D}_2'$ with coefficients α and $1 - \alpha$, resp.

Observe that the only possible concave angle of the quadrilateral $Q = v_1v_2v_3v_4$ in $\mathcal{D}_1'$ and $\mathcal{D}_2'$ is at v_3 and at v_2, resp. In $\mathcal{D}_\alpha'$, Q has at most one concave vertex.

Since the shape of Q changes continuously with α, w eobtain that there is a value of α for which Q is a *convex* quadrilateral in $\mathcal{D}_\alpha$. Let $\mathcal{D}'$ be the straight-line drawing obtained from $\mathcal{D}'_\alpha$ b y adding v at the unique point of the segment $v_1 v_4$, whose x-coordinate is $x(v)$, and connect it to $v_1, \ldots, v_5$. Clearly, $\mathcal{D}'$ meets the requirements of Theorem 2.

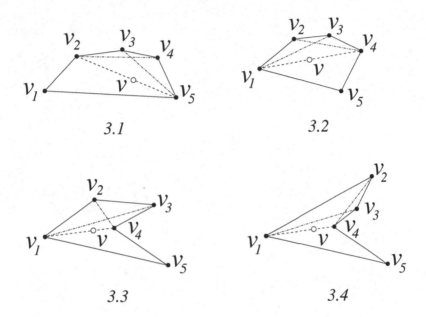

Fig. 3. CASE 3.

References

[Ch34] Ch. Chojnacki (A. Hanani), Über w esertlich unpl¨attbareKurven im drei-dimensionalen Raume, *Fund. Math.* **23** (1934), 135–142.

[CN00] G. Cairns and Y. Nikolayevsky ,Bounds for generalized thrackles, *Discr ete Comput. Geom.* **23** (2000), 191–206.

[DETT99] G. Di Battista, P. Eades, R. Tamassia, and I. G. T ollis, *Graph Drawing,* Prentice Hall, Upper Saddle River, NJ, 1999.

[EET76] G. Ehrlich, S. Even, and R. E. Tarjan, In tersection graphs of curⱱs in the plane, *Journal of Combinatorial Theory, Series B* **21** (1976), 8–20.

[F48] I. F´ary, On straigh line representation of planar graphs, *Acta Univ. Szeged. Sect. Sci. Math.* **11** (1948), 229–233.

[T70] W. T. Tutte, Tow ard a theory of crossing mmbers, *J. Combinatorial Theory* **8** (1970), 45–53.

[W36] K. Wagner, Bemerkungen zum Vierfarbenproblem, *Jber. Deutsch. math. Vere-inigung* **46** (1936), 26–32.

Author Index

Lecture Notes in Computer Science

For information about Vols. 1–2446

please contact your bookseller or Springer-Verlag

Vol. 2487: D. Batory, C. Consel, W. Taha (Eds.), Generative Programming and Component Engineering. Proceedings, 2002. VIII, 335 pages. 2002.

Vol. 2488: T. Dohi, R. Kikinis (Eds), Medical Image Computing and Computer-Assisted Intervention – MICCAI 2002. Proceedings, Part I. XXIX, 807 pages. 2002.

Vol. 2489: T. Dohi, R. Kikinis (Eds), Medical Image Computing and Computer-Assisted Intervention – MICCAI 2002. Proceedings, Part II. XXIX, 693 pages. 2002.

Vol. 2490: A.B. Chaudhri, R. Unland, C. Djeraba, W. Lindner (Eds.), XML-Based Data Management and Multimedia Engineering – EDBT 2002. Proceedings, 2002. XII, 652 pages. 2002.

Vol. 2491: A. Sangiovanni-Vincentelli, J. Sifakis (Eds.), Embedded Software. Proceedings, 2002. IX, 423 pages. 2002.

Vol. 2492: F.J. Perales, E.R. Hancock (Eds.), Articulated Motion and Deformable Objects. Proceedings, 2002. X, 257 pages. 2002.

Vol. 2493: S. Bandini, B. Chopard, M. Tomassini (Eds.), Cellular Automata. Proceedings, 2002. XI, 369 pages. 2002.

Vol. 2495: C. George, H. Miao (Eds.), Formal Methods and Software Engineering. Proceedings, 2002. XI, 626 pages. 2002.

Vol. 2496: K.C. Almeroth, M. Hasan (Eds.), Management of Multimedia in the Internet. Proceedings, 2002. XI, 355 pages. 2002.

Vol. 2497: E. Gregori, G. Anastasi, S. Basagni (Eds.), Advanced Lectures on Networking. XI, 195 pages. 2002.

Vol. 2498: G. Borriello, L.E. Holmquist (Eds.), UbiComp 2002: Ubiquitous Computing. Proceedings, 2002. XV, 380 pages. 2002.

Vol. 2499: S.D. Richardson (Ed.), Machine Translation: From Research to Real Users. Proceedings, 2002. XXI, 254 pages. 2002. (Subseries LNAI).

Vol. 2501: D. Zheng (Ed.), Advances in Cryptology – ASIACRYPT 2002. Proceedings, 2002. XIII, 578 pages. 2002.

Vol. 2502: D. Gollmann, G. Karjoth, M. Waidner (Eds.), Computer Security – ESORICS 2002. Proceedings, 2002. X, 281 pages. 2002.

Vol. 2503: S. Spaccapietra, S.T. March, Y. Kambayashi (Eds.), Conceptual Modeling – ER 2002. Proceedings, 2002. XX, 480 pages. 2002.

Vol. 2504: M.T. Escrig, F. Toledo, E. Golobardes (Eds.), Topics in Artificial Intelligence. Proceedings 2002. XI, 432 pages. 2002. (Subseries LNAI).

Vol. 2506: M. Feridun, P. Kropf, G. Babin (Eds.), Management Technologies for E-Commerce and E-Business Applications. Proceedings, 2002. IX, 209 pages. 2002.

Vol. 2507: G. Bittencourt, G.L. Ramalho (Eds.), Advances in Artificial Intelligence. Proceedings, 2002. XIII, 418 pages. 2002. (Subseries LNAI).

Vol. 2508: D. Malkhi (Ed.), Distributed Computing. Proceedings, 2002. X, 371 pages. 2002.

Vol. 2509: C.S. Calude, M.J. Dinneen, F. Peper (Eds.), Unconventional Models in Computation. Proceedings, 2002. VIII, 331 pages. 2002.

Vol. 2510: H. Shafazand, A Min Tjoa (Eds.), EurAsia-ICT 2002: Information and Communication Technology. Proceedings, 2002. XXIII, 1020 pages. 2002.

Vol. 2511: B. Stiller, M. Smirnow, M. Karsten, P. Reichl (Eds.), From QoS Provisioning to QoS Charging. Proceedings, 2002. XIV, 348 pages. 2002.

Vol. 2513: R. Deng, S. Qing, F. Bao, J. Zhou (Eds.), Information and Communications Security. Proceedings, 2002. XII, 496 pages. 2002.

Vol. 2514: M. Baaz, A. Voronkov (Eds.), Logic for Programming, Artificial Intelligence, and Reasoning. Proceedings 2002. XIII, 465 pages. 2002. (Subseries LNAI).

Vol. 2515: F. Boavida, E. Monteiro, J. Orvalho (Eds.), Protocols and Systems for Interactive Distributed Multimedia. Proceedings, 2002. XIV, 372 pages. 2002.

Vol. 2516: A. Wespi, G. Vigna, L. Deri (Eds.), Recent Advances in Intrusion Detection. Proceedings, 2002. X, 327 pages. 2002.

Vol. 2517: M.D. Aagaard, J.W. O'Leary (Eds.), Formal Methods in Computer-Aided Design. Proceedings, 2002. XI, 399 pages. 2002.

Vol. 2518: P. Bose, P. Morin (Eds.), Algorithms and Computation. Proceedings, 2002. XIII, 656 pages. 2002.

Vol. 2519: R. Meersman, Z. Tari, et al. (Eds.), On the Move to Meaningful Internet Systems 2002: CoopIS, DOA, and ODBASE. Proceedings, 2002. XXIII, 1367 pages. 2002.

Vol. 2521: A. Karmouch, T. Magedanz, J. Delgado (Eds.), Mobile Agents for Telecommunication Applications. Proceedings 2002. XII, 317 pages. 2002.

Vol. 2522: T. Andreasen, A. Motro, H. Christiansen, H. Legind Larsen (Eds.), Flexible Query Answering. Proceedings 2002. XI, 386 pages. 2002. (Subseries LNAI).

Vol. 2525: H.H. Bülthoff, S.-Whan Lee, T.A. Poggio, C. Wallraven (Eds.), Biologically Motivated Computer Vision. Proceedings 2002. XIV, 662 pages. 2002.

Vol. 2526: A. Colosimo, A. Giuliani, P. Sirabella (Eds.), Medical Data Analysis. Proceedings 2002. IX, 222 pages. 2002.

Vol. 2527: F.J. Garijo, J.C. Riquelme, M. Toro (Eds.), Advances in Artificial Intelligence – IBERAMIA 2002. Proceedings 2002. XVIII, 955 pages. 2002. (Subseries LNAI).

Vol. 2528: M.T. Goodrich, S.G. Kobourov (Eds.), Graph Drawing. Proceedings 2002. XIII, 384 pages. 2002.

Vol. 2529: D.A. Peled, M.Y. Vardi (Eds.), Formal Techniques for Networked and Distributed Sytems – FORTE 2002. Proceedings 2002. XI, 371 pages. 2002.

Vol. 2534: S. Lange, K. Satoh, C.H. Smith (Ed.), Discovery Science. Proceedings 2002. XIII, 464 pages. 2002.

Vol. 2535: N. Suri (Ed.), Mobile Agents. Proceedings 2002. X, 203 pages. 2002.

Vol. 2536: M. Parashar (Ed.), Grid Computing – GRID 2002. Proceedings 2002. XI, 318 pages. 2002.

Vol. 2540: W.I. Grosky, F. Plášil (Eds.), SOFSEM 2002: Theory and Practice of Informatics. Proceedings 2002. X, 289 pages. 2002.